TECHNICAL DRAWING 101

With AutoCAD 2024

A Multidisciplinary Guide to Drafting Theory and Practice

DOUGLAS SMITH

Professor
Architectural and Engineering Computer Aided Design
Austin Community College

ANTONIO RAMIREZ

Professor Emeritus
Architectural and Engineering Computer Aided Design
Austin Community College

ASHLEIGH CONGDON-FULLER

Associate Professor and Program Coordinator
Architectural and Engineering Computer Aided Design
Austin Community College

SDC Publications
P.O. Box 1334
Mission, KS 66222
913-262-2664
www.SDCpublications.com
Publisher: Stephen Schroff

ISBN-13: 978-1-63057-601-1
ISBN-10: 1-63057-601-8
Printed and bound in the United States of America.

ABOUT THIS BOOK

In the Architectural and Engineering Computer-Aided Design program at Austin Community College (ACC), we take our Introduction to Technical Drawing class very seriously. In fact, although we offer forty-three CAD-related courses, we believe Introduction to Technical Drawing is the most important class. Our reasoning is simple: Many years of experience tell us that students will decide whether to major in our department based on their success and satisfaction with their first technical drawing course. So, if we don't do a great job teaching this course, students don't enroll for the other forty-two. You may have found this to be true in your department as well.

Each semester, when a new crop of students enrolls in our Introduction to Technical Drawing classes, our department has a unique opportunity to recruit and retain these students, and we feel that the best way to accomplish this is to offer them a course that is challenging, interesting, and supportive. With this philosophy in mind, the authors began in the spring of 2000 to create a curriculum system (text and supporting materials) that would assist both students and faculty through the process of successfully learning and teaching the basics of technical drawing, including AutoCAD fundamentals, in a one-semester course (about 80 class hours). We have worked to improve this system every semester since. To date more than 3,500 students have successfully completed this curriculum in our CAD labs at Austin Community College. The best proof we can offer of the efficacy of this system with regard to student retention is that in the 2022/2023 academic year our department will have approximately 1,000 enrollments.

THE CURRICULUM

Technical Drawing 101 covers topics ranging from the most basic, such as making freehand, multiview sketches of machine parts, to the advanced—creating an AutoCAD dimension style containing the style settings defined by the ASME Y14.5 Dimensioning and Tolerancing standard. But unlike the massive technical drawing reference texts on the market, Technical Drawing 101 aims to present just the right mix of information and projects that can be reasonably covered by faculty, and assimilated by students, in one semester.

The CAD portion of the text incorporates drafting theory whenever possible and covers the basics of drawing setup (units, limits, and layers), the tools of the Draw, Modify, and Dimension toolbars, and the fundamentals of 3D modeling. By focusing on the fundamental building blocks of CAD, Technical Drawing 101 provides a solid foundation for students going on to learn advanced CAD concepts and techniques (xrefs, annotative text, etc.) in intermediate CAD courses.

In recognition of the diverse career interests of our students, Technical Drawing 101 includes projects in which students create working drawings for a mechanical assembly as well as for an architectural project. We include architectural drawing because our experience has shown that many (if not most) first-semester drafting students are interested in careers in the architectural design field, and that a traditional technical drawing text, which focuses solely on mechanical drawing projects, often holds limited interest for these students. The multidisciplinary approach of this text and its supporting materials is intended to broaden the appeal of the curriculum and increase student interest and, it is hoped, future enrollments.

To be sure, this is not intended to be an architectural drafting text, but the steps involved in drawing the elevation views of a building dovetail nicely with the concepts and techniques of multiview drawing. Most CAD tools and techniques are basically the same whether applied to a mechanical or an architectural project.

HIGH SCHOOL STEM AND TECH PREP PROGRAMS

Our program at Austin Community College has articulation of college credit agreements with many Austin-area high schools, and each year our department provides faculty development training to the CAD faculty from these schools. By requiring that our high school partners adopt the Technical Drawing 101 curriculum, both programs can be confident that our curricula are aligned and equivalent in scope and quality, thus facilitating the awarding of college credit to students for the work completed at the high school level. By building relationships with high school faculty in our area, we are maintaining an important pipeline for recruiting well-prepared students into our program at Austin Community College.

FEATURES NEW TO THE THIRTEENTH EDITION

The design layout of the thirteenth edition of Technical Drawing 101 is intended to highlight the step-by-step instruction of drafting and CAD techniques.

The twelfth edition has been updated to reflect the latest ASME (and other) standards applying to the creation of engineering drawings including the ASME Y14.5 Dimensioning and Tolerancing standard.

SUPPLEMENTS AND ONLINE RESOURCES FOR INSTRUCTORS AND STUDENTS

Technical Drawing 101 is intended to be a complete turnkey curriculum that provides supporting materials for both full-time and adjunct faculty, thus improving the consistency and quality of instruction across the curriculum. Faculty members using this text can access supplementary instructional materials at:
https://www.sdcpublications.com/Resources/978-1-63057-601-1

Faculty members should create an instructor account in order to access the instructor material.

Supplements available to instructors include:

- A comprehensive Instructor's Manual that explains how to present the material in each chapter, and the resources available to both teachers and students.
- Lecture materials for each chapter in PowerPoint format.
- Check prints consisting of AutoCAD dwg files with solutions to all drawing problems.
- Prototype drawings for each CAD assignment in AutoCAD DWG file format.
- Syllabus, tests, quizzes, answer keys, and sketching and traditional drafting files in doc and PDF format.

Students can download supplementary materials including prototype drawings and instructional videos by redeeming the access code that comes with this book. Instructions for redeeming the access code can be found inside the front cover of the text.

Supplements available for download by students include:

- Video tutorials for AutoCAD's Draw, Modify, Dimension and Object Snap commands (students can refer to these videos as needed for help outside of class).
- Video tutorials to guide students through creation of CAD assignments.
- AutoCAD prototype drawings for each CAD assignment.
- Lettering practice sheets and multiview sketching grid sheets in PDF format.

A FINAL WORD

At the end of registration for the spring semester of 2023, our program at ACC had enrolled over 100 beginning students in Introduction to Technical Drawing. From our program's perspective, this is 100-plus opportunities to cast our department in the best possible light to learners who are making important decisions about their future educational (and career) paths. We feel that the curriculum presented in this book, along with the supplementary materials, positions our department to deliver the best training possible to these students. Ultimately, the future of our entire department depends on the success and retention of these students because not only are they why our program exists, they are the life-blood of our advanced courses.

ACKNOWLEDGMENTS

Doug, Tony, and Ashleigh would like to thank their families for their encouragement, understanding, and support during the development of the thirteenth edition.

The authors would like to express our sincere thanks to Brian Hagerty, Jennifer Black, and Gabe Zaldivar for their expert advice and guidance, our students and colleagues at Austin Community College, and our reviewers and adopters whose suggestions, expertise, and encouragement help us to constantly improve this curriculum.

Douglas Smith, Tony Ramirez, and Ashleigh Congdon-Fuller

Austin, Texas - April 2023

TABLE OF CONTENTS

CHAPTER

1

TECHNICAL DRAWING

CHAPTER OBJECTIVES:

After studying the material in this chapter, you should be able to:

1. Explain what technical drawings are.
2. Explain the terminology used to describe the process of creating technical drawings.
3. Explain how technical drawings are produced.
4. Explain the training needed to become an engineer, architect, designer, or drafter.
5. Describe the process of obtaining employment in the technical drawing field and the qualities that employers seek.
6. Describe what career prospects and opportunities, including salary ranges, are available in the field of technical drawing.

CHAPTER OVERVIEW

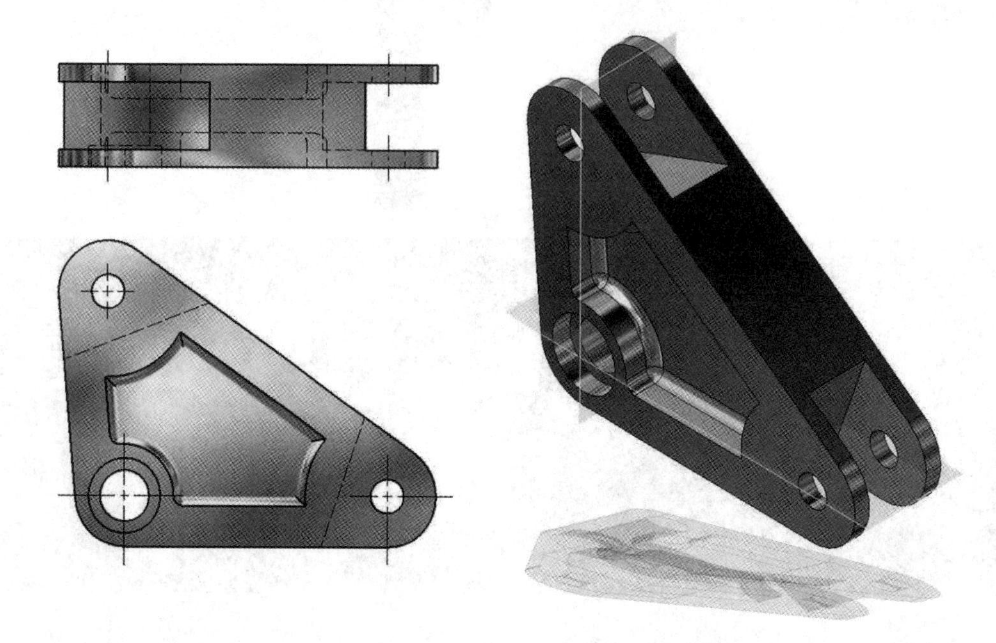

TECHNICAL DRAWING:

Technical drawings are the graphics and documentation (including notes and specifications) used by manufacturers to fabricate electronic and mechanical products and by construction professionals to produce houses, commercial buildings, roads, bridges, and water and wastewater systems. In fact, from the electronic devices inside your cell phone to the handle of your toothbrush, technical graphics are produced before almost all products are manufactured.

Refer to the following websites:	
American Design Drafting Association:	www.adda.org
American Institute of Architects:	www.aia.org
American Society for Engineering Education:	www.asee.org
National Society of Professional Engineers:	www.nspe.org
U.S. Department of Labor:	www.bls.gov

1.1 THE ORIGINS OF TECHNICAL DRAWING

Technical drawing is not a new concept; archaeological evidence suggests that humans first began creating crude technical drawings several thousand years ago. Through the ages, architects and designers, including Leonardo da Vinci, created technical drawings. However, a French mathematician, Gaspard Monge, is considered by many to be the founder of modern technical drawing. Monge's thoughts on the subject, *Géométrie Descriptive* (Descriptive Geometry), published around 1799, became the basis for the first university courses. In 1821, the first English-language text on technical drawing, *Treatise on Descriptive Geometry*, was published by Claudius Crozet, a professor at the U.S. Military Academy. Other terms often used to describe the creation of technical drawings are ***drafting, engineering graphics, engineering drawings, and computer-aided design (CAD).***

1.2 THE ROLE OF TECHNICAL DRAWING IN THE DESIGN PROCESS

To appreciate technical drawing's role in the design process, you must first understand some basics about the design process itself. For most projects, the first phase of a design project is to define clearly the design criteria that the finished design must meet to be considered a success. Many designers refer to this phase in the design process as problem identification. For example, before designing a house, an architectural designer needs to know the size and style of home the client wants, the number of bedrooms and baths, and the approximate budget for the project. The designer also needs information about the site where the house will be built. Is it hilly or flat? Are there trees, and if so, where are they located? What is the orientation of the site relative to the rising and setting of the sun? These concerns represent just a few of many design parameters that the designer needs to define before beginning the design process.

Once the design problem is clearly defined, the designer begins preparing preliminary designs that can meet the parameters defined during the problem identification phase. During this step, multiple solutions to the design problem may be generated in the form of freehand sketches, formal CAD drawings, or even rendered three-dimensional (3D) models. Designers refer to this process of generating many possible solutions to the design problem as the ideation, or brainstorming, phase of the process.

The preliminary designs are shown to the client to determine whether the design is in line with the client's expectations. This step allows the designer to clarify the client's needs and expectations. It also is an opportunity for a designer to educate the client about other, possibly better, solutions to the design problem.

After the client decides on a preliminary design that meets the criteria established in the first problem identification phase, the designer begins preparing design inputs that more clearly define the details of the design project. Design inputs may include freehand sketches with dimensional information, detailed notes, or even CAD models. Figure 1.1 shows an example of an architectural designer's sketch of a foundation detail for a house.

When the design inputs are finished, they are given to the drafter(s) responsible for preparing the technical drawings for the project. ***Drafters*** are individuals who have received specialized training in the creation of technical drawings.

One of the most important skills that drafters must acquire during their training is the ability to interpret design inputs and transform them into technical drawings. Drafters usually work closely with other members of the design team, which may include designers, checkers, engineers, architects, and other drafters during the creation of technical drawings.

Most drafters use CAD software to prepare technical drawings. CAD allows drafters to produce drawings much more quickly than traditional drafting techniques. Figure 1.2 shows a CAD drawing prepared from the designer's sketch shown in Figure 1.1.

Popular CAD programs include Autodesk® AutoCAD®, Autodesk® Revit®, Autodesk® Inventor®, SOLIDWORKS®, and PTC Creo®.

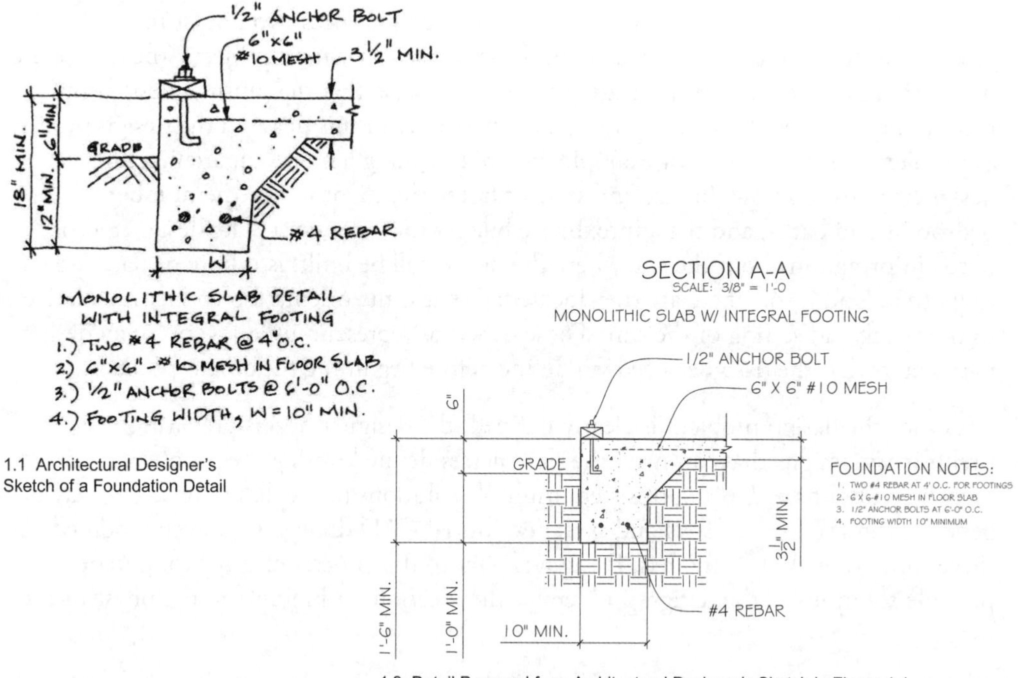

1.1 Architectural Designer's Sketch of a Foundation Detail

1.2 Detail Prepared from Architectural Designer's Sketch in Figure 1.1

JOB SKILL

Although modern drafters use CAD tools to create drawings, traditional drafting skills such as sketching and blueprint reading are still very important for facilitating communication between drafters and designers.

When the drafter is finished preparing the technical drawings, the designer, or in some cases a ***checker***, reviews the drawings carefully for mistakes. If mistakes are found, or if the design has been revised, the drafter will make the necessary corrections or revisions to the drawings. This process is repeated until the construction drawings are considered to be complete. When the entire set of construction drawings is finalized, the drafter and designer(s) put their initials in an area of the drawing called the ***title block***.

The finished construction drawings represent the master plan for the project. Everything required to complete the project, from applying for a building permit to securing financing for the project, revolves around the construction drawings. Building contractors use the construction documents to prepare bids for the project, and the winning bidders will use them to construct the building.

Engineering designers follow a similar process when designing products. Most engineering projects begin with a definition of initial design criteria and progress through the phases of preliminary design, design refinement, preparation of technical drawings, manufacturing, and inspection.

The trend in modern design, whether architectural or engineering, is to use CAD tools to create a dynamic, often three-dimensional, database that can be shared by all members of the design team. Increasingly, others in the organization, such as those involved in marketing, finance, or service and repair, will access information from the CAD database to accomplish their jobs.

JOB SKILL

Catching problems and mistakes during the design and drafting stages of the project can result in huge savings versus correcting mistakes on the job site or after the project has been built or manufactured. An example is the enormous cost incurred by an automobile manufacturer who has to recall thousands of cars to correct a design problem versus the cost of catching the problem on the technical drawing before the cars are manufactured.

1.3 TRAINING FOR CAREERS IN TECHNICAL DRAWING

Most drafters acquire their training by attending community college or technical school programs that lead to a certificate or associate's degree in drafting and design or CAD. These programs usually take from one to two years to complete and focus on the skills necessary to work as a drafter in industry, such as drafting techniques, knowledge of drafting standards, and the use of CAD programs to create drawings. Although most employers do not require that drafters be certified, the American Design Drafting Association (ADDA) has established a certification program for drafters. Individuals seeking certification must pass a test, which is administered periodically at ADDA-authorized sites. Some publishers of CAD software also offer certification on their products through authorized training sites.

Most drafters are full-time employees of architectural and engineering firms. Usually, drafters qualify for overtime pay when they work in excess of 40 hours per week. However, some drafters prefer to work as non-employee contractors. Contractors are usually very experienced drafters who often earn higher salaries than direct employees but have less job stability. Some organizations allow drafters to telecommute and transfer drawing files to the office via the Internet. ***Designers*** are often former drafters who have proven their ability to take on more responsibility and decision-making duties. Designers usually earn higher salaries than drafters because they are charged with more responsibility for the design—and even the successful completion—of the project.

To become an engineer or architect, an individual must first earn a bachelor's degree in engineering or architecture from a university program. Bachelor's degree programs generally take four to five years to complete and usually require a mastery of higher level courses in mathematics and physics.

After earning a degree, an engineer may become a ***Professional Engineer (P.E.)***, and an architect may become licensed, through a process involving both work experience and strenuous professional exams. The accreditation of architects is regulated by the ***American Institute of Architects (AIA)*** and by individual state statutes. The accrediting agency for engineers is the ***National Society of Professional Engineers***.

QUICK TIP

You can learn more about the American Design Drafting Association by visiting its website at www.adda.org. You can learn more about the American Institute of Architects and the National Society of Professional Engineers by visiting their websites at www.aia.org and www.nspe.org, respectively.

Career Paths in Technical Drawing

Architectural drafters work with architects and designers to prepare the drawings used in construction projects. These drawings may include floor plans, elevations, and construction details. Study of construction techniques and materials, as well as building codes, is important to the education of an architectural drafter. Some architectural drafters specialize in residential architecture (houses), whereas others may specialize in commercial architecture (buildings and apartments) or structural drafting (steel buildings or concrete structures). Figures 1.3 and 1.4 show details from a set of architectural drawings prepared by an architectural design drafter using Autodesk® Revit® CAD software.

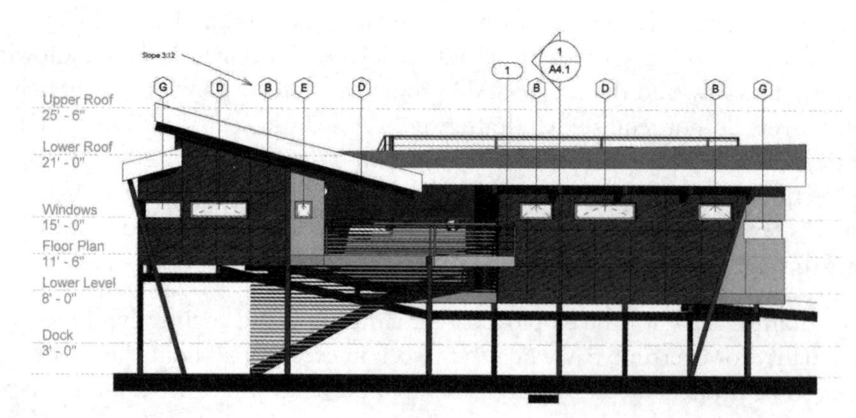

1.3 Elevation View of a Custom Home Created with 3D Modeling Software (Image courtesy of David Naumann)

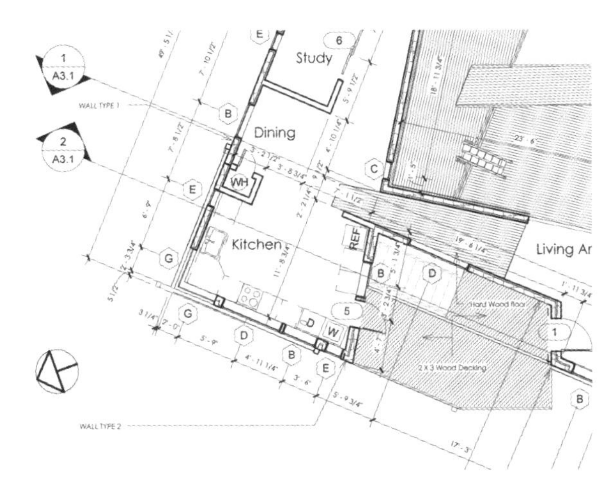

1.4 Detail from the Floor Plan of the Home Shown in Figure 1.3 (Image courtesy of David Naumann)

Aeronautical or aerospace drafters prepare technical drawings used in the manufacture of spacecraft and aircraft. These drafters often split their duties between mechanical drafting and electrical/electronic drafting and are sometimes referred to as electro/mechanical drafters.

Civil drafters and design technicians prepare construction drawings and topographical maps used in civil engineering projects. Civil projects may include roads, bridges, and water and wastewater systems. Civil drafters may also work for surveying companies to create site plans and plats for new subdivisions. Figure 1.5 shows an image of a subdivision plat created with Autodesk® AutoCAD® Civil 3D® CAD software.

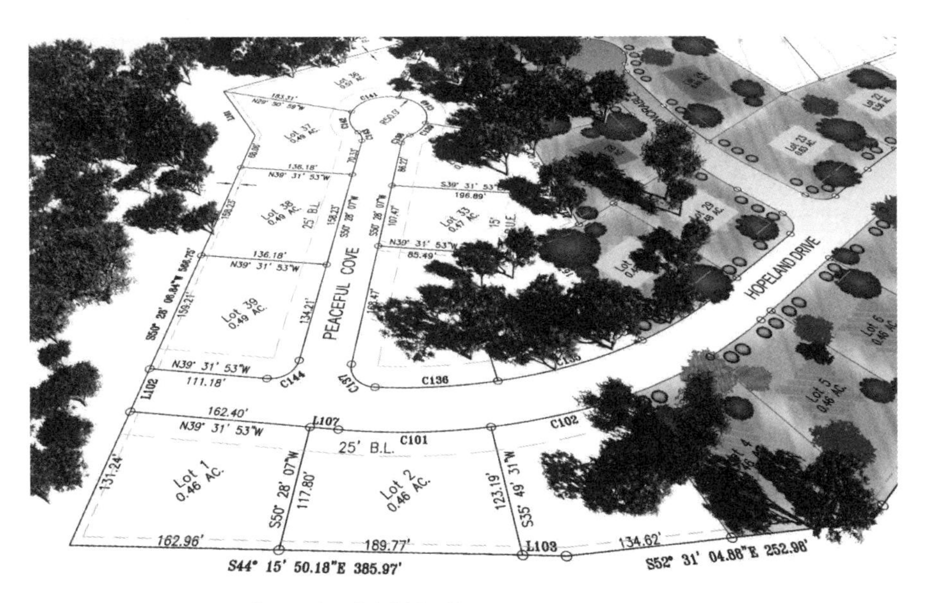

1.5 Civil Engineering Image Created for a Subdivision Plat. (Image courtesy of Jeffrey B. Muhammad)

Electrical drafters prepare diagrams used in the installation and repair of electrical equipment and the building wiring. Electrical drafters create documentation for systems ranging from low-voltage fire and security systems to high-voltage electrical distribution networks.

Electronics drafters create schematic diagrams, printed circuit board (PCB) artwork, integrated circuit layouts, and other graphics used in the design and maintenance of electronic (semiconductor) devices. Figure 1.6 shows a detail from an electronics schematic drawing. Figure 1.7 shows a detail of a PCB prepared from the schematic shown in Figure 1.6.

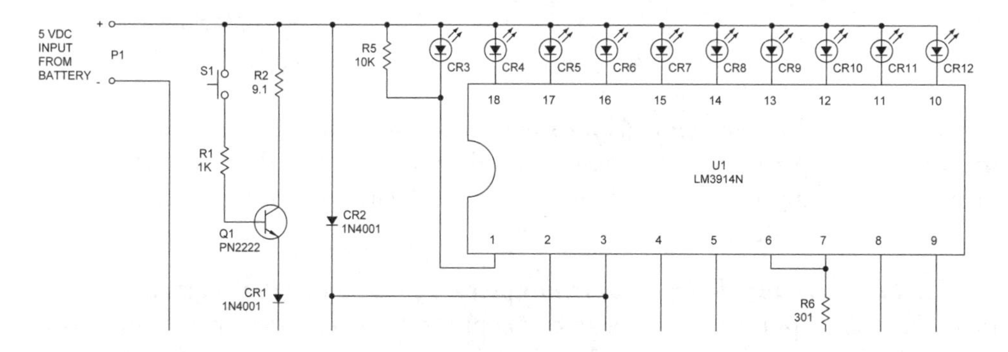

1.6 Detail from an Electronics Schematic

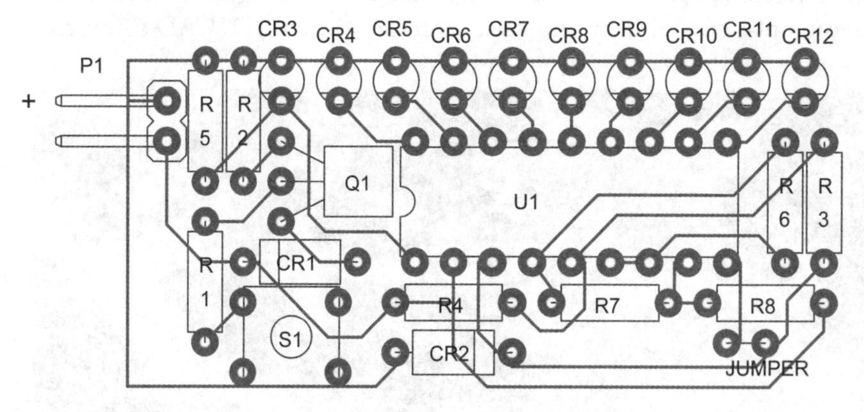

1.7 Detail of Printed Circuit Board Prepared from the Schematic Shown in Figure 1.7

Pipeline drafters and ***process piping drafters*** prepare drawings used in the construction and maintenance of oil refineries, oil production and exploration industries, chemical plants, and process piping systems such as those used in the manufacture of semiconductor devices. Figure 1.8 shows a 3D CAD model of a process piping system created using CADWorx® Plant software.

1.8 3D CAD Model of a Process Piping Design.
(Image Courtesy of Mustang Engineering L.P.)

Mechanical drafters work with mechanical engineers and designers to prepare detail and assembly drawings of machinery and mechanical devices. Mechanical drafters are usually trained in basic engineering theory as well as drafting standards and manufacturing techniques. They may be responsible for specifying items on a drawing such as the types of fasteners (nuts, bolts, and screws) needed to assemble a mechanical device or the fit between mating parts. Figure 1.9 shows a 3D model of a bellcrank created with Autodesk® Inventor® CAD software.

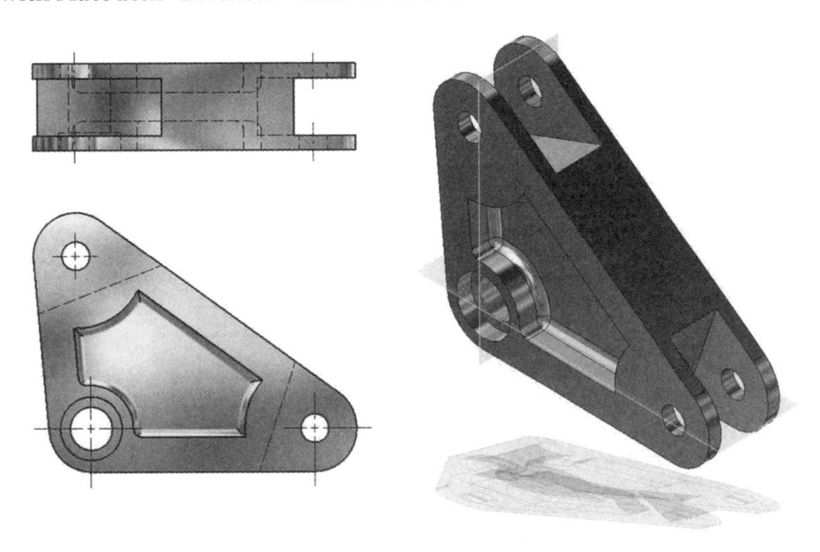

1.9 Model of a Machine Part Created with 3D Modeling Software

Mechanical, Electrical, Plumbing (MEP) drafters prepare drawings used in the design, fabrication, and installation of heating, ventilating, and air conditioning (HVAC), electrical, and plumbing systems in houses and commercial buildings.

Qualities That Employers Look for in Drafters

Most employers are very careful in making hiring decisions. Interviews are usually very thorough; in extreme cases, the interview process may take several hours. Candidates may be called back for more than one interview. When interviewing for a job, the candidate may interview with one person or with the entire design team. In the modern engineering or architectural office, the candidate's attitude regarding work and his or her ability to learn quickly and contribute to the team immediately is factored into the hiring decision. Often, tests are administered during the interview to measure proficiency with CAD or the candidate's understanding of necessary concepts that should have been mastered while in college or during other training.

During the job interview, employers may ask questions to help them determine the following:

- Did your previous training in the creation of technical drawings prepare you for this job?
- Are you a quick learner (will you contribute to the organization quickly)?
- Are you intelligent, competent, positive, and energetic?
- Will you fit into their team (do you get along well with your co-workers)?
- Will you require a lot of supervision (can you work independently)?
- Did you make good grades in your major (would your instructors give you good recommendations)?
- Are you able to meet deadlines?
- Do you communicate well with others (both verbally and in writing)?
- Do you have good work habits?
- Are you dependable?
- If you are hired, will the employer profit from your efforts?

JOB SKILL

PREPARING FOR THE WORKPLACE

If you are interested in working in the technical drawing field, develop good work habits while you are in school. Set high standards for your work, strive to create outstanding drawings for your portfolio, and be able to explain how the drawings in your portfolio were created and why. Come to class prepared and on time, meet your deadlines, and try to impress your instructors with your work habits and attitude, because often the recommendation of a current or former instructor will determine whether you get a job.

Salary Information for Drafters, Architects, and Engineers

Earnings for drafters vary depending on the specialty (architecture, mechanical, electrical/electronics, etc.), level of responsibility, and geographic area. According to U.S. Department of Labor statistics (2021), median annual earnings for architectural and civil drafters were $60,340, median annual earnings for mechanical drafters were $60,200, and median annual earnings for electrical and electronics drafters were $61,510.

By comparison, median annual earnings for architects were $80,180, median annual earnings for mechanical engineers were $95,300, median annual earnings for civil engineers were $88,050, and median annual earnings for electrical engineers were $101,780.

QUICK TIP

Median earnings refer to the middle of a distribution; that is, half the earnings are above the median, and half are below the median.
Salary data taken from the Bureau of Labor Statistics, U.S. Department of Labor, *Occupational Outlook Handbook*, on the Internet at: http://www.bls.gov/ooh/architecture-and-engineering/drafters.htm.

Job Prospects for Drafters

According to the U.S. Department of Labor (2021), about 19,000 new openings for drafters are projected each year through 2031. Overall employment for drafters varies nationwide by discipline and geographic region. For example, mechanical drafters will find more employment opportunities near large manufacturing hubs while architectural and civil drafters may find more job openings in regions experiencing rapid growth in population.

QUICK TIP

You can find out more about careers in drafting by visiting the Bureau of Labor Statistics, U.S. Department of Labor, *Occupational Outlook Handbook*, on the Internet at: http://www.bls.gov/ooh/architecture-and-engineering/drafters.htm.

CHAPTER REVIEW

The curriculum of this course is designed to introduce students to the field of technical drawing. This course is a good way to explore whether you possess the interests and aptitudes to pursue a career in which technical drawings are created or interpreted. If you wish to pursue more training, most community colleges and many universities offer specialized courses in engineering or architectural drawing. Your instructor may also be able to advise you on training opportunities and possible career paths.

KEY WORDS

American Institute of Architects
Architect
American Society of Mechanical Engineers
Computer-Aided Design (CAD) Designers
Design Technicians
Designers
Checkers
Drafters
- Aerospace
- Architectural
- Civil
- Electrical
- Mechanical
- Piping
- MEP

Drafting
Engineering Drawings
Engineering Graphics
Technical Drawings
National Society of Professional Engineers
Professional Engineer

REVIEW

Short Answer

1. Name three terms that are used to describe the creation of technical drawings.
2. In which field must a drafter be familiar with floor plans, elevations, and construction details?
3. How can an employer find out about a student's CAD skills, work habits, and dependability?
4. A drafter who divides his or her duties between mechanical drafting and electrical/electronics drafting is known as what kind of drafter?
5. Name the job titles of people who might constitute a design team.

REVIEW

Matching

Column A	Column B
1. Mechanical	1. Type of college degree held by most engineers and architects
2. Process piping	2. Drafting field concerned with chemical plants and oil refineries
3. Bachelor's	3. Type of college degree held by many drafters
4. Electronics	4. Drafting field concerned with screw threads and fasteners
5. Associate's	5. Drafting field concerned with the design of semiconductor devices

REVIEW

Multiple Choice

1. The acronym AIA stands for what?
    a. Architectural Institute of America
    b. American and International Architects
    c. American Institute of Architects
    d. Association of International Architects
2. According to the U.S. Department of Labor Statistics, which field of drafting has the highest median salary?
    a. Architectural
    b. Mechanical
    c. Electronics
    d. Radiological
3. In which field must a drafter be familiar with roads, bridges, water and wastewater systems, and surveying techniques?
    a. Architectural
    b. Civil
    c. Electronics
    d. Mechanical
4. Which of the following attributes do most employers value?
    a. Ability to fit into a team
    b. Ability to work independently
    c. Ability to communicate clearly
    d. All the above
5. Designers have a higher salary than drafters because:
    a. They usually have advanced technical degrees.
    b. They are licensed.
    c. They work longer hours than drafters.
    d. They have a higher degree of responsibility for the success of the project.

EXERCISE

Locate Bachelor Degree Programs in Architecture or Engineering

Conduct an internet search for university programs in your geographic area offering a bachelor's degree in architecture and/or engineering. Try to locate the degree plans posted at the site for each major, and compare the courses required for each degree.

EXERCISE

Locate Associate Degree Programs in CAD

Conduct an internet search for community college and technical school programs in your geographic area offering an associate's degree or certificate program in CAD or a drafting-related field. Try to locate the degree plan posted at the site for the program. Compare the courses required for the associate degree with the degree plan for a bachelor's degree program in architecture or engineering (see previous exercise).

CHAPTER

2

MULTIVIEW DRAWING

CHAPTER OBJECTIVES:

After studying the material in this chapter, you should be able to:

1. Explain what multiview drawings are and their importance to the field of technical drawing.
2. Define the terminology used to describe the geometry of multiview drawings.
3. Describe points, lines, planes, and angles.
4. Explain how views are chosen and aligned in a multiview drawing.
5. Describe projection planes.
6. Describe normal, inclined and oblique surfaces.
7. Describe orthographic projection including the miter line technique.
8. Describe the line types and line weights used in technical drawings as defined by the ASME Y14.2 standard.
9. Explain the difference between drawings created with First Angle and Third Angle projection techniques.

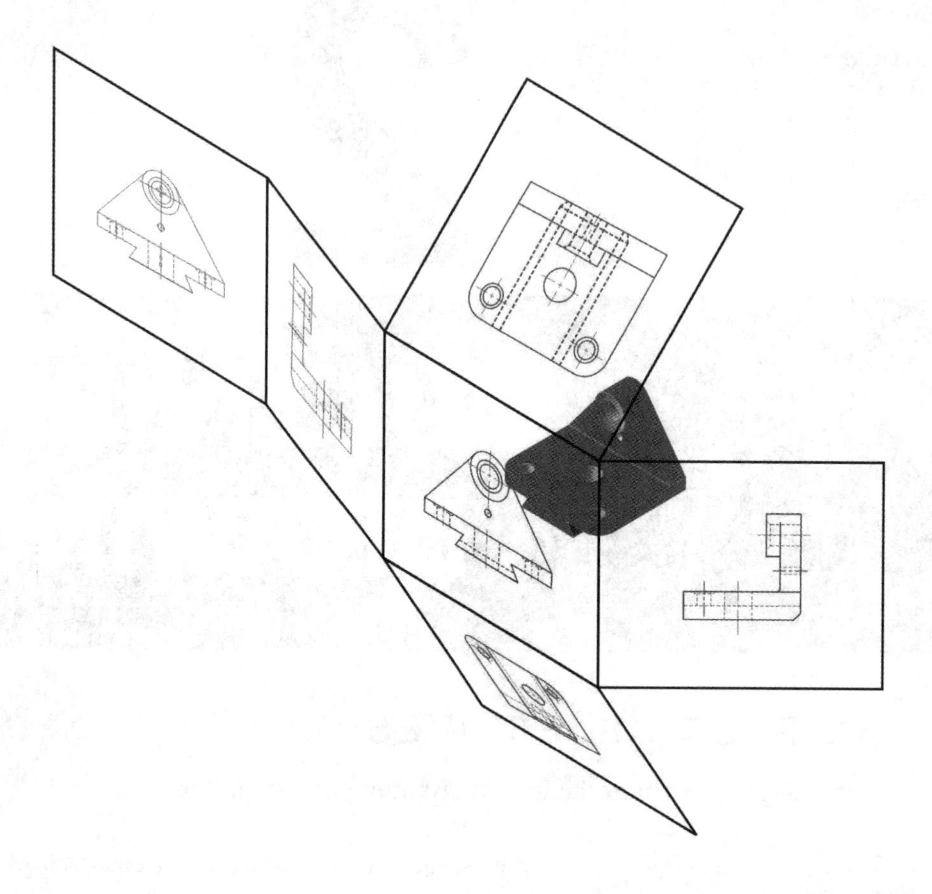

MULTIVIEW DRAWING:

Multiview drawing is a technique used by drafters and designers to depict a ***three-dimensional (3D) object*** (an object having height, width, and depth) as a group of related ***two-dimensional (2D) objects*** (having only width and height, or width and depth, or height and depth). A person trained in interpreting multiview drawings can visualize an object's 3D shape by studying the 2D multiview drawings of the object.

Refer to the following standard(s):
ASME Y14.2 Line Conventions and Lettering
ASME Y14.1 Decimal Inch Drawing Sheet Size and Format
ASME Y14.3 Multiview and Sectional View Drawings
ASME Y14.1M Metric Drawing Sheet Size and Format

2.1 THE TERMINOLOGY OF MULTIVIEW DRAWING

Multiview drawings are created by configuring the points, lines and planes of an object to create ***views*** that represent the object's features as they would appear if viewed from different points of view. For example, a mechanical drafter might construct the ***front, top,*** and ***side*** views of a machine part. An architectural drafter may draw the ***front, sides,*** and ***rear*** views of a building.

Before learning the techniques involved in the creation of the multiviews of an object, students should be familiar with the terminology used to describe the features of an object.

2.2 POINTS, PLANES, COORDINATE SYSTEMS, LINES, AND ANGLES

Points

A point is an exact "location" in space that is defined by coordinates that are located relative to a known origin point.

Points are often represented in technical drawings by a visible "dot," and while the dot representing a point is visible, the point has no dimensional size.

Locating Points in Two Dimensional (2D) Coordinate Systems

In a ***two dimensional (2D) coordinate system***, points are defined on a 2D flat surface that represents a plane. The coordinates of the point are located by measuring from two perpendicular lines that represent the ***X*** (horizontal) and ***Y*** (vertical) ***axes***. The intersection where the X and Y axes meet is called the ***origin***. The flat surface the X and Y axes lie on represents a 2D ***planar*** space referred to as the ***XY plane***.

QUICK TIP

Coordinates that are defined relative to a 0,0 origin point are referred to as ***absolute coordinates***.

In Figure 2.1, the origin point's value would be stated as "zero comma zero," or 0,0, which means the location of the origin is zero units (0) on the X axis and zero units (0) on the Y axis. A point, represented by a dot, is located at coordinate 2,3. This means that the point's location is 2 units to the right of the origin on the X axis and 3 units above the origin on the Y axis. Because points 0,0 and 2,3 are both located on the XY plane they are considered to be ***coplanar***.

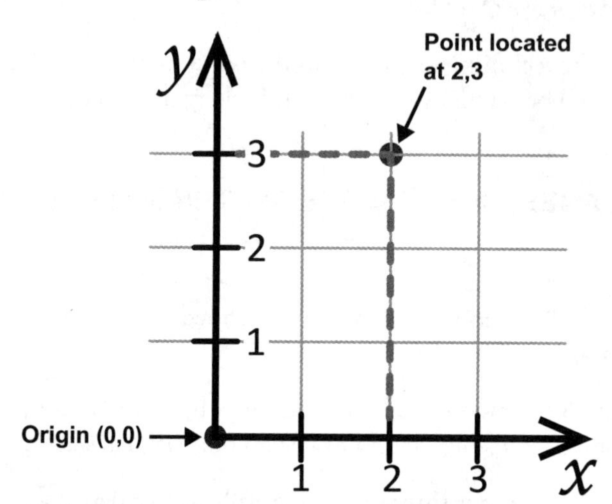

2.1 Two Dimensional Coordinate System with X and Y Axes Noted

In Figure 2.2, two large dots have been defined on the XY plane at coordinates 1,2 and 4,3 relative to the origin point. Points that lie on the same plane are referred to as ***coplanar***. Points that share the same location are referred to as ***coincident***.

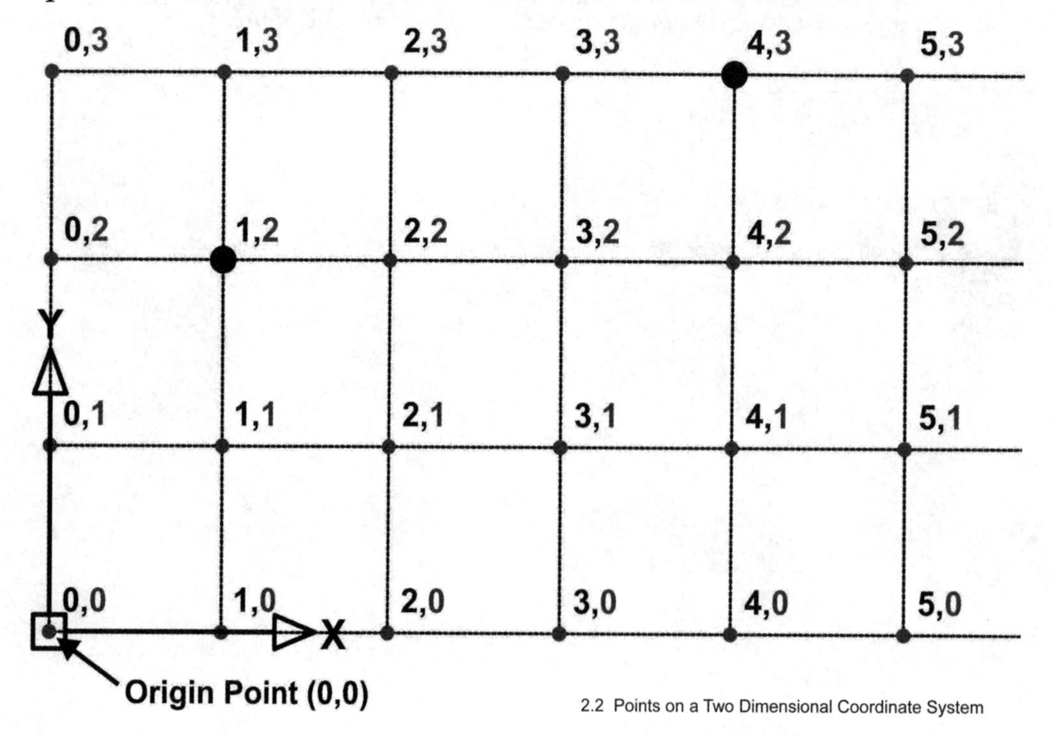

2.2 Points on a Two Dimensional Coordinate System

Three Dimensional (3D) Coordinate Systems

In a ***three dimensional (3D) coordinate system***, a Z axis is added to the X and Y axes. Using this system, points can be located relative to the origin along the X, Y, and Z axes. The Z axis represents the height of the point above or below the XY plane (see Figure 2.3). For example, a 3D coordinate might be defined with the coordinates 1,1,1. This coordinate would lay one unit to the right of the origin along the X axis, one unit from the origin along the Y axis, and one unit *above* the XY plane.

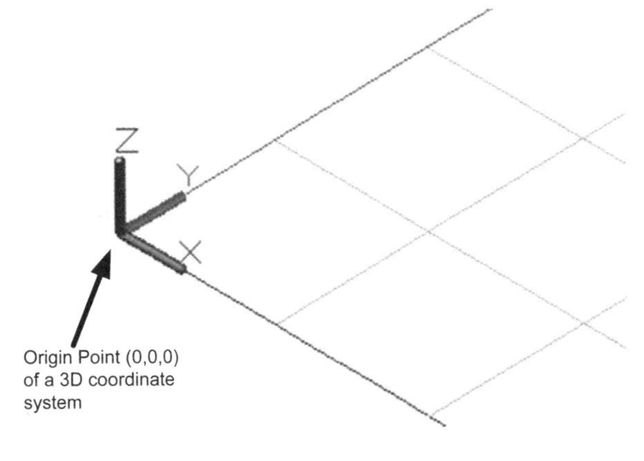

2.3 Three Dimensional Coordinate System

Lines

In the field of mathematics, a line is defined as a set of continuous points that extend indefinitely in either direction. In technical drawing terminology, a line is a segment defined by two points, the start point and the *endpoint*. These endpoints are defined with coordinates.

In Figure 2.4, a line begins at a start point located at coordinate 2,2 and ends at a point located at coordinate 8,7 (relative to the origin 0,0). Points that lie on the same line are referred to as ***collinear***. Points 2,2 and 8,7 are collinear. Non-collinear points do not lie on the same line.

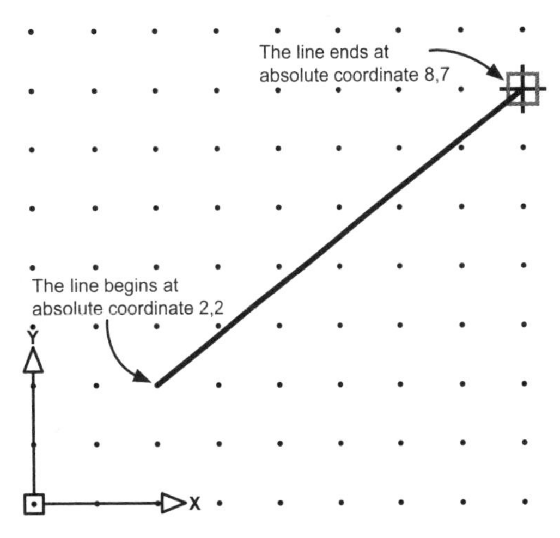

2.4 Coordinates for the Start and End Points of a Two-dimensional Line Drawn on the XY Plane

A collinear point located exactly halfway between the start and end points of the line in Figure 2.4 would be its *midpoint*.

Angles

An angle is formed when two, non-collinear lines share the same endpoint. The angle in Figure 2.5 is formed by sides BA and BC. The angle formed by these lines is referred to as angle ABC.

QUICK TIP

Parallel lines run side by side at a uniform distance and never intersect, even if extended.

Two or more planes may also be parallel relative to each other. For example, the floor and ceiling planes of a room may be parallel. Lines can also be parallel to planes.

The point where two lines ***cross*** is referred to as an ***intersection*** - as opposed to a vertex.

Vertex

In 2D space, the common point where two lines meet is called a ***vertex***. The plural of vertex is ***vertices***.

In Figure 2.5, angle ABC's vertex is point B and its sides are lines AB and CB. When specifying an angle using letters, or numbers, the vertex should be the middle letter or number in the series.

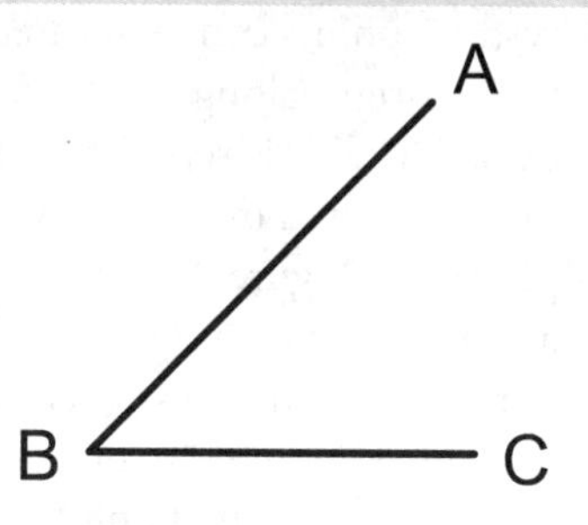

2.5 Two Lines Meeting to Form an Angle

In three dimensional objects, a vertex is where the object's edges meet. In the object shown in Figure 2.6, each of the vertices has been assigned a number. The object in Figure 2.6 has a total of 17 ***vertices***.

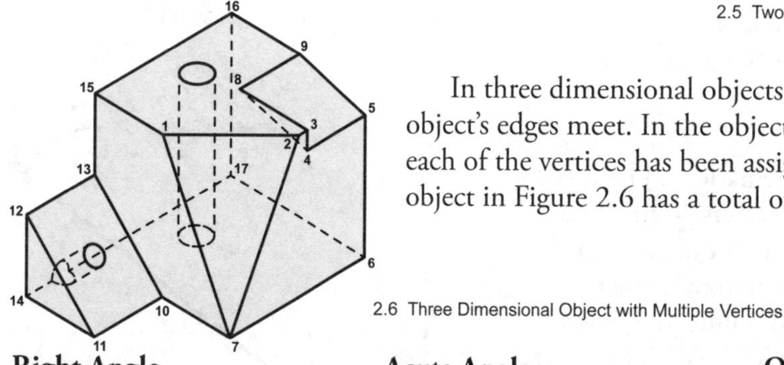

2.6 Three Dimensional Object with Multiple Vertices

Right Angle

In a right angle, the angle between the sides measures exactly 90 degrees as shown in Figure 2.7(a).

Acute Angle

In an acute angle, the angle between the sides measures less than 90 degrees as shown in Figure 2.7(b).

Obtuse Angle

In an obtuse angle, the angle between the sides measures greater than 90 degrees but less than 180 degrees as shown in Figure 2.7(c).

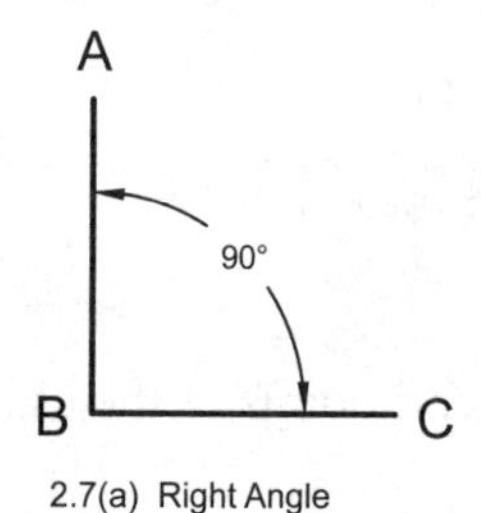

2.7(a) Right Angle

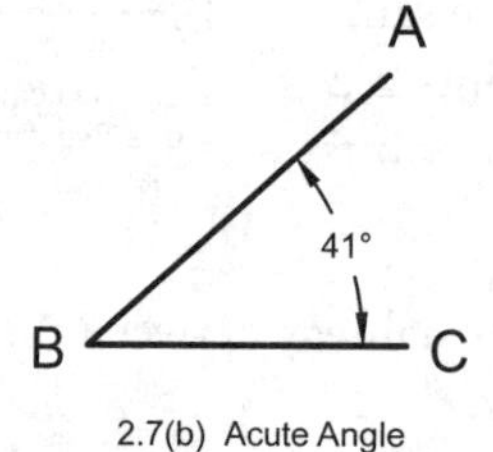

2.7(b) Acute Angle

A

126°

B C

2.7(c) Obtuse Angle

When two lines form a right angle they are ***perpendicular***. In Figure 2.7(a), line AB is perpendicular to line CB.

Complementary Angles

When two angles that share a common side have a total measurement of 90 degrees, they are ***complementary*** angles. See Figure 2.8(a).

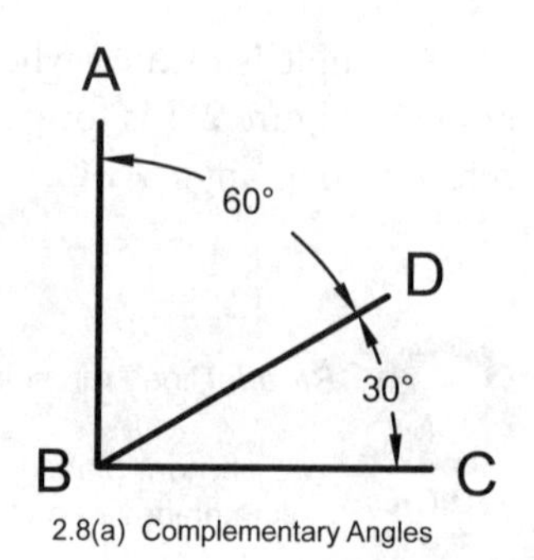

2.8(a) Complementary Angles

Supplementary Angles

If two angles that share a common side have a total measurement of 180 degrees they are ***supplementary*** angles. See Figure 2.8(b).

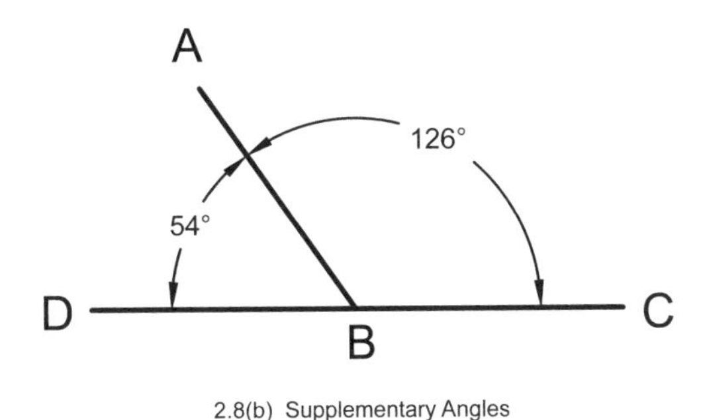

2.8(b) Supplementary Angles

Opposite Angles

When two lines cross, they form 4 angles. The angles ***opposite*** each other have the same measurements.

For example, in Figure 2.9(a), Angle A = Angle B and Angle C = Angle D.

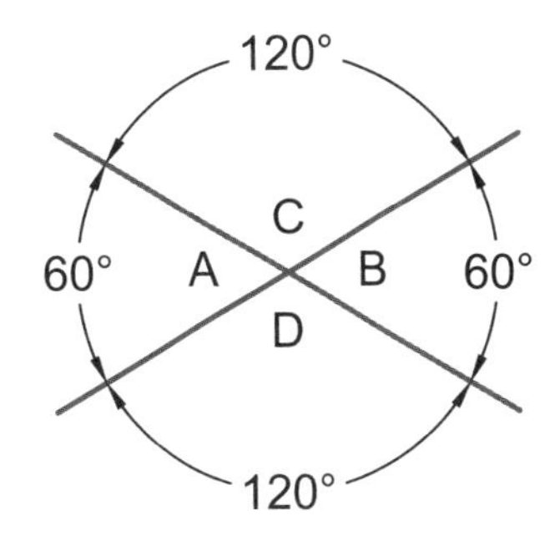

2.9(a) Opposite Angles

Adjacent Angles

Where two lines cross, angles that share a common side and common vertex are called ***adjacent*** angles. The sum of any two adjacent angles equals 180 degrees.

Therefore, in Figure 2.9(b), Angles A + C = 180 degrees and Angle C + B = 180 degrees, Angle B + D = 180 degrees and Angle D + A = 180 degrees.

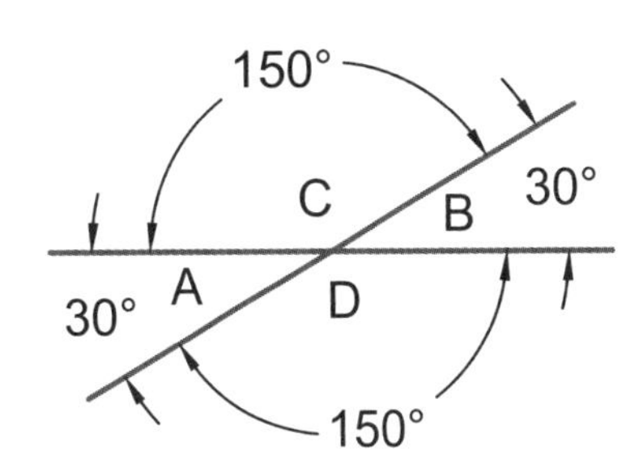

2.9(b) Adjacent Angles

2.3 TERMINOLOGY OF GEOMETRIC SHAPES

Circles

A circle can be defined by the locations of its ***center point*** and either its ***diameter*** or ***radius***. The radius of a circle is its diameter divided by 2.

In Figure 2.10, note how the radius and diameter of the circle are labeled.

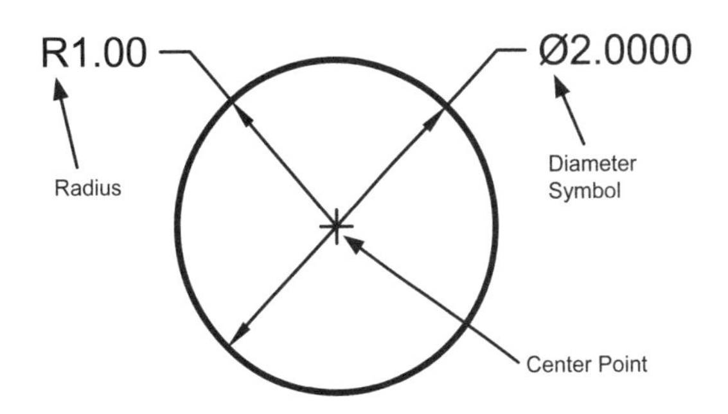

2.10 Circle Terminology

Tangency

When a line makes contact with a circle at only one point, but does not intersect the circle, it is ***tangent*** to the circle. See Figure 2.11(a). A line that is tangent to a circle must originate at a point outside of the circle. Circles that make contact at only one point can also be tangent. See Figure 2.11(b).

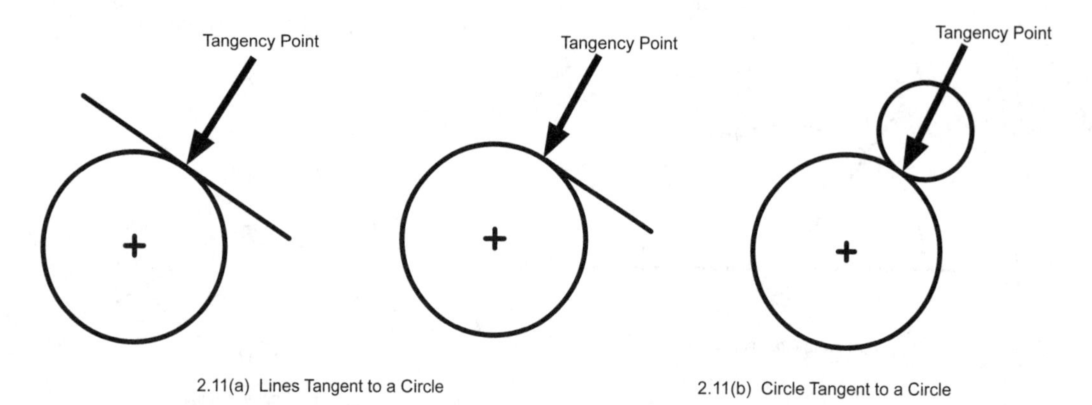

2.11(a) Lines Tangent to a Circle

2.11(b) Circle Tangent to a Circle

Concentricity and Eccentricity of Circles

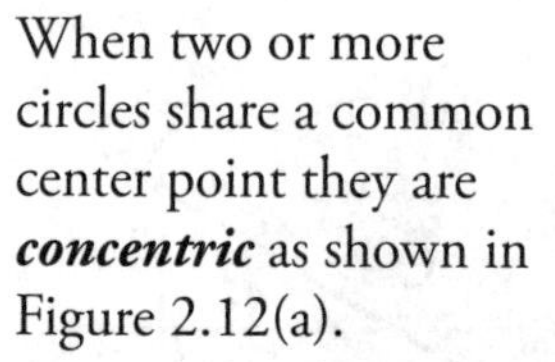

When two or more circles share a common center point they are ***concentric*** as shown in Figure 2.12(a).

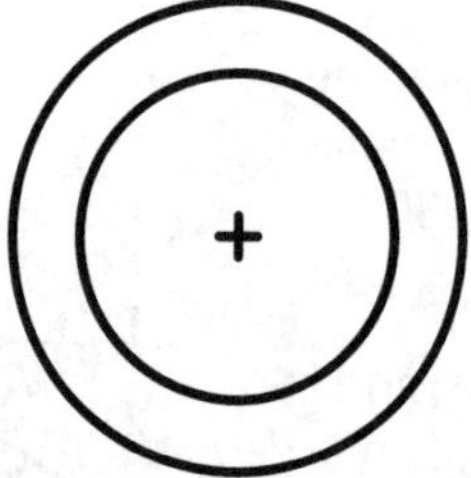

2.12(a) Concentric Circles

When two or more circles do not share a common center point they are ***eccentric*** as shown in Figure 2.12(b).

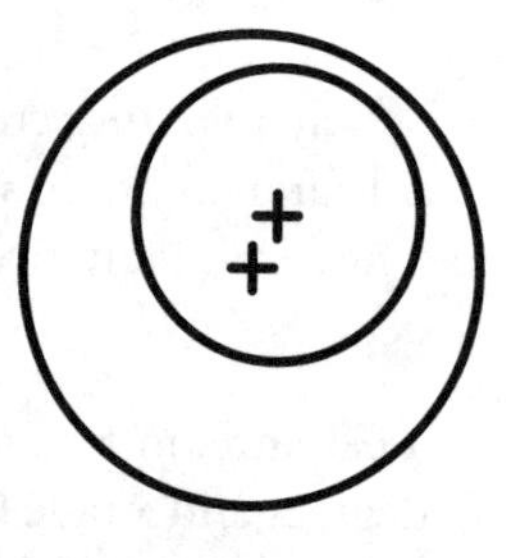

2.12(b) Eccentric Circles

Polygons

Polygons are multi-sided, 2D figures composed of straight line segments. The start and end points of a polygon's lines meet at the same point which creates a ***closed*** figure. Polygons are classified by the number of sides that define them. Examples of triangles, quadrilaterals, and hexagons are shown in Figures 2.13(a), 2.13(b), and 2.13(c).

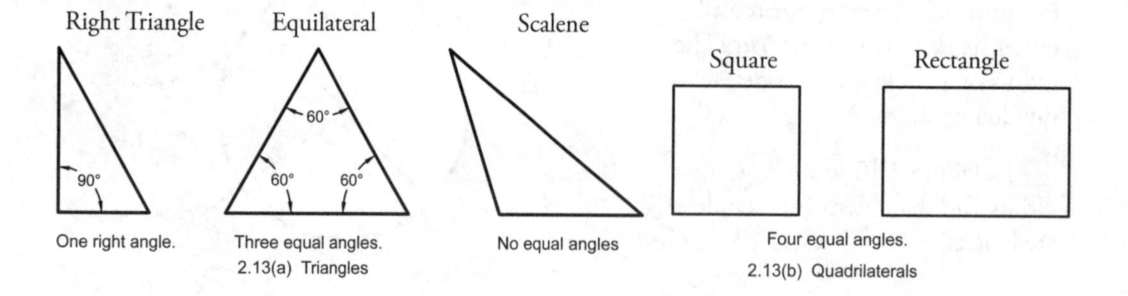

2.13(a) Triangles

2.13(b) Quadrilaterals

Hexagons are six-sided polygons. Hexagons may be constructed by ***inscribing*** them inside of circles or ***circumscribing*** them around a circle. See Figure 2.13(c).

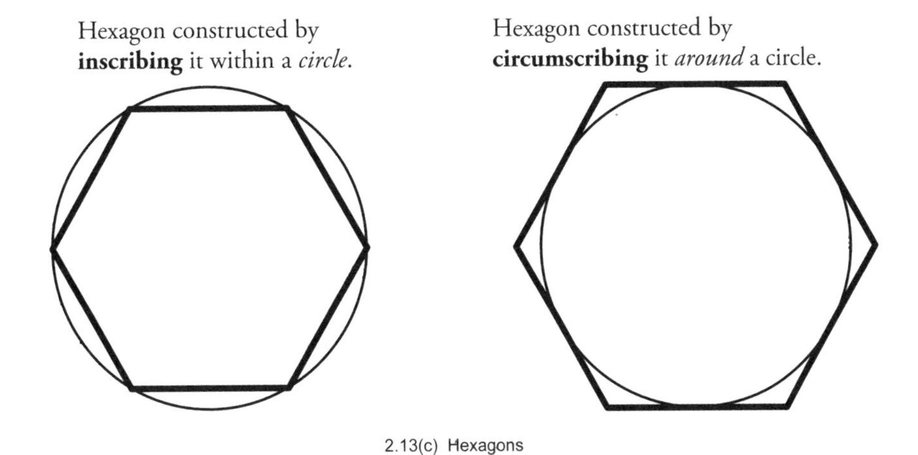

Hexagon constructed by **inscribing** it within a *circle.*
Hexagon constructed by **circumscribing** it *around* a circle.

2.13(c) Hexagons

NOTE

Pentagon	Five-sided polygon
Heptagon	Seven-sided polygon
Octagon	Eight-sided polygon
Nonagon	Nine-sided polygon
Decagon	Ten-sided polygon
Dodecagon	Twelve-sided polygon

Cylinders

A cylinder is a three-dimensional object that is defined by the diameter or radius, its length, and the location of its ***center axis***. See Figure 2.14(a). When two or more cylinders are aligned along the same center axis they are ***coaxial***. The two cylinders shown in Figure 2.14(b) are coaxial.

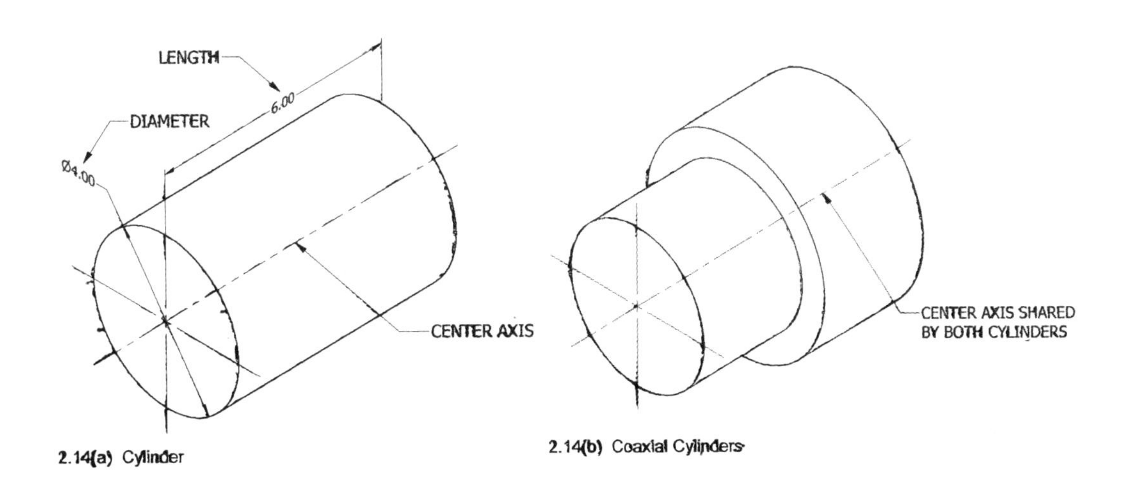

2.14(a) Cylinder

2.14(b) Coaxial Cylinders

2.4 MULTIVIEW DRAWINGS

Figure 2.15 represents a 3D image of a school bus. The 3D image of the bus is very helpful in visualizing its shape, because the viewer can quickly get an idea of the overall height, width, and depth of the bus. However, the 3D view cannot show the viewer all the sides of the bus, or its true length, width, or height. On the other hand, a multiview drawing (Figure 2.16) can provide the viewer with all the sides of a bus, represented in its true proportions: width, height, and depth. The six views representing the bus in Figure 2.16—the front, top, right side, left side, back, and bottom—are referred to in technical drawing terminology as the six ***regular views***.

2.15 Three Dimensional Image of a School Bus

JOB SKILL

Although a total of six views are possible using the multiview drawing technique, drafters only draw the views necessary to show clearly all the features of the object. Dimensional information for the features is added to these views. If no dimensions are placed on a view, the view is probably unnecessary.

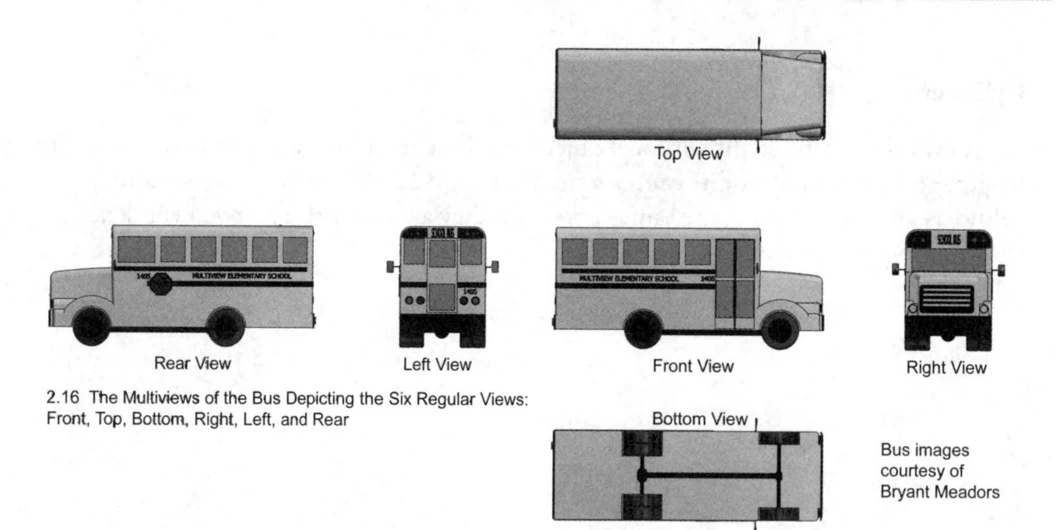

2.16 The Multiviews of the Bus Depicting the Six Regular Views: Front, Top, Bottom, Right, Left, and Rear

When the multiview drawing of the bus is created, the front, or principal, view is drawn first. The top, bottom, right, and left-side views are drawn by rotating the bus at 90° intervals relative to the front view. For example, the top view is created by rotating the front view 90° toward the top; the right-side view is created by rotating the front view 90° to the right.

2.5 VIEW SELECTION AND ALIGNMENT OF MULTIVIEW DRAWINGS

In Figure 2.16, the *side* view of the bus was chosen to be the front, or principal, view because it provides the viewer with the most information about the shape of the bus. The actual *front* of the bus is drawn as the right-side view. The top view is a bird's-eye view from directly above the front view. The left-side view shows the back of the bus, and the rear view of the bus is projected from this view. The bottom view is drawn directly below the front view. Features of the bus, such as the headlights, tires, and windows, are aligned in all views.

Because the front view is the principal view of the bus, it is drawn first. The other views are created by projecting from the geometry of the front view. For example, after the width of the front view has been measured and drawn, it can be projected to the top and bottom views, so the drafter doesn't need to remeasure the width in those views. Likewise, the height of the front view can be projected to the right and left-side views. Because drafters can avoid remeasuring features by projecting between views, multiview drawing is an efficient way to create technical drawings.

2.6 USING PROJECTION PLANES TO VISUALIZE MULTIVIEWS

If a house were placed inside a glass box as in Figure 2.17, the glass sides of the box would create ***projection planes*** (also referred to as *viewing planes*). If the 3D geometry of the front, side, and top of the house were projected onto the corresponding 2D projection plane, the resulting 2D image would represent a front, top, or side view as shown in Figures 2.18, 2.19, and 2.20, respectively. In Figure 2.18 the front of the house is shown as it would appear if projected onto a *frontal projection plane* that is placed between the viewer and the house. Likewise, in Figures 2.19 and 2.20, the top and right views are shown as they would appear if projected onto the *horizontal* and *profile projection planes* respectively.

In Figure 2.21, the front, top, left, and right-side views of the house are shown as they would be arranged in a multiview drawing. In the creation of this drawing, the front view was drawn first. Then, the top view was drawn directly above the front view. Next, the right and left views were drawn directly to the right and left of the front view, respectively. Architects refer to multiview drawings of the exterior of a house as ***elevation drawings***.

Notice that a feature in the front view, such as the peak of the roof, is exactly in line with the top of the roof in both the left and right views. Also note how the features of the chimney are aligned in each of the views.

The planes representing the roof in the right, left, and top views appear as rectangles in the multiviews, but by studying them in relation to the front view, you will see that they actually represent the sloping of the roof. Because the planes of the roof, as projected to the top and side viewing planes, are sloping, they are not drawn actual, or true size. In technical drawing, this phenomenon is referred to as ***foreshortening***.

QUICK TIP

When choosing the front, or principal, view of an object, select the view you would choose if you could show the viewer only one view to describe the object.

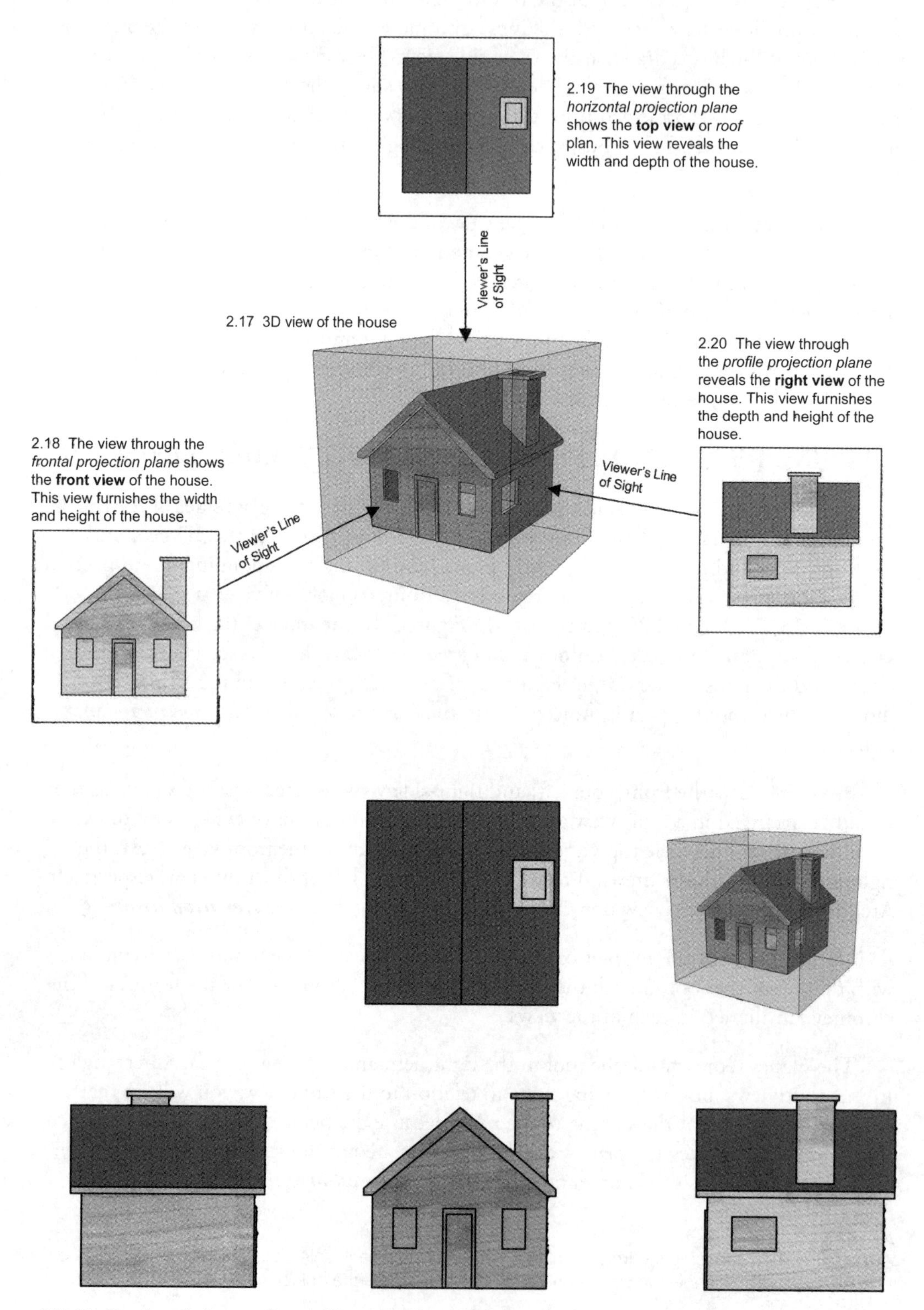

2.19 The view through the *horizontal projection plane* shows the **top view** or *roof* plan. This view reveals the width and depth of the house.

2.17 3D view of the house

2.20 The view through the *profile projection plane* reveals the **right view** of the house. This view furnishes the depth and height of the house.

2.18 The view through the *frontal projection plane* shows the **front view** of the house. This view furnishes the width and height of the house.

2.21 The Elevations of a house as they would be arranged in a Multiview Drawing.

2.7 PROJECTION PLANES

The projection plane technique of visualizing multiviews can also be applied to visualizing machine parts like the one shown in Figure 2.22. The arrows indicate the desired viewing position for the front, top and side views.

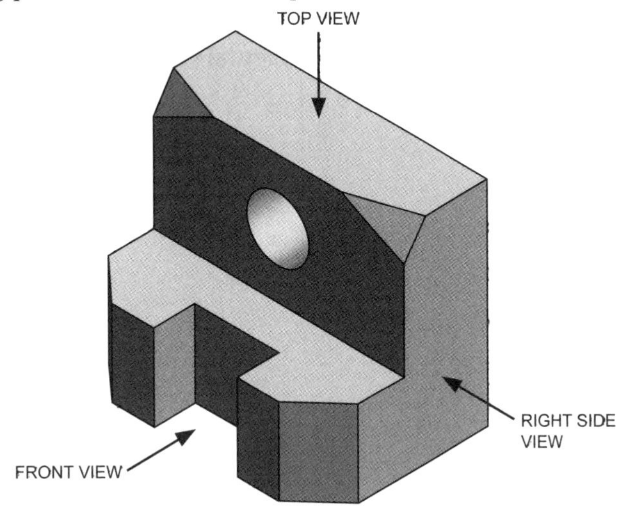

2.22 Machine Part with Desired Views Labeled

Visualizing the Front View

In Figure 2.23, a frontal projection plane has been placed between the viewer and the object and the ***front view*** of the object has been projected onto the projection plane. The viewer's line of sight is perpendicular to the projection plane in this example.

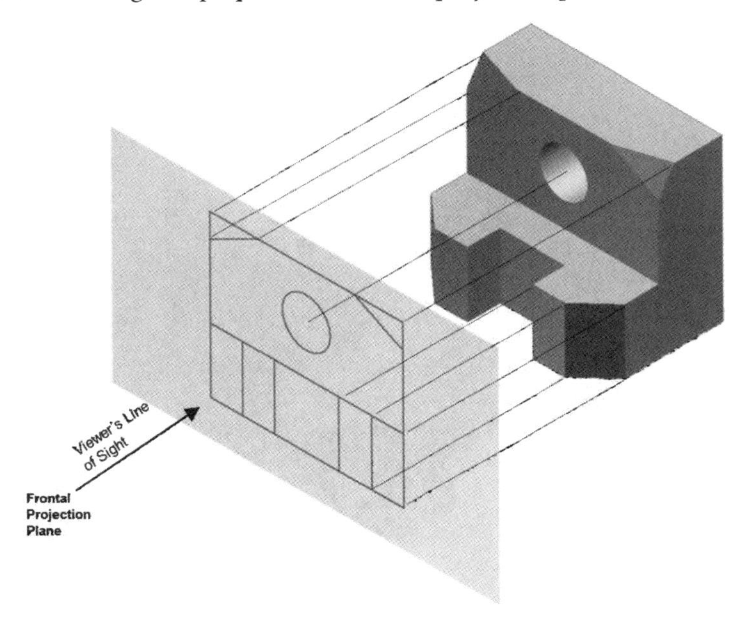

2.23 Projecting the Front View of an Object to a Frontal Projection Plane

Visualizing the Top View

In Figure 2.24, a horizontal projection plane has been placed between the viewer and the object and the ***top view*** of the object has been projected onto the projection plane. The viewer's line of sight is perpendicular to the projection plane in this example.

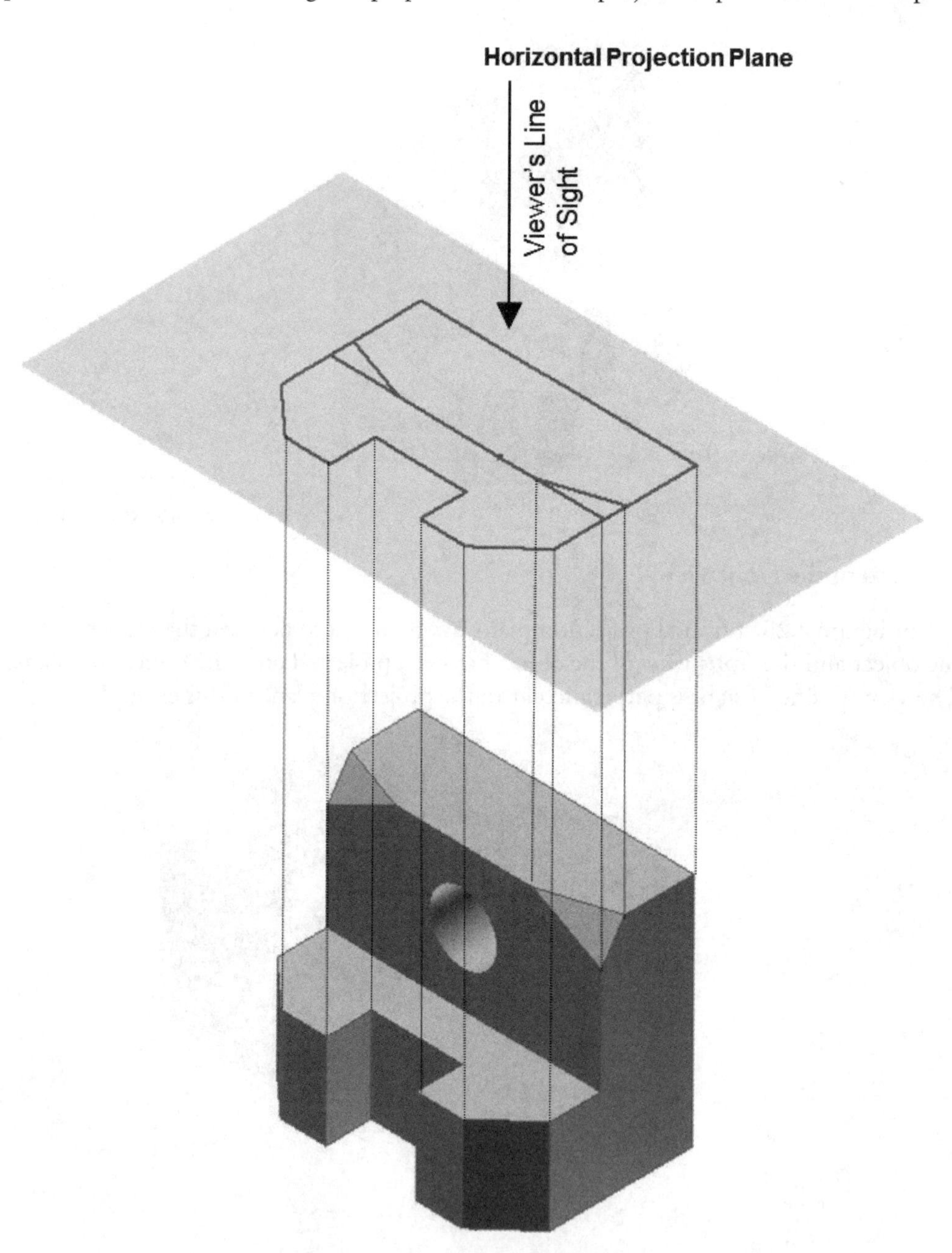

2.24 Projecting the Top View of an Object to a Horizontal Projection Plane

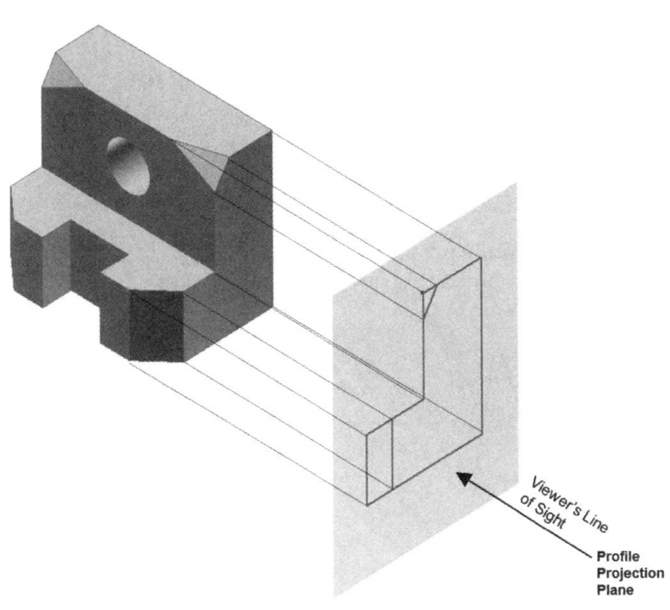

2.25 Projecting the Side View of an Object to a Profile Projection Plane

Visualizing the Side View

In Figure 2.25, a profile projection plane has been placed between the viewer and the object and the ***right side view*** of the object has been projected onto the projection plane. The viewer's line of sight is perpendicular to the projection plane in this example.

2.8 NORMAL, INCLINED, AND OBLIQUE SURFACES

Normal Surfaces

Normal surfaces are parallel to the projection plane. As a result, these surfaces appear ***true size*** and ***true shape*** on the projection plane that they are parallel with. The viewer's line of sight is perpendicular to the surface when the surface is normal. See Figure 2.26.

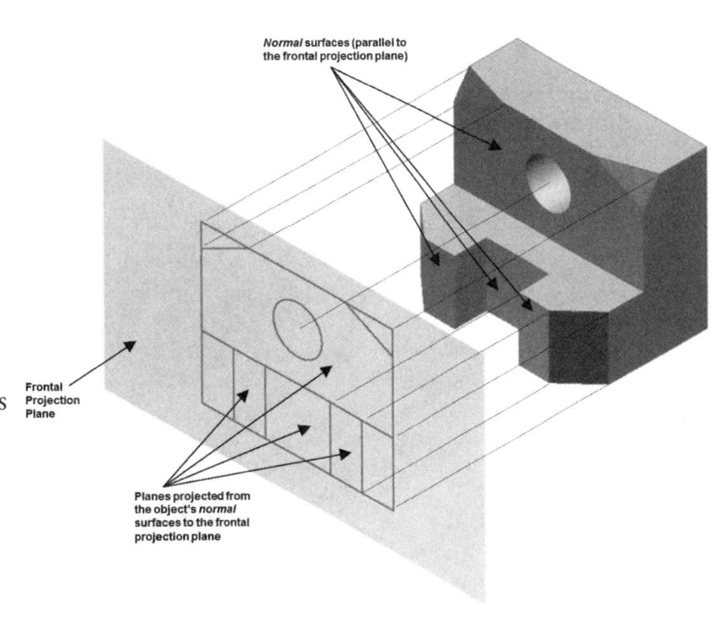

2.26 Projecting Normal Surfaces to a Frontal Projection Plane

Projecting Normal Surfaces to the Projection Plane

In Figure 2.27, the object has two surfaces that are parallel (normal) to the Horizontal Projection Plane. These surfaces appear true size and shape on the horizontal projection plane.

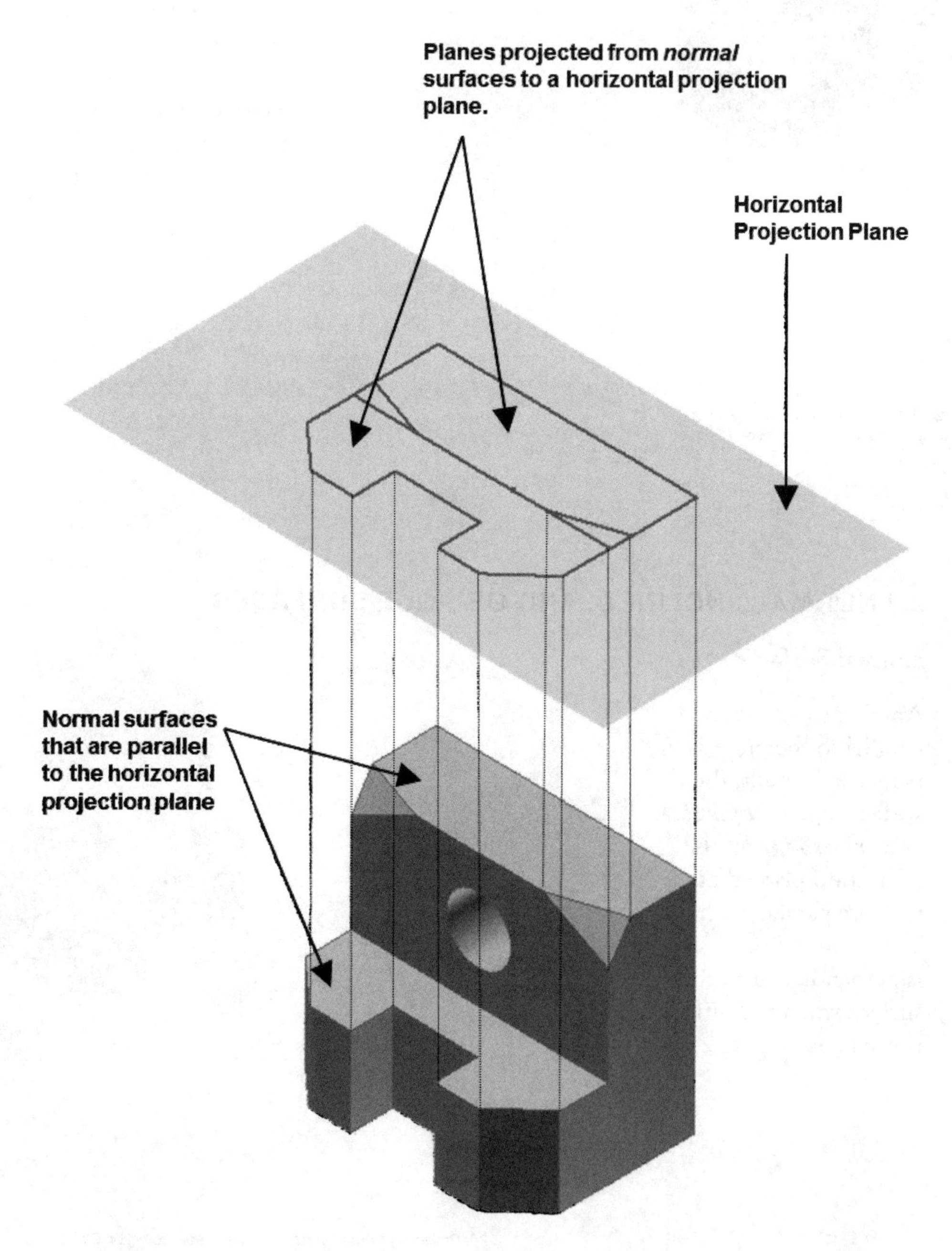

2.27 Projecting Normal Surfaces to a Horizontal Projection Plane

Inclined Surfaces

Inclined surfaces are slanted, or inclined, relative to the projection plane. As a result, these surfaces appear ***foreshortened*** on the projection plane and are not true size. See Figure 2.28.

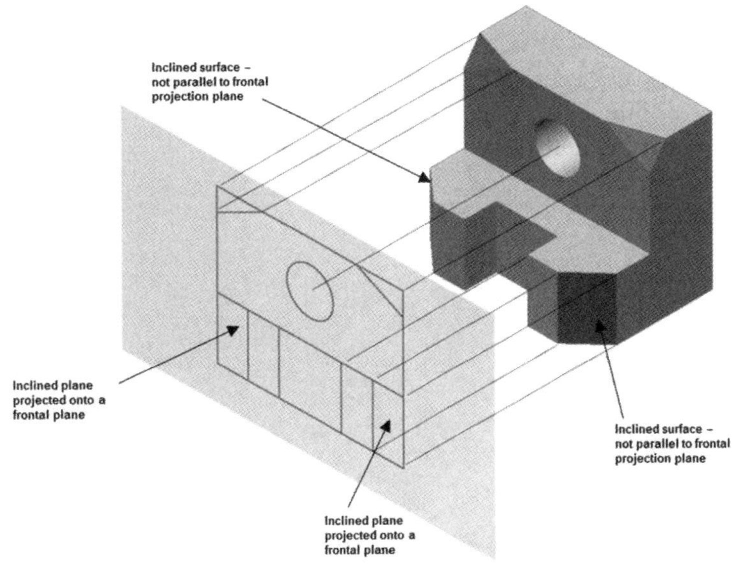

2.28 Projecting Inclined Surfaces to a Frontal Projection Plane

Projecting Inclined Surfaces to the Projection Plane

In Figure 2.29, the object's two inclined surfaces project to the top view as edges that are ***true length***.

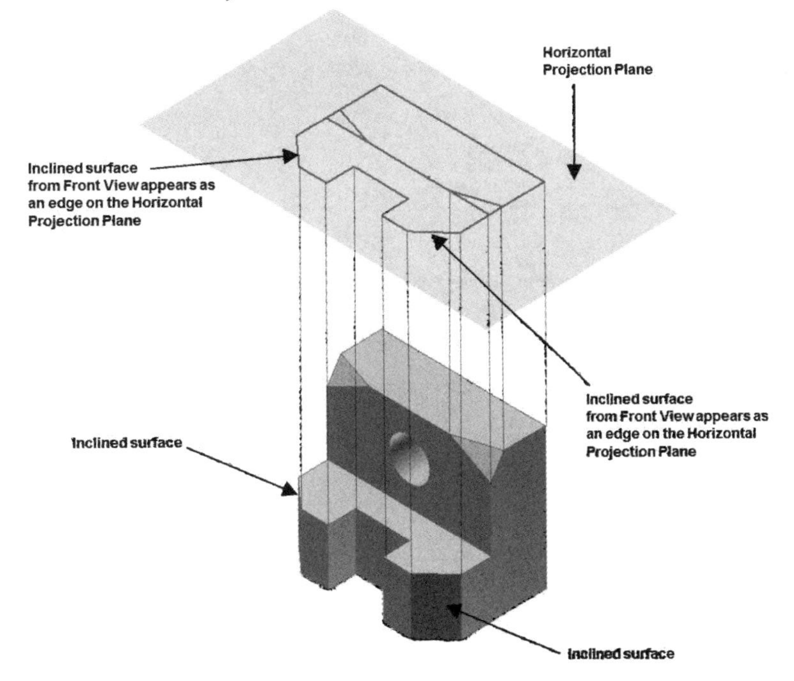

2.29 Projecting Inclined Surfaces to a Horizontal Projection Plane

Oblique Surfaces

Oblique surfaces are both ***inclined*** and ***rotated*** relative to the frontal, horizontal, and profile projection planes. As a result, these surfaces never appear true size or shape in any of these views. In Figure 2.30, the object's two oblique surfaces project to the front view as oblique planes that are not true size.

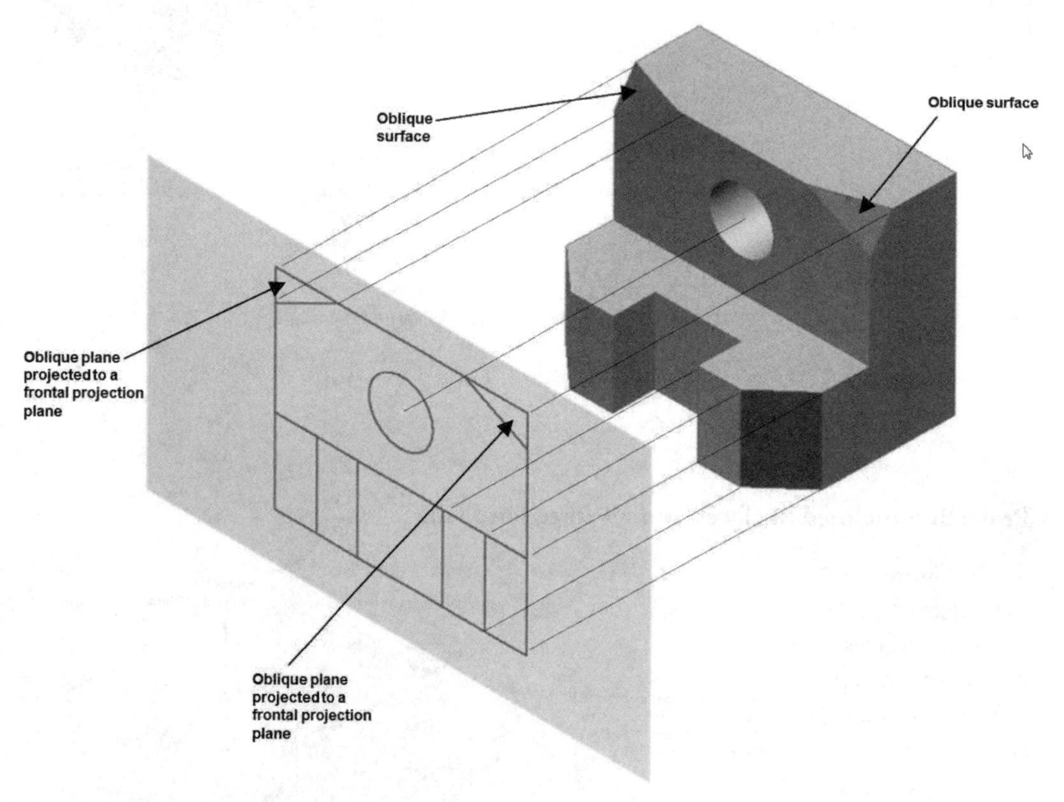

2.30 Projecting Oblique Surfaces to a Frontal Projection Plane

Projecting Oblique Surfaces to Projection Planes

In Figure 2.31, the object's two oblique surfaces project to the top view as oblique planes that are not true size.

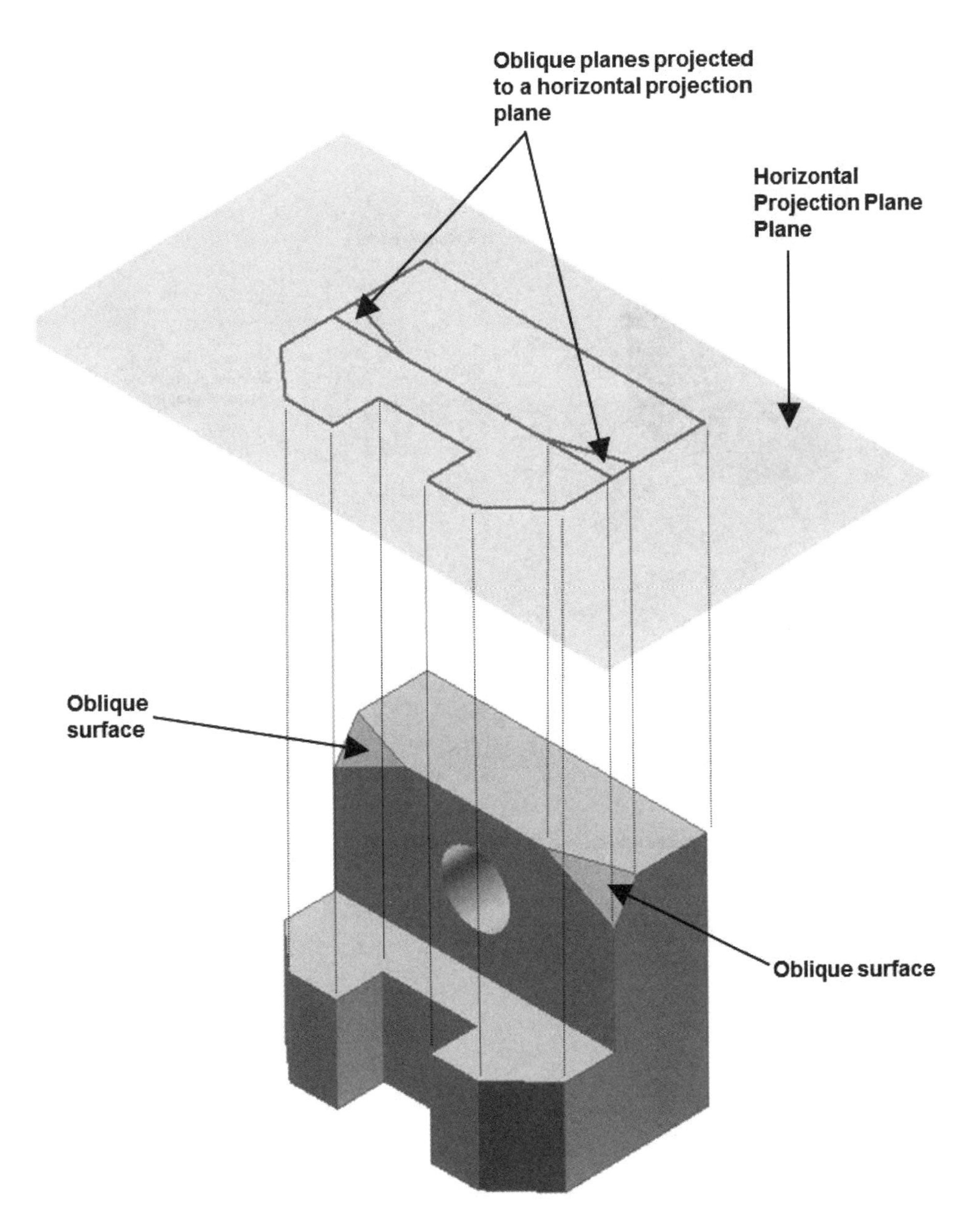

2.31 Projecting Oblique Surfaces to a Horizontal Projection Plane

In Figure 2.32, the object's normal, inclined, and oblique planes are projected to the profile projection plane.

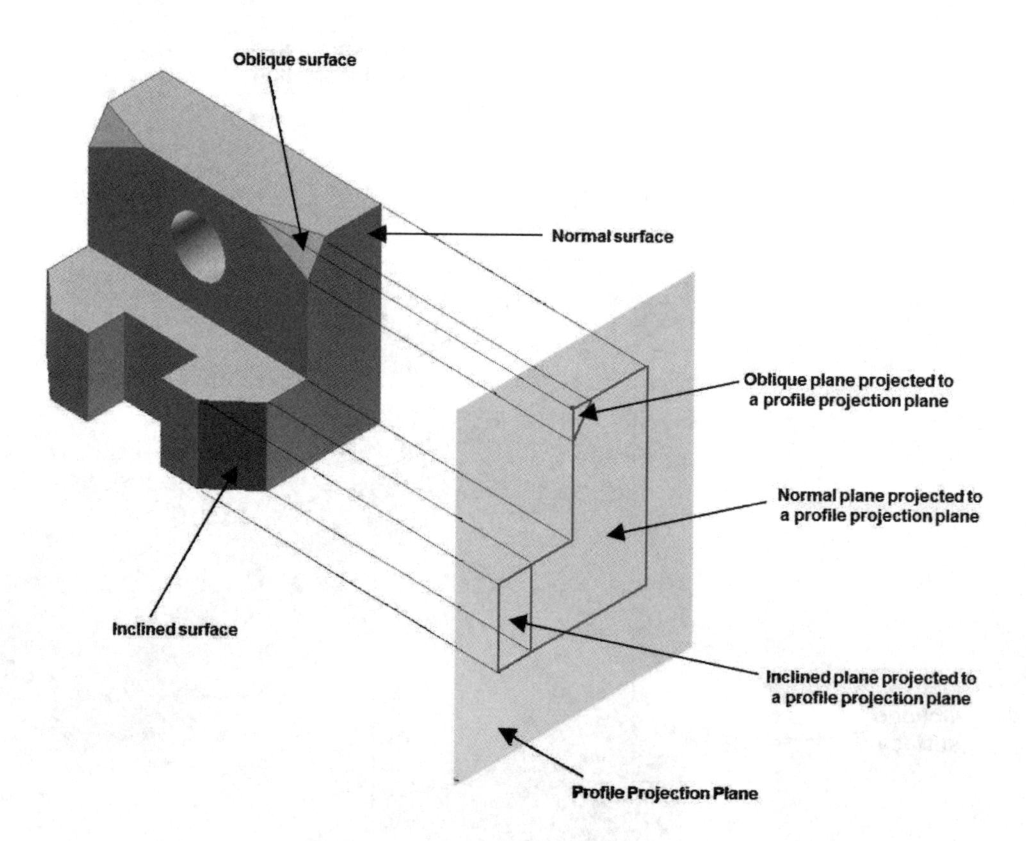

2.32 Normal, Inclined, and Oblique Surfaces Projected to a Profile Projection Plane

Visualizing the True Size and Shape of an Oblique Surface

In order to create a true shape view of an oblique surface, the surface must be projected onto a projection plane that is parallel to the oblique surface. When the viewer's line of sight is perpendicular to the projection plane, the surface will appear as true size and shape. See Figure 2.33. These types of drawings are known as *auxiliary views*. Creation of Auxiliary Views is presented in Appendix E.

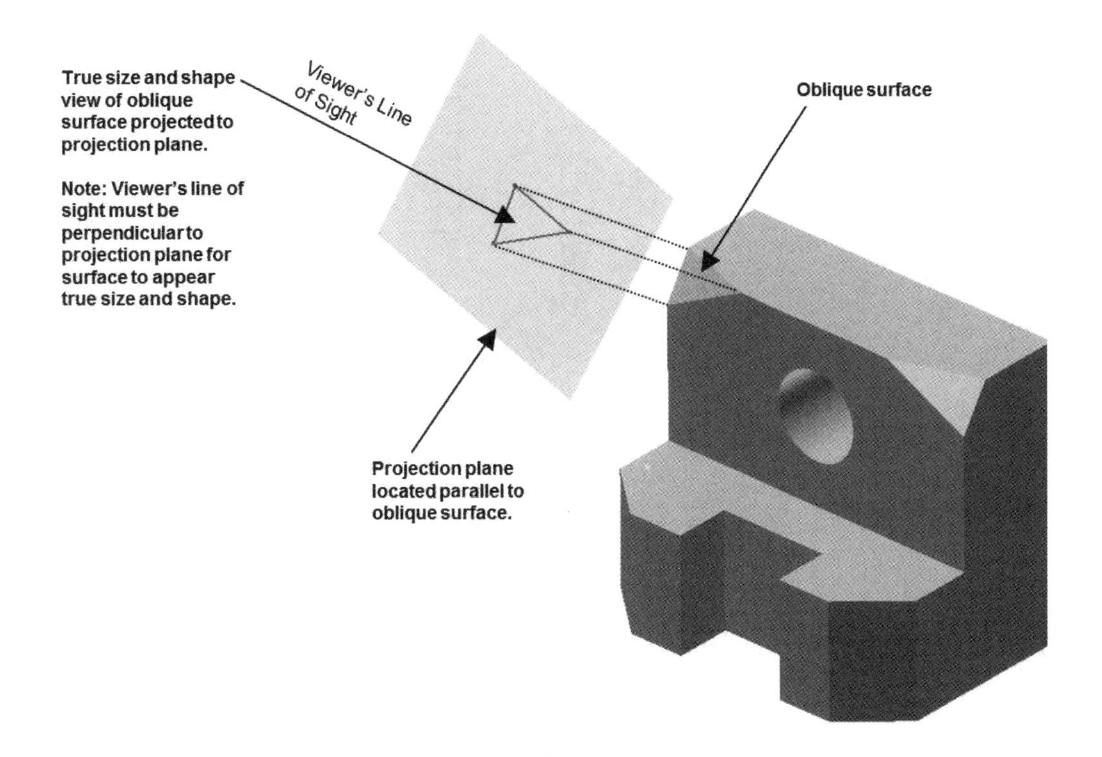

2.33 Oblique Surface Projected to a Parallel Projection Plane to Reveal Its True Size and Shape

Labeling Points and Vertices of Features to Aid in Visualizing the Multiviews of Objects

Sometimes it is helpful to number the corresponding point or vertex of a feature in its regular views to assist with visualizing the object. In Figure 2.34 a number has been assigned to each vertex and each plane has been labeled as normal, inclined or oblique.

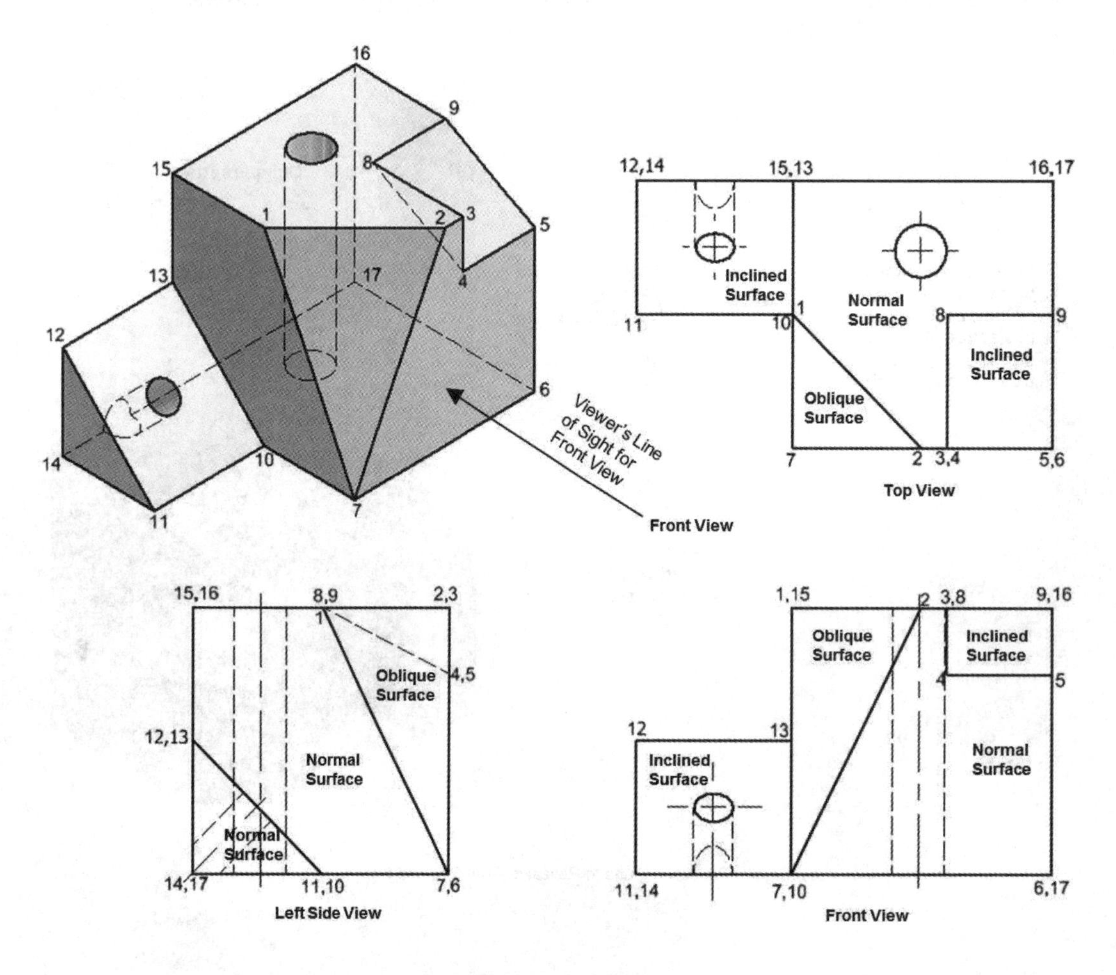

2.34 Labeling Points and Vertices of Features to Assist in Visualizing Views

STEP-BY-STEP

USING THE GLASS BOX TECHNIQUE OF VISUALIZING MULTIVIEWS

Figure 2.17 introduced the concept of placing an object inside a glass box to create viewing planes. This method of visualizing multiviews is known as the "Glass Box" technique. This technique is often helpful for beginners who are learning the process of visualizing an object's multiviews. The following steps detail the process of using the glass box technique to visualize the multiviews of the object in Figure 2.35.

Step 1. Imagine the object shown in Figure 2.35 is centered inside a glass box and its six regular views are projected out to the glass planes surrounding the object (see Figure 2.36). The plane that the front view is projected onto is called the *frontal* projection plane.

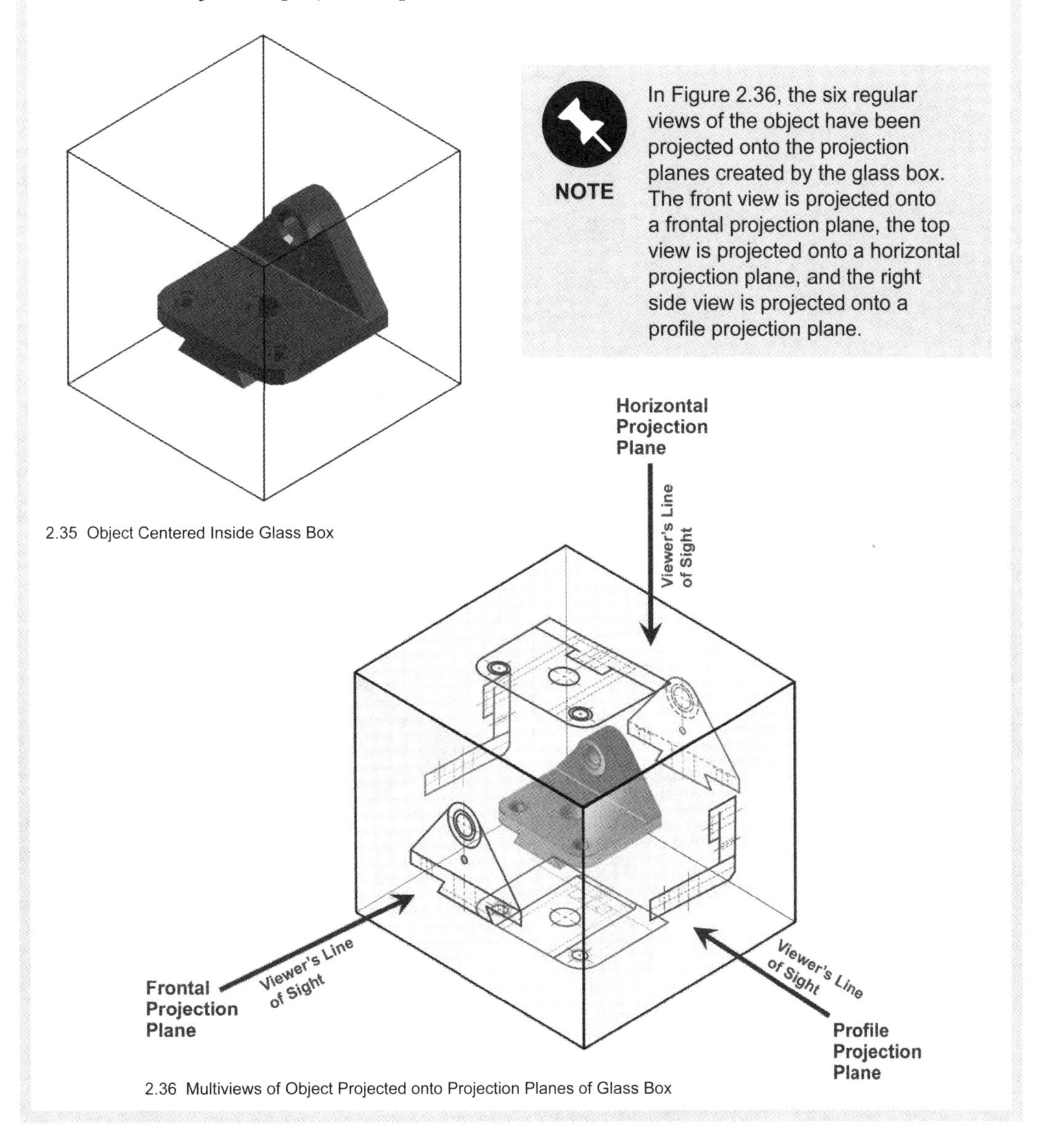

NOTE In Figure 2.36, the six regular views of the object have been projected onto the projection planes created by the glass box. The front view is projected onto a frontal projection plane, the top view is projected onto a horizontal projection plane, and the right side view is projected onto a profile projection plane.

2.35 Object Centered Inside Glass Box

2.36 Multiviews of Object Projected onto Projection Planes of Glass Box

STEP–BY–STEP

Step 2. Unfold the glass box as if the four sides of the frontal projection plane were hinged as shown in Figures 2.37 and 2.38.

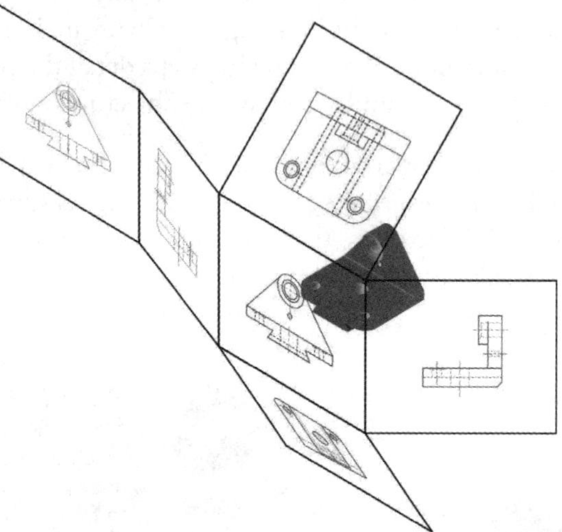

2.37 Unfolding the Glass Box

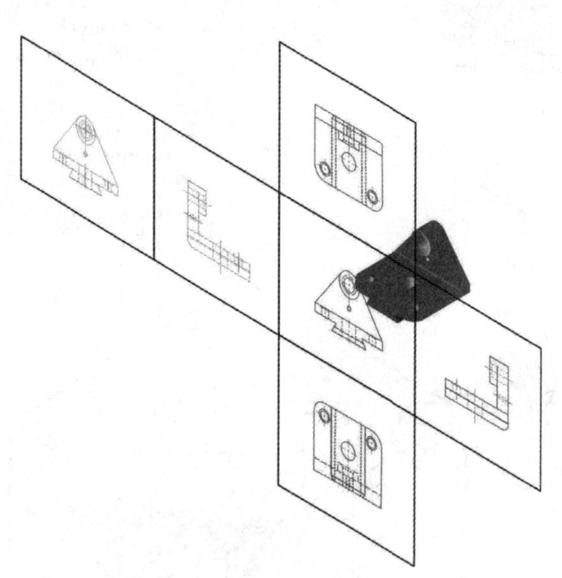

2.38 The Glass Box Unfolded

NOTE

When Step 2 is completed, the projection planes of the glass box will be coplanar as shown in Figure 2.38, revealing the six regular views of the object.

STEP-BY-STEP

Step 3. After the sides of the glass box are unfolded, the six regular multiviews of the object (front, top, bottom, right, left, and rear) are displayed in their "projected" positions as shown in Figure 2.39. Note that the front, right, left and rear views are aligned horizontally, and the front, top, and bottom views are in vertical alignment.

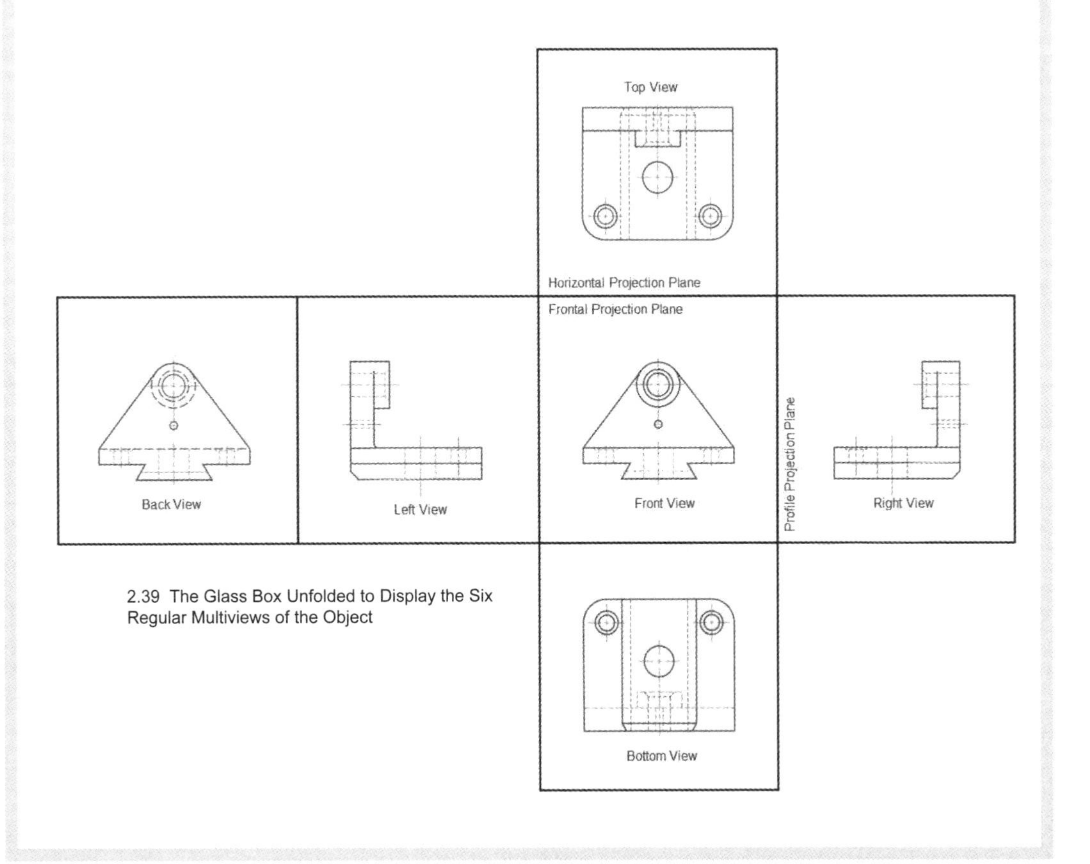

2.39 The Glass Box Unfolded to Display the Six Regular Multiviews of the Object

2.9 LINETYPES AND LINEWEIGHTS IN MULTIVIEW DRAWINGS

The features of an object are shown with differing ***linetypes*** and ***lineweights***. Commonly used linetypes include *visible lines*, which show the visible edges and features of an object; *hidden lines*, which represent features that would not be visible; and *centerlines*, which locate the centers of features such as holes and arcs. The terminology used for the various linetypes is shown in Figure 2.40.

Lineweight refers to the width of the lines in a technical drawing. Standard linetypes and lineweights have been established by the *American Society of Mechanical Engineers (ASME)*. The ASME standard for line conventions and lettering is *ASME Y14.2-2014*. The lineweights for lines specified by this standard for use on technical drawings are shown in Table 2.1.

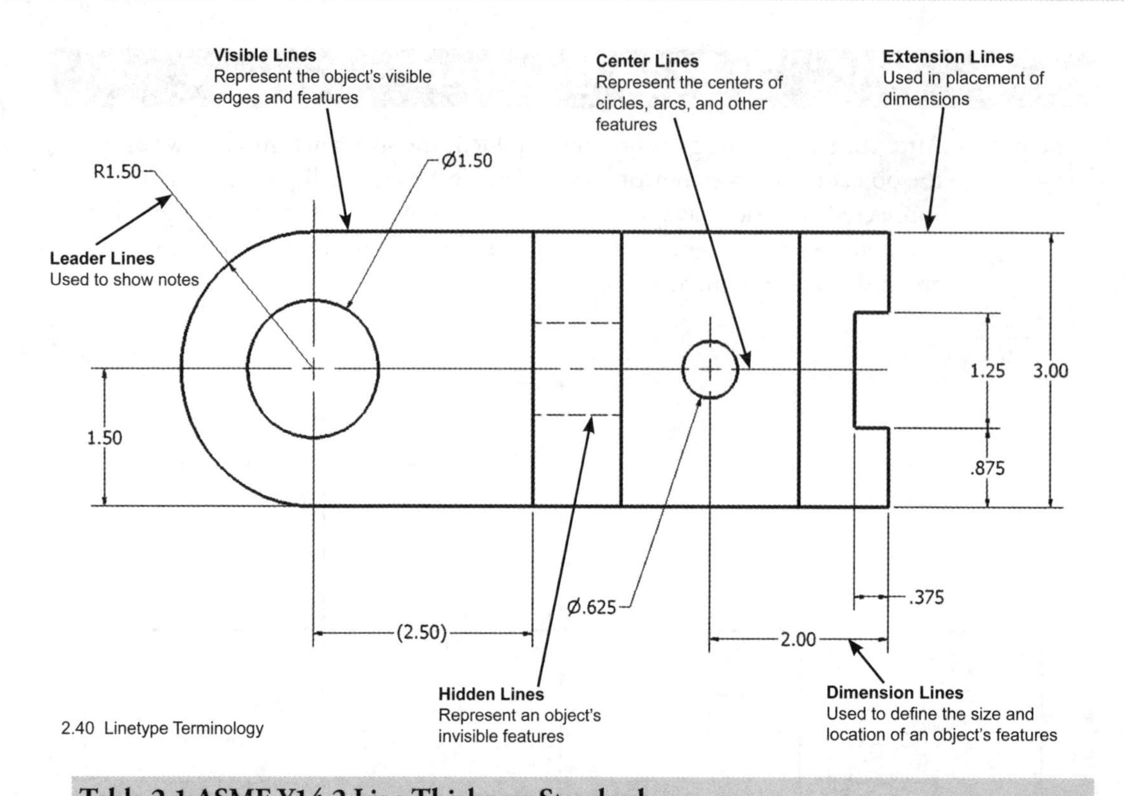

2.40 Linetype Terminology

Table 2.1 ASME Y14.2 Line Thickness Standard

Line Name	*ASME Line Thickness*	*Precedence*
Visible line	.6mm thick	Visible lines cover Hidden and Center lines
Hidden line	.3mm thick	Hidden lines cover Center lines
Centerline	.3mm thick	Center lines have the lowest precedence
Dimension line	.3mm thick	
Extension line	.3mm thick	
Cutting plane line	.6mm thick	
Section line	.3mm thick	

NOTE ASME describes these line thicknesses as the *approximate* widths.

2.10 HIDDEN FEATURES AND CENTERLINES IN MULTIVIEW DRAWINGS

Figure 2.42 shows the six regular views of the object shown in Figure 2.41. Study these examples and note how features that would otherwise be invisible in a view (such as the edges of the hole and slot in the side views) are depicted with hidden lines. Also, note the different ways that the centerlines representing the center of the hole are drawn in each view.

2.11 USE YOUR IMAGINATION!

As a drafter-in-training, you should develop the ability to use your imagination to visualize the multiviews of an object. Engineering and architecture are fields in which a powerful imagination is an important tool for success because most of the objects being designed exist first only in the imagination of the designer(s).

2.41 Object to Be Visualized with Multiviews

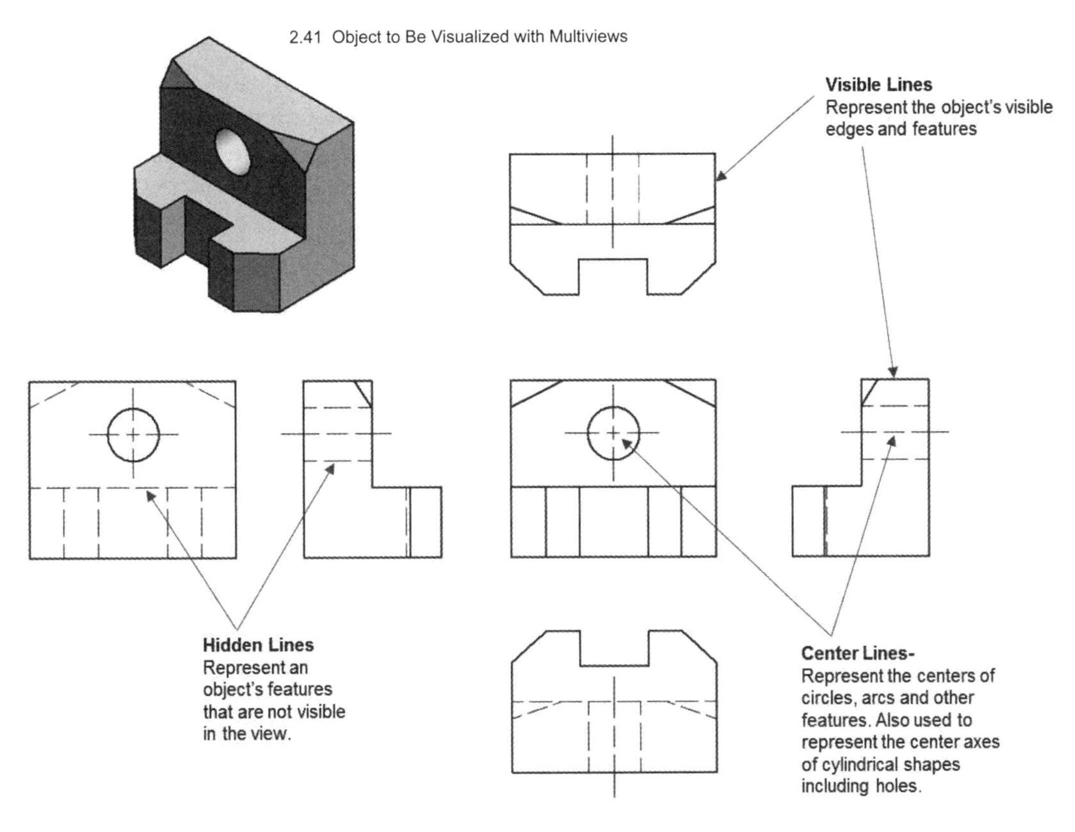

2.42 The Six Regular Views of the Object in Figure 2.14 with Visible, Hidden Lines and Centerlines Displayed

The challenge for the design team is to take the design from the imagination stage and turn it into a set of drawings that can be used to make the design a reality.

The following steps document the process of creating a multiview drawing for the object shown in the designer's sketch in Figure 2.43.

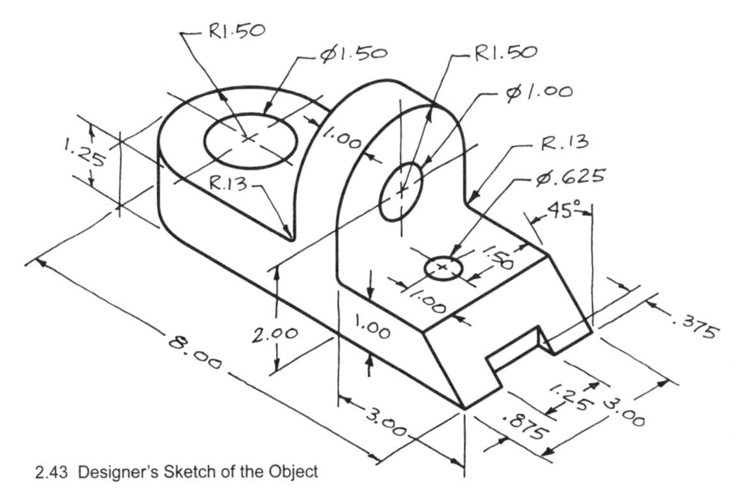

2.43 Designer's Sketch of the Object

MATERIAL-ALUMINUM 6061

STEP–BY–STEP

Step 1. The drafter studies the sketch of the object in Figure 2.43 and imagines it as a 3D object as in Figure 2.44. Next, the drafter determines the front, or principal, view of the object and imagines it positioned as shown in Figure 2.45.

- Then, the drafter rotates the object in his "mind's eye" toward the top (Figure 2.46) until the top, or "bird's eye" view, of the principal view has been rotated a full 90 degrees relative to the front view as shown in Figure 2.47.
- Next, the drafter imagines the front view of the object rotated toward the right (Figure 2.48) until the right-side view is rotated a full 90 degrees relative to the front view (Figure 2.49).
- This process could be likened to creating a 3D movie of the object in one's imagination to facilitate the visualization of the desired views. Note that because the top and right views (Figure 2.47 and Figure 2.49) are drawn at right angles (90 degrees) to the front view, the viewer's line of sight in these views is *perpendicular* to the front view (Figure 2.45).

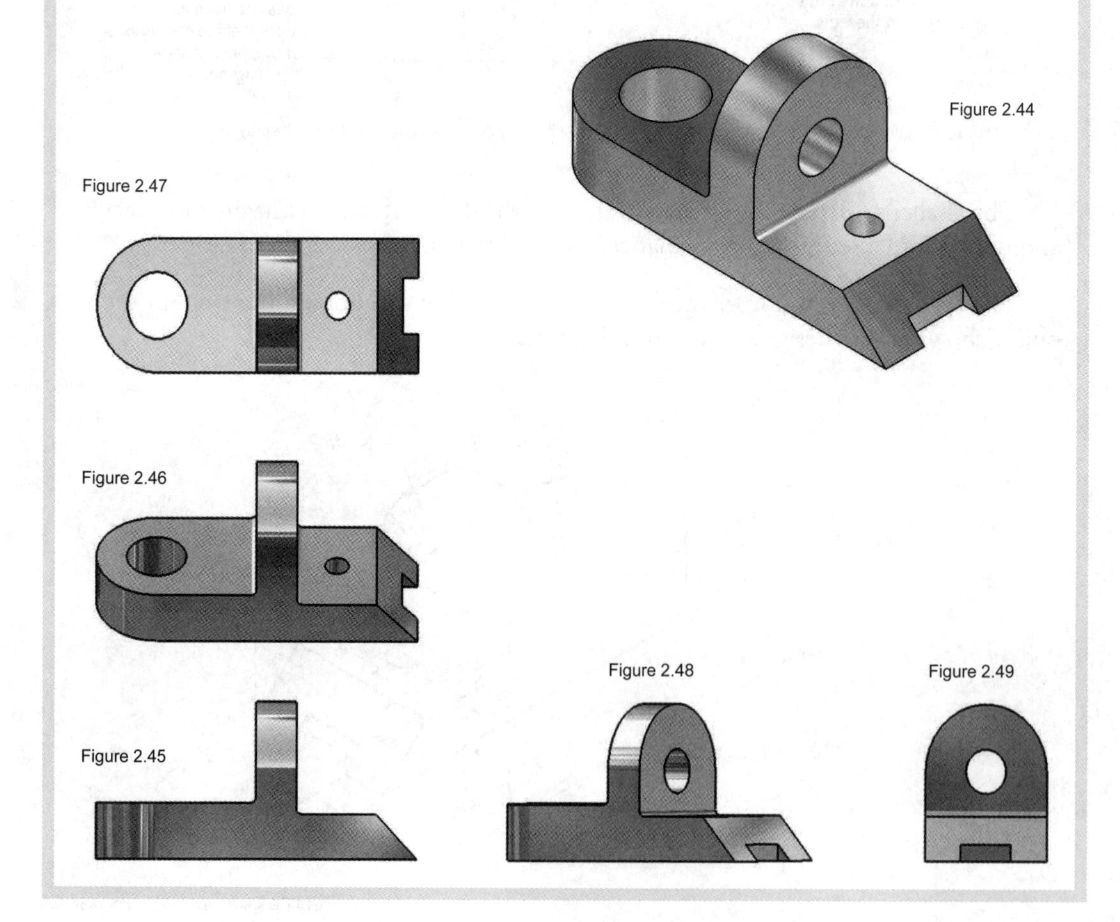

Figure 2.44

Figure 2.47

Figure 2.46

Figure 2.45

Figure 2.48

Figure 2.49

STEP–BY–STEP

Step 2. The drafter continues the process begun in Step 1, rotating the object until the bottom, left, and rear views of the object have been visualized as shown in Figure 2.50. All six views together make up what drafters refer to as the six ***regular*** views of the object.

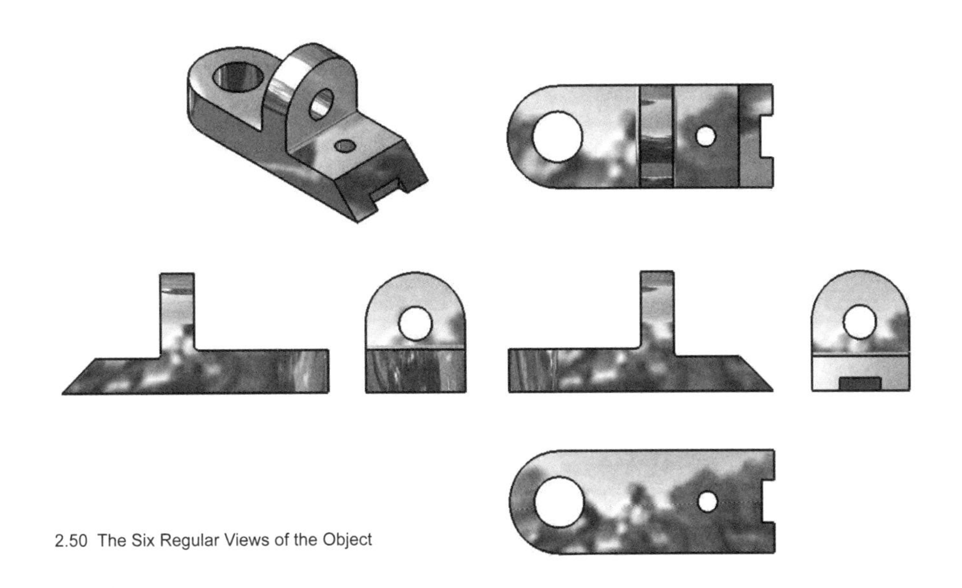

2.50 The Six Regular Views of the Object

Step 3. The drafter visualizes which lines in the regular views will represent the ***visible*** lines of the object (Figure 2.51).

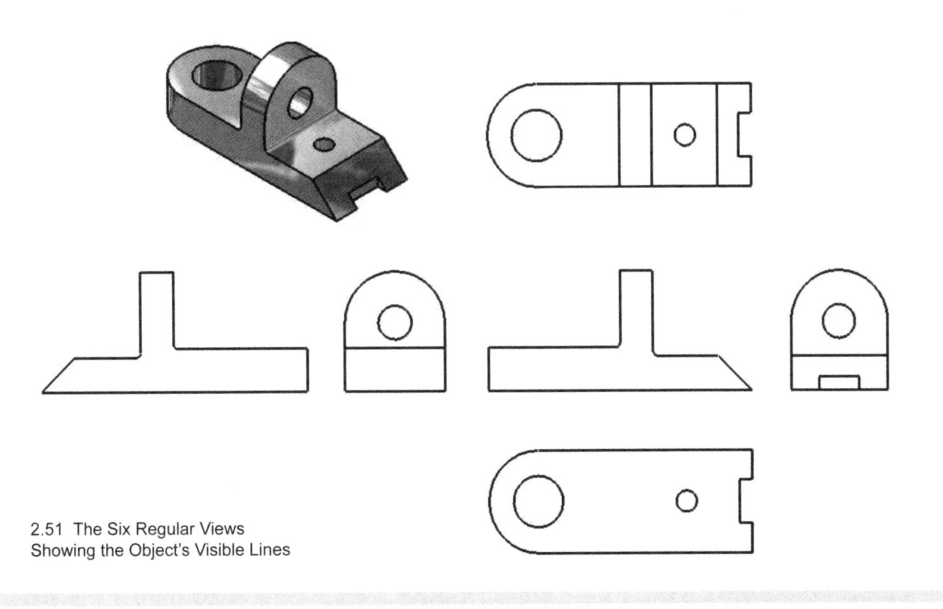

2.51 The Six Regular Views Showing the Object's Visible Lines

STEP-BY-STEP

Step 4. Next, the drafter visualizes the location of the object's hidden lines and centerlines (Figure 2.52).

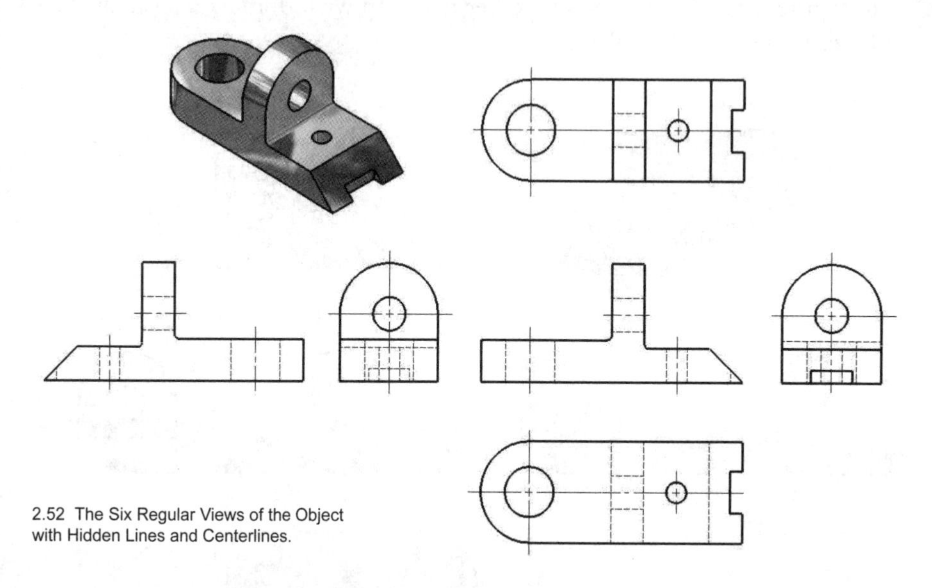

2.52 The Six Regular Views of the Object with Hidden Lines and Centerlines.

Step 5. In the last step, the drafter determines which of the six views are ***necessary*** in order to describe the object. Views portraying the object's features as visible lines or in their ***profile*** views are selected. Views that are redundant (in this case the left and rear views), or that portray the object's features as hidden lines (the bottom view), are usually not drawn. All of the dimensions necessary to manufacture the object will be placed on these views (Figure 2.53).

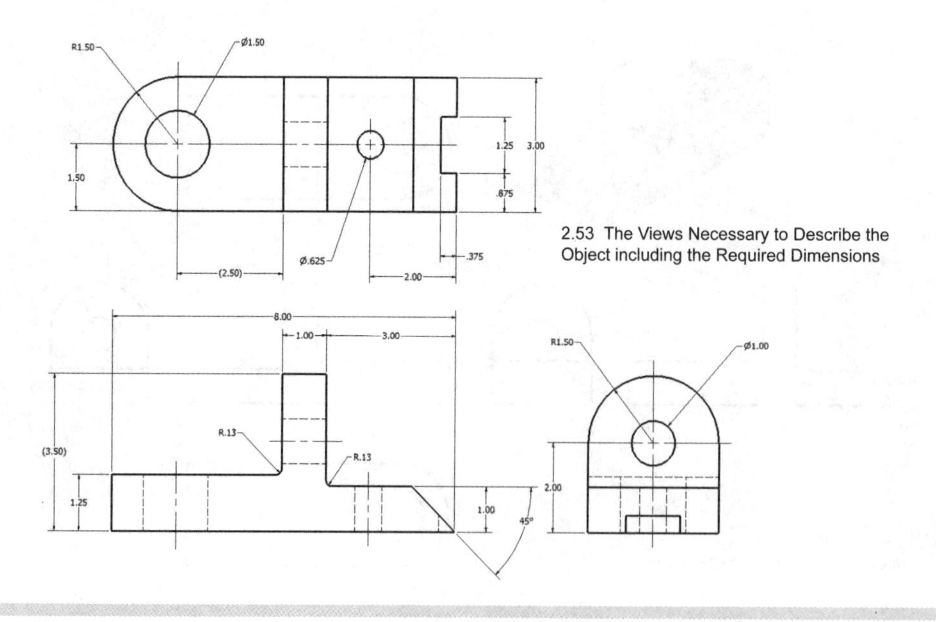

2.53 The Views Necessary to Describe the Object including the Required Dimensions

2.12 VISUALIZING THE MULTIVIEWS OF BASIC GEOMETRIC SHAPES

The multiview representations of some basic geometric shapes are shown in Figures 2.54 through 2.67. Shapes such as boxes (rectangular prisms), cylinders, cones, spheres, wedges, and prisms are often referred to as ***graphic primitives***, because by combining, or unioning, these shapes, or in some cases *subtracting* the geometry of one shape from another one, more complicated shapes can be formed. Graphic primitives can be considered as building blocks used to construct more complex objects. Students who learn to visualize the multiviews of graphic primitives will find it easier to visualize the multiviews of the more complicated shapes formed when they are combined.

Study Figures 2.54 through 2.67 and familiarize yourself with how the front, top, and side views of the graphic primitives and their combinations are drawn, including how hidden and centerlines are placed.

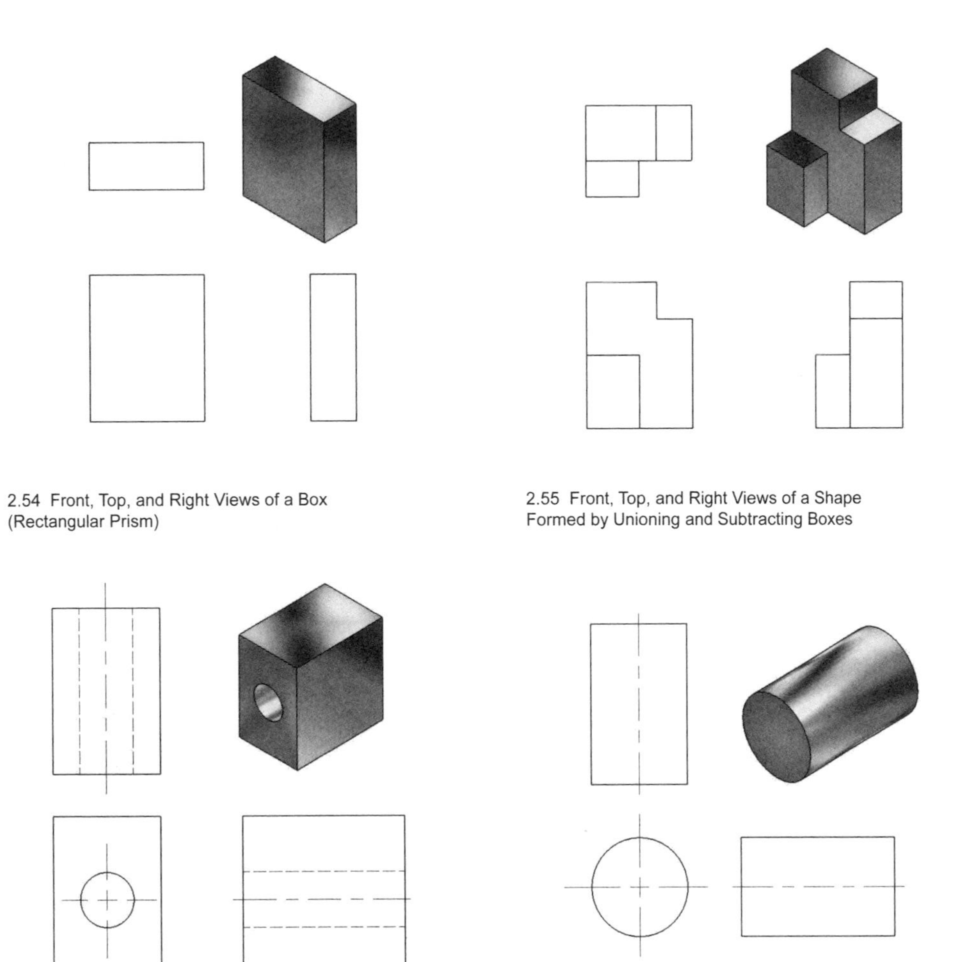

2.54 Front, Top, and Right Views of a Box (Rectangular Prism)

2.55 Front, Top, and Right Views of a Shape Formed by Unioning and Subtracting Boxes

2.56 Front, Top, and Right Views of a Shape formed by Subtracting a Cylinder from a Box

2.57 Front, Top, and Right Views of a Cylinder

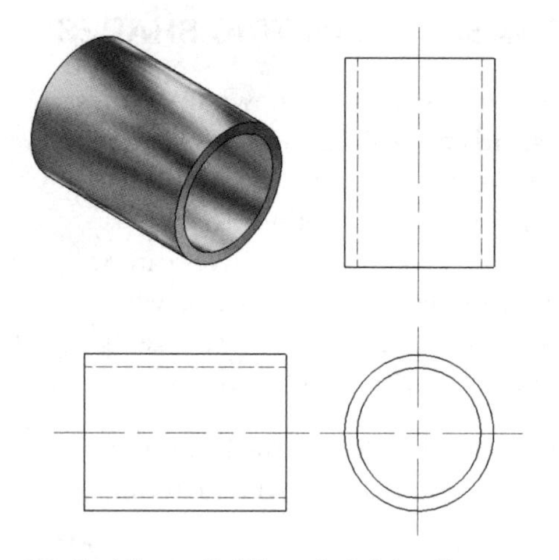

2.58 Front, Top, and Left Views of a Cylinder with a Smaller Cylinder Subtracted from Its Center

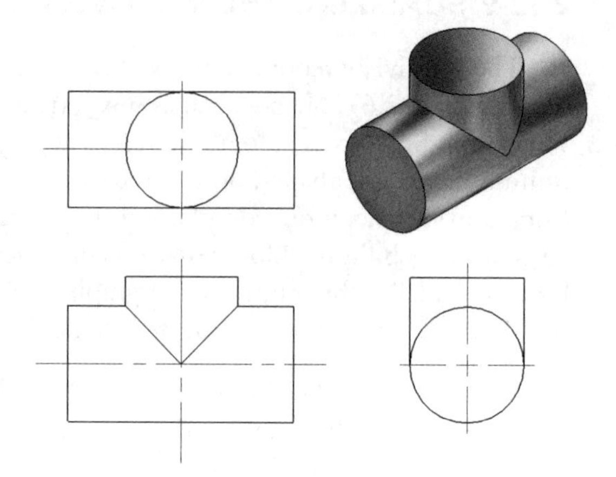

2.59 Front, Top, and Right Views of a Shape Formed by the Intersection of Two Cylinders of Equal Diameter

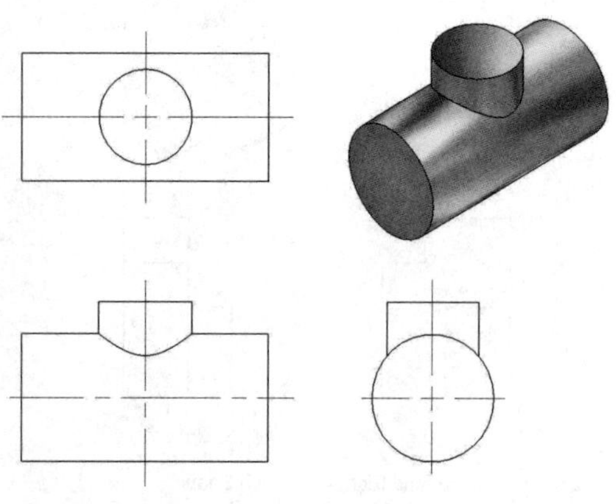

2.60 Front, Top, and Right Views of a Shape Resulting from the Intersection of Two Cylinders of Unequal Diameters

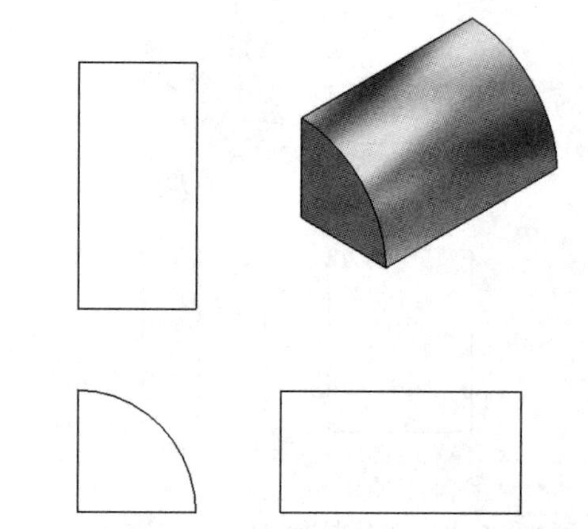

2.61 Front, Top, and Right Views of a Quarter-Round Shape

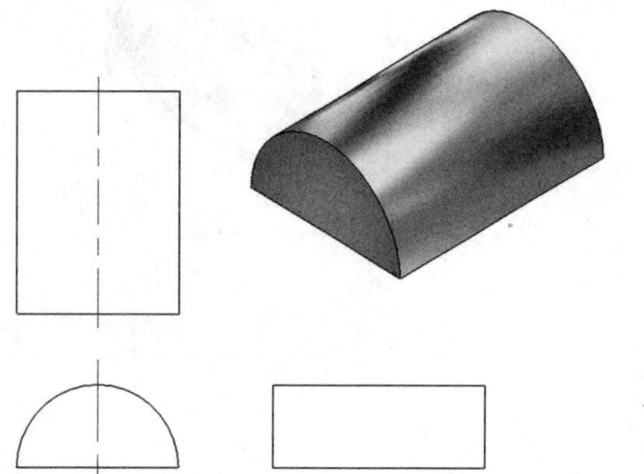

2.62 Front, Top, and Right Views of a Half-Round Shape

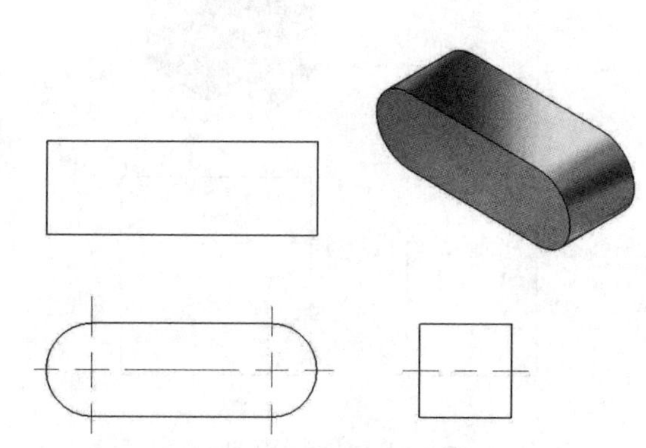

2.63 Front, Top, and Right Views of a Stadium Shape Resulting from the Union of a Box and Two Half-Rounds

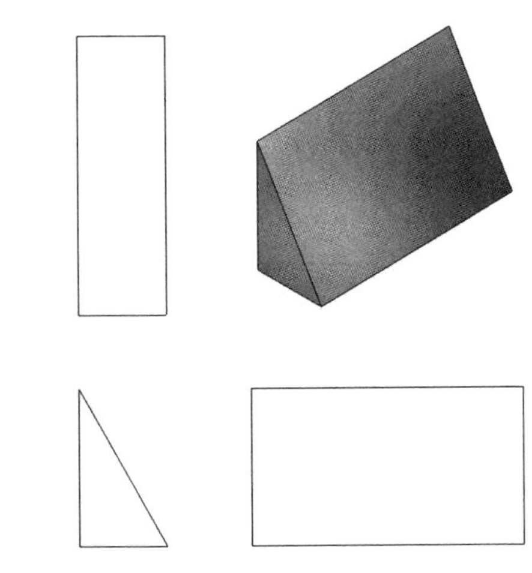

2.64 Front, Top, and Right Views of a Wedge

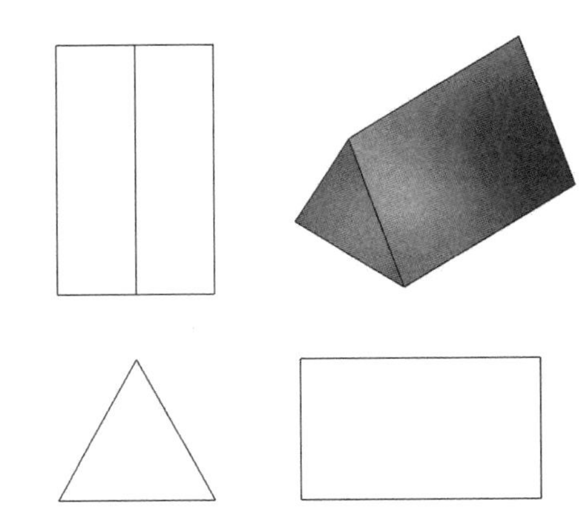

2.65 Front, Top, and Right Views of a Prism

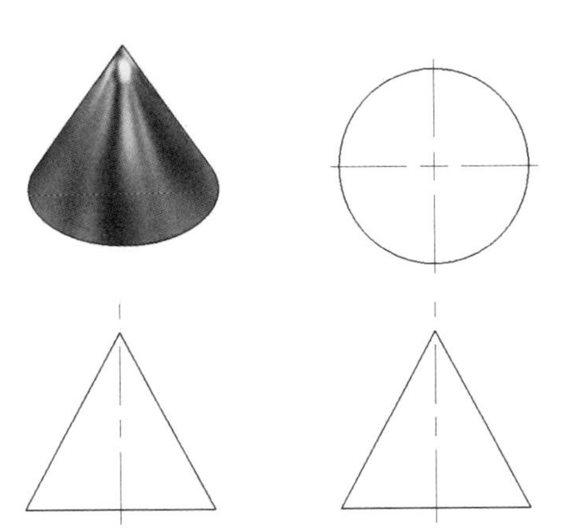

2.66 Front, Top, and Left Views of a Cone

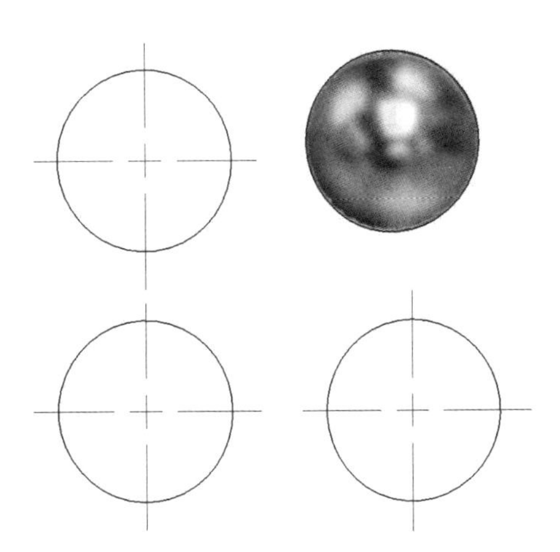

2.67 Front, Top, and Right Views of a Sphere

2.13 ORTHOGRAPHIC PROJECTION

Orthographic projection is the technique employed in the creation of multiview drawings to project the size and location of geometric features (points, lines, planes, or other features) from one view to another. Light construction lines are usually drawn between views to facilitate the transfer of this information. Projecting information in this manner is a more efficient way to construct technical drawings than by remeasuring the features in each view.

The orthographic projection technique also utilizes a ***miter line*** drawn at 45°, which allows geometric information to be projected between the top and side views.

Figure 2.68 shows an example of this technique. Dotted lines illustrate how the size and location of the object's features can be projected from view to view. Note how the 45° miter line is used to facilitate the transfer of information between the top and side views.

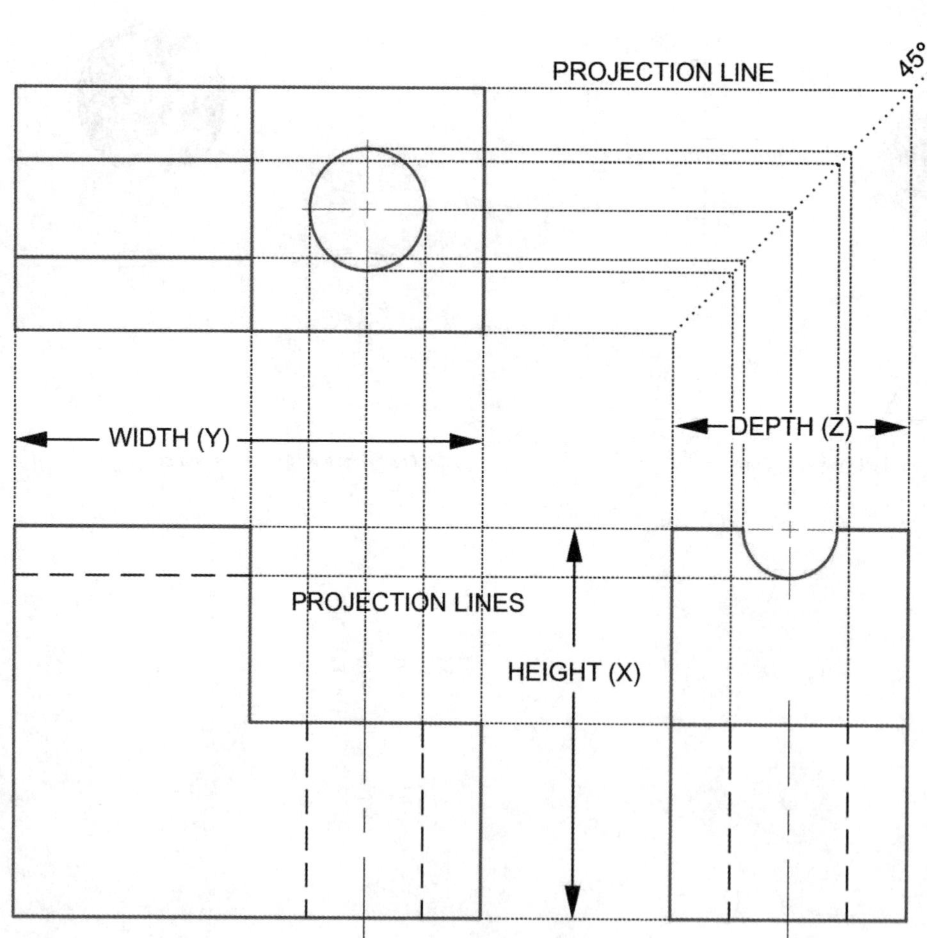

2.68 Multiview Drawing of an Object Using Orthographic Projection Techniques

STEP-BY-STEP

Step 1. Study the sketch of the part shown in Figure 2.69 and try to imagine it as a 3D object. With this image in mind, visualize the front, top, and right-side views of the object.

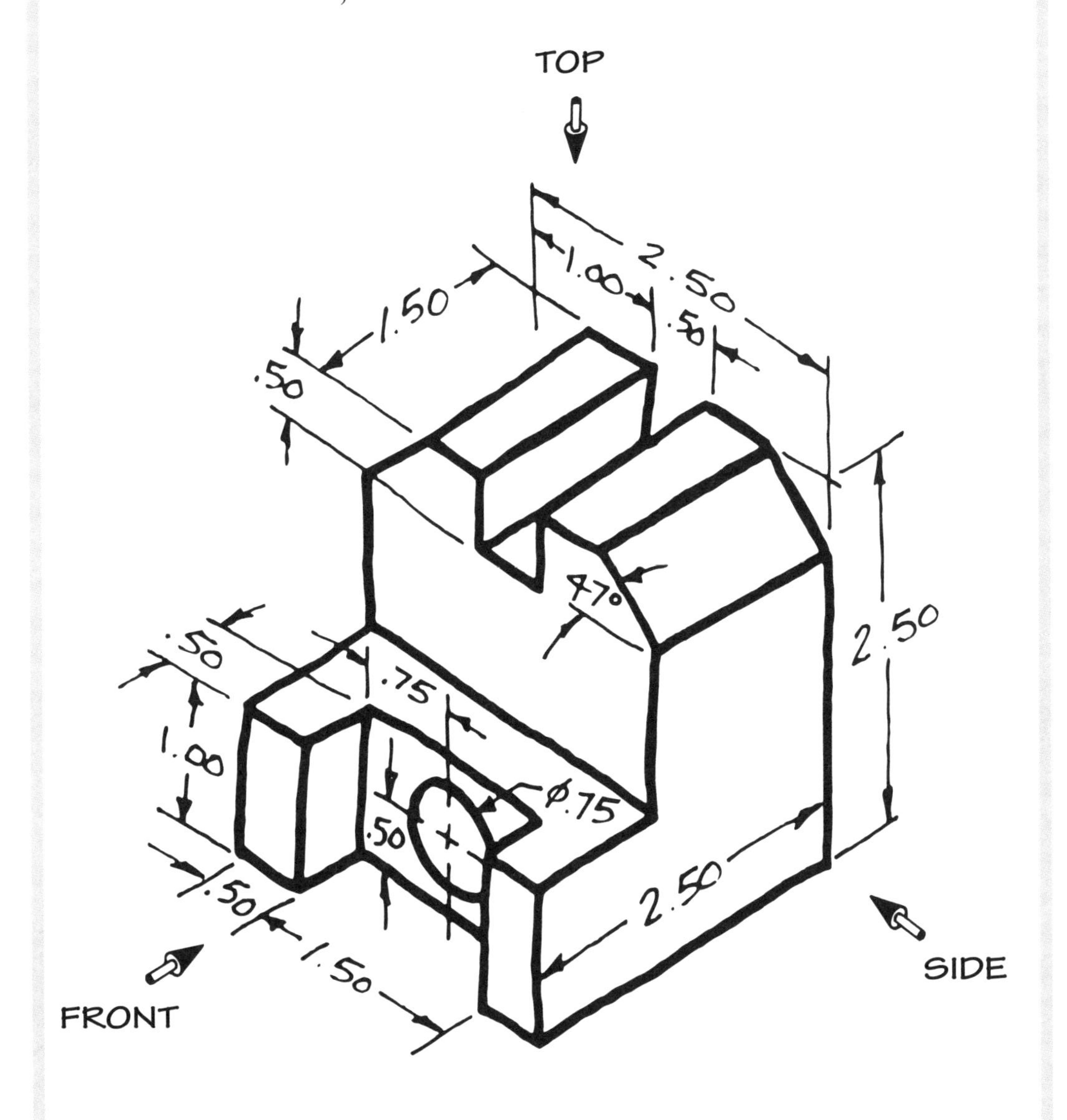

2.69 Sketch of Object to Be Drawn as a Multiview Drawing

JOB SKILL

The American Society of Mechanical Engineers Standard for creating multiview drawings is ASME Y14.3-2012.

STEP–BY–STEP

Step 2. Sketch the front view of the object. Try to sketch the part proportionally to the dimensions specified on the sketch. Extend light construction lines out from the features of the front view to the top and right sides and place a 45° miter line as shown in Figure 2.70.

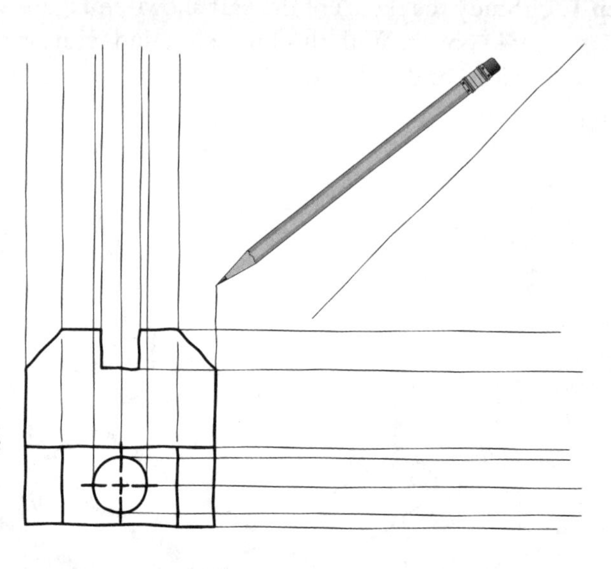

2.70 Extending Construction Lines from the Features of the Front View to the Side and Top

Step 3. Sketch the top and right-side views of the object as shown in Figure 2.71. Use the construction lines projected from the front view, and construction lines projected through the miter line, to locate the features of each view. Darken the visible, hidden, and centerlines as needed. Erase construction lines that appear too dark.

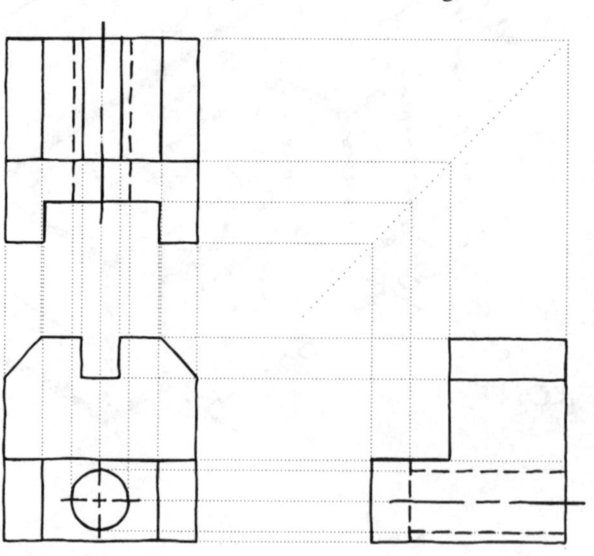

2.71 Sketching the Top and Right-Side Views

2.14 DRAWING OBJECTS TO SCALE

In technical drawings, objects are often drawn to scale. This term refers to the relationship between the size of the object in the drawing and the actual size of the object after it is manufactured. Following are four of the scales most commonly used in the creation of mechanical drawings:

- **Full Scale** This means that the size of the object in the drawing will be the same size as the object after it is manufactured. This is usually only feasible on smaller objects such as machine parts (to draw an average-size house at full scale, you might need a sheet of paper that is 136' long by 88' wide). When noting on a drawing that the object is drawn full scale, the drafter may write 1 = 1, 1/1, or 1:1.
- **Half Scale** This means that the size of the object in the drawing is half the size of the object after it is manufactured. The drafter will still place the full-size dimensions on the views of the object so that even though the drawing is half size, the part will be manufactured full size. When noting on a drawing that the object is drawn half scale, the drafter may write 1 = 2, 1/2, .5X, or 1:2.
- **Quarter Scale** This means that the size of the object in the drawing is one fourth the size of the object after it is manufactured. The drafter will still place the full-size dimensions on the views of the object so that even though the drawing is one-fourth size, the part will be manufactured full size. When noting on a drawing that the object is drawn quarter scale, the drafter may write 1 = 4, 1/4, .25X, or 1:4.
- **Double Scale** This means that the size of the object in the drawing is twice the size of the object after it is manufactured. The drafter will still place the full-size dimensions on the views of the object so that even though the drawing is twice the size, the part will be manufactured full size. This scale is used for smaller objects that would be difficult to dimension if drawn at actual size. When noting on a drawing that the object is drawn double scale, the drafter may write 2 = 1, 2/1, 2X, or 2:1.

2.15 DRAWING ARCHITECTURAL PLANS TO SCALE

Following are two of the scales most commonly used in the creation of architectural drawings:

- **Quarter Inch Equals One Foot** This means that every 1/4" on the plotted drawing represents a measurement of 1' on the actual construction project. For example, a wall that is to be built 16' in length will measure 4" on the drawing. This allows a drafter to fit a house that is 100' long and 50' wide on a sheet of paper measuring only 34" by 24". The 100' distance will measure only 25" on the drawing sheet (100 X 1/4"= 25"), and 50' will measure 12.5" on the sheet (50 X 1/4"= 12.5"). The dimensions on the drawing will be labeled at the actual distance (in feet and inches) required to construct the building full size. When noting on a drawing that the object is drawn to this scale, the drafter would write 1/4"= 1'-0".
- **Eighth Inch Equals One Foot** This means that every 1/8" on the plotted drawing will represent a measurement of 1' on the actual construction project. For example, a wall that is to be built 16' in length will measure 2" on the drawing. This allows a drafter to fit a house that is 200' long and 100' wide on a sheet of paper measuring only 34" by 24". The distance of 200' will measure only 25" on the drawing sheet (200 X 1/8" = 25"), and 100' will measure 12.5" on the sheet (100 X 1/8"= 12.5"). The dimensions on the drawing will be labeled at the actual distance (in feet and inches) required to construct the building full size. When noting on a drawing that the object is drawn to this scale, the drafter would write 1/8"= 1'-0".

2.16 DRAWING SHEET SIZES

ASME and other standards organizations have defined standardized ***sheet sizes*** for the preparation of technical drawings. These sheet sizes vary depending on the type of drawing and/or the unit of measurement used to create the drawing.

The ASME standard for decimal sheet sizes is *ASME Y14.1-2012*. The sheet sizes defined in this standard begin with an A sheet, which is 11" X 8.5". A B sheet's dimensions are 17" X 11", which is the equivalent of two A sheets laid side by side. A C sheet is 22" X 17", which is the equivalent of two B sheets laid side by side. A D sheet is 34" X 22", which is the equivalent of two C sheets laid side by side. Figure 2.72 illustrates the sheet sizes used in mechanical drawings.

The ASME standard for metric sheet sizes is *ASME Y14.1M-2012*. In this standard, an A4 sheet measures 297 X 210 millimeters (mm), an A3 sheet measures 420 mm X 297 mm, an A2 sheet measures 594 mm X 420 mm, an A1 sheet measures 841 mm X 594 mm, and an A0 sheet measures 1189 mm X 841 mm.

For architectural drawings in which inches are used as the unit of measurement, an A sheet measures 12" X 9", a B sheet is 18" X 12", a C sheet is 24" X 18", and a D sheet is 36" X 24".

A high-quality paper known as vellum, or tracing paper, is used to plot drawings that are made to be reproduced by blueprinting. Vellum is a strong, thin paper that allows light to pass through it relatively easily. Vellum can be purchased in rolls 24" to 36" in width or in standard sheet sizes. Vellum can also be purchased with preprinted title blocks.

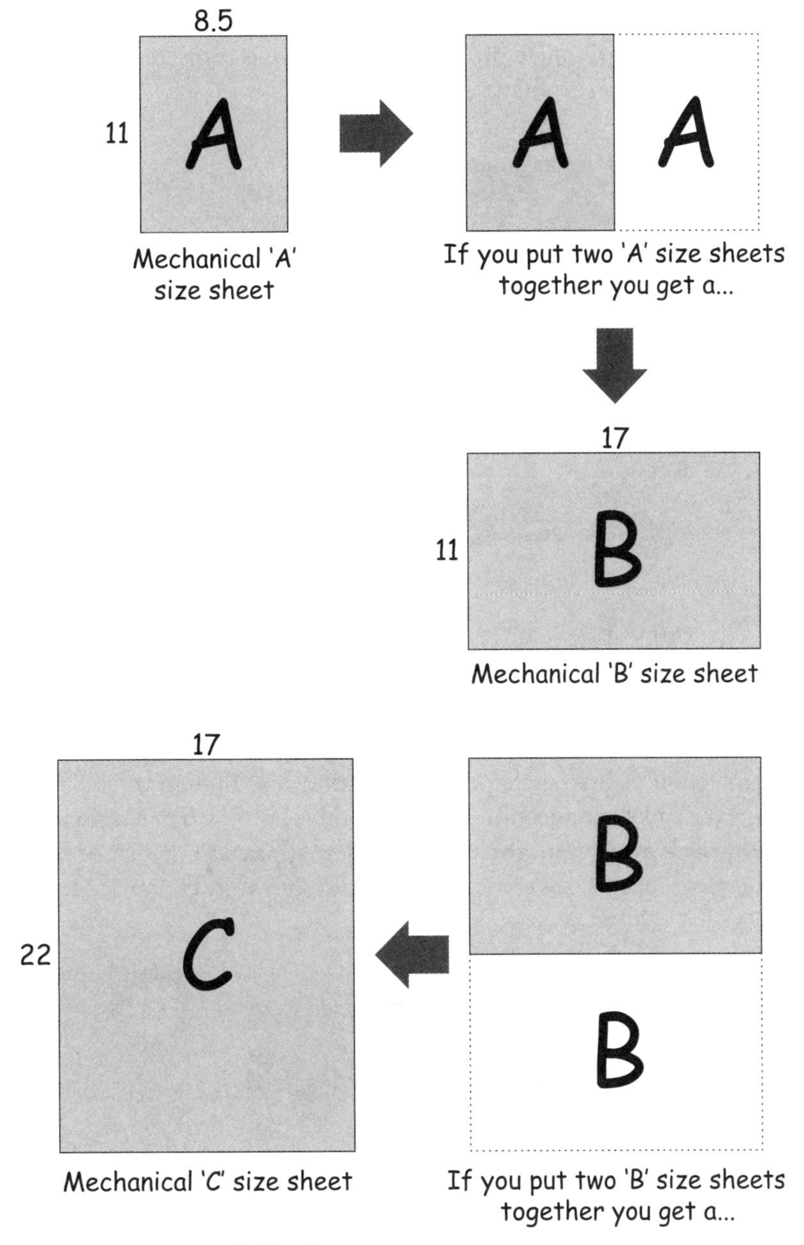

2.72 Sheet Sizes for a Mechanical Drawing

2.17 THIRD-ANGLE PROJECTION VERSUS FIRST-ANGLE PROJECTION

The method of arranging multiviews shown in Figure 2.73, with the top view drawn above the front view and the right-side view drawn to the right of the front view, is called ***third-angle projection***. This method is widely used in technical drawings created in the United States. In a third-angle projection, the image is projected onto a viewing plane that is located between the object and the viewer.

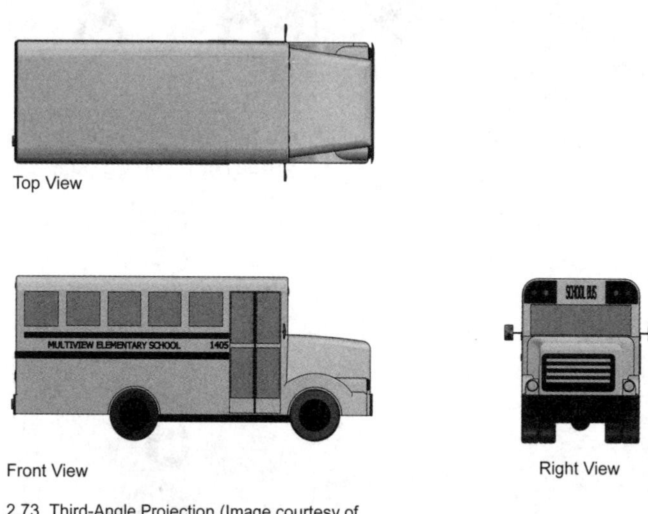

2.73 Third-Angle Projection (Image courtesy of Bryant Meadors)

In many parts of the world, multiviews are arranged using first-angle projection instead of third-angle projection. A first-angle projection is drawn as though the object is between the observer and the projection plane. For this reason, when a drawing is created with ***first-angle projection***, the right-side view appears to the *left* of the front view, and the top view appears *below* the front view, as shown in Figure 2.74.

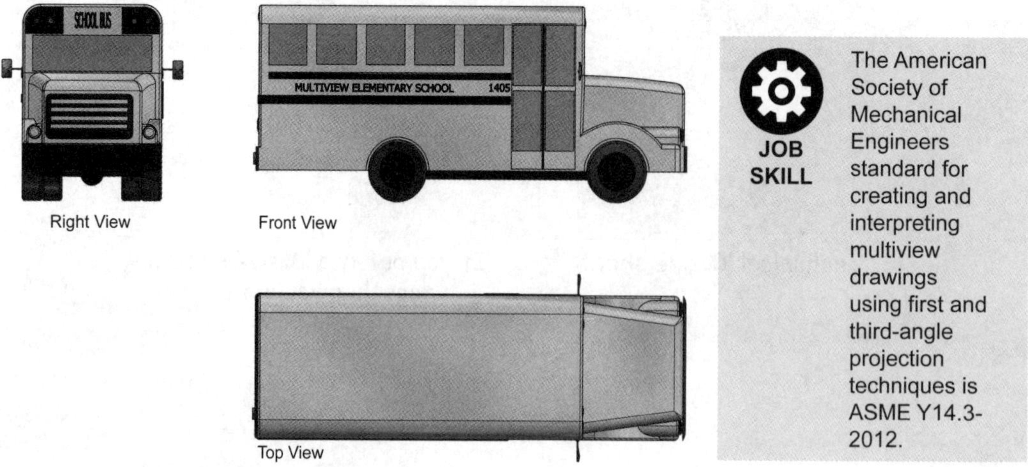

2.74 Arrangement of Views in First-Angle Projection (Image courtesy of Bryant Meadors)

JOB SKILL

The American Society of Mechanical Engineers standard for creating and interpreting multiview drawings using first and third-angle projection techniques is ASME Y14.3-2012.

To avoid confusion, it may be necessary to note on the drawing whether first or third-angle projection was used to create the drawing. For this reason, symbols have been developed to distinguish between the two types of projection techniques.

Third-angle projection can be noted on drawings by placing the symbol shown in Figure 2.75 in, or near, the title block.

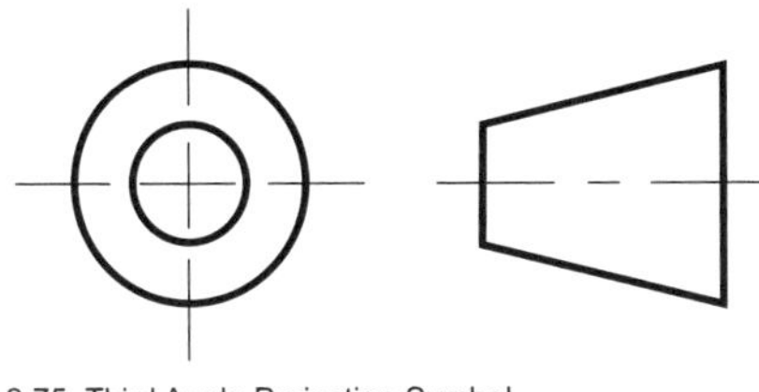

2.75 Third Angle Projection Symbol

JOB SKILL

The American Society of Mechanical Engineers standard governing the symbols used for first and third-angle projection is ASME Y14.1-2012.

The first-angle projection technique can be noted on drawings with the symbol shown in Figure 2.76. The letters *SI* (International System of Units) indicate that the drawing was prepared using metric units. The unit of measurement commonly used in the creation of mechanical engineering drawings is the millimeter.

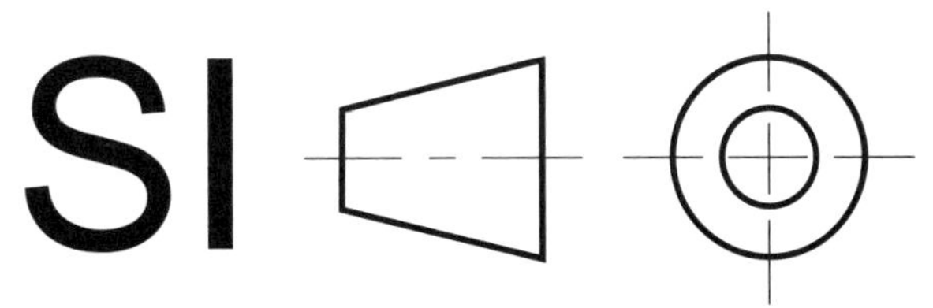

2.76 First-Angle Projection Symbol. SI signifies that the drawing was created in Metric Units

CHAPTER REVIEW

The ability to visualize and create multiview drawings, as well as the ability to interpret multiview drawings produced by others, is an essential job skill that every successful architect, engineer, designer, and drafter must possess. Some students may find that to develop this skill they will need to devote a significant amount of time practicing the visualization and sketching techniques presented in this unit.

Developing a solid understanding of the multiview drawing techniques presented in this unit is essential to mastering the concepts and drawing assignments you will encounter later in this text.

KEY WORDS

- Elevation Drawings
- First-Angle Projection
- Foreshortening
- Graphic Primitives
- Inclined Plane
- Linetype
- Lineweight
- Miter Line
- Multiview Drawing
- Orthographic Projection
- Projection Planes
- Regular Views
- Sheet Sizes
- Third-Angle Projection
- Three-Dimensional (3D) Object
- Two-Dimensional (2D) Object

REVIEW

Short Answer

1. Name the six regular views of an object.
2. How does a drafter determine which view will be the front view?
3. How many views of an object should a drafter draw?
4. What does quarter scale mean?
5. Name the standard that controls line thickness in an ASME drawing.

REVIEW

Matching

Column A	Column B
a. 2 = 1	1. Ratio indicating a drawing is half scale
b. .3mm	2. ASME line thickness for a visible line
c. .3mm	3. Ratio indicating a drawing is double scale
d. 1 = 2	4. ASME line thickness for a centerline
e. .6mm	5. ASME line thickness for a hidden line

Multiple Choice

1. What is the sheet size for an architectural D sheet?
   a. 12" X 9"
   b. 297 mm X 210 mm
   c. 36" X 24"
   d. 18" X 12"
2. What is the sheet size for a mechanical B size sheet?
   a. 17" X 11"
   b. 18" X 12"
   c. 841 mm X 594 mm
   d. 36" X 24"
3. Which ratio indicates that a drawing is full size?
   a. 1 = 1
   b. 1/1
   c. 1:1
   d. All the above
4. A 'C' size sheet is equal in size to two of these sheets laid side by side:
   a. A
   b. B
   c. D
   d. None of the above
5. What does 1/4"= 1'-0" represent on an architectural drawing of a house?
   a. The scale of the drawing is 1 = 4.
   b. One foot on the construction site equals 1/4 inch on the drawing.
   c. The house should be built one-quarter scale.
   d. A distance of 100 feet will measure 12.5 inches on the drawing.

EXERCISE 2.1

On the grid sheet located at the back of the text, sketch the front, top, and right-side views of the objects shown on the following page. Begin views in the darkened corners shown in each grid.

The following sketching exercises are designed to help you develop multiview sketching and visualization skills.

Directions

1. On the grid sheets located at the back of the text, sketch the front, top, and side views of the objects in Exercises 2.1 through 2.6. The black arrows included on each object establish the viewer's line of sight for the object's front view.
2. Begin each sketch by counting the number of grids that define the features of the front view and transfer these distances to the grid sheet. Start the front view in the darkened corner located in the lower left corner of each numbered grid box (see the example shown on the upper right grid box of Exercise 2.1's grid sheet). Begin the top and right views in the darkened corners above, or to the right, of the front view. **Note:** Assume parts that appear to be symmetrical are symmetrical. If it is unclear whether or not a hole or other cut-out feature shown on a sketch goes through the part, assume that it does.
3. Take advantage of the miter line drawn in each grid box to transfer information between the top and right views whenever possible (refer to Figures 2.69 through 2.71).

If you have trouble with a sketching problem, you may find referring to Figures 2.54 through 2.67 helpful. Also, do not hesitate to ask your instructor for assistance. This activity may seem difficult at first, but keep working at it, because through practice it is possible for you to develop this important drafting skill.

NOTE

Remove the grid sheets at the back of the book by carefully tearing along the perforation. If you need extra grid sheets, download and print the *Multiview Grid Sheet.pdf* file located in the *Sketching and Lettering Plates* folder of this book's file downloads. These downloads are available by redeeming the access code located on the inside front cover of this book.

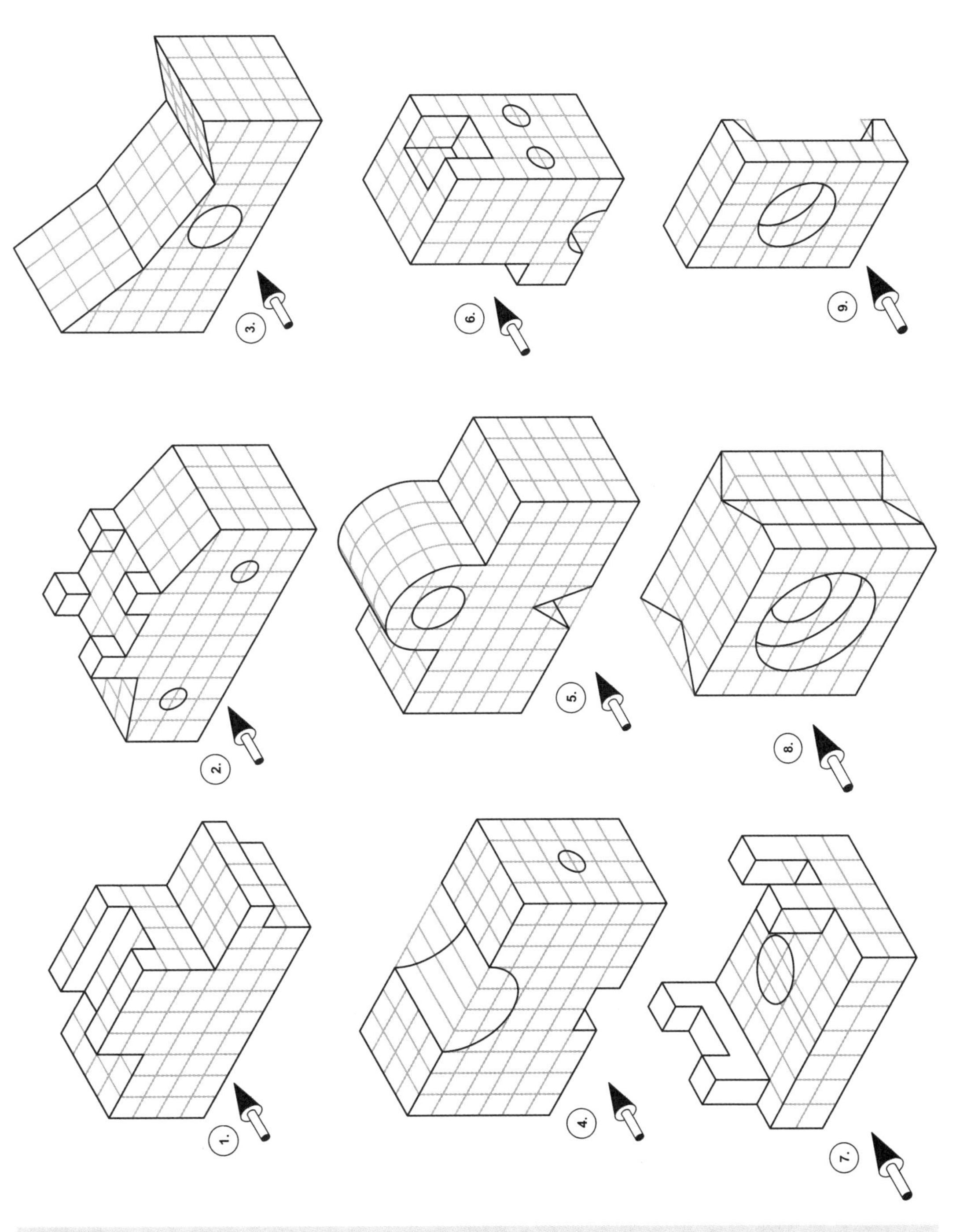

QUICK TIP

For the purposes of this exercise, assume that objects that appear to be symmetrical are symmetrical. If it is unclear whether or not a hole or other cut-out feature shown on a sketch goes all the way through the part, assume that it does.

EXERCISE 2.2

On the grid sheet located at the back of the text, sketch the front, top, and right-side views of the objects shown on the following page. Begin views in the darkened corners shown in each grid.

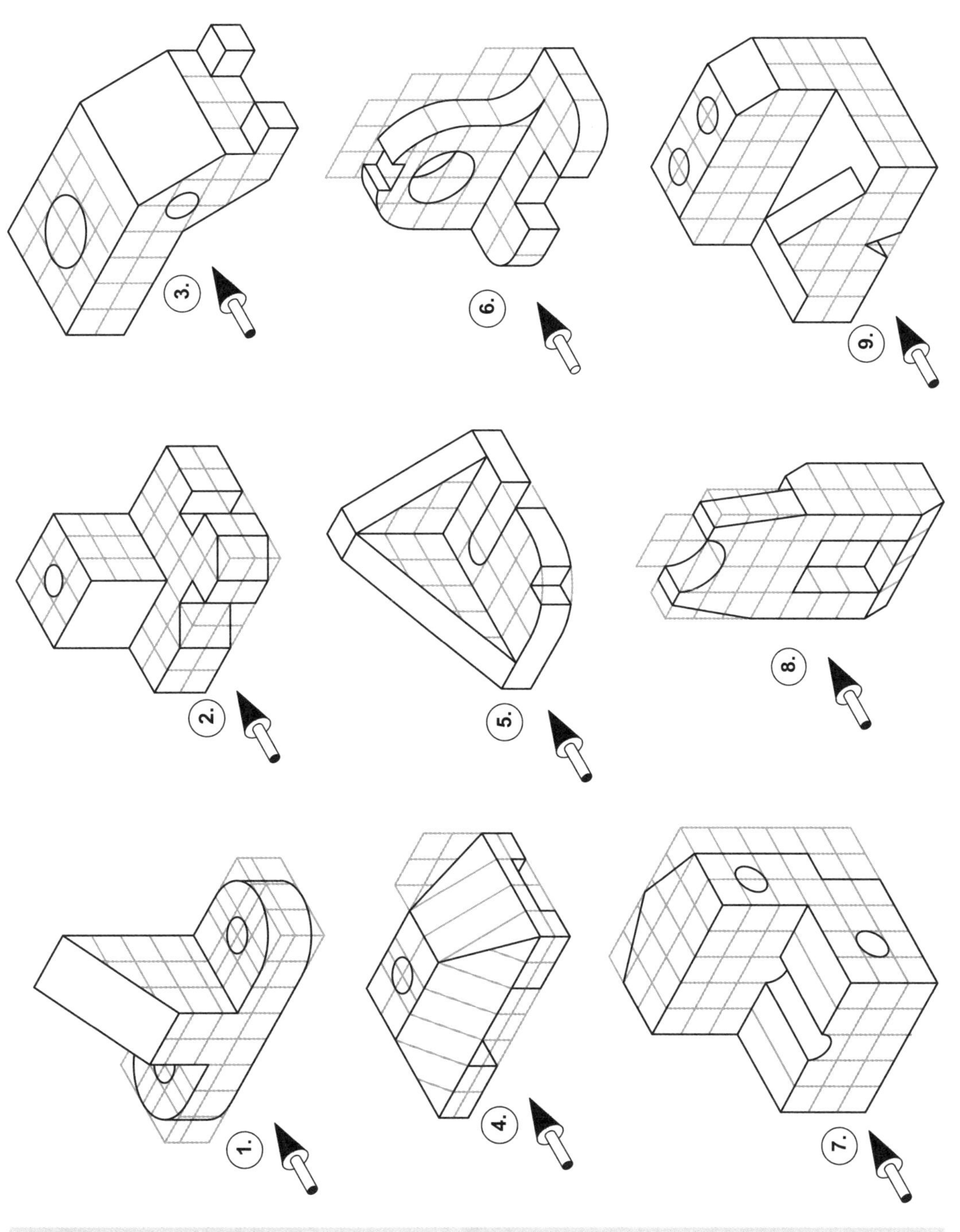

QUICK TIP

For the purposes of this exercise, assume that objects that appear to be symmetrical are symmetrical. If it is unclear whether or not a hole or other cut-out feature shown on a sketch goes all the way through the part, assume that it does.

EXERCISE 2.3

On the grid sheet located at the back of the text, sketch the front, top, and right-side views of the objects shown on the following page. Begin views in the darkened corners shown in each grid.

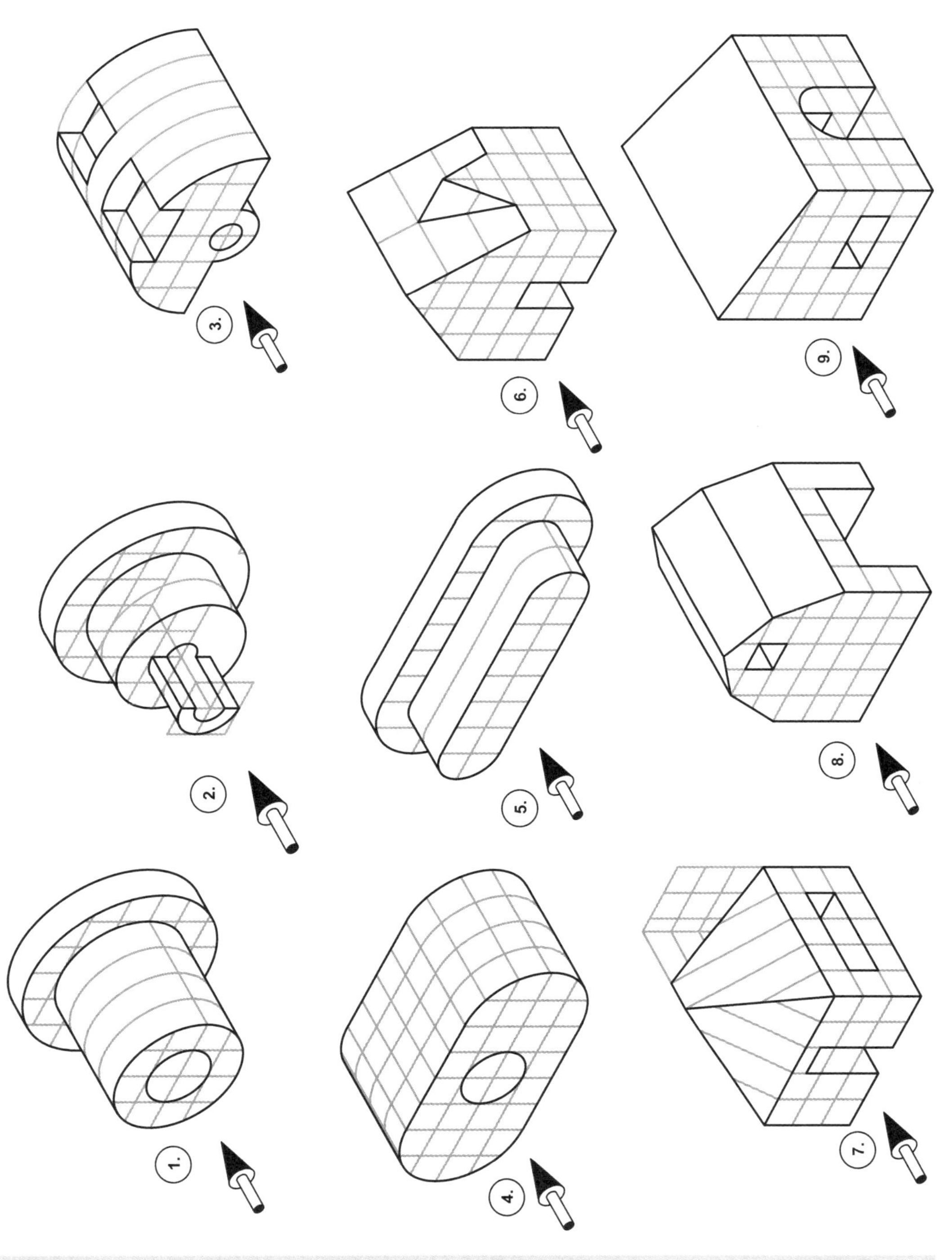

QUICK TIP

For the purposes of this exercise, assume that objects that appear to be symmetrical are symmetrical. If it is unclear whether or not a hole or other cut-out feature shown on a sketch goes all the way through the part, assume that it does.

EXERCISE

EXERCISE 2.4

On the grid sheet located at the back of the text, sketch the front, top, and right-side views of the objects shown on the following page. Begin views in the darkened corners shown in each grid.

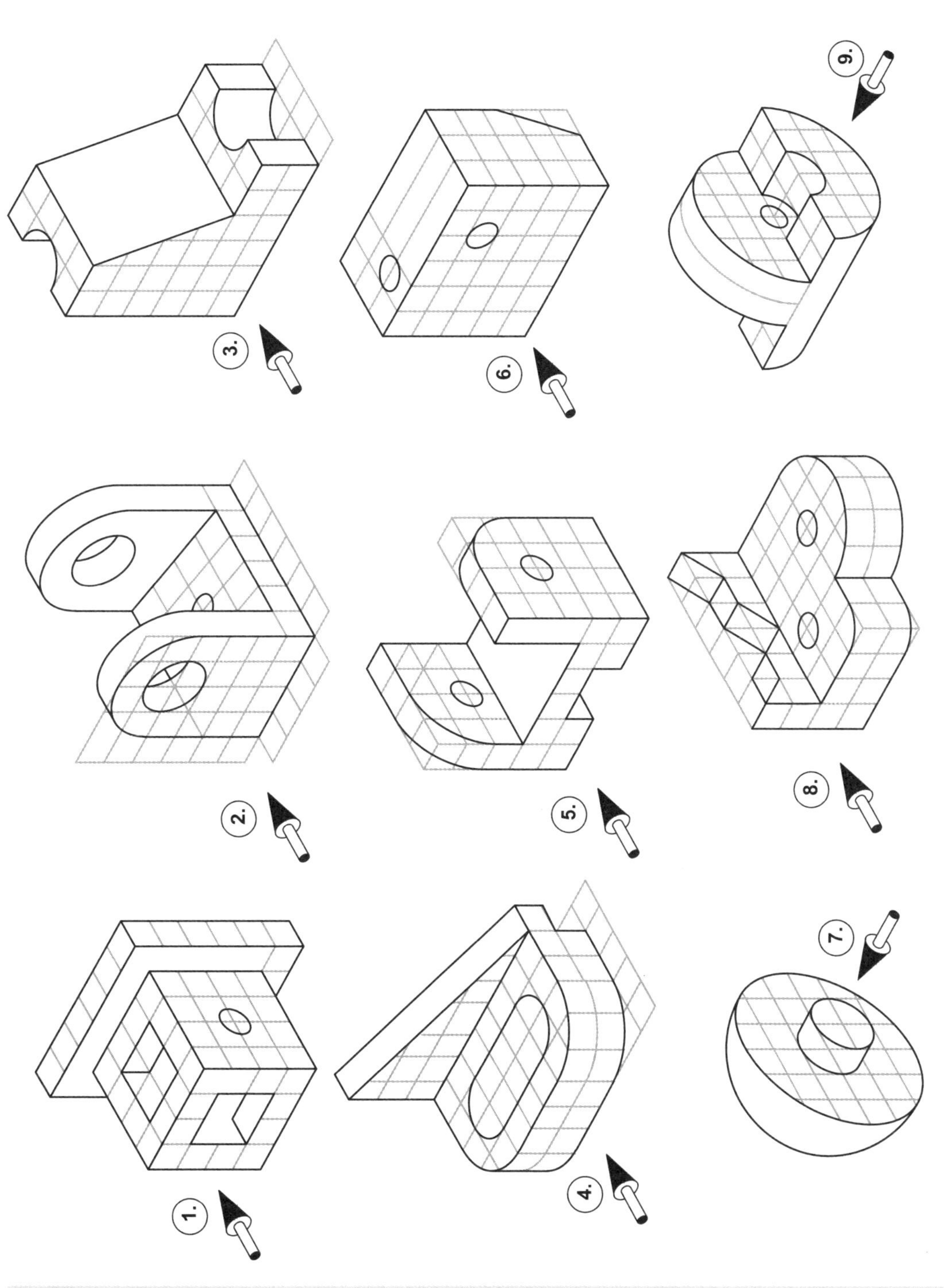

QUICK TIP

For the purposes of this exercise, assume that objects that appear to be symmetrical are symmetrical. If it is unclear whether or not a hole or other cut-out feature shown on a sketch goes all the way through the part, assume that it does.

EXERCISE 2.5

On the grid sheet located at the back of the text, sketch the front, top, and right-side views of the objects shown on the following page. Begin views in the darkened corners shown in each grid.

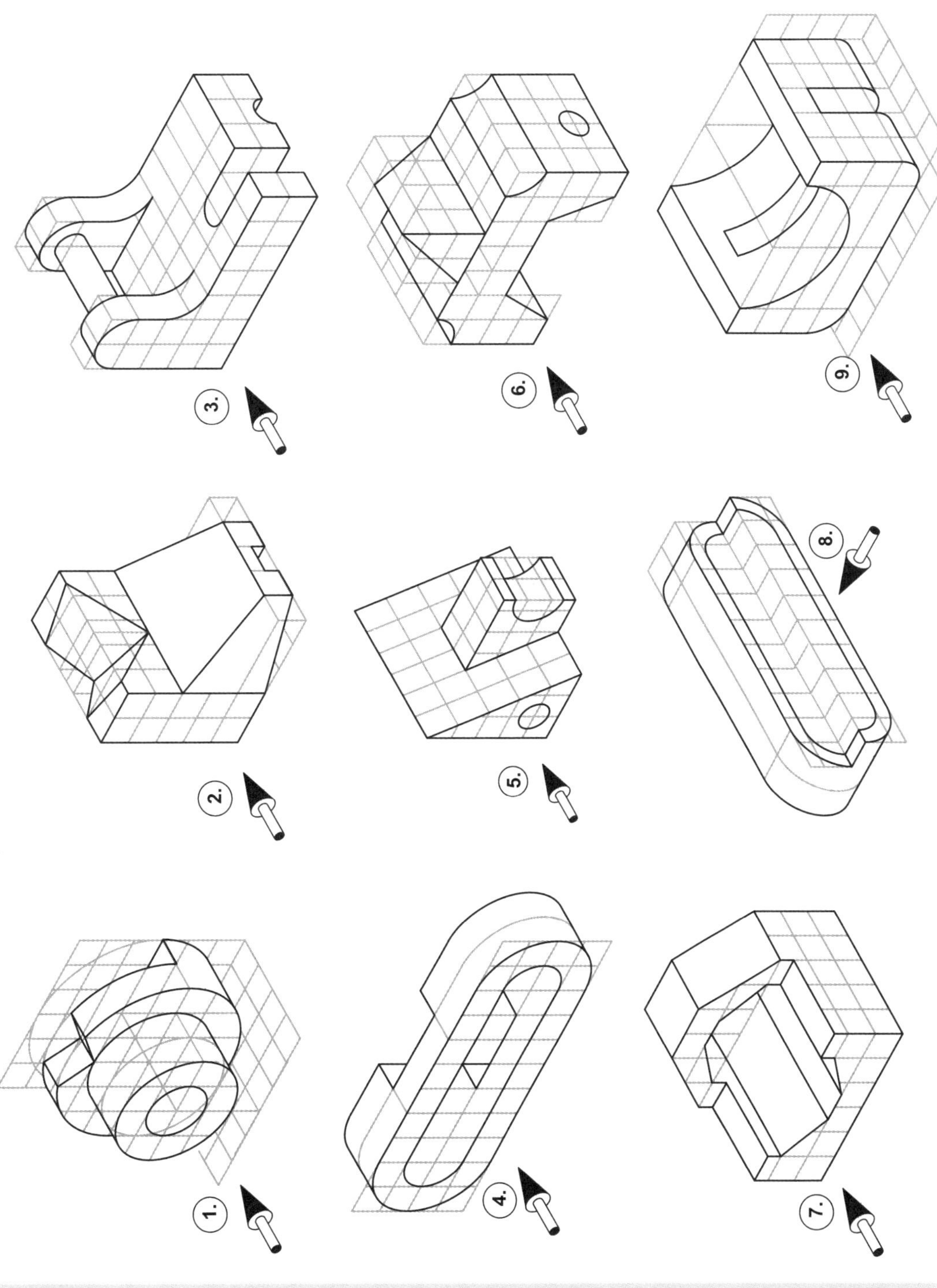

QUICK TIP

For the purposes of this exercise, assume that objects that appear to be symmetrical are symmetrical. If it is unclear whether or not a hole or other cut-out feature shown on a sketch goes all the way through the part, assume that it does.

EXERCISE 2.6

On the grid sheet located at the back of the text, sketch the front, top, and right-side views of the objects shown on the following page. Begin views in the darkened corners shown in each grid.

EXERCISE

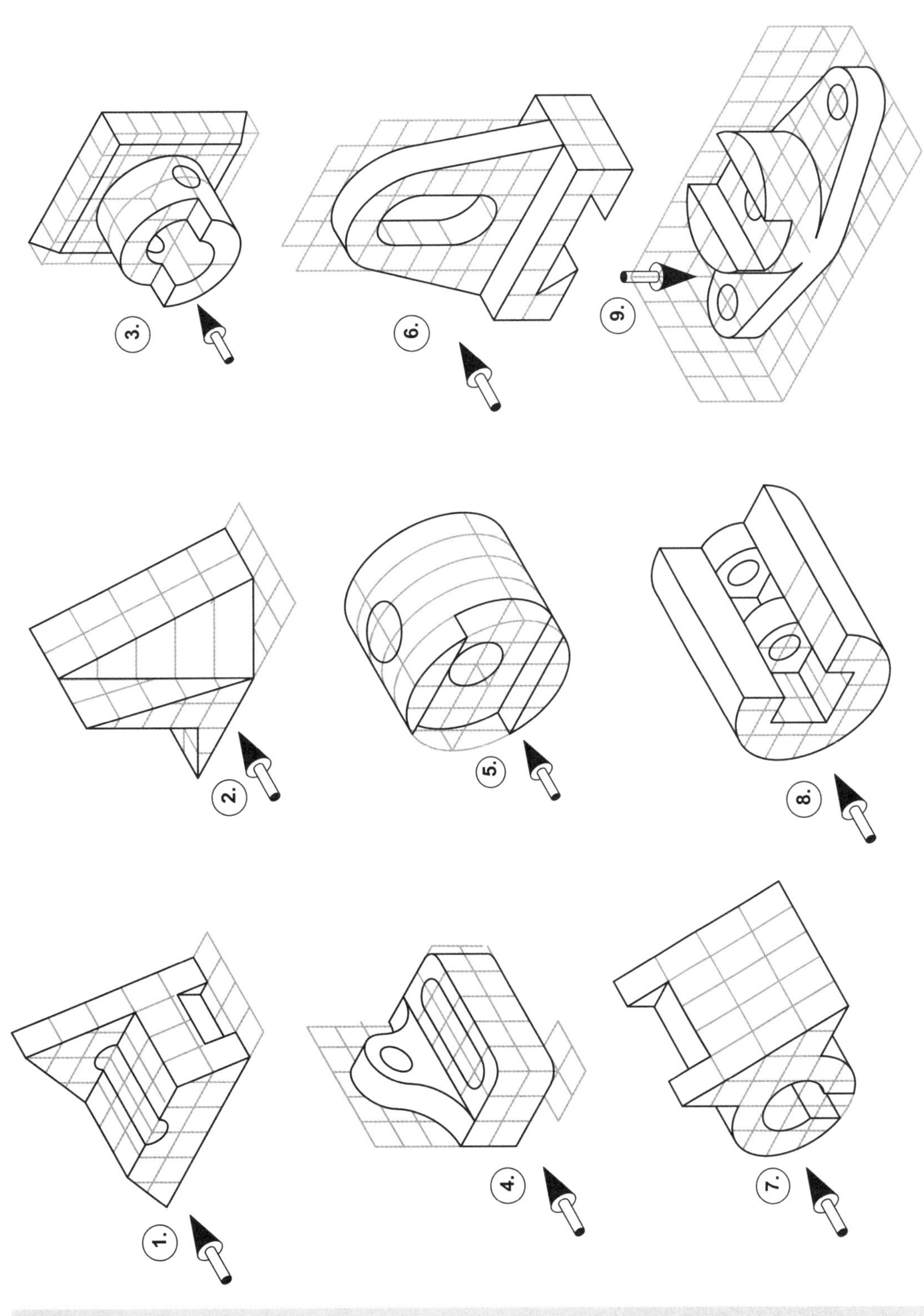

QUICK TIP

For the purposes of this exercise, assume that objects that appear to be symmetrical are symmetrical. If it is unclear whether or not a hole or other cut-out feature shown on a sketch goes all the way through the part, assume that it does.

CHAPTER

3

TRADITIONAL DRAFTING TOOLS AND TECHNIQUES

CHAPTER OBJECTIVES:

After studying the material in this chapter, you should be able to:

1. Describe the tools and techniques used in traditional drafting.
2. Use technical pencils, straightedges, triangles, scales, protractors, and templates to construct the geometry of technical drawings.
3. Read a conversion table to convert between decimal, fractional, and metric units.
4. Use traditional drafting tools to create multiview drawings of objects including correctly located and depicted visible, hidden, and centerlines.
5. Hand-letter notes and dimensions on technical drawings that are clear and legible.

CHAPTER OVERVIEW

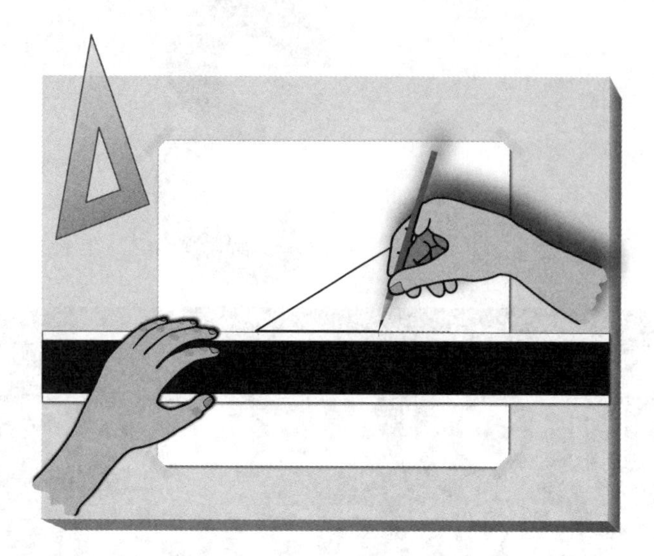

TRADITIONAL DRAFTING TOOLS AND TECHNIQUES

Before CAD revolutionized the way technical drawings are produced, drafters and designers sat at drawing tables and used ***traditional drafting tools*** like technical pens, straightedges, triangles, scales, protractors, and templates to draw on sheets of vellum or Mylar (a thin sheet of plastic with a matte surface). In today's engineering or architectural office, it would be rare for a drafter to create a drawing in the traditional way, but many of the techniques developed by traditional drafters, such as orthographic projection, are still used to create 2D CAD drawings. The traditional tools discussed in this chapter can be purchased through drafting supply outlets.

MORE INFO

Refer to the following websites:

Koh-I-Noor:	www.kohinoorusa.com
Alvin:	www.alvinco.com
Pentel:	www.pentel.com
Staedtler:	www.staedtler.com

3.1 TRADITIONAL DRAFTING TOOLS AND TECHNIQUES

Shown in Figures 3.1 through 3.3 are examples of traditional drafting equipment. A drafting machine, or a parallel straightedge, attached to the top of a drawing table allows a drafter to draw horizontal lines that are parallel to each other. Another tool that can be used to draw parallel horizontal lines is a T-square.

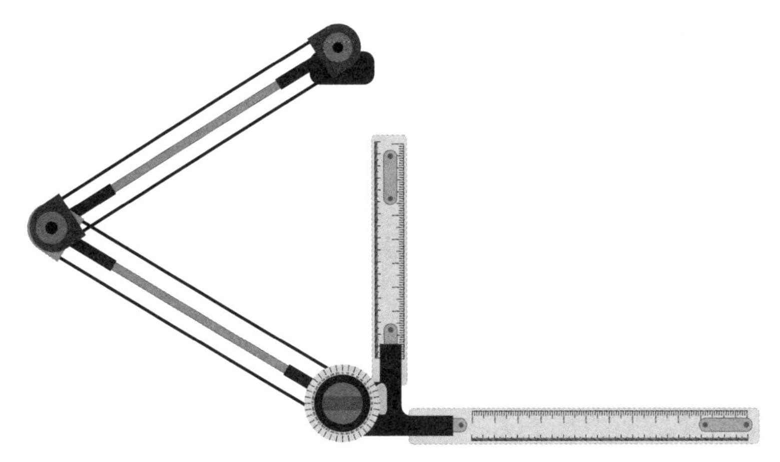

3.1 Drafting machines can easily be adjusted for drawing variable angles. Drafting machines were once common in engineering offices.

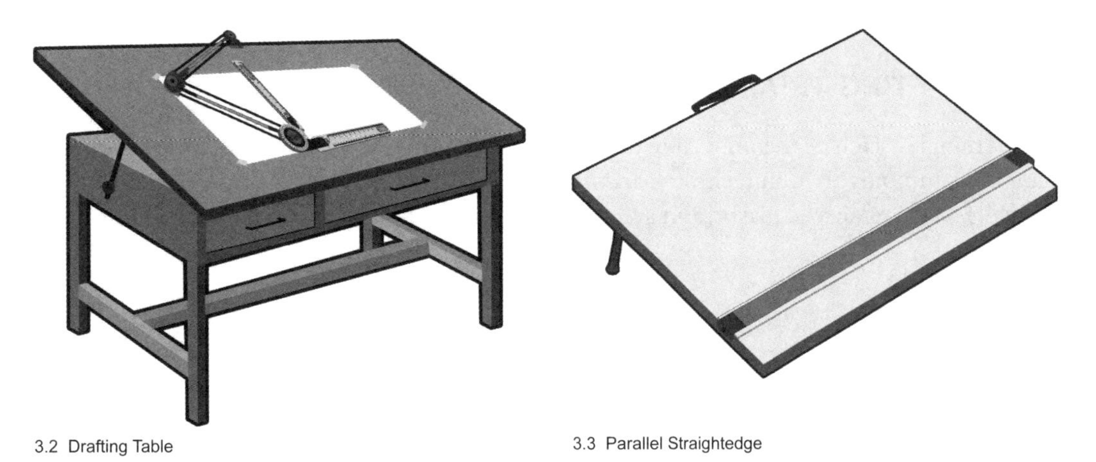

3.2 Drafting Table

3.3 Parallel Straightedge

3.2 TECHNICAL PENCILS AND PENS

Professional-grade technical pens and pencils are often used by design professionals to create technical drawings and sketches (see Figure 3.4).

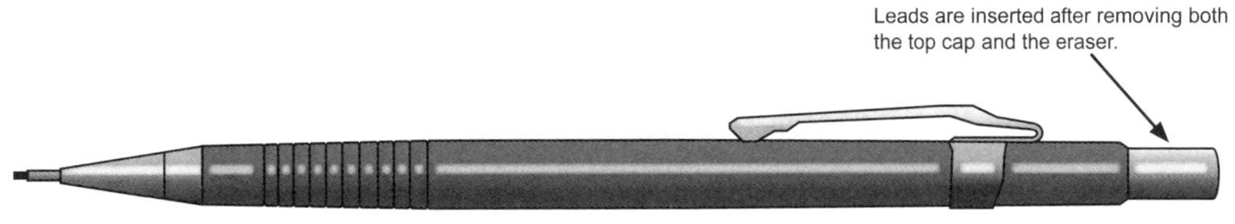
Leads are inserted after removing both the top cap and the eraser.

3.4 Technical Pencil

Technical pens and pencils come in differing line widths (.3mm, .5mm, .7mm, or .9mm). Leads for technical pencils are available in a variety of hardness grades, defined by the **lead hardness grade**, depending on the type of work to be performed. The info box below shows the lead (graphite) hardness grades and the appropriate application for the leads in each hardness range.

MORE INFO

Hard leads (4H–9H) are useful for laying out construction lines on a technical drawing.

Medium leads (3H, 2H, H, F, HB, and B) are often used for general drafting and sketching (2H is a popular medium lead).

Soft leads (2B–7B) are often used for shading and rendering drawings

3.3 BEGINNING A TRADITIONAL DRAFTING PROJECT

Figure 3.5 illustrates the proper technique for attaching a sheet of vellum to the top of a drawing table. Align the bottom edge of the sheet with the top edge of the drafting machine arm, parallel straightedge, or T-square, and tape all four corners to the table top.

Horizontal lines are drawn along the top edge of the straightedge as shown in Figure 3.6. Right-handed drafters would hold the straightedge in place with their left hand when drawing a horizontal line.

3.4 DRAFTING TRIANGLES

Triangles, such as the ones shown in Figures 3.7 and 3.8, provide drafters with angles commonly used in technical drawings: 30°, 45°, 60°, and 90°. Triangles are usually made of acrylic plastic and are available in a variety of sizes.

3.5 Align the bottom edge of a sheet to the top edge of a parallel straightedge and tape the corners

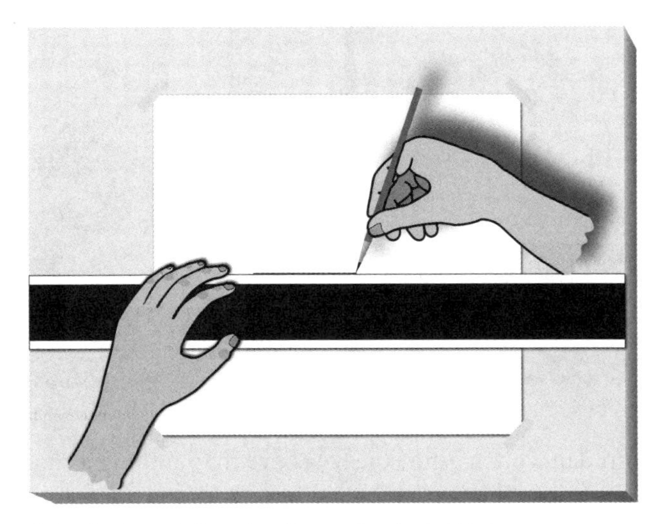

3.6 Drawing horizontal lines along the top edge of a straightedge

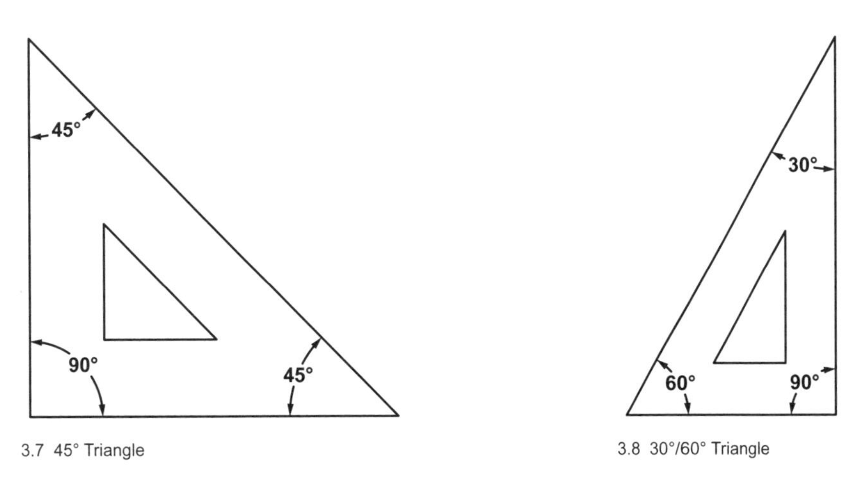

3.7 45° Triangle

3.8 30°/60° Triangle

JOB SKILL

Some triangles are designed for drawing with graphite, and others for drawing with pen and ink. Inking triangles have a beveled, or stepped, edge to prevent the ink from running under the edge of the triangle and smearing.

3.5 DRAWING LINES WITH TRIANGLES AND PARALLEL STRAIGHTEDGES

To draw horizontal and vertical lines that are mutually perpendicular (90°), drafters use a triangle in conjunction with a parallel straightedge.

To draw vertical lines, place the triangle on top of the straightedge, as in Figure 3.9. Hold both the straightedge and triangle with your left hand while drawing the vertical line. In this way you are assured that a vertical line will be drawn at a 90° angle relative to lines drawn along the top of the parallel straightedge. You can use either the 30°/60° triangle (as shown) or the 45° triangle.

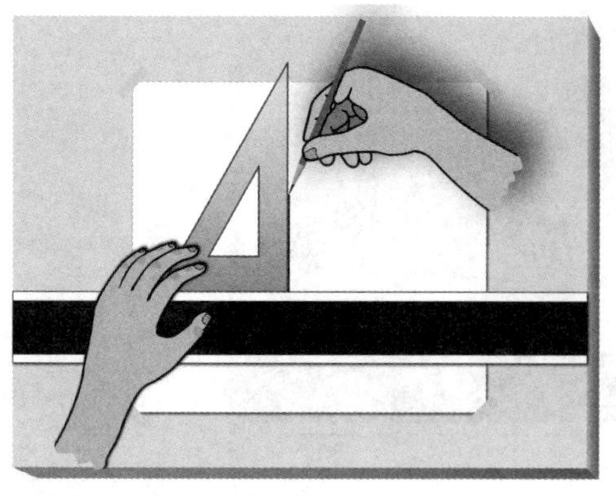

3.9 Drawing a Vertical Line

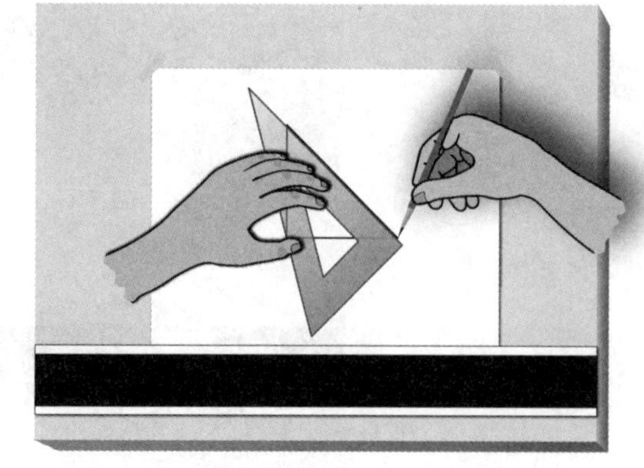

3.10 Drawing a Line between the Endpoints of Two Lines

Beginning students often erroneously believe that holding a triangle as vertical as their eyes can position it (known as "eyeballing") and drawing a line will give them a perpendicular vertical line, when in fact the only way to draw an accurate vertical line is to make sure the triangle rests on the straightedge as shown in Figure 3.9. However, there are occasions when the triangle is not positioned on the straightedge, for example, when connecting two points or the ends of two lines, when the angle of the resulting line is not equal to 30°, 45°, 60°, or 90°.

Figure 3.10 illustrates how you can draw a line connecting the endpoints of vertical and horizontal lines. The desired angle does not match one of the triangle's normal angles (30°, 45°, 60°, or 90°). In this case, the triangle would be floated and aligned with the ends of the vertical and horizontal lines. The line is drawn between the points along the edge of the triangle.

Placing the 30°/60° triangle on the top of the straightedge as in Figure 3.11 allows you to draw lines 30° from horizontal. You can draw these lines sloping either to the right (as shown) or to the left by flipping the triangle over.

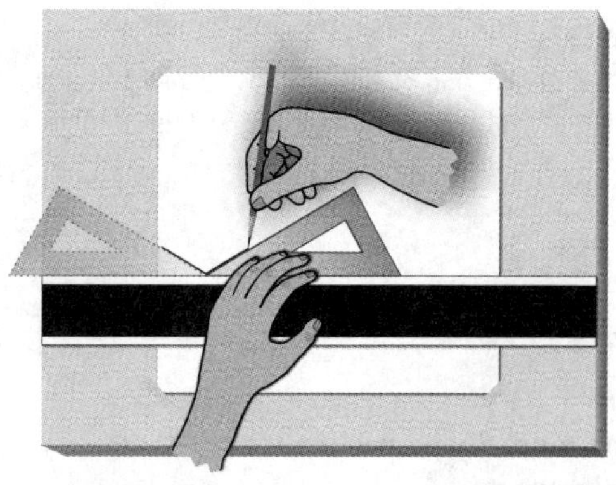

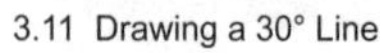

3.11 Drawing a 30° Line

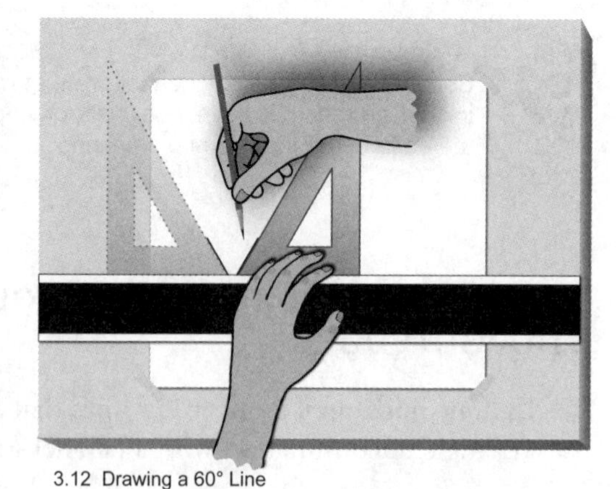

3.12 Drawing a 60° Line

Figure 3.12 shows lines being drawn at a 60° angle from horizontal. Flipping the triangle over allows you to draw lines sloping to the left or right as needed. The 45° triangle can be used to draw 45° lines sloping to either the left or right, as in Figure 3.13.

3.13 Drawing a 45° Line and 165° Line

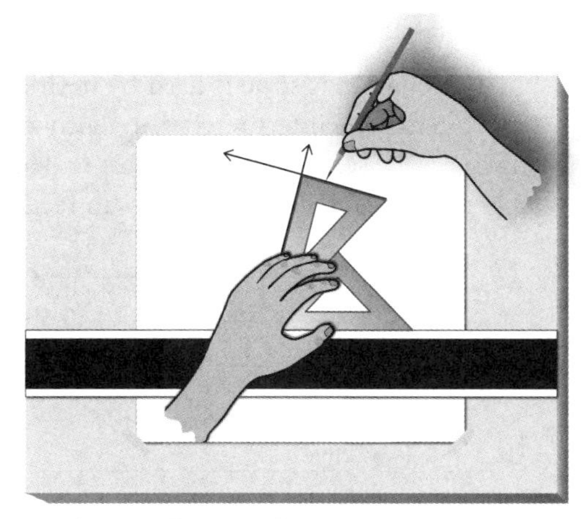

3.14 Combining Triangles to Produce 75° Lines

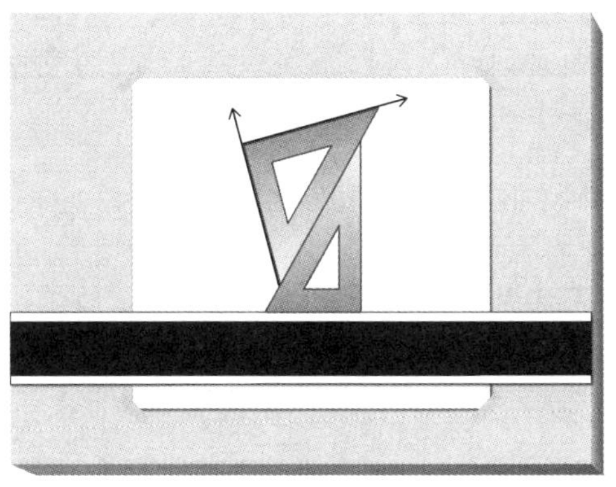

3.15 Combining Triangles to Produce 15° and 105° Lines

Using two triangles in combination allows you to draw lines at 15° increments.

- In Figure 3.14, the 45° and 30°/60° triangles are combined to draw lines at 75° and 165°.
- In Figure 3.15, the 45° and 30°/60° triangles are combined to draw lines at 105° and 15°.

3.6 MAKING MEASUREMENTS WITH THE ENGINEER'S, ARCHITECT'S, AND METRIC SCALES

In engineering and architectural offices, designers and drafters use scales in two ways. The first is to take measurements from existing drawings or plots; the second is to layout distances when constructing a drawing using traditional drafting techniques.

Depending on the type of drawing being created, a designer may choose the **engineer's**, **architect's**, or **metric scale** to measure distances (a combination scale is also available that has a mix of the most commonly used scales).

Reading the Engineer's Scale

The engineer's scale is used by both mechanical and civil engineers. The marks on the scale may be interpreted differently depending on the discipline. In Figure 3.16, the engineer's 10 scale is used to measure decimal inches. In Figure 3.17, the engineer's 10 scale is used to measure distances in feet.

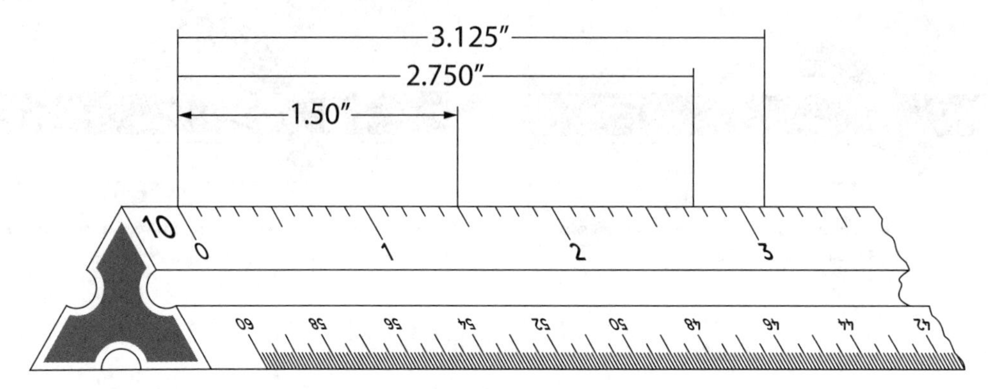

3.16 The Engineer's 10 Scale Showing Decimal Inches

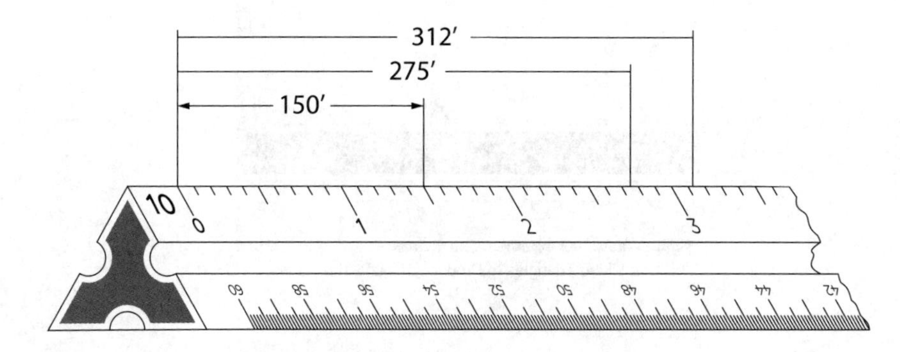

3.17 The Engineer's 10 Scale Showing Feet. This Scale would be Interpreted as 1" = 100'

Reading the Metric Scale

Mechanical engineers can use the metric scale shown in Figure 3.18 to measure distances in millimeters. In Figure 3.18, each small mark on the scale equals 1 millimeter.

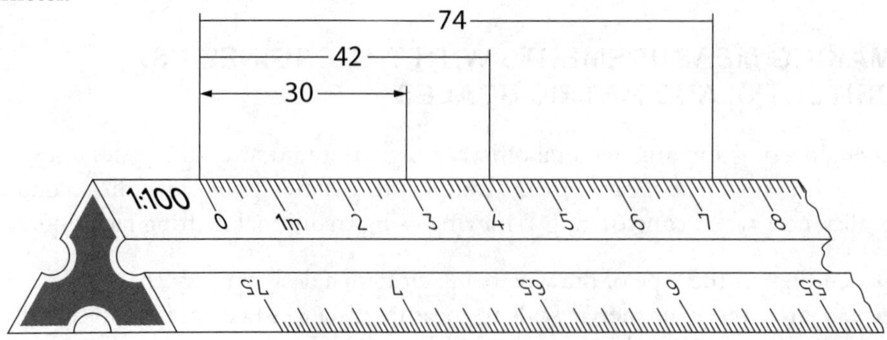

3.18 The Metric Scale Used to Measure Millimeters

Reading the Architect's Scale

The architect's scale is used to measure feet and inches on a floor plan. The architect's 1/4 scale shown in Figure 3.19 would be interpreted as 1/4" = 1'-0". This means that 1/4 inch measured on the drawing would equal 1 foot at the construction site. Figures 3.20, 3.21, and 3.22 show examples of other commonly used architectural scales.

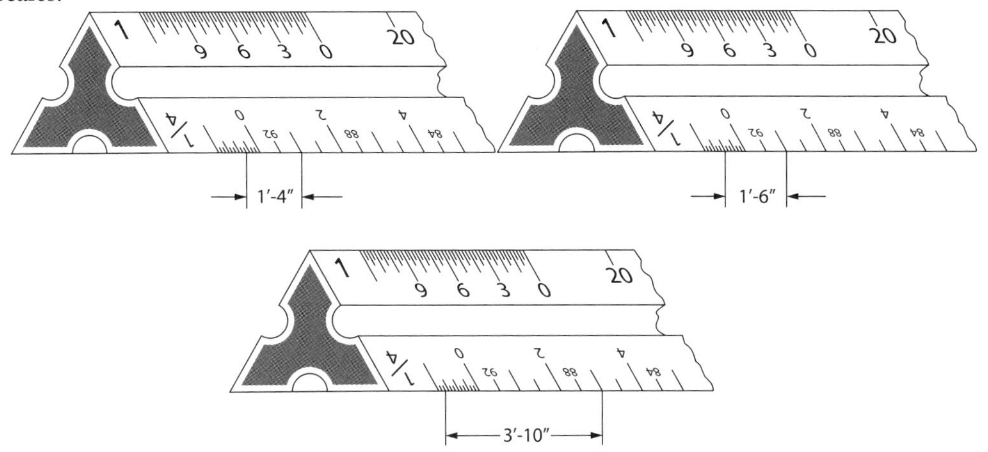
1'-4"
1'-6"
3'-10"

3.19 The Architect's 1/4 Scale Illustrating Several Different Measurements of Feet and Inches

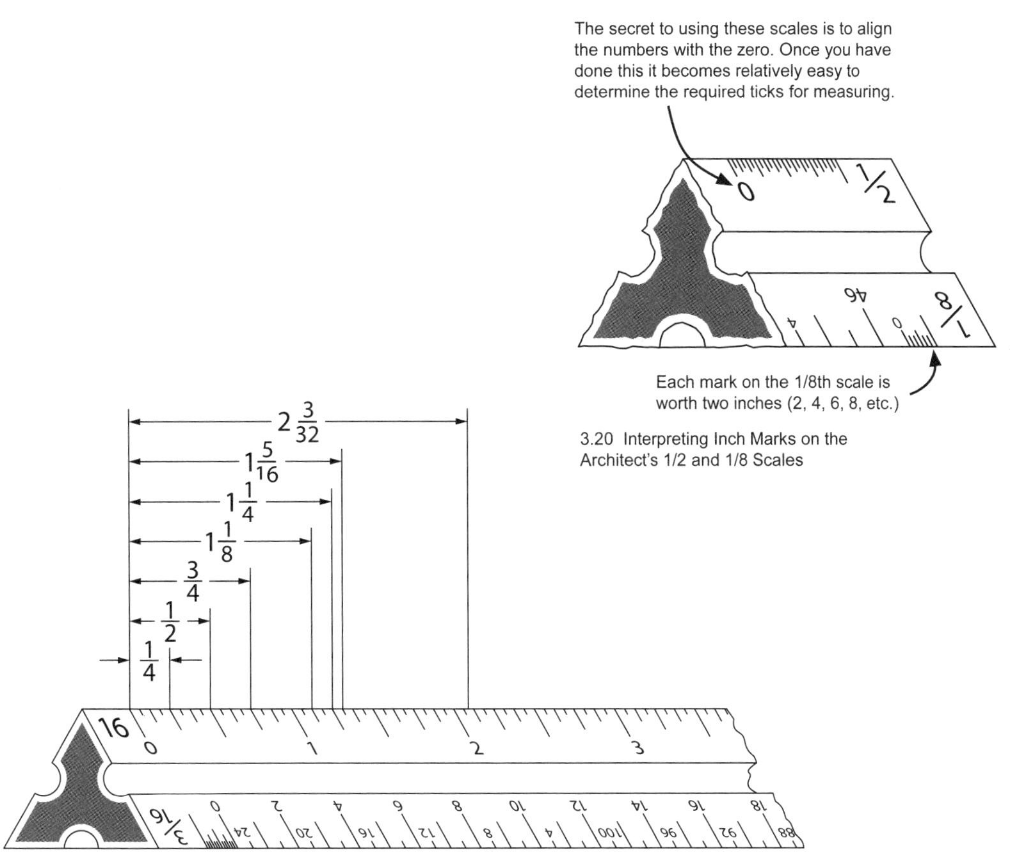
The secret to using these scales is to align the numbers with the zero. Once you have done this it becomes relatively easy to determine the required ticks for measuring.
Each mark on the 1/8th scale is worth two inches (2, 4, 6, 8, etc.)
2 3/32
1 5/16
1 1/4
1 1/8
3/4
1/2
1/4

3.20 Interpreting Inch Marks on the Architect's 1/2 and 1/8 Scales

3.21 The Architect's 16 Scale Can Be Used to Measure Fractional Inches

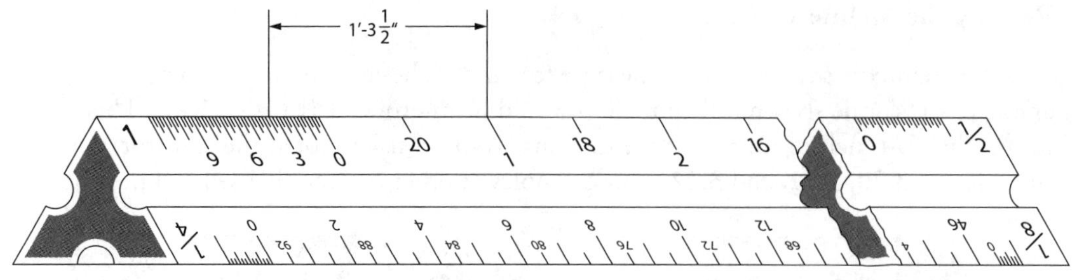

The scale along this edge - if read from the right - would be 1/8"=1'0"

3.22 The Architect's 1 Scale Can Be Used to Measure Feet and Inches on a Floor Plan. The scale along the top edge, if read from the left, would be interpreted as 1" = 1'-0".

QUICK TIP

To avoid having to search for a desired scale on a triangular scale, use a binder clip to mark your scale. For example, if using the 10 scale on an engineer's scale, attach the binder clip to the 50 scale. When you pick up your scale, the binder clip will orient you to the 10 scale.

3.7 CONVERTING UNITS OF MEASUREMENT

Drafters may sometimes need to convert from one unit of measure to another. Some commonly used conversion factors are as follows:

- Fractional inches can be converted to decimal inches by dividing the numerator (the top number) by the denominator (the bottom number).
- Decimal inches can be converted to millimeters by multiplying them by 25.4.
- Millimeters can be converted to decimal inches by dividing them by 25.4.

Table 3.1 is useful for quickly finding the equivalent value among the various units listed.

3.8 READING THE PROTRACTOR

A protractor is a tool used to lay out angles on a drawing. Full-circle protractors are divided into 360°, and half-moon protractors (see Figure 3.23) are divided into 180°. Both are divided in 10° increments. Note that the protractor in Figure 3.23 is divided from 0° to 180° in both the clockwise and counterclockwise directions.

Figures 3.24 through 3.27 illustrate some of the ways the protractor can be used to measure angles.

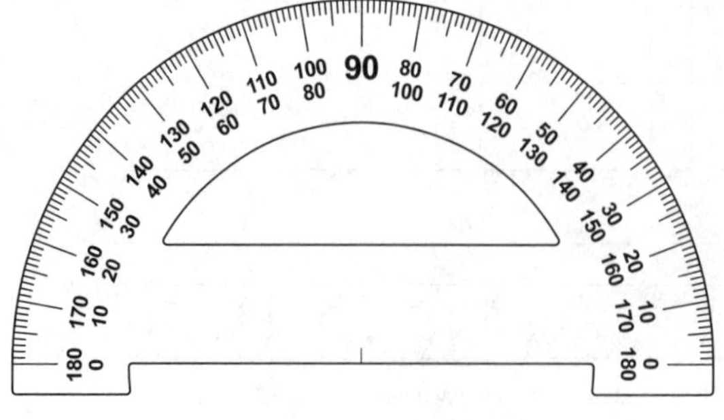

3.23 A Half-Moon, or 180° Protractor

Fractional Inch	Decimal Inch	Metric (mm)	Fractional Inch	Decimal Inch	Metric (mm)
1/64	.015625	0.3969	33/64	.515625	12.0969
1/32	.03125	0.7938	17/32	.51325	13.4938
3/64	.046875	1.1906	35/64	.546875	13.8906
1/16	.0625	1.5875	9/16	.5625	14.2875
5/64	.078125	1.9844	37/64	.578125	14.6844
3/32	.09375	2.3813	19/32	.59375	15.0813
7/64	.109375	2.7781	39/64	.609375	14.4781
1/8	.1250	3.1750	5/8	.6250	15.8750
9/64	.140625	3.5719	41/64	.640625	16.2719
5/32	.15625	3.9688	21/32	.65625	16.6688
11/64	.171875	4.3656	43/64	.671875	17.0656
3/16	.1875	4.7625	11/16	.6875	17.4625
13/64	.203125	5.1594	45/64	.703125	17.8594
7/32	.21875	5.5563	23/32	.71875	18.2563
15/64	.234375	5.9531	47/64	.734375	18.6531
1/4	.250	6.3500	3/4	.750	19.0500
17/64	.265625	6.7469	49/64	.765625	19.4469
9/32	.28125	7.1438	25/32	.78125	19.8438
19/64	.296875	7.5406	51/64	.796875	20.2406
5/16	.3125	7.9375	13/16	.8125	20.6375
21/64	.328125	8.3344	53/64	.828125	21.0344
11/32	.34375	8.7313	27/32	.84375	21.4313
23/64	.359375	9.1281	55/64	.859375	21.8281
3/8	.375	9.5250	7/8	.8750	22.2250
25/64	.390625	9.9219	57/64	.890625	22.6219
13/32	.40625	10.3188	29/32	.90625	23.0188
27/64	.421875	10.7156	59/64	.921875	23.4156
7/16	.4375	11.1125	15/16	.9375	23.8125
29/64	.453125	11.5094	61/64	.953125	24.2094
15/32	.46875	11.9063	31/32	.96875	24.6063
31/64	.484375	12.3031	63/64	.984375	25.0031
1/2	.500	12.700	1	1.000	25.4000

Table 3.1 Conversions among Fractional Inches, Decimal Inches, and Millimeters for Values Up to One Inch

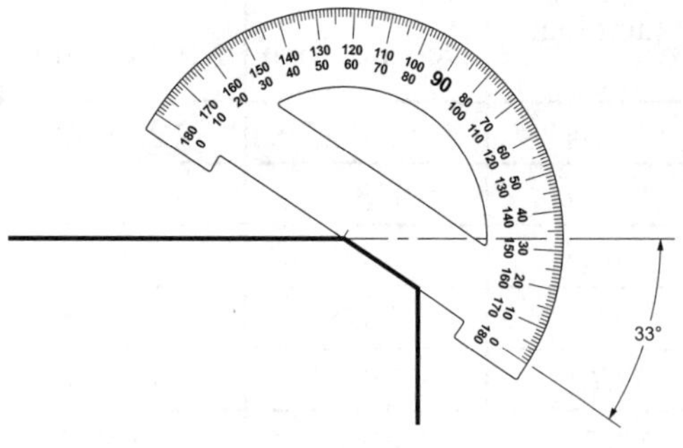

3.24 A 33° Angle Using the Inner Dial

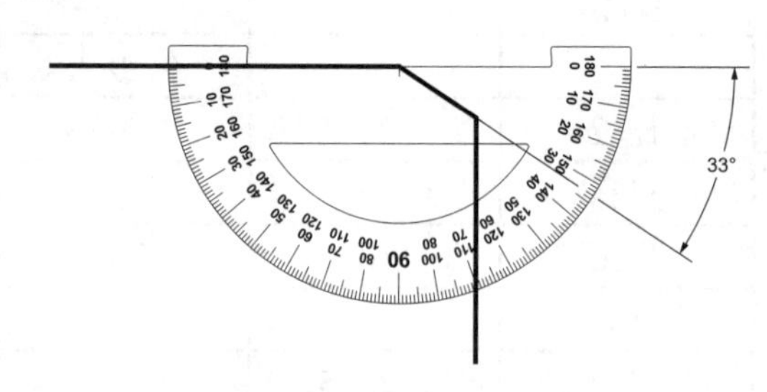

3.25 A 33° Angle Using the Outer Dial

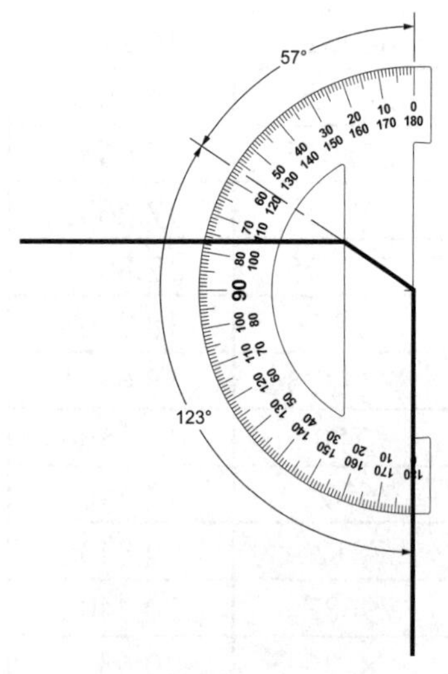

3.26 A 57° Angle Measured from Vertical

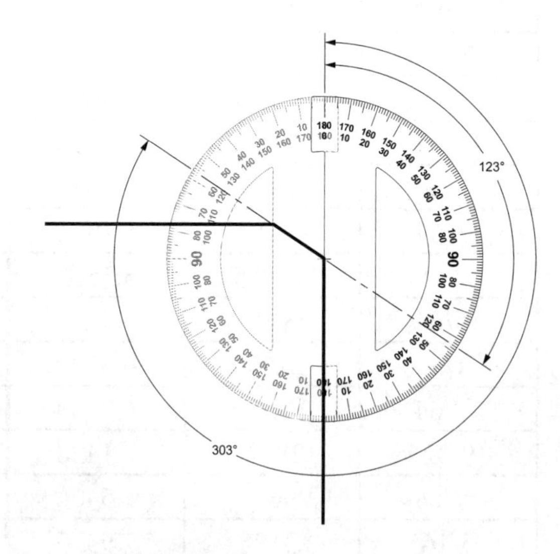

3.27 Measuring an Angle Greater Than 180° Using a Half-Moon Protractor

3.9 CIRCLE TEMPLATE

Circle templates come in a wide range of units (decimal inches, fractional inches, and millimeters) and diameters (see Figure 3.29). Figure 3.29 illustrates the steps in drawing a circle with the circle template.

3.10 ISOMETRIC ELLIPSE TEMPLATE

On isometric drawings, circles are represented as ellipses. Isometric ellipse templates allow drafters to place ellipses on isometric drawings quickly. Figure 3.30 illustrates the steps in drawing an ellipse with this template.

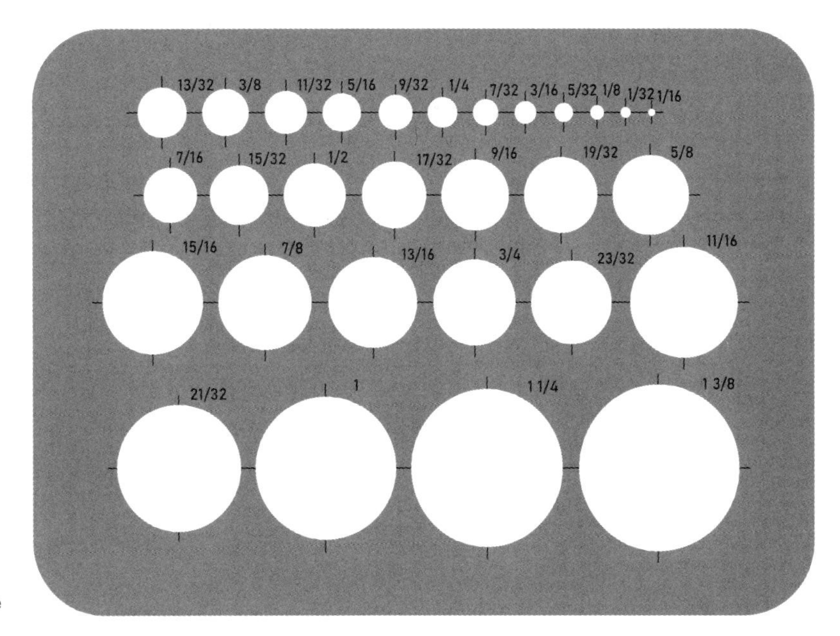
13/32 3/8 11/32 5/16 9/32 1/4 7/32 3/16 5/32 1/8 1/32 1/16
7/16 15/32 1/2 17/32 9/16 19/32 5/8
15/16 7/8 13/16 3/4 23/32 11/16
21/32 1 1 1/4 1 3/8

3.28 A Circle Template

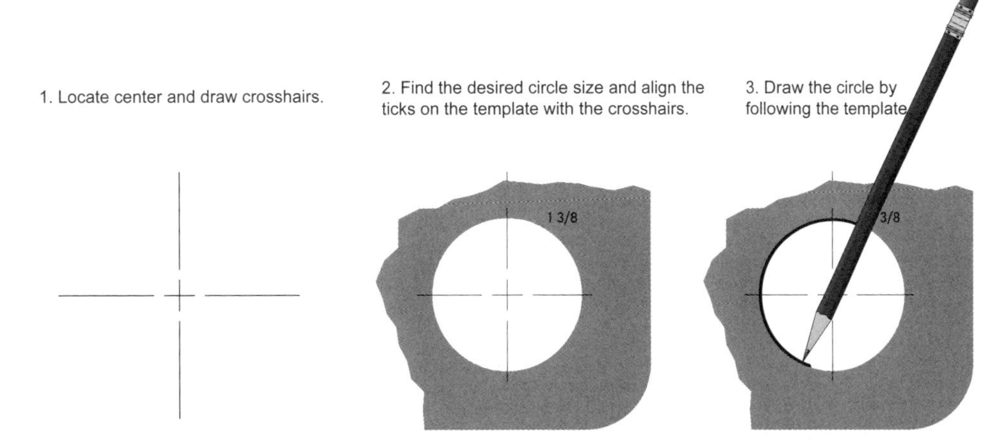
1. Locate center and draw crosshairs.
2. Find the desired circle size and align the ticks on the template with the crosshairs.
3. Draw the circle by following the template.
1 3/8
3/8

3.29 Steps in using the Circle Template

1. Locate center and draw crosshairs at 30° angles.

2. Find the desired ellipse size and align the ticks on the template with the crosshairs.

3. Draw the isometric ellipse by following the template.

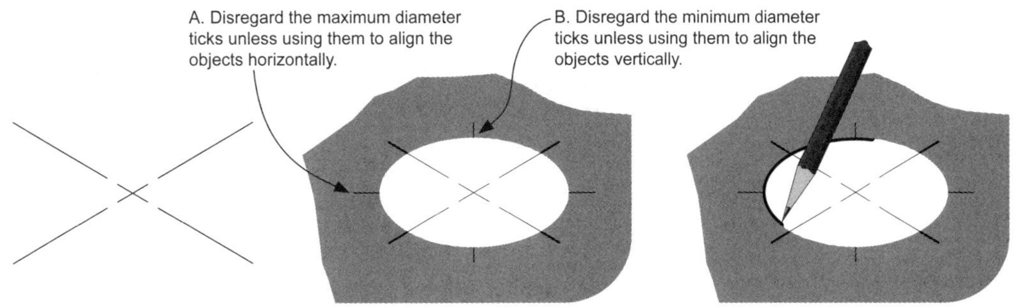
A. Disregard the maximum diameter ticks unless using them to align the objects horizontally.
B. Disregard the minimum diameter ticks unless using them to align the objects vertically.

3.30 Steps in using the Isometric Ellipse Template

STEP–BY–STEP

STEPS IN CONSTRUCTING A SIMPLE DRAWING

Steps 1 through 8 illustrate how a drafter may use the triangle, scale, and parallel straightedge to construct a simple technical drawing. At the end of this chapter are drafting projects in which you will have an opportunity to apply the techniques shown in these steps.

Step 1. Draw a light 30° line (Figure 3.31)

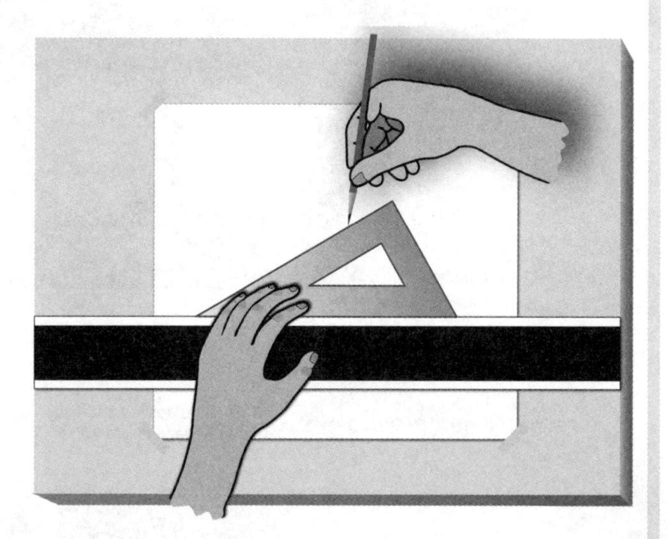

3.31 Drawing a 30° Construction Line

Step 2. Align the scale along the construction line and place light tick marks to denote the desired measurement (Figure 3.32).

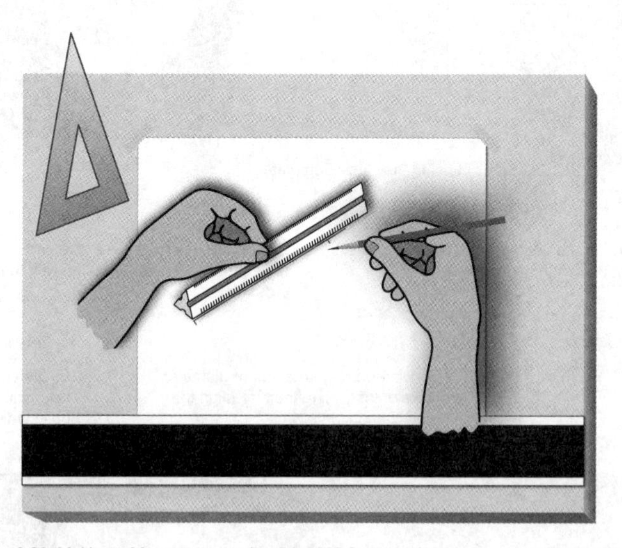

3.32 Making a Measurement along the 30° Construction Line

STEP–BY–STEP

Step 3. Align the edge of the 30°/60° triangle with the construction line and draw a dark line along the top edge of the triangle between the tick marks located in Step 2 (Figure 3.33).

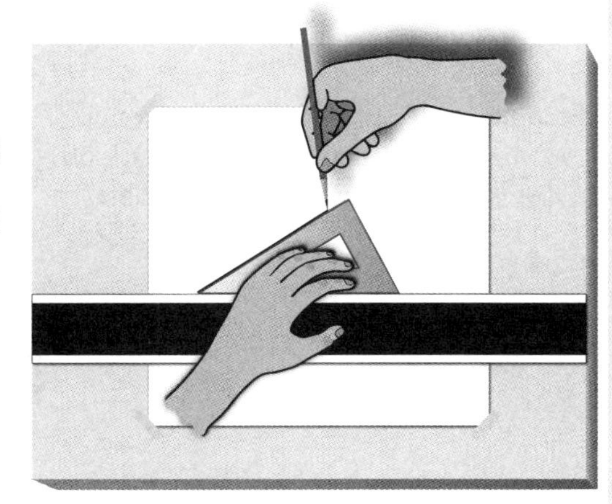

3.33 Darkening the Line

Step 4. Draw the darkened visible line the desired distance (Figure 3.34).

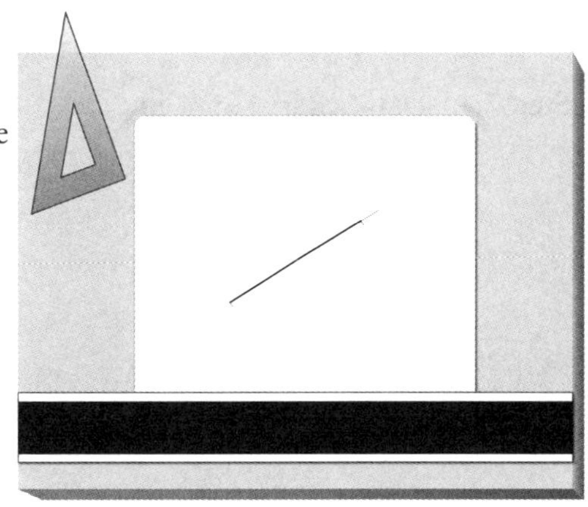

3.34 The Line Drawn to the Desired Length

Step 5. Slide the horizontal bar until it is aligned with the lower tick mark and lightly draw a horizontal construction line (Figure 3.35).

3.35 Drawing a Horizontal Construction Line

STEP–BY–STEP

Step 6. Measure along the horizontal construction line and mark off the required distance with a small tick mark (Figure 3.36).

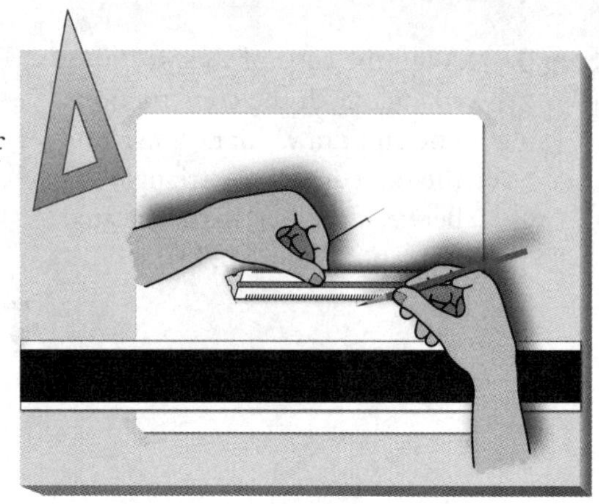

3.36 Making a Measurement along the Horizontal Construction Line

Step 7. Draw a dark visible line between the tick marks (Figure 3.37).

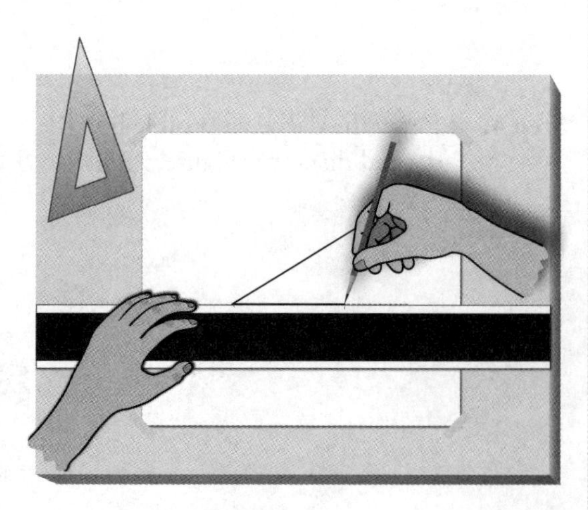

3.37 Darkening the Horizontal Line

STEP-BY-STEP

Step 8. Because the angle between the ends of the lines does not match an angle on either the 45° or 30°/60° triangle, move the triangle until it is aligned with the endpoints of each line and draw a dark line connecting them (Figure 3.38). The completed drawing is shown in Figure 3.39.

3.38 Connecting the Endpoints of the Lines

3.39 The Finished Drawing

3.11 SPACING VIEWS EQUALLY ON A SHEET

The multiviews of an object are often equally spaced both horizontally and vertically on the sheet as shown in Figure 3.40. One method for determining the horizontal spacing between views is to take the overall width of the sheet and subtract the width of the front view and the depth of the side view and divide the remainder by 3. To find the vertical spacing between views, take the overall height of the sheet and subtract the height of the front view and the depth of the top view and divide the remainder by 3.

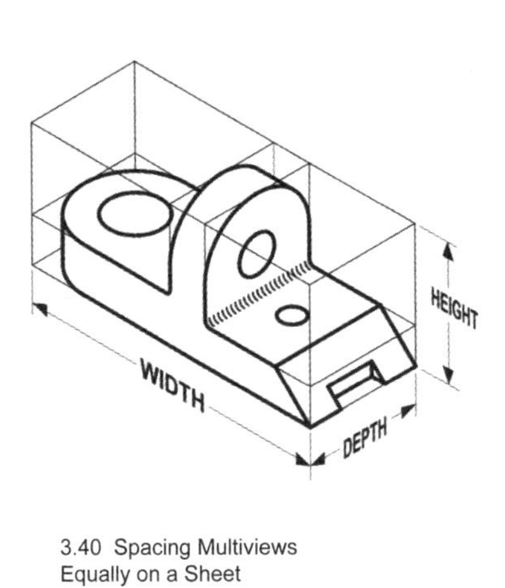

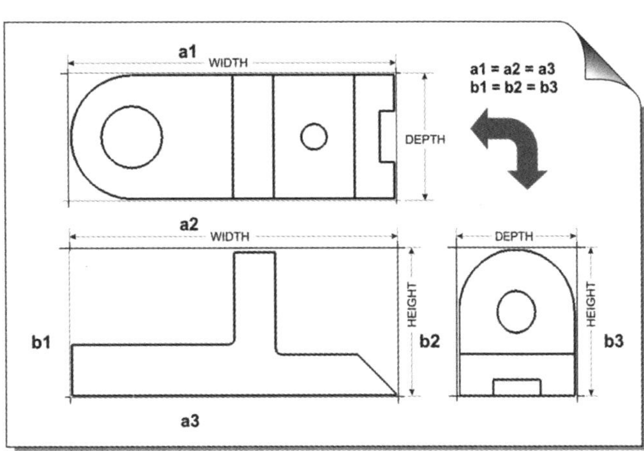

3.40 Spacing Multiviews Equally on a Sheet

STEP–BY–STEP

STEPS IN CONSTRUCTING A MULTIVIEW DRAWING

The steps below show how to construct the views for the object in the Traditional Drafting Project 3.1 presented later in this chapter. These views are drawn on the tear-out sheet located at the back of the textbook. Refer to Figure 3.47 for the dimensions and features of the object in the examples below.

Step 1. Study Figure 3.47 and locate the dimensions for the width of the front view (63mm), height of the front view (47.6mm), and depth of the top view (47.6mm), and measuring with the Metric 1:100 scale, draw light construction lines to "box in" the front, top, and side views as shown in Figure 3.41. Use construction lines to project the width of the object's front view to the top view. Using light lines, project the height of the front view to the side view and project the depth of the top view to the side view using the miter line.

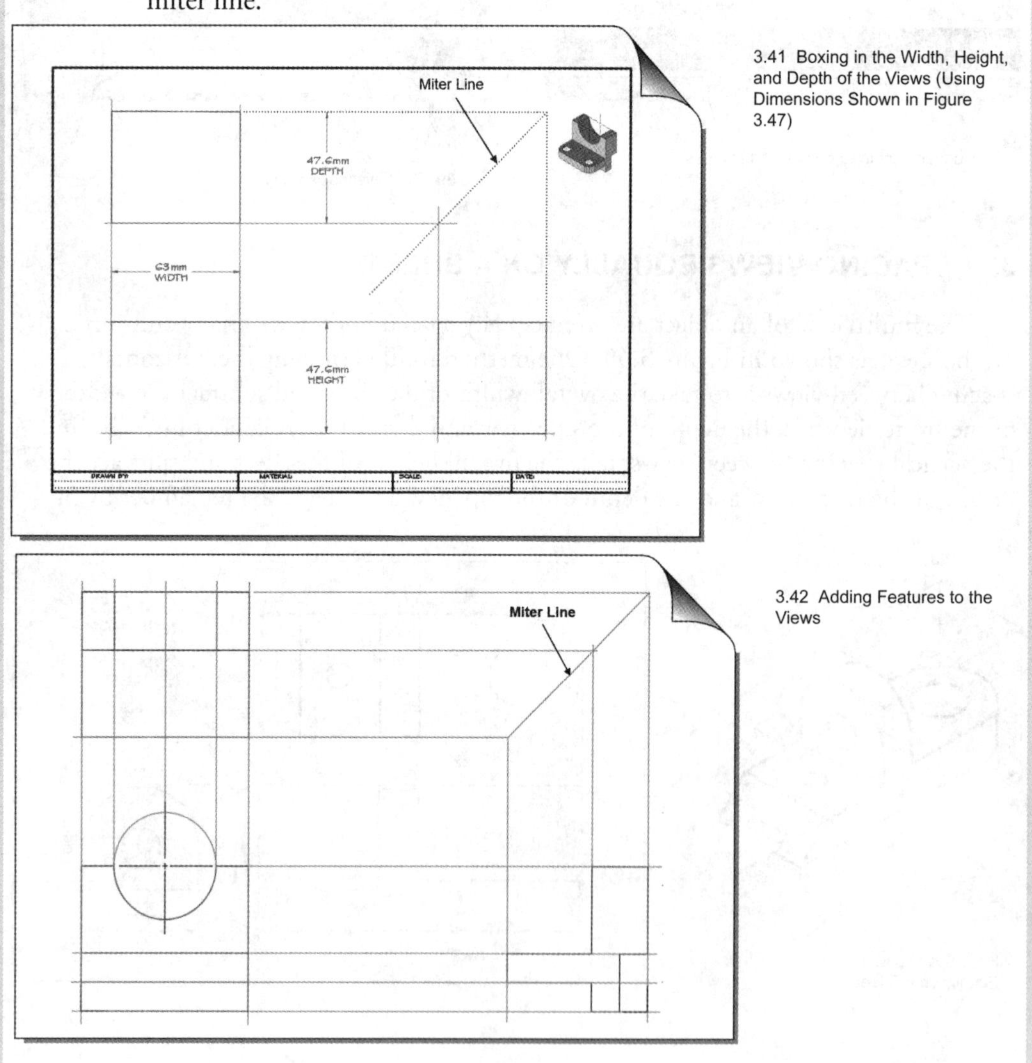

3.41 Boxing in the Width, Height, and Depth of the Views (Using Dimensions Shown in Figure 3.47)

3.42 Adding Features to the Views

STEP–BY–STEP

Step 2. Using light construction lines, continue to add basic features to the views as shown in Figure 3.42. The 38mm diameter circle (R19 x 2) in the front view is drawn in lightly using a circle template.

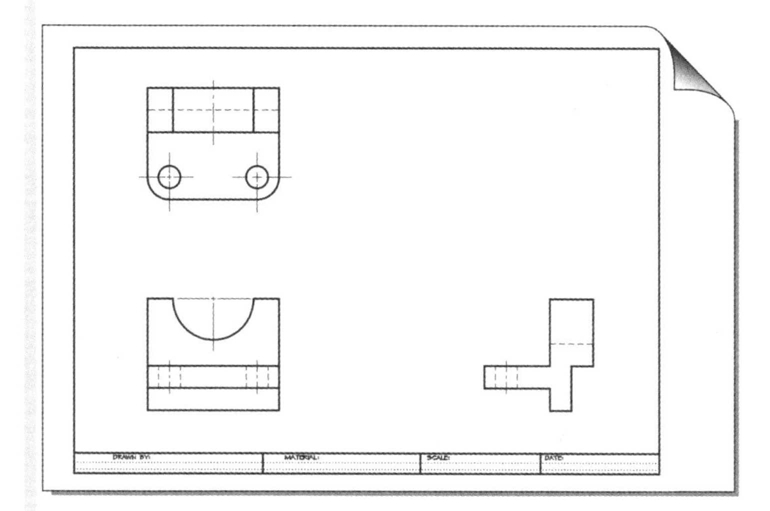

3.43 Projecting Hidden and Center Lines of Holes

Step 3. Using light construction lines, add other features as needed, such as the fillets (rounded corners) and holes to the top view. Project the hidden and center lines for holes and other features between the views using light lines as shown in Figure 3.43.

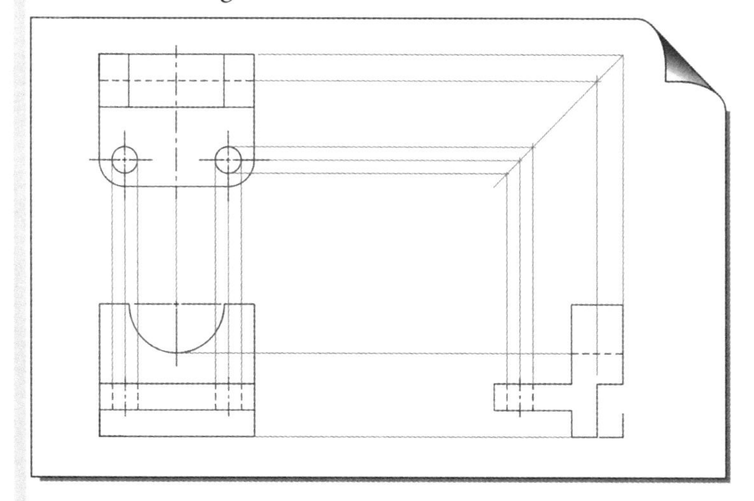

3.44 The Views of the Object with Darkened Visible, Hidden, and Center Lines

Step 4. Erase the construction lines and darken the visible, hidden, and center lines to complete the views as shown in Figure 3.44.

JOB SKILL

It is best to draw all the views using light construction lines. If you make a mistake, it is easier to erase a construction line than a darkened visible, hidden, or centerline. After you have drawn the views with construction lines, darken the top view first, then the front view, and the side view last. Darkening your lines in this order allows you to work "away" from your darkened views and minimizes smudging.

3.12 ADDING TEXT TO DRAWINGS AND SKETCHES

During the design process, drafters and designers often make sketches or drawings that contain handwritten notes and dimensions. The act of placing handwriting on a technical drawing is referred to as "lettering" the drawing. It is very important that any hand-printed text on a drawing be neat, uniform, and legible. In fact, *lettering* that does not have these qualities would be considered unprofessional in most engineering or architectural offices because illegible handwriting on a design sketch could cause dimensions or notes to be incorrectly translated to a technical drawing which could lead to a costly mistake on a construction site or manufacturing center – it could even cost a designer his or her job. It is hoped that last sentence caught your attention and that you realize that developing good lettering skills is very important to your success as a drafter/designer.

Good lettering is a skill that is mastered only through practice. Over time, most drafters and designers develop their own unique lettering style. Some drafters, especially in the architectural field, develop an artistic lettering style (see Figure 3.45 top) while engineers often adopt a simpler lettering style (Fig 3.45 bottom). Although you may never develop a style that qualifies as "art," with practice you can develop a style that is neat, legible, and uniform.

ABCDEFGHIJKLMNOPQRSTUVWXYZ
1234567890 1'-0" 1/4"=1'0"

ABCDEFGHIJKLMNOPQRSTUVWXYZ
1234567890 1'-0" 1/4"= 1'0"

3.45 Examples of Technical Lettering Styles (Top: *Architectural* Lettering Style; Bottom: *Engineering* Lettering Style)

JOB SKILL

Before placing a line of text on a technical drawing, it is a good practice to use a straightedge to draw two, very light, horizontal lines called guidelines about .125 inches apart and then letter the text between these guidelines. This helps ensure that the lettering is uniform height and orientation.

Developing a Technical Lettering Style

Study the examples of the two lettering styles shown in Figure 3.45. In each of the styles, notice how the characters are uniform with regard to height, width, and style. The angle of vertical strokes should be consistent, and each character should be clear and legible.

In the *guidelines* provided below the lettering examples in Figure 3.45, practice lettering the alphabet and numerals. Try to match the lettering style shown in the example. Lettering should be dark, so press down hard with the pencil when making the strokes that form a letter or numeral.

Although traditional drafting tools are rarely used in modern engineering and architectural offices, many of the same techniques developed by traditional drafters are still used in the creation of drawings with CAD. For example, the location of points and planes is still projected between multiviews; however, instead of using a drafting triangle, a drafter working with AutoCAD® uses a drafting setting called **Ortho** to draw perfectly straight horizontal and vertical lines. An understanding of how angles are measured with a protractor facilitates drawing angles with AutoCAD®.

Drafters and designers sometime use scales to take measurements from plotted drawings (which is not always a good idea, by the way), and an understanding of how to interpret an engineer's, architect's, or metric scale is useful in understanding how scaling applies to CAD drawings.

The creation of freehand sketches with legible lettering remains an important skill in today's design office. Every professional drafter/designer, whether traditional or CAD, should develop a freehand lettering style that is neat, uniform, and legible to facilitate communication among designers, or between designers and their clients.

Lead Hardness Grades
Architect's, Engineer's and Metric Scales
Tick Marks
Construction Lines
45° Triangle
30°/60° Triangle
Circle Template
Ellipse Template
Protractor
Technical Lettering
Lettering Guidelines

Short Answer

1. Name three tools used by traditional drafters to draw parallel horizontal lines.
2. Drafting triangles come in what angles?
3. How is a triangle used with a parallel straightedge to draw a vertical line that is perpendicular to a horizontal line?
4. What is the range of lead hardness for general drafting and sketching?
5. Name three qualities that technical lettering should possess.
6. What unit of measurement is represented by each small increment on the 1:100 metric scale as interpreted by a mechanical engineer?
7. Which scale is used by mechanical engineers to measure drawings in decimal inches?
8. How many degrees is a half-moon protractor divided into?
9. Name three lead widths in which technical pencils can be purchased.
10. What is the multiplier used to convert decimal inches to millimeters?

TECHNICAL LETTERING PRACTICE

The two drafting lettering exercises below are designed to give you additional practice in developing your technical lettering style. Follow the directions below to print the *Architectural Style Lettering Sheet.pdf* and *Mechanical Style Lettering Sheet.pdf* located in the file downloads available for *Technical Drawing 101 with AutoCAD®*. Practice lettering the alphabet and numerals in the guidelines as shown in Figures 3.46(a) and 3.46(b). The goal of this exercise is to help you develop a lettering style that is neat, legible, and uniform.

EXERCISE

EXERCISE 3.1 ARCHITECTURAL LETTERING

Print the Architectural Style Lettering Sheet.pdf located in the file downloads available for Technical Drawing 101 with AutoCAD. Practice lettering the alphabet and numerals in the guidelines as shown in Figures 3.46(a) using the architectural lettering style shown on the sheet as an example.

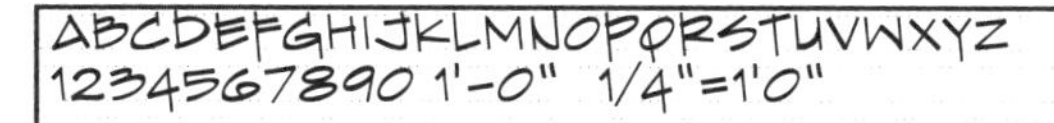

ABCDEFGHIJKLMNOPQRSTUVWXYZ 1234567890 1/8"=1'0"
abcdefghijklmnopqrstuvwxyz

3.46(a) Architectural Lettering Exercise

EXERCISE

EXERCISE 3.2 MECHANICAL LETTERING

Print the Mechanical Style Lettering Sheet.pdf located in the file downloads available for Technical Drawing 101 with AutoCAD. Practice lettering the alphabet and numerals in the guidelines as shown in Figures 3.46(b) using the mechanical lettering style shown on the sheet as an example.

ABCDEFGHIJKLMNOPQRSTUVWXYZ
1234567890 1'-0" 1/4"= 1'0"

ABCDEFGHIJKLMNOPQRSTUVWXYZ 1234567890 1/8"= 1'0"

3.46(b) Mechanical Lettering Exercise

DOWNLOAD

The PDF files for the lettering sheets used in Exercise 3.1 and 3.2 are located inside the *Sketching and Lettering Plates* folder of the book's file downloads. These downloads are available by redeeming the access code that comes with this book. Please see the inside front cover of the book for further details.

Step 1. Carefully remove the sheet for Traditional Drafting Project 1 (refer to Figure 3.48) located at the back of this book and tape it to the top of your drafting board as shown in Figure 3.5.

Step 2. Draw the front, top, and right views of the object in Figure 3.47. The dimensions in Figure 3.47 are given in millimeters, so use the **Metric scale marked 1:100** (refer to Figure 3.18) to draw the views full scale (1 = 1). Refer to the steps shown in Figures 3.41 through 3.44 when constructing the views. When complete, the views should resemble the ones shown in Figure 3.48.

Step 3. Add dimensions to the views as shown in Figure 3.48. Letter your name, the material of the object, today's date, and the drawing scale in the guidelines provided (refer to Figure 3.48).

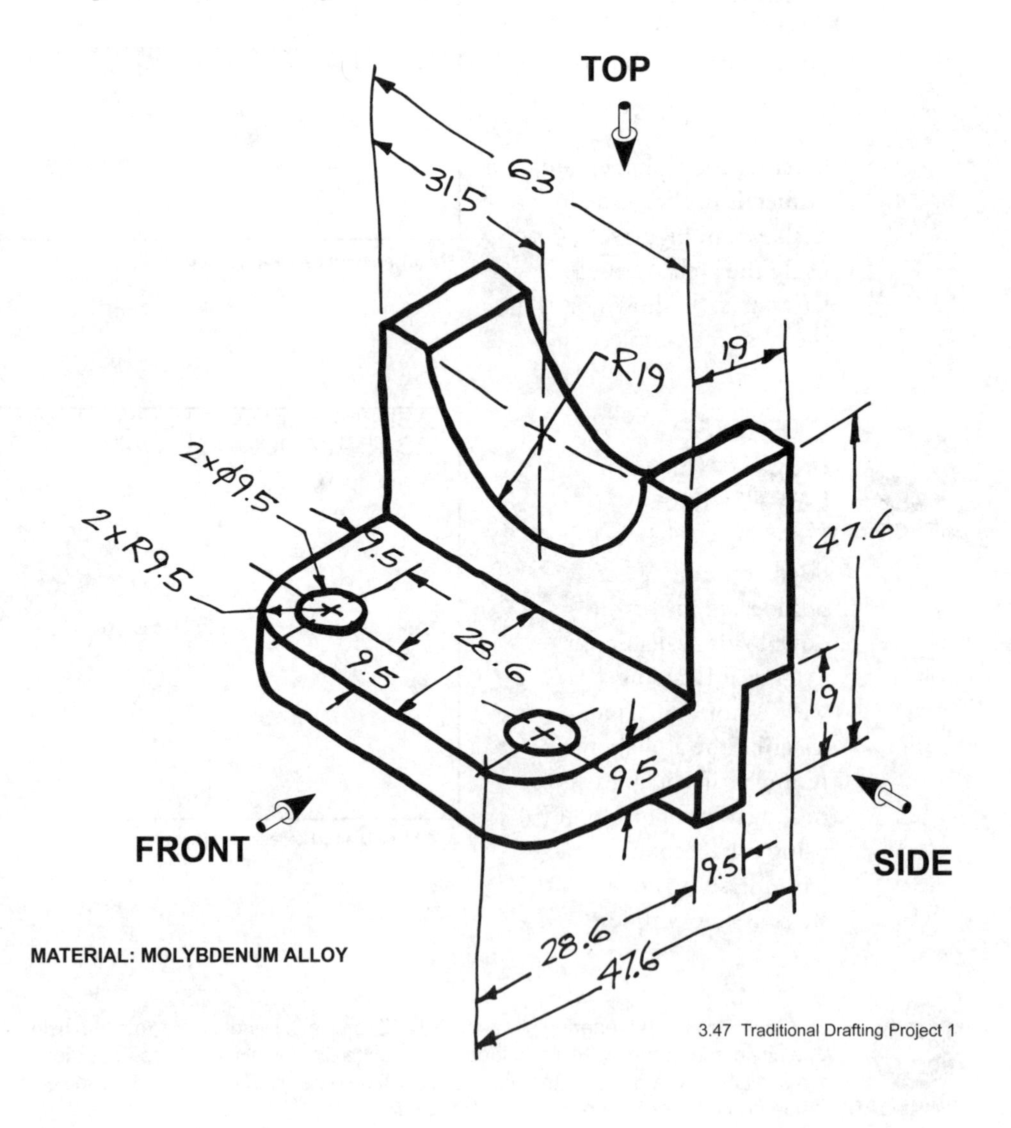

3.47 Traditional Drafting Project 1

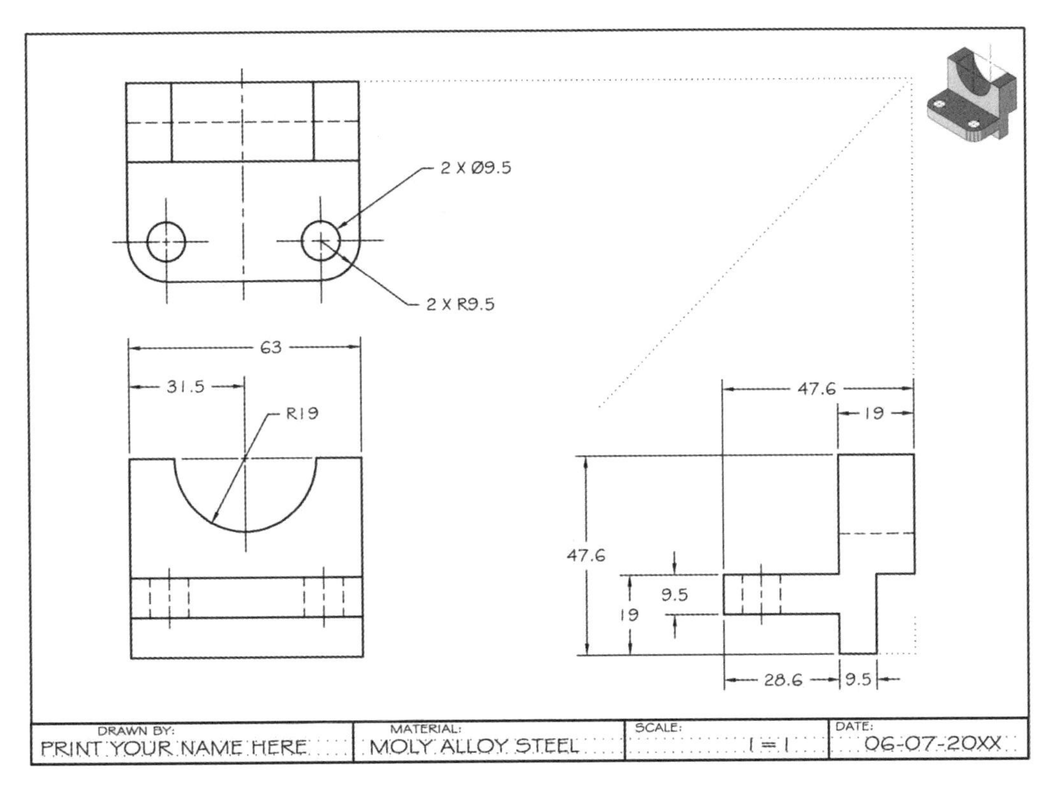

3.48 Traditional Drafting Project 1 with Views, Dimensions, and Notes

Step 1. Carefully remove the sheet for Traditional Drafting Project 2 (refer to Figure 3.50(a)) located at the back of this book and tape it to the top of your drafting board as shown in Figure 3.5.

Step 2. Draw the front, top, and right views of the object in Figure 3.49. Study Figure 3.49 and locate the dimensions for the width of the front and top views (2.50"), height of the front and side views (2.50"), and depth of the top and side views (2.625"), and measuring with the **Engineer 10 scale**, draw light construction lines to "box in" the front, top, and side views as shown in Figure 3.50(a). Using light construction lines add other features as needed, such as the circles and fillet (rounded corner) to the front view. Project the visible, hidden and center lines for holes and other features between the views using light lines as shown in Figure 3.50(b).

Step 3. Add dimensions to the views as shown in Figure 3.48. Letter your name, the material of the object, today's date, and the drawing scale in the guidelines provided (refer to Figure 3.48).

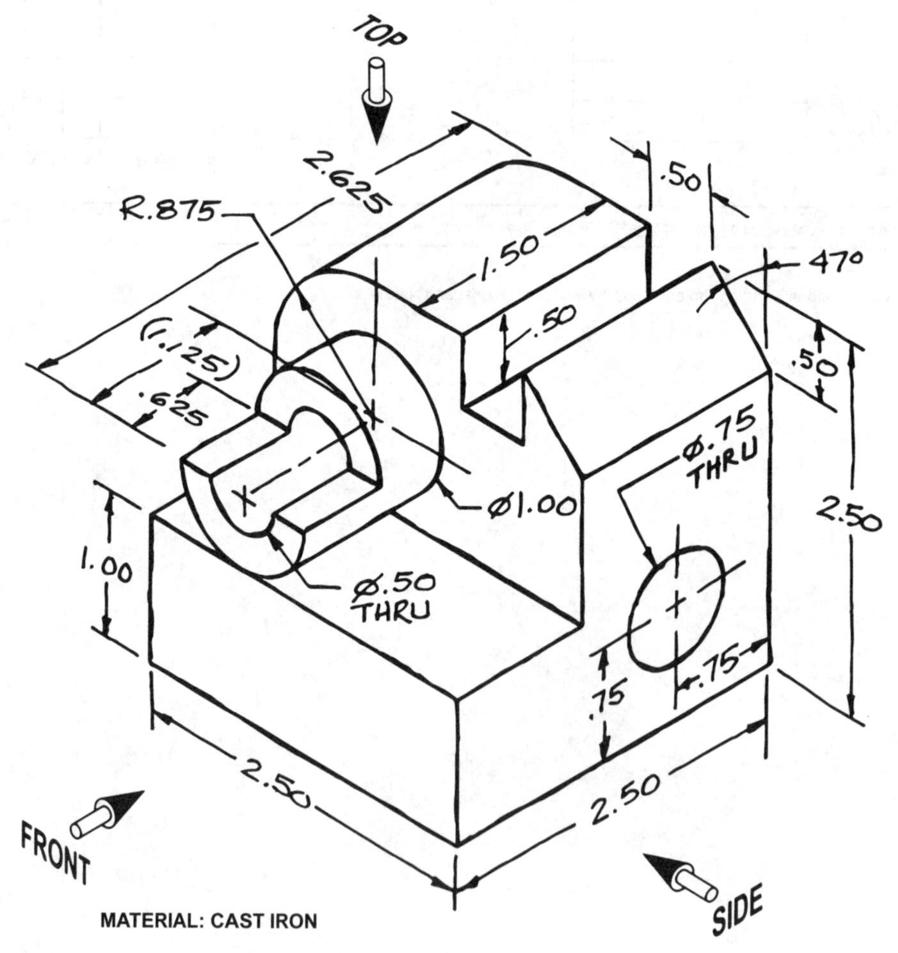

3.49 Traditional Drafting Project 2

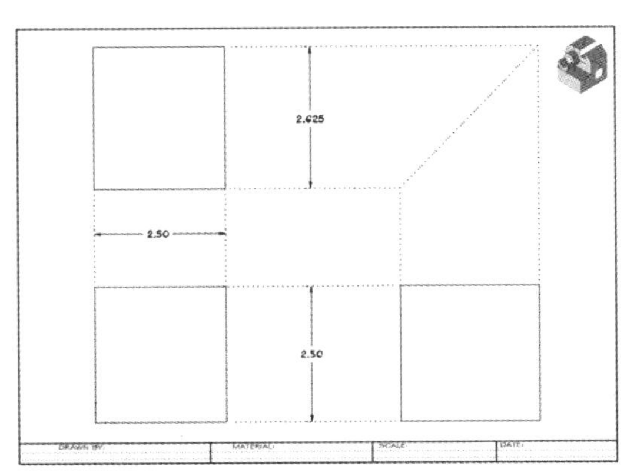

3.50(a) Boxing in the Width, Height, and Depth of the Views (Using Dimensions Shown in Figure 3.49)

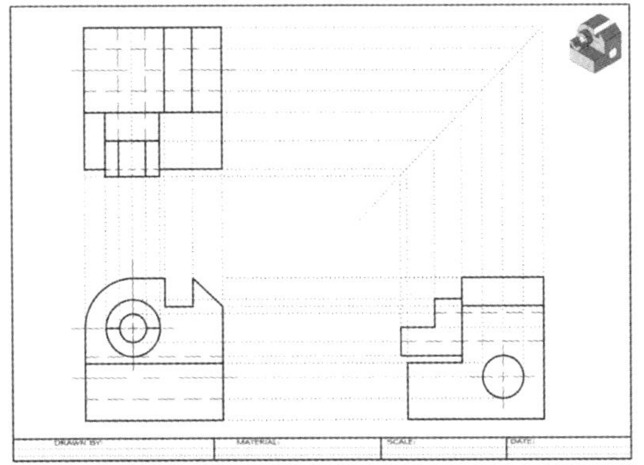

3.50(b) Projecting the Geometry of the Object's Features between the Views

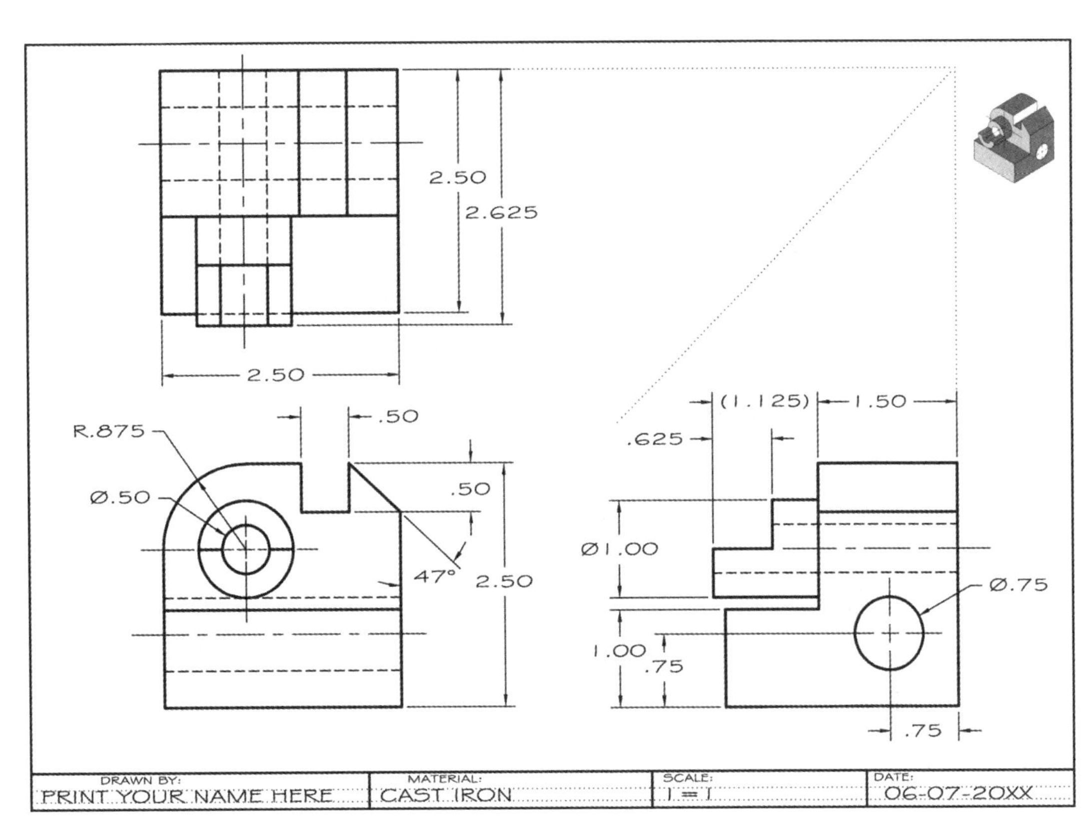

3.50(c) Traditional Drafting Project 2 showing completed Views, Dimensions, and Notes

The three projects that follow are designed to give you additional practice applying the tools and techniques of traditional drafting.

Directions for Optional Projects

Follow your instructor's directions to print copies of the *Traditional Drafting Sheet.pdf* file located inside the *Sketching and Lettering Plates* folder of this book's file downloads. Then, draw the front, top, and right views of the objects in Figures 3.51 through 3.53.

Add dimensions to the views as instructed by your teacher. Letter your name, the material of the part, the date the drawing was created, and the drawing scale in the guidelines provided.

- Use the Engineer's 20 scale to draw the half-scale views of the object in Figure 3.51. Label the drawing scale in the title block as 1=2.
- Use the Metric 1:100 scale to draw the full-scale views of the object in Figure 3.52. Label the drawing scale in the title block as 1=1.
- Use the Engineer's 20 scale to draw the half-scale views of the object in Figure 3.53. Label the drawing scale in the title block as 1=2.

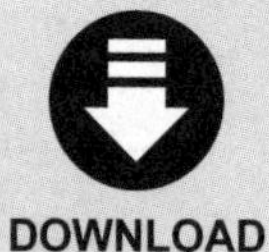

The *Traditional Drafting Sheet.pdf* file used for the optional traditional drafting projects is located inside the *Sketching and Lettering Plates* folder of this book's file downloads. These downloads are available by redeeming the access code that comes with this book. Please see the inside front cover of this book for further details.

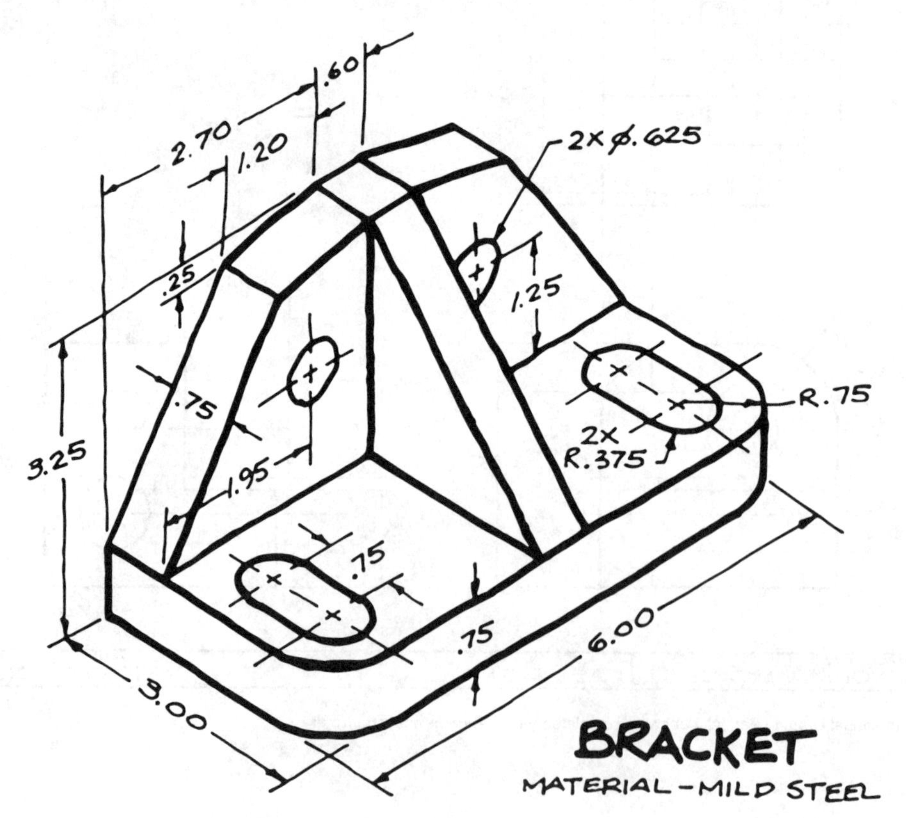

3.51 Traditional Drafting Project 3.3 (Engineer 20 Scale)

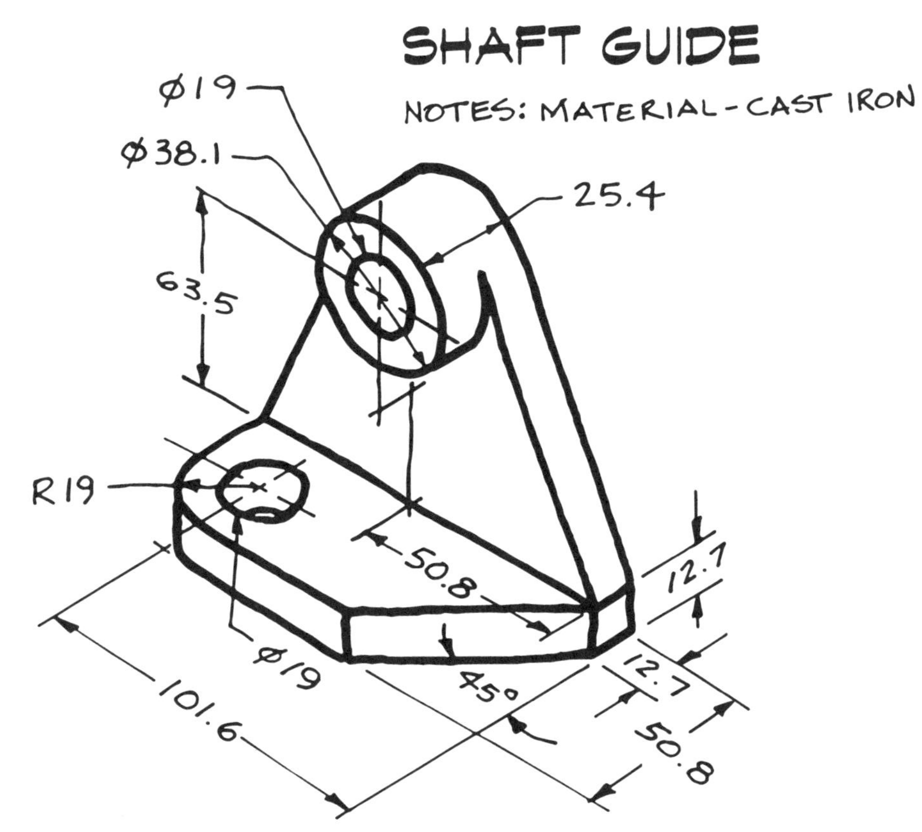

3.52 Traditional Drafting Project 3.4 (Metric 1:100 Scale)

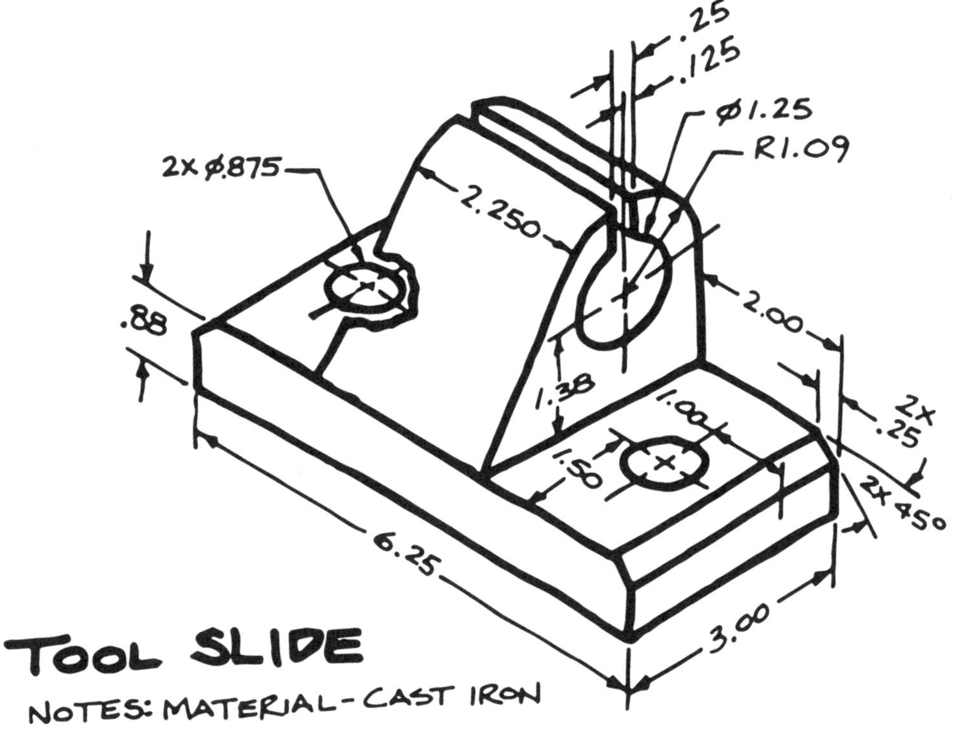

3.53 Traditional Drafting Project 3.5 (Engineer 20 Scale)

CHAPTER

4

COMPUTER - AIDED DESIGN BASICS

CHAPTER OBJECTIVES:

After studying the material in this chapter, you should be able to:

1. Describe the AutoCAD user interface and screen layout.
2. Add toolbars and menus to the AutoCAD user interface.
3. Create, open, and save AutoCAD drawing files.
4. Explain the coordinate system used to create AutoCAD drawings.
5. Draw horizontal and vertical lines using AutoCAD's direct entry method.
6. Draw lines with AutoCAD using absolute, relative, and polar coordinates.
7. Set the units, limits, layers, linetypes, and lineweights for an AutoCAD drawing.
8. Create and edit AutoCAD drawings using AutoCAD's Draw and Modify commands.
9. Employ Object Snap tools to facilitate accurate construction of AutoCAD drawings.
10. Use the Properties tool to inquire about, or to change the properties of, an entity.
11. Add text to a drawing and manage the text's height, font, and justification settings.
12. Create a new text style and edit text on a drawing.
13. Create a floor plan for a small cottage.
14. Create multiview drawings of machine parts.
15. Plot AutoCAD drawings.

CHAPTER OVERVIEW

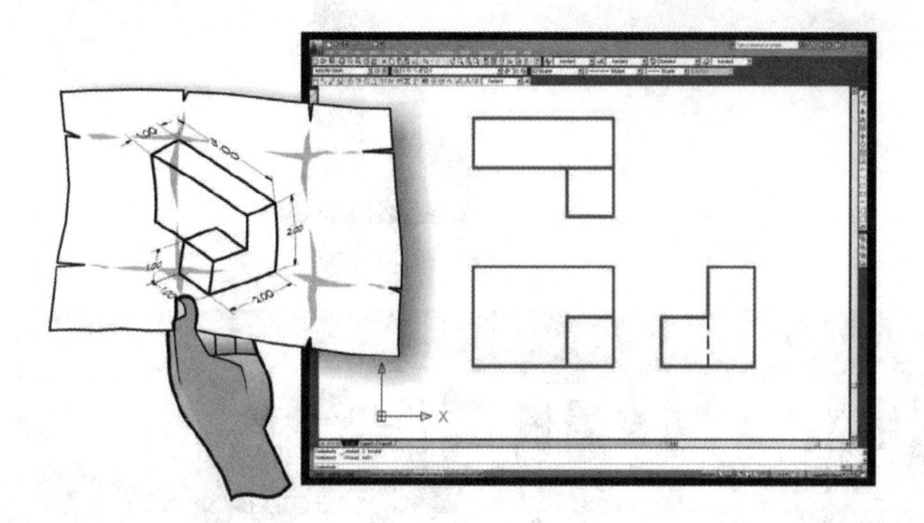

COMPUTER-AIDED DESIGN BASICS:

In most engineering and architectural offices, drafters and designers produce technical drawings using CAD systems. A CAD system consists of a personal computer (PC) or workstation coupled with a CAD software program. One of the most widely used CAD software programs is called AutoCAD®. When AutoCAD was introduced in 1982, it was one of the first CAD programs that could operate on a PC. Autodesk, the parent company that publishes AutoCAD software, reports that there are now millions of AutoCAD users worldwide. The price for a single station of AutoCAD for a professional user is about $4,350, but at the present time, Autodesk offers free downloads of the latest release of AutoCAD (and other CAD products) to students, educators, and educational institutions at its website.

There are many other CAD programs on the market as well. Some CAD programs are designed to perform work in a specialized area. In mechanical design, Autodesk® Inventor®, PTC Creo®, and SOLIDWORKS® are three of the principal CAD programs. In electronics design, Cadence Virtuoso®, Cadence Allegro®, and Mentor Graphics Calibre® are widely used. In the civil and architectural fields Autodesk® AutoCAD® Civil 3D®, Bentley MicroStation®, and Autodesk® Revit® are popular CAD programs.

- Autodesk® (AutoCAD®, Revit®, Inventor®): www.autodesk.com
- Bentley (MicroStation®): www.bentley.com
- Cadence (Allegro®, Virtuoso®): www.cadence.com
- Mentor Graphics (Calibre®): www.mentor.com
- PTC (Creo®): www.ptc.com
- SOLIDWORKS®: www.solidworks.com

4.1 BEGINNING AN AUTOCAD DRAWING

Use the mouse's left-click button to double-click on the AutoCAD icon located on the desktop of your computer. This will launch the AutoCAD program.

When the program opens, the **Create Page** will appear similar to the one shown in Figure 4.1. From this page, the user can start a new drawing, open AutoCAD template files, open existing drawing files, and open recently used drawing files. The Create Page also displays notifications regarding updates and other information from Autodesk. Clicking on the **Learn Page** button at the bottom of the page opens a page containing learning resources including online videos, tips, and other content. (Note: If there is no internet, the Learn Page is not displayed.)

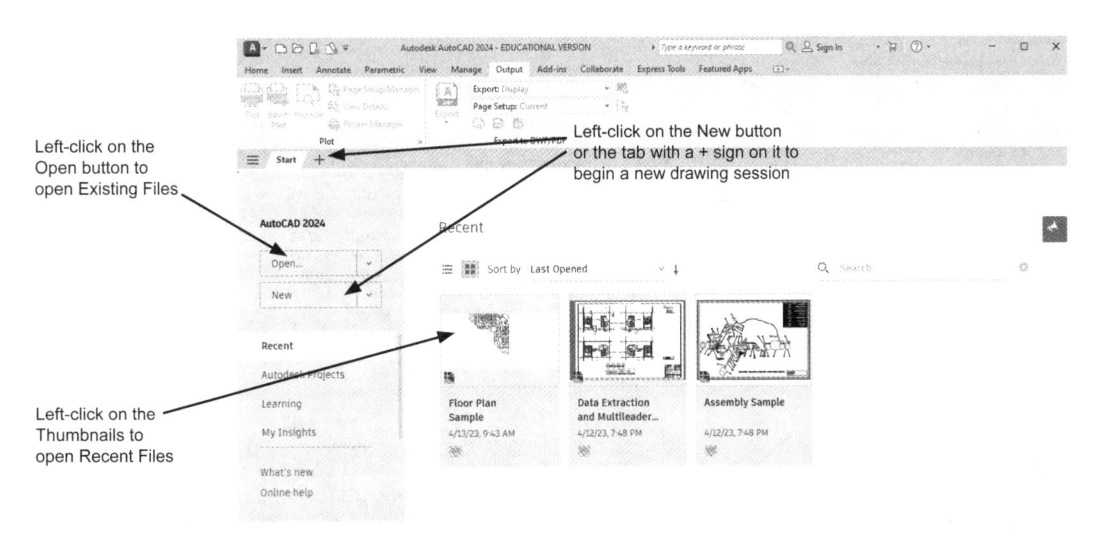

4.1 The AutoCAD Create Page

Begin a new drawing session, by left-clicking the mouse on the Start Drawing button shown in Figure 4.1. When the drawing session opens, the AutoCAD user interface will appear (see Figure 4.2). Study the AutoCAD interface shown in Figure 4.2 and acquaint yourself with the features. Your instructor may call your attention to these features as you proceed with your CAD training. This textbook often refers to these features as well.

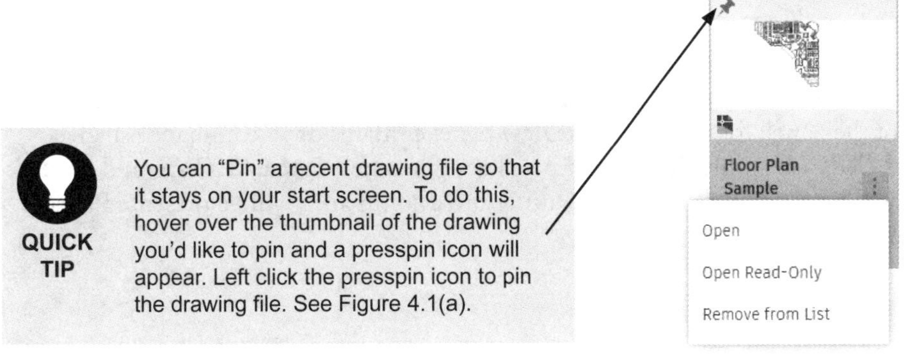

QUICK TIP

You can "Pin" a recent drawing file so that it stays on your start screen. To do this, hover over the thumbnail of the drawing you'd like to pin and a presspin icon will appear. Left click the presspin icon to pin the drawing file. See Figure 4.1(a).

4.1(a) Pin Existing Drawing to Start menu

Find the ***command line*** located along the bottom edge of Figure 4.2; it is very important for beginners to refer to the command line frequently because it offers important prompts and cues necessary to successfully complete AutoCAD commands.

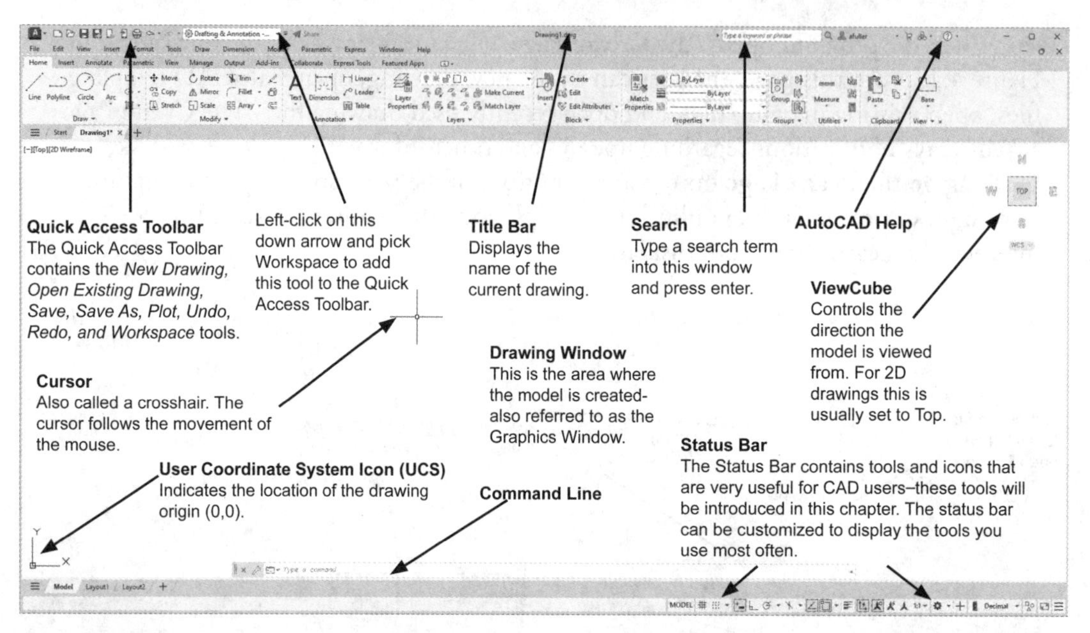
Quick Access Toolbar
The Quick Access Toolbar contains the *New Drawing, Open Existing Drawing, Save, Save As, Plot, Undo, Redo, and Workspace* tools.
Left-click on this down arrow and pick Workspace to add this tool to the Quick Access Toolbar.
Title Bar
Displays the name of the current drawing.
Search
Type a search term into this window and press enter.
AutoCAD Help
ViewCube
Controls the direction the model is viewed from. For 2D drawings this is usually set to Top.
Drawing Window
This is the area where the model is created- also referred to as the Graphics Window.
Cursor
Also called a crosshair. The cursor follows the movement of the mouse.
User Coordinate System Icon (UCS)
Indicates the location of the drawing origin (0,0).
Command Line
Status Bar
The Status Bar contains tools and icons that are very useful for CAD users–these tools will be introduced in this chapter. The status bar can be customized to display the tools you use most often.

4.2 The AutoCAD Screen Layout in the Default Workspace

JOB
SKILL

Whenever you need help or information regarding a command or feature of the AutoCAD program, press the <F1> function key located on the keyboard (or by clicking on the AutoCAD *Help* icon shown in Figure 4.2). This will open AutoCAD's Help system window. You can also search for information by entering a search term in the *Search* window shown in Figure 4.2.

Locate the **Ribbon** noted in the AutoCAD user interface in Figure 4.3; in Windows-based programs like AutoCAD, the ribbon provides a framework of *tabs* and *panels* for organizing AutoCAD's commands, tools and menus. As you proceed through this textbook, you will often be directed to select (or *left-click* on) an icon located on this ribbon in order to launch one of AutoCAD's commands.

In Figure 4.3 the **Home** tab of the ribbon is selected, which makes the **Draw**, **Modify**, **Annotation**, **Layers**, **Block**, **Properties**, **Groups**, **Utilities**, **Clipboard** and **View** panels visible. Each panel contains a set of commands that are related to the function of the panel; for example, the Draw panel contains draw commands such as **LINE** and **CIRCLE**, while the **Modify** panel contains the **ERASE**, **MOVE**, and **COPY** commands. The techniques involved in using these commands will be presented in this chapter.

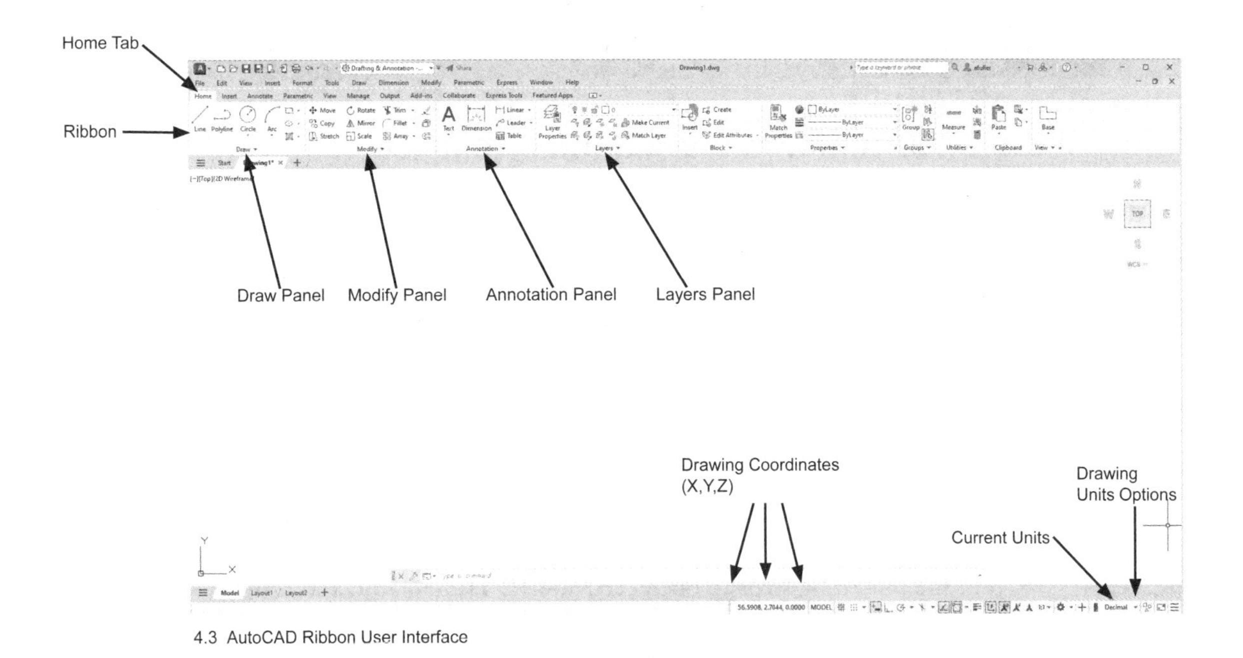

4.3 AutoCAD Ribbon User Interface

4.2 ADDING TOOLBARS TO THE RIBBON INTERFACE

To make the drafting environment easier to use for beginners, and more efficient for experienced users, the authors recommend the following additions to the user interface:

Turn on the **menu bar**. The menu bar contains 13 drop-down menus that provide AutoCAD users easy access to many important commands and settings (see Figure 4.4 for the menu bar's location). The **Format** and **Tools** menus are especially useful to users of AutoCAD and are referred to later in this chapter.

Make the **Draw**, **Modify**, **Dimension**, and **Object Snap** toolbars visible. These four toolbars contain almost all the commands necessary to produce 2D multiview drawings with AutoCAD - and although most of these commands are also located on the ribbon panels, it is often quicker to initiate a command from a toolbar because only *one* mouse left-click is required - versus *two* clicks to open a command whose icon is not initially visible in the panel (and thus requires a second click on the panel's down arrow in order to expand the panel and display the icon so that it can be selected).

QUICK TIP

FLOATING DRAWING WINDOW—IDEAL FOR DUAL MONITORS
AutoCAD offers an option to have floating drawing windows, which allows the user to have two drawings visible at one time. This is ideal for users working with two monitors.

- Left-click the tab of the drawing you'd like to move, then drag and drop it away from the default location (i.e.: drag it to your second monitor).
- To redock the drawing, select the floating tab and drag and drop it back to the original location with the other drawing tabs.

STEP–BY–STEP

STEPS IN ADDING THE MENU BAR AND TOOLBARS TO THE RIBBON INTERFACE

Step 1. To turn on the Menu Bar: type **MENUBAR** on the command line and press **<Enter>**. At the prompt type **1** and press **<Enter>**. When this step is complete, the menu bar shown near the top of the screen in Figure 4.4 will be visible.

Step 2. Move the mouse cursor to the word **Tools** located on the menu bar and left-click, when the submenu drops down, left-click on the word **Toolbars**.

Step 3. When the next submenu appears, left-click on the word **AutoCAD**; this will open the **Toolbar** menu. Move the mouse onto the word **Draw** and left-click; this will open the **Draw** toolbar.

Step 4. Repeat the preceding steps to open the **Modify**, **Dimension**, and **Object Snap** toolbars. By left-clicking on the dark gray bars at either end of a toolbar and holding down the left-click button, you can drag the toolbar to the side of the screen and dock it as shown in Figure 4.4.

Once a toolbar has been opened, additional toolbars can be opened quickly by right-clicking the mouse on the open toolbar and selecting from the menu of toolbars that appears.

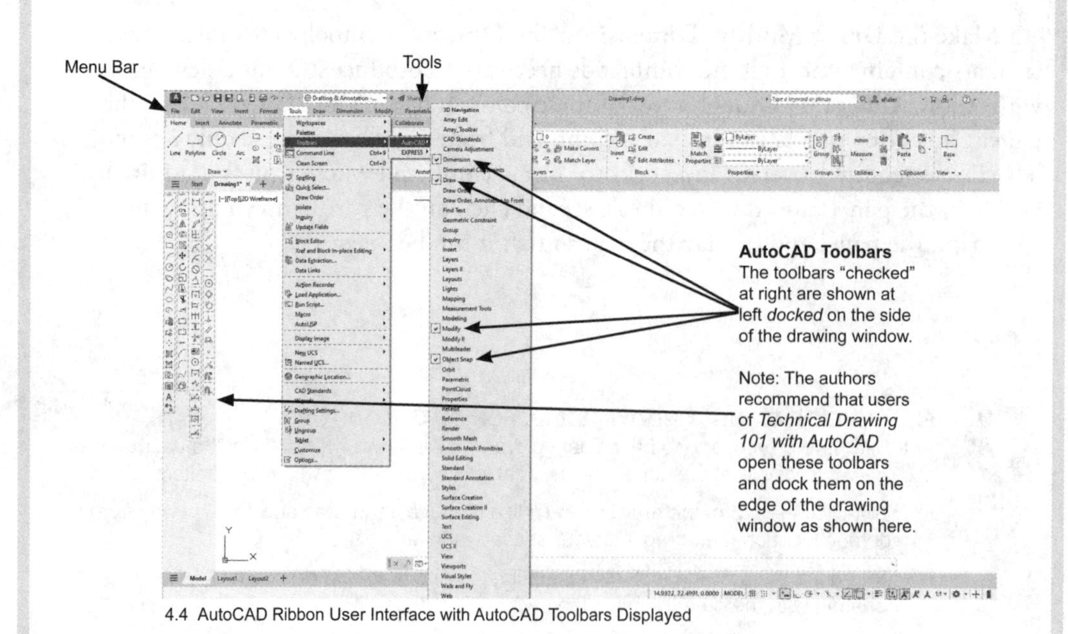

Menu Bar
Tools
AutoCAD Toolbars
The toolbars "checked" at right are shown at left *docked* on the side of the drawing window.
Note: The authors recommend that users of *Technical Drawing 101 with AutoCAD* open these toolbars and dock them on the edge of the drawing window as shown here.

4.4 AutoCAD Ribbon User Interface with AutoCAD Toolbars Displayed

JOB SKILL

By having the toolbars shown in Figure 4.4 displayed, new users of AutoCAD may become familiar with the icons associated with these commands more quickly. Experienced users may instead prefer to open commands from the ribbon, or by typing command aliases. For example, to begin the **LINE** command, an experienced user might type its alias, **L**, and press **<Enter>**. This would launch the Line command. The aliases for many of AutoCAD's commands are presented later in this chapter.

4.3 CREATING, OPENING, AND SAVING AUTOCAD DRAWING FILES

Locate the **Application Button** noted in Figure 4.5 (the red **A** in the upper left corner of the user interface). By left-clicking on this button and choosing from the application menu, users can create a new drawing, open an existing drawing, save the current drawing, or save a copy of the current drawing in a new location or with a different drawing title.

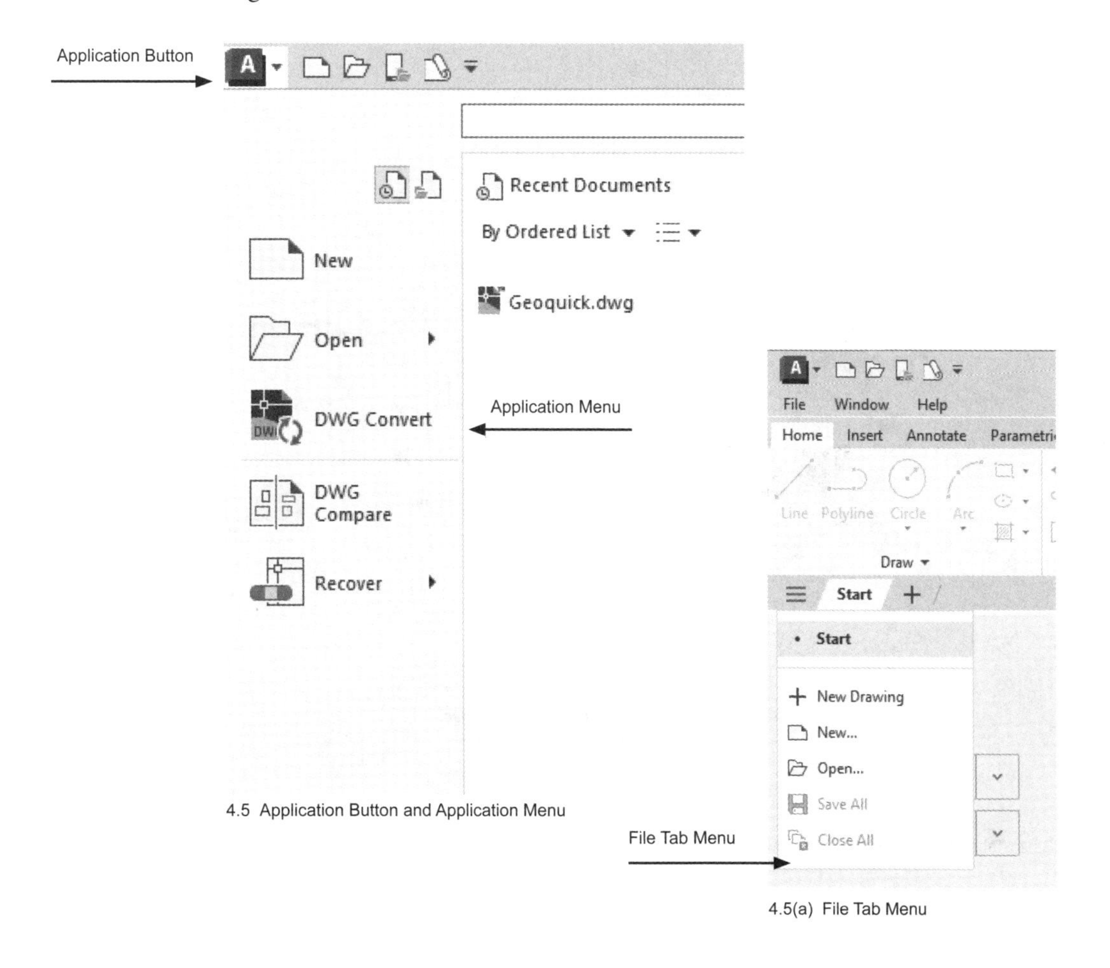

4.5 Application Button and Application Menu

4.5(a) File Tab Menu

NEW

NEW IN RELEASE 2024: FILE TAB MENU

AutoCAD replaced the previous 'Overflow Menu' with a new 'File Tab Menu'.

- Left-click the icon with the 3 horizontal lines (known in computing as a "Hamburger Menu" or "Hamburger Icon" to choose from the following options: New Drawing, New..., Open..., Save All, and Close All

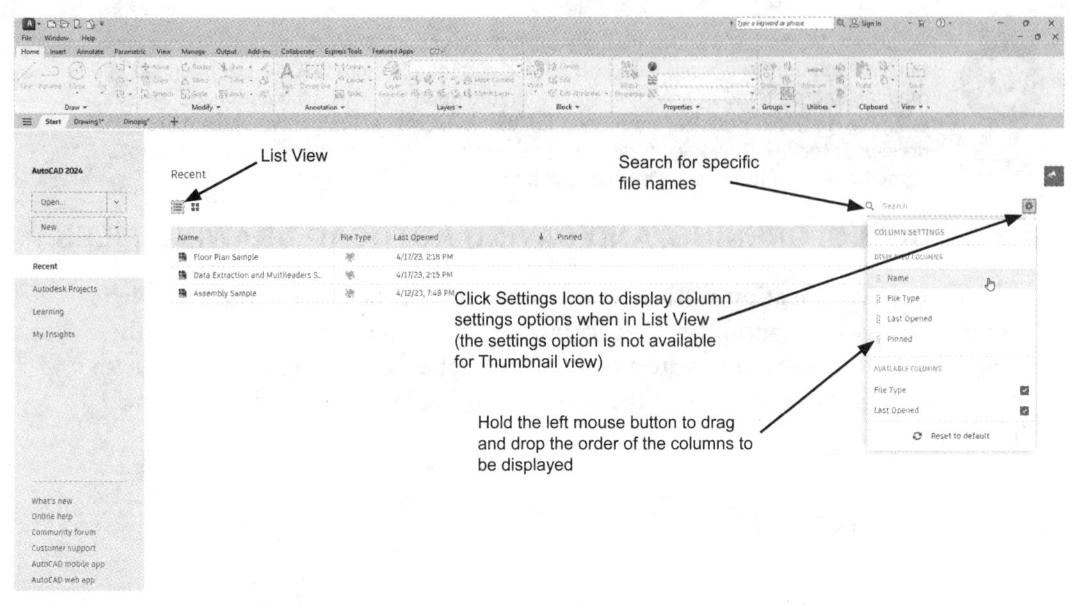

4.6 Smart Tab: Recent Drawings List View

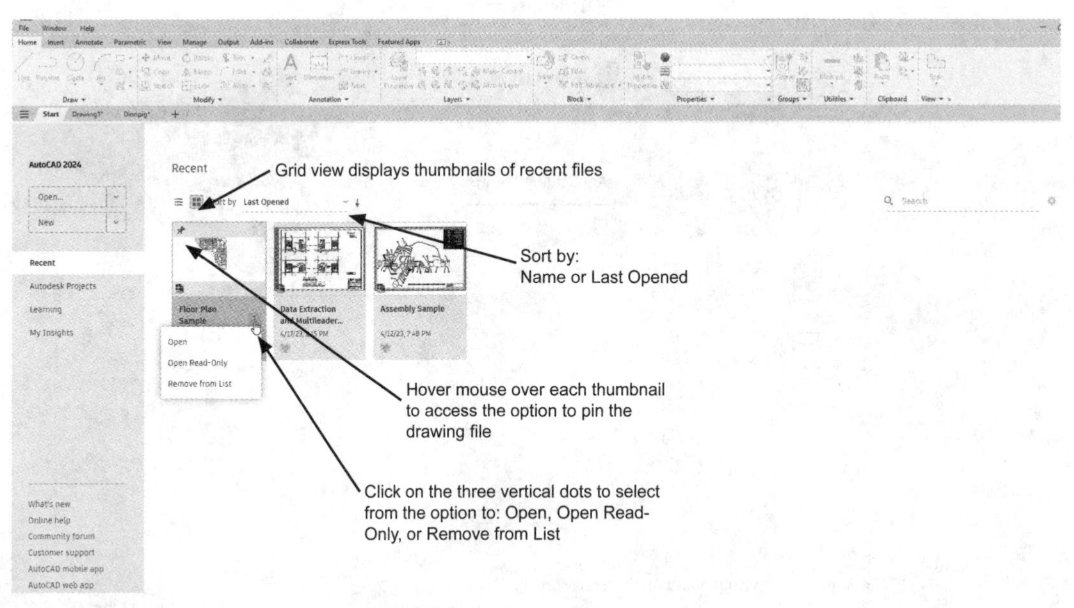

4.7 Smart Tab: Recent Drawings Grid View

NEW

NEW IN RELEASE 2024: SMART TAB UPDATES

AutoCAD updated the Startup screen

- Choose between “List View” or “Grid View”
- When viewing the List view, users have the option to customize column settings
- In Grid view, thumbnails of recent drawings are smaller to allow for more drawings to be listed
- Hover over the thumbnails to view more options

Beginning a New Drawing from the Application Menu

Left-clicking the **New** option from the application menu (see Figure 4.5) opens the **Select Template** window. You can choose from the list of template files (**dwt** file extension) displayed in this box, or by selecting the **Open** button, you can default to the **acad.dwt** template.

After you have selected the **Open** button, a new drawing session will begin. This drawing is automatically named **Drawing1.dwg.** This drawing title will also appear in the title bar at the top of the user interface (see Figure 4.2 for help in locating the title bar).

Opening an Existing Drawing

Left-clicking on the **Open** option from the application menu (see Figure 4.5) opens the **Select File** window. In this window, you can select the file name and location (the drive and folder) of the drawing file to be opened. After making these selections, select the **Open** button to open the file.

Saving a Drawing

To save a new drawing, left-click on the **Save** option located on the application menu (see Figure 4.5) and when the **Save Drawing As** window opens select the drive and folder in which the drawing should be saved and change the file name if desired. After making these changes in the **Save Drawing As** window, select the **Save** button to complete the operation. Once a drawing has been saved, you can continue to save changes made to the drawing by left-clicking on the **Save** option or by picking on the diskette icon located to the right of the **Workspace Switching** drop-down menu (see Figure 4.5).

AutoCAD drawing file names end with a dwg extension. For example, a drawing named **House Plan 1** will appear in a Windows Explorer folder as **House Plan 1.dwg**. Two things occur each time the drawing is saved: first, the dwg file is saved reflecting the most recent edits to the drawing, and second, the extension of the previously saved version of the drawing is changed from dwg to *bak*. The bak extension indicates that the file has been converted to a *backup* copy. In the event that the most recent drawing file is lost due to a software or hardware malfunction (or user error), you can rename the backup copy of the file by replacing the *bak* in the extension with *dwg*. You may then open the newly renamed drawing file and resume working on the drawing. Unfortunately, the drawing will have lost any edits made between the last save and the malfunction, so these edits will have to be redone.

Performing a Save As

Save As is used to rename a current drawing and/or save the drawing to a new location. Left-clicking on the **Save As** option from the application menu (see Figure 4.5) opens the **Save Drawing As** window. In this window, you can select the drive and folder in which the drawing should be saved and change the file name as desired. After making the desired changes, select the **Save** button to complete the operation. The newly named drawing file will become the current drawing.

STEP–BY–STEP

CUSTOMIZING THE AUTOCAD STATUS BAR

The AutoCAD user interface includes a status bar along its bottom-right border providing convenient access to basic drawing tools and settings. Icons representing these settings can be added or removed from the status bar based on the user's preference. Use the steps listed below to customize the icons on the status bar:

Locate the status bar and left-click the *status bar customization* button identified in Figure 4.8.

Step 1. From the *status bar customization menu*, select the settings you wish to add (or remove) from the bar. If using Technical Drawing 101 with AutoCAD curriculum, add the settings "checked" in Figure 4.8.

Step 2. Close the customization menu by left clicking in the graphics window. The icons selected now appear on the status bar. Note that when a setting is *ON* the icon is illuminated but the icon is not illuminated when *OFF*.

NOTE

IMPORTANT NOTE TO USERS OF TECHNICAL DRAWING 101 WITH AutoCAD
The instructions described in this textbook are written with the assumption that users have the Dynamic Input icon shown in Figures 4.8 and 4.9 turned, or *"toggled"*, *ON*. Dynamic Input can be toggled *ON* (or *OFF*) by left clicking on the Dynamic Input icon on the status bar or by pressing the <F12> key located on the keyboard.

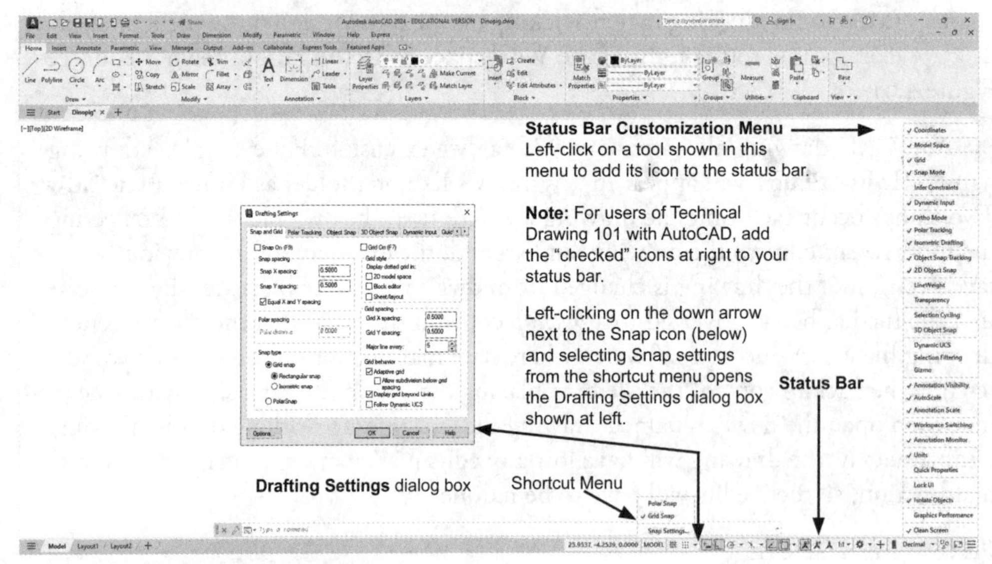

4.8 AutoCAD Status Bar Customization

JOB SKILL

For quick access to the Drafting Settings dialog box shown in Figure 4.8, left click the down arrow next to the Snap icon shown in Figure 4.8 and select Snap Settings on the shortcut menu. The AutoCAD Drafting Settings dialog box is covered in detail in Section "4.12 Drafting Settings Dialog Box" of this chapter.

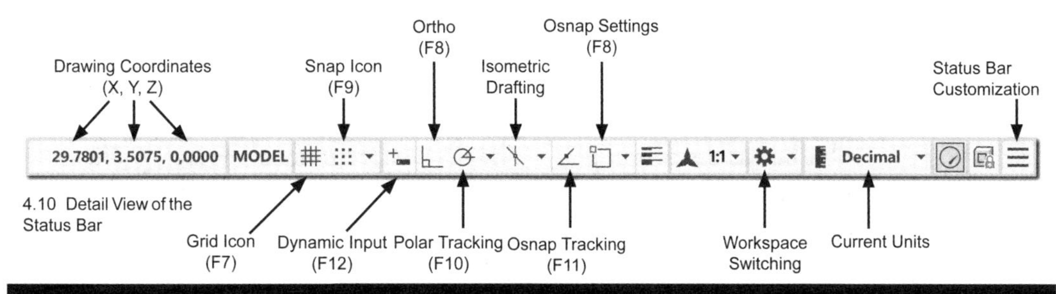

4.10 Detail View of the Status Bar

STEP–BY–STEP

DRAWING YOUR FIRST LINE WITH AUTOCAD

Step 1. Use the left-click button of your mouse to select the icon from the toolbar, or select the **LINE** icon from the **Draw** panel located on the ribbon's **Home** tab, or type the letter **L** (the Line command's *alias*) at the command line and press **<Enter>**.

Step 2. Move the cursor into the graphics window and *pick* a random location inside the window with the left-click button, then move the mouse to a new position and pick again. Congratulations, you've drawn your first line! By continuing to pick points in the graphics area, you can add to the line (note the *Dynamic prompt* and *Input* fields shown in Figure 4.10 – you'll need these later). When you are finished, press the **<Esc>** (escape) or **<Enter>** key to end the **LINE** command. Drawing lines in this manner to *random* points is easy; drawing lines to exact points is a bit more complicated. For this, you'll need to understand how to draw lines with *absolute coordinates*, *direct entry*, *relative coordinates*, and *polar coordinates*. These techniques are presented in Sections 4.5-4.8.

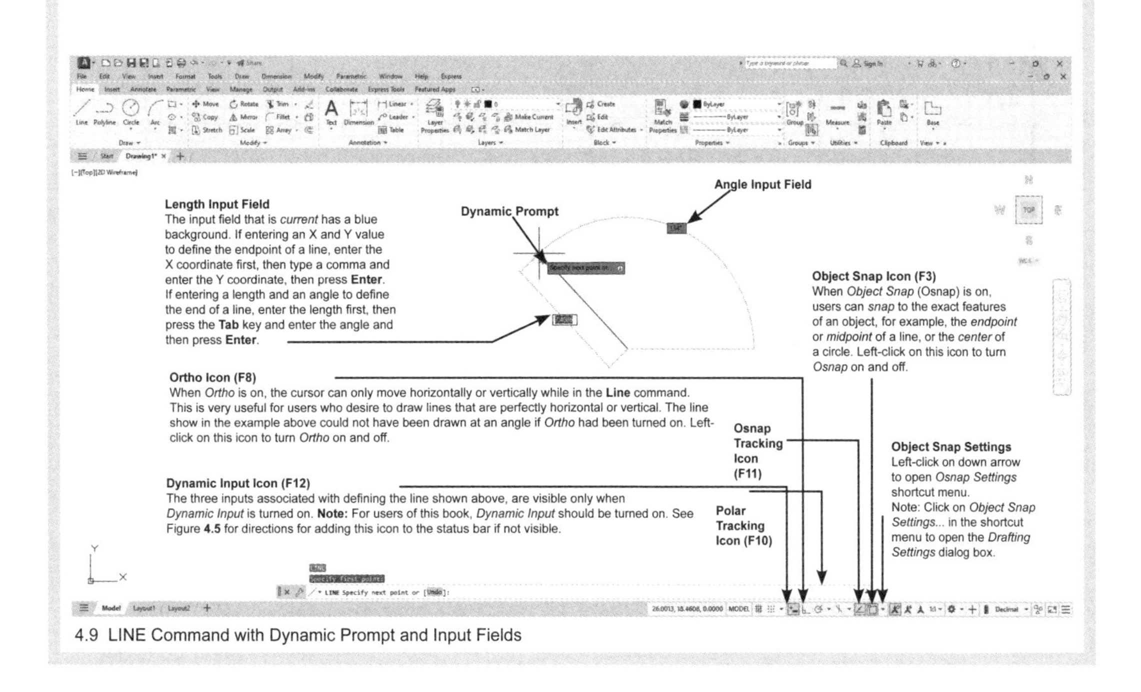

4.9 LINE Command with Dynamic Prompt and Input Fields

4.4 LOCATING POINTS ON THE CARTESIAN COORDINATE SYSTEM

AutoCAD employs the ***Cartesian coordinate system*** to define the exact location of points in the graphics window. In the Cartesian coordinate system, a **0,0** (zero, zero) point is established as the **origin** point. The first zero represents the start point of measurements along the *X*-axis (horizontal), and the second zero represents the start point of measurements along the *Y*-axis (vertical). All other points are located along the *X*- and *Y*-axes using **0,0** as the starting point. In Figure 4.11 the **0,0** point is located in the lower left corner. The coordinates of the other points labeled on the grid refer to each point's location measured along the *X*- and *Y*-axes relative to **0,0**.

Locate the point labeled with the coordinates **1,2** in Figure 4.11. This point is located on the grid by starting at the **0,0** origin in the lower left corner of the grid and measuring 1 unit to the right along the *X*-axis and up 2 units along the *Y*-axis. The *X*- and *Y*-values are separated with a comma. CAD drafters refer to this point as **1,2.**

Next, locate the point labeled with the coordinates **4,3** in Figure 4.11. This point is found by starting at the **0,0** origin in the lower left corner of the grid and measuring 4 units to the right along the *X*-axis and up 3 units along the *Y*-axis. CAD drafters refer to this point as **4,3.** Lines drawn in two dimensions have a start and an endpoint. Both points are defined by their respective *X*- and *Y*-coordinates.

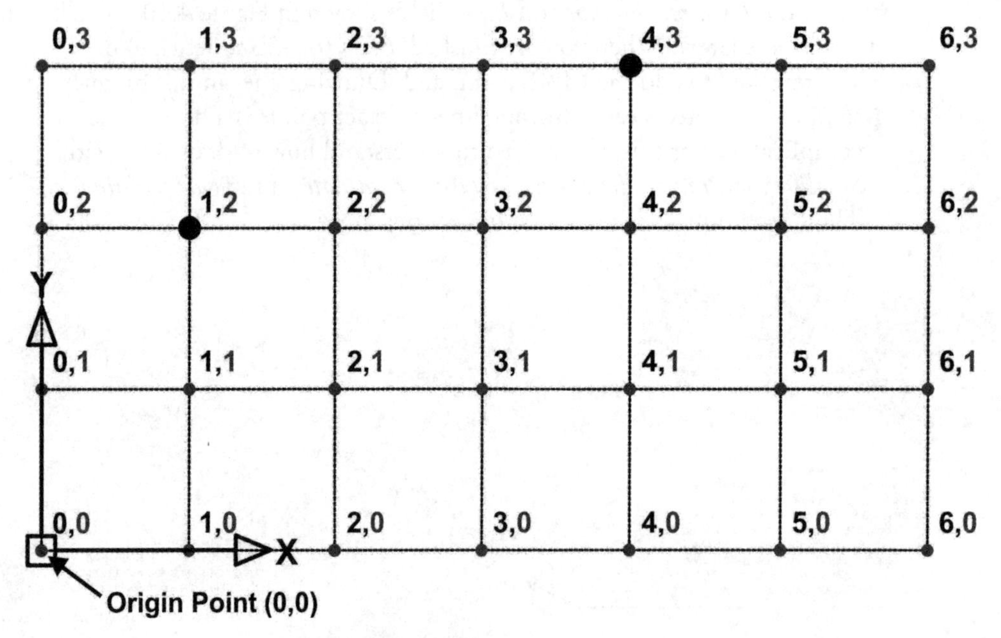

4.11 Location of Points on the Cartesian Coordinate System

The User Coordinate System (UCS) Icon

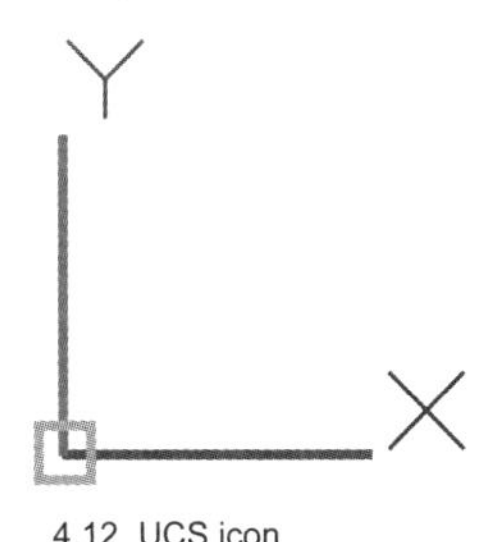

4.12 UCS icon

In AutoCAD the **0,0** (or *origin*) point in the graphics window is represented in the lower left corner of the screen by the icon shown in Figure 4.12. This icon is called the ***user coordinate system (UCS)*** icon. The visibility of this icon can be controlled by typing **UCSICON** at the command line, pressing **<Enter>**, and selecting **On** or **Off** from the settings listed. This icon orients the CAD operator to AutoCAD's **0,0** point.

AutoCAD employs several methods to specify the location of points, although each method has its basis in Cartesian coordinates. In AutoCAD terminology these other coordinate systems are referred to as *absolute coordinates*, *relative coordinates*, and *polar coordinates*. CAD drafters must be familiar with each system.

Absolute Coordinates

In AutoCAD terminology, points that are relative to origin point **0,0**—usually identified by the UCS icon in the lower-left corner of the AutoCAD screen (see Figure 4.12)—are referred to as ***absolute coordinates***. When drawing a line, a drafter often defines the line's start point by typing absolute coordinates – for example **2,2** and pressing **<Enter>**). The drafter could continue the line to another point by typing in the absolute coordinates of the point preceded by the # symbol – for example **#7,3** and pressing **<Enter>**. Using this method, *both* endpoints of the line are defined relative to **0,0**. It is important to note, however, that while it is possible to locate the starting and ending points of a line by entering absolute coordinates, this method of drawing a line is seldom used as in most cases only the starting point of a line is defined with absolute coordinates and the following points are located using either the *direct input*, *polar*, or *relative coordinates* methods discussed later in this chapter.

4.5 DRAWING LINES USING ABSOLUTE COORDINATES

The line shown in Figure 4.13(b) was defined using absolute coordinates. This line begins at absolute coordinates **2,2** and ends at absolute coordinates **8,7** (both points are relative to **0,0**). The steps involved in drawing this line are shown on the next page.

JOB SKILL

While it is possible to define lines by the absolute coordinates of their endpoints, this method of drawing lines is rarely used. It is more common to define only the start point of a line using absolute coordinates and then define the next point using either the *direct input*, *polar*, or *relative coordinates* methods discussed later in this chapter.

VIDEO

Video tutorials for this chapter, including the **Line** command, are located inside the *Chapter 4* folder of the book's *Video Training* downloads. These video downloads are available by redeeming the access code that comes with this book. Please see the inside front cover for further details.

STEP–BY–STEP

STEPS IN DRAWING A LINE WITH ABSOLUTE COORDINATES

Step 1. To draw the line shown in Figure 4.13(b), select the **LINE** command icon as shown in Figure 4.13(a) and at the *Specify the first point*: prompt, type **2,2** and press **<Enter>**. Notice you are still in the **Line** command and the command prompt has directed you to *Specify next point*.

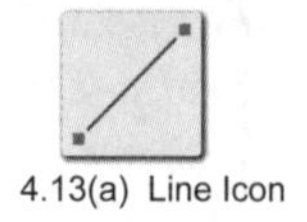

4.13(a) Line Icon

Step 2. At the *Specify the next point*: prompt, type **#8,7** and press **<Enter>**. Press **<Esc>** or **<Enter>** to end the command. The resulting line begins at absolute coordinates **2,2** and ends at absolute coordinates **8,7**. Press **<Esc>** or **<Enter>** to end the **Line** command. For more information on drawing lines with *absolute coordinates*, see the **Line** command tutorial video in the *Chapter 4* folder located inside the *Video Training* downloads that accompany this text.

NOTE

Absolute Coordinates are relative to 0,0 which is located in the lower left corner of the graphics window and is represented by the ***User Coordinate System (UCS)*** icon.

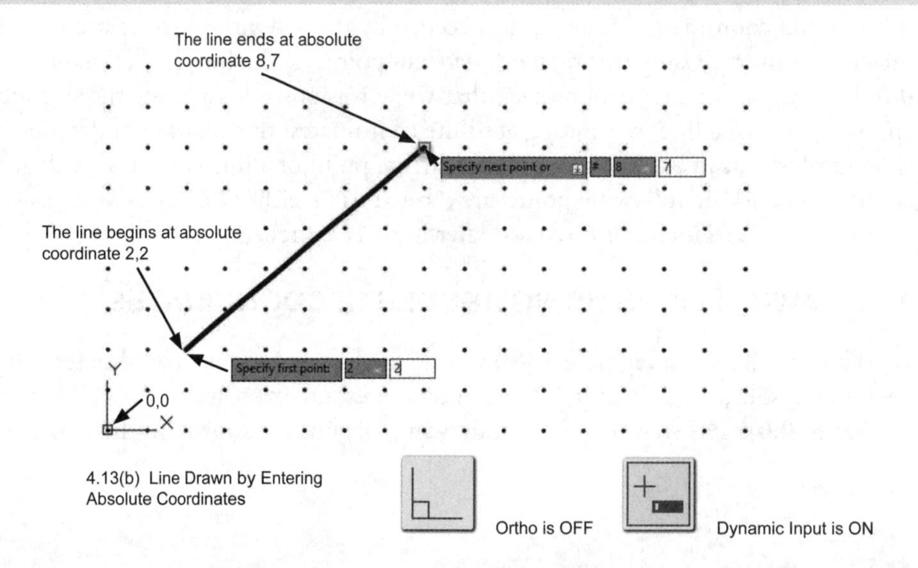

4.13(b) Line Drawn by Entering Absolute Coordinates

JOB SKILL

The table below shows how to enter coordinates with Dynamic Input turned ON and OFF.

	Dynamic Input On (after release 2006)	Dynamic Input Off (before release 2006)
Absolute Coordinate (coming from origin)	#x,y	x,y (assumed)
Relative Coordinate (coming from last point)	x,y (assumed)	@x,y
Polar Coordinate (angled line)	length ***TAB*** angle	@10<45 (@length<angle)

4.6 DRAWING HORIZONTAL AND VERTICAL LINES

The quickest and easiest way to draw horizontal and vertical lines is by *direct entry*. To use this method, first turn *ON* the **ORTHOMODE** and **Dynamic Input** toggles located on the status bar detail shown in Figure 4.9 (or by pressing **F8** and **F12** respectively). Turning **Ortho** *ON* ensures that the lines will be drawn exactly horizontally and/or vertically. Follow the steps shown below to draw lines using the *direct entry* technique.

STEP–BY–STEP

STEPS IN DRAWING LINES WITH DIRECT ENTRY

Step 1. To draw the bold horizontal line shown in Figure 4.14(a), turn **Orthomode** (**F8**) and **Dynamic Input** (**F12**) on, then select the **LINE** command icon. At the *Specify the first point:* prompt, type **1,1** and press **<Enter>**. At the *Specify the next point:* prompt, move the mouse cursor toward the right (any distance), type **1.75**, and press **<Enter>**. A horizontal line is drawn **1.75** units to the right of the start point.

Step 2. To draw the bold vertical line shown in Figure 4.14(b), at the *Specify the next point:* prompt, move the mouse cursor up (any distance), type **1.25** and press **<Enter>**. The resulting horizontal line is drawn **1.25** units above the start point.

Step 3. To draw the new horizontal line shown in Figure 4.14(c), at the *Specify the next point:* prompt, move the cursor any distance to the left, type **1.75** and press **<Enter>**. The resulting horizontal line is drawn **1.75** units to the left of the start point.

Step 4. To draw the new vertical line shown in Figure 4.14(d), at the *Specify the next point:* prompt, move the cursor down (any distance), type **1.25** and press **<Enter>**. The resulting vertical line is drawn **1.25** units below the start point. Press **<Esc>** or **<Enter>** to end the **Line** command. For more information on drawing lines with *direct entry*, see the **Line** command tutorial video in the *Chapter 4* folder located inside the *Video Training* downloads that accompany this book.

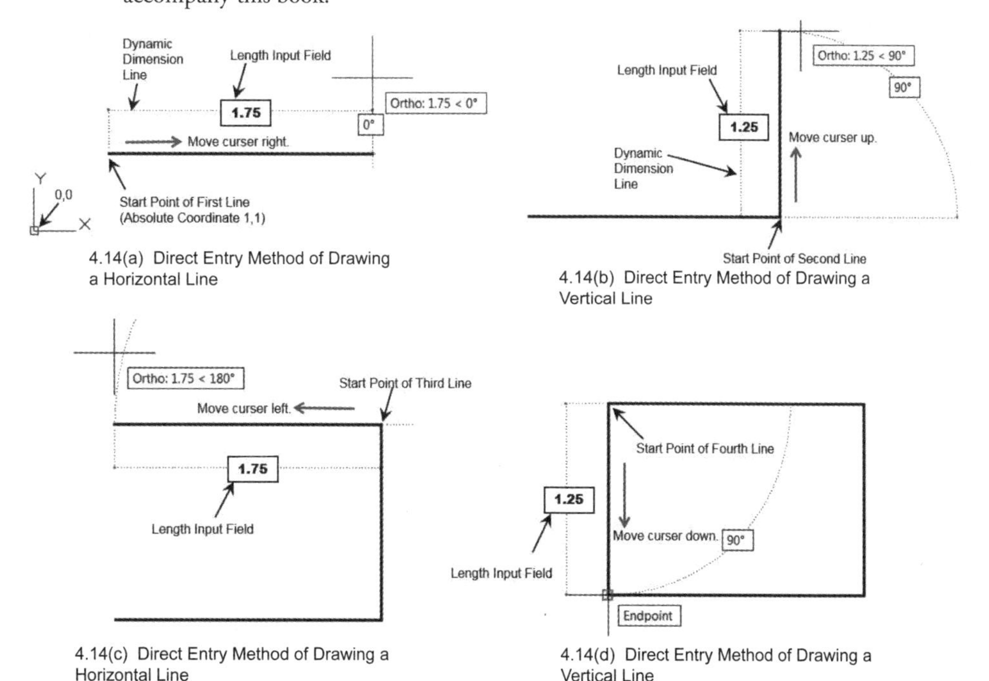

4.14(a) Direct Entry Method of Drawing a Horizontal Line

4.14(b) Direct Entry Method of Drawing a Vertical Line

4.14(c) Direct Entry Method of Drawing a Horizontal Line

4.14(d) Direct Entry Method of Drawing a Vertical Line

4.7 DRAWING LINES WITH RELATIVE COORDINATES

Relative coordinates employ *X-* and *Y-* coordinates to draw a line to a point located relative to the start point of the line. For example, in Figure 4.15, the start point of a line begins at absolute coordinate **1,1** and is drawn to a second point located **4** units to the right (along the *X*-axis) and **1** unit above (along the *Y*-axis) the starting point. The line continues to a third point located **-2** units to the left (along the *X*-axis) and **2** units above (along the *Y*-axis) relative to the second start point. The line continues to a fourth point located **-1** units to the left (along the *X*-axis) and **-1** unit below (along the *Y*-axis) relative to the third start point. The line continues in this fashion until it returns to the original start point. With the exception of the start point of the first line, which is an absolute coordinate, each endpoint that defines a line in Figure 4.15 is located relative to the previous point.

JOB SKILL

When defining a point with a relative coordinate that is located to the left of, or below, the start point, it is necessary to enter a negative coordinate value. This is done by typing a minus sign (-) before the coordinate value. For example, typing -3,-2 draws a line to a point -3 units to the left (on the negative *X*-axis) and -2 units below (on the negative *Y*-axis) the point previously defined.

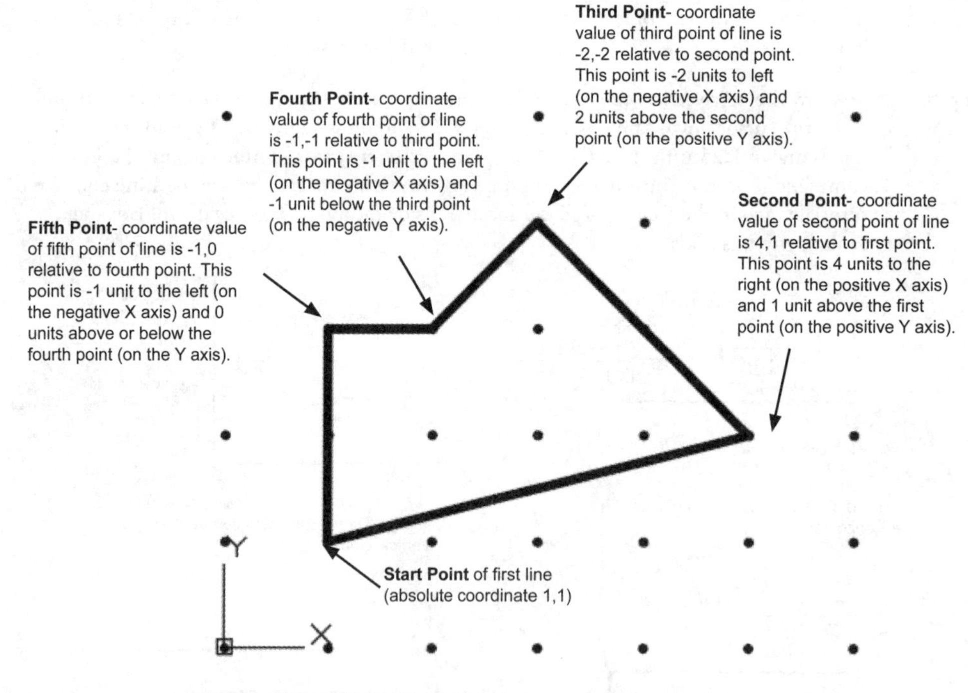

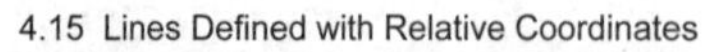

4.15 Lines Defined with Relative Coordinates

JOB SKILL

To enter relative coordinate values in releases of AutoCAD prior to Release 2006, or when drawing with the Dynamic Input setting toggled *OFF* in newer releases, you must first type the @ symbol before entering the *X* and *Y* distances. For example: typing @6,5 at the *Specify next point:* prompt draws a line to a point located 6 units to the right (on the positive *X*-axis) and 5 units above (on the positive *Y*-axis) the start point of the line. Typing @-6,-5 at the *Specify next point:* prompt draws a line to a point located 6 units to the left (on the negative *X*-axis) and 5 units below (on the negative *Y*-axis) the start point of the line.

STEPS IN DRAWING LINES WITH RELATIVE COORDINATES

Step 1. To draw the angled line shown in Figure 4.16(a), turn **Orthomode** (**F8**) *OFF* and **Dynamic Entry** (**F12**) *ON*, then select the **LINE** command icon and at the *Specify the first point:* prompt, type **1,1** and press **<Enter>**.

Step 2. At the *Specify the next point:* prompt, type **2,3** and press **<Enter>**. An angled line is drawn **2** units to the right (positive X), and **3** units above (positive Y) the start point.

Step 3. To draw the second angled line shown in Figure 4.16(b), at the *Specify the next point:* prompt, type **-2,-1** and press **<Enter>**. The resulting angled line is drawn **2** units to the left (negative X), and one unit below (negative Y) the start point. Press **<Esc>** or **<Enter>** to end the **Line** command. For more information on drawing lines with *relative coordinates*, see the **Line** command tutorial video in the *Chapter 4* folder located inside the *Video Training* downloads that accompany this book.

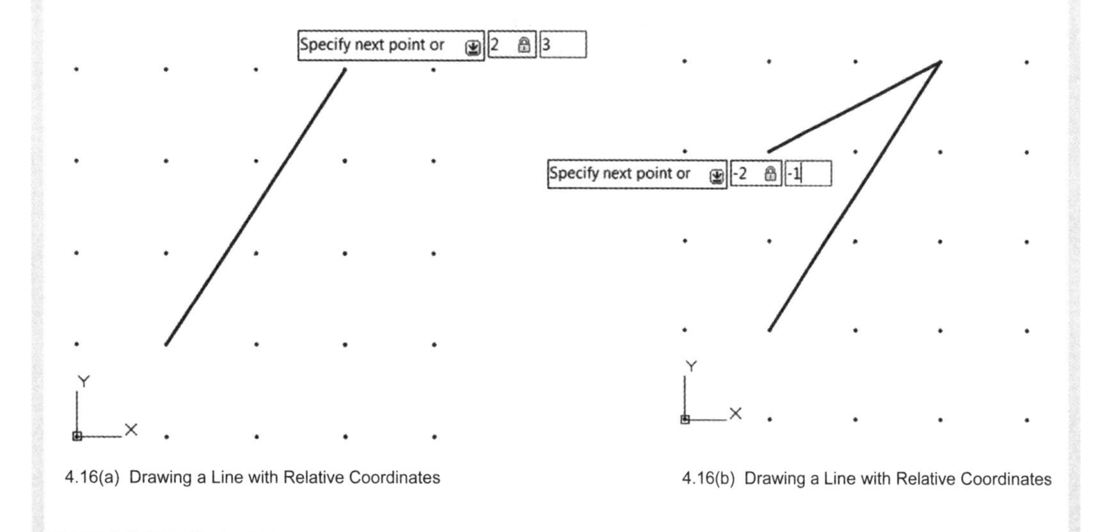

4.16(a) Drawing a Line with Relative Coordinates

4.16(b) Drawing a Line with Relative Coordinates

4.8 DRAWING LINES WITH POLAR COORDINATES

Polar coordinates are used to draw lines to exact lengths and angles. To understand polar coordinates, you first need to understand that in AutoCAD, *East* (as marked on a compass) is considered 0°, so angles entered in AutoCAD are measured above or below the eastern horizon.

For example, the angled line in Figure 4.17 is oriented **45°** above the eastern horizon. The line could also have been defined **45°** below the eastern horizon (later in this chapter you will learn that when using AutoCAD's **ROTATE** command to *rotate* an object at a *positive* angle, the object rotates *counterclockwise* relative to east and a *negative* rotation angle results in *clockwise* rotation).

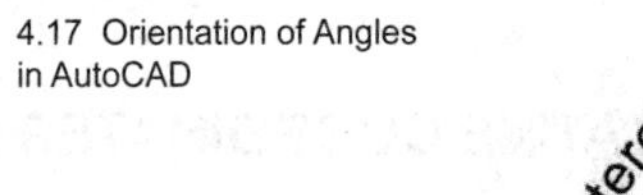

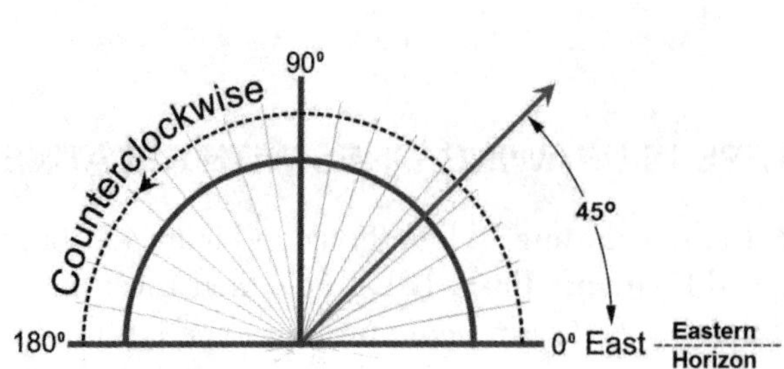

Lines drawn with polar coordinates are defined with a length *and* an angle and are defined relative to the last point entered. When specifying a polar coordinate in AutoCAD, it is necessary to first move the cursor in the desired direction of the angled line (relative to the start point of the line) and then type the length of the line, press the **<Tab>** key, and enter the desired angle. Pressing **<Tab>** switches AutoCAD's direct entry mode from length units to angular values (degrees). For example, entering **10 <Tab> 30** would draw a line **10** units long at a **30°** angle relative to the previously defined point (remember that AutoCAD defines East as 0°). Moving the cursor above or below the eastern horizon determines the direction of the angle above or below the horizon.

In Figure 4.16, the start point of the first line begins at absolute coordinate **1,1** and is drawn with polar coordinates to a second point **6** units in length along a **0°** angle (**6 <Tab> 0**). The second line begins at the second point and is drawn to a third point **2.25** units in length at a **60°** angle (**2.25 <Tab> 60**). The third line is drawn **2** units in length at a **120°** angle (**2 <Tab> 120**). The line continues in this fashion until it ends at the seventh point.

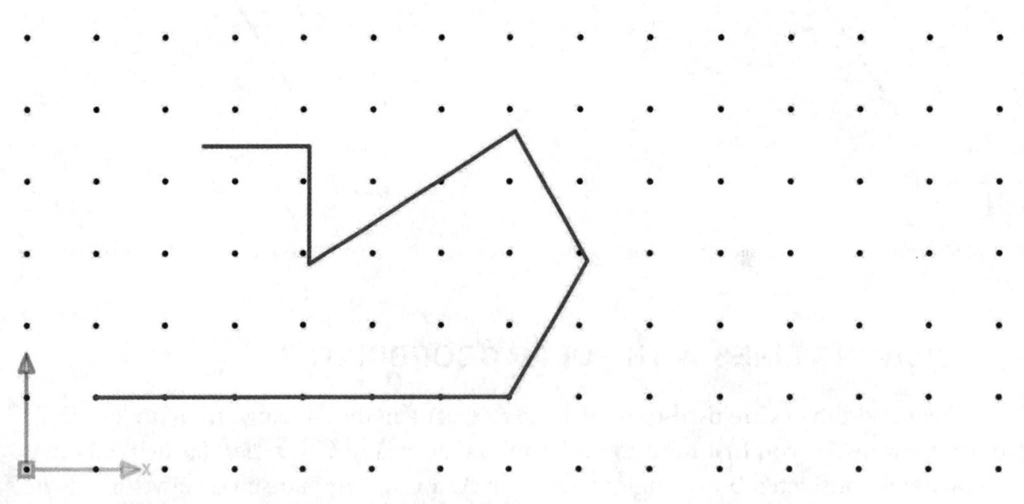

4.18 Lines Drawn with Polar Coordinates

When entering the *polar* coordinates value for a line with Dynamic Input toggled *OFF*, you must type the @ symbol before entering the length value and instead of pressing <Tab>, type the < (less than) symbol; for example, @4<45.

STEPS IN DRAWING LINES WITH POLAR COORDINATES

Step 1. To draw the angled line in Figure 4.19(a) turn **Orthomode** (**F8**) *OFF*, then select the **Line** command icon and type **1,1** for the location of the start point and press **<Enter>**. At the *Specify the next point:* prompt, move the mouse above and to the right of the start point (the desired direction of this line) and type **1.50** and press the **<Tab>** key and type **45** and press **<Enter>**. This will result in a line beginning at absolute coordinates **1,1** that is drawn **1.50** units in length at a **45°** angle *above* the eastern horizon. See *Angle Input* field in Figure 4.19(a).

Step 2. At the *Specify the next point:* prompt, move the mouse above and to the left of the second point and type **1.00** and press **<Tab>** and type **135** and press **<Enter>**. The resulting line is drawn **1** unit in length and at an angle of **135°** above the eastern horizon. See Figure 4.19(b).

Step 3. At the *Specify the next point:* prompt, move the mouse below and to the left of the third point and type **1.00** and press **<Tab>** and type **160** and press **<Enter>**. This line is drawn **1** unit in length and at an angle of **160°** relative to East but is measured below the eastern horizon as shown in Figure 4.19(c).

Step 4. At the *Specify the next point:* prompt, move the mouse below and to the right of the fourth point and type **1.00** and press **<Tab>** and type **45** and press **<Enter>**. This line is drawn **1** unit in length and at an angle of **45°** below the eastern horizon. See Figure 4.19(d). Press **<Esc>** or **<Enter>** to end the **Line** command. For more information on drawing lines with *polar coordinates*, see the **Line** command tutorial video in the *Chapter 4* folder located inside the *Video Training* downloads that accompany this book.

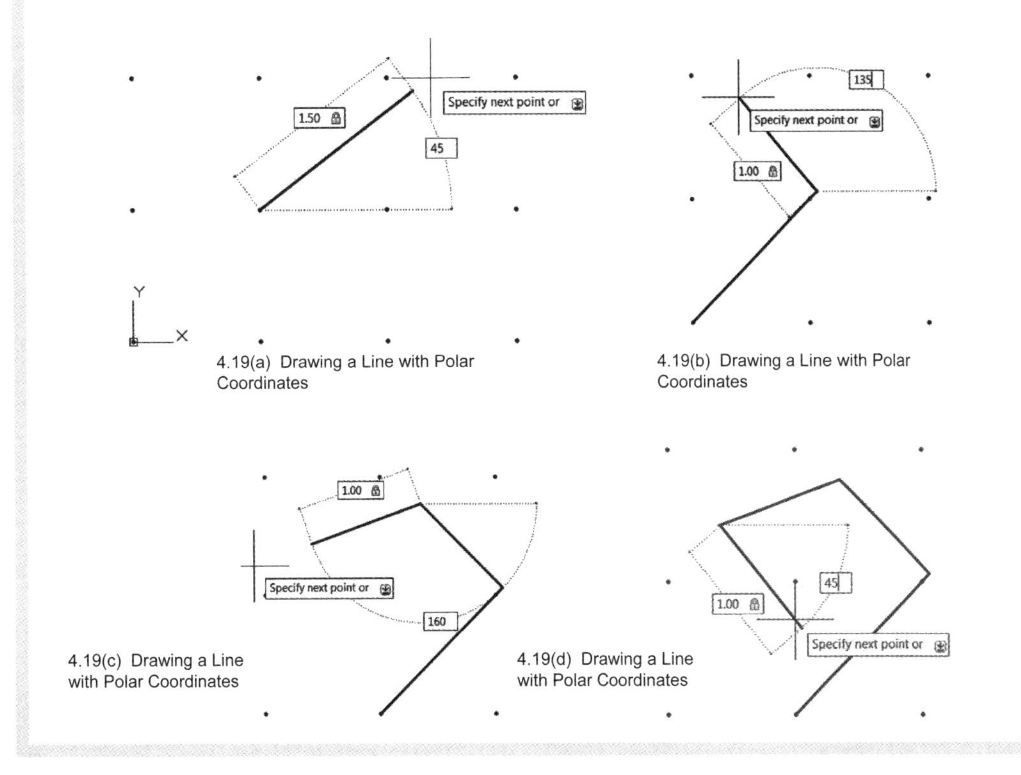

4.19(a) Drawing a Line with Polar Coordinates

4.19(b) Drawing a Line with Polar Coordinates

4.19(c) Drawing a Line with Polar Coordinates

4.19(d) Drawing a Line with Polar Coordinates

Exercise 4.1

Begin a new AutoCAD drawing (default to the **acad.dwt** template). Using the **Line** command and the *direct entry*, *relative coordinates* and *polar coordinates* methods described earlier in the chapter, draw the object shown in Figure 4.20. Begin the bottom left corner of the object at absolute coordinate **2,2** and draw the first line in the positive-*X* direction using direct entry. In fact, the first 10 lines can be drawn using direct entry, then switch to polar coordinates for the twelfth and twelfth lines. Continue drawing lines until you are unable to continue due to lack of coordinate information (21 contiguous lines total). At this point, start a new line at absolute coordinate **2,2** and draw in the positive-*Y* direction. Continue drawing lines in this manner until you are unable to continue drawing due to lack of coordinate information (4 contiguous lines total). To complete the drawing, draw a line connecting the endpoints of the two sets of lines to *close* the shape. If you need assistance with this exercise, ask your instructor for help, or refer to the video available for download from the publisher.

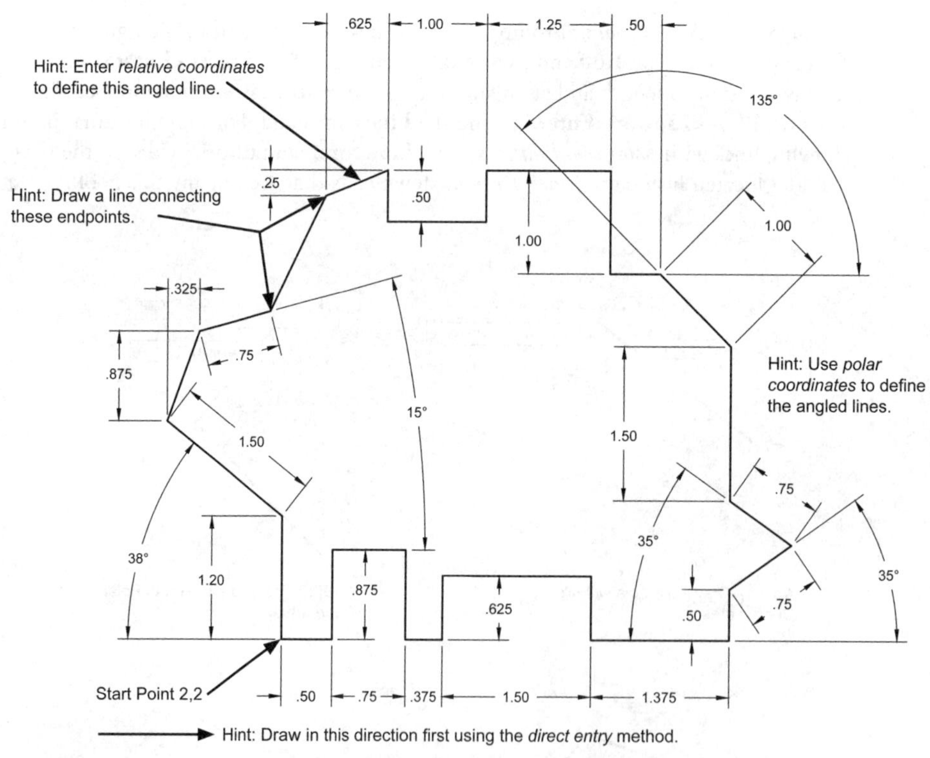

4.20 Exercise 4.1

VIDEO

There is a video tutorial available for this project inside the Chapter 4 folder of the book's Video Training downloads. The video downloads are available by redeeming the access code that comes with this book. Please see the inside front cover for further details.

4.9 SETTING THE ENVIRONMENT FOR AUTOCAD DRAWINGS

Before beginning an AutoCAD drawing, a drafter must first determine the appropriate ***drawing units***, or units of measurement, for the type of drawing being created. For example, for architectural drawings, architectural units (feet and fractional inches) would be appropriate. For civil engineering drawings, engineering units (feet and decimal inches) would be appropriate; for mechanical engineering drawings, decimal units would be chosen.

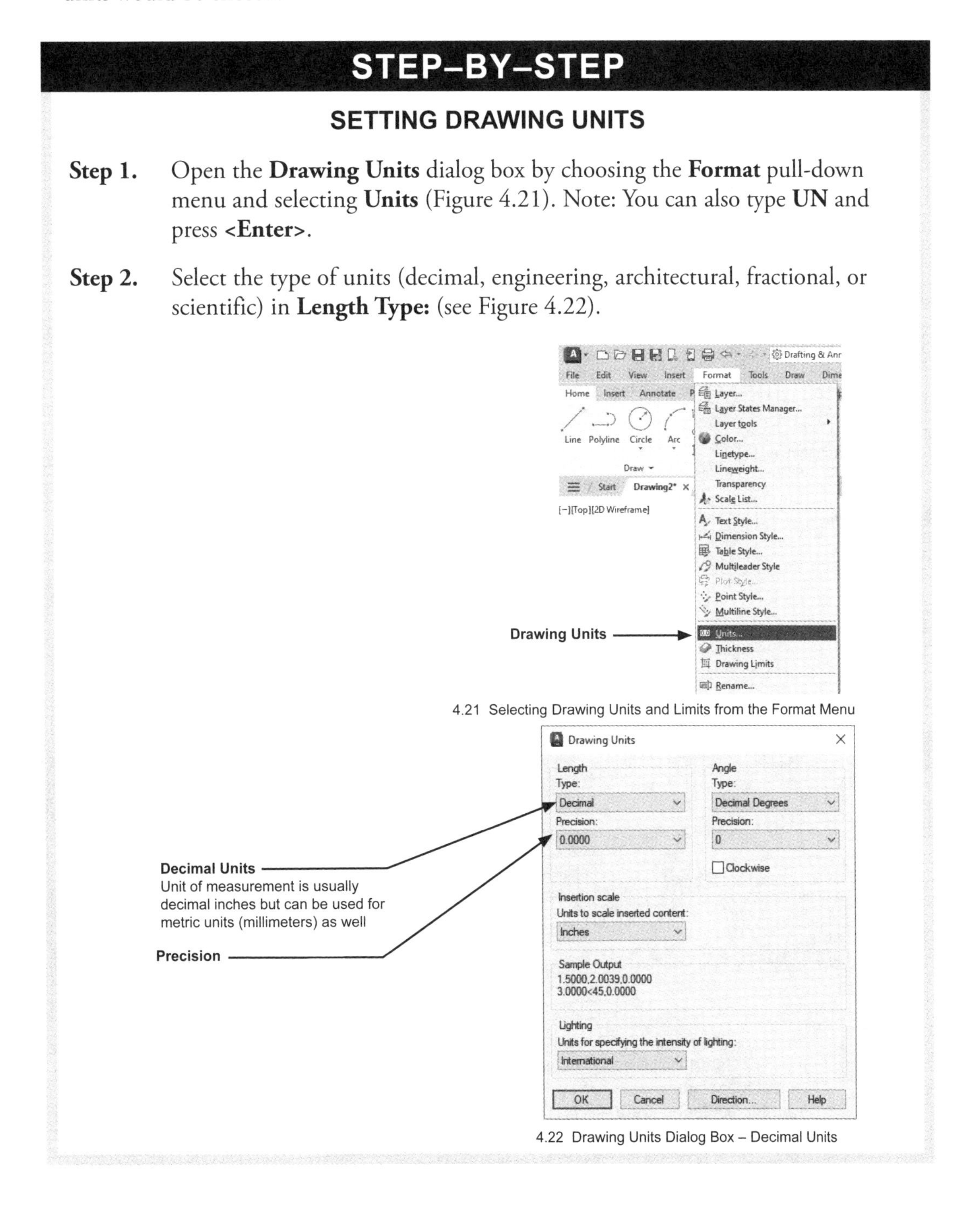

STEP–BY–STEP

SETTING DRAWING UNITS

Step 1. Open the **Drawing Units** dialog box by choosing the **Format** pull-down menu and selecting **Units** (Figure 4.21). Note: You can also type **UN** and press **<Enter>**.

Step 2. Select the type of units (decimal, engineering, architectural, fractional, or scientific) in **Length Type:** (see Figure 4.22).

4.21 Selecting Drawing Units and Limits from the Format Menu

4.22 Drawing Units Dialog Box – Decimal Units

STEP–BY–STEP

Step 3. Select the level of precision (the number of decimal places or fractional precision) for entering units in **Precision:** below **Length Type:**

In Figure 4.22, the drawing units are set to **Decimal**, which means that coordinates will be entered and displayed in decimal units. Precision for entering and displaying data is set to four decimal places.

In Figure 4.23, the drawing units are set to **Architectural**, which means that coordinates will be entered and displayed in feet and fractional inches. Precision for entering and displaying data is set to 1/16".

In Figure 4.24, the drawing units are set to **Engineering**, which means that coordinates will be entered and displayed in feet and decimal inches. Precision for entering and displaying data is set to four decimal places.

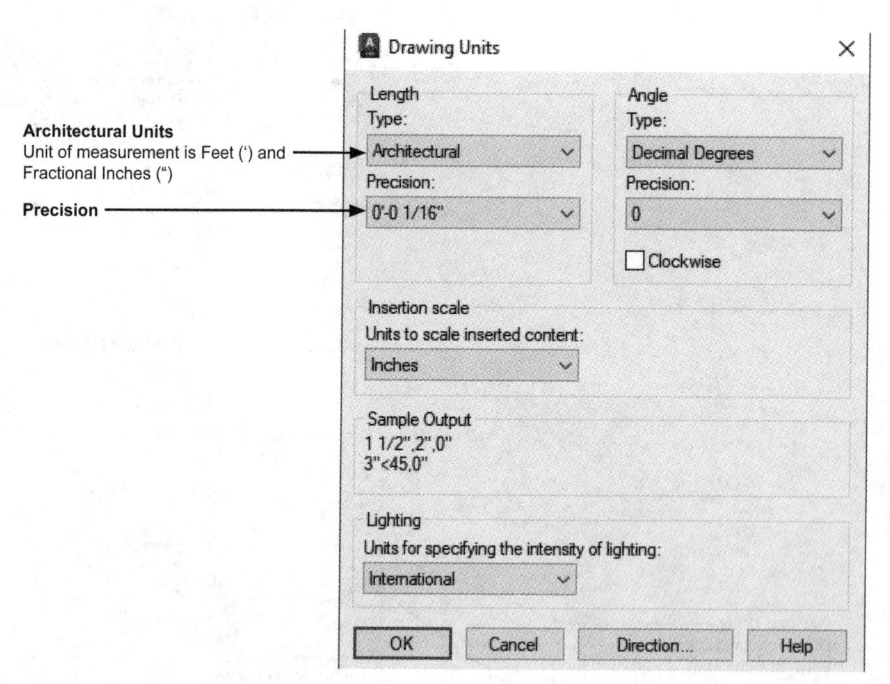

4.23 Drawing Units Dialog Box – Architectural Units

QUICK TIP

For quick reference, the drawing's *Current Units* are displayed on the Status bar (see Figure 4.3 and 4.9). Drawing Units can also be changed by clicking on the Drawing Units Options drop-down arrow located on the Status Bar and selecting different units from the list (see Figure 4.3 for the location of this tool).

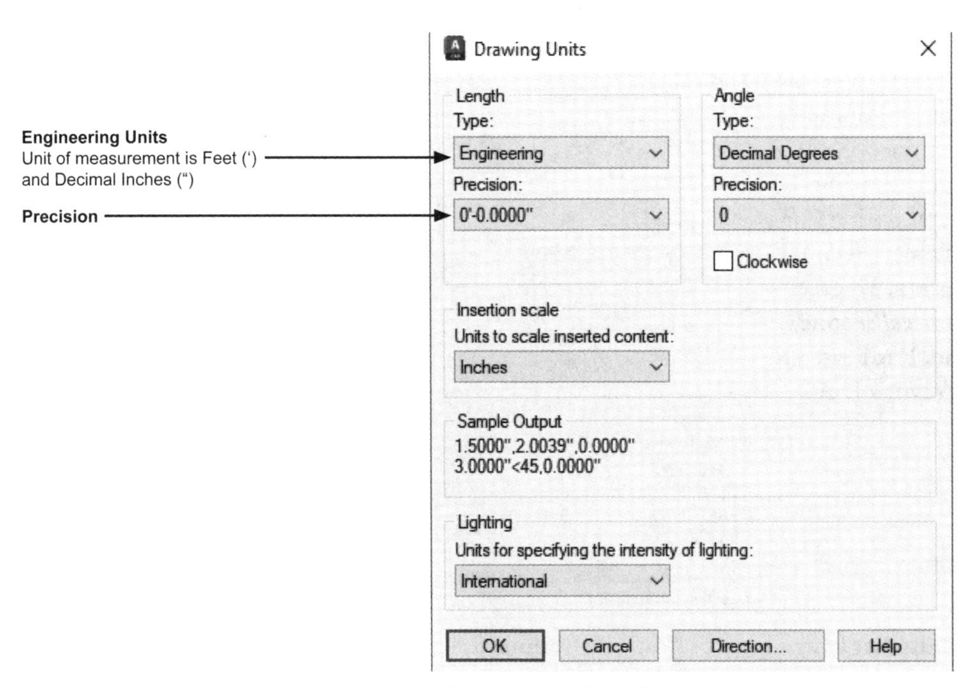

4.24 Drawing Units Dialog Box – Engineering Units

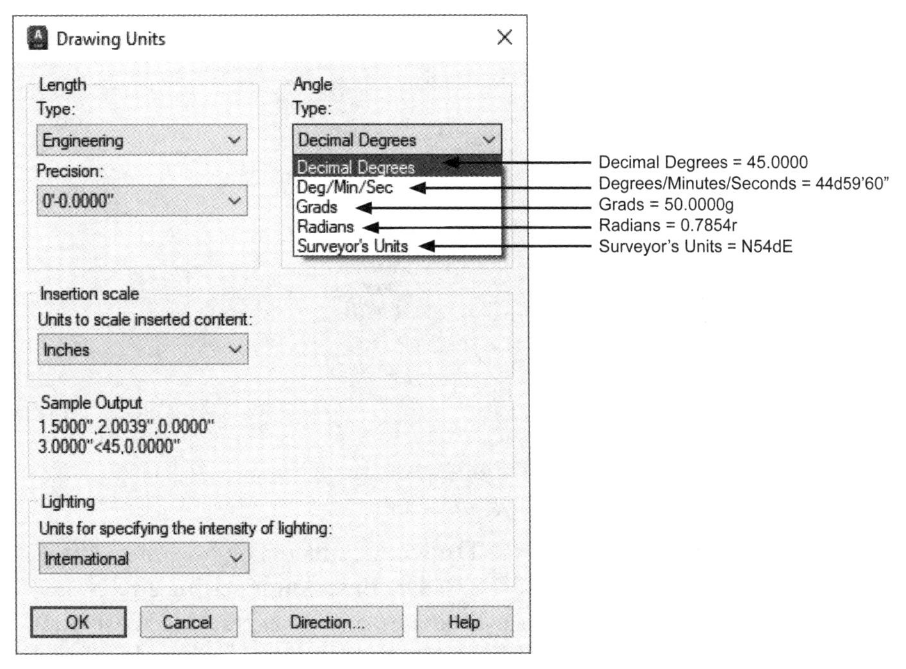

4.25 Drawing Units Dialog Box – Setting Angle Type

Setting Angle Type

After selecting the length type and precision, drafters select the angle type for the drawing. Several options are available: **Decimal Degrees**, **Degrees/Minutes/Seconds**, **GradsRadians**, and **Surveyor's Units**.

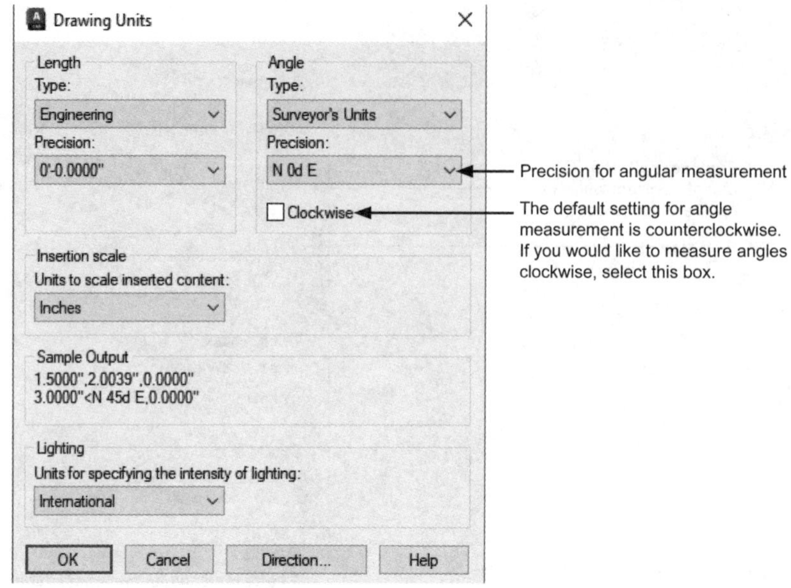

4.26 Drawing Units Dialog Box – Setting Precision and Direction for Angular Measurement

1. In **Angle Type:** select from the options (decimal, deg/min/sec, grads, radians, or surveyor's units) (see Figure 4.25)
2. In **Precision:** select the measurement of angles (below **Angle Type:**) (see Figure 4.26).

Setting the Direction of Angle Measurement

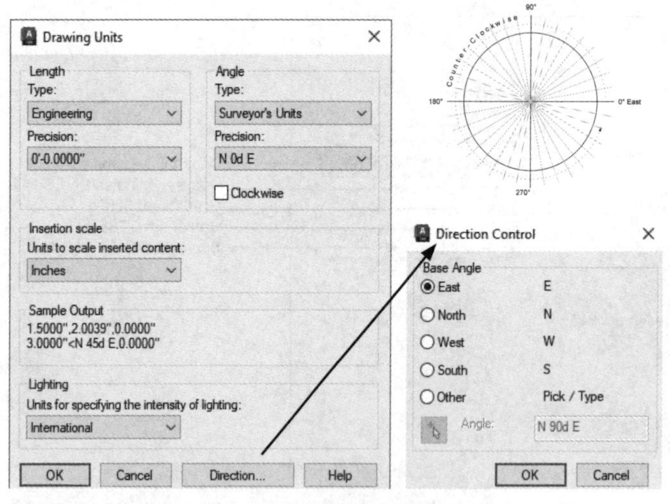

4.27 Setting Direction for Measuring Angles in AutoCAD

Selecting the **Direction…** button in the **Drawing Units** dialog box opens the **Direction Control** dialog box (see Figure 4.27). The **Base Angle** setting affects the starting point for measuring angles, polar coordinates, and polar tracking. **East** (the default direction), is usually assigned as the base angle; however, a base angle other than East can be set as the direction for 0° by selecting either **North**, **West**, **South**, or **Other**.

Drawing Limits

Setting the *drawing limits* defines a drawing area; this is similar to defining a sheet size for the drawing (note: defining drawing limits does not limit drafters to drawing *inside* of those limits). Limits should be set *after* the units of the drawing have been set because the value for the limits will be displayed in the current units. When setting limits, you are prompted to specify the lower left and upper right corners of the drawing area. In most cases, the lower left corner is *defaulted* to **0,0** and the upper right corner is defined by typing in the coordinates of the corresponding sheet size. For example, if using decimal units, an A-size sheet limits are **0,0** and **12,9**; a B-size sheet limits are **0,0** and **17,11**; a C-size sheet limits are **0,0** and **22,17**; and a D-size sheet limits are **0,0** and **34,22**.

STEP-BY-STEP

SETTING DRAWING LIMITS

Step 1. Open the **Drawing Limits** dialog box by choosing the **Format** pull-down menu and selecting **Drawing Limits**, as shown in Figure 4.21.
Note: You can also type **LIM** and press **<Enter>**.

```
Command: Limits
Reset Model space limits:
Specify lower left corner or [ON/OFF] <0.0000,0.0000>:
Specify upper right corner <12.0000,9.0000>:
```

4.28 Command Line Displaying Default Limits When Decimal Units are in Effect

Step 2. When prompted to *Specify lower left corner:* (The default limits **0,0** will be displayed as shown in Figure 4.28). Press **<Enter>** to accept **0,0** as the lower left limit.

```
Command: Limits
Reset Model space limits:
Specify lower left corner or [ON/OFF] <0.0000,0.0000>:
Specify upper right corner <12.0000,9.0000>:
```

4.29 Command Line Displaying Limits of 0,0 and 24,18

Step 3. When prompted to *Specify upper right corner:* (the default limits **12,9** will be displayed as shown in Figure 4.28). Type the coordinates for a different sheet size, for example, **24,18**, and press **<Enter>** (see Figure 4.29).

The drawing area will update to the new limits, but a **VIEW** command named **Zoom All** must be performed to see the new limits displayed in the graphics window. To perform a **Zoom All**, type **Z**, press **<Enter>**, and then type **A** and press **<Enter>**. **Note:** If your mouse has a scroll wheel, you can also *double-click* the wheel.

QUICK TIP

In AutoCAD terminology, to *default* means that you accept the initial setting offered by AutoCAD. Default settings are often framed in brackets (<>). For example, drafters usually accept <0.0000,0.0000> as the default setting for the lower left corner of the Limits setting (see Figure 4.29). As a new user of AutoCAD, you may find that you will often accept AutoCAD's default settings, but as your knowledge of AutoCAD grows you will become more comfortable with making changes to the AutoCAD environment.

The limits of a drawing are dependent on the units of measurement assigned to the drawing. Therefore, the limits assigned are based on the sheet sizes that are appropriate for drawings created with the assigned units. Table 4.1, Table 4.2, and Table 4.3 show the **Limits** settings for various sheet sizes relative to the units of measurement for the drawing (decimal, architectural, or engineering).

Limits Based on ASME Y14.1 Decimal Sheet Sizes
A Sheet Limits = 0,0 and 11,8.5
B Sheet Limits = 0,0 and 17,11
C Sheet Limits = 0,0 and 22,17
D Sheet Limits = 0,0 and 34,22
E Sheet Limits = 0,0 and 44,32
Limits Based on ASME Y14.1 Decimal Sheet Sizes
A4 Sheet Limits = 0,0 and 297,210
A3 Sheet Limits = 0,0 and 420,297
A2 Sheet Limits = 0,0 and 594,420
A1 Sheet Limits = 0,0 and 841,594
A0 Sheet Limits = 0,0 and 1189,841

Table 4.1 Limits Settings for Decimal and Metric Units

Limits or Scale of: 1/4" = 1'-0"
A Sheet Limits = 0',0' and 48',36'
B Sheet Limits = 0',0' and 72',48'
C Sheet Limits = 0',0' and 96',72'
D Sheet Limits = 0',0' and 144',96'
Limits or Scale of: 1/8" = 1'-0"
A Sheet Limits = 0',0' and 96',72'
B Sheet Limits = 0',0' and 144',96'
C Sheet Limits = 0',0' and 192',144'
D Sheet Limits = 0',0' and 288',192'

Table 4.2 Limits Settings for Architectural Units

Limits or Scale of: 1" = 100'
A Sheet Limits = 0',0' and 1200',900'
B Sheet Limits = 0',0' and 1800',1200'
C Sheet Limits = 0',0' and 2400',1800'
D Sheet Limits = 0',0' and 3600',2400'

Table 4.3 Limits Settings for Engineering Units

Layers

In AutoCAD drawings, lines and other entities are drawn on ***layers***. Think of layers as sheets of clear glass layered one on top of the other. A layer can have its own color, linetype, or lineweight assigned to it.

When you begin an AutoCAD drawing from scratch, it contains only one layer, layer **0** (zero). If more layers are needed, they must be created.

STEPS INVOLVED IN CREATING NEW LAYERS

Step 1. Click on the **Layer Properties Manager** icon located in the **Layers** panel of the **Home** tab on the ribbon (see Figure 4.30).

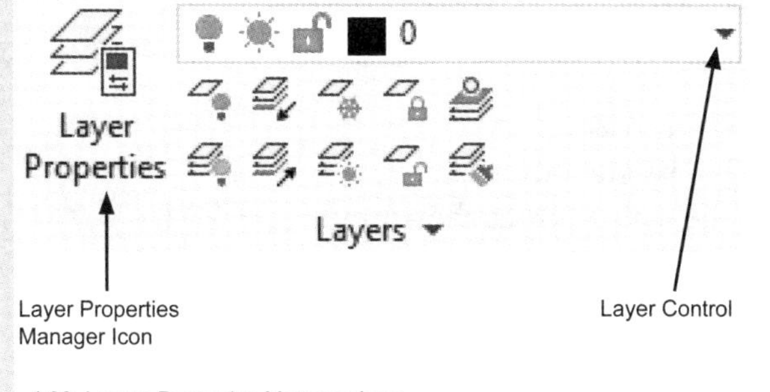

4.30 Layers Properties Manager Icon

There is a video tutorial available for Layer Properties tool inside the Chapter 4 folder of the book's Video Training downloads. The video downloads are available by redeeming the access code that comes with this book. Please see the inside front cover for further details.

Step 2. When the **Layer Properties Manager** palette shown in Figure 4.31 opens, click the **New** button.

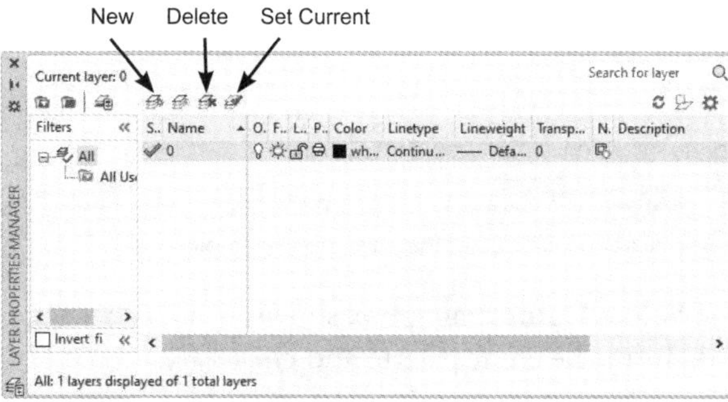

4.31 Layers Properties Manager Palette

Step 3. Select the new layer and replace its default name, **Layer 1**, with the new layer name.

Step 4. Repeat Step 3 to create other layers. When all the new layers have been created, click **OK**. Figure 4.32 shows the layers created for a mechanical drawing.

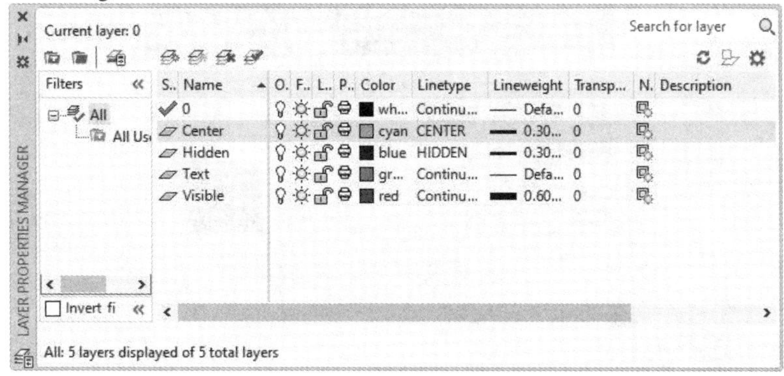

4.32 Layer Examples Shown in the Layers Properties Manager Palette

STEP–BY–STEP

SETTING LAYER COLOR

Step 1. Click on the **Layer Properties Manager** icon located in the upper left corner of the **Layers** panel of the **Home** tab on the ribbon (see Figure 4.33).

4.33 Layer Properties Manager Icon

Step 2. Select the layer to which you want to assign a new color and click on the color assigned to the layer in the **Color** column in the dialog box as shown in Figure 4.34.

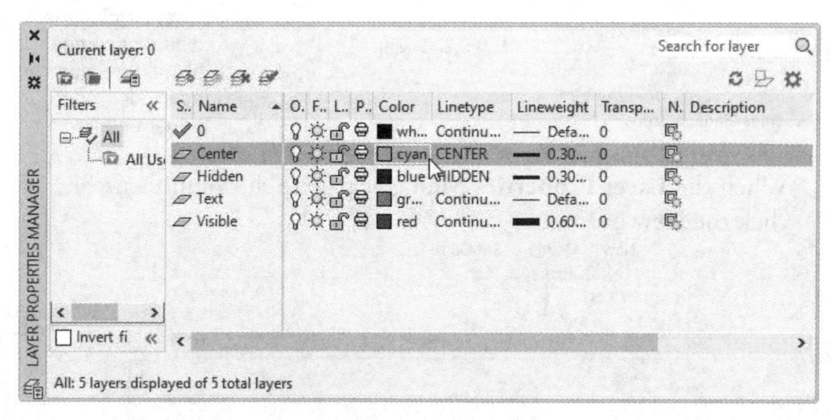

4.34 Assigning Color to a Layer in the Layer Properties Manager Palette

Step 3. When the **Select Color** dialog box shown in Figure 4.35 opens, select the desired tile from the color palette and click **OK**.

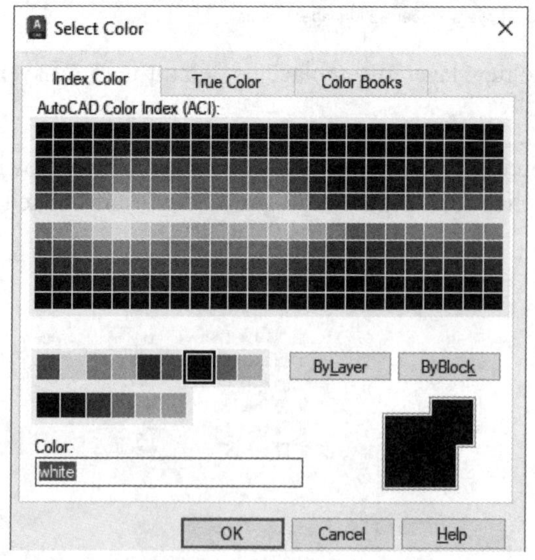

4.35 Select Color Dialog Box

SETTING LAYER LINETYPE

Step 1. Click on the **Layer Properties Manager** icon located in the upper left corner of the **Layers** panel of the **Home** tab on the ribbon (see Figure 4.36).

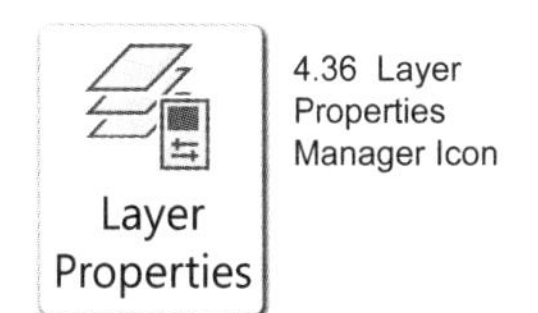

4.36 Layer Properties Manager Icon

Step 2. Select the layer to which you want to assign a new linetype and click on its linetype name in the **Linetype** column (see Figure 4.37).

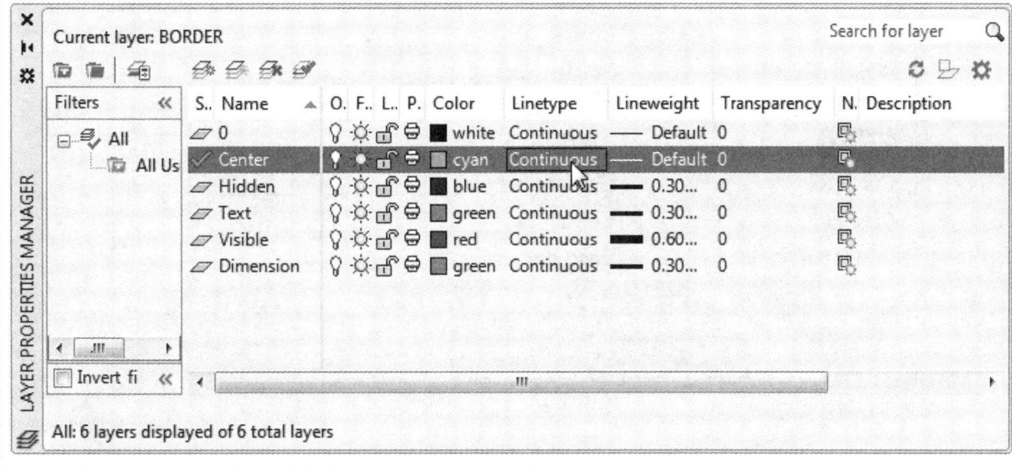

4.37 Assigning Linetype to a Layer in the Layer Properties Manager Palette

Step 3. The **Select Linetype** dialog box shown in Figure 4.38 will open. If you do not see the desired linetype listed, click the **Load...** button.

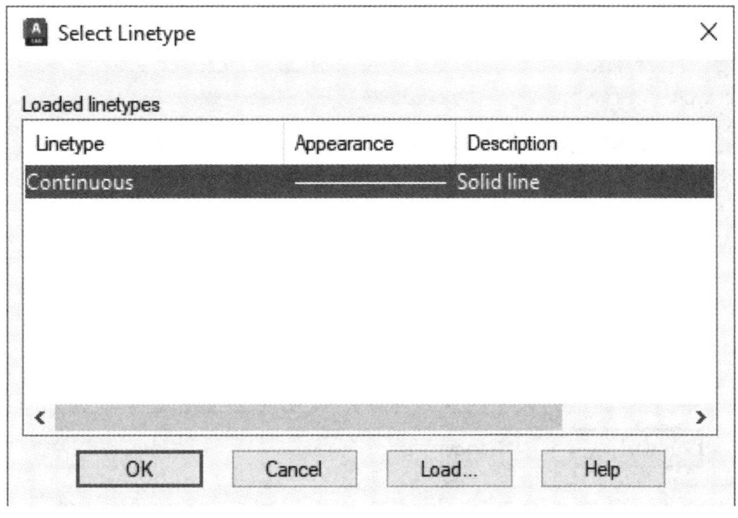

4.38 Select Linetype Dialog Box

STEP–BY–STEP

Step 4. The **Load or Reload Linetypes** dialog box shown in Figure 4.39 will open. Scroll through the linetypes. Select the linetype you wish to load and click **OK**.

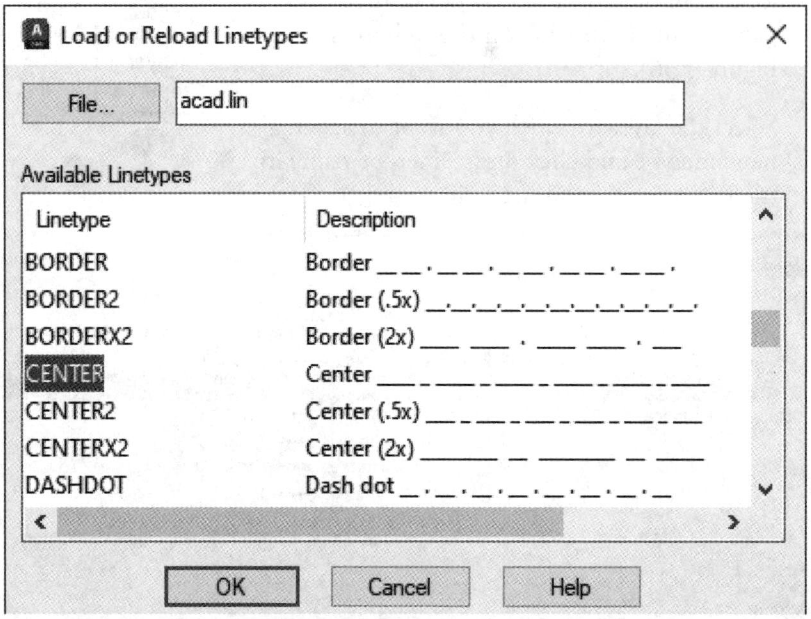

4.39 Load or Reload Linetypes Dialog Box

Step 5. Select the newly loaded linetype from the **Select Linetype** dialog box and click **OK**. The new linetype will be assigned to the layer selected in Step 2 (see Figure 4.40).

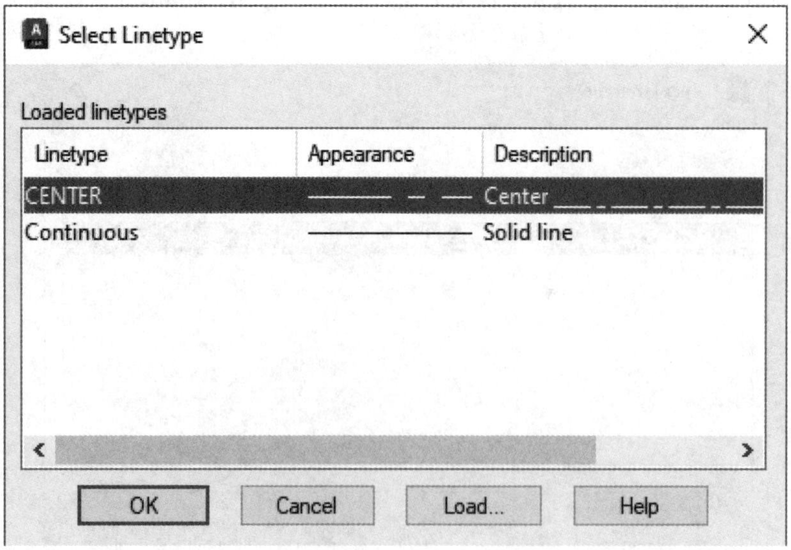

4.40 Select Linetype Dialog Box

SETTING LAYER LINEWEIGHT

Step 1. Click on the **Layer Properties Manager** icon located in the upper left corner of the **Layers** panel of the **Home** tab the ribbon (see Figure 4.41).

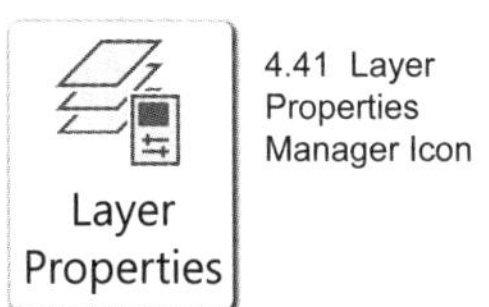

4.41 Layer Properties Manager Icon

Step 2. When the **Layer Properties Manager** palette opens, select the layer to which you want to assign a new lineweight, and click on the lineweight setting in the **Lineweight** column (see Figure 4.42).

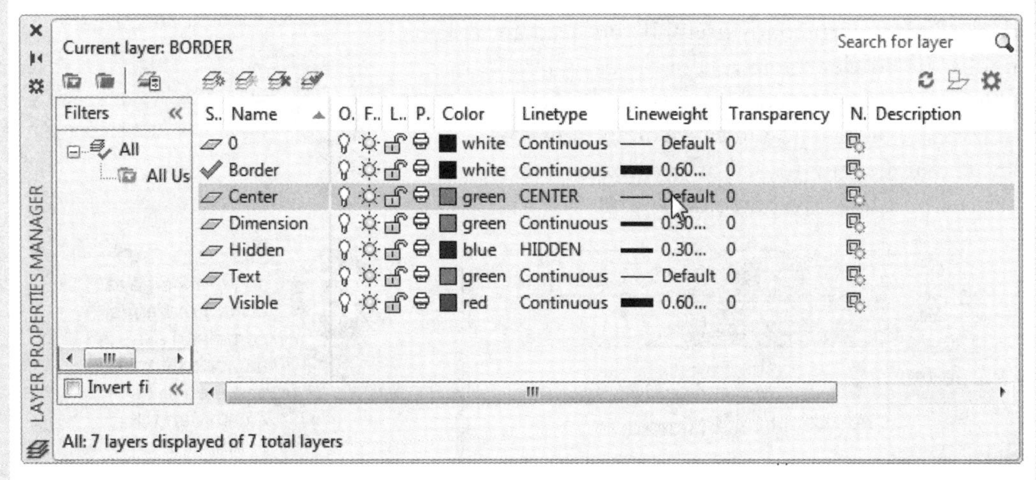

4.42 Assigning Lineweight to a Layer in the Layer Properties Manager Palette

Step 3. When the **Lineweight** dialog box opens, scroll through and select the desired line thickness in which you want the layer to be printed and click **OK** (see Figure 4.43).

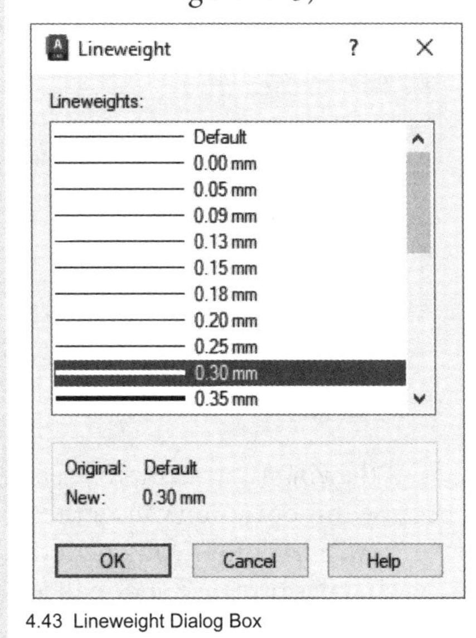

4.43 Lineweight Dialog Box

The **LTSCALE** (*linetype scale*) command controls the spacing of the breaks and dashes in non-continuous linetypes such as centerlines and hidden lines. Sometimes a non-continuous line will appear to be a continuous line in the drawing window because its spaces are too small or too large. To change the linetype scale of all the non-continuous lines in the drawing, type **LTS** and press **<Enter>** and change the default value (which is **1**) to a larger or smaller value. Settings like **LTS**, which affect every non-continuous line in the drawing, are referred to as *global* settings.

Setting the Current Layer

In an AutoCAD drawing, you can draw only on the current layer. To set a different layer current, select the down arrow in the **Layer Control** window located in the **Layers** panel of the **Home** tab on the ribbon and left-click on the layer you want to make current from the list of layers shown (see Figure 4.44).

JOB SKILL

An entity drawn on one layer can be moved to a different layer simply by left-clicking on the entity in the drawing window, selecting the down arrow in the **Layer Control** window (see Figure 4.44) located in the **Layers** panel of the **Home** tab on the ribbon, and picking the layer to which you want to move the entity.

Controlling Layer Visibility

Visibility of a drawing's layers can be controlled in two ways: either turning the layers off or freezing them. This is particularly useful if you need an unobstructed view of an area of the drawing, or if you are working in detail on a particular layer or set of layers. Construction lines are often drawn on layers that are later turned off, or frozen, because entities on these layers are not plotted.

QUICK TIP

The *current layer* can be turned off, but it cannot be frozen.

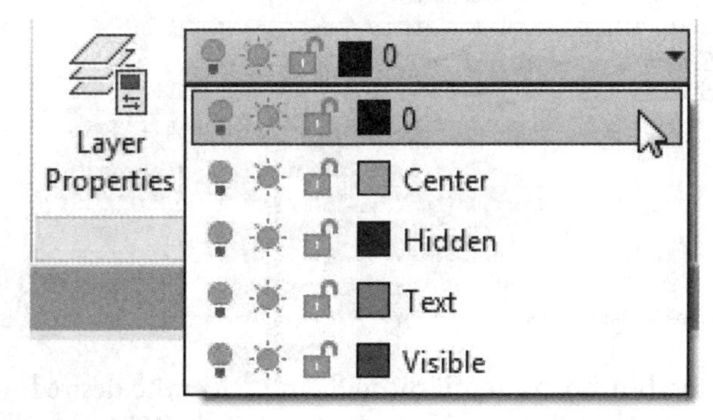

4.44 Layer Control Drop-down Menu

JOB SKILL

Many AutoCAD commands have a ***command alias***. A command alias is a shortcut that can be typed to begin a command. For example, a quick way to perform a **Zoom Window** is to type the **ZOOM** command alias, **Z** (upper- or lowercase), and press **<Enter>**. Then, type **W** (for Window), press **<Enter>**, and pick two points to define the area to be zoomed into. Likewise, a quick way to perform a **Zoom All** is to type **Z**, press **<Enter>**, type **A**, and press **<Enter>**. Command aliases for the commands on the **Draw** and **Modify** toolbars are included later in this chapter.

Turning Layers Off

Select the down arrow in the **Layer Control** window and turn a layer off by left-clicking the yellow light bulb next to the layer name. Layers that are off will display the darkened light bulb symbol (see Figure 4.44).

Freezing Layers

Select the down arrow in the **Layer Control** window and freeze a layer by clicking on the sun symbol next to its name. When the layer is frozen, the sun symbol will be replaced with a snowflake. Freezing, and thawing (unfreezing), layers takes a little more time than turning layers on and off because this operation causes the drawing to be regenerated. If you do not intend to use, or view, a layer for a long period of time, the layer can be frozen. Doing so will increase drawing performance and reduce the drawing calculation time.

4.10 ZOOM AND PAN COMMANDS

The **ZOOM** command allows you to view a drawing up close or far away. Zooming does not actually change the scale of entities in the drawing (this is accomplished with the **SCALE** command), just their magnification in the graphics window.

The **PAN** command allows you to reposition the view of the drawing in the graphics window. Panning does not change the location of entities in the drawing (this is accomplished with the **MOVE** command), just the viewer's point of view. The **PAN** and **ZOOM** commands are located on the **View** tab on the ribbon in the Navigate panel. To see all the options for the **ZOOM** command, left-click on the down arrow next to **Extents** in the **Navigate** panel. For a detailed explanation of these important viewing tools, see Figure 4.45.

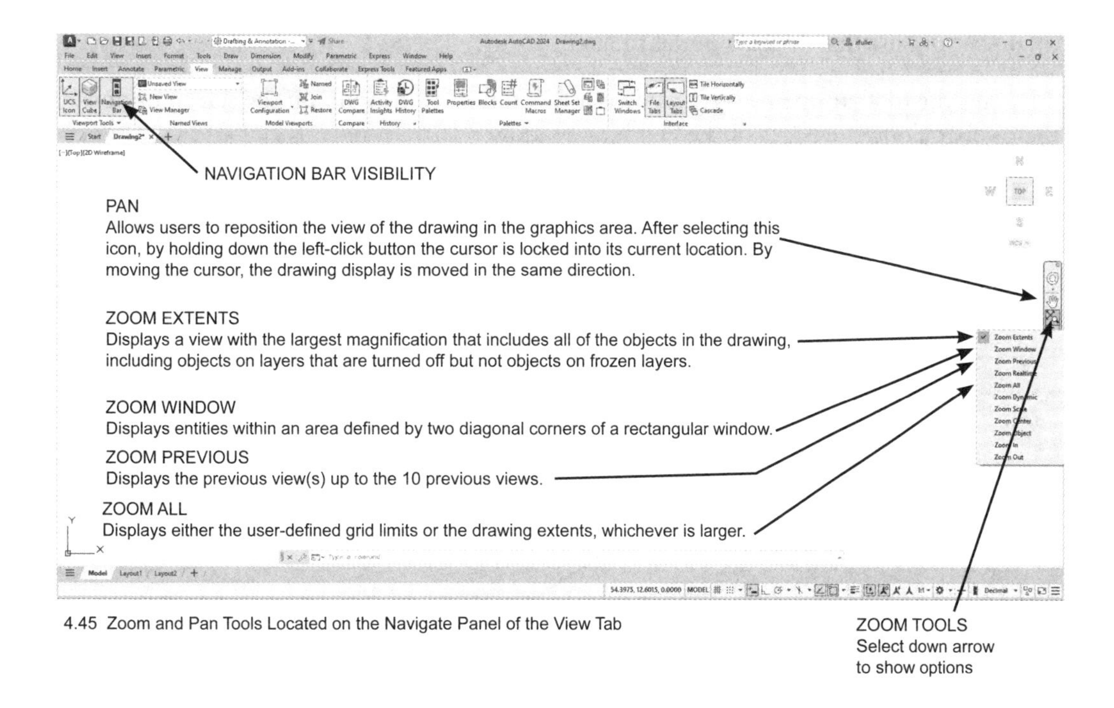

4.45 Zoom and Pan Tools Located on the Navigate Panel of the View Tab

QUICK TIP

AutoCAD's **Zoom** icons are also located on the Zoom toolbar.

4.11 AUTOCAD COMMANDS

AutoCAD commands can be invoked by choosing the appropriate icon from a panel on the ribbon, from command toolbars such as **Draw**, **Modify**, or **Dimension**, from a **menu bar** pull-down menu, or by typing the command's name or alias on the command line. The quickest way to open an AutoCAD toolbar is to right-click on any open toolbar. This will open a list of available toolbars, as shown in Figure 4.46. From this list, select the name of the toolbar that you wish to open by left-clicking on it.

Another way to open a toolbar when no toolbars are open is to left-click on the **Tools** menu located on the menu bar. When the submenu drops down, left-click on the word **Toolbars**. When the next submenu appears, left-click on the word **AutoCAD** and select the name of the toolbar from the **Toolbar** menu. In AutoCAD, toolbars can also be opened by selecting the ribbon's **View** tab and picking on the **Toolbars** tool located on the **Windows** panel.

When the drop-down list appears, click on the word AutoCAD and select the toolbars from list that appears in the toolbar menu.

After the toolbar opens, you can drag it to a different location on the screen or dock the toolbar along the edges of the graphics window.

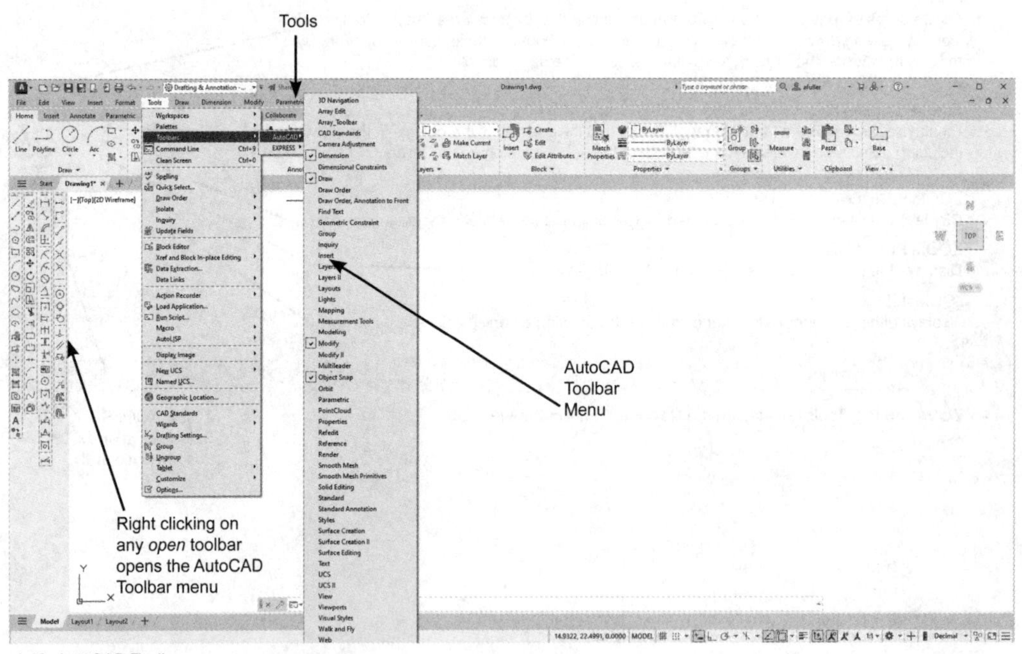

4.46 AutoCAD Toolbars

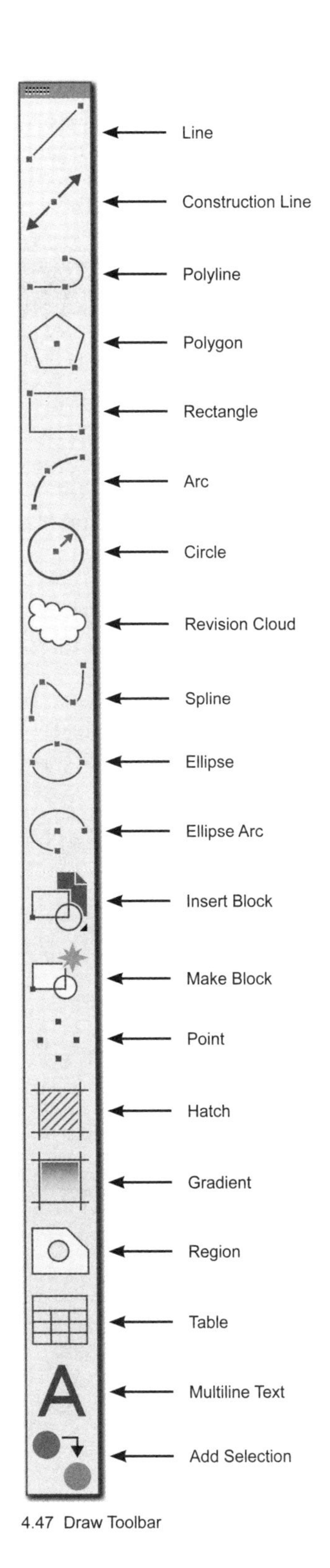

4.47 Draw Toolbar

Essential Draw Commands

The icons for AutoCAD's **Draw** commands are shown as they appear on the AutoCAD ***Draw*** toolbar as in Figure 4.47. Some of the commands located on this toolbar, such as **Line**, **Circle**, **Arc**, and **Multiline Text**, are used more frequently than others; however, all the commands on this toolbar are useful, and you should familiarize yourself with each of them. These command icons can also be found in the **Draw** panel of the **Home** tab of the ribbon.

There are video tutorials available for the Draw commands inside the Chapter 4 folder of the book's Video Training downloads. The video downloads are available by redeeming the access code that comes with this book. Please see the inside front cover for further details.

LINE COMMAND			
	Ribbon	Home	The icon for the **LINE** command is shown in Figure 4.48(a). As stated earlier in the chapter, lines can be drawn by using *absolute coordinates* [see Figure 4.13(b)], *direct entry* [see Figures 4.14(a) through 4.14(d)], *relative coordinates* [see Figure 4.15], or *polar coordinates* [see Figures 4.19(a) through 4.19(d)].
	Panel	Draw	
	Command Line	Line	
4.48(a)	Alias	L	

STEP–BY–STEP

LINE COMMAND TUTORIAL

Step 1. Select the **Line** icon from the **Draw** toolbar or the **Draw** panel of the **Home** tab.

Step 2. When prompted to *Specify start point*, define the start point of the line by left-clicking to select a point in the graphics window or by entering an absolute coordinate and pressing **<Enter>**.

Step 3. When prompted to *Specify next point*, define the next point of the line by left-clicking to select a point in the graphics window or by entering absolute, relative, or polar coordinates and pressing **<Enter>**. You can continue to define lines in this manner or end the command by pressing **<Esc>** or **<Enter>**. See Figure 4.48(b).

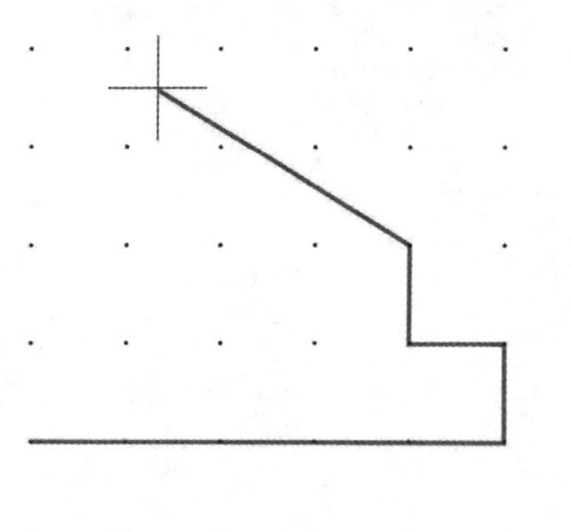

Example: Line command is selected and the object is drawn by snapping to the grid points.

4.48(b) LINE Command

CONSTRUCTION LINE COMMAND			
	Ribbon	Home	The icon for the **Construction Line** command is shown in Figure 4.49(a). This command creates lines that extend to infinity that can be placed on a drawing to facilitate the construction of other objects.
	Panel	Draw	
	Command Line	Xline	
4.49(a)	Alias	XL	

STEP–BY–STEP

CONSTRUCTION LINE COMMAND TUTORIAL

Step 1. Select the **Construction Line** icon from the **Draw** toolbar or **Draw** panel of the **Home** tab.

Step 2. When prompted to *Specify a point,* type **H** and press **<Enter>** to place a horizontal construction line, type **V** and press **<Enter>** to place a vertical construction line, or type **A** and press **<Enter>** and an angle value at the *Enter angle of xline:* prompt to place a construction line at an angle. Press **<Enter>** after entering the value for an angle.

Step 3. At the *Specify through point:* prompt, select a point on the screen through which the construction line is to be drawn. You can continue to pick points for placement of other construction lines or end the command by pressing **<Esc>** or **<Enter>**. See Figure 4.49(b).

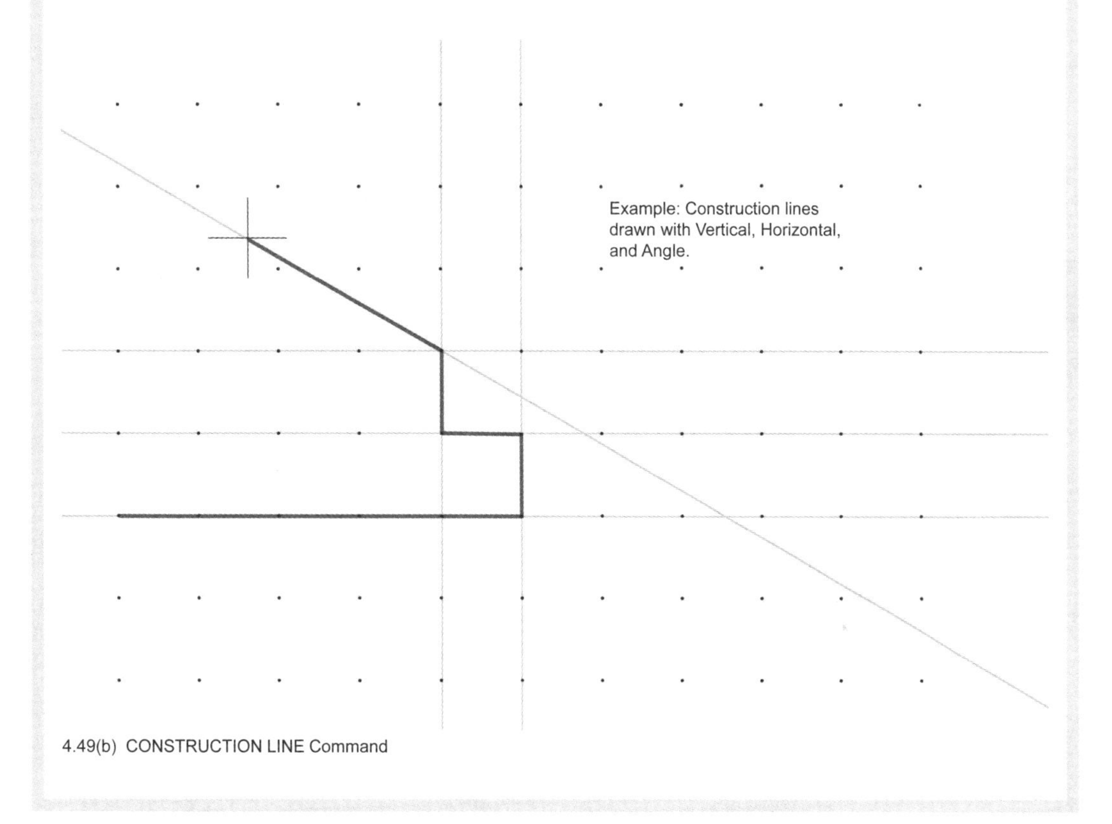

Example: Construction lines
drawn with Vertical, Horizontal,
and Angle.

4.49(b) CONSTRUCTION LINE Command

POLYLINE COMMAND			
	Ribbon	Home	The icon for the **Polyline** command is shown in Figure 4.48(a). This command creates continuous lines that may vary in width and shape.
	Panel	Draw	
	Command Line	Polyline	
4.50(a)	Alias	PLINE	

STEP–BY–STEP

POLYLINE COMMAND TUTORIAL

Step 1. Select the **Polyline** icon from the **Draw** toolbar or the **Draw** panel of the **Home** tab.

Step 2. When prompted to *Specify start point*, type absolute coordinates **1,8** and press **<Enter>**.

Step 3. When prompted to *Specify next point or [Arc/Halfwidth/Undo/Width]*, move your mouse to the right (with **Polar Tracking** turned on), type **4.5**, and press **<Enter>**. See Figure 4.50(b).

Step 4. When prompted to *Specify next point or Arc/Halfwidth/Undo/Width]*, type **A** for **Arc** and press **<Enter>**.

Step 5. When prompted to *Specify endpoint of arc or [Angle/Center/Close/Direction/Halfwidth/Line/Radius/Undo/Width]*, type **W** for **Width** and press **<Enter>**.

Step 6. When prompted to *Specify starting width (0.0000)*, type **.06** and press **<Enter>**. When prompted to *Specify ending width <0.0600>*, press **<Enter>** to accept the default lineweight. Notice that the lineweight has changed.

Step 7. When prompted to *Specify endpoint of arc or [Angle/Center/Close/Direction/Halfwidth/Line/Radius/Undo/Width]*, ensure that **Polar Tracking** is tracking at **270°** and type **2** for the distance and **<Enter>**.

Step 8. When again prompted to *Specify endpoint of arc or [Angle/Center/Close/Direction/Halfwidth/Line/Radius/Undo/Width]*, type **L** for **Line** and press **<Enter>**.

Step 9. When prompted to *Specify next point or [Arc/Halfwidth/Undo/Width]*, type **W** for **Width** and press **<Enter>**.

Step 10. When prompted to *Specify starting width (0.0600)* type **0**. When prompted with *Specify ending width (0.0000):* press **<Enter>**.

Step 11. Continue drawing polylines at either a 0 width or changing to different widths. Press **<Enter>** to end the command.

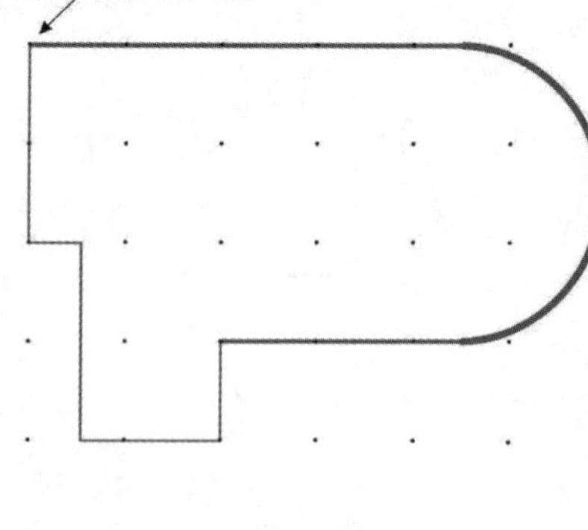

Example: POLYLINE drawn with different line weights and a transition from line to arc and back to line (select option Arc to draw arc and then Line to resume straight lines). POLYLINES are one continuous line. You can turn a POLYLINE into LINES with the EXPLODE command.

Draw an Arrowhead with the POLYLINE command. Select POLYLINE and pick a point in the drawing area. Type W for width and default the starting width of 0.0000. Enter .10 for the ending width and pick a point about .250 inches either vertically or horizontally from the beginning point.

4.50(b) POLYLINE Command

POLYGON COMMAND			
	Ribbon	Home	The icon for the **POLYGON** command is shown in Figure 4.51(a). This command is used to create multi-sided shapes with sides of equal length.
	Panel	Draw	
	Command Line	Polygon	
4.51(a)	Alias	POL	

STEP–BY–STEP

POLYGON COMMAND TUTORIAL

Step 1. Select the **Polygon** icon from the **Draw** toolbar or the **Draw** panel of the **Home** tab.

Step 2. When prompted to *Enter number of sides*, enter a value and press **<Enter>**.

Step 3. At the *Specify center of polygon* prompt, select a point on the screen by left-clicking or typing coordinate values. See Figure 4.51(b).

Step 4. When prompted to *Enter an option [Inscribed in circle/Circumscribed about circle]:* type either an **I** for inscribed or a **C** for circumscribed, and press **<Enter>**.

Step 5. When prompted to *Specify radius of circle* enter a value for the radius of the circle in which the polygon will be inscribed inside or circumscribed about, and press **<Enter>**.

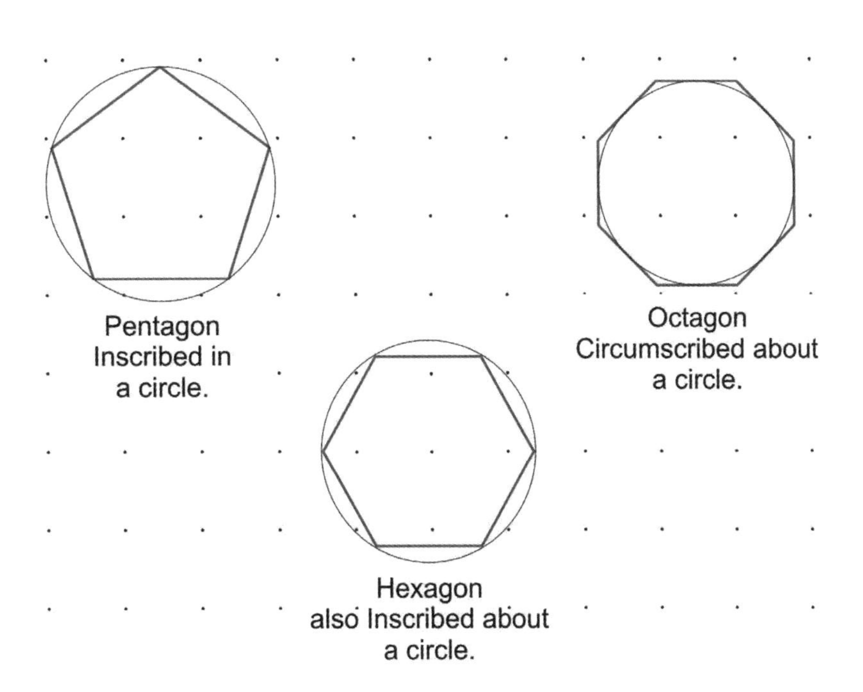

4.51(b) POLYGON Command

RECTANGLE COMMAND			
	Ribbon	Home	The icon for the **RECTANGLE** command is shown in Figure 4.52(a). This command is used to create continuous-line rectangles.
	Panel	Draw	
	Command Line	Rectangle	
4.52(a)	Alias	REC	

STEP–BY–STEP

RECTANGLE COMMAND TUTORIAL

This tutorial presents the steps for drawing a rectangle that begins at the point with absolute coordinates **1,2** and has an X-value of 6 units and a Y-value of 2 units.

Step 1. Select the **Rectangle** icon from the **Draw** toolbar or the **Draw** panel of the **Home** tab.

Step 2. When prompted to *Specify first corner point or [Chamfer/Elevation/Fillet/Thickness/Width]*, type **1,2** and press **<Enter>**. See Figure 4.52(b).

Step 3. When prompted to *Specify other corner point or [Area/Dimensions/Rotation]*, type **6,2** (with dynamic input on), or **@6,2** and press **<Enter>**.

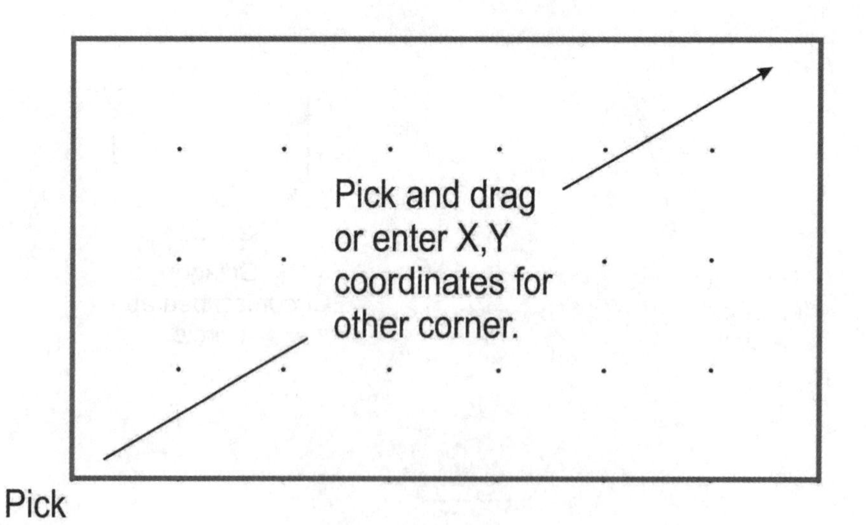

4.52(b) RECTANGLE Command

ARC COMMAND			
	Ribbon	Home	The icon for the **ARC** command is shown in Figure 4.53(a). This command can create arcs using 11 different methods.
	Panel	Draw	
	Command Line	Arc	
4.53(a)	Alias	A	

STEP–BY–STEP

ARC COMMAND TUTORIAL (START, END, RADIUS METHOD)

This tutorial presents the steps for drawing an arc that begins at absolute coordinates 3,5 and ends at absolute coordinates 7,5 and has a radius of 2 units. Note: Arcs are drawn counter-clockwise.

Step 1. From the **Draw** panel of the ribbon, highlight the **Arc** icon and pick on the icon's down arrow and pick the **Start**, **End**, **Radius** option.

Step 2. When prompted to *Specify start point of arc or [Center]*, type **3,5** and press **<Enter>**. See Figure 4.53(b).

Step 3. When prompted to *Specify end point of arc:* type **#7,5** (with dynamic input on) or **4,0** (with dynamic input off) and press **<Enter>**.

Step 4. When prompted to *Specify radius of arc*, type **2** and press **<Enter>**. See Figure **4.51(b)**.

Step 5. Draw another **Start**, **End**, **Radius** arc by typing **7,5** for the start and **#3,5** (with dynamic input on) or **-4,0** (with dynamic input off) for the end point and enter a radius of **2**.

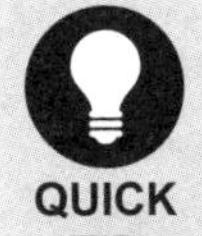

QUICK TIP

The ***Start, Center, End*** option for drawing an arc is also commonly used because it is similar to the way an arc is drawn with a compass in traditional drafting techniques.

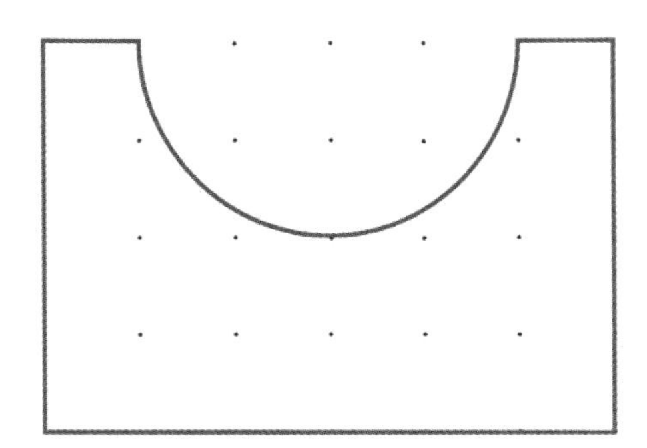

4.53(b) ARC Command

NOTE

EXAMPLE
In this example, pick Point 1 to define the *Start* point of the arc and select Point 2 to define the *End* point of the arc, then type 2 and press *Enter* to define the arc's radius.

CIRCLE COMMAND			
	Ribbon	Home	The icon for the **CIRCLE** command is shown in Figure 4.54(a). This command is used to draw a circle after you are prompted to select (or enter) the center point and the radius or diameter.
	Panel	Draw	
	Command Line	Circle	
4.54(a)	Alias	C	

STEP–BY–STEP

CIRCLE COMMAND TUTORIAL

Step 1. Select the **Circle** icon from the **Draw** toolbar or the **Draw** panel of the **Home** tab.

Step 2. When prompted to *Specify center point of circle or [3P/2P/Ttr (tan tan radius)]*, type **2,7** and press **<Enter>**.

Step 3. When prompted to *Specify radius of circle or [Diameter]*: type **1** and press **<Enter>**. See Figure 4.54(b).

Step 4. Try it again except type **6,7** for the center and press **<Enter>**. When prompted to *Specify radius of circle or [Diameter]:* type **D** for **Diameter**.

Step 5. When prompted to *Specify diameter of circle <2.0000>:* type **1** and press **<Enter>**.

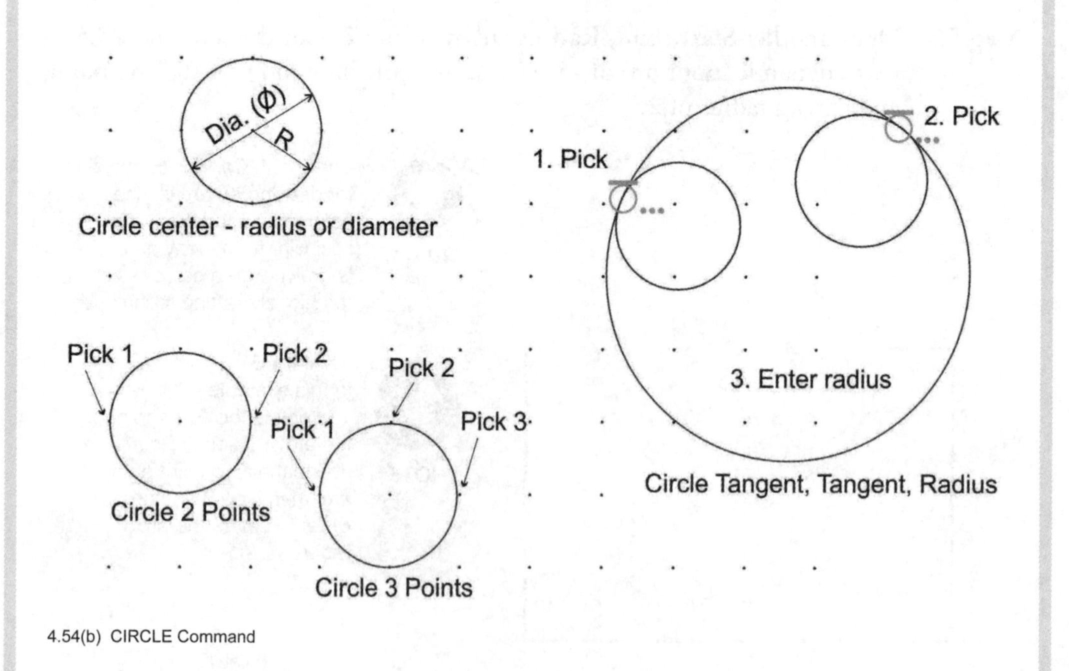

4.54(b) CIRCLE Command

REVISION CLOUD COMMAND			
	Ribbon	Home	The icon for the Revision Cloud command is shown in Figure 4.55(a). This command creates a cloud shape made of polyline arcs. Revision clouds are placed on drawings to draw attention to an area of the drawing.
	Panel	Draw	
	Command Line	Revcloud	
4.55(a)	Alias	-	

STEP–BY–STEP

REVISION CLOUD COMMAND TUTORIAL

Step 1. Select the **Revision Cloud** icon from the **Draw** toolbar or the **Draw** panel of the **Home** tab.

Step 2. When prompted to *Specify start point or [Arc length/Object/Style]*, pick a point on the screen and draw the revision cloud either clockwise or counterclockwise. Selecting a point close to the beginning point will close the cloud. See Figure 4.55(b).

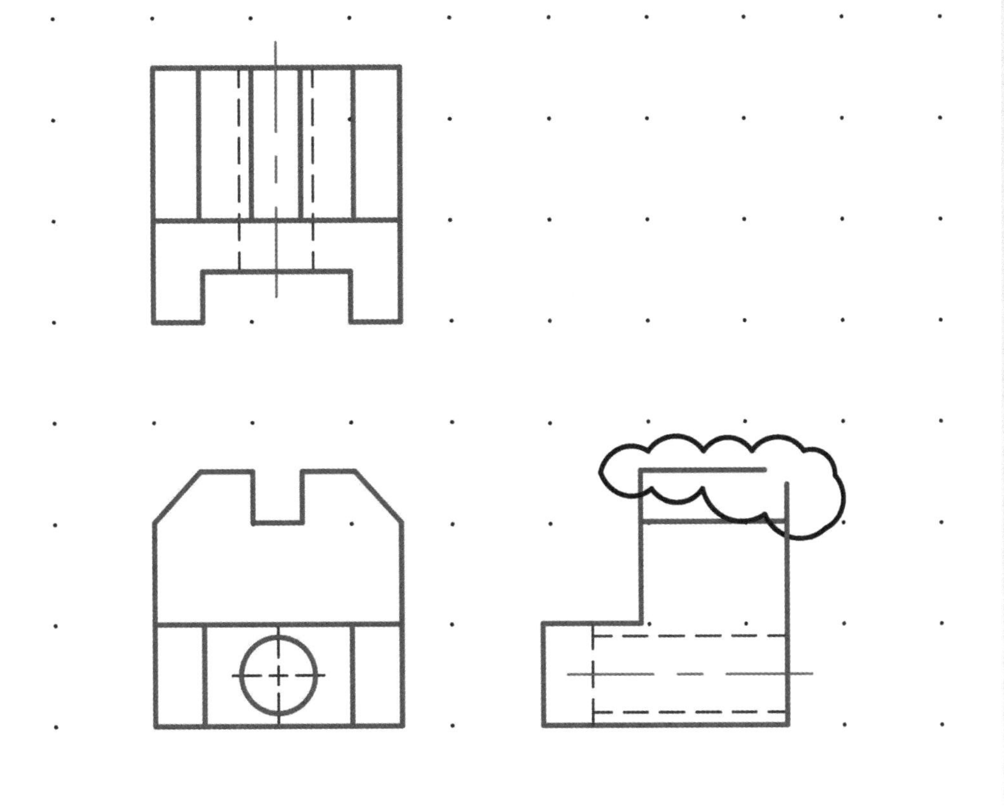

4.55(b) REVCLOUD Command

SPLINE COMMAND			
	Ribbon	Home	The icon for the **SPLINE** command is shown in Figure 4.56(a). This command creates a non-uniform spline curve.
	Panel	Draw	
	Command Line	Spline	
4.56(a)	Alias	SPL	

STEP–BY–STEP

SPLINE COMMAND TUTORIAL

Step 1. Select the **Spline** icon from the **Draw** toolbar or **Draw** panel of the **Home** tab.

Step 2. When prompted to *Specify first point or [Object]*, type **2,2** and press **<Enter>**.

Step 3. When prompted to *Specify next point*, type **#4,5** (with dynamic input on) and press **<Enter>**.

Step 4. When prompted to *Specify next point*, type **#6,2** and press **<Enter>**.

Step 5. When prompted to *Specify next point*, type **#8,5** and press **<Enter>**.

Step 6. When prompted to *Specify next point*, type **#10,2** and press **<Enter>**.

Step 7. When prompted to *Specify next point*, press **<Enter>**. See Figure 4.56(b).

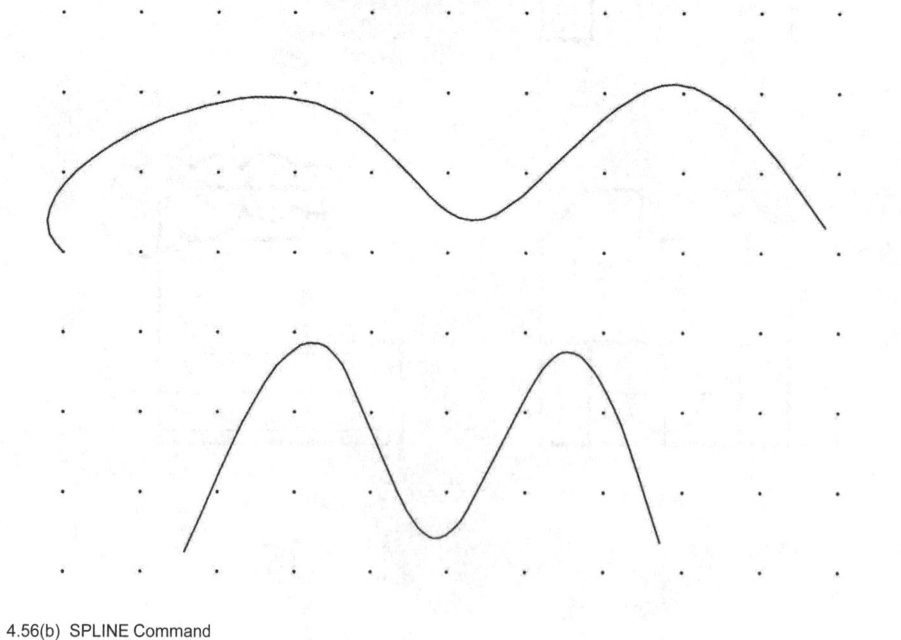

4.56(b) SPLINE Command

ELLIPSE COMMAND			
	Ribbon	Home	The icon for the **ELLIPSE** command is shown in Figure 4.57(a). This command is used to create an ellipse. An ellipse is defined by two axes: a long axis that defines its length (called the major axis) and a shorter axis that defines its width (called the minor axis).
	Panel	Draw	
	Command Line	Ellipse	
4.57(a)	Alias	EL	

STEP-BY-STEP

ELLIPSE COMMAND TUTORIAL

Step 1. Select the **Ellipse** icon from the **Draw** toolbar or the Draw panel of the **Home** tab.

Step 2. At the *Specify axis endpoint of ellipse* prompt, pick the start point of the major axis of the ellipse. See Figure 4.57(b).

Step 3. At the *Specify other endpoint of axis* prompt, pick the location of the endpoint of the major axis, or define the endpoint with coordinates, and press **<Enter>**.

Step 4. At the *Specify distance to other axis* prompt, drag the mouse in the direction desired for the minor axis and enter a distance for half the length of the minor axis and press **<Enter>**.

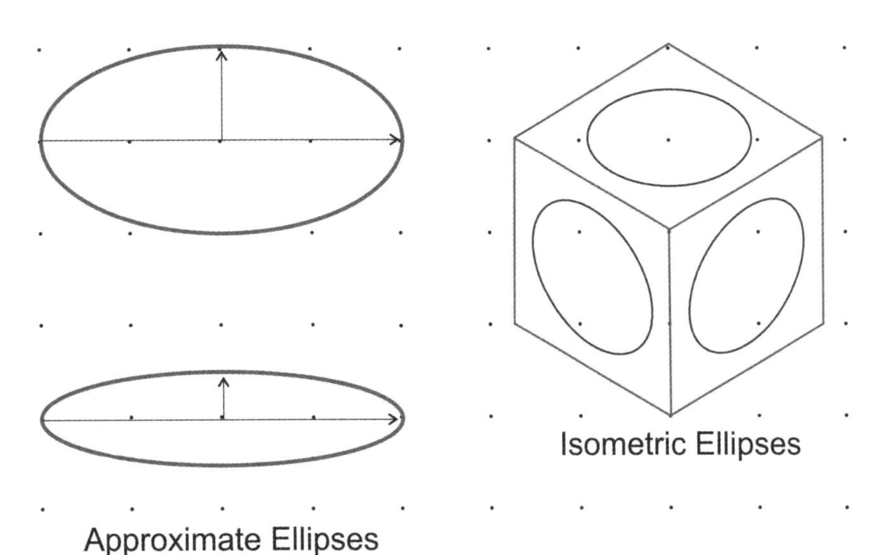

4.57(b) ELLIPSE Command

ELLIPTICAL ARC COMMAND			
	Ribbon	Home	The icon for the **Elliptical Arc** command is shown in Figure 4.58(a). This command is used to draw an elliptical arc. An elliptical arc is defined by the length of its major axis and the endpoints of the break in the ellipse.
	Panel	Draw	
	Command Line	Ellipse	
4.58(a)	Alias	EL	

STEP-BY-STEP

ELLIPTICAL ARC COMMAND TUTORIAL

Step 1. Select the **Elliptical Arc** icon from the **Draw** toolbar or the **Draw** panel of the **Home** tab (select the down arrow next to the **Ellipse** icon and choose from the drop-down list).

Step 2. At the *Specify axis endpoint of elliptical arc* prompt, pick the start point of the major axis of the ellipse. See Figure 4.58(b).

Step 3. At the *Specify other endpoint of axis* prompt, pick the location of the endpoint of the major axis, or define the endpoint with coordinates, and press **<Enter>**.

Step 4. At the *Specify distance to other axis* prompt, drag the mouse in the direction desired for the minor axis and enter a distance equal to half the length of the minor axis and press **<Enter>**.

Step 5. At the *Specify start angle* prompt, enter an angle relative to the point defined in Step 4 where the elliptical arc should begin.

Step 6. At the *Specify end angle* prompt, enter an angle relative to the point defined in Step 5 where the elliptical arc should end.

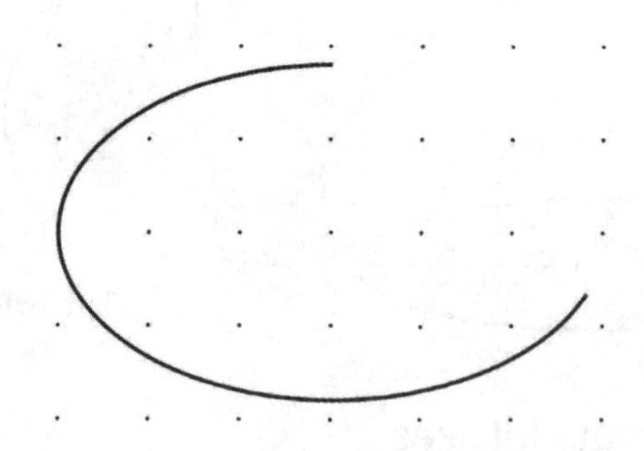

4.58(b) ELLIPTICAL ARC Command

QUICK TIP

Consulting the value noted in the *Angle Input* field of the **Dynamic Input** prompt is helpful when defining the start and end angles for an elliptical arc (see Figure 4.7).

INSERT BLOCK COMMAND			
	Ribbon	Home	The icon for the **Insert Block** command is shown in Figure 4.59. This command is used to place a named block, or a drawing, into the current drawing. The steps involved in using the **Insert Block** command are presented in Chapter 10 of this text.
	Panel	Block	
	Command Line	Insert	
4.59	Alias	I	

BLOCK COMMAND			
	Ribbon	Home	The icon for the **BLOCK** command is shown in Figure 4.60. A *block* is a named image defined within a drawing file that can be *inserted* into the drawing whenever it is needed. An example of this is a symbol for a door that is used multiple times in the creation of a floor plan. The steps involved in using the **BLOCK** command are presented in Chapter 10 of this text.
	Panel	Block	
	Command Line	Block	
4.60	Alias	B	

REGION COMMAND			
	Ribbon	Home	The icon for the **REGION** command is shown in Figure 4.61. Regions are two-dimensional closed shapes or loops. Regions are often created with closed lines or polylines. The steps involved in using the **REGION** command are presented in Chapter 13 of this text.
	Panel	Block	
	Command Line	Region	
4.61	Alias	REG	

POINT COMMAND			
	Ribbon	Home	The icon for the **POINT** command is shown in Figure 4.62(a). This command places a single point on a drawing. The point style is defined by selecting the **Format** pull-down menu and selecting **Point Style**. When the **Point Style** dialog box opens, select the tile containing the desired style and click **OK**.
	Panel	Draw	
	Command Line	Point	
4.62(a)	Alias	PO	

STEP-BY-STEP

POINT COMMAND TUTORIAL

Step 1. Select the **Point** icon from the **Draw** toolbar or **Draw** panel of the **Home** tab.

Step 2. When prompted to *Specify a point*, type **4.5,4.5** and press **<Enter>**. A point (a dot) will be placed at the point defined by the coordinates.

Step 3. To change the point style, select the **Format** pull-down menu and select **Point Style**. See Figure 4.62(b).

Step 4. When the Point Style dialog box opens, pick the tile whose symbol looks like a box with a center mark inside it and click **OK**. The original point will change to the new point style definition.

Step 5. Select **Format**, **Point Style** and practice placing another point with a different style.

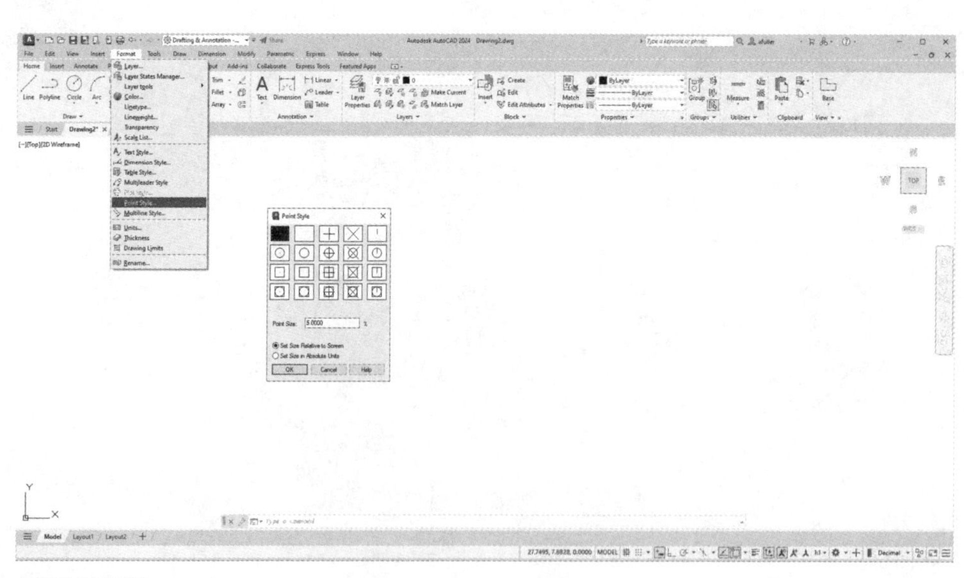

4.62(b) POINT Command

HATCH COMMAND			
	Ribbon	Home	The icon for the **HATCH** command is shown in Figure 4.63. This command fills an enclosed area, or selected objects, with a hatch pattern, a solid fill, or a gradient fill. The steps involved in using the **HATCH** command are presented in Chapter 8 of this text.
	Panel	Draw	
	Command Line	Hatch	
4.63	Alias	H	

GRADIENT COMMAND			
	Ribbon	Home	The icon for the **GRADIENT** command is shown in Figure 4.64(a). This command specifies a fill that allows a smooth transition from a darker color to a lighter one.
	Panel	Draw	
	Command Line	Gradient	
4.64(a)	Alias	GD	

STEP-BY-STEP

GRADIENT COMMAND TUTORIAL

Step 1. Draw a rectangle.

Step 2. Select the **Gradient** icon from the **Draw** toolbar or the **Draw** panel of the **Home** tab.

Step 3. Choose from the gradient styles shown in the **Pattern** panel of the **Hatch Creation** tab of the ribbon. See Figure 4.64(a).

Step 4. Select the **Add: Pick Points** button in the dialog box.

Step 5. When prompted to *Pick internal point or [Select objects/remove Boundaries]*, pick a point inside the rectangle and press **<Enter>**.

Step 6. Click **OK**.

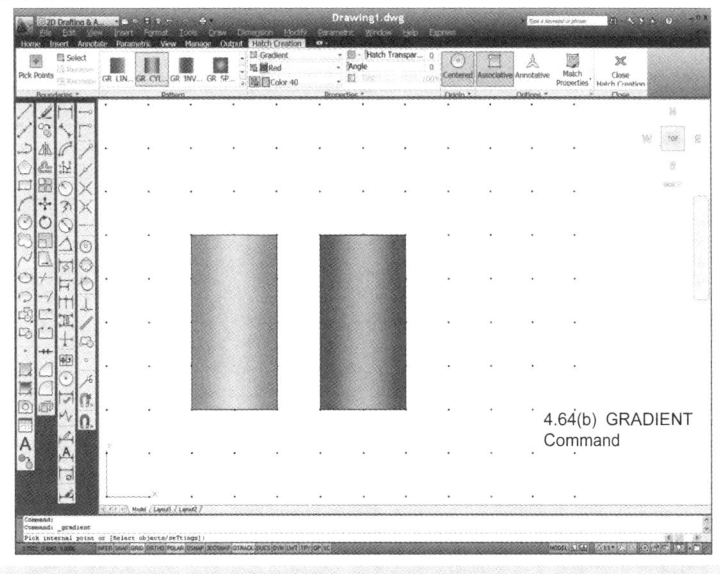

4.64(b) GRADIENT Command

TABLE COMMAND

	Ribbon	Home	The icon for the **TABLE** command is shown in Figure 4.65(a). A table is an object that displays data in rows and columns.
	Panel	Draw	
	Command Line	Table	
4.65(a)	Alias	TB	

STEP–BY–STEP

TABLE COMMAND TUTORIAL

Step 1. Select the **Table** icon from the **Draw** toolbar or the **Annotation** panel of the **Home** tab.

Step 2. When the **Insert Table** dialog box opens, enter the number of columns and rows to be created, as well as their respective widths and heights, in the **Column & Row Settings area**. Click **OK** when these values have been entered.

Step 3. At the *Specify insertion point:* prompt, select a point in the graphics window to insert the table. When the table appears, you will see the cursor blinking in the top cell. This indicates that text can be entered into this cell using the **Text Editor**. Content for other cells can be added or edited by double-clicking inside a cell with the left-click button of the mouse and using the options available in the **Text Editor** tab. Click the **Close Text Editor** tool on the **Text Editor** tab after entering the content. See Figure 4.65(b).

	A	B	C	D
1	TABLE WITH 4 COLUMNS AND 22 ROWS			
2				
3				
4				
5				
6				
7				
8				
9				
10				
11				
12				
13				
14				
15				
16				
17				
18				
19				
20				
21				
22				

4.65(b) TABLE Command

MULTILINE TEXT COMMAND

	Ribbon	Home	The icon for the **MTEXT** command is shown in Figure 4.66. This command is used to place one or more paragraphs of multiline text into the field of a drawing. Saved text from other formats can also be inserted into a drawing using this command. The procedures involved in placing text with the **MTEXT** command are described later in this chapter.
	Panel	Annotation	
	Command Line	Mtext	
4.66	Alias	T	

ADD SELECTED COMMAND

	Ribbon	Home	The icon for the **Add Selected** command is shown in Figure 4.67. This command allows you to initiate a drawing command by selecting an object that was created using that command. For example, selecting **Add Selected** and selecting an arc will open the **ARC** command.
	Panel	Draw	
	Command Line	Mtext	
4.67	Alias	T	

STEP-BY-STEP

ADD SELECTED COMMAND TUTORIAL

Step 1. Select the **Add Selected** icon from the **Draw** toolbar.

Step 2. Select an object in the graphics window, for example, a line, and the **LINE** command is invoked beginning with the *Specify the first point prompt.* If the selected entity is a circle, the **CIRCLE** command is invoked beginning with the *Specify the center point for the circle:* prompt.

QUICK TIP

When the **Add Selected** command is selected, and the user selects an object created with the RECTANGLE or POLYGON commands, the Polyline command opens.

NEW

ACTIVITY INSIGHTS

New in AutoCAD 2024: The Activity Insights command tracks events and lets the user know when changes have been made to the current drawing:

When the drawing is opened, worked on, renamed, copied, or saved in a new location

This command can be found on the History panel of the View ribbon.

4.68 Activity Insights Icon

4.69 Activity Insights Panel

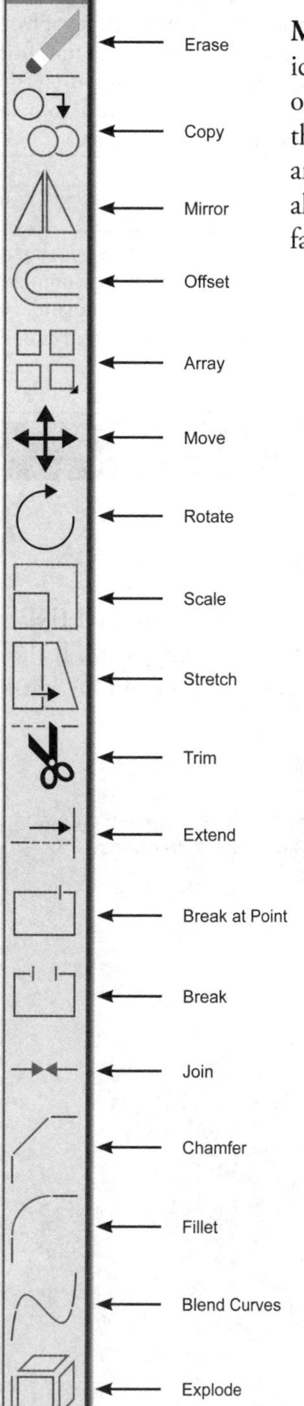

4.70 AutoCAD Modify Toolbar

Essential Modify Commands

The icons for the ***Modify*** commands located on the **Modify** toolbar are shown in Figure 4.70. These command icons can also be found in the **Modify** panel of the **Home** tab of the ribbon. Although you may find yourself using a few of the commands on this toolbar, such as **Move**, **Copy**, **Trim**, and **Offset**, much more frequently than some of the others, all the commands on this toolbar are useful, and you should familiarize yourself with each of them.

VIDEO

There are video tutorials available for the Modify commands inside the Chapter 4 folder of the book's Video Training downloads. The video downloads are available by redeeming the access code that comes with this book. Please see the inside front cover for further details.

JOB SKILL

There's a saying among experienced CAD drafters, "Never draw anything twice." What they mean by this is that drafters should use commands like **Copy**, **Move**, and **Rotate** to create technical drawings more quickly and efficiently. The **Modify** toolbar has many tools that help *speed up* the drafting process.

ERASE COMMAND			
	Ribbon	Home	The icon for the **ERASE** command is shown in Figure 4.71(a). This command is used to remove objects from a drawing. *Objects can also be removed from the drawing by selecting them and pressing the Delete <DEL> key on your keyboard
	Panel	Modify	
	Command Line	Erase	
4.71(a)	Alias	E	

STEP–BY–STEP

ERASE COMMAND TUTORIAL

Step 1. Select the **Erase** icon from the **Modify** toolbar or the **Modify** panel of the **Home** tab.

Step 2. When prompted to *Select objects*, you can select them by left-clicking and selecting in the following ways: **Window** (picking left to right), **Crossing Window** (picking right to left), or by typing **Crossing Polygon** (type **CP**), **Fence** (type **F**), and **Lasso** (click and hold down the left mouse button, then drag to select objects). When you are finished selecting the objects to erase, press **<Enter>** to complete the command. See Figure 4.71(b).

QUICK TIP

The **OOPS** command can be used to restore erased objects (just type **OOPS** and press **<Enter>**). This is a useful tool whenever an **Undo** would reverse steps that were performed following the last erase that you do not wish to be undone.

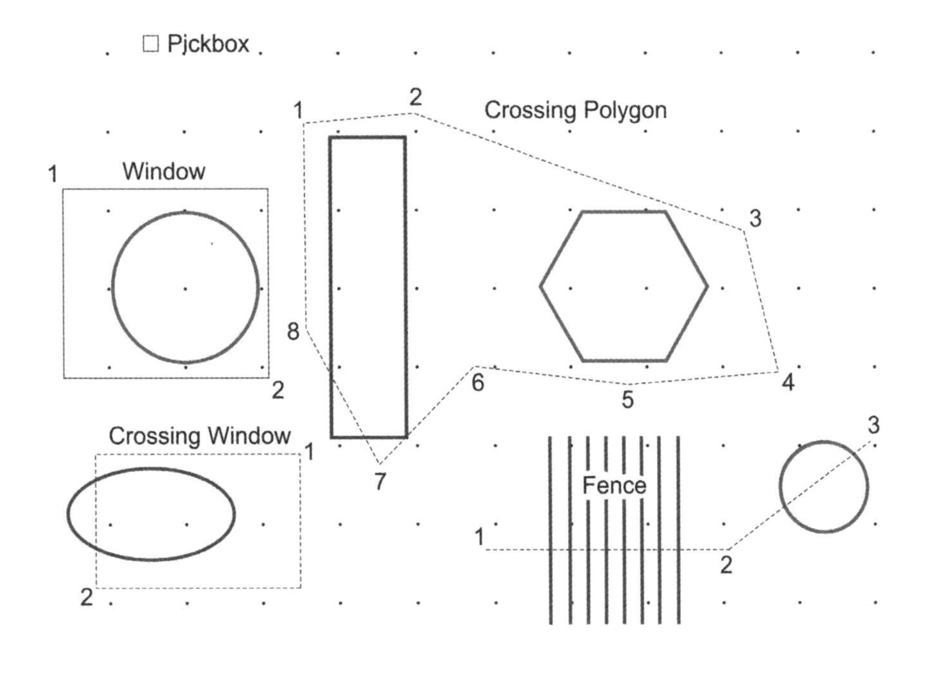

4.71(b) ERASE Command

COPY COMMAND			
	Ribbon	Home	The icon for the **COPY** command is shown in Figure 4.72(a). This command is used to create a copy of a selected object or objects.
	Panel	Modify	
	Command Line	Copy	
4.72(a)	Alias	CO	

STEP-BY-STEP

COPY COMMAND TUTORIAL

Step 1. Select the **Copy** icon from the **Modify** toolbar or the **Modify** panel of the **Home** tab.

Step 2. When prompted to *Select objects*, pick the objects you would like to copy by selecting them either individually or with a window and press **<Enter>**. See Figure 4.72(b).

Step 3. When prompted to *Specify the Base point*, pick a point on the object. The base point can be thought of as a handle on the object to be copied.

Step 4. When prompted to *Define the Displacement*, use the mouse to pick a point in the drawing window located at a specific distance from the base point of the original object, or enter absolute, relative, or polar coordinates, to define the point where you want to place the copy. You can continue to place copies by defining more points. Press **<Esc>** or **<Enter>** to end the command.

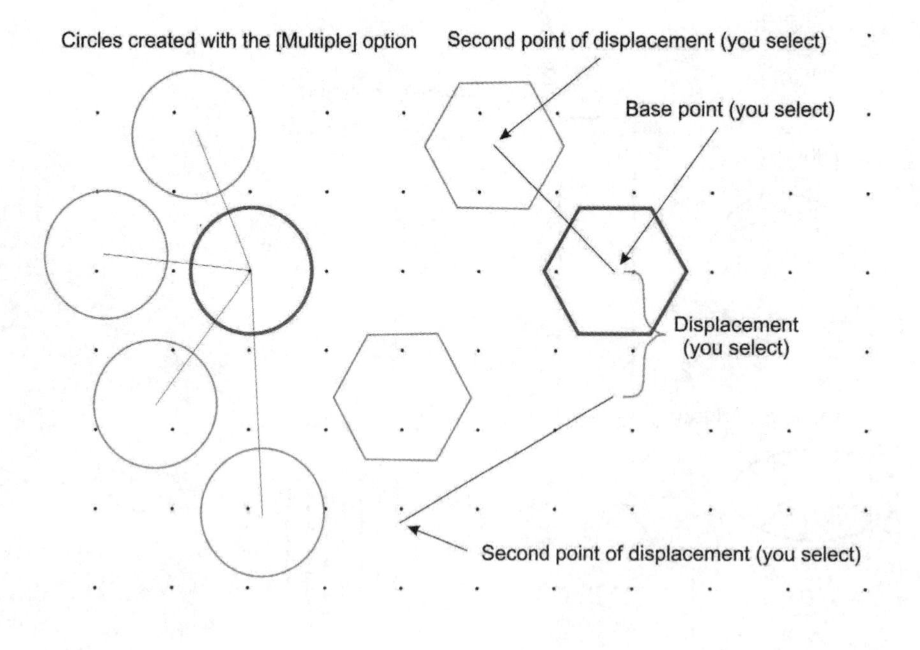

4.72(b) COPY Command

MIRROR COMMAND			
	Ribbon	Home	The icon for the **MIRROR** command is shown in Figure 4.73(a). This command is used to create a mirror image of an object around an axis called a *mirror line.*
	Panel	Modify	
	Command Line	Mirror	
4.73(a)	Alias	MI	

STEP–BY–STEP

MIRROR COMMAND TUTORIAL

Step 1. Select the **Mirror** icon from the **Modify** toolbar or the **Modify** panel of the **Home** tab.

Step 2. At the *Select objects* prompt, pick the objects you would like to mirror by selecting them either individually, or with a window, and press **<Enter>**. See Figure 4.73(b).

Step 3. When prompted to *Specify first point of mirror line*, define the first point of the mirror axis by selecting a point in the graphics window where you want the mirror axis to begin.

Step 4. When you are prompted to *Select the second point of the mirror line*, select a second point in the graphics window defining the other end of the mirror axis.

Step 5. When prompted to *Erase source objects? [Y/N]*, press **<Enter>** to retain the source object (the prompt's default is **No**), or type a **Y** and press **<Enter>** to erase the object being mirrored.

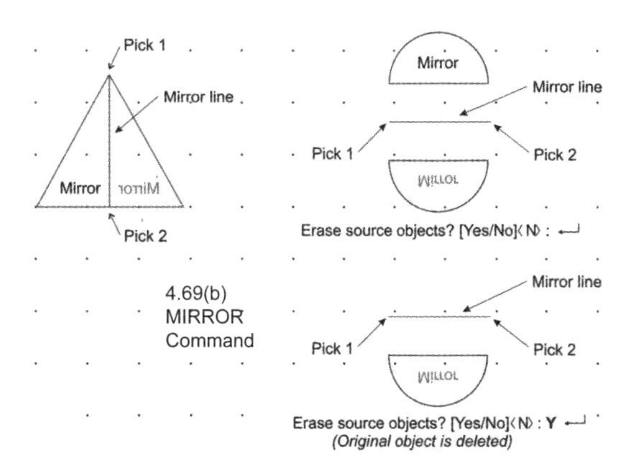

4.73(b) MIRROR Command

QUICK TIP

To prevent text from being mirrored, type **MIRRTEXT** on the command line, press **<Enter>**, set the value to 0, and press **<Enter>**.
Turn **ORTHOMODE <F8>** *ON* when defining a vertical or horizontal mirror line.

OFFSET COMMAND			
	Ribbon	Home	The icon for the **OFFSET** command is shown in Figure 4.70(a). The **OFFSET** command is used to create a new object whose shape parallels the shape of the selected object. Offsetting a circle or an arc creates a larger or smaller circle or arc (depending on which side you specify for the offset) that is concentric with the original circle or arc.
	Panel	Modify	
	Command Line	Offset	
4.74(a)	Alias	O	

STEP-BY-STEP

OFFSET COMMAND TUTORIAL

Step 1. Select the **Offset** icon from the **Modify** toolbar or the **Modify** panel of the **Home** tab.

Step 2. When prompted to *Specify the offset distance*, type the value of desired offset distance and press **<Enter>**.

Step 3. When prompted to *Select object to offset:* select the object by left-clicking.

Step 4. At the *Specify point on side to offset prompt*, pick a point on the side of the object where you want the new object to be created. See Figure 4.74(b).

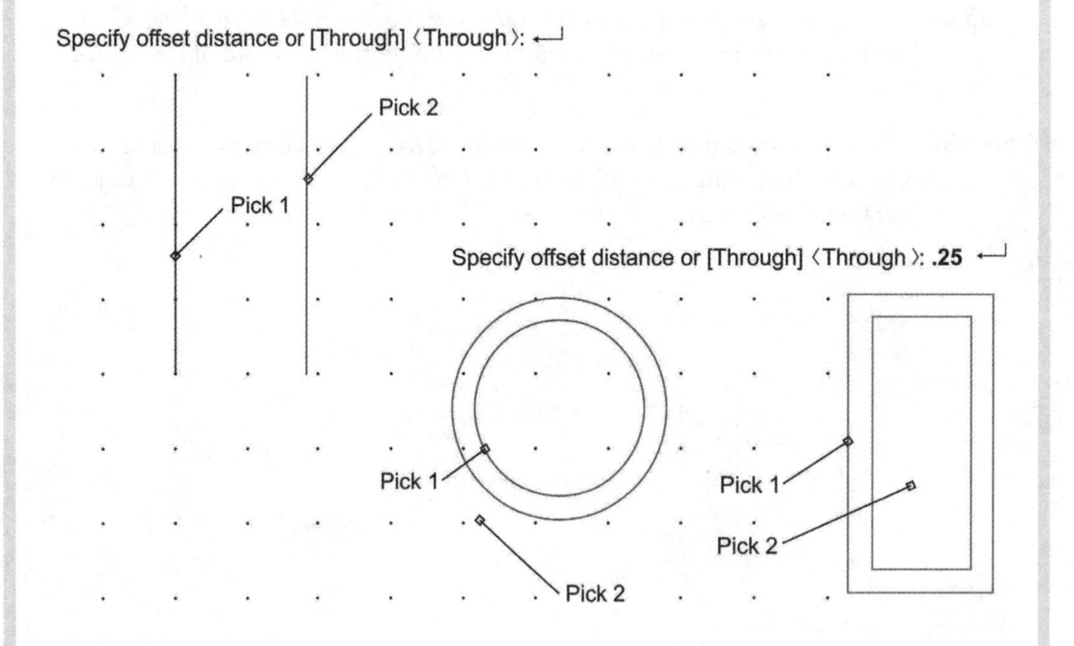

4.74(b) OFFSET Command

ARRAY COMMAND

4.75(a)			The icon for the **ARRAY** command is shown in Figure 4.75(a). This command is used to create multiple copies of objects in either a *rectangular* or *polar* array, see Figure 4.75(c). **Note:** Items arrayed in a circle are referred to as a *polar* array.
	Ribbon	Home	
	Panel	Modify	
	Command Line	Array	
	Alias	-	

STEP–BY–STEP

STEPS IN CREATING A RECTANGULAR ARRAY

Step 1. Select the **RECTANGULAR ARRAY** command from the **Modify** toolbar or the **Modify** panel of the **Home** tab. See Figure 4.75(a) and select object(s) to be arrayed and press **<Enter>**.

Step 2. Enter the number of columns and rows desired into the **Columns** and **Rows** fields of the contextual ribbon. **Note:** Columns run vertically and rows run horizontally.

Step 3. Next, enter the desired spacing between rows and columns into the **Between** fields of the contextual ribbon. Note: Entering positive values creates an array of the selected object that is above and to the right of the object. Entering negative values (**-4** for example) in the *Between* fields creates an array that is to the left and below. Press **<Enter>** or the **Close Array** button to complete the command. A rectangular array is created based on the settings defined in steps 3 and 4. An **Array** can be edited by *double-clicking* it with the left click mouse button and changing the settings in the contextual ribbon, see 4.75(d).

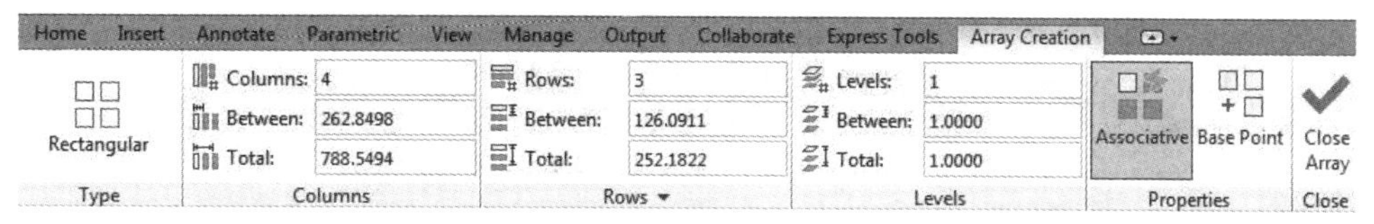

4.75(b) Detail of the Array command's contextual ribbon. This ribbon provides tools that are in context to the Array command.

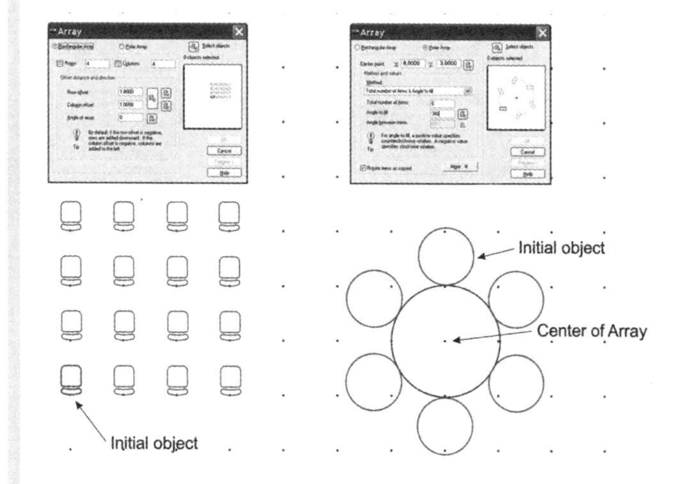

Inputs for the **Array** command are entered into the fields of the *contextual* Array ribbon. This ribbon opens when the Array command has been selected, see Figure 4.75(b).

4.75(c) Examples of Rectangular and Polar Arrays.

STEP-BY-STEP

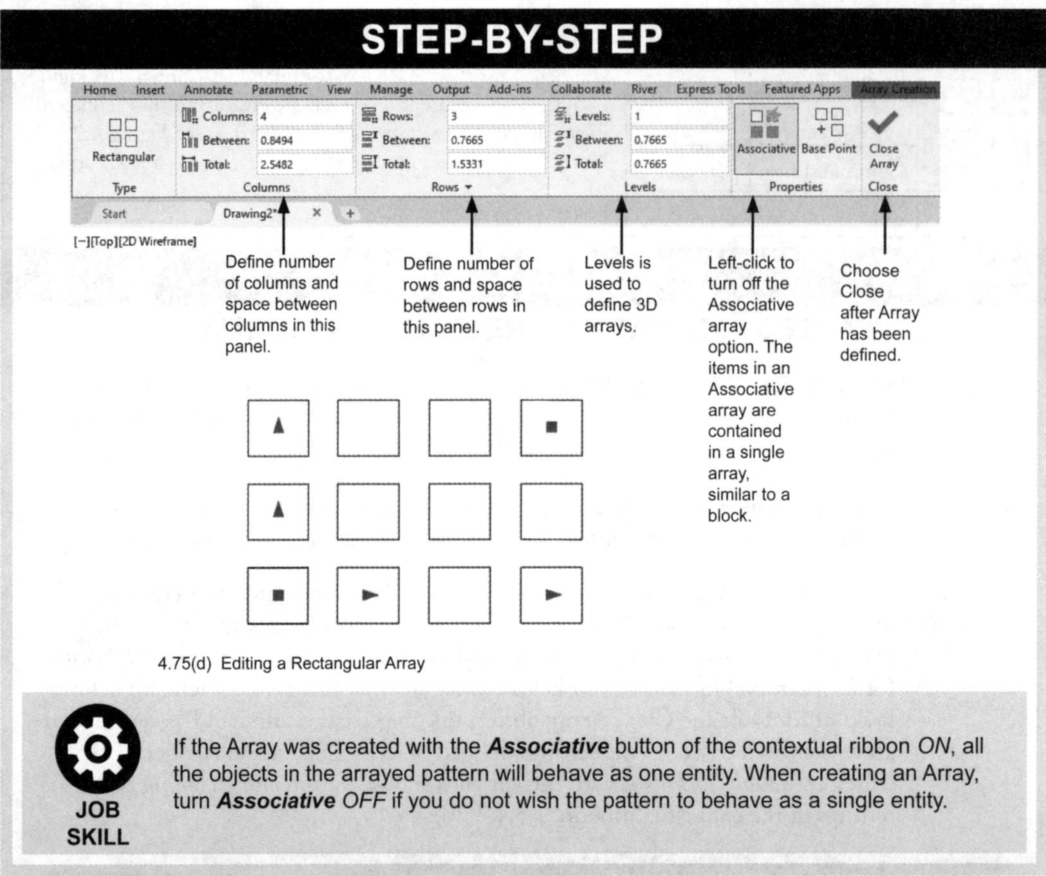
Home Insert Annotate Parametric View Manage Output Add-ins Collaborate River Express Tools Featured Apps Array Creation
Rectangular
Type
Columns: 4
Between: 0.8494
Total: 2.5482
Columns
Rows: 3
Between: 0.7665
Total: 1.5331
Rows
Levels: 1
Between: 0.7665
Total: 0.7665
Levels
Associative
Base Point
Properties
Close Array
Close
Start
Drawing2*
[−][Top][2D Wireframe]
Define number of columns and space between columns in this panel.
Define number of rows and space between rows in this panel.
Levels is used to define 3D arrays.
Left-click to turn off the Associative array option. The items in an Associative array are contained in a single array, similar to a block.
Choose Close after Array has been defined.

4.75(d) Editing a Rectangular Array

JOB SKILL

If the Array was created with the ***Associative*** button of the contextual ribbon *ON*, all the objects in the arrayed pattern will behave as one entity. When creating an Array, turn ***Associative*** *OFF* if you do not wish the pattern to behave as a single entity.

STEP-BY-STEP

STEPS IN CREATING A POLAR ARRAY

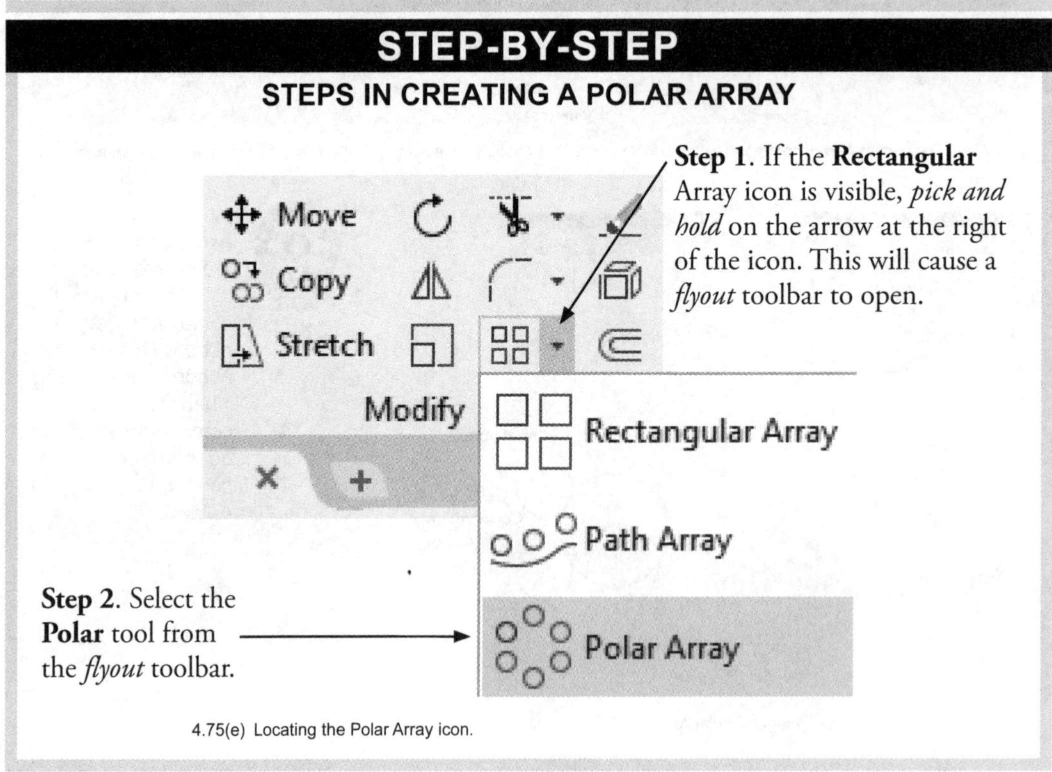
Step 1. If the **Rectangular** Array icon is visible, *pick and hold* on the arrow at the right of the icon. This will cause a *flyout* toolbar to open.
Move
Copy
Stretch
Modify
Rectangular Array
Path Array
Polar Array
Step 2. Select the **Polar** tool from the *flyout* toolbar.

4.75(e) Locating the Polar Array icon.

Step 3. Select the object(s) to be arrayed and press **<Enter>**. See Figure 4.75(f).

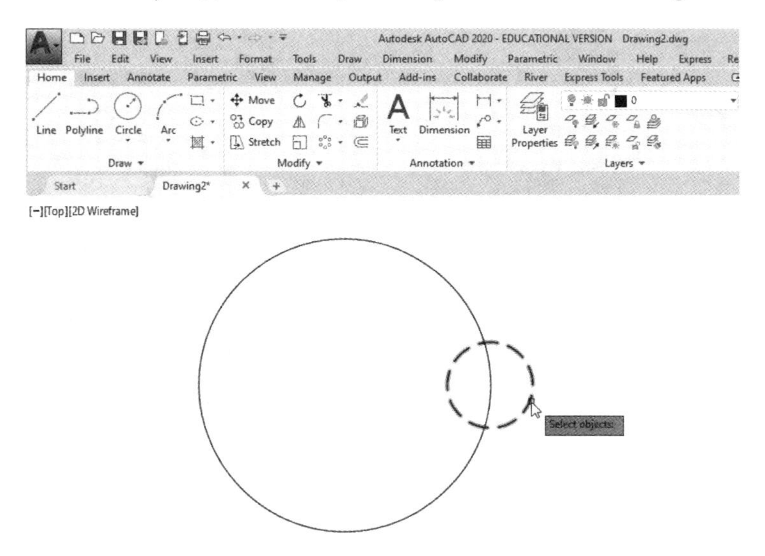

4.75(f) Selecting Object(s) to be Arrayed

Step 4. Define the center point of the array by picking a point or typing X,Y coordinates (then pressing **<Enter>**). See Figure 4.75(g). In this figure, the center of the larger circle will be the center of the array.

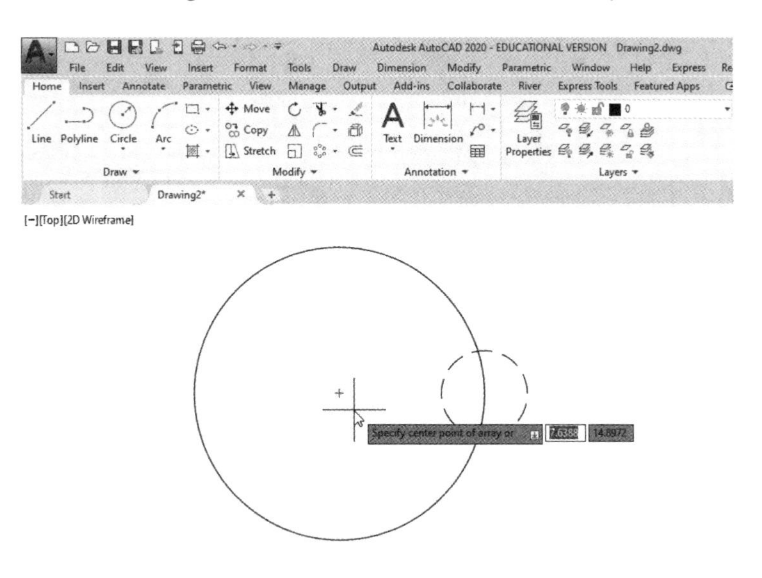

4.75(g) Defining the Center of the Arrayed Pattern

STEP-BY-STEP

Step 5. Enter the information shown in Figure 4.75(h).

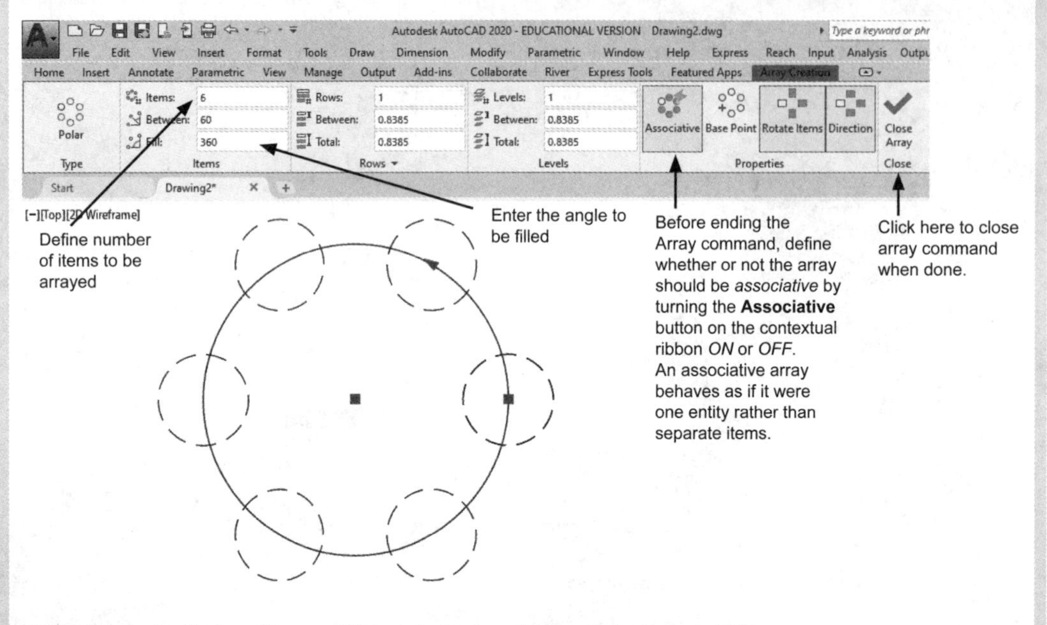
Autodesk AutoCAD 2020 - EDUCATIONAL VERSION Drawing2.dwg
Polar
Items: 6
Between: 60
Fill: 360
Rows: 1
Between: 0.8385
Total: 0.8385
Levels: 1
Between: 0.8385
Total: 0.8385
Associative
Base Point
Rotate Items
Direction
Close Array
Type
Items
Rows
Levels
Properties
Close
Define number of items to be arrayed
Enter the angle to be filled
Before ending the Array command, define whether or not the array should be *associative* by turning the **Associative** button on the contextual ribbon *ON* or *OFF*. An associative array behaves as if it were one entity rather than separate items.
Click here to close array command when done.

4.75(h) Defining the Number of Items and Fill Angle for the Arrayed Pattern in the Contextual Ribbon

STEPS TO EDITING AN ASSOCIATIVE POLAR ARRAY

Step 1. Select one of the arrayed entities. Blue boxes and a triangle (called *grips*) will appear along with an Array *properties* box, see Figure 4.75(i). By selecting the *grips*, or changing the values in the properties box, the array can be edited, see Figures 4.75(j) and 4.75(k).

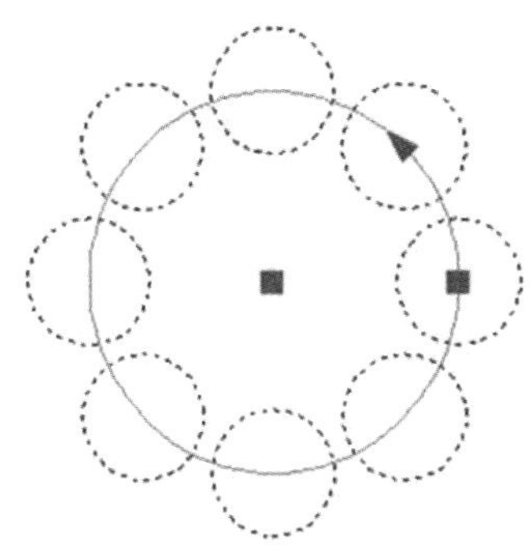

4.75(i) Editing a Polar Array

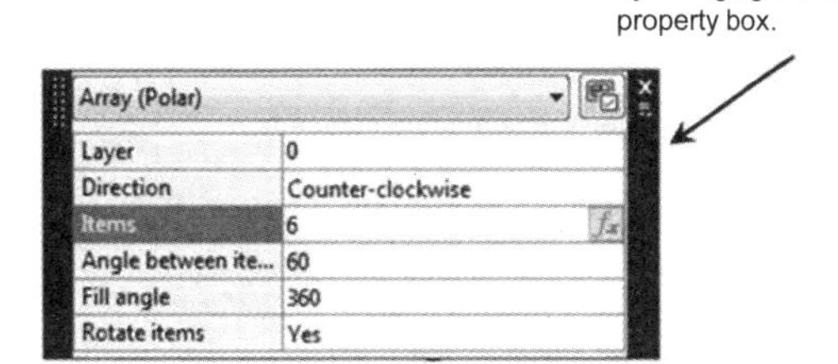

In addition to the contextual ribbon, the Array can be edited by changing the values in the property box.

Step 2. By selecting this grip, and dragging it to a different location, you can redefine the radius of the pattern; as in Figure 4.75(j).

4.75(j) Editing the radius of a Polar Array

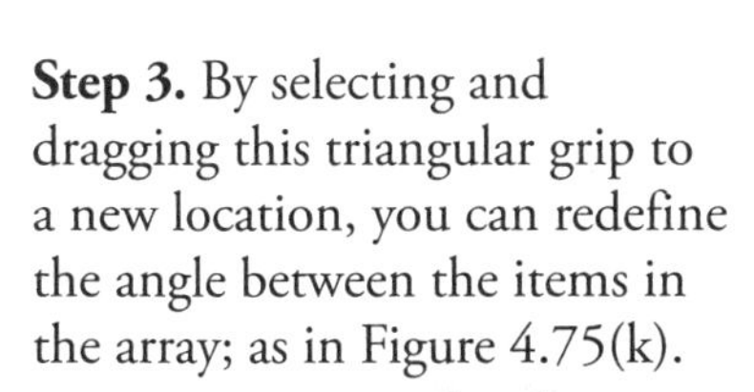

Step 3. By selecting and dragging this triangular grip to a new location, you can redefine the angle between the items in the array; as in Figure 4.75(k).

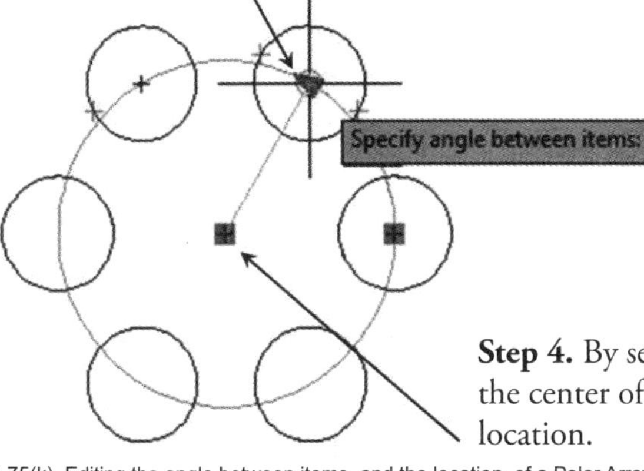

Step 4. By selecting this grip you can drag the center of the arrayed pattern to a new location.

4.75(k) Editing the angle between items, and the location, of a Polar Array

MOVE COMMAND			
	Ribbon	Home	The icon for the **MOVE** command is shown in Figure 4.76(a). This command is used to move existing objects to a new location.
	Panel	Modify	
	Command Line	Move	
4.76(a)	Alias	M	

STEP–BY–STEP

MOVE COMMAND TUTORIAL

Select the Move icon from the Modify panel of the Home tab.

Step 1. When prompted to *Select objects*, pick the objects you would like to move by selecting them either individually or with a window and press **<Enter>**.

Step 2. When prompted to *Specify the base point:*, pick a point on the object. The base point is like a handle on the object to be moved.

Step 3. When prompted to *Define the Displacement*, pick a point at a specific distance from the original object or enter absolute, relative, or polar coordinate values. See Figure 4.76(b).

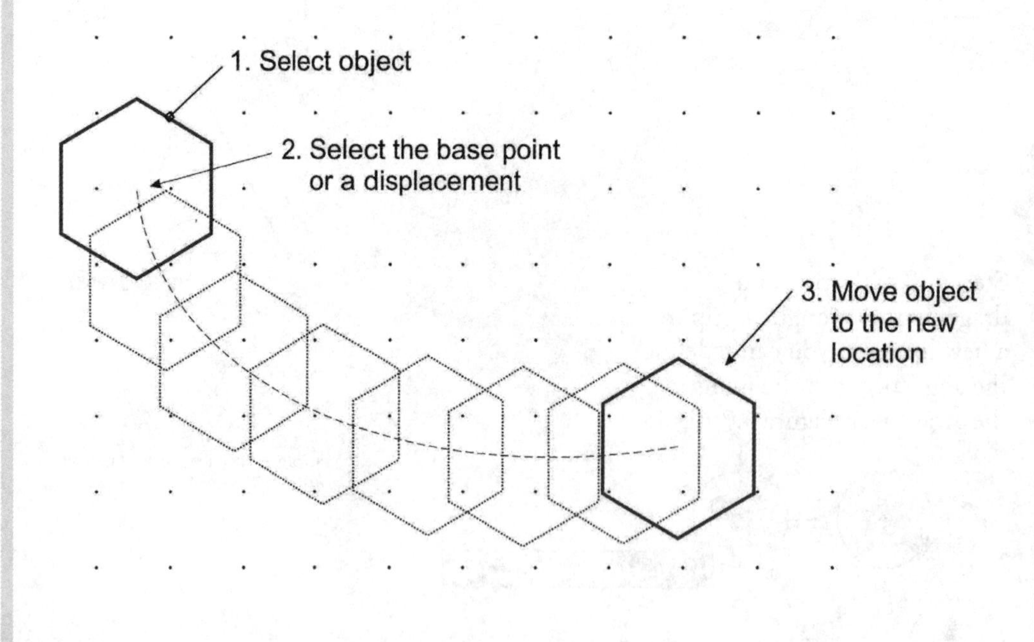

4.76(b) MOVE Command

ROTATE COMMAND			
	Ribbon	Home	The icon for the **ROTATE** command is shown in Figure 4.77(a). This command is used to rotate objects about a base point. The rotation angle is based on the initial position of the object to be rotated and the selected base point. With **Ortho** on, rotations are limited to increments of 90° angles only.
	Panel	Modify	
	Command Line	Rotate	
4.77(a)	Alias	RO	

STEP–BY–STEP

ROTATE COMMAND TUTORIAL

Step 1. Select the **Rotate** icon from the **Modify** toolbar or the **Modify** panel of the **Home** tab.

Step 2. When prompted to *Select objects*, pick the objects you would like to rotate by selecting them either individually or with a window, and press **<Enter>**.

Step 3. When prompted to *Specify the base point*, select a point on the object. The base point is the pivot point around which the rotation will occur.

Step 4. When prompted to *Specify the rotation angle*, enter a value and press **<Enter>**. See Figure 4.77(b).

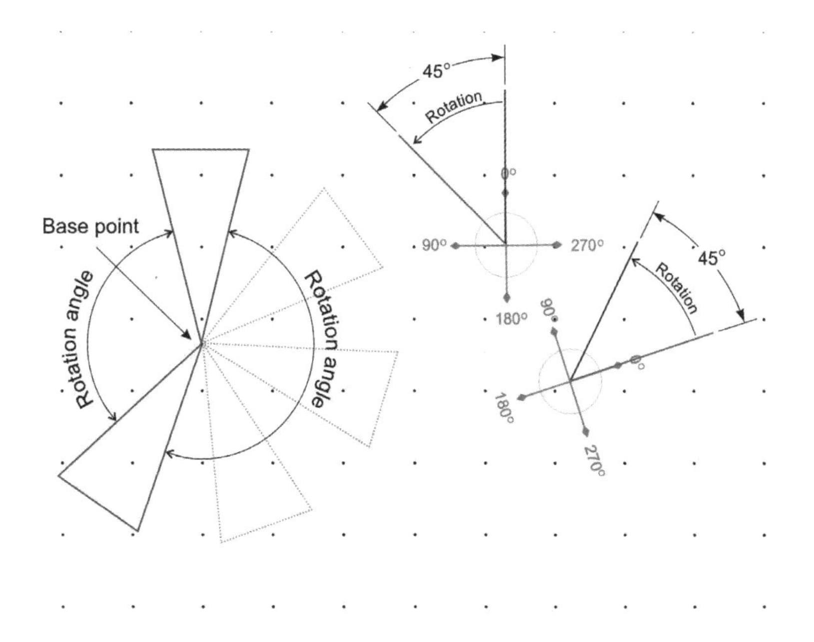

4.77(b) ROTATE Command

QUICK TIP

Objects are rotated *counterclockwise* when positive rotation values are entered and *clockwise* when negative values are entered.

SCALE COMMAND			
	Ribbon	Home	The icon for the **SCALE** command is shown in Figure 4.78(a). This command is used to change the size of objects.
	Panel	Modify	
	Command Line	Scale	You can scale up or down by either dragging or entering a scale factor. Scale factors are as follows: .5 = half size, 2 = twice the size, .75 = 3/4 size, and so on.
4.78(a)	Alias	SC	

STEP–BY–STEP

SCALE COMMAND TUTORIAL

Step 1. Select the **Scale** icon from the **Modify** toolbar or the **Modify** panel of the **Home** tab.

Step 2. When prompted to *Select objects*, pick the objects you would like to scale by selecting them either individually or with a window, and press **<Enter>**.

Step 3. When prompted to *Specify the base point*, pick a point on the object. The objects that are scaled will expand or contract relative to this point.

Step 4. When prompted to *Specify scale factor*, enter a value (for example: .5 = half size, 2 = twice the size, .75 = 3/4 size) and press **<Enter>**. See Figure 4.78(b).

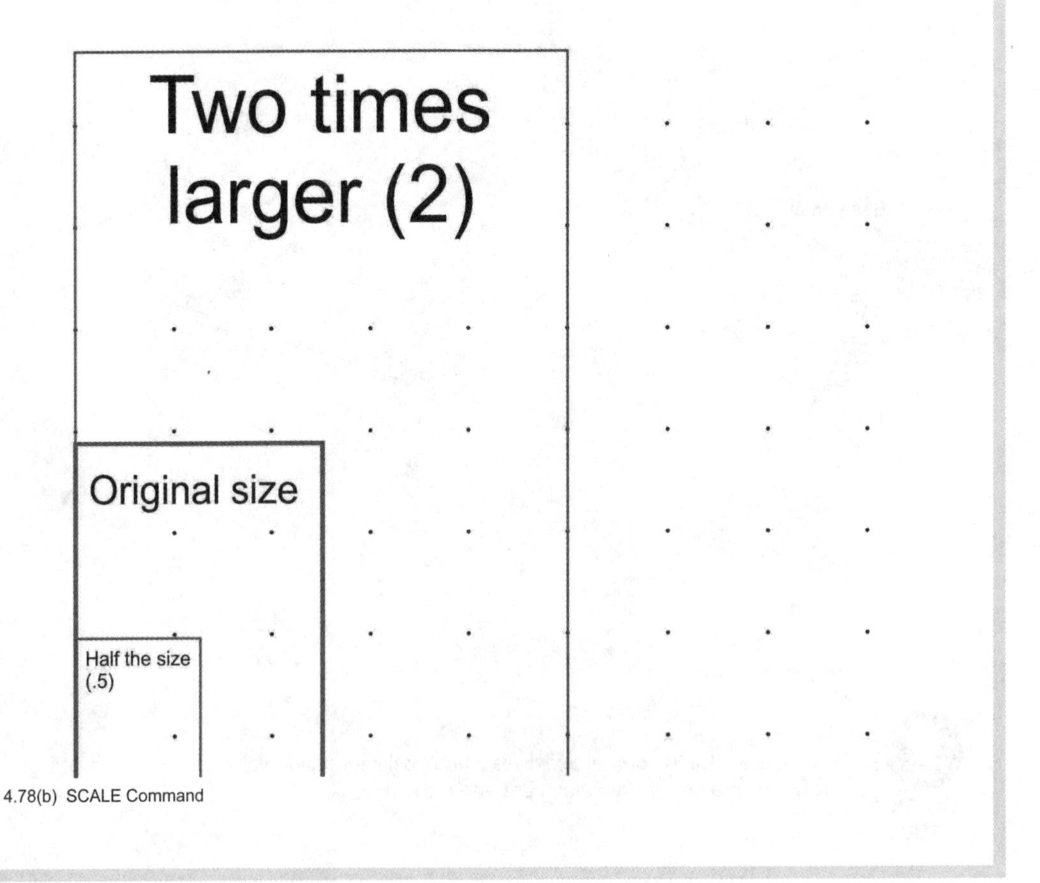

4.78(b) SCALE Command

STRETCH COMMAND			
	Ribbon	Home	The icon for the **STRETCH** command is shown in Figure 4.79(a). This command can be used to lengthen or shorten objects. It can also distort objects. To stretch an object, you can either drag the base point to a new location or enter coordinates.
	Panel	Modify	
	Command Line	Stretch	
4.79(a)	Alias	S	

STEP–BY–STEP

STRETCH COMMAND TUTORIAL

Step 1. Select the **Stretch** icon from the **Modify** toolbar or the **Modify** panel of the **Home** tab.

Step 2. When prompted to *Select objects*, pick the objects you would like to stretch with a *crossing window* (define the window by picking from right to left) and press **<Enter>**.

Step 3. When prompted to *Specify base point*, select a point located on the object to be stretched.

Step 4. When prompted to *Specify second point*, define the point you wish the object to stretch to by picking a point with the mouse, or entering absolute, relative, or polar coordinate values. See Figure 4.79(b).

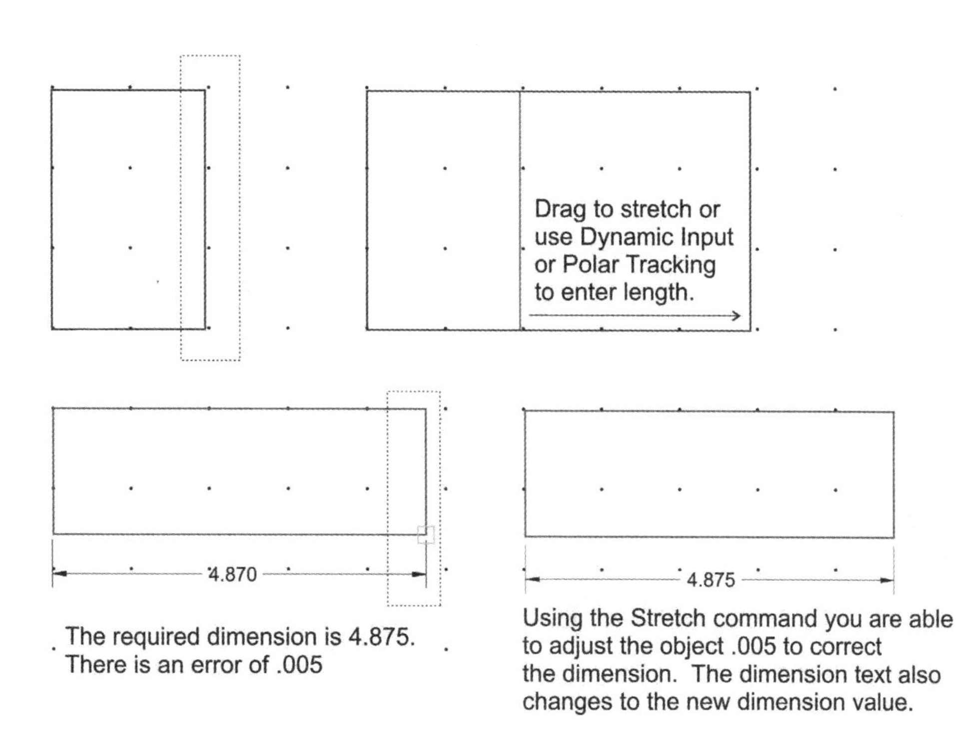

Drag to stretch or use Dynamic Input or Polar Tracking to enter length.
4.870
The required dimension is 4.875. There is an error of .005
4.875
Using the Stretch command you are able to adjust the object .005 to correct the dimension. The dimension text also changes to the new dimension value.

4.79(b) STRETCH Command

TRIM COMMAND			
	Ribbon	Home	The icon for the **TRIM** command is shown in Figure 4.80(a). This command is used to trim objects to an intersection or cutting edge defined by another object. Two trim methods are available: *quick trim* and *regular trim.*
	Panel	Modify	
	Command Line	Trim	
4.80(a)	Alias	TR	

STEP–BY–STEP

TRIM COMMAND TUTORIAL (QUICK TRIM OPTION)

Step 1. Select the **Trim** icon from the **Modify** toolbar or the **Modify** panel of the **Home** tab.

Step 2. When prompted to *Select object to trim*, pick the object to be trimmed on the side of the cutting edge you want to trim to.

Step 3. Press **<Enter>** to end the command. See Figure 4.80(b).

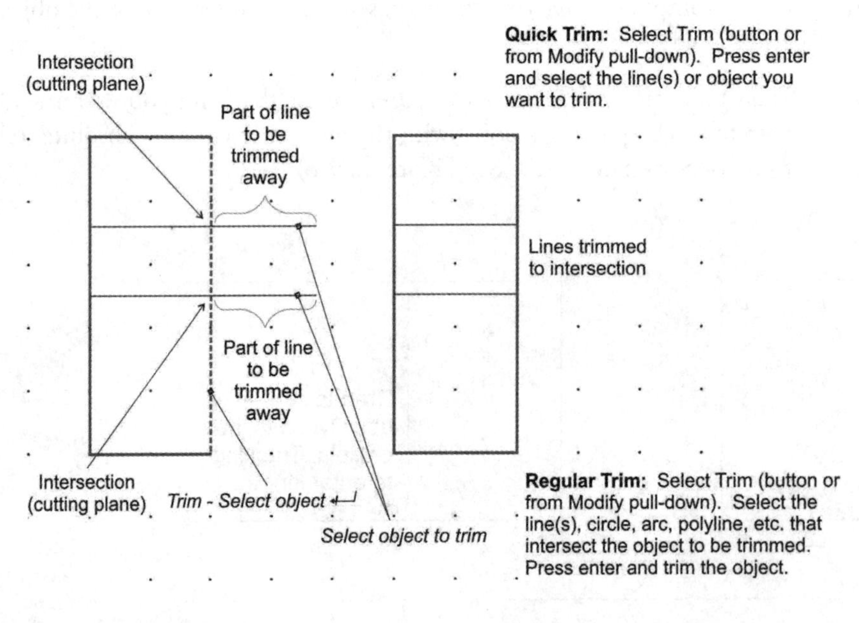
Intersection
(cutting plane)
Part of line
to be
trimmed
away
Part of line
to be
trimmed
away
Intersection
(cutting plane)
Trim - Select object
Select object to trim
Quick Trim: Select Trim (button or from Modify pull-down). Press enter and select the line(s) or object you want to trim.
Lines trimmed
to intersection
Regular Trim: Select Trim (button or from Modify pull-down). Select the line(s), circle, arc, polyline, etc. that intersect the object to be trimmed. Press enter and trim the object.

4.80(b) TRIM Command

QUICK TIP

TRIM COMMAND UPDATE
The trim command no longer prompts for a cutting edge as the command begins. Simply select the object to trim. The trim command will now also erase lines that don't cross a cutting edge.

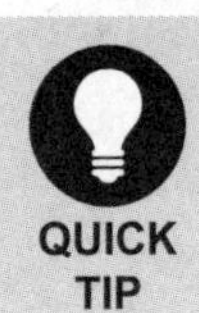
QUICK TIP

Cutting edges can be selected individually with the mouse instead, if you prefer. This is known as a *regular* trim.

EXTEND COMMAND			
	Ribbon	Home	The icon for the **EXTEND** command is shown in Figure 4.81(a). This command is used to extend a line or object to meet another object or line (called a *boundary edge*). Objects that can be extended are arcs, elliptical arcs, lines, polylines (2D and 3D), and rays.
	Panel	Modify	
	Command Line	Extend	
4.81(a)	Alias	EX	

STEP–BY–STEP

EXTEND COMMAND TUTORIAL

Step 1. Select the **Extend** icon from the **Modify** toolbar or the **Modify** panel of the **Home** tab. (Select the down arrow next to the **Trim** icon and choose from the drop-down list.)

Step 2. When prompted to *Select object to extend*, select the object to be extended near the end that you would like to extend, and the object will be extended to the next boundary edge.

Step 3. Press **<Esc>** or **<Enter>** to end the command. See Figure 4.81(b).

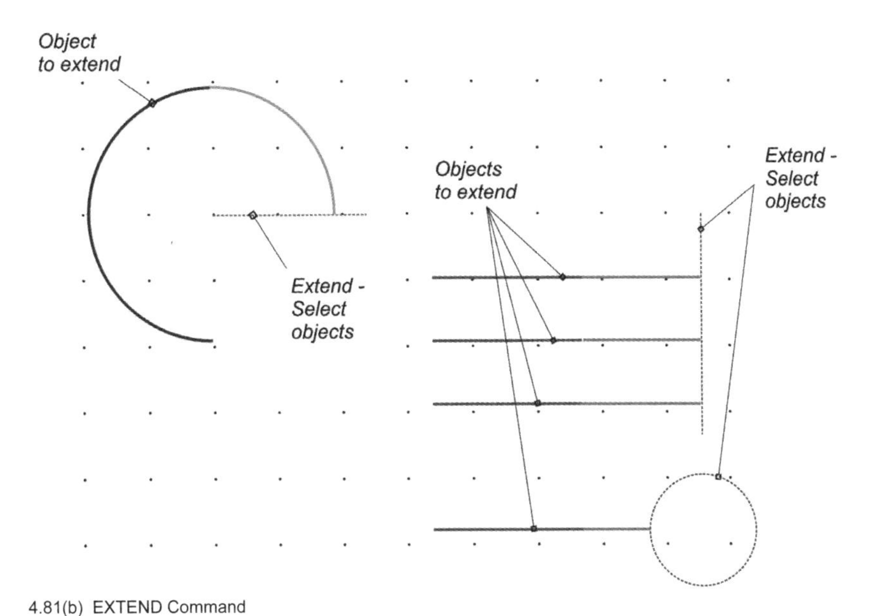

4.81(b) EXTEND Command

QUICK TIP

EXTEND COMMAND UPDATE

The extend command no longer prompts for a boundary edge as the command begins. Simply select the object to extend.

BREAK AT POINT COMMAND			
	Ribbon	Home	The icon for the **BREAK AT POINT** command is shown in Figure 4.82. This command is used to break a line into two separate but collinear lines at the selected point.
	Panel	Modify	
	Command Line	Break	
4.82	Alias	BR	

STEP-BY-STEP

BREAK AT POINT COMMAND TUTORIAL

Step 1. Select the **Break at Point** icon from the **Modify** toolbar or the **Modify** panel of the **Home** tab.

Step 2. When prompted to *Select object to break*, select the line to be broken.

Step 3. When prompted to *Select first break point*, select the point on the line where you would like the break to occur.

BREAK COMMAND			
	Ribbon	Home	The icon for the **BREAK** command is shown in Figure 4.83(a). This command is used to break an object between two points.
	Panel	Modify	
	Command Line	Break	
4.83(a)	Alias	BR	

STEP–BY–STEP

BREAK COMMAND TUTORIAL

Step 1. Select the **Break** icon from the **Modify** toolbar or the **Modify** panel of the **Home** tab.

Step 2. When prompted to *Select object to break*, select the object to be broken.

Step 3. Type **F** for **First Point** and press **<Enter>**.

Step 4. When prompted to *Select first break point*, select the object to be broken at the start point of the desired break.

Step 5. When prompted to *Select second break point*, select the endpoint of the desired break. See Figure 4.83(b).

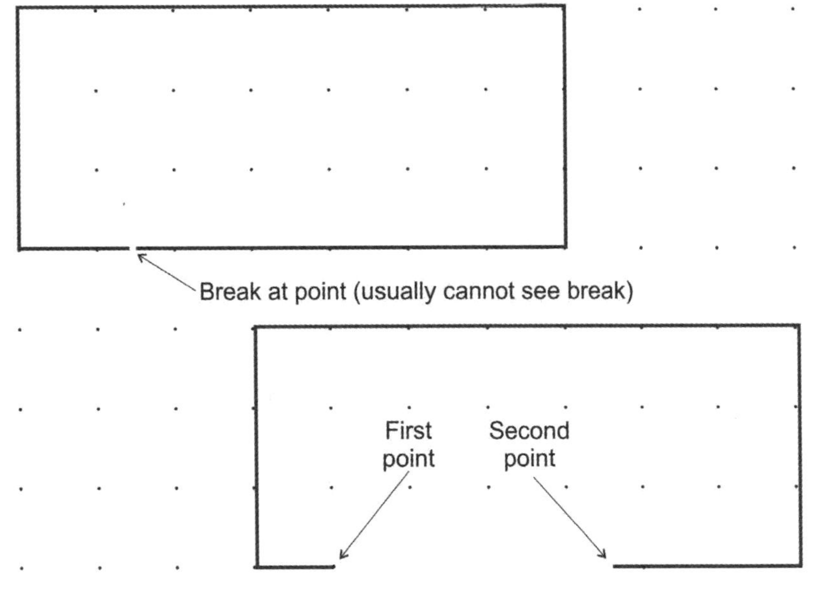

4.83(b) BREAK Command

JOIN COMMAND			
	Ribbon	Home	The icon for the **JOIN** command is shown in Figure 4.84(a). This command is used to join objects (collinear lines, for example, can have a gap between them) into one object.
	Panel	Modify	
	Command Line	Join	
4.84(a)	Alias	J	

STEP-BY-STEP

JOIN COMMAND TUTORIAL

Step 1. Select the **Join** icon from the **Modify** toolbar or the **Modify** panel of the **Home** tab.

Step 2. When prompted to *Select source object*, select the first line of a set of collinear lines.

Step 3. When prompted to *Select lines to join to source*, select the collinear line(s) that you would like to join to the source line. Press **<Enter>** to complete the command. See Figure 4.84(b).

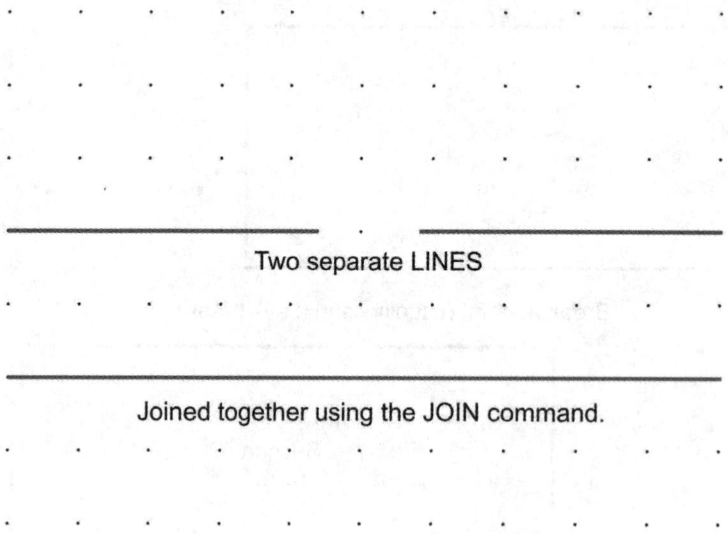

4.84(b) JOIN Command

CHAMFER COMMAND

	Ribbon	Home	The icon for the **CHAMFER** command is shown in Figure 4.85(a). This command is used to bevel the corners of objects and lines. The **Distance** option allows you to enter a chamfer distance for both sides of the beveled corner. The **Angle** option allows you to enter a distance and an angle for the beveled corner. The **Polyline** option will bevel the corners of the polyline.
	Panel	Modify	
	Command Line	Chamfer	
4.85(a)	Alias	CHA	

STEP–BY–STEP

CHAMFER COMMAND TUTORIAL

Step 1. Select the **Chamfer** icon from the **Modify** toolbar or the **Modify** panel of the **Home** tab (select the down arrow next to the **Fillet** icon and choose from the drop-down list).

Step 2. Type **D** (for distance) and press **<Enter>**.

QUICK TIP

To enter a distance and an angle for a chamfer, type **A** (for angle) instead of **D** and enter the desired chamfer distance and angle in response to the prompts.

Step 3. When prompted to *Specify first chamfer distance*, enter the distance to bevel the first edge and press **<Enter>**.

Step 4. When prompted to *Specify second chamfer distance*, enter the distance to bevel the second edge and press **<Enter>**.

Step 5. When prompted to *Select first line*, pick the first line to be beveled near the end to be beveled.

Step 6. When prompted to *Specify second line*, pick the second line to be beveled. See Figure 4.85(b).

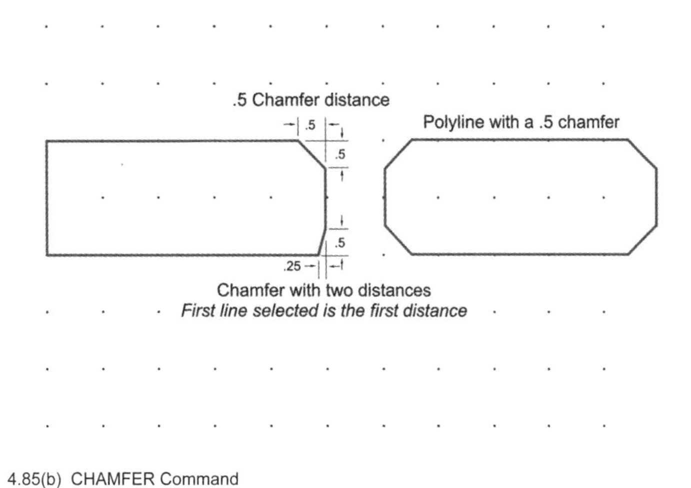

4.85(b) CHAMFER Command

FILLET COMMAND			
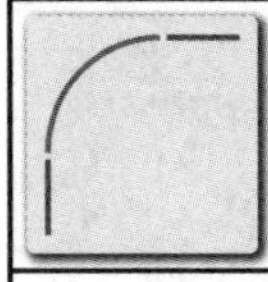	Ribbon	Home	The icon for the **FILLET** command is shown in Figure 4.86(a). This command is used to create fillets (rounded *inside* corners) and rounds (rounded *outside* corners) on objects. The **Polyline** option creates fillets and rounds on the corners of a polyline.
	Panel	Modify	
	Command Line	Fillet	
4.86(a)	Alias	F	

STEP-BY-STEP

FILLET COMMAND TUTORIAL

Step 1. Select the **Fillet** icon from the **Modify** toolbar or the **Modify** panel of the **Home** tab.

Step 2. Type **R** (for radius) and press **<Enter>**.

Step 3. When prompted to *Specify fillet radius*, enter the radius of the fillet and press **<Enter>**.

Step 4. When prompted to *Select first object*, pick the first line to be filleted near the end to be filleted.

Step 5. When prompted to *Select second object*, pick the second line to be filleted. See Figure 4.86(b).

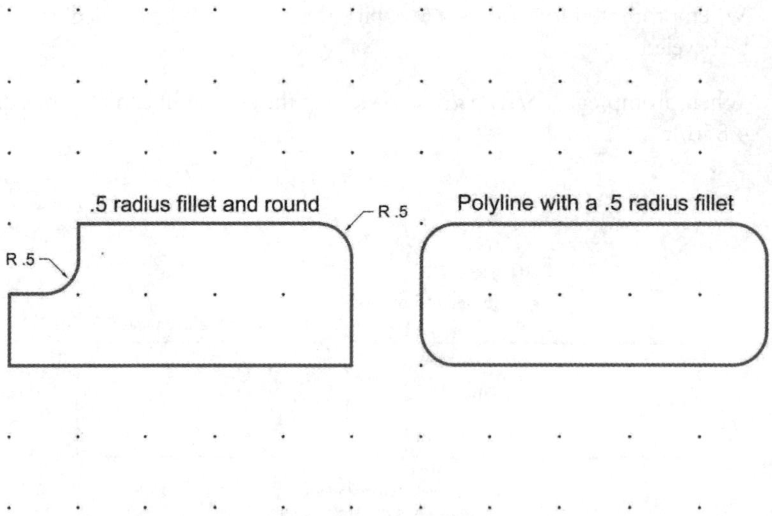

4.86(b) FILLET Command

EXPLODE COMMAND			
	Ribbon	Home	The icon for the **EXPLODE** command is shown in Figure 4.87(a). This command breaks compound objects, such as rectangles, polygons, polylines, and blocks, into separate entities—usually for the purpose of editing them.
	Panel	Modify	
	Command Line	Explode	
4.87(a)	Alias	X	

STEP–BY–STEP

EXPLODE COMMAND TUTORIAL

Step 1. Select the **Explode** icon from the **Modify** toolbar or the **Modify** panel of the **Home** tab.

Step 2. When prompted to *Select objects*, select the object(s) to be exploded.

Step 3. Press **<Enter>** to complete the command. See Figure 4.87(b).

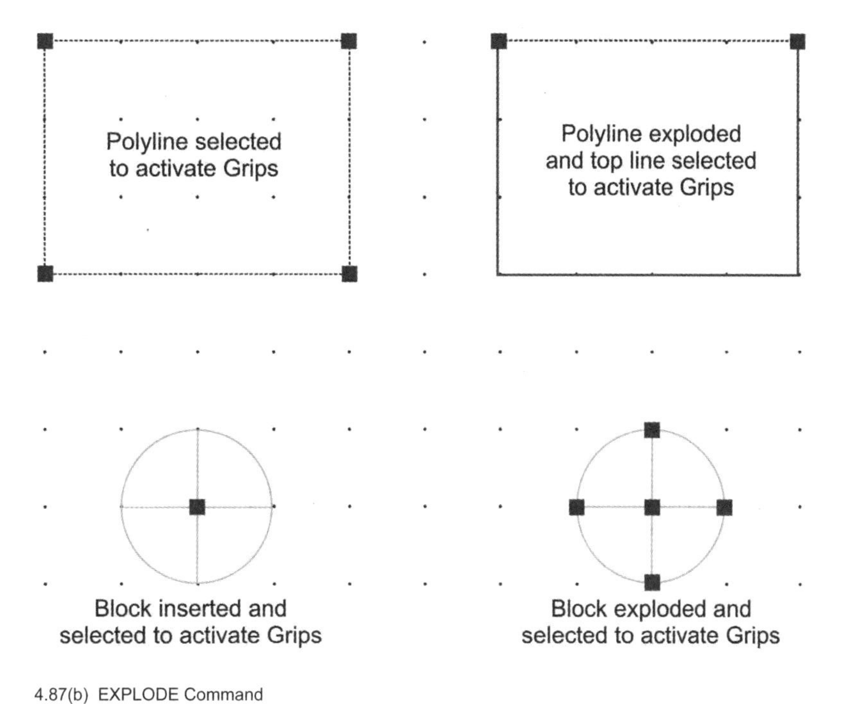

4.87(b) EXPLODE Command

Placing and Editing Text

There are two commands commonly used to place text in the field of a drawing, **MTEXT** (*multiline* text) and **DTEXT** (also called *single-line* or *dynamic* text). The **MTEXT** icon is the bold **A** located on the **Annotation** panel of the **Home** tab, as shown in Figure 4.88(a), and on the **Draw** toolbar, as shown in Figure 4.47. The **DTEXT** command can be found by selecting the down arrow located beneath the words **Multiline Text** in the **Annotation** panel and clicking on **Single Line**. See Figure 4.88(b).

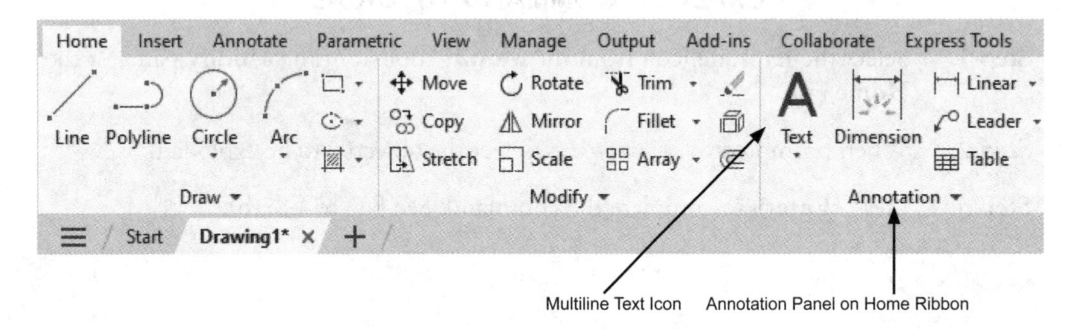

4.88(a) Location of the Multiline Text Icon in the Annotation Panel

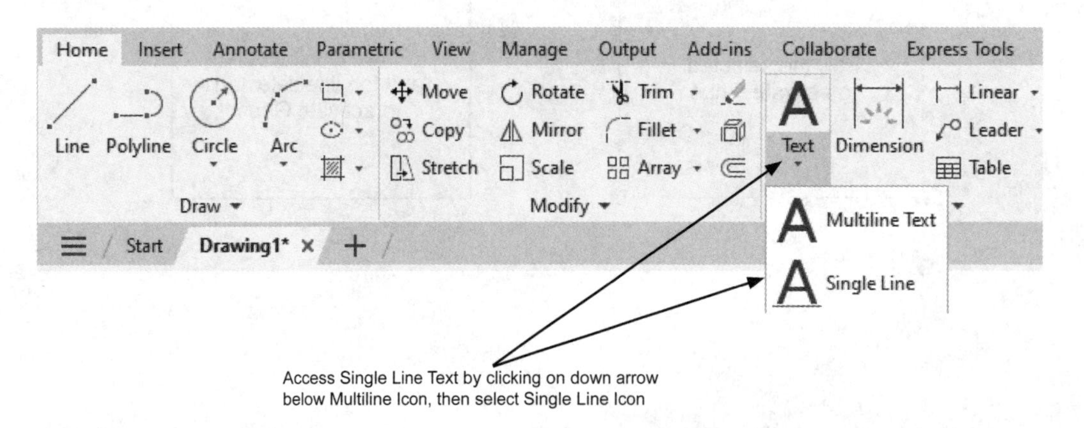

4.88(b) Location of the Single Line Text Command in the Annotation Panel

MULTILINE TEXT COMMAND

4.89(a)		
	Ribbon	Home
	Panel	Annotate
	Command Line	Mtext
4.89(a)	Alias	T

Left-click on the **Multiline Text** icon located on the Annotation panel, see Figure 4.88(a), or type **MTEXT** and press **<Enter>** to start the command. You will be prompted to *Specify the first and Opposite corner of a text box.* By picking two points in the drawing window, you can define the width of the text box. After you have selected the two points defining the corners of the text box, a text box containing a blinking cursor will appear on the screen. See Figure 4.89(b). Simply type the desired text in the text box. If you would like to edit the properties of the text, such as the text height, font style, or justification, highlight the text by holding down the left-click button, and change the appropriate setting(s) in the panels of the **Text Editor** ribbon located above the drawing window, see Figure 4.89(b), and press **<Enter>**. To end the command, left-click the mouse on a point located anywhere outside the text box or select the **Close Text Editor** tool located on the ribbon.

STEP-BY-STEP

EDITING MULTILINE TEXT

Step 1. Use the mouse's left-click button to double-click on the annotation to be edited or type **ED** and press **<Enter>** and select the text to be edited.

Step 2. When the **Text Editor** box appears, see Figure 4.89(b), highlight the text to be edited with the mouse's left-click button and make the desired changes in the box. When you've completed the edit(s), left-click the mouse on a point located anywhere outside the text box. You can also change settings such as text height by highlighting the text and entering a new value in the **Text Editor** and pressing **<Enter>.** See Figure 4.89(b).

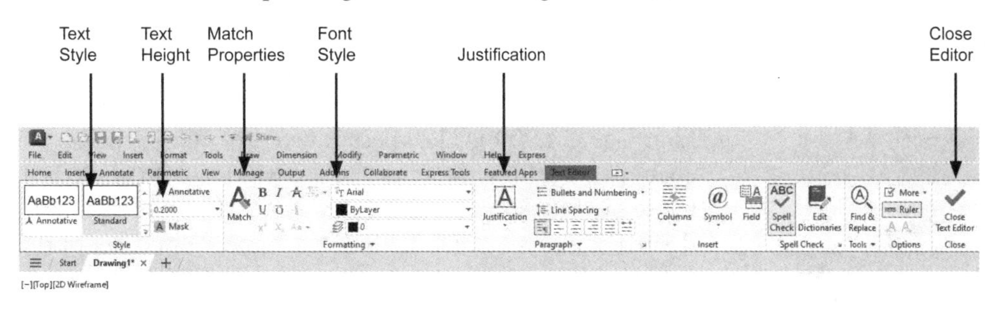

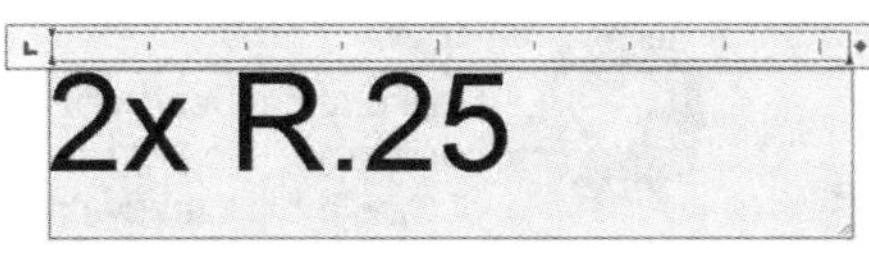

4.89(b) The Multiline Text Box and Text Editor Panel

SINGLE LINE TEXT COMMAND

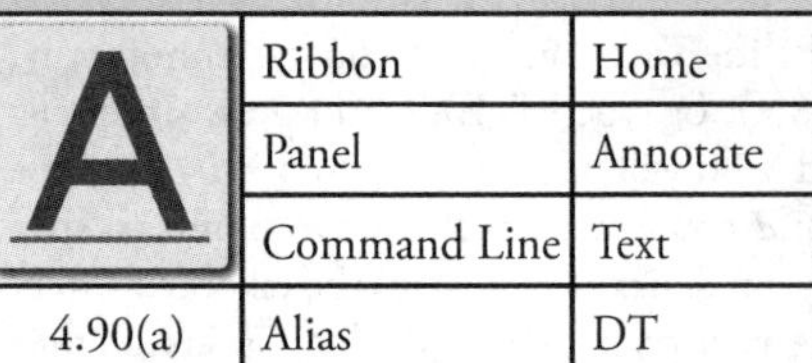

4.90(a)		
Ribbon	Home	
Panel	Annotate	
Command Line	Text	
Alias	DT	

4.90(b) Single Line Text Edit Box

To start the command select on the down arrow located below the words **Multiline Text** in the **Annotation** panel and select **Single Line**, as shown in Figure 4.90(b), or type **DTEXT** or **TEXT** and press **<Enter>**. You will be prompted to select the text's start point, height, and rotation angle (these prompts appear in the dynamic prompts next to the cursor and on the command line). When you have defined all three text specifications, a cursor will appear in the drawing window. At the cursor, you can type the desired text. When you are finished entering the text you can end the command by pressing **<Enter>** twice, or you can continue to place text at a different location in the drawing window by left-clicking in the new location and entering text at the cursor.

STEP-BY-STEP

EDITING SINGLE-LINE TEXT

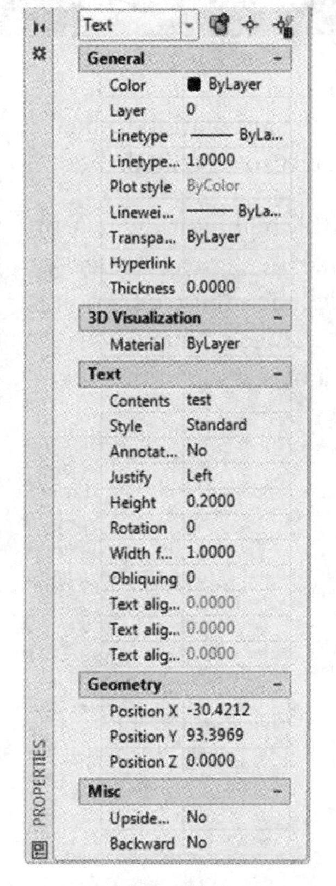

4.90(c) Properties Palette Displaying the Properties of the Single Line Text Shown in Figure 4.90(a)

Step 1. Use the mouse's left-click button to double-click on the annotation to be edited or type ED and press <Enter> and select the text to be edited.

Step 2. When the Text Editor box appears, move the cursor into the box and highlight the text to be edited with the mouse's left-click button, and enter the changes to the text; as in Figure 4.90(b). When you have completed the edit(s), left-click the mouse on a point located anywhere outside the text box.

Step 3. To edit the height, rotation, or style of text placed with DTEXT, select the text by left-clicking on it, then right-click the mouse and select Properties from the shortcut menu that appears. This will open the Properties palette, as in Figure 4.90(c). The selected text's properties can be changed by editing the values in the windows of the palette. Close the Properties palette when you have completed the desired edits by picking the X in the upper-left corner of the palette and press <Esc> to exit the command.

NOTE

Unlike text placed with the Multiline Text command, it is not possible to edit the properties of Single Line text with the Text Editor.

JOB SKILL

One advantage of **DTEXT** (or **Single Line Text**) over **Multiline Text** is that the former can be used to place text in multiple locations in the drawing window without having to restart the command each time. The properties of **DTEXT**, however, are more complicated to edit, since the **Text Editor** does not work with **DTEXT**.

Controlling Text Style

The characteristics of text used in a drawing, such as font name, height, width factor, and oblique angle, are determined by the values set in the **Text Style** dialog box. You can choose to default to the settings of the **Standard** style, edit the **Standard** style, or create a new ***text style***.

STEP–BY–STEP

CHANGE TEXT STYLE SETTINGS

Step 1. Left-click on the down arrow to the right of the word **Annotation** in the **Annotation** panel of the **Home** tab of the ribbon and select the **Text Style** icon as shown in Figure 4.91, or pick **Text Style...** from the **Format** pull-down menu of the menu bar, or type **Style** and press **<Enter>**. The **Text Style** dialog box will open, as shown in Figure 4.92.

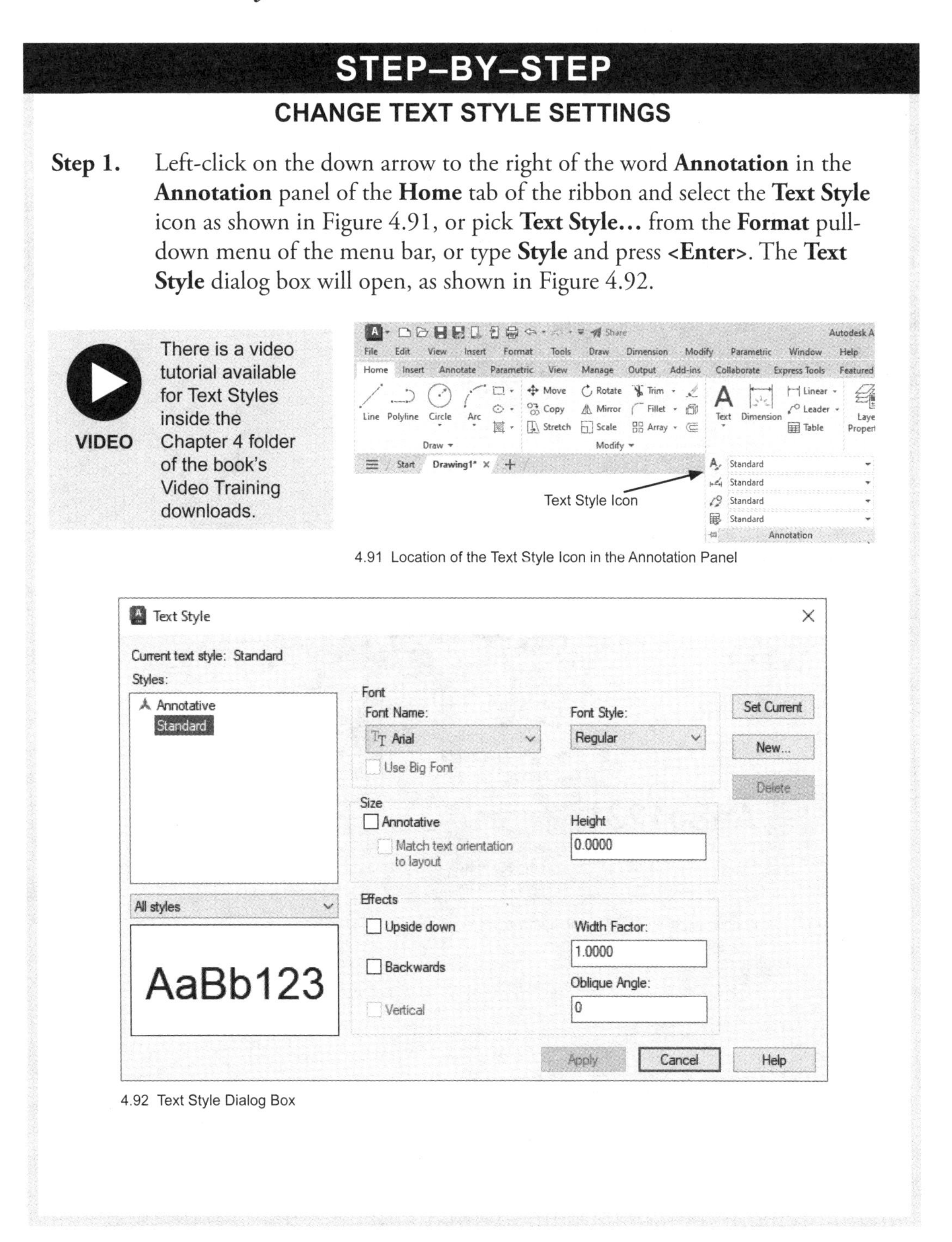

4.91 Location of the Text Style Icon in the Annotation Panel

4.92 Text Style Dialog Box

STEP–BY–STEP

Step 2. Select the **Standard** style name in the **Styles** window or select the **New...** button if you wish to create a new text style.

Step 3. To change the font, left-click the down arrow in the **Font Name** window (the default font name is **txt.shx**), and select the desired font from the list of font styles (see Figure 4.93). For mechanical drawings, a Gothic font, such as Arial, is appropriate. For architectural drawings, an architectural font, such as Stylus BT or City Blueprint, may be appropriate. The value for the height, width factor, and oblique angle can also be changed by editing the values in the appropriate window of the **Text Style** dialog box. After making the desired edits to the text style, select **Apply**. Click the **Set Current** button to set the new style as the current text style for the drawing. The new text style can also be assigned as the text used for dimensions by changing the **Text Style** setting in the **Text** tab of the **Dimension Style Manager** (see Figure 5.56 on page 286).

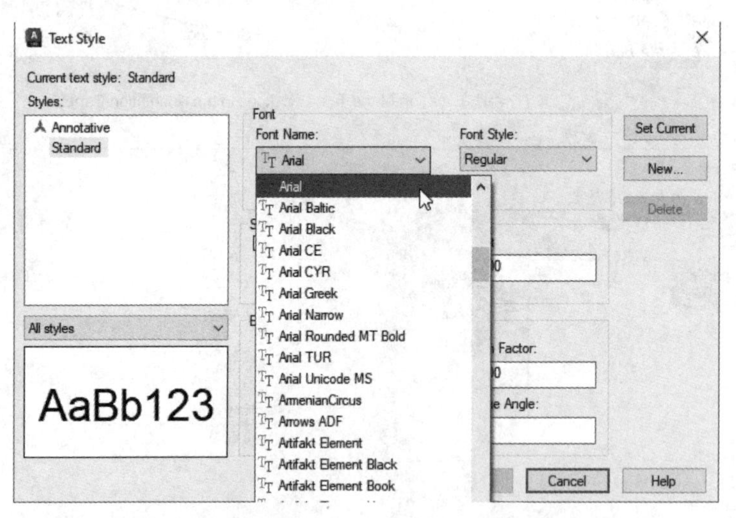

4.93 Selecting a New Font Name from the List in the Text Style Dialog Box

JOB SKILL

If a desired font cannot be located in the drop-down list in the **Font Name** window of the **Text Style** dialog box (see Figure 4.93), the organization's computer systems administrator may need to install it. When creating a new text style, it is best to leave the text height set to **0.0000**; otherwise, all text will default to the height defined in the **Text Style** dialog box.

4.12 DRAFTING SETTINGS DIALOG BOX

Drafting settings are a group of drawing aids that can help you produce accurate drawings more quickly. These drawing aids include Grid, Snap, Polar Tracking, Running Object Snaps, Dynamic Input, 3D Object Snap, Quick Properties, and Selection Cycling.

Open the Drafting Settings dialog box by selecting Drafting Settings from the Tools pull-down menu of the menu bar (see Figure 4.94), or by left-clicking on the down arrow near the Snap icon on the status bar and selecting Snap Settings (see Figure 4.8).

As you can see in Figure 4.95, the Drafting Settings dialog box has seven tabs along its top edge, but this chapter focuses on four tabs: Snap and Grid, Polar Tracking, Object Snap, and Dynamic Input.

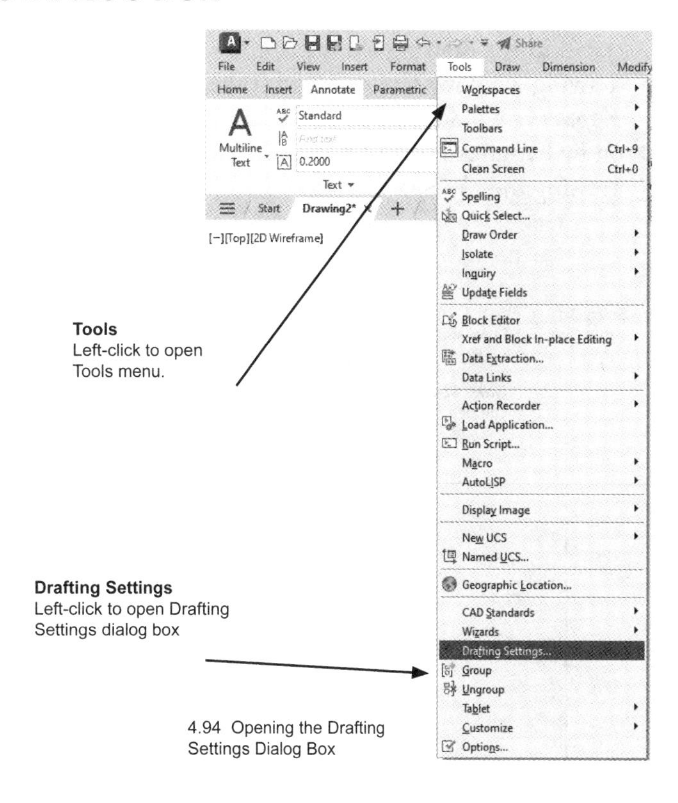

4.94 Opening the Drafting Settings Dialog Box

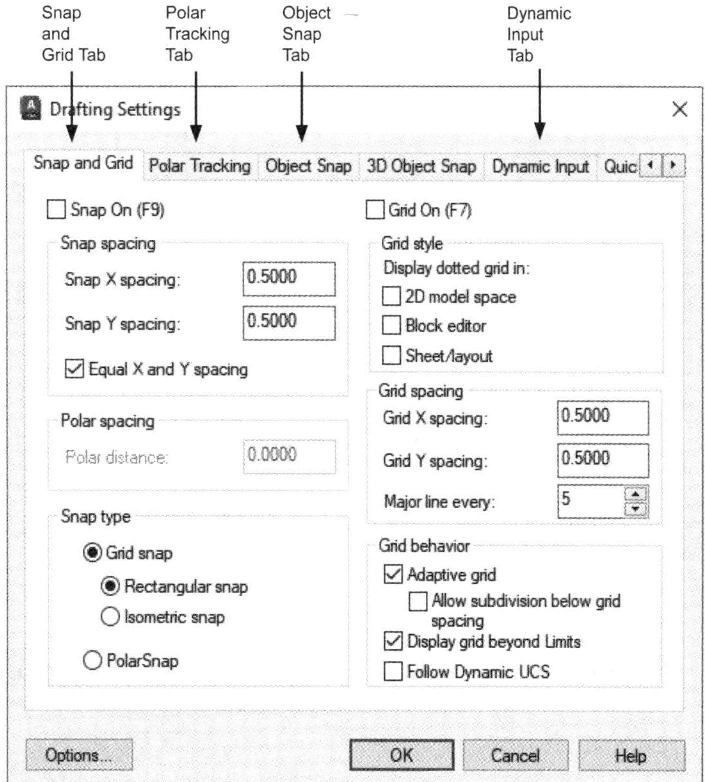

4.95 Drafting Settings Dialog Box

Snap and Grid Tab

Grid is a drawing aid that displays a rectangular pattern of dots or lines in the drawing window. The spacing for the grid is defined in the ***Grid X spacing*** and ***Grid Y spacing*** windows of the **Snap and Grid** tab (see Figure 4.96). The grid can be turned on by left-clicking the **Gridmode** icon located on the status bar (see Figure 4.8) or by pressing the **<F7>** key.

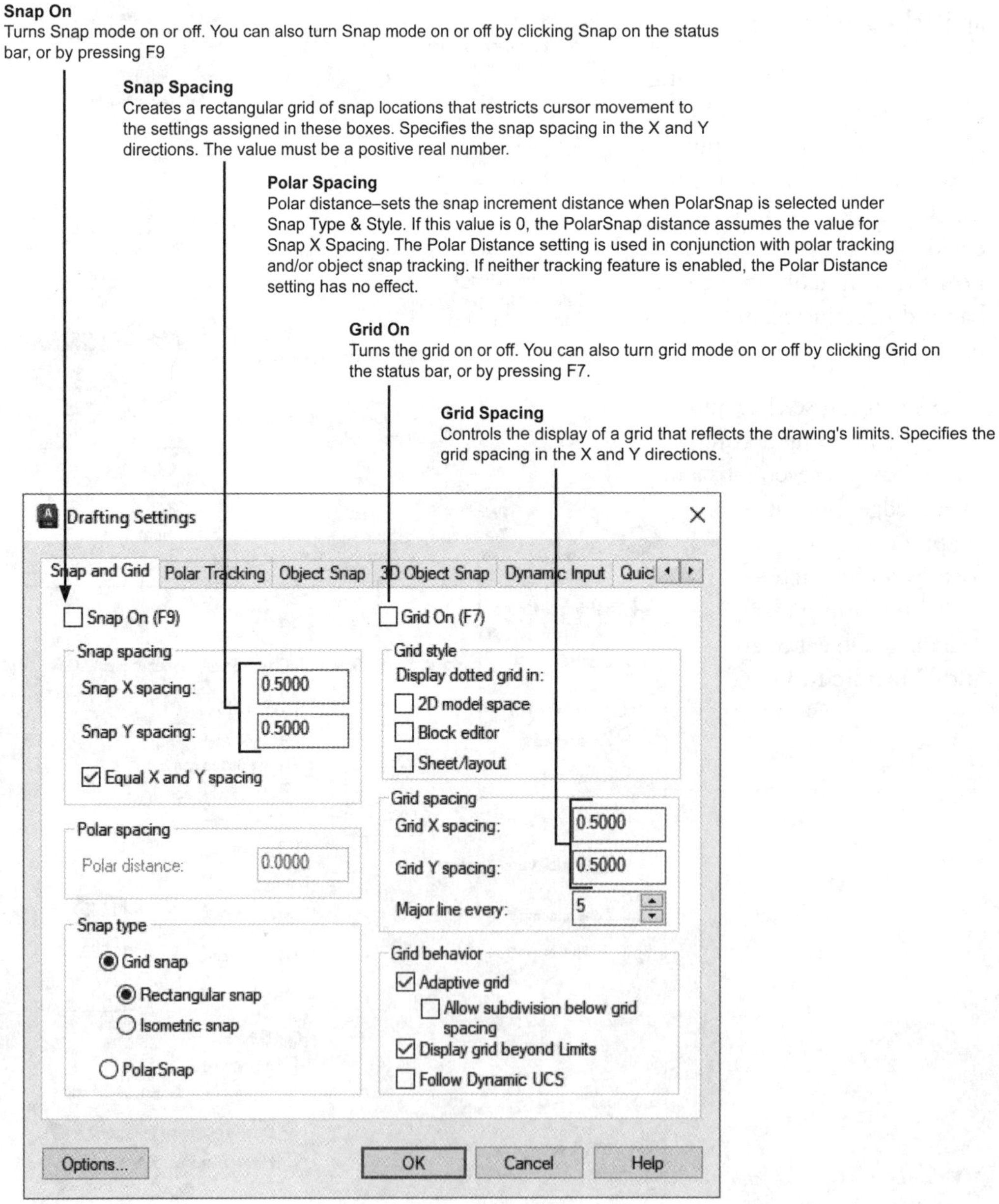

Snap On
Turns Snap mode on or off. You can also turn Snap mode on or off by clicking Snap on the status bar, or by pressing F9

Snap Spacing
Creates a rectangular grid of snap locations that restricts cursor movement to the settings assigned in these boxes. Specifies the snap spacing in the X and Y directions. The value must be a positive real number.

Polar Spacing
Polar distance–sets the snap increment distance when PolarSnap is selected under Snap Type & Style. If this value is 0, the PolarSnap distance assumes the value for Snap X Spacing. The Polar Distance setting is used in conjunction with polar tracking and/or object snap tracking. If neither tracking feature is enabled, the Polar Distance setting has no effect.

Grid On
Turns the grid on or off. You can also turn grid mode on or off by clicking Grid on the status bar, or by pressing F7.

Grid Spacing
Controls the display of a grid that reflects the drawing's limits. Specifies the grid spacing in the X and Y directions.

Drafting Settings
Snap and Grid | Polar Tracking | Object Snap | 3D Object Snap | Dynamic Input | Quic
Snap On (F9)
Snap spacing
Snap X spacing: 0.5000
Snap Y spacing: 0.5000
Equal X and Y spacing
Polar spacing
Polar distance: 0.0000
Snap type
Grid snap
Rectangular snap
Isometric snap
PolarSnap
Grid On (F7)
Grid style
Display dotted grid in:
2D model space
Block editor
Sheet/layout
Grid spacing
Grid X spacing: 0.5000
Grid Y spacing: 0.5000
Major line every: 5
Grid behavior
Adaptive grid
Allow subdivision below grid spacing
Display grid beyond Limits
Follow Dynamic UCS
Options... | OK | Cancel | Help

4.96 Snap and Grid Tab

In Figure 4.97 other settings that can affect the snap and grid drawing aids are shown. *Snap* is a drawing aid that restricts the cursor's movement to the increments defined in the **Snap X spacing** and **Snap Y spacing** windows of the **Snap and Grid** tab (see Figure 4.96). Snap increments can be set to the same increments as the grid spacing or they can be set to different values. **Snap** can be turned on by left-clicking the **Object Snap** icon located on the status bar (see Figure 4.9) or by pressing the **<F9>** key.

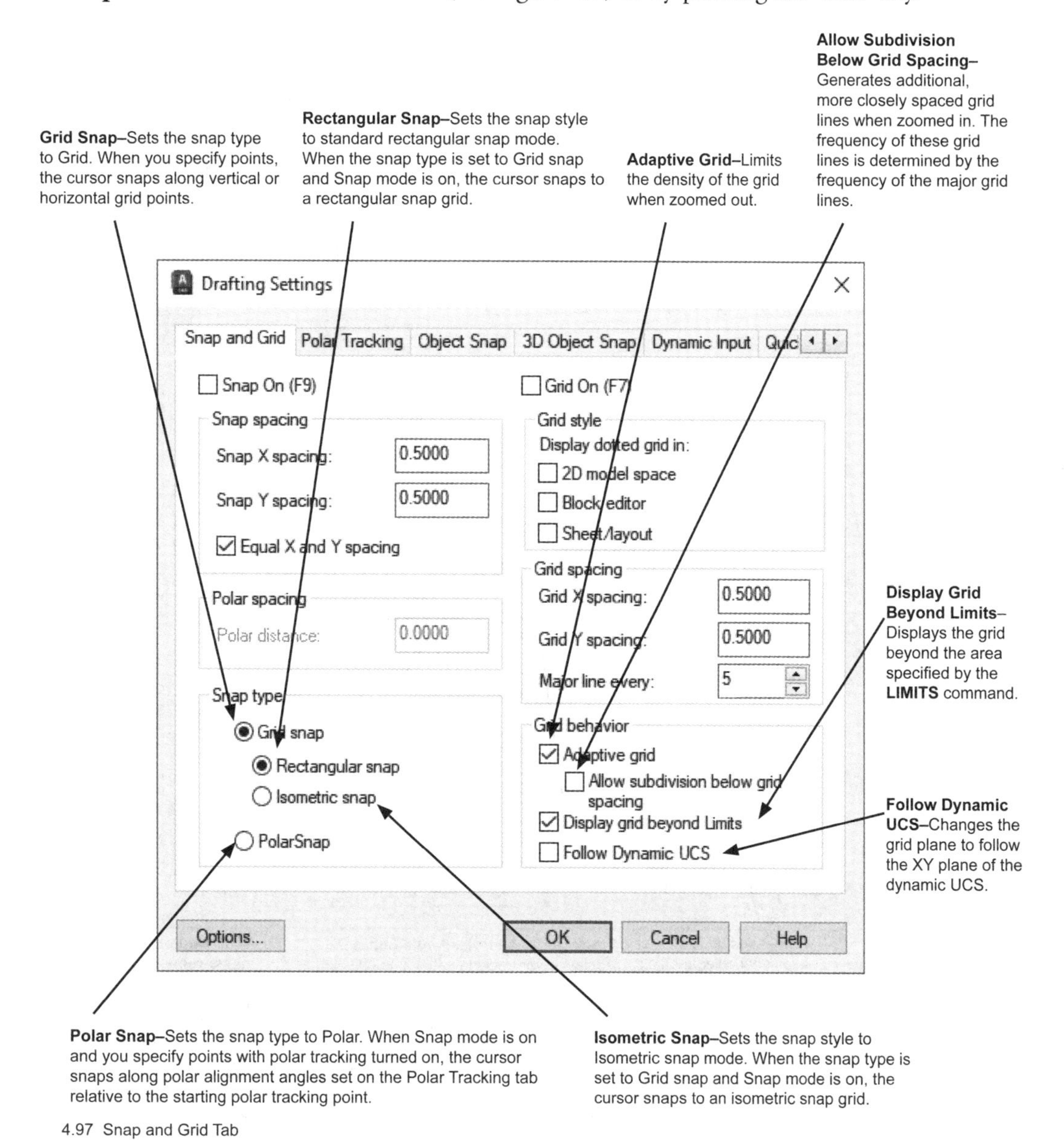

Grid Snap–Sets the snap type to Grid. When you specify points, the cursor snaps along vertical or horizontal grid points.
Rectangular Snap–Sets the snap style to standard rectangular snap mode. When the snap type is set to Grid snap and Snap mode is on, the cursor snaps to a rectangular snap grid.
Adaptive Grid–Limits the density of the grid when zoomed out.
Allow Subdivision Below Grid Spacing–Generates additional, more closely spaced grid lines when zoomed in. The frequency of these grid lines is determined by the frequency of the major grid lines.
Display Grid Beyond Limits–Displays the grid beyond the area specified by the **LIMITS** command.
Follow Dynamic UCS–Changes the grid plane to follow the XY plane of the dynamic UCS.
Polar Snap–Sets the snap type to Polar. When Snap mode is on and you specify points with polar tracking turned on, the cursor snaps along polar alignment angles set on the Polar Tracking tab relative to the starting polar tracking point.
Isometric Snap–Sets the snap style to Isometric snap mode. When the snap type is set to Grid snap and Snap mode is on, the cursor snaps to an isometric snap grid.
Drafting Settings
Snap and Grid
Polar Tracking
Object Snap
3D Object Snap
Dynamic Input
Snap On (F9)
Snap spacing
Snap X spacing:
0.5000
Snap Y spacing:
0.5000
Equal X and Y spacing
Polar spacing
Polar distance:
0.0000
Snap type
Grid snap
Rectangular snap
Isometric snap
PolarSnap
Grid On (F7)
Grid style
Display dotted grid in:
2D model space
Block editor
Sheet/layout
Grid spacing
Grid X spacing:
0.5000
Grid Y spacing:
0.5000
Major line every:
5
Grid behavior
Adaptive grid
Allow subdivision below grid spacing
Display grid beyond Limits
Follow Dynamic UCS
Options...
OK
Cancel
Help

4.97 Snap and Grid Tab

Polar Tracking Tab

Polar tracking is a drawing aid that restricts the cursor's movement to specified increments along a polar angle. The user specifies the angle(s) by choosing an angle from the drop-down menu located next to the **Increment angle** window located under the **Polar Tracking** tab (see Figure 4.98). Additional angles can be defined by checking the **Additional Angles** box and picking the **New** button.

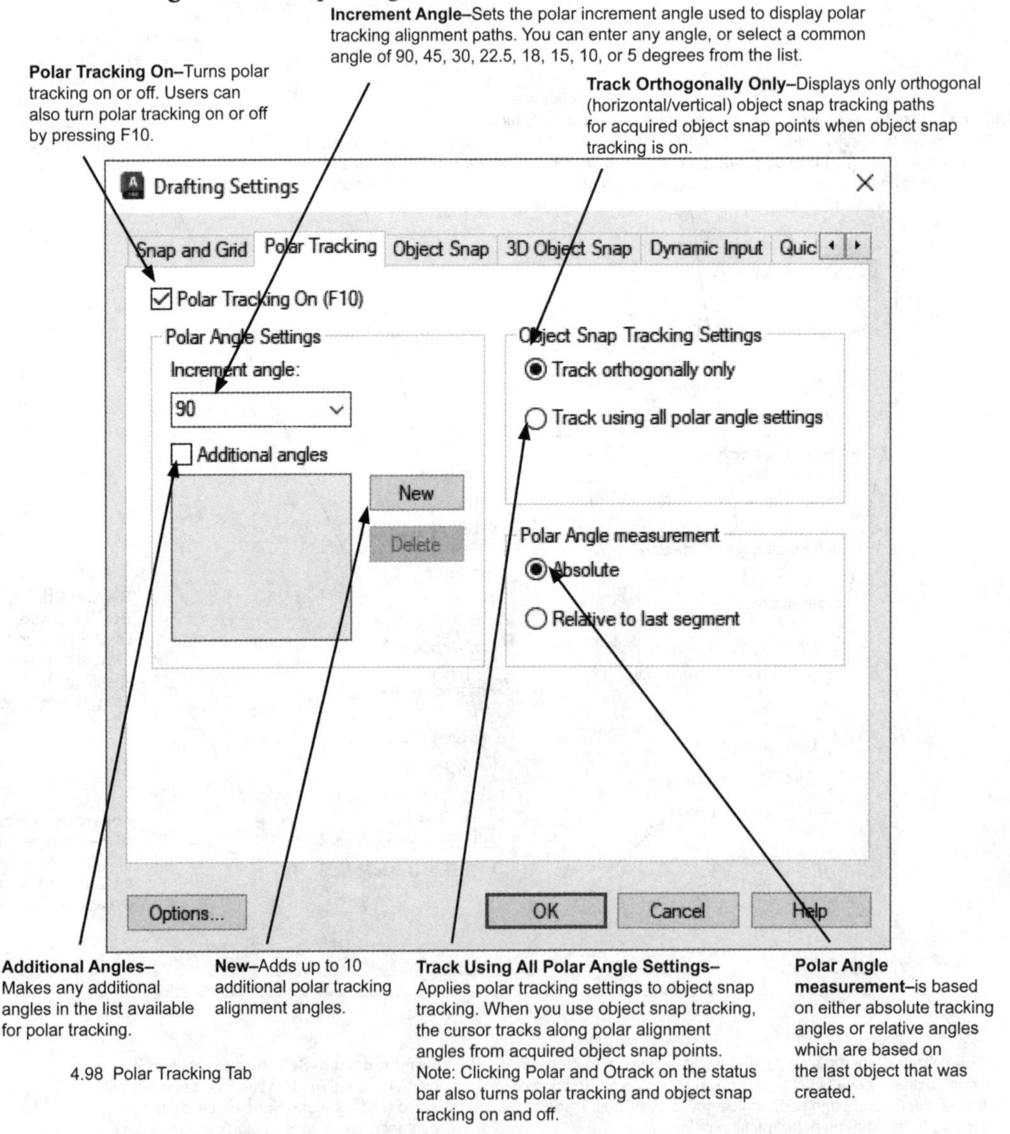

Increment Angle–Sets the polar increment angle used to display polar tracking alignment paths. You can enter any angle, or select a common angle of 90, 45, 30, 22.5, 18, 15, 10, or 5 degrees from the list.
Polar Tracking On–Turns polar tracking on or off. Users can also turn polar tracking on or off by pressing F10.
Track Orthogonally Only–Displays only orthogonal (horizontal/vertical) object snap tracking paths for acquired object snap points when object snap tracking is on.
Drafting Settings
Snap and Grid
Polar Tracking
Object Snap
3D Object Snap
Dynamic Input
Quic
Polar Tracking On (F10)
Polar Angle Settings
Increment angle:
90
Additional angles
New
Delete
Object Snap Tracking Settings
Track orthogonally only
Track using all polar angle settings
Polar Angle measurement
Absolute
Relative to last segment
Options...
OK
Cancel
Help
Additional Angles–Makes any additional angles in the list available for polar tracking.
New–Adds up to 10 additional polar tracking alignment angles.
Track Using All Polar Angle Settings–Applies polar tracking settings to object snap tracking. When you use object snap tracking, the cursor tracks along polar alignment angles from acquired object snap points. Note: Clicking Polar and Otrack on the status bar also turns polar tracking and object snap tracking on and off.
Polar Angle measurement–is based on either absolute tracking angles or relative angles which are based on the last object that was created.

4.98 Polar Tracking Tab

Polar tracking can be turned on by putting a check in the **Polar Tracking On** box of the tab, by left-clicking on the **Polar Tracking** icon located on the status bar (see Figure 4.9), or by pressing the **<F10>** key.

NOTE

The **Polar Tracking** and **Orthomode** icons of the status bar cannot both be on at the same time (when one is selected the other is automatically turned *OFF*).

Dynamic Input Tab

The **Dynamic Input** tab is divided into three panes: **Pointer Input**, **Dimension Input**, and **Dynamic Prompts** (see Figure 4.99).

The Dimension Input Pane

Selecting the **Settings** button in the **Dimension Input** pane (Figure 4.99) controls the dimension settings shown at the bottom right of Figure 4.100 (it is recommended that this setting be set to show two dimension input fields).

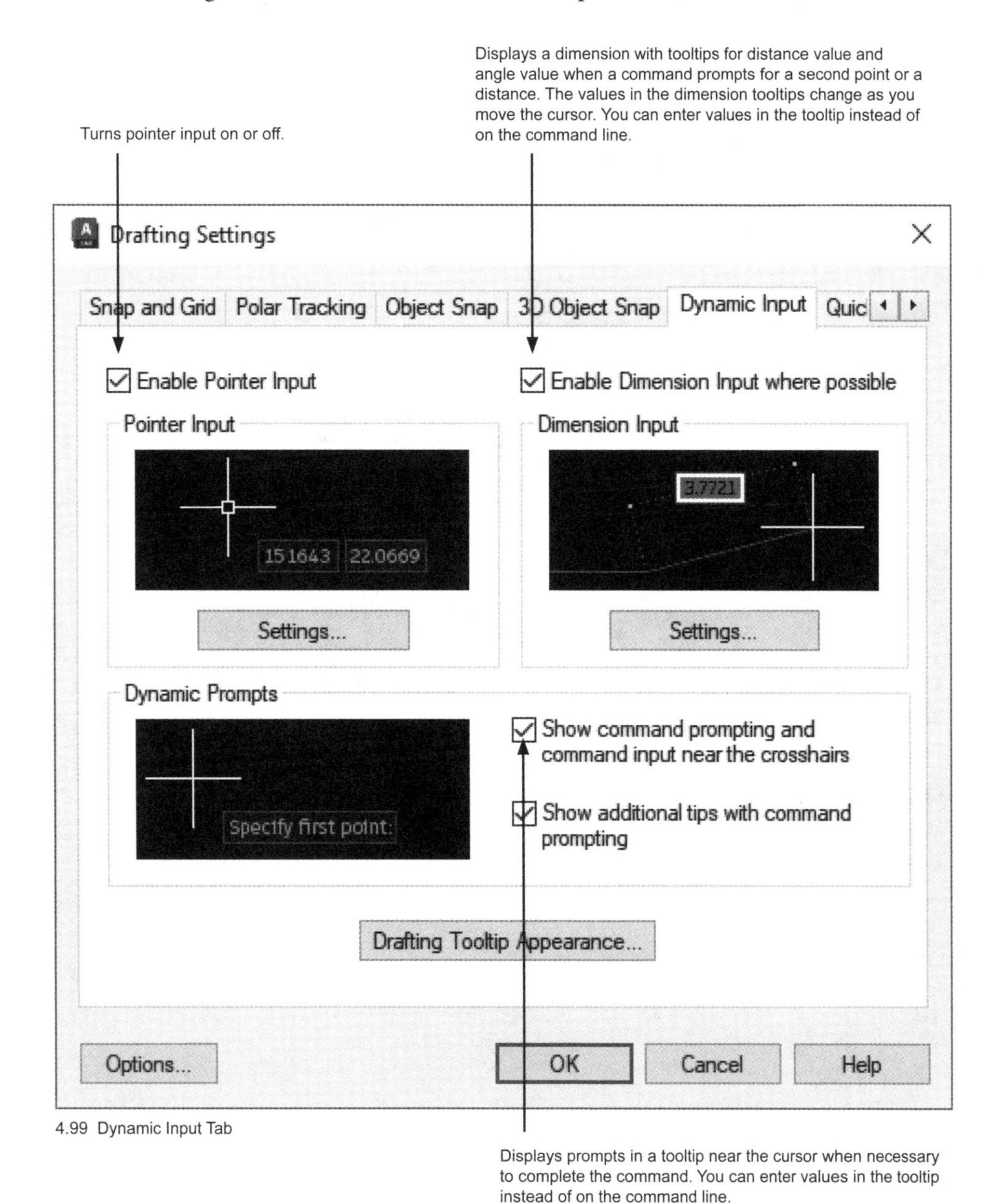
Turns pointer input on or off.
Displays a dimension with tooltips for distance value and angle value when a command prompts for a second point or a distance. The values in the dimension tooltips change as you move the cursor. You can enter values in the tooltip instead of on the command line.
Drafting Settings
Snap and Grid
Polar Tracking
Object Snap
3D Object Snap
Dynamic Input
Quic
Enable Pointer Input
Enable Dimension Input where possible
Pointer Input
Dimension Input
15.1643
22.0669
3.7721
Settings...
Settings...
Dynamic Prompts
Specify first point:
Show command prompting and command input near the crosshairs
Show additional tips with command prompting
Drafting Tooltip Appearance...
Options...
OK
Cancel
Help
Displays prompts in a tooltip near the cursor when necessary to complete the command. You can enter values in the tooltip instead of on the command line.

4.99 Dynamic Input Tab

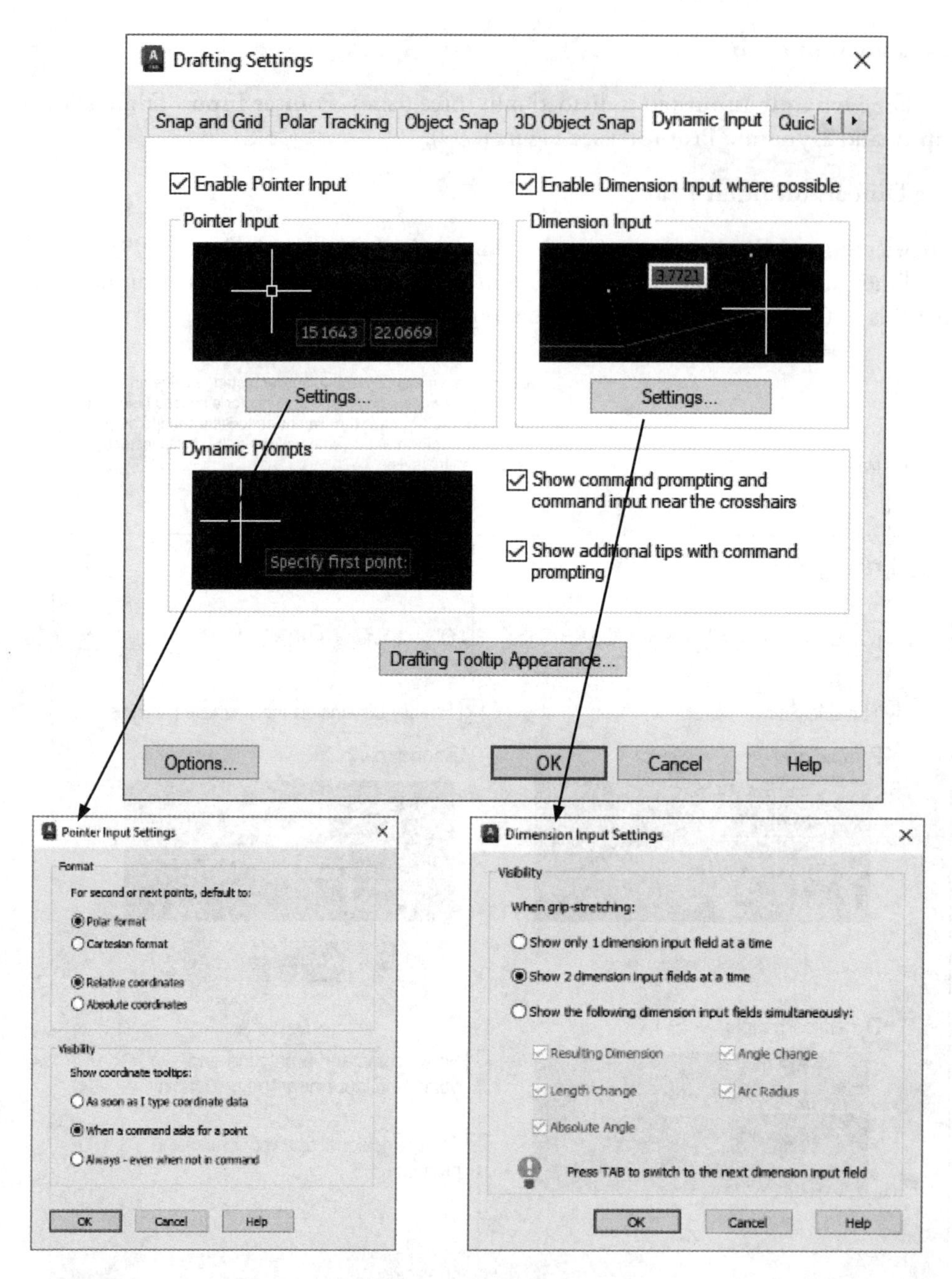

4.100 Dynamic Input Tab Settings

The Dynamic Prompts Pane

Selecting the **Drafting Tooltip Appearance** button in the **Dynamic Prompts** pane (Figure 4.99) controls the color, size, and transparency of dynamic tooltips.

Object Snap Tab

Object snaps (also called *Osnaps*) are drafting aids that allow you to snap to exact points on an object, such as the *endpoint* or *midpoint* of a line, or the *center* of a circle, when an AutoCAD command has prompted you to select a point. For example, if the **Endpoint** object snap is on, and you used the **LINE** command to place the first point of a line and are being prompted to *Select next point*, moving the cursor over, or near, the endpoint of a line will cause a marker to appear at the endpoint. If you left-click the mouse near the marked endpoint, the new line will snap to the exact endpoint of the existing line. The **Object Snap** function can be turned on by left-clicking the **Object Snap** icon located on the status bar (see Figure 4.9), or by pressing the **<F3>** key, or by putting a check in the **Object Snap On** box of the **Object Snap** tab.

JOB SKILL

Using the keyboard's function keys (<F> keys) to speed up the process of creating drawings is mentioned numerous times in this chapter. Here's a complete list of the function keys and what they do:
<F1> AutoCAD Help
<F2> Displays the AutoCAD text window
<F3> Object snap on/off
<F4> 3D Object snap on/off
<F5> Isoplane toggle
<F6> Dynamic UCS on/off
<F7> Grid on/off
<F8> Ortho on/off
<F9> Snap on/off
<F10> Polar Tracking on/off
<F11> Osnap tracking on/off
<F12> Dynamic input on/off

By checking the boxes next to the ***Object Snap settings*** listed on the **Object Snap** tab of the **Drafting Settings** dialog box, you can set one or more ***running object snaps*** (see Figure 4.101). When **Object Snap** is on, the running object snap(s) will be active whenever you are prompted to select a point. When multiple object snaps have been selected in the **Object Snap** tab, you can press **<Tab>** to cycle through the available snap options before selecting the desired point.

JOB SKILL

The quickest method to select running object snaps is to right-click on the **Object Snap** icon located on the status bar and select from the list of object snaps available on the pop-up menu. (See Figure 4.9)

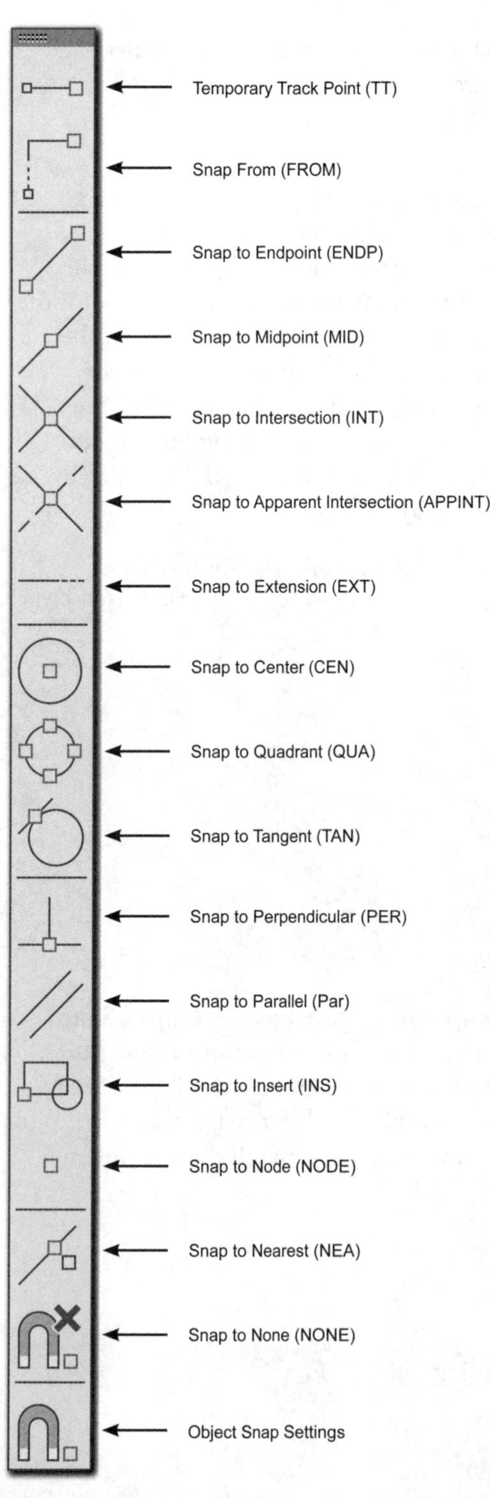

4.101 Object Snap Toolbar, Tools, Aliases, and Snap and Grid tab of Drafting Settings Dialog Box

4.13 OBJECT SNAP TOOLS

When an object snap is needed for a one-time selection (versus a *running* object snap), the appropriate Osnap icon can be selected from the Object Snap toolbar or by typing its alias (shown in parentheses in Figure 4.101) and pressing **<Enter>**. Selecting a single object snap while a command is active is referred to as using an *Osnap interrupt*. Follow the numbered steps shown in Figures 4.102 through 4.111 to see how object snaps are used to simplify drawing.

There are video tutorials available for the Object Snap tools inside the Chapter 4 folder of the book's Video Training downloads. The video downloads are available by redeeming the access code that comes with this book. Please see the inside front cover for further details.

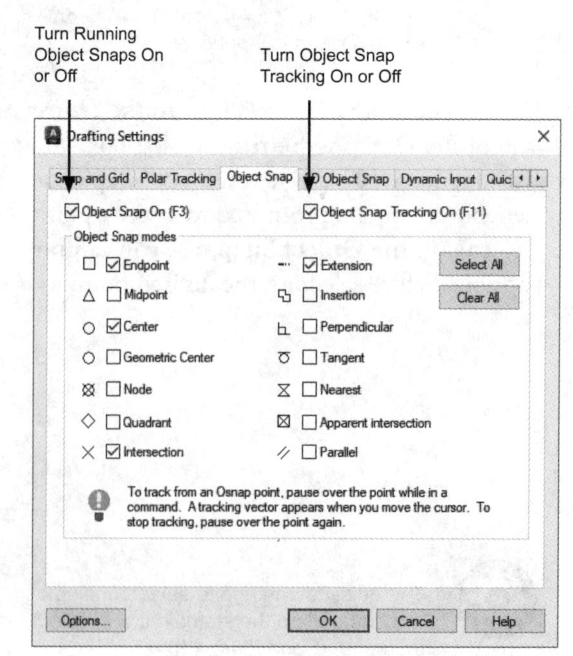

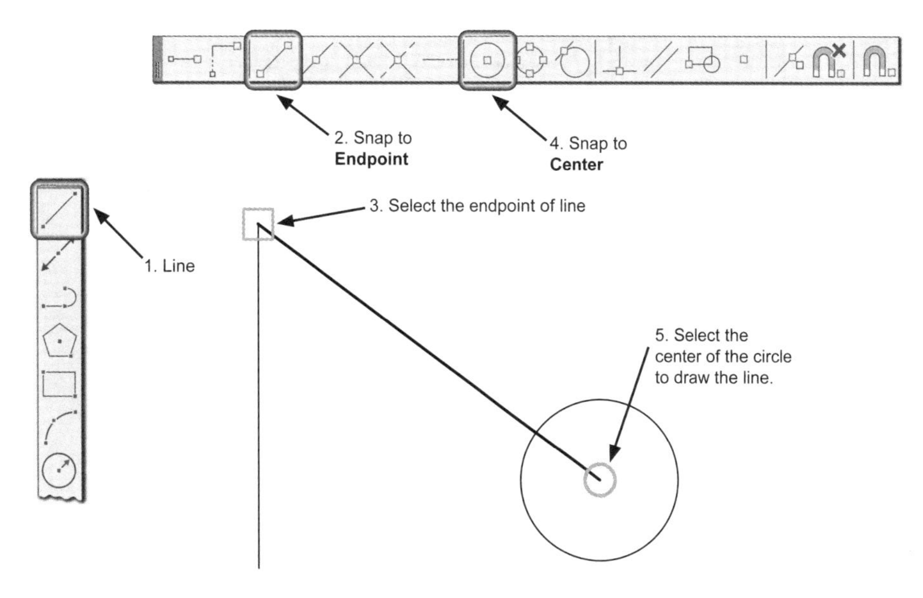

4.102 Steps in Drawing a Line from an Endpoint to a Center Point

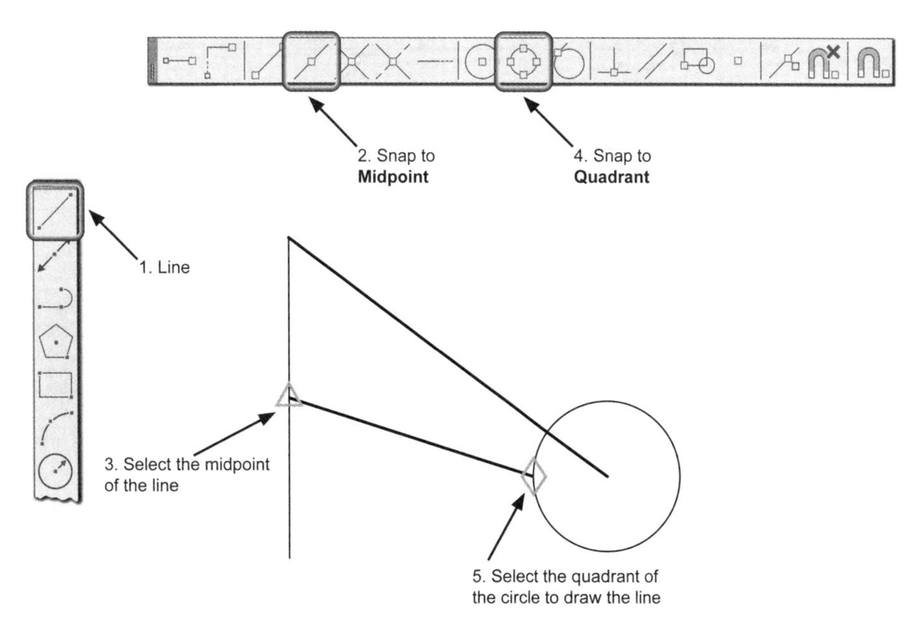

4.103 Steps in Drawing a Line from a **Midpoint** to a **Quadrant**

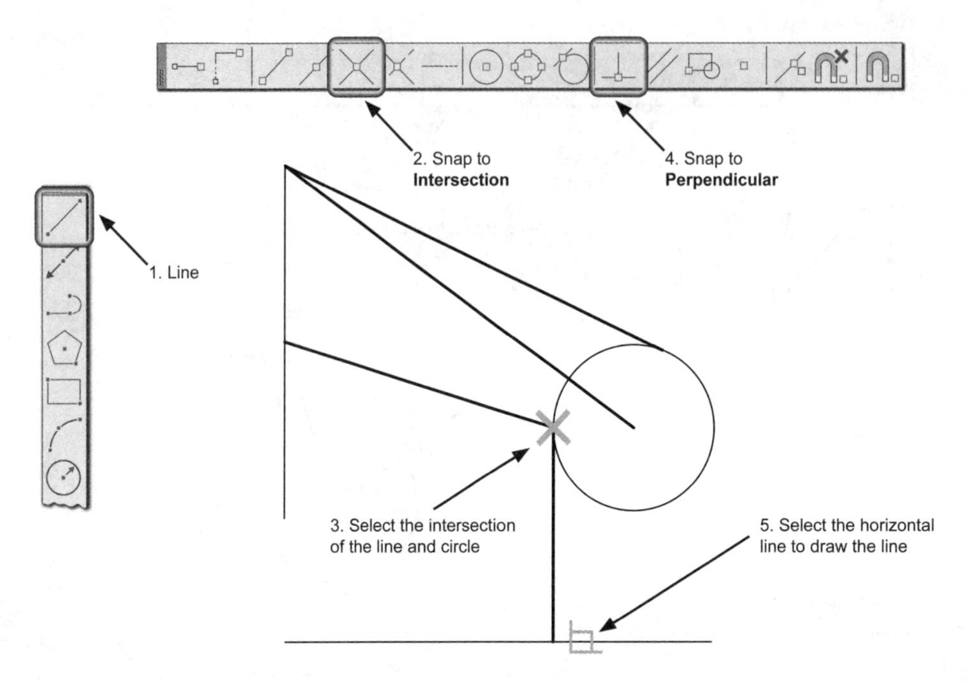

4.104 Steps in Drawing a Line from an Intersection that is Perpendicular to another Line

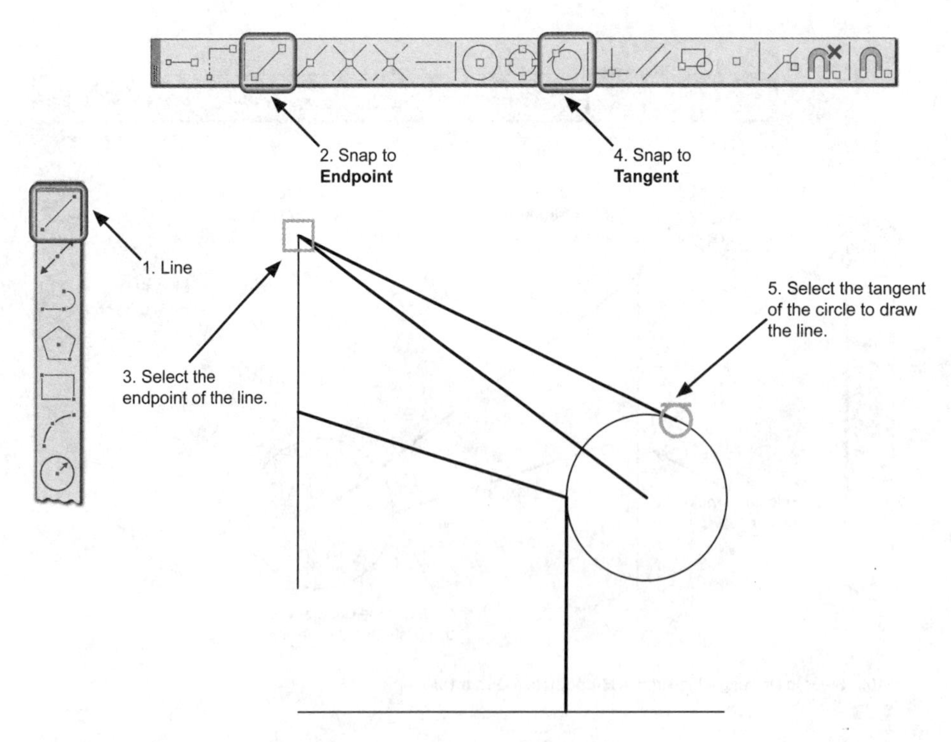

4.105 Steps in Drawing a Line from an **Endpoint** to a **Tangency** Point on a Circle

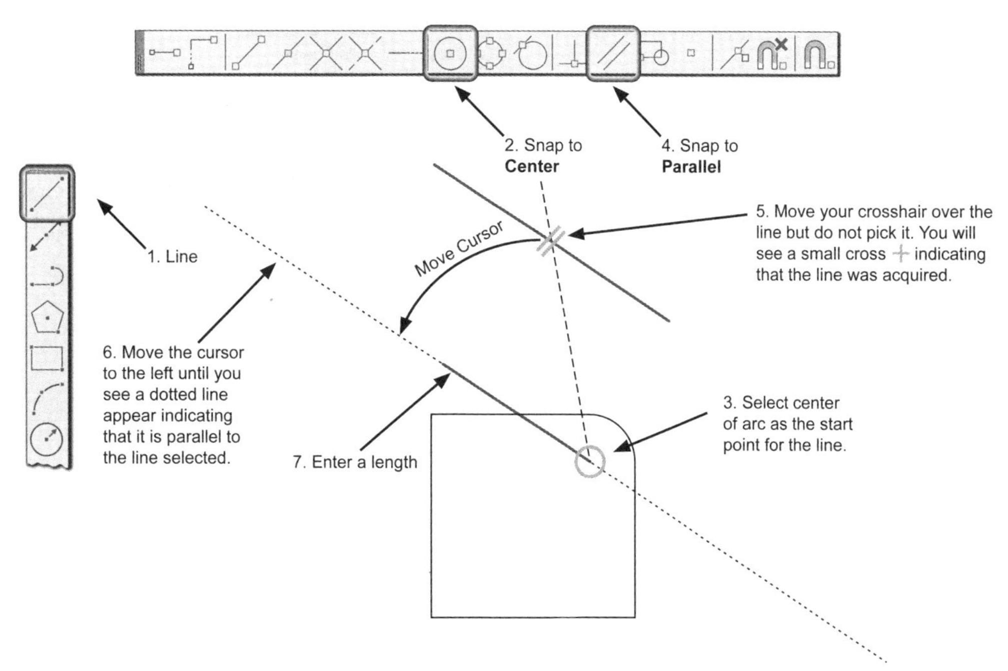
2. Snap to
Center
4. Snap to
Parallel
1. Line
Move Cursor
5. Move your crosshair over the line but do not pick it. You will see a small cross + indicating that the line was acquired.
6. Move the cursor to the left until you see a dotted line appear indicating that it is parallel to the line selected.
7. Enter a length
3. Select center of arc as the start point for the line.

4.106 Steps in Drawing a Line Parallel to another Line Using the Parallel Osnap Option

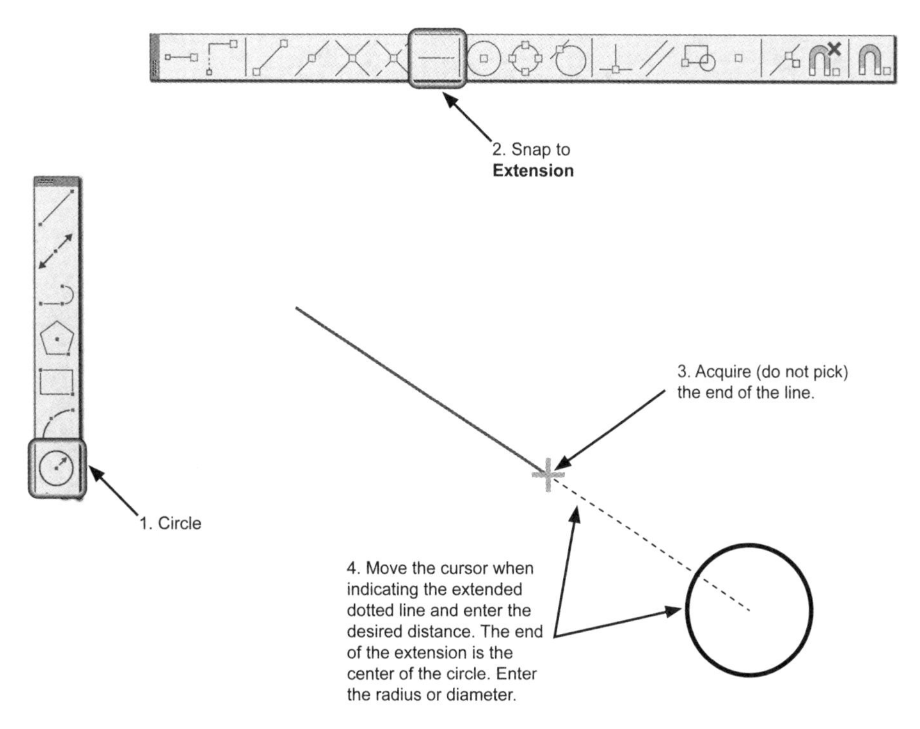
2. Snap to
Extension
3. Acquire (do not pick) the end of the line.
1. Circle
4. Move the cursor when indicating the extended dotted line and enter the desired distance. The end of the extension is the center of the circle. Enter the radius or diameter.

4.107 Steps in Defining the Center of a Circle Using the **Extension** Osnap Option

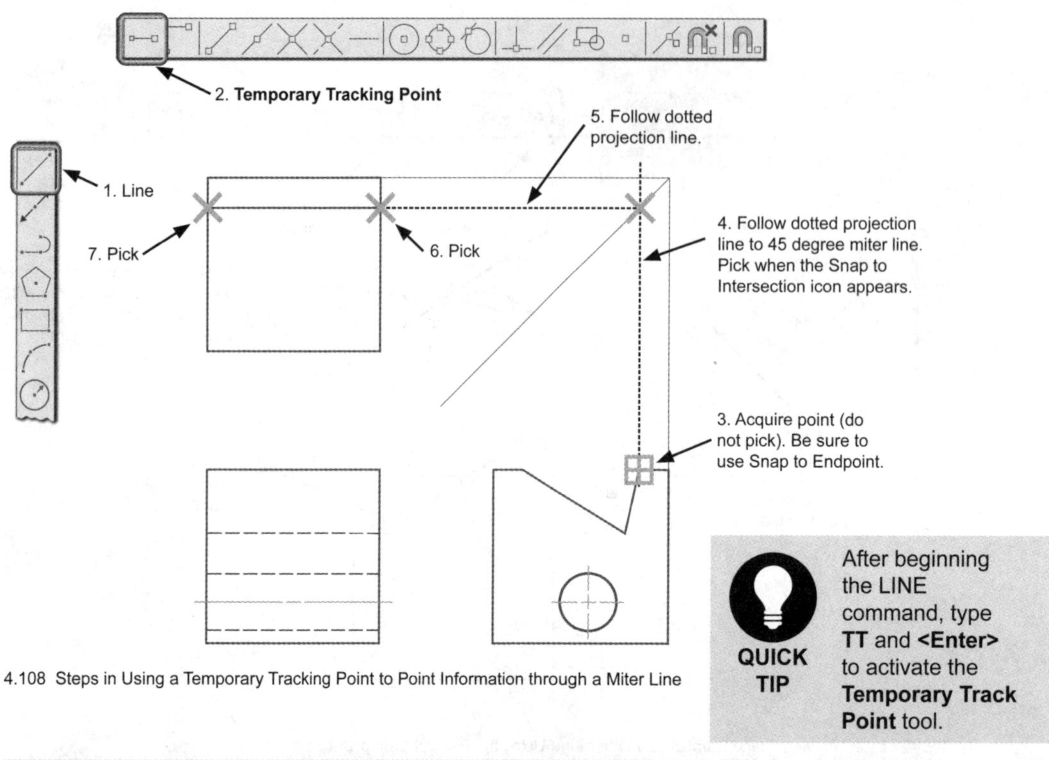

2. Temporary Tracking Point
5. Follow dotted projection line.
1. Line
7. Pick
6. Pick
4. Follow dotted projection line to 45 degree miter line. Pick when the Snap to Intersection icon appears.
3. Acquire point (do not pick). Be sure to use Snap to Endpoint.

4.108 Steps in Using a Temporary Tracking Point to Point Information through a Miter Line

QUICK TIP

After beginning the LINE command, type **TT** and **<Enter>** to activate the **Temporary Track Point** tool.

QUICK TIP

Polar Tracking, **Osnap** and **Otrack** should be turned *ON* (see Figure 4.9 to locate these icons on the status bar).

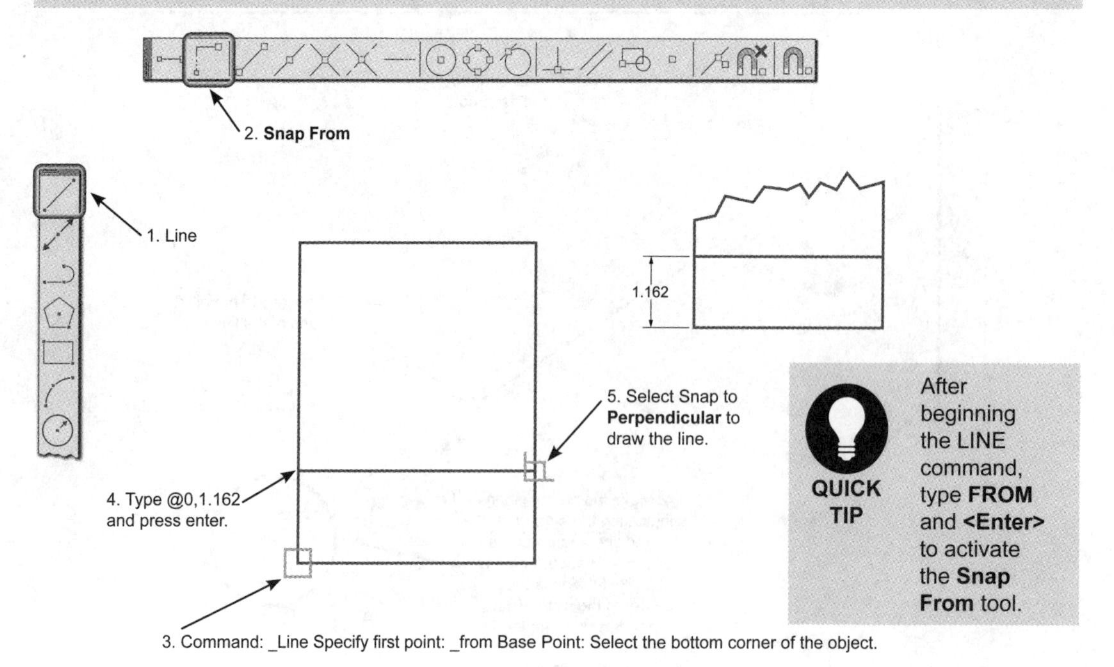

2. Snap From
1. Line
1.162
5. Select Snap to Perpendicular to draw the line.
4. Type @0,1.162 and press enter.
3. Command: _Line Specify first point: _from Base Point: Select the bottom corner of the object.

QUICK TIP

After beginning the LINE command, type **FROM** and **<Enter>** to activate the **Snap From** tool.

4.109 Steps in Drawing a Line Using the **Snap From** Osnap Option

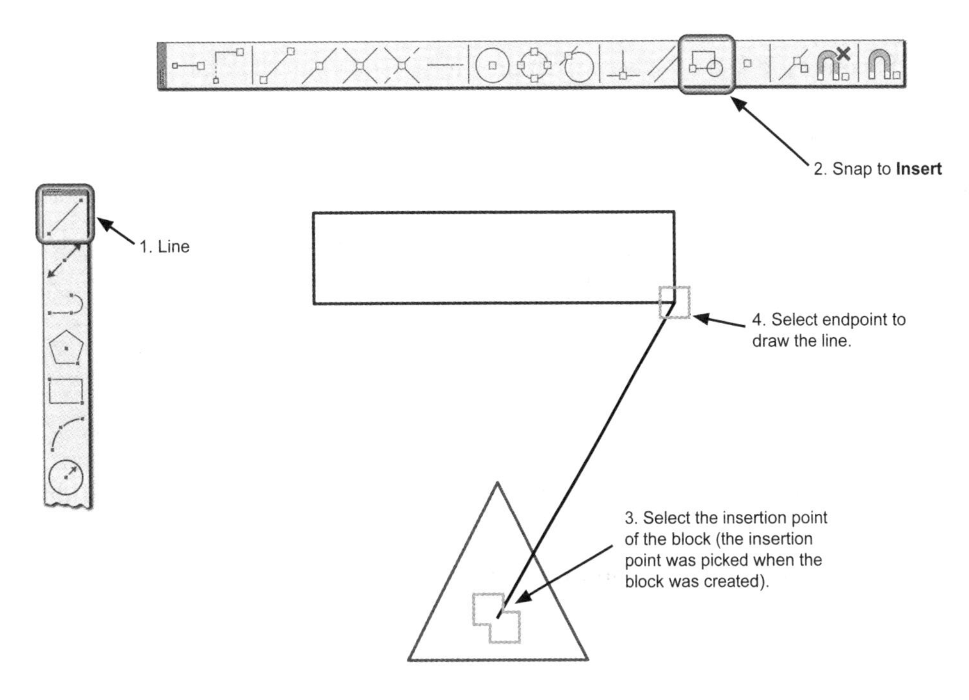
2. Snap to **Insert**
1. Line
4. Select endpoint to draw the line.
3. Select the insertion point of the block (the insertion point was picked when the block was created).

4.110 Steps in Drawing a Line from the Intersection Point of a Block Using the Snap to Insert Option

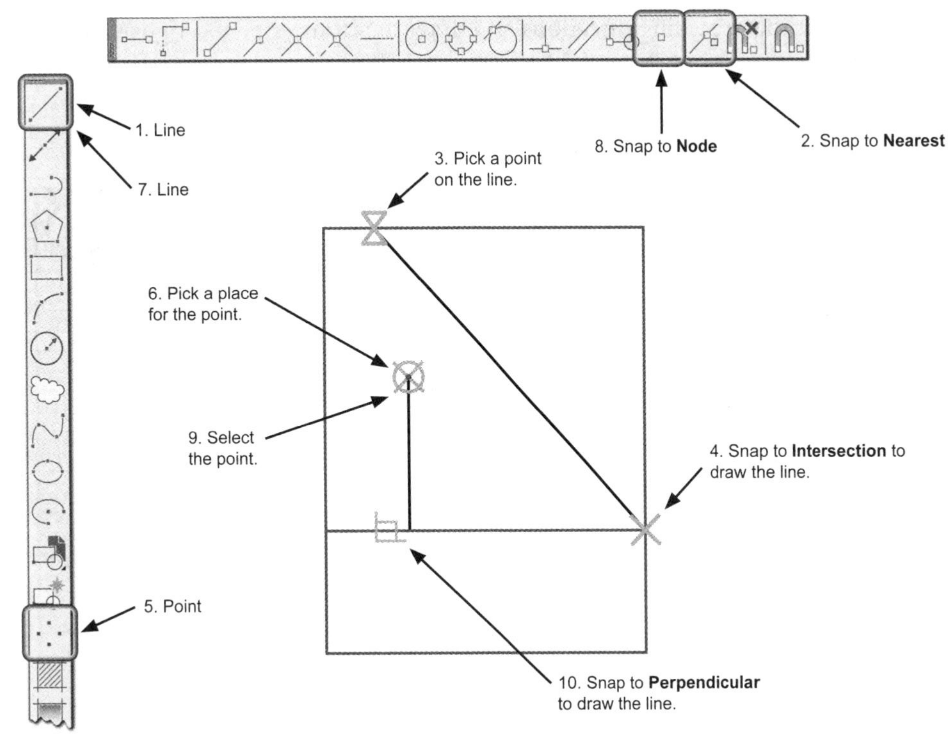
1. Line
7. Line
3. Pick a point on the line.
8. Snap to **Node**
2. Snap to **Nearest**
6. Pick a place for the point.
9. Select the point.
4. Snap to **Intersection** to draw the line.
5. Point
10. Snap to **Perpendicular** to draw the line.

4.111 Drawing Lines Using the **Snap to Nearest** and **Snap to Node** Options

EXERCISE

Open the **Object Snap** toolbar by right-clicking on any open toolbar and selecting the **Object Snap** toolbar with the left-click mouse button. Also, whenever you are in an AutoCAD command and have been prompted to select a point, if you hold down the **<Shift>** key and right-click, you can select an object snap from the pop-up menu that appears next to the cursor.

JOB SKILL

When using the object snap technique shown in Figure 4.108 to project a point in the side view through a miter line to the top view (or vice versa), the **Snap to Endpoint** and **Snap to Intersection** running object snaps must be turned *ON*. The **Polar Tracking**, **Object Snap Tracking**, and **Object Snap** icons located on the status bar must be turned *ON* as well (see Figure 4.9).

Directions

1. Begin a new AutoCAD drawing (default to the acad.dwt template).
2. Set the grid spacing to **1.00**. Make the following layers: **Visible**, **Hidden**, and **Center**. Assign a color to each layer and set the Hidden layer's linetype to **Hidden** and the Center layer's linetype to **Center**.
3. Set running osnaps for **Endpoint**, **Quadrant**, **Tangent**, **Perpendicular**, and **Intersection.**
4. Draw the front, top, and right-side views of the object shown in Figure 4.112(a). Begin the views at the absolute coordinates shown in Figure 4.112(a). Count the grids to determine the size of the object's features.
5. Save the drawing as Figure 4.112(a).
6. Repeat the preceding directions and draw the front, top, and right-side views of the objects shown in Figures 4.112(b) and 4.112(c). Save the drawings as Figure 4.112(b) and Figure 4.112(c), respectively.

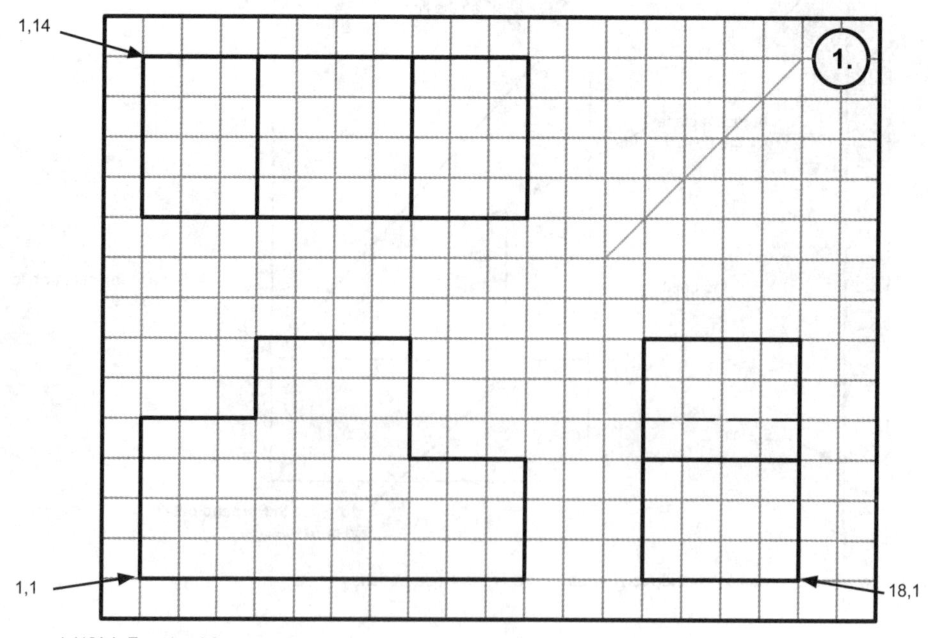

4.112(a) Exercise 4.2

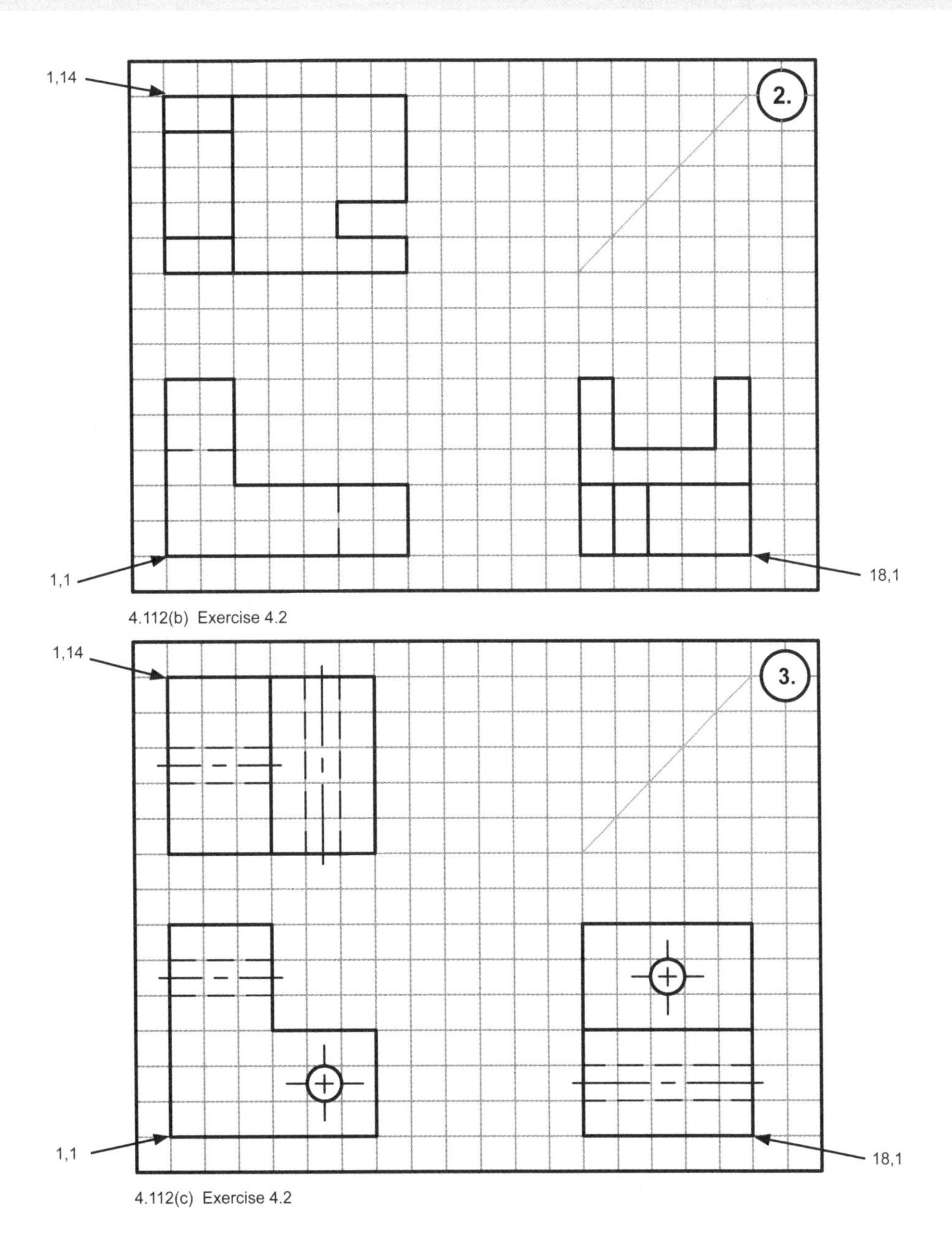

4.112(b) Exercise 4.2

4.112(c) Exercise 4.2

4.14 PROPERTIES COMMAND

The **PROPERTIES** command is used to display, or change, the ***properties*** of an object or group of objects, such as color, lineweight, layer, linetype, and linetype scale. This is a very useful feature of AutoCAD. Begin this command by choosing the arrow next to **Properties** at the bottom of the **Properties** panel of the **Home** tab of the ribbon, see Figure 4.113(a), or by selecting the **Properties** icon from the **Palettes** panel of the **View** tab of the ribbon, see Figure 4.113(b). Next, select an object in the drawing window and a **Properties** palette will open that displays the object's properties. In the example in Figure 4.114, the circle has been selected, and its properties are displayed in the **Properties** palette.

To change the properties of the circle, select the field in the palette next to the property to be changed. For example, the circle shown in Figure 4.114 is on the **VISIBLE** layer. Clicking on the field next to the **Layer** property displays a list of available layers. Selecting a different layer from the list moves the circle to the selected layer. The circle's other properties, such as lineweight, or linetype scale, can be changed in a similar fashion. Other information about the circle, such as its diameter, radius, circumference, area, and center location, can be located or changed in the **Properties** palette under the **Geometry** heading.

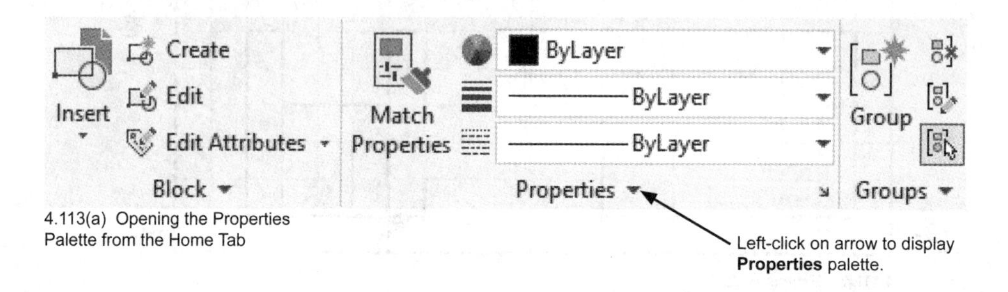

4.113(a) Opening the Properties Palette from the Home Tab

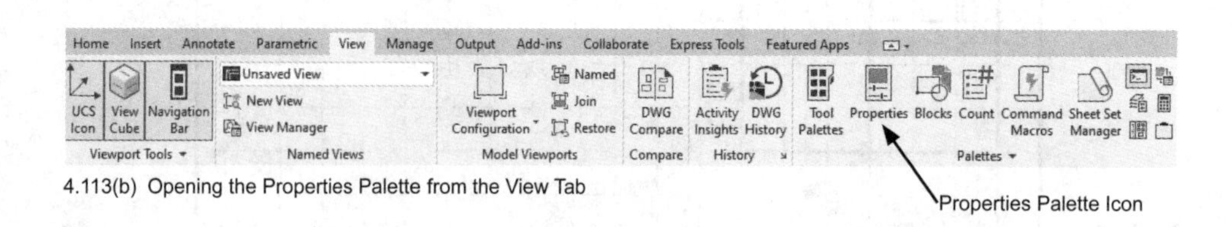

4.113(b) Opening the Properties Palette from the View Tab

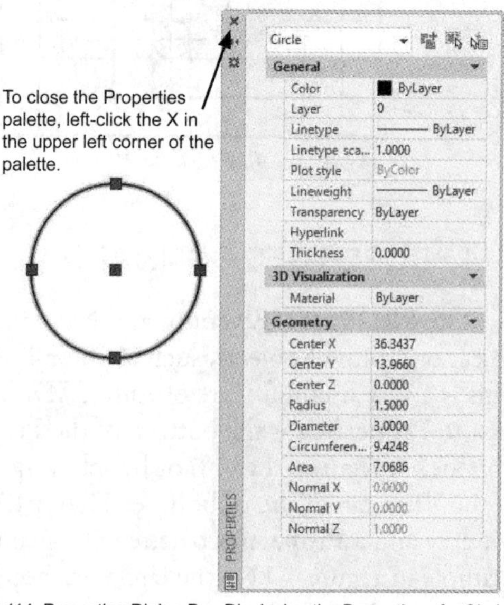

4.114 Properties Dialog Box Displaying the Properties of a Circle

NOTE

Changing the linetype scale value of every non-continuous line in the drawing by changing the **LTS** value to a larger or smaller value was discussed earlier in this chapter. Sometimes you may want to change the linetype scale of only one line. To do this, select the line, right-click, pick **Properties**, and change the **Linetype** scale value in the **Properties** palette.

4.15 INQUIRY TOOLBAR

The commands on the **Inquiry** toolbar are used to display information about AutoCAD entities, such as the distance between two points, the coordinates of a single point, the area of a closed figure like a rectangle or circle, or the volume of a 3D object. The icons and functions of the commands on this toolbar are identified in Figure 4.115(a). Open the **Inquiry** toolbar by placing the cursor on any open AutoCAD toolbar and right-clicking the mouse and selecting **Inquiry** from the list of toolbars.

QUICK TIP

Many of the **Inquiry** commands can be accessed by picking on the down arrow under the word **Measure** located on the **Utilities** panel of the **Home** tab of the ribbon.

Left-click and hold down arrow in the corner of the icon to see all of the **Inquiry** commands.

REGION/MASS PROPERTIES–produces information about 3D objects such as volume and center of gravity.

LIST–produces the properties of an object.

LOCATE POINT–displays X, Y, and Z coordinates of a selected point.

DISTANCE–measures distance between 2 selected points.

RADIUS–calculates the radius of an arc or circle.

ANGLE–calculates the angle.

AREA–calculates the are of a selected perimeter.

VOLUME–calculates the volume of a selected 3D object.

4.115(a) The Inquiry Toolbar

4.115(b) Quick Measure

QUICK MEASURE–Shows dimensions, distances, angles and other measurements dynamically as you mouse over the objects.

Quick measure can be found in the Utilities panel of the Home ribbon.

*This command was introduced in AutoCAD 2020

4.16 PREPARING TO PLOT

Before plotting in AutoCAD, drafters must first set up a **Layout**. A ***Layout*** is a convenient way to set up your final drawing sheet and set the drawing scale.

In the lower left corner of the AutoCAD screen, there are three default tabs: *Model*, *Layout1*, and *Layout2* (see Figure 4.116).

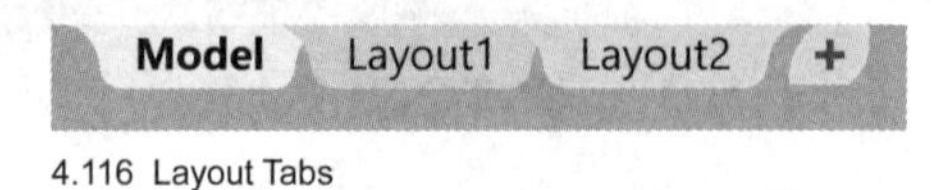
Model
Layout1
Layout2
+

4.116 Layout Tabs

To view a **Layout**, left click on the tab named *Layout1* at the bottom left of the screen. Once this tab is opened, you will see a visual representation of the final sheet to be plotted (see Figure 4.117).

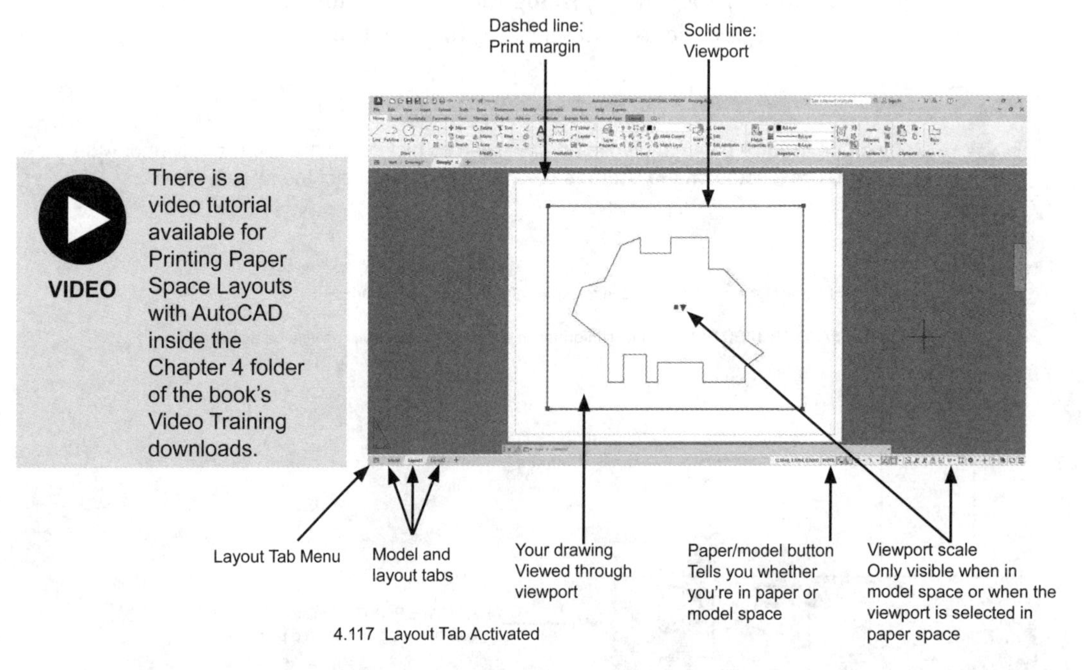
Dashed line:
Print margin
Solid line:
Viewport
There is a video tutorial available for Printing Paper Space Layouts with AutoCAD inside the Chapter 4 folder of the book's Video Training downloads.
VIDEO
Layout Tab Menu
Model and layout tabs
Your drawing Viewed through viewport
Paper/model button Tells you whether you're in paper or model space
Viewport scale Only visible when in model space or when the viewport is selected in paper space

4.117 Layout Tab Activated

After clicking on the **Layout** tab, you will see a white rectangle against a gray background that represents the sheet to be plotted. AutoCAD's default **Layout** is set up to print an A sized sheet of paper (8.5"x11").

In Figure 4.117, the rectangle with dashed lines represents the print margins. The size of the rectangle represented by the dashed line may change as different printers are assigned because printers have varying print margin capabilities. Anything drawn outside of the dashed line will not plot.

QUICK TIP

While AutoCAD provides two Layout tabs by default, more can be added as needed. To add a new layout, simply left click on the + sign to the right of the tab named Layout2. AutoCAD will allow up to 255 layouts.

The rectangle shown by a solid line in Figure 4.117 is called a **Viewport**. A ***Viewport*** is like a window into where you have originally drawn your views. The area where your views reside is called ***Model Space***. (Note: By clicking on the *Model* tab in the lower left corner, you will return to **Model Space**.) Just as a window in a door allows you to view what's inside a room, the viewport you see is actually just window-like rectangle that allows you to see into **Model Space**. The scale of the objects visible inside the Viewport can be adjusted by left clicking on the rectangle representing the Viewport and selecting a new scale from the list on the status bar. (See Figure 4.119.)

QUICK TIP

NOTE
Any *closed* object, such as a rectangle or circle, can be converted to a viewport. It is possible to have multiple viewports on a layout and the viewports can be set to display different scales.

When a layout tab is activated, you are working in an area that is called **Paper Space** (versus **Model Space** where you drew your views). ***Paper Space*** is where you will place entities that should not be affected by the viewport scale. Entities commonly placed in **Paper Space** include Borders, Title Blocks, Notes, Parts Lists and Schedules. The rectangle representing the Viewport is located in paper space.

Toggling Between Paper Space and Model Space

To edit the views, you need to be in Model Space, but to edit objects like the Title Block, you need to be in Paper Space. AutoCAD allows drafters several ways to *toggle* between Paper Space and Model Space. For instance, when in a Layout, model space can be accessed by either double clicking inside the Viewport, by clicking on the "Paper" button on the status bar at the bottom of the screen (thus changing it to read *Model*, letting you know that you're now in Model Space), or by left clicking on the Model tab, which exits the Layout completely. To enter Paper Space when you are working inside the viewport, you can double click outside of the viewport, or click on the "Model" button on the status bar (thus changing it to read *Paper*, letting you know that you're now in Paper Space.)

QUICK TIP

To completely exit a layout and return to Model Space, left click on the model tab in the lower left hand corner.

NOTE

BENEFITS OF PLOTTING WITH LAYOUTS
When plotting with layouts, the user has a visual representation of what the final plotted page will look like. One can easily and accurately adjust the drawing scale within a Viewport to ensure the drawing will fit within the page size.

QUICK TIP

ADJUSTING THE VIEWPORT SCALE
Two common ways to set the Viewport Scales are to either left click the edge of the Viewport while in Paper Space, or to double click inside the viewport, and select the scale from the Viewport Scale list located in the Status Bar. (See Figure 4.119)

STEP–BY–STEP

STEPS IN SETTING UP A LAYOUT

Follow the steps below to create a layout and set the Viewport scale once you've completed your drawing.

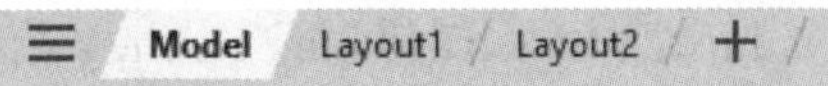

4.118 Layout Tabs

Step 1. Select the *Layout1* tab in the lower left corner of your screen, as shown in Figure 4.118.

Step 2. To adjust what's visible inside the **Viewport**, enter **Model Space** by either double clicking inside the **Viewport**, or clicking on the word PAPER in the status bar at the bottom of your screen. Pan as needed inside the Viewport to center objects as needed. (Note: Once you're in Model Space, your Viewport will be bold and you will see the word MODEL instead of PAPER in the status bar.)

Step 3. Adjust your Viewport Scale by clicking on the down arrow next to the current Viewport Scale in the status bar at the bottom right of your screen, as shown in Figure 4.119. You will see a list of possible scales to choose from, as shown in Figure 4.119. Choose the desired scale to make your objects fit on your page (pan again to center the drawing on the page as desired). (Note: With the Viewport selected, you can drag the grips on the corners of the Viewport to enlarge or shrink it as needed.)

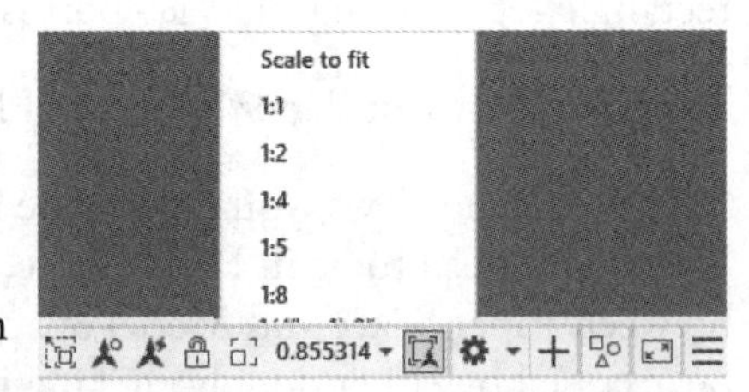

4.119 Viewport Scale

Step 4. Once your Viewport Scale is set, enter Paper Space by either double-clicking outside the Viewport, or clicking on the word MODEL in the status bar at the bottom of your screen.

Step 5. Lock the Viewport by clicking the image of the Padlock icon to the left of the new Viewport scale (see Figure 4.119). This will lock in your drawing at the desired scale.

Step 6. Edit the Title Block, Parts List, Drawing Notes and any other Paper Space objects, as needed.

QUICK TIP

PAPER SPACE OBJECTS
Users place objects in Paper Space when they don't want the size of that object to change as the Viewport Scale changes. Examples of things to put in Paper Space include Borders, Title Blocks, Notes, Parts Lists and Schedules.

4.17 PLOTTING WITH AUTOCAD

Left click on the **Output** tab of the ribbon and from the **Plot** panel select the **PLOT** icon (see Figure 4.120). This opens the **Plot** dialog box shown in Figure 4.121.

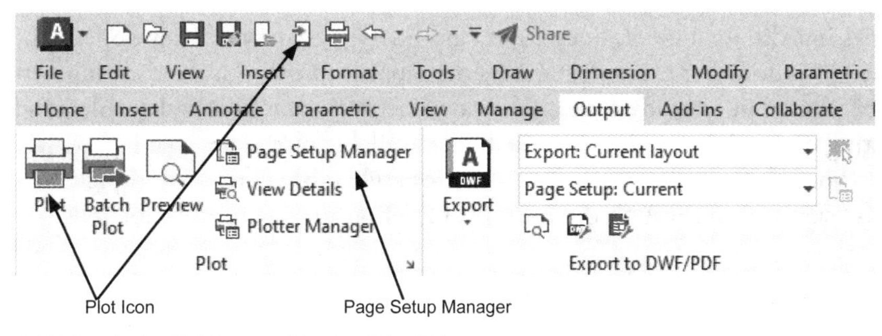

4.120 Opening the Plot Command from the Output Tab

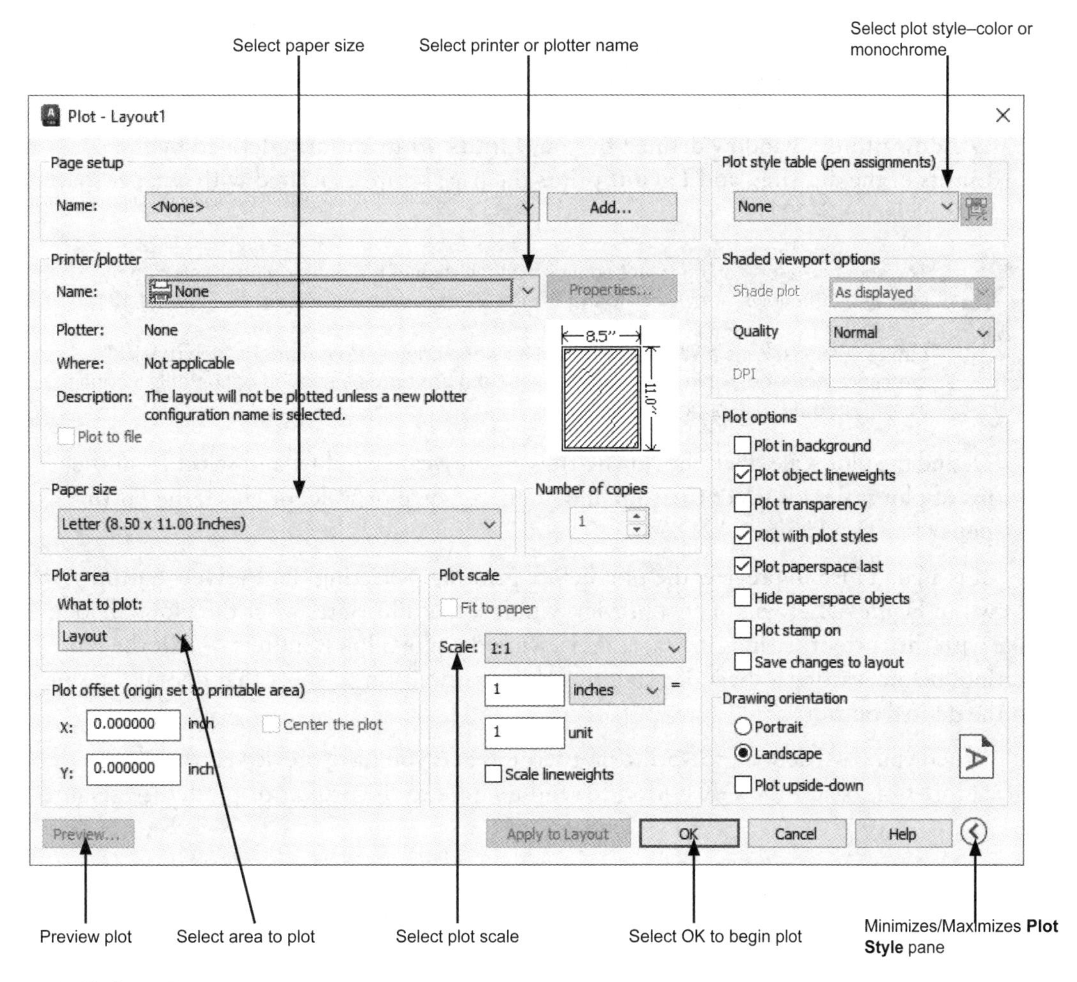

4.121 Plot Dialog Box

The options in the **Plot** dialog box allow you to do the following (see Figure 4.121):

- Select the name of the printer/plotter where you want to send the plot. Pick the down arrow in the **Printer/plotter** window and select from the list of available printers and/or plotters (see Figure 4.121).
- Change the **Plot style table.** By changing the **Plot style table** settings, you can plot entities in a different color, lineweight, and linetype than what they appear in the drawing window. For example, entities that appear in color in the drawing window can be plotted in black by picking the down arrow in the **Plot style table** window and selecting **monochrome.ctb** from the list of available plot styles (see Figure 4.121). When you select the down arrow in the **Plot style table** window and pick **New** from the list, you are led through the prompts to define a new, custom plot style.

NOTE

Under **Plot style table**, select **acad.ctb** for color printing and plotting, or **monochrome.ctb** for black-and-white plots. Selecting **acad.ctb** from the list in the plot style table ensures that entities are printed in the color they are assigned in the drawing.

- Select the paper size. Pick the down arrow in the **Paper size** window and select from the list of available paper sizes.
- Define the plot area. Pick the down arrow in the **What to plot** window and select from the list of available choices. **Display** prints the image currently visible in the drawing window, **Extents** prints to the extents of all entities placed in the drawing, **Window** prints a window defined by you, **Limits** prints the area defined by the **Limits** of the drawing, and **Layout** prints the page setup associated with a paper space tab (see Figure 4.121).

NOTE

The sheet size for plotting is limited to the capabilities of the printer/plotter selected by picking on the down arrow next to the **Printer/Plotter *name*** window (see Figure 4.121).

Limits is available as a **What to Plot** option when printing from model space (basically, the area encompassed within the drawing limits), but **Layout** is a printing option only if printing from a **Layout** tab in paper space.

- Define the plot scale. Pick the down arrow in the **Scale** window and select from the list of plot scales, or select **Custom** and define a new plot scale, or check the **Fit to paper** box (see Figure 4.121).

It is a good idea to preview the plot before printing. Selecting the **Preview** button shown in Figure 4.121 will open a preview window that illustrates what the plot will look like with the current settings in place. Previewing the plot allows drafters to edit the plot settings before wasting a sheet of paper by printing or plotting a sheet that does not result in the desired output.

When you have defined the print/plot settings and you have previewed the plot to check for errors, select the **OK** button to send the plot to the printer or plotter (see Figure 4.121).

NOTE

Fit to paper is a **Plot scale** option, but the object will not be printed to an exact scale if this is selected because the plot will probably be scaled up or down to fit the paper size. Select the down arrow in the window next to **Scale** to see a list of available plot scales.

4.18 CREATING A PAGE SETUP FOR PLOTTING

Creating a page setup for an AutoCAD project streamlines the process of plotting drawings. A **Page Setup** is a named set of predefined print/plot settings. When you wish to plot the project, you simply select the name of the page setup from the list in the **Plot** dialog box rather than redefining the plot settings each time.

STEP-BY-STEP

STEPS IN CREATING A PAGE SETUP

Follow the steps below to create a page setup for a plot named Cottage.

Step 1. Left click on the **Layout1** tab at the bottom left of the graphics window. Select the **Page Setup Manager** from the **Plot** panel of the **Output** tab of the ribbon (see Figure 4.120). The **Page Setup Manager** dialog box opens, as shown in Figure 4.122.

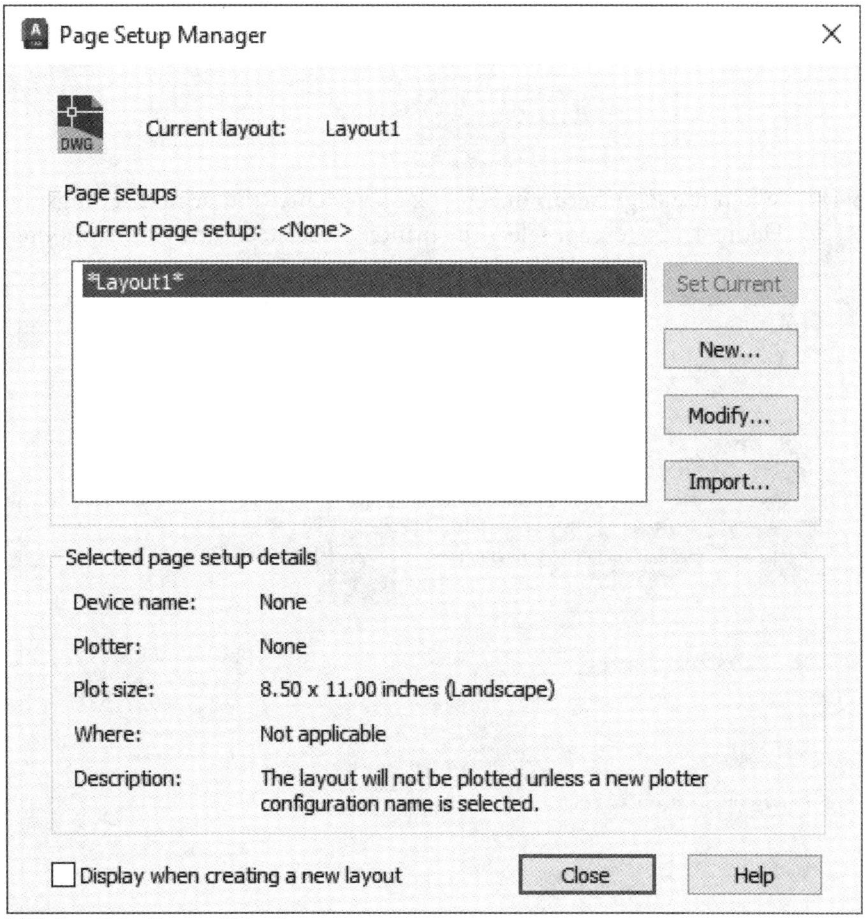

4.122 Page Setup Manager Dialog Box

STEP–BY–STEP

Step 2. Select the **New...** button from the **Page Setup Manager** dialog box shown in Figure 4.122.

Step 3. When the **New Page Setup** dialog box opens, type **Cottage** (or other desired name) into the **New page setup name** field as shown in Figure 4.123 and click **OK**.

New Page Setup

New page setup name:

Cottage

Start with:

<None>

<Default output device>

Layout1

OK Cancel Help

4.123 New Page Setup Dialog Box

Step 4. When the **Page Setup** dialog box opens, enter the printer settings shown in Figure 4.124 (except select the printer associated with your computer).

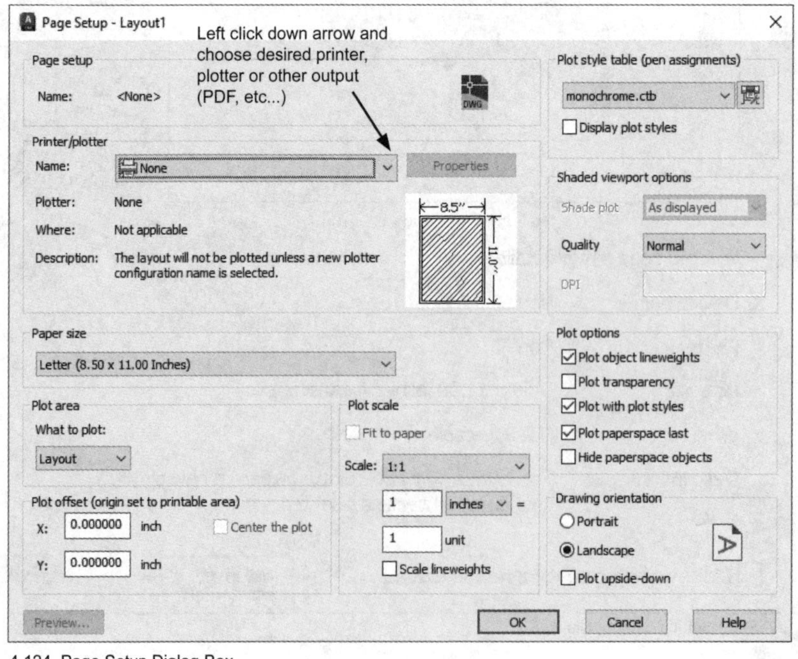

4.124 Page Setup Dialog Box

STEP–BY–STEP

When the desired settings are in place, click **OK** to return to the **Page Setup Manager** dialog box, then click **Close** to return to your drawing.

Step 5. To plot the page setup, select the **Plot** icon from the **Plot** panel of the **Output** tab of the ribbon (see Figure 4.120), and when the **Plot** dialog box opens, select the down arrow in the **Page Setup** window and select **Cottage** from the list, as shown in Figure 4.125. Click **OK** to send the drawing file to the plotter or printer.

QUICK TIP

It is sometimes necessary to adjust the settings in the **Page Setup** dialog box through trial and error until the desired plot settings are in place. Selecting the **Preview...** button allows you to see the settings and adjust accordingly. Previewing the plot before selecting OK allows you to confirm that the settings in the page setup will result in the desired plot.

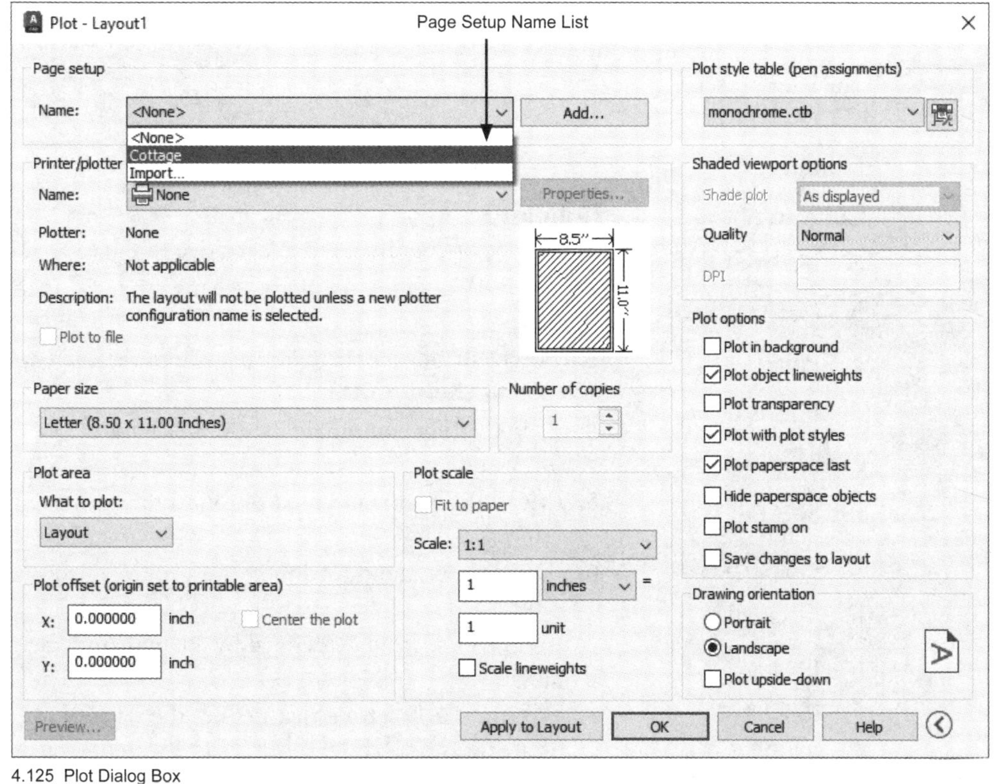

4.125 Plot Dialog Box

CHAPTER REVIEW

With each new software release, computer-aided design tools become more powerful, allowing CAD users to accomplish higher-level design and drafting tasks. However, as CAD software evolves, it also drives changes in how technical drawings are created, viewed, shared, and managed. Persons employed in the CAD field must be committed to the principle of life-long learning to keep up with the dynamic technology of this field.

- Absolute Coordinates
- Cartesian Coordinate System
- Draw Commands
- Drawing Limits
- Drawing Units
- Inquiry Commands
- Layers
- Modify Commands
- Object Snap Settings
- Polar Coordinates
- Properties
- Relative Coordinates
- Running Object Snaps
- Text Style
- User Coordinate System

Short Answer

1. How is the point with coordinates 0,0 represented in the graphics window of an AutoCAD drawing?
2. What is the direction (North, South, East, or West) of the default Base Angle setting in AutoCAD?
3. What setting should be defined before the limits of an AutoCAD drawing are set?
4. Which of the following is not an AutoCAD Object Snap setting: Node, Tangent, Quadrant, or Startpoint?
5. What is the name of the tool on the Inquiry toolbar that is used to calculate the distance between two points?

Matching

Column A	Column B
a. POLYGON	1. A type of coordinate that is relative to 0,0
b. CHAMFER	2. MODIFY command used to change the size of an object
c. Polar	3. A type of coordinate that is relative to the last point defined
d. Limits	4. Units setting commonly used in mechanical drawings
e. Absolute	5. MODIFY command used to round the corners of objects
f. SCALE	6. DRAW command used to create a multi-sided object
g. RECTANGLE	7. A type of coordinate defined by a length and an angle
h. Decimal	8. An AutoCAD setting that defines the size of the drawing area
i. FILLET	9. DRAW command that requires you to define two diagonal points
j. Relative	10. MODIFY command used to bevel the corners of objects

Step 1. Getting Started

a. Download the **Imperial Prototype** drawing file located in the *Prototype Drawing Files* folder associated with this book.

b. **SAVE AS Geoquick** to save the drawing to your **Home** directory.

c. Activate the **MODEL** tab in the lower-left corner of the graphics window (if it is not already highlighted) by left-clicking on the tab.

d. Set Units to **Decimal.**

e. Set **Text Style** *font* to **Arial**.

f. Create a Visible and a Center layer (Center should be set to the "Center" linetype)

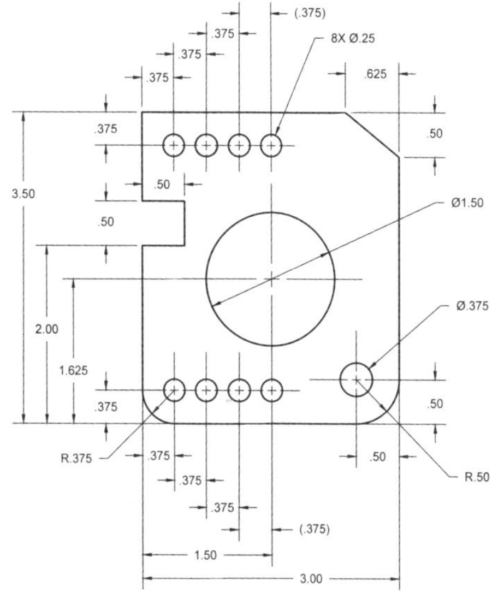

4.126 Gasket with Dimensions

g. Set the Visible layer current

h. Follow the steps below to draw the Gasket shown in Figure 4.126.

Note: The Gasket will be drawn inside the magenta rectangle located in model space.

There is a video tutorial available for this project inside the Chapter 4 folder of the book's Video Training downloads.

Step 2. Begin Drawing

a. Select the **RECTANGLE** command from the **Draw** panel of the ribbon.

b. When asked to *specify the first corner* type **4,5** and press **Enter**.

c. When prompted to *specify the other corner* type **3.00,3.50** and press **<Enter>**, as in Figure 4.127(a).

d. Select the **EXPLODE** command from the **Modify** panel of the ribbon and when prompted to *select objects* left-click on the edge of the rectangle and press **<Enter>**. *Exploding* the rectangle breaks it into four separate lines.

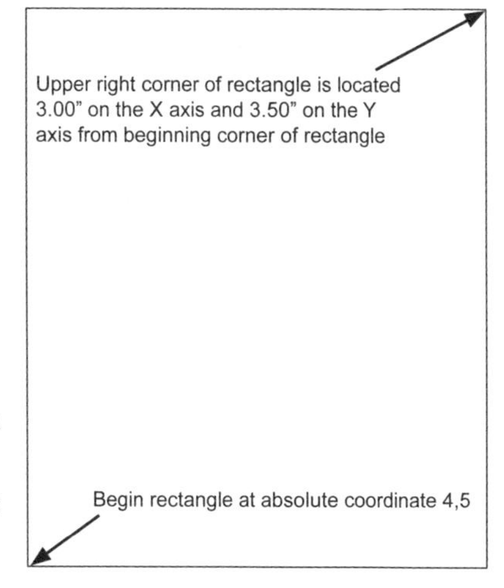

4.127(a) Drawing the 3.00 X 3.50 rectangle

Step 3. Select the **OFFSET** command from the **Modify** panel

a. When asked to *specify the offset distance* type **.50** and press **<Enter>**.

b. When prompted to *select object to be offset* left-click the vertical line on the left side of the rectangle.

c. When prompted to *specify point on side to offset* left-click in the graphics window *anywhere* to the right of the selected line and a second line will be created that is parallel to the selected line and located at a distance of .50" to the right of the line. See Figure 4.127(b).

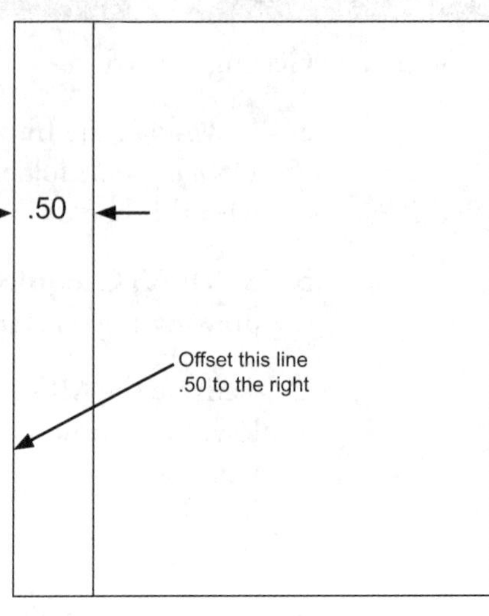

4.127(b) Offsetting line .50

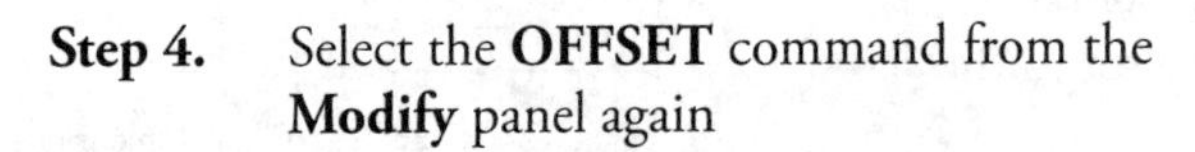

Step 4. Select the **OFFSET** command from the **Modify** panel again

a. When asked to *specify the offset distance* type **2.00** and press **<Enter>**.

b. When asked to *select object to be offset* left-click the line on the bottom edge of the rectangle.

c. When prompted to *specify point on side to offset* left-click in the graphics window above the selected line and a second line will be created that is parallel to the selected line and 2.00" above the line. See Figure 4.127(c).

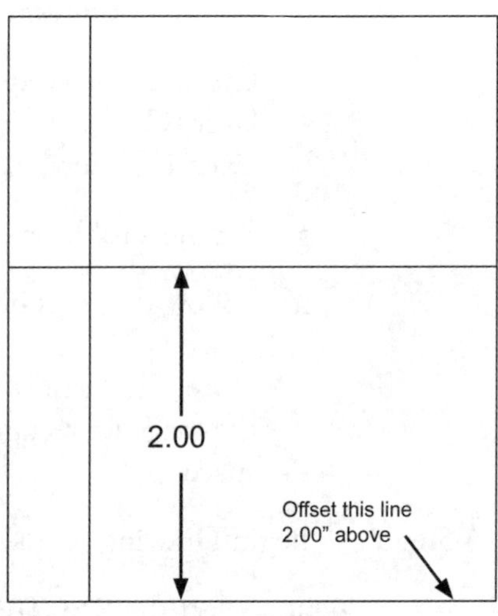

4.127(c) Offsetting line 2.00

Step 5. Select the **OFFSET** command again

a. When asked to *specify the offset distance* type **.50** and press **<Enter>**.

b. When asked to *select object to be offset* left-click the line created in Step 4.

c. When prompted to *specify point on side to offset* left-click in the graphics window above the selected line and a second line will be created that is parallel to the selected line and .50" above the line.
See Figure 4.127(d).

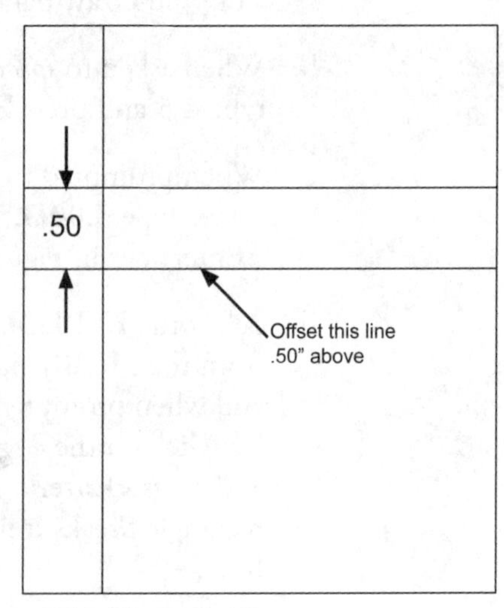

4.127(d) Offsetting line .50

Step 6. Select the **TRIM** command from the **Modify** panel and press **<Enter>**.

a. Left-click on the lines to be trimmed in the order shown in Figure 4.128(a).

b. When you are finished with this step, your drawing should resemble the object in Figure 4.128(b).

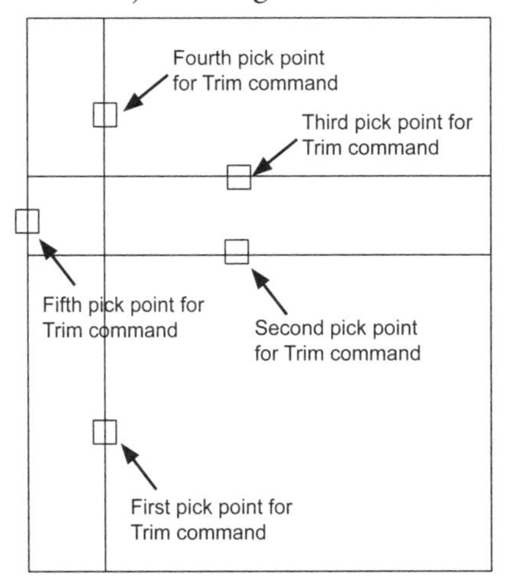

4.128(a) Trimming lines to create slot.

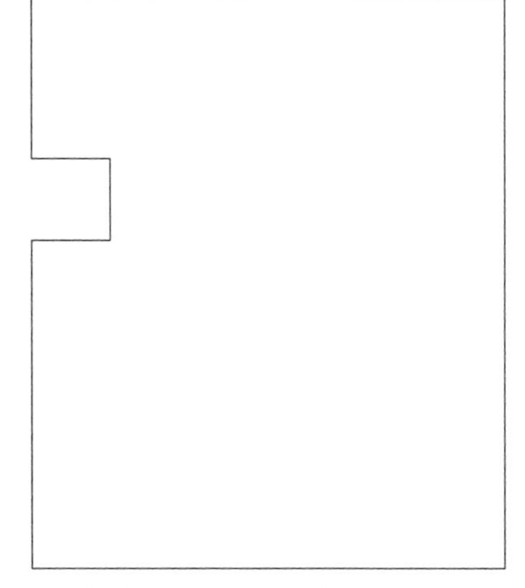

4.128(b) Gasket with slot on left side.

QUICK TIP

Pressing <Enter> immediately after selecting the Trim command causes AutoCAD to recognize every line as a ***cutting edge.***

Step 7. Select the **CIRCLE/Diameter** command from the **Draw** panel

a. When prompted to *specify the center point* for circle, type **from**, or select the **Snap From** icon on the Object Snap toolbar (See Figure 4.109), and press **<Enter>**.

b. When prompted to *specify base point* left-click on the lower-left corner of the rectangle drawn in Figure 4.129, then type **@1.50,1.625** and press **<Enter>**.

c. When prompted to *specify diameter of circle* type **1.50** and press **<Enter>**.

d. This step creates a 1.50" diameter circle that is located 1.50" on the X-axis and 1.625" on the Y-axis from the lower-left corner of the rectangle. (See Figure 4.129.)

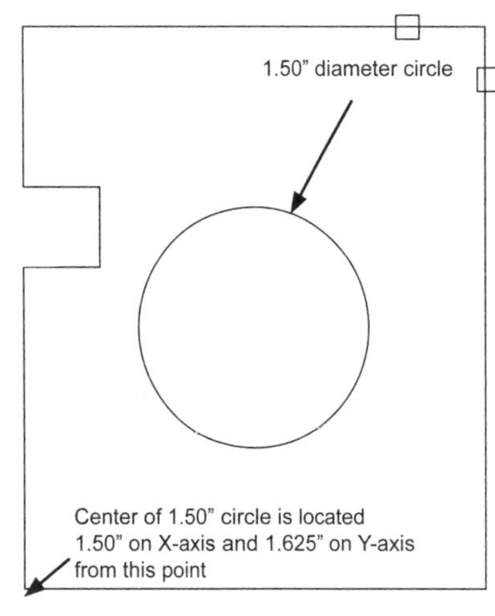

4.129 Adding the 1.50 Diameter Circle

PROJECT

Step 8. Select the **CHAMFER** command from the **Modify** panel

a. Type **D** and enter (or click on the word **Distance** on the command line).

b. When prompted to *specify first chamfer distance* type **.625** and press **<Enter>**.

c. When prompted to *specify second chamfer distance* type **.50** and press **<Enter>**.

d. When prompted to *select first line* left-click the top line near the first pick point shown in Figure 4.130.

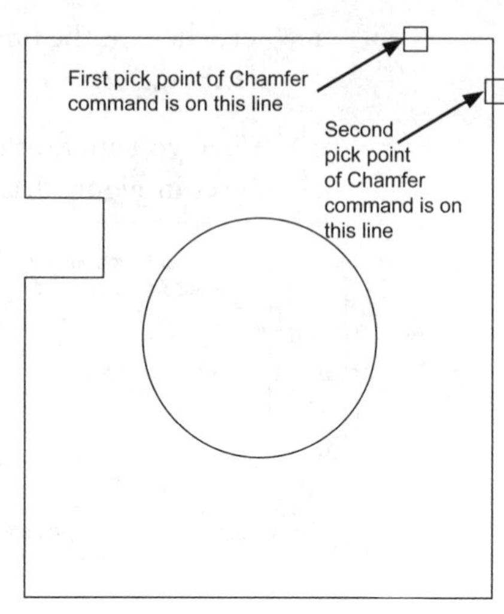
First pick point of Chamfer command is on this line
Second pick point of Chamfer command is on this line

4.130 Adding the top right Chamfer

e. When prompted to *select second line* left-click the line on the right side of the rectangle near the second pick point shown in Figure 4.130.

f. When this step is complete, the corner will be chamfered as shown in Figure 4.131.

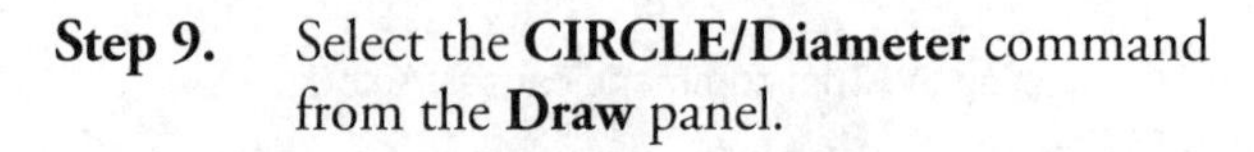

Step 9. Select the **CIRCLE/Diameter** command from the **Draw** panel.

a. When prompted to *specify the center point for circle*, type **from**, or select the **Snap From** icon on the Object Snap toolbar, and press **<Enter>**.

b. When prompted to *specify base point* left-click on the lower-right corner of the rectangle drawn in Step 1, then type **@-.50,.50** and press **<Enter>**.

c. When prompted to specify diameter of circle type **.375** and press **<Enter>**.

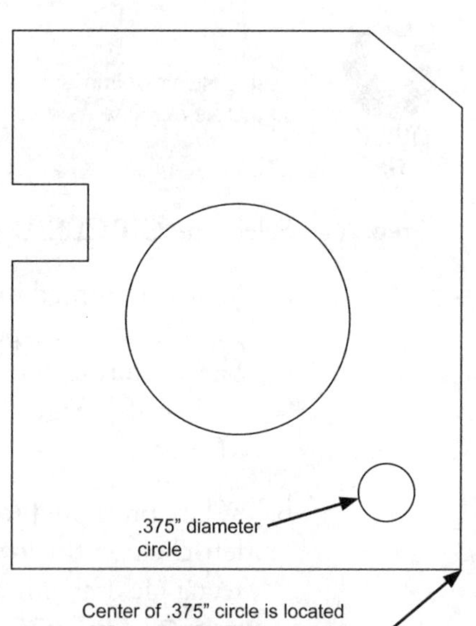
.375" diameter circle
Center of .375" circle is located .50" on negative X axis and .50" on Y axis from this point

4.131 Gasket with Chamfered Corner and .375 Diameter Circle

d. This draws a **.375"** diameter circle that is located **.50"** on the *(negative) X-axis* and **.50"** on the *Y-axis* from the *lower-right corner of the rectangle.* (See Figure 4.131).

Step 10. Select the **FILLET** command from the **Modify** panel

a. Type **R** and **<Enter>** (or click on the word Radius on the command line).

b. When prompted to *specify radius* type **.50** and press **<Enter>**.

c. When prompted to *select first object* left-click the line on the right side of the rectangle near the first pick point shown in Figure 4.132.

d. When prompted to *select second object* left-click the line on the bottom edge of the rectangle near the second pick point shown in Figure 4.132.

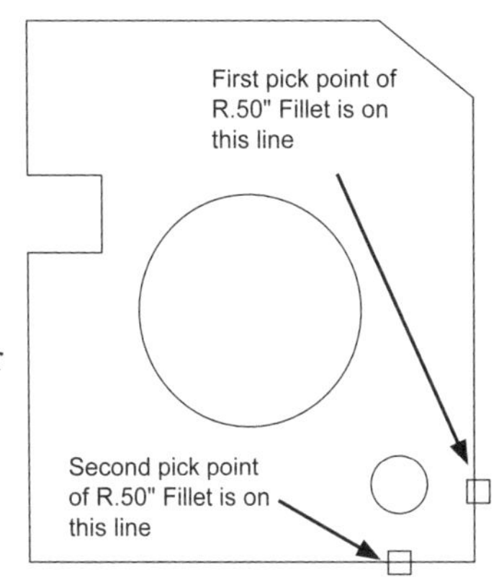
First pick point of R.50" Fillet is on this line
Second pick point of R.50" Fillet is on this line
4.132 Adding R.50 Fillet

Step 11. Select the **CIRCLE/Diameter** command from the **Draw** panel

a. When prompted to *specify center point* for circle, type **from**, or select the **Snap From** icon on the Object Snap toolbar, and press **<Enter>**.

b. When prompted for the *specify base point* pick the lower-left corner of the rectangle you drew in Step 1, then type **@.375,.375** and press **<Enter>**.

c. When prompted to *specify diameter* of the circle type **.25** and press **<Enter>**.

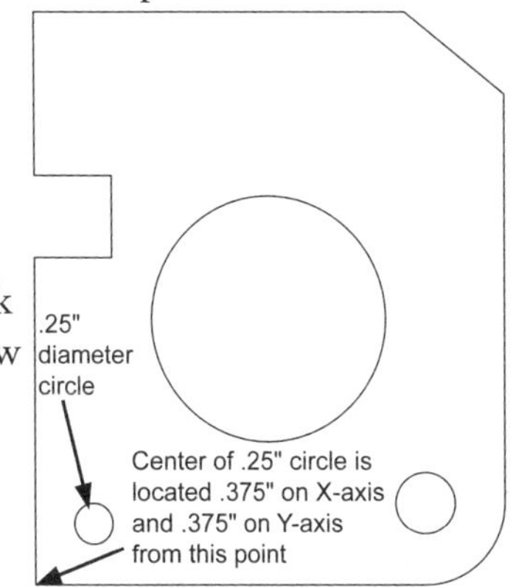
.25" diameter circle
Center of .25" circle is located .375" on X-axis and .375" on Y-axis from this point
4.133 Adding the .25 Diameter Circle

d. This draws a .25" diameter circle that is located .375" on the X-axis and .375" on the Y-axis from the lower-left corner of the rectangle. (See Figure 4.133.)

Step 12. Select the **FILLET** command from the **Modify** panel

a. Type **R** and enter (or click on the word **Radius** on the command line).

b. When prompted to *specify radius* type **.375** and press **<Enter>**.

c. When prompted to *select first object* left-click the line on the left side of the rectangle near the first pick point shown in Figure 4.134.

d. When prompted to *select second object* left-click the line on the bottom edge of the rectangle near the second pick point shown in Figure 4.134.

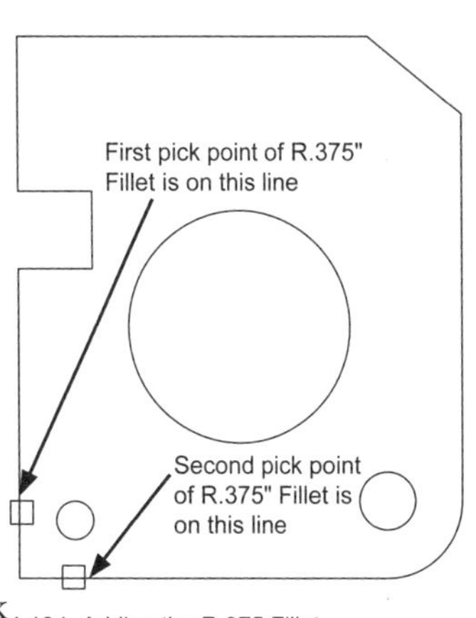
First pick point of R.375" Fillet is on this line
Second pick point of R.375" Fillet is on this line
4.134 Adding the R.375 Fillet.

PROJECT

Step 13. Select the **RECTANGULAR ARRAY** command from the **Modify** panel.

a. When prompted to *select objects* left-click on the .25" diameter circle drawn in Step 11 and press **<Enter>**.

b. When the **Array Contextual Tab** ribbon opens:

c. Set the number of **Columns** (||||) to **4**, and the distance **Between** columns to **.375**.

d. Set the number of **Rows** (----) to **2** and the distance **Between** rows to **2.75** (See Figure 4.135).

e. Click the **Close Array** button on the ribbon to finish the command.

f. The completed Gasket should look like the object in Figure 4.136.

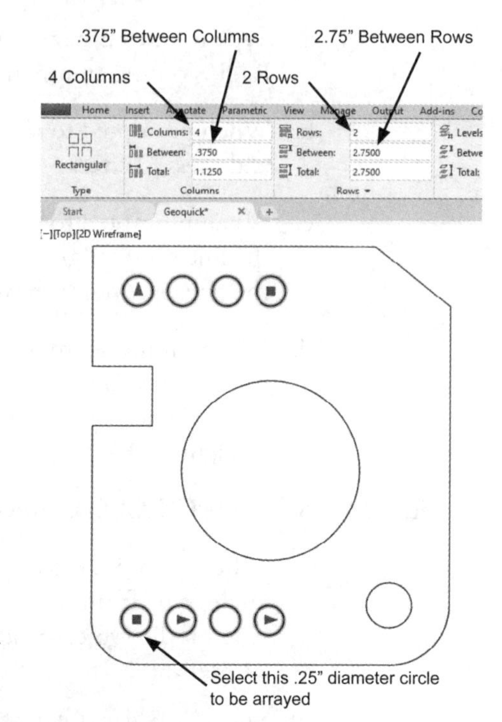

4.135 Arraying the .25 Diameter Circle

Step 14. When finished drawing the Gasket, left-click on the **Layout1** tab in the lower-left corner of the graphics window

a. Edit the text in the title block by double-clicking on it to enable the text edit option.

b. Save the file and plot the drawing as instructed by your instructor.

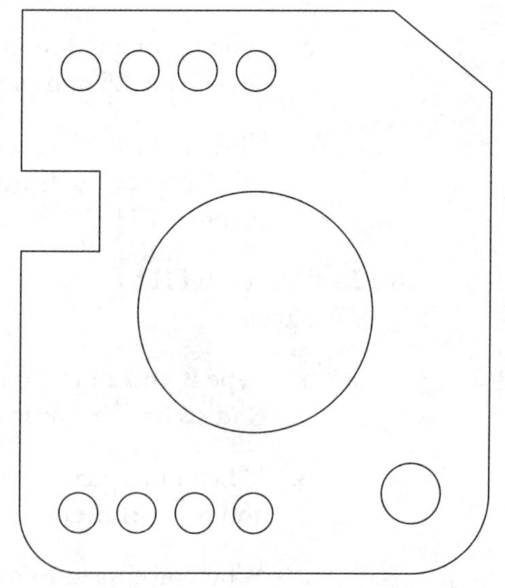

4.136 Completed Gasket Drawing.

VIDEO

There is a video tutorial available for this project inside the Chapter 4 folder of the book's Video Training downloads.

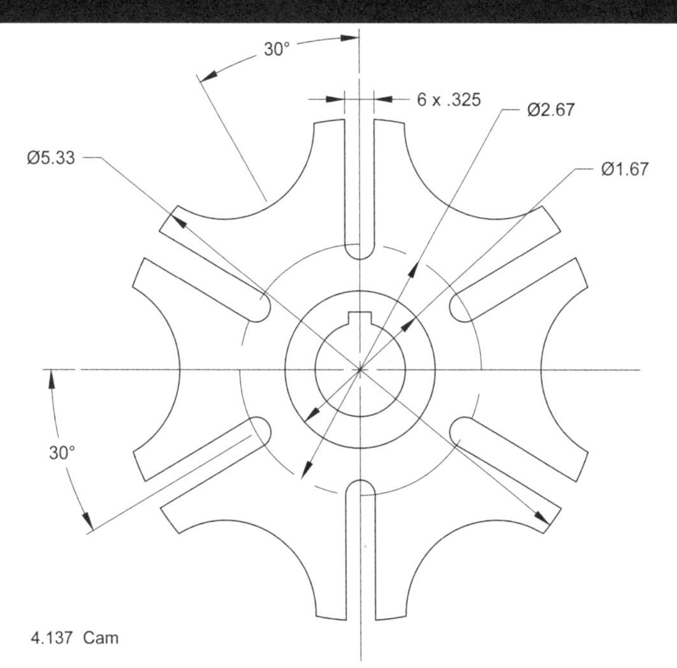

4.137 Cam

Step 1. Select File and open the GEOQUICK drawing you created earlier.

Step 2. Activate the Model tab before beginning this step.

a. The Cam will be drawn in model space inside the same rectangle that the object in Exercise 4.1 is drawn.

b. Draw two concentric circles

c. With a center point at absolute coordinates 14,11.

d. Draw one circle with a 6" diameter. Draw the second circle with a 5.33" diameter.

e. Set the Center layer current.

f. Choose the "Center Mark" icon on the "Annotate" ribbon, then click on the outer circle and a center mark will appear (see Figure 4.138).

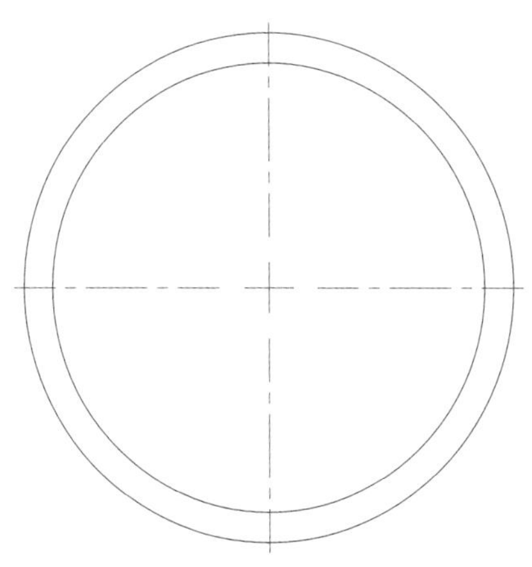

4.138 Concentric Circles

Step 3. Draw a circle 2" in diameter at the right quadrant of the largest circle, as shown in Figure 4.139.

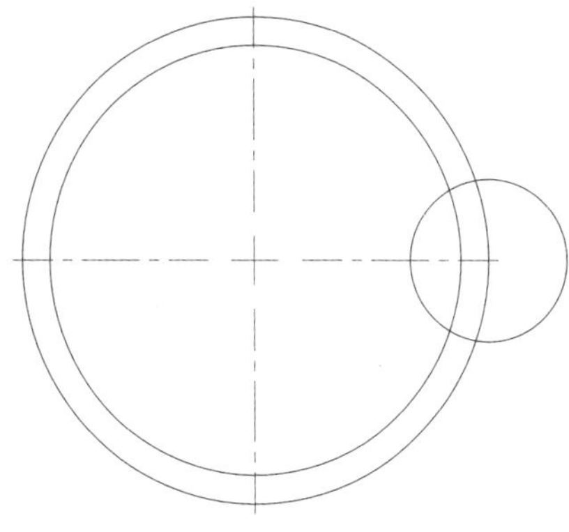

4.139 Adding 2" Circle

Step 4. Use the ARRAY command's Polar option (refer to page 173)

a. Array 6 circles around the center of the cam

b. Fill an angle of 360° as shown in Figure 4.140.

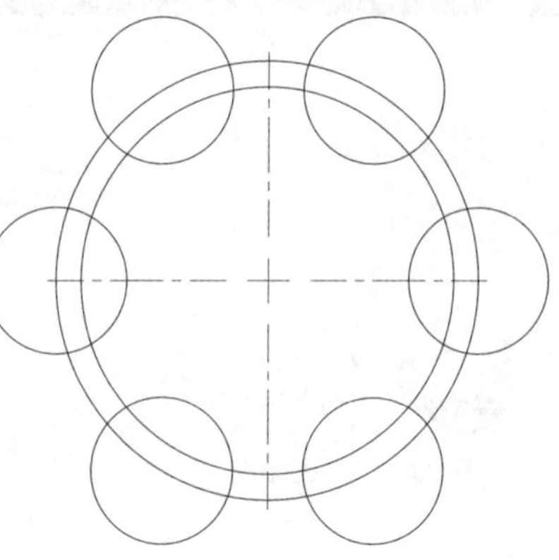

4.140 Polar Array

Step 5. Erase and Trim

a. Erase the larger-diameter circle

b. Trim the object to create the shape shown in Figure 4.141.

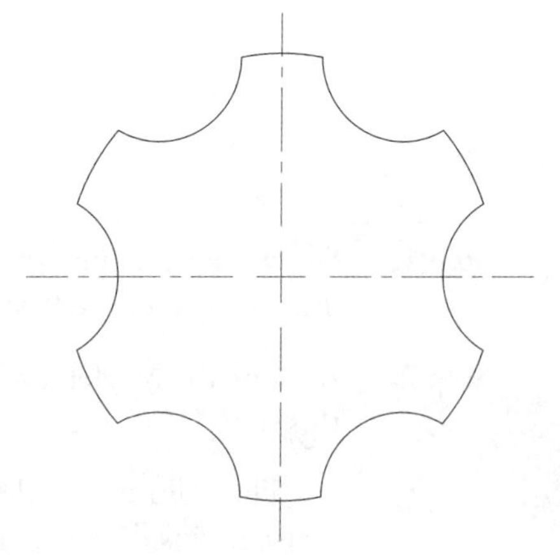

4.141 Erase and Trim

Step 6. To create the slot:

a. Draw a circle with a diameter of 2.67", centered on the existing CAM.

b. Draw another circle with a diameter of .325, using the point where the vertical centerline and the new circle intersect as the center point as shown in Figure 4.142.

c. With Ortho on, draw two vertical lines starting from the quadrants on each side of the small circle as shown in Figure 4.142.

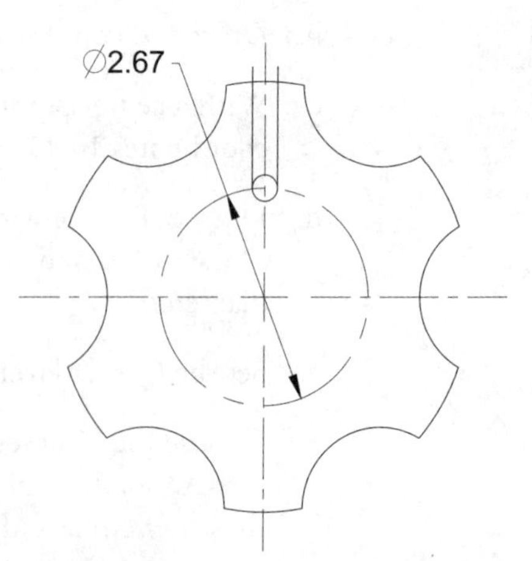

4.142 Creating a Slot

Step 7. Use **Polar Array** to array the slot created in Step 5.

a. Pick the center of the cam for the center of the array

b. Rotate 6 objects to fill an angle of 360°, rotating the objects as they are arrayed.

c. Your cam drawing should look like the object shown in Figure 4.143.

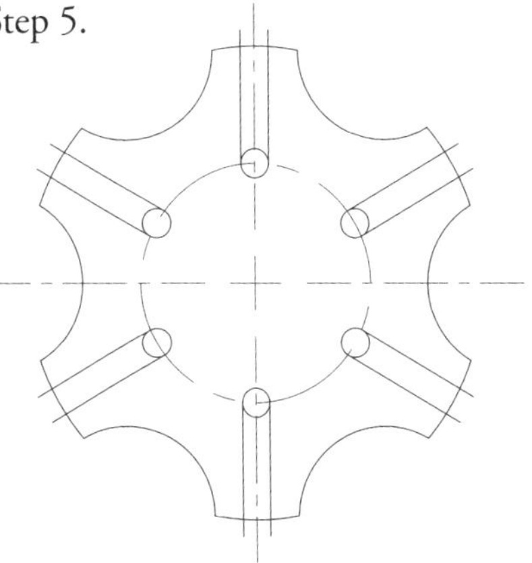

4.143 Polar Array of Slot

Step 8. Erase and Trim

a. Erase the 2.67 diameter circle created in step 6

b. Trim the slots as shown in Figure 4.144.

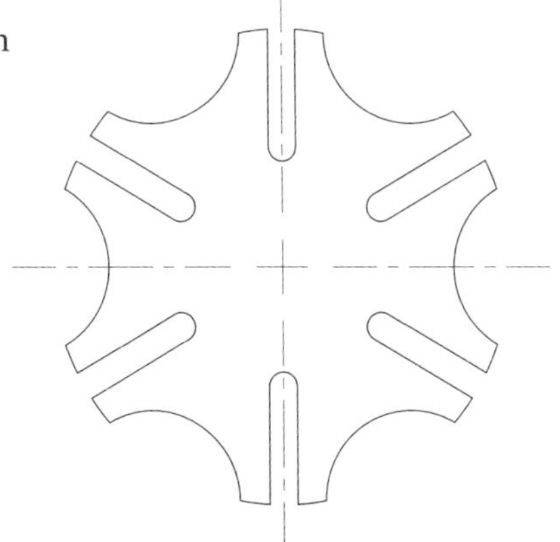

4.144 Trimmed Slots

Step 9. Draw two concentric circles to add the hub to the cam.

a. Draw a circle with a diameter of 1.67", centered on the existing CAM

b. Draw another circle with a diameter of 1.00" to create the hub.

Step 10. Add a keyway to the CAM

a. Draw a .25" wide and .15" tall rectangle at the top of the 1.00" circle.

b. Trim the top of the 1.00 circle to open the keyway. See Figure 4.145.

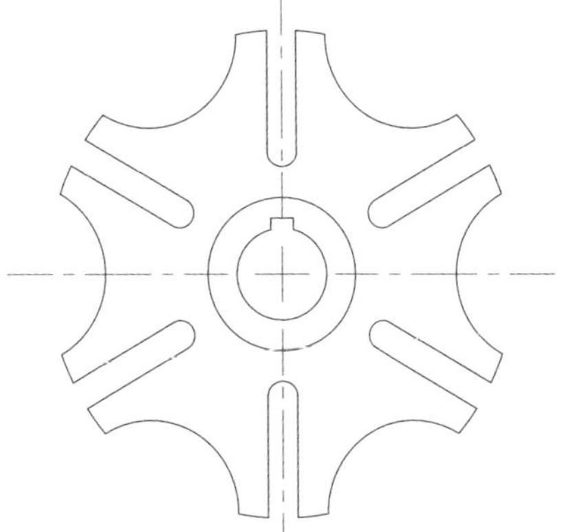

4.145 Adding the Keyway

This completes the cam construction (see Figure 4.146). Follow your instructor's directions to plot/print the drawing. Be sure to save the drawing file before you close AutoCAD.

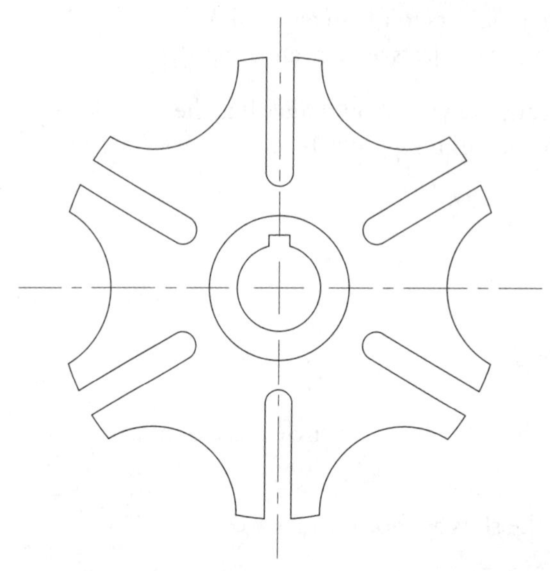

4.146 Completed Cam Exercise

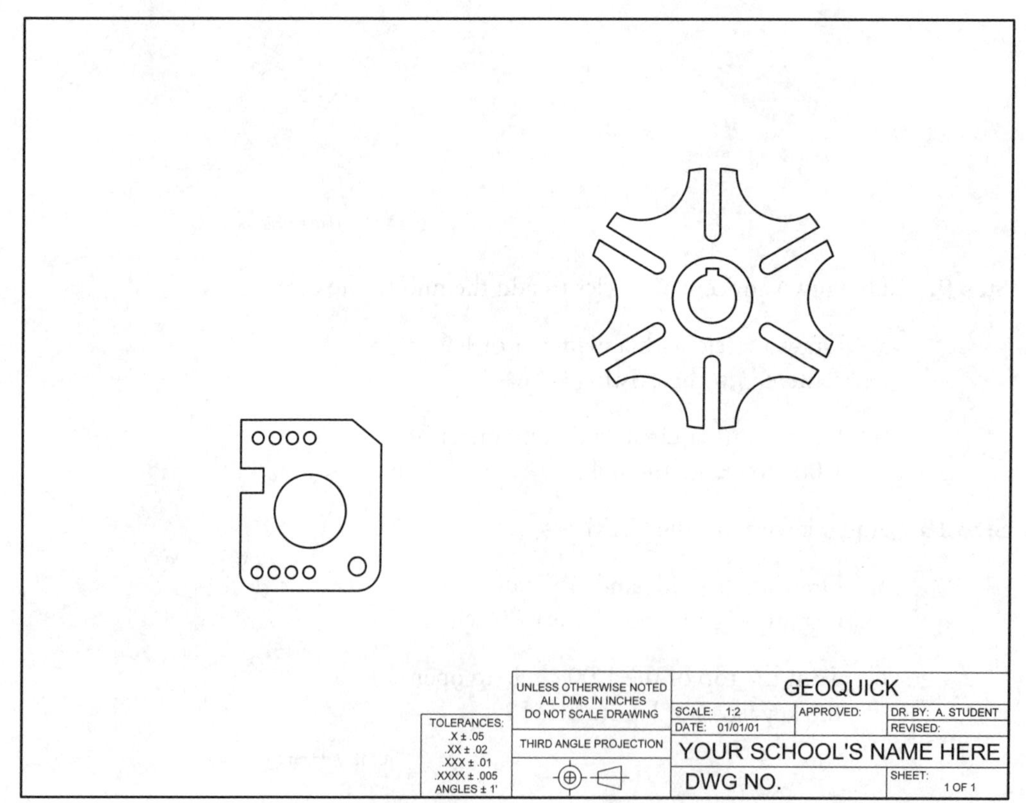

4.147 Completed Geoquick drawing with Gasket and Cam.
No dimensions or Notes Needed

In this project you will draw the floor plan of the Guest Cottage shown in the designer's sketch in Figure 4.148.

In creating this drawing, you will be required to use many of the **Draw** and **Modify** tools that you learned about earlier in this chapter. Your instructor will guide you through this project, so do not hesitate to ask for help if you get stuck.

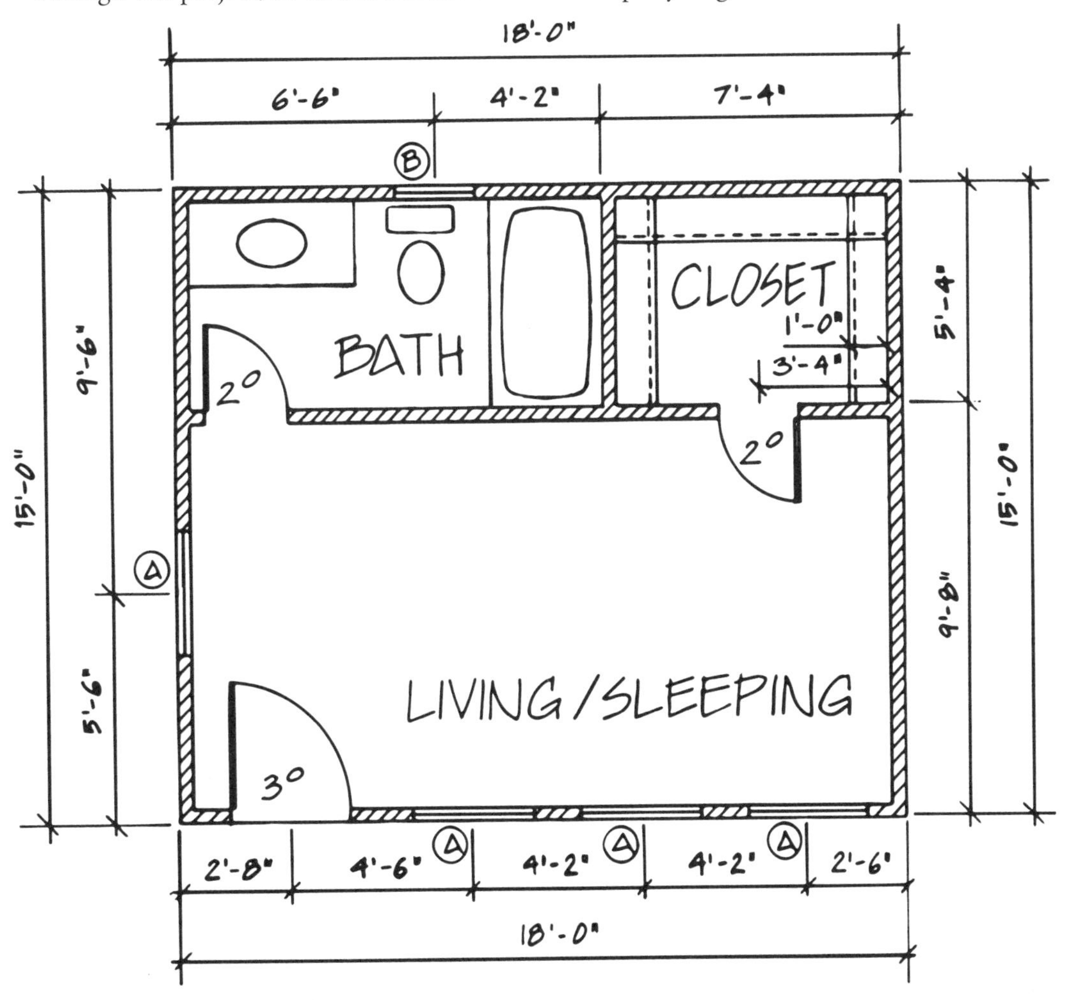

4.148 Designer's Sketch of the Guest Cottage

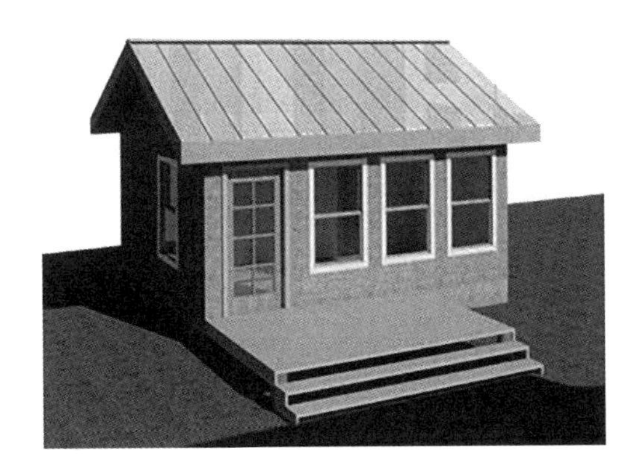

WINDOW SCHEDULE			
MARK	SIZE	TYPE	QTY
A	$3^0 \times 5^0$	S.H.	4
B	$2^0 \times 2^0$	S.H.	1

VIDEO

There is a video tutorial available for this project inside the Chapter 4 folder of the book's Video Training downloads.

PROJECT

Step 1. Download and open the **Cottage Prototype** drawing file located in the *Prototype Drawing Files* folder associated with this book.

a. Use **SAVE AS** to save the drawing to your **Home** directory

b. Rename the drawing **GUEST COTTAGE.**

Step 2. Activate the **Model** tab in the lower-left corner of the graphics window (if it is not already highlighted) by left-clicking on the tab.

- **Note:** The floor plan will be drawn inside the magenta rectangle in model space.

Step 3. Drawing Setup

a. Set units to **Architectural** and precision to **1/16"**.

b. Set the font to **Stylus BT** in the *Standard* Text Style dialog box.

c. Set the following running object snaps: Endpoint, Midpoint, Quadrant, Intersection, and Perpendicular.

QUICK TIP

If you neglect to include the foot mark (') when entering the drawing limits, the drawing limits for the upper right corner will default to inches and be set to 48",38".

Step 4. Create the following layers:

- To draw the exterior walls, set **Walls** as the current layer.

LAYER NAME	COLOR	LINETYPE	LINEWEIGHT
Clothes Rod	Magenta	Dashed	Default- Draw Closet Rods on this layer
Doors	Blue	Continuous	Default- Draw doors on this layer
Hatch	Cyan	Continuous	Default- Hatch walls on this layer
Shelves	Yellow	Continuous	Default- Draw closet shelves on this layer
Text	Magenta	Continuous	Default- Place all text on this layer
Walls	Red	Continuous	Default- Draw walls on this layer
Windows	Green	Continuous	Default- Draw windows on this layer

NOTE

ENTERING ARCHITECTURAL UNITS OF MEASUREMENT
When architectural units are assigned to the drawing, lengths and distances will default to inches unless you enter a foot mark (') symbol. For example, to draw a line 24' 6" in length, type **24'-6** (you do not need to include an inch mark after the **6** because AutoCAD defaults to inches).

To enter lengths containing fractions, type a dash between the inch value and the fractional value - for example, type **15'9-1/2**.

Step 5. Exterior walls

a. Use the **RECTANGLE** command to draw a rectangle
- Begin the rectangle at absolute coordinates **10',12'** (see Figure 4.149).
- Rectangle is **18' X 15'**

b. Use the **OFFSET** command to offset the rectangle **4"** to the inside.
- The two rectangles represent the exterior walls of the guest cottage.

4.149 Exterior Walls

Step 6. **EXPLODE** both rectangles so that they can be edited.

Step 7. Interior walls

a. Use the **OFFSET** command—refer to Figure 4.74(b)—with the dimensions shown in Figure 4.150 to place the lines for the interior walls.
- Be careful to offset the wall to the side shown in the example.

b. Use the **TRIM** command to trim the offset lines at the intersections of the walls as shown in Figure 4.151.

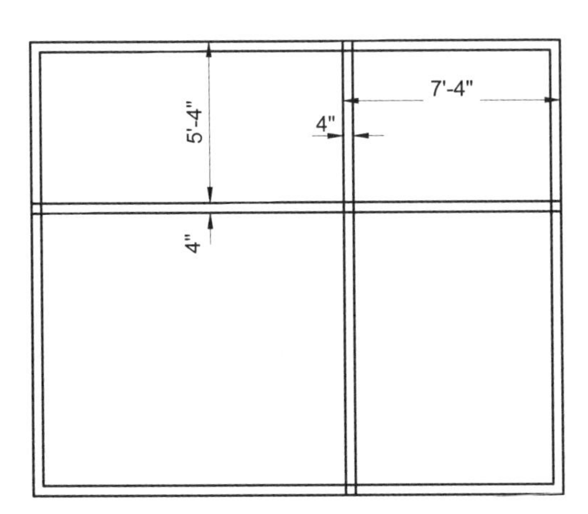

4.150 Placing Interior Walls in the Floor Plan

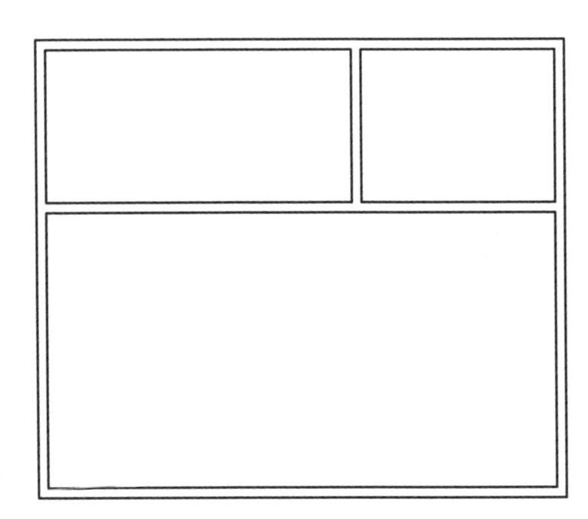

4.151 Trimmed Wall Openings Where Walls Intersect

QUICK TIP

When architectural units are in effect, values will default to inches unless you enter a foot mark ('). For example, to enter the length of a line 5 feet 4 inches long, type 5'-4 (you do not need to type an inch mark after the 4 because AutoCAD defaults to inches).

Step 8. Create the front door opening:

a. Offset the left outside wall line 2'8" to the right to locate the center of the front door opening.

b. Offset this line 1'6" to both its right and left sides to define the edges of an opening 3'-0" wide (as shown in Figure 4.152).

c. Trim the offset lines as needed to create the front door opening as shown in Figure 4.153.

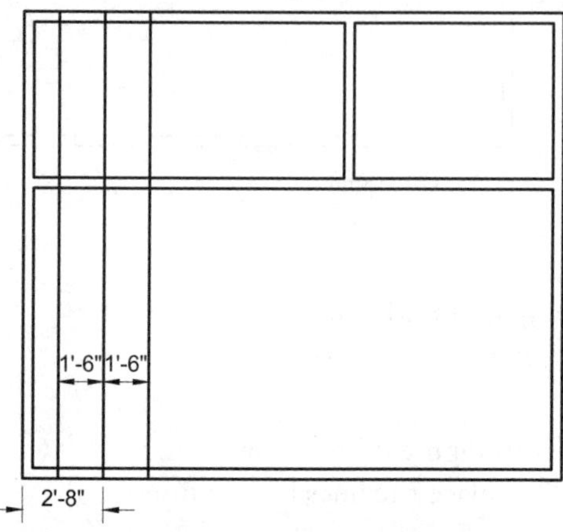

4.152 Offsetting Walls to Locate the Front Door

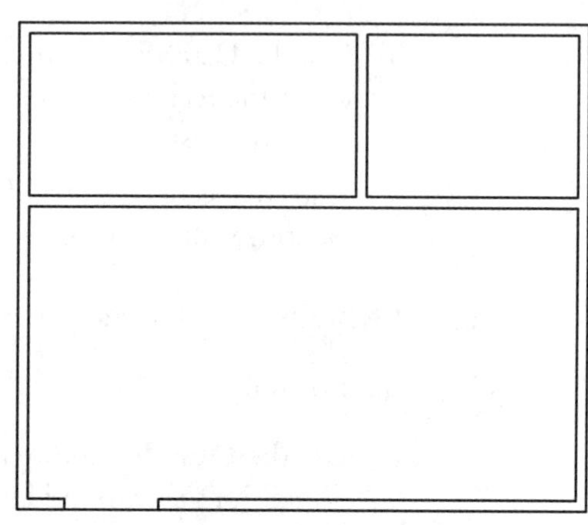
4.153 Trimming the Front Door Opening

Step 9. Create the interior door openings for the bath and closet

a. Offset the walls as shown in Figure 4.154 to define the sides of two **2'-0"** wide openings.

b. Trim the offset lines to create the door openings as shown in Figure 4.155.

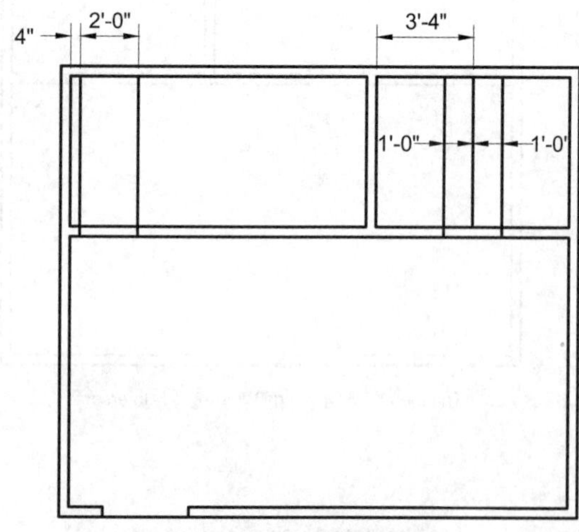

4.154 Constructing Interior Door Openings

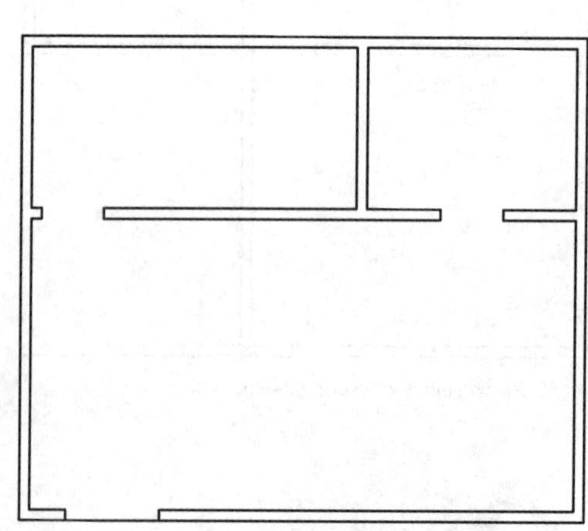
4.155 Trimming Interior Door Openings

Step 10. Windows

a. Set **Windows** as the current layer.

b. Offset the walls as shown in Figure 4.156(a) to locate the centers of the windows.

c. Use the **RECTANGLE** command—refer to Figure 4.52(b)—to draw two rectangles:

d. One **36" X 4"**, and another **24" X 4"**.

e. Draw a horizontal line through each rectangle from the midpoints of each vertical side as shown in Figures 4.156(c) and 4.156(d).

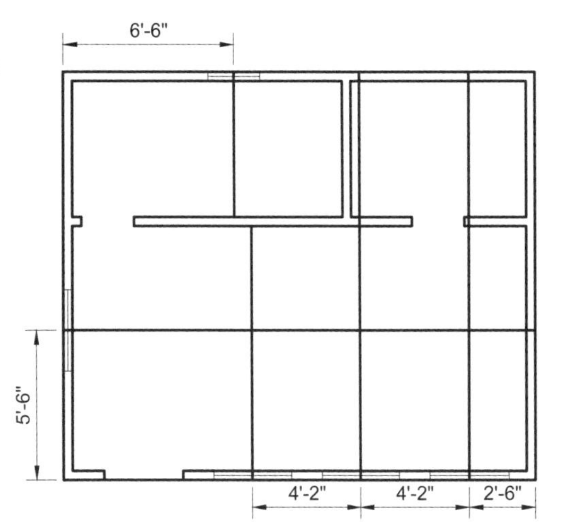

4.156(a) Locating the Centers of the Windows

f. Use **COPY** to make three copies of the larger window

g. Use the **MOVE** command—refer to Figure 4.76(b)—to place these copies in their positions along the front wall of the cottage as in Figure 4.156(a).

h. Next, move the smaller window into position along the back wall.

i. Finally, **ROTATE**—refer to Figure 4.77(b)—the remaining window 90° and move it to its place in the left side wall.

4.156(b) Drawing a 36" X 4" Rectangle

4.156(c) Horizontal Line Drawn from Midpoint to Midpoint of the Vertical Sides of the Rectangle

4.156(d) 24" X 4" Window

QUICK TIP

When interpreting door notes, the first number (usually largest) represents the width of the door in feet. The second number (usually smaller, or "superscript") represents inches. For example, 3^0 means that the door is 3'-0" wide. When a door height isn't given, it is assumed to be 6'-8", which is the standard height for a residential door.

QUICK TIP

Use **Object Snap** settings to move the window from its midpoint of the window to the intersection of the offset line and the outside edge of the wall to facilitate the exact placement of the windows.

Step 11. Doors

a. Set **Doors** as the current layer.

b. Use **RECTANGLE** to draw the three doors.

c. Begin each rectangle at the corner of the door opening where the hinges would be attached (referred to as the hinge jamb) as shown in Figure 4.157(a), and use relative coordinates to define the door.

d. The doors are drawn **1"** thick.

e. For example, in Figure 4.157(a) the rectangle for the **3'-0"** front door starts at the top of the left inside corner of the door opening and then is drawn at a distance of **1"** along the X-axis and **36"** along the Y-axis.

4.157(a) Creating and Locating the Doors

f. The other two interior doors are **1" X 24"** rectangles and are placed as in Figure 4.157(a).

g. To draw the arcs that represent the door swings, select the down arrow below the **ARC** command icon located on the **Draw** panel of the ribbon

h. Select the **Start**, **Center**, **End** option—refer to Figure 4.54(a)—to see the location and orientation of the door swings.

i. Next, select the *start point* of the arc for the front door swing by picking at the point shown in Figure 4.157(b).

j. Then define the *center point* of the arc by selecting at the point where the door hinge is attached to the door as shown in Figure 4.157(b).

k. Lastly, define the *end point* of the arc by selecting the top right corner of the door as shown in Figure 4.157(b).

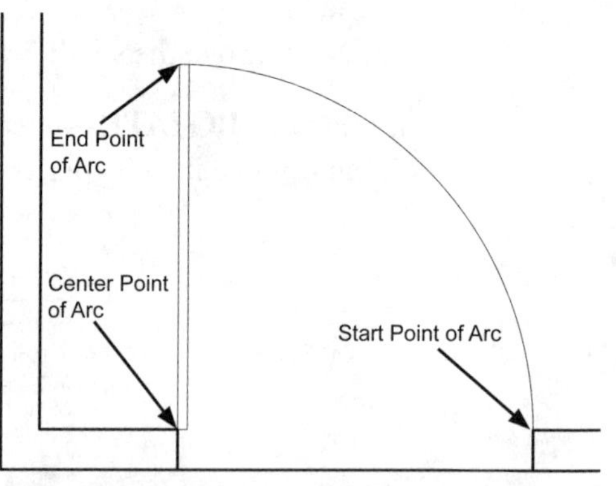

4.157(b) Selecting the Start, Center, and End Points in a Counterclockwise Order to Draw the Arc Representing the Front Door Swing.

l. The doors to the bath and closet areas will be drawn in a similar fashion but because arcs are drawn counterclockwise, the order that the start, center, and end points are selected must be considered when defining the orientation of the arc.

NOTE

The side of the door frame where the hinges are attached is referred to as the *hinge jamb*. The side of the door frame where the latching mechanism attaches is referred to as the *strike jamb*.

Step 12. Closet Rods and Shelves

a. Set the Shelves layer current

b. Offset the inside walls 12" into the closet to add shelves to the closet (see Figure 4.158).

c. Set the **Clothes Rod** layer current

d. Offset the inside walls 10" into the closet to add the clothes rods to the closet (see Figure 4.158).

4.158 Closet Shelves and Clothes Rods

QUICK TIP

The **Offset** command makes a copy of an existing object at a preset distance, so you may notice the shelf offset will still be on the Floor Plan layer. In order to automatically make an offset line on the current layer, type **L** for layer immediately after starting the Offset command (before entering the offset distance). You can then choose **Source** (which is the original layer of the object being offset) or **Current** (which is the layer set current in your drawing at the moment). Once you select Current, it will stay active throughout the rest of the drawing.

Step 13. Adding text

a. To add room labels to the floor plan, set the **Text** layer current.

b. Use the **SINGLE LINE TEXT** command to add the room names shown in Figure 4.158.

c. The text height for room names should be **6"**.

d. Use **MULTILINE TEXT** command to add the door callouts.

e. The text height for door callouts should be **4"**.

f. To add the superscript to the 3^0 door callout, use the **Multiline Text** command and type *30*; see Figure 4.88(a). Then highlight the *0* and select the **Superscript** option located in the **Formatting** panel of the **Text Editor** ribbon.

QUICK TIP

Single Line Text will allow a user to justify the text placement. Start the Single Line Text command. Instead of specifying the start point of text immediately after starting the command, type J (for justify), then type MC (for middle center). It will then prompt you for the start point (select the center of the circle you drew), text height and rotation.

Step 14. Window Markers

Window marks are used to identify each window in the floor plan (see Figure 4.159).

a. Create the window mark by drawing a circle that is **9"** in diameter

b. Use Single Line Text and center the letter **A** or **B** inside the circle.

A

4.159 Window Mark

c. Use **4"** text height for the letters.

d. Place window marks next to each window as shown in the designer's sketch in Figure 4.148.

Step 15. Hatching the Walls

a. To hatch the walls, set the **Hatch** layer current

b. Use the **HATCH** command on the home ribbon and refer to Figure 4.160 to apply the settings listed below.

- Select the **Net** hatch pattern
- Set the hatch scale to **18** and the hatch angle to **45** degrees

c. Use the **Pick Points** option to select points inside the walls.

d. After hatching, the plan should look like the one shown in Figure 4.161.

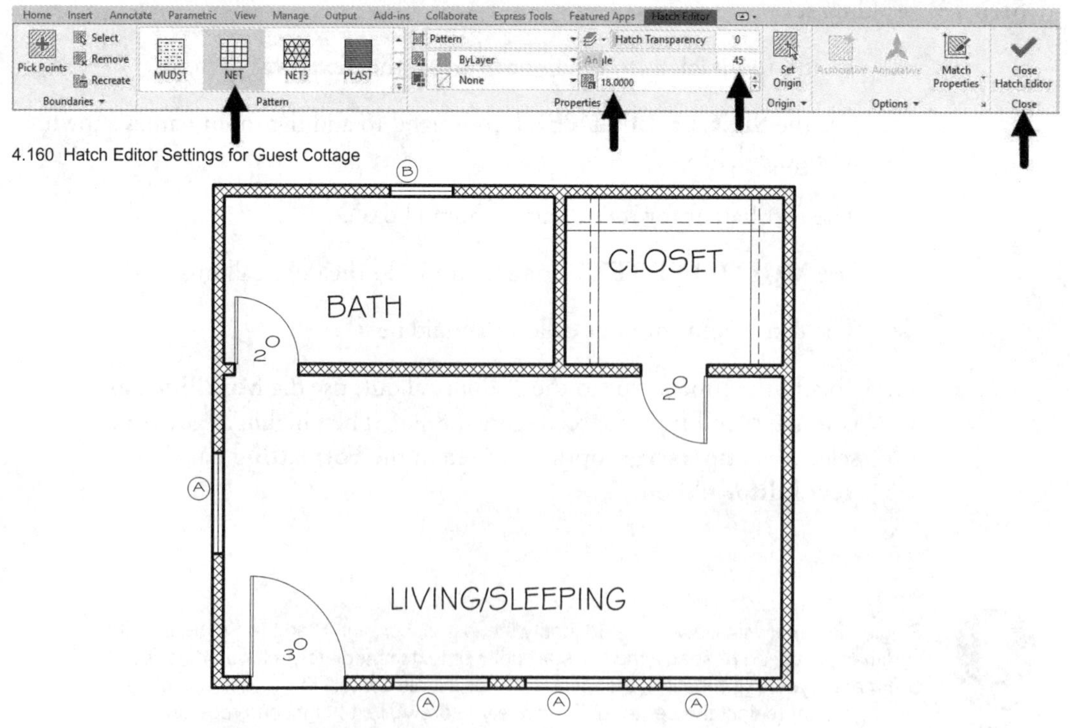

4.160 Hatch Editor Settings for Guest Cottage

4.161 Walls with Hatch Pattern Applied

Step 16. Window Schedule

a. When finished drawing the Guest Cottage floor plan, left-click on the **Layout1** tab in the lower-left corner of the graphics window.

b. Complete the window schedule in **PAPER SPACE** by using **Text Edit** to change the placeholders (the X's) in the schedule to the values shown in Figure 4.162.

WINDOW SCHEDULE			
MARK	SIZE	TYPE	QTY
A	3^0 X 5^0	S.H.	4
B	2^0 X 2^0	S.H.	1

4.162 Window Schedule

c. In the Window Schedule, the **3^0 X 5^0** label refers to a window that is **3'-0"** wide by **5'-0"** tall.

d. The **2^0 X 2^0** label refers to a window that is **2'-0"** wide by **2'-0"** tall.

e. The S.H. entry in the TYPE column indicates that the window type is *single hung* which means in this case that the bottom half of the window can slide up and the top half is fixed.

QUICK TIP

When editing the text, it is common to accidentally double-click empty space next to the text, instead of on the text itself. If this happens, you may inadvertently end up in MODEL SPACE. You can tell your drawing is in Model Space when you see the word MODEL in your Drafting Settings bar at the bottom of your screen. Simply click on the button that says MODEL to toggle to Paper Space (this button will now say PAPER if done correctly).

Step 17. In paper space, edit the text in the title block by double-clicking on it to enable the text edit option.

a. Follow the steps presented in "4.17 Plotting With AutoCAD" to create a **Page Setup** for the cottage.

b. Plot the project by selecting the **Cottage** page setup from the **Plot** dialog box. Be sure to save the drawing file before closing AutoCAD.

PROJECT

BRACKET PROJECT 4.4

DIRECTIONS:

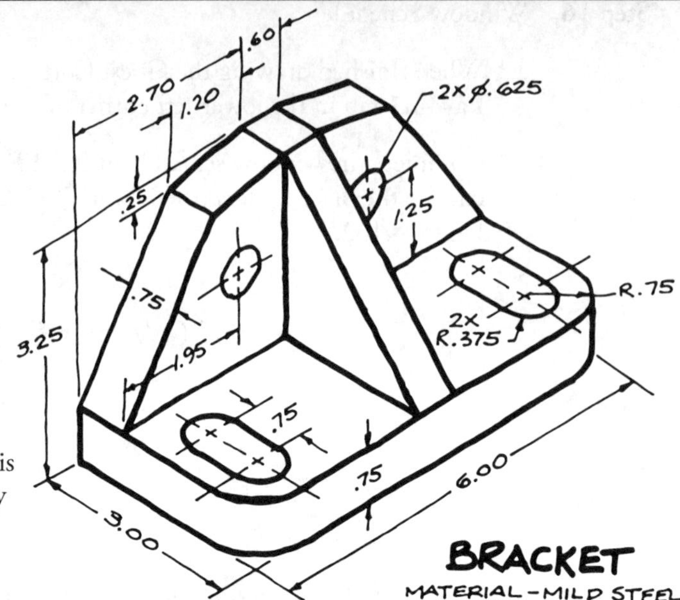

4.163 Designer's Sketch of the Bracket

1. Download the **Imperial Prototype** drawing file located in the *Prototype Drawing Files* folder associated with this book.
2. Use **SAVE AS** to save the drawing to your **Home** directory and rename the drawing **Bracket**.
3. Activate the **MODEL** tab in the lower-left corner of the graphics window (if it is not already highlighted) by left-clicking on the tab.
4. Set **Units** to *decimal* and *precision* to 0.000.
5. Set **Text Style** *font* to **Arial**.
6. Create and assign layers as follows and set the **Visible** layer current:

LAYER NAME	COLOR	LINETYPE	LINEWEIGHT
Visible	Red	Continuous	.60 mm - draw visible lines on this layer
Hidden	Blue	Hidden	Default - draw hidden lines on this layer
Center	Green	Center	Default - draw center lines on this layer
Text	Magenta	Continuous	Default - place text on this layer

7. Draw the front, top, and right side views of the Bracket shown in the designer's sketch in Figure 4.163.
   - **Note:** The views of the Bracket will be drawn inside the magenta rectangle located in model space. Turn on the Miter Box layer.
   - Adjust the Linetype Scale (type **LTS** and **<Enter>**) as needed to properly display breaks and dashes in non-continuous lines.
8. When finished drawing the object, left-click on the **Layout1** tab in the lower-left corner of the graphics window and edit the text in the title block by double-clicking on it to enable the text editing option.
9. Follow your instructor's directions to print the drawing.
10. Be sure to save the drawing file before closing AutoCAD.

VIDEO

There is a video tutorial available for this project inside the Chapter 4 folder of the book's Video Training downloads.

QUICK TIP

When editing the text, it is common to accidentally double-click empty space next to the text, instead of on the text itself. If this happens, you may inadvertently end up in MODEL SPACE. You can tell your drawing is in Model Space when you see the word MODEL in your Status Bar at the bottom of your screen. Simply click on the button that says MODEL to toggle to Paper Space (this button will now say PAPER if done correctly).

DIRECTIONS:

1. Download the **Metric Prototype** drawing file located in the *Prototype Drawing Files* folder associated with this book.
2. Use **SAVE AS** to save the drawing to your **Home** directory and rename the drawing **Shaft Guide**.
3. Activate the **MODEL** tab in the lower-left corner of the graphics window (if it is not already highlighted) by left-clicking on the tab.
4. Set **Units** to *decimal* and *precision* to 0.000.
5. Set **Text Style** *font* to **Arial**.
6. Create and assign layers as follows and set the **Visible** layer current:

4.164 Designer's Sketch of the Shaft Guide

LAYER NAME	COLOR	LINETYPE	LINEWEIGHT
Visible	Red	Continuous	.60 mm- draw visible lines on this layer
Hidden	Blue	Hidden	Default- draw hidden lines on this layer
Center	Green	Center	Default- draw center lines on this layer
Text	Magenta	Continuous	Default- place text on this layer

7. Draw the front, top, and right side views of the Shaft Guide shown in the designer's sketch in Figure 4.164.
   - **Note:** The views of the Shaft Guide will be drawn inside the magenta rectangle located in model space. Turn on the Miter Box layer.
   - Set the **LTS** (Linetype Scale) to **.375**. (**Note:** Experiment with other LTS values if non-continuous lines do not display correct sizes for dashes and breaks in lines.)
8. When finished drawing the object, left-click on the **Layout1** tab in the lower-left corner of the graphics window and edit the text in the title block by double-clicking on it to enable the text editing option.
9. Follow your instructor's directions to print the drawing.
10. Be sure to save the drawing file before closing AutoCAD.

VIDEO There is a video tutorial available for this project inside the Chapter 4 folder of the book's Video Training downloads.

QUICK TIP When editing the text, it is common to accidentally double-click empty space next to the text, instead of on the text itself. If this happens, you may inadvertently end up in MODEL SPACE. You can tell your drawing is in Model Space when you see the word MODEL in your Status Bar at the bottom of your screen. Simply click on the button that says MODEL to toggle to Paper Space (this button will now say PAPER if done correctly).

TOOL HOLDER
PROJECT 4.6

DIRECTIONS:

1. Download the **Imperial Prototype** drawing file located in the *Prototype Drawing Files* folder associated with this book.
2. Use **SAVE AS** to save the drawing to your **Home** directory and rename the drawing **Tool Holder**.
3. Activate the **MODEL** tab in the lower-left corner of the graphics window (if it is not already highlighted) by left-clicking on the tab.

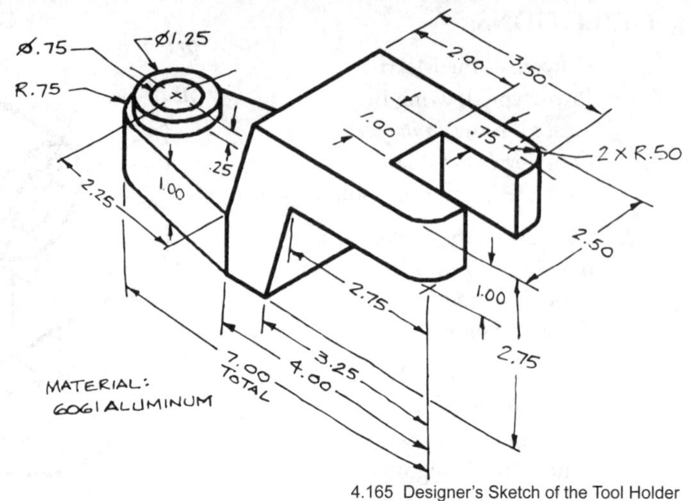

4.165 Designer's Sketch of the Tool Holder

4. Set **Units** to *decimal* and *precision* to 0.000.
5. Set **Text Style** *font* to **Arial**.
6. Create and assign layers as follows and set the **Visible** layer current:

LAYER NAME	COLOR	LINETYPE	LINEWEIGHT
Visible	Red	Continuous	.60 mm- draw visible lines on this layer
Hidden	Blue	Hidden	Default- draw hidden lines on this layer
Center	Green	Center	Default- draw center lines on this layer
Text	Magenta	Continuous	Default- place text on this layer

7. Draw the front, top, and right side views of the Tool Holder shown in the designer's sketch in Figure 4.165.
    - **Note:** The views of the Tool Holder will be drawn inside the magenta rectangle located in model space. Turn on the Miter Box layer.
    - Adjust the Linetype Scale (type **LTS** and **<Enter>**) as needed to properly display breaks and dashes in non-continuous lines.
8. When finished drawing the object, left-click on the **Layout1** tab in the lower-left corner of the graphics window and edit the text in the title block by double-clicking on it to enable the text editing option.
9. Follow your instructor's directions to print the drawing.
10. Be sure to save the drawing file before closing AutoCAD.

VIDEO

There is a video tutorial available for this project inside the Chapter 4 folder of the book's Video Training downloads.

QUICK TIP

When editing the text, it is common to accidentally double-click empty space next to the text, instead of on the text itself. If this happens, you may inadvertently end up in MODEL SPACE. You can tell your drawing is in Model Space when you see the word MODEL in your Status Bar at the bottom of your screen. Simply click on the button that says MODEL to toggle to Paper Space (this button will now say PAPER if done correctly).

DIRECTIONS:

1. Download the **Imperial Prototype** drawing file located in the *Prototype Drawing Files* folder associated with this book.
2. Use **SAVE AS** to save the drawing to your **Home** directory and rename the drawing **Tool Slide**.
3. Activate the **MODEL** tab in the lower-left corner of the graphics window (if it is not already highlighted) by left-clicking on the tab.

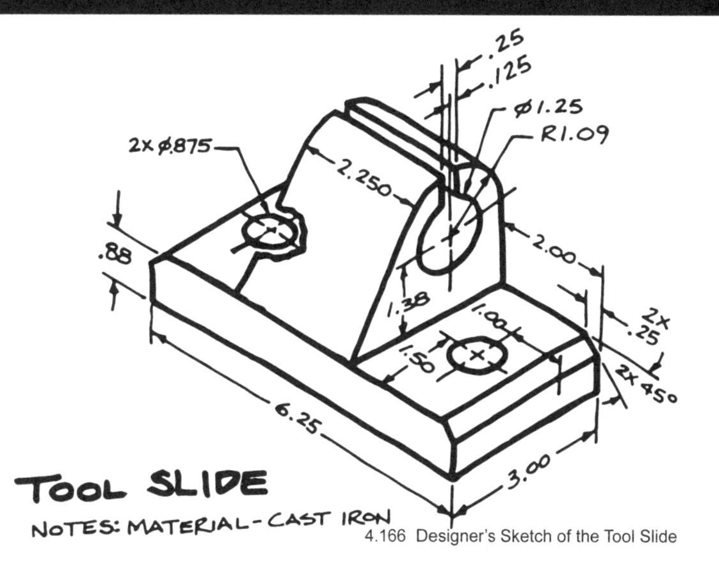

4.166 Designer's Sketch of the Tool Slide

4. Set **Units** to *decimal* and *precision* to 0.000.
5. Set **Text Style** *font* to **Arial**.
6. Create and assign layers as follows and set the **Visible** layer current:

LAYER NAME	COLOR	LINETYPE	LINEWEIGHT
Visible	Red	Continuous	.60 mm- draw visible lines on this layer
Hidden	Blue	Hidden	Default- draw hidden lines on this layer
Center	Green	Center	Default- draw center lines on this layer
Text	Magenta	Continuous	Default- place text on this layer

7. Draw the front, top, and right side views of the Tool Slide shown in the designer's sketch in Figure 4.166.
   - **Note:** The views of the Tool Slide will be drawn inside the magenta rectangle located in model space. Turn on the Miter Box layer.
   - Adjust the Linetype Scale (type **LTS** and **<Enter>**) as needed to properly display breaks and dashes in non-continuous lines.
8. When finished drawing the object, left-click on the **Layout1** tab in the lower-left corner of the graphics window and edit the text in the title block by double-clicking on it to enable the text editing option.
9. Follow your instructor's directions to print the drawing.
10. Be sure to save the drawing file before closing AutoCAD.

VIDEO

There is a video tutorial available for this project inside the Chapter 4 folder of the book's Video Training downloads.

QUICK TIP

When editing the text, it is common to accidentally double-click empty space next to the text, instead of on the text itself. If this happens, you may inadvertently end up in MODEL SPACE. You can tell your drawing is in Model Space when you see the word MODEL in your Status Bar at the bottom of your screen. Simply click on the button that says MODEL to toggle to Paper Space (this button will now say PAPER if done correctly).

OFFSET FLANGE (OPTIONAL)
PROJECT 4.8

DIRECTIONS:

1. Download the **Metric Prototype** drawing file located in the *Prototype Drawing Files* folder associated with this book.
2. Use **SAVE AS** to save the drawing to your **Home** directory and rename the drawing **Offset Flange**.
3. Activate the **MODEL** tab in the lower-left corner of the graphics window
(if it is not already highlighted) by left-clicking on the tab.
4. Set **Units** to *decimal* and *precision* to 0.000.
5. Set **Text Style** *font* to **Arial**.
6. Create and assign layers as follows and set the **Visible** layer current:

4.167 Designer's Sketch of the Offset Flange

LAYER NAME	COLOR	LINETYPE	LINEWEIGHT
Visible	Red	Continuous	.60 mm- draw visible lines on this layer
Hidden	Blue	Hidden	Default- draw hidden lines on this layer
Center	Green	Center	Default- draw center lines on this layer
Text	Magenta	Continuous	Default- place text on this layer

7. Draw the front and right side views of the Offset Flange shown in the designer's sketch in Figure 4.167.
    - **Note:** The views of the Offset Flange will be drawn inside the magenta rectangle located in model space. Turn on the Miter Box layer.
    - Set the **LTS** (Linetype Scale) to **.375**. (**Note:** Experiment with other LTS values if non-continuous lines do not display correct sizes for dashes and breaks in lines.)
8. When finished drawing the object, left-click on the **Layout1** tab in the lower-left corner of the graphics window and edit the text in the title block by double-clicking on it to enable the text editing option.
9. Follow your instructor's directions to print the drawing.
10. Be sure to save the drawing file before closing AutoCAD.

QUICK TIP

When editing the text, it is common to accidentally double-click empty space next to the text, instead of on the text itself. If this happens, you may inadvertently end up in MODEL SPACE. You can tell your drawing is in Model Space when you see the word MODEL in your Status Bar at the bottom of your screen. Simply click on the button that says MODEL to toggle to Paper Space (this button will now say PAPER if done correctly).

DIRECTIONS:

1. Download the **Metric Prototype** drawing file located in the *Prototype Drawing Files* folder associated with this book.
2. Use **SAVE AS** to save the drawing to your **Home** directory and rename the drawing **Angle Stop**.
3. Activate the **MODEL** tab in the lower-left corner of the graphics window (if it is not already highlighted) by left-clicking on the tab.
4. Set **Units** to *decimal* and *precision* to 0.000.
5. Set **Text Style** *font* to **Arial**.
6. Create and assign layers as follows and set the **Visible** layer current:

4.168 Designer's Sketch of the Angle Stop

LAYER NAME	COLOR	LINETYPE	LINEWEIGHT
Visible	Red	Continuous	.60 mm- draw visible lines on this layer
Hidden	Blue	Hidden	Default- draw hidden lines on this layer
Center	Green	Center	Default- draw center lines on this layer
Text	Magenta	Continuous	Default- place text on this layer

7. Draw the front, top, and right side views of the Angle Stop shown in the designer's sketch in Figure 4.168.
   - **Note:** The views of the Angle Stop will be drawn inside the magenta rectangle located in model space. Turn on the Miter Box layer.
   - Set the **LTS** (Linetype Scale) to **.375**. (**Note:** Experiment with other LTS values if non-continuous lines do not display correct sizes for dashes and breaks in lines.)
8. When finished drawing the object, left-click on the **Layout1** tab in the lower-left corner of the graphics window and edit the text in the title block by double-clicking on it to enable the text editing option.
9. Follow your instructor's directions to print the drawing.
10. Be sure to save the drawing file before closing AutoCAD.

QUICK TIP

When editing the text, it is common to accidentally double-click empty space next to the text, instead of on the text itself. If this happens, you may inadvertently end up in MODEL SPACE. You can tell your drawing is in Model Space when you see the word MODEL in your Status Bar at the bottom of your screen. Simply click on the button that says MODEL to toggle to Paper Space (this button will now say PAPER if done correctly).

SWIVEL STOP (OPTIONAL) PROJECT 4.10

DIRECTIONS:

1. Download the **Imperial Prototype** drawing file located in the *Prototype Drawing Files* folder associated with this book.
2. Use **SAVE AS** to save the drawing to your **Home** directory and rename the drawing **Swivel Stop.**
3. Activate the **MODEL** tab in the lower-left corner of the graphics window (if it is not already highlighted) by left-clicking on the tab.

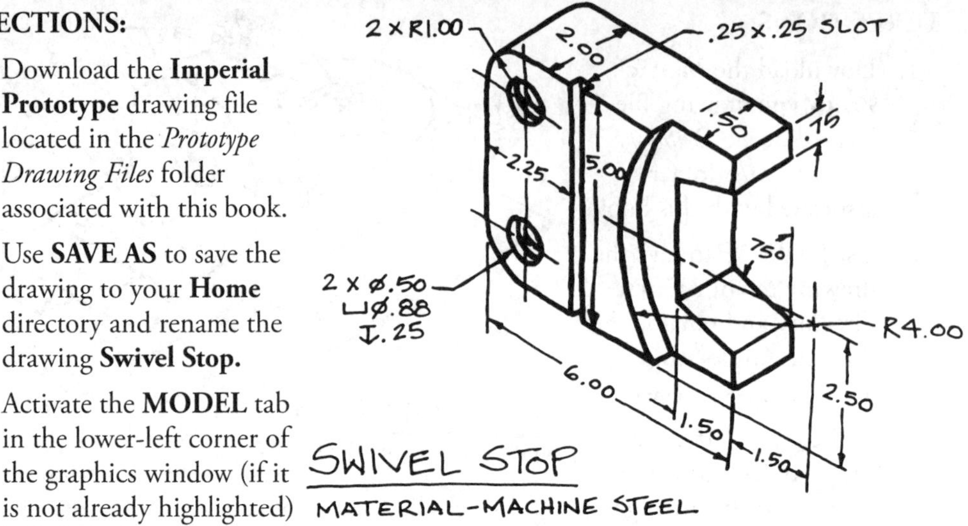

4.169 Designer's Sketch of the Swivel Stop

4. Set **Units** to *decimal* and *precision* to 0.000.
5. Set **Text Style** *font* to **Arial**.
6. Create and assign layers as follows and set the **Visible** layer current:

LAYER NAME	COLOR	LINETYPE	LINEWEIGHT
Visible	Red	Continuous	.60 mm- draw visible lines on this layer
Hidden	Blue	Hidden	Default- draw hidden lines on this layer
Center	Green	Center	Default- draw center lines on this layer
Text	Magenta	Continuous	Default- place text on this layer

7. Draw the front, top, and right side views of the Swivel Stop in Figure 4.169.
   - **Note:** The views of the Swivel Stop will be drawn inside the magenta rectangle located in model space. Turn on the Miter Box layer.
   - Adjust the Linetype Scale (type **LTS** and **<Enter>**) as needed to properly display breaks and dashes in non-continuous lines.
8. When finished drawing the object, left-click on the **Layout1** tab in the lower-left corner of the graphics window and edit the text in the title block by double-clicking on it to enable the text editing option.
9. Follow your instructor's directions to print the drawing.
10. Be sure to save the drawing file before closing AutoCAD.

QUICK TIP

When editing the text, it is common to accidentally double-click empty space next to the text, instead of on the text itself. If this happens, you may inadvertently end up in MODEL SPACE. You can tell your drawing is in Model Space when you see the word MODEL in your Status Bar at the bottom of your screen. Simply click on the button that says MODEL to toggle to Paper Space (this button will now say PAPER if done correctly).

DIRECTIONS:

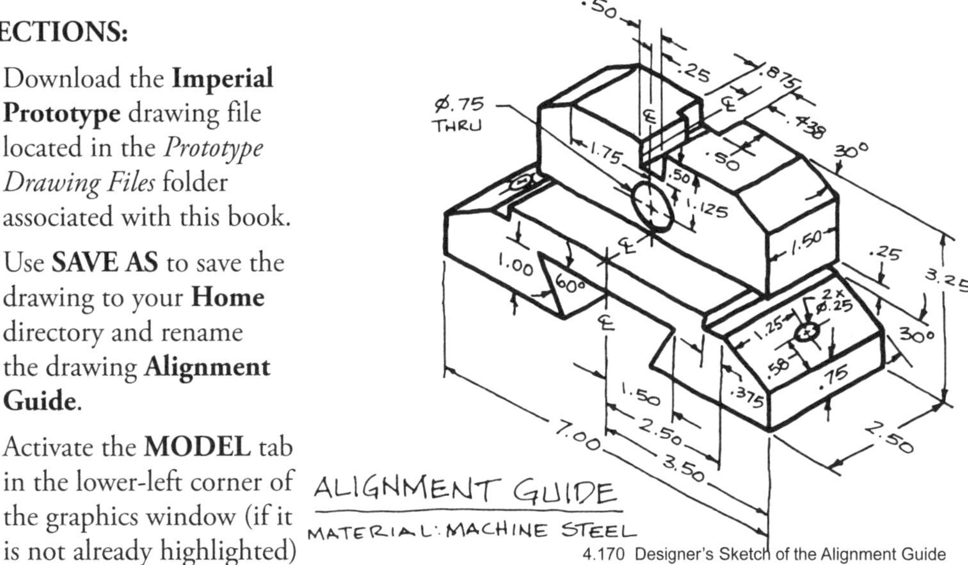

4.170 Designer's Sketch of the Alignment Guide

1. Download the **Imperial Prototype** drawing file located in the *Prototype Drawing Files* folder associated with this book.
2. Use **SAVE AS** to save the drawing to your **Home** directory and rename the drawing **Alignment Guide**.
3. Activate the **MODEL** tab in the lower-left corner of the graphics window (if it is not already highlighted) by left-clicking on the tab.
4. Set **Units** to *decimal* and *precision* to 0.000.
5. Set **Text Style** *font* to **Arial**.
6. Create and assign layers as follows and set the **Visible** layer current:

LAYER NAME	COLOR	LINETYPE	LINEWEIGHT
Visible	Red	Continuous	.60 mm- draw visible lines on this layer
Hidden	Blue	Hidden	Default- draw hidden lines on this layer
Center	Green	Center	Default- draw center lines on this layer
Text	Magenta	Continuous	Default- place text on this layer

7. Draw the front, top, and right side views of the Alignment Guide shown in the designer's sketch in Figure 4.170.
    - **Note:** The views of the Alignment Guide will be drawn inside the magenta rectangle located in model space. Turn on the Miter Box layer.
    - Adjust the Linetype Scale (type **LTS** and **<Enter>**) as needed to properly display breaks and dashes in non-continuous lines.
8. When finished drawing the object, left-click on the **Layout1** tab in the lower-left corner of the graphics window and edit the text in the title block by double-clicking on it to enable the text editing option.
9. Follow your instructor's directions to print the drawing.
10. Be sure to save the drawing file before closing AutoCAD.

QUICK TIP

When editing the text, it is common to accidentally double-click empty space next to the text, instead of on the text itself. If this happens, you may inadvertently end up in MODEL SPACE. You can tell your drawing is in Model Space when you see the word MODEL in your Status Bar at the bottom of your screen. Simply click on the button that says MODEL to toggle to Paper Space (this button will now say PAPER if done correctly).

FLANGE #1105 (OPTIONAL) PROJECT 4.12

DIRECTIONS:

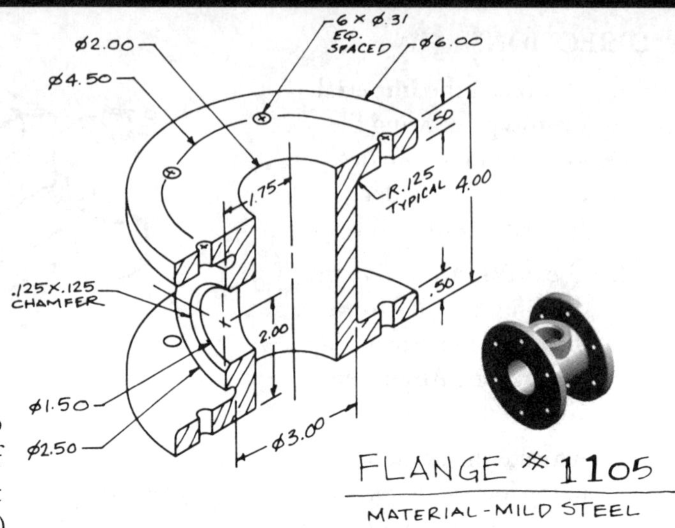

4.171 Designer's Sketch of the Flange #1105

1. Download the **Imperial Prototype** drawing file located in the *Prototype Drawing Files* folder associated with this book.
2. Use **SAVE AS** to save the drawing to your **Home** directory and rename the drawing **Flange 1105**.
3. Activate the **MODEL** tab in the lower-left corner of the graphics window (if it is not already highlighted) by left-clicking on the tab.
4. Set **Units** to *decimal* and *precision* to 0.000.
5. Set **Text Style** *font* to **Arial**.
6. Create and assign layers as follows and set the **Visible** layer current:

LAYER NAME	COLOR	LINETYPE	LINEWEIGHT
Visible	Red	Continuous	.60 mm- draw visible lines on this layer
Hidden	Blue	Hidden	Default- draw hidden lines on this layer
Center	Green	Center	Default- draw center lines on this layer
Text	Magenta	Continuous	Default- place text on this layer

7. Draw the front, top, and right side views of the Flange 1105 shown in the designer's sketch in Figure 4.171.
   - **Note:** The views of the Flange 1105 will be drawn inside the magenta rectangle located in model space. Turn on the Miter Box layer.
   - Adjust the Linetype Scale (type **LTS** and **<Enter>**) as needed to properly display breaks and dashes in non-continuous lines.
8. When finished drawing the object, left-click on the **Layout1** tab in the lower-left corner of the graphics window and edit the text in the title block by double-clicking on it to enable the text editing option.
9. Follow your instructor's directions to print the drawing.
10. Be sure to save the drawing file before closing AutoCAD.

QUICK TIP

When editing the text, it is common to accidentally double-click empty space next to the text, instead of on the text itself. If this happens, you may inadvertently end up in MODEL SPACE. You can tell your drawing is in Model Space when you see the word MODEL in your Status Bar at the bottom of your screen. Simply click on the button that says MODEL to toggle to Paper Space (this button will now say PAPER if done correctly).

CHAPTER

5

DIMENSIONING MECHANICAL DRAWINGS

CHAPTER OBJECTIVES:

After studying the material in this chapter, you should be able to:

1. Define what dimensions are and explain the difference between size and location dimensions.
2. Describe the terminology associated with dimensioning.
3. Apply ASME Y14.5 standards when dimensioning machine parts.
4. List the dos and don'ts of dimensioning mechanical drawings.
5. Describe how dimensions are determined by designers.
6. Describe the importance of tolerances to the dimensioning process.
7. Calculate a fit between two simple parts.
8. Describe how ASME and ISO standards affect the creation of mechanical drawings.
9. Describe the role of drafters in the dimensioning process.
10. Use the commands on AutoCAD's **Dimension** toolbar.
11. Create and modify mechanical dimension styles with AutoCAD's **Dimension Style Manager.**
12. Add dimensions to mechanical engineering drawings.

CHAPTER OVERVIEW

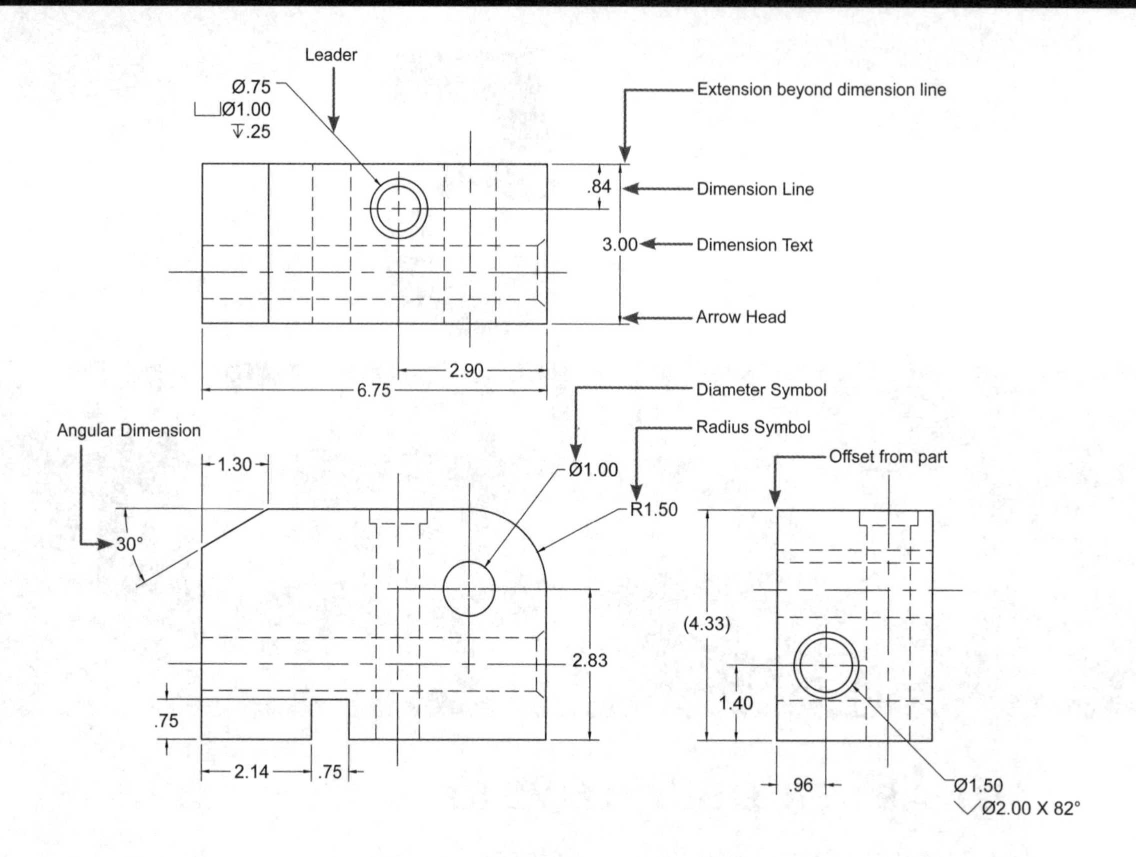

DIMENSIONING MECHANICAL DRAWINGS:

After the views necessary to describe the geometrical features of a machine part have been drawn, dimensions and notes—which fully describe the part—are added to the views. Dimensions communicate vital manufacturing information about the features of an object, for example, the diameter and location of a hole through a machine part. Designers and engineers carefully calculate the values of the dimensions included on their design inputs, and these values must be accurately reproduced, or the resulting part may not fit or function as intended. It is the drafter's responsibility to ensure that the dimensions and notes included on the designer's sketch (or other form of design input) are accurately represented on the finished mechanical drawing.

In Chapter 4 you created multiview drawings for machine parts by using the dimensional information presented on designer's sketches. The dimensions included on each sketch provided all the information that was needed to create the views of the part. In theory, if the designer's sketch contained 15 dimensions, all the drafter would need to dimension the part would be to add the 15 dimensions to the views of the object. However, in practice, it is a little more complicated than this, and here are several reasons why:

- The drafter must ensure that the value of each of the 15 dimensions is represented on the drawing with the exact precision (number of decimal places) shown on the designer's sketch. Each dimension must also reference the same **datum** shown on the designer's sketch. A datum is a theoretically perfect element (an edge, plane, axis, point, or other geometric feature) from which a dimension is placed. The features of finished parts are inspected by measuring the distance between the datums and the features described on the drawing. The use of datums will be presented in more detail later in this chapter.
- Before placing a dimension, the drafter must choose the view where the feature will most clearly be described by the dimension. For inexperienced drafters, this is sometimes a difficult decision to make.
- There are industry-wide guidelines for spacing and formatting of dimensions. Drafters must be familiar with these guidelines and know how to control the AutoCAD dimension settings necessary to comply with the guidelines.

5.1 DIMENSIONING FUNDAMENTALS

Essentially, there are two types of dimensions: *size* and *location*. In Figure 5.1 the overall width and height of the part are size dimensions. The diameter of the hole through the part is also a size dimension, but the location of the center point of the hole is defined with location dimensions. Generally, the features of an object are located on the object with location dimensions and described with size dimensions.

Dimensional information may also include notes that provide information necessary to manufacture a machine part. For example, notes that specify the type of material from which the part is manufactured, or special processes to be performed on the part during manufacture (e.g., heat treating or polishing), are included in the field of the drawing. If the notes included on the designer's input are omitted from the technical drawing, the part may not be manufactured as the designer intended.

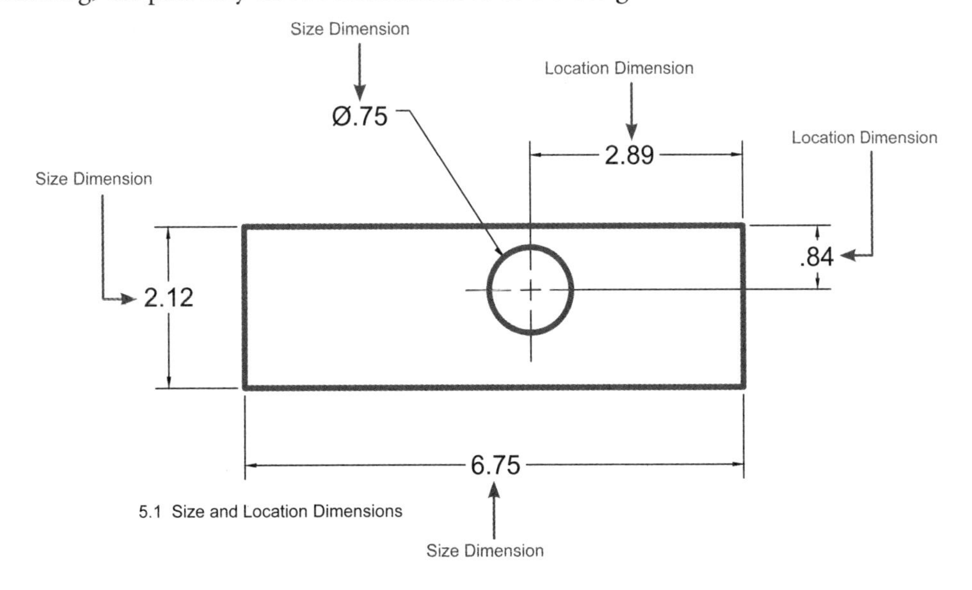

5.1 Size and Location Dimensions

Dimensioning Terminology

Figure 5.2 shows the terminology used to refer to the elements of dimensioning. Locate the *dimension* and *extension* lines noted in Figure 5.2. Extension lines extend out from the features of the object that are being dimensioned. Dimension lines are drawn between the extension lines and typically terminate in arrowheads that point to the extension lines. Dimension lines contain text that denotes the distance, or angle, between the extension lines. Now locate the *leader*, shown in Figure 5.2. Leaders point to the features of the part and are used in notations that describe the feature, like the diameter of a hole or the radius of an arc. Later in this chapter you will be required to relate the terminology shown in Figure 5.2 to the corresponding settings in AutoCAD's **Dimension Style Manager** dialog box to create an AutoCAD dimension style.

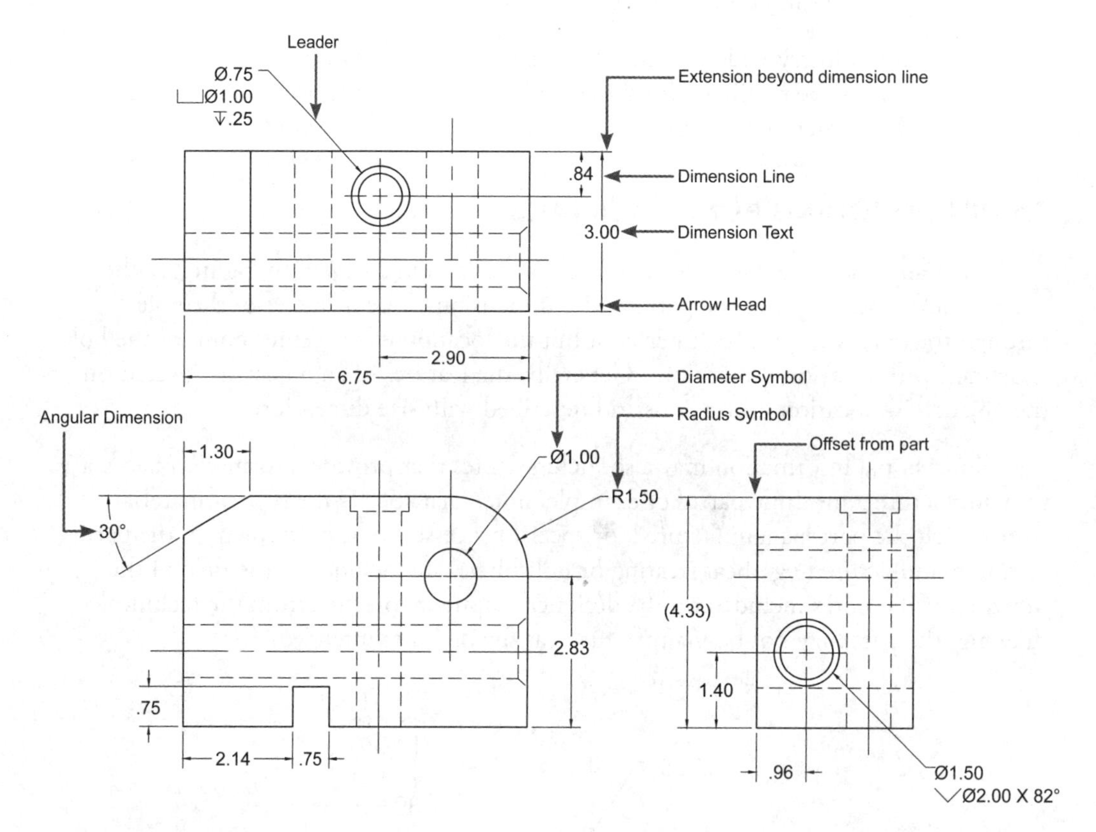

5.2 Dimensioning Terminology

Guidelines for Dimension Placement

Dimensions placed by the drafter on the multiviews of an object should be represented exactly as they are shown on the designer's sketch. For example, if a dimension on the sketch is shown at three decimal places of precision, for instance .625", it should be depicted with three decimal places on the drawing and not rounded to .63".

Dimensions should also be referenced from the same datum geometry indicated on the designer's sketch. This is because the designer may have designed a mating part that is intended to fit exactly in the space between the datum and the feature defined by the dimension. If the drafter references the dimension from a different datum than one used by the designer, the mating part may not fit.

Drafters are responsible for choosing the best placement of the dimensions on the views of the object. Drafters should use the following guidelines when determining the best location and placement of dimensions on a drawing:

- Place dimensions on the profile view of the feature. For example, if you were trying to describe the length and angle of someone's nose, it would be best to show this information in a profile, or side, view rather than head-on. Conversely, the width of the nose and the distance between the eyes would best be described with a head-on view.
- Avoid dimensioning to hidden lines or centerlines of hidden holes.
- Whenever possible, group dimensions and place them between the views of the object.
- Avoid drawing extension lines and leaders through dimension lines.
- Avoid placing dimensions on the object unless it is absolutely necessary.

Figure 5.3 provides an example of a multiview drawing in which the preceding dimensioning guidelines were ignored. Can you find the mistakes? Compare this drawing with Figure 5.4. In Figure 5.4 the dimensions were placed following the guidelines. As a result, the dimensions in this figure are much better organized and easier to interpret than in Figure 5.3.

5.2 DIMENSION STANDARDS FOR MECHANICAL DRAWINGS

Dimension standards define the rules and guidelines for the preparation of technical drawings. In the United States, the industry-wide standard for dimensioning machine parts is *ASME Y14.5 Dimensioning and Tolerancing*, published by the **American Society of Mechanical Engineers (ASME)**. According to ASME's website, "This standard establishes uniform practices for stating and interpreting dimensioning, tolerancing, and related requirements for use on engineering drawings."

ASME also publishes other standards concerning engineering drawing practices. These standards are available for purchase from the ASME. In the United States, standards of this type are created under the aegis of the ***American National Standards***

Institute (ANSI). The ***International Organization for Standardization (ISO)*** also publishes drawing standards for the preparation of technical drawings including dimensioning. ANSI represents the United States as its delegate to ISO. The ISO dimensioning standard is very similar to the ASME dimensioning standard.

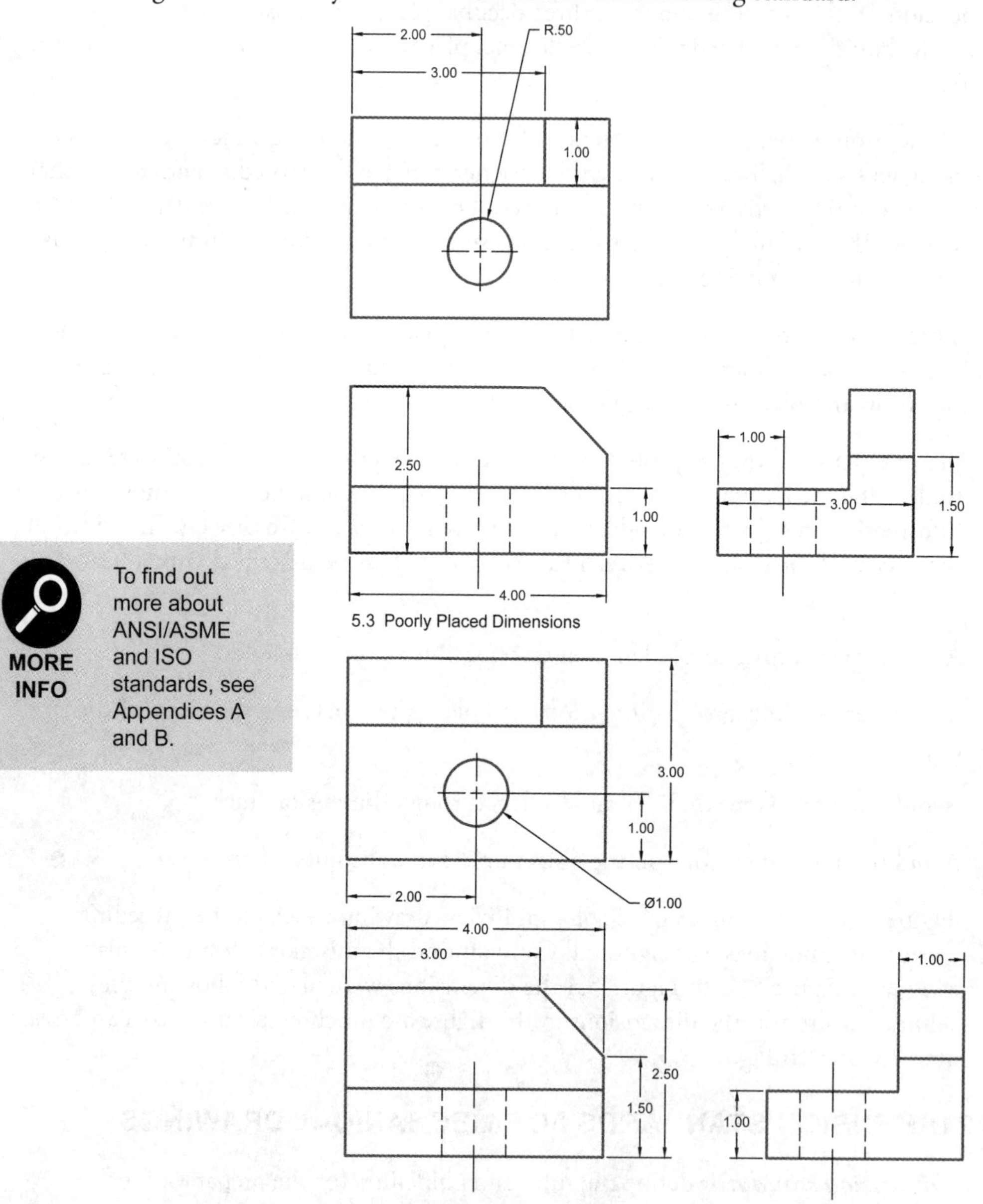

5.3 Poorly Placed Dimensions

5.4 Dimensions Placed Following the Recommended Guidelines

MORE INFO

To find out more about ANSI/ASME and ISO standards, see Appendices A and B.

Recommended Size and Spacing of Dimension Features

Figure 5.5 shows recommended size and spacing of common dimension features. The *ASME Y14.5* standard specifies that the space between the part outline and the first dimension should not be less than .40" (10mm), and the spacing between succeeding parallel dimension lines should not be less than .25" (6mm). Dimension text height should not measure less than .12" (3mm) when the drawing is printed at its intended scale and length of arrowheads is usually set to match text height.

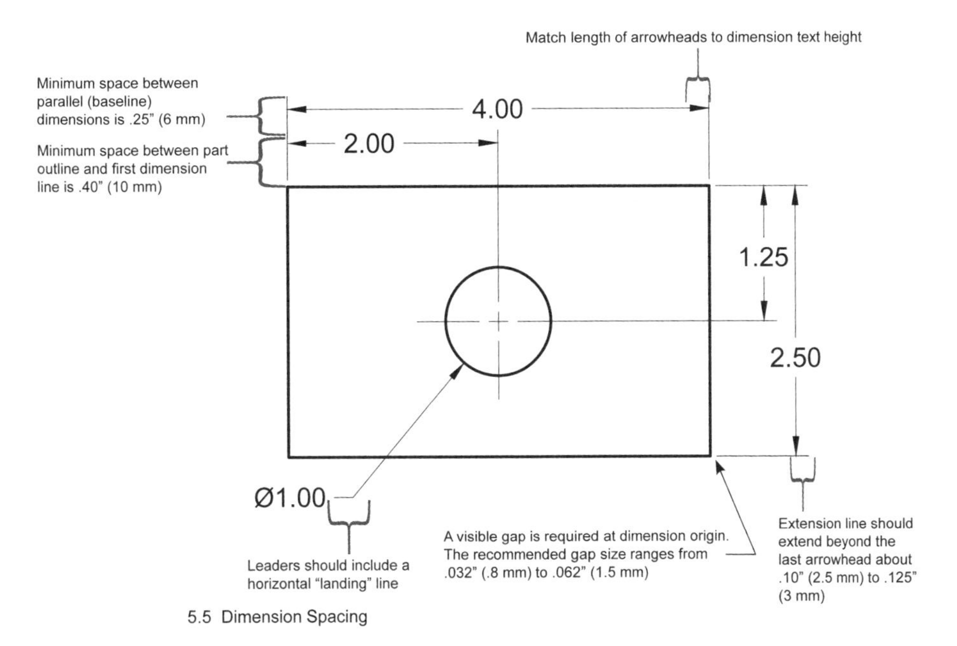

5.5 Dimension Spacing

There should be a visible gap between the outline of the part and the beginning of an extension line. Usually, this gap is set to between .032" and .062" (1 to 1.5mm). Extension lines should extend about .10" to .125" (2 to 3mm) beyond the last arrowhead.

JOB SKILL

The ASME Y14.5 standard states that its recommendations for dimension spacing are "intended as guides only. If the drawing meets the reproduction requirements of the accepted industry or military reproduction specification, nonconformance to these spacing requirements is not a basis for rejection of the drawing."

QUICK TIP

In AutoCAD drawings, the recommended size and spacing for dimensions are assigned in the settings of the Dimension Style Manager dialog box. This dialog box and the steps involved in entering these settings are presented in detail later in this chapter.

Text Height and Style

When placing text on a technical drawing, legibility is the primary concern. The ASME standard governing text height and style on engineering drawings is *ASME Y14.2-2008 Line Conventions and Lettering*. This standard states that text used for titles and for denoting special characters, such as section view labels, should be no less than .24" (6mm). All other characters should have a minimum text height of .12" (3mm). Uppercase letters should be used unless lowercase letters are required. Single-stroke Gothic-style letters are recommended. Gothic characters do not have serifs at the ends of the strokes—*serifs* are the small flourishes found at the ends of the main strokes of characters in some font styles. Fonts without serifs are referred to as *sans serif* fonts.

Alignment of Dimension Text

The ASME standard for text states that text should face the bottom of the sheet. This is known as ***unidirectional text***. An example of unidirectional text is shown in Figure 5.6.

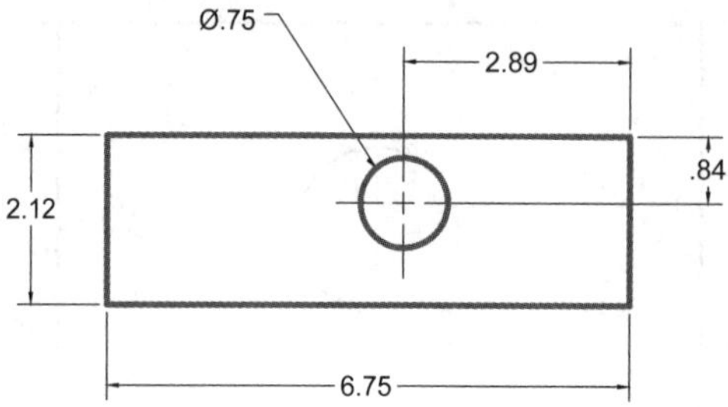

5.6 Unidirectional Text

Aligned text is aligned to dimension lines and may face the bottom and the right side of the sheet. This style is allowed on mechanical drawings prepared following the ISO standard but is not allowed on drawings employing the ASME dimensioning standard. An example of aligned text is shown in Figure 5.7.

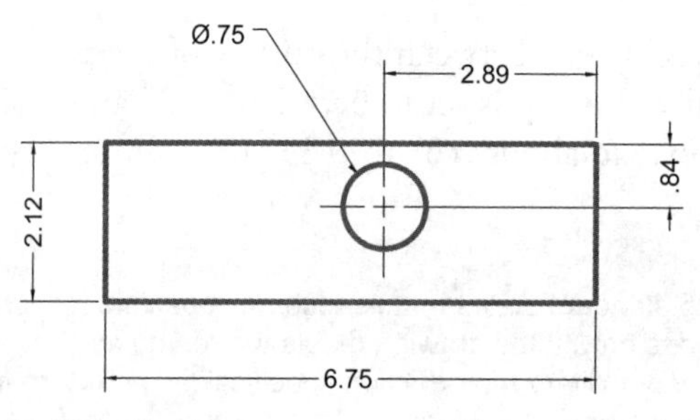

5.7 Aligned Text

Notating Holes and Arcs

When holes are dimensioned, the *X*- and *Y*-values to the center of the hole from the designer's datums should be provided and the hole's diameter specified. An arc should be described as a radius.

Figure 5.8 illustrates how leaders for small arcs and diameters are depicted. Figure 5.9 illustrates how leaders and notes are represented for large diameters and radii.

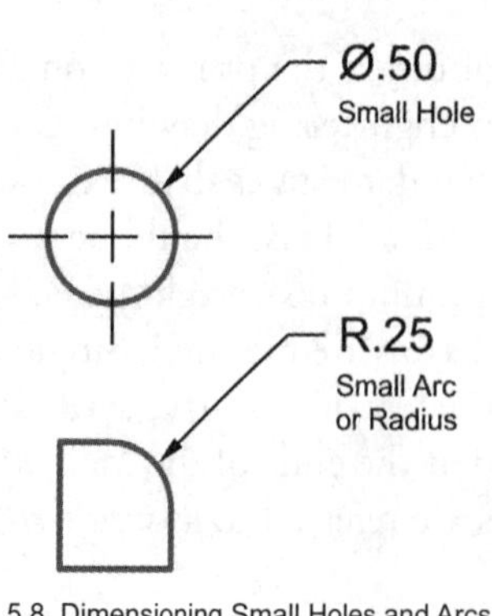

5.8 Dimensioning Small Holes and Arcs

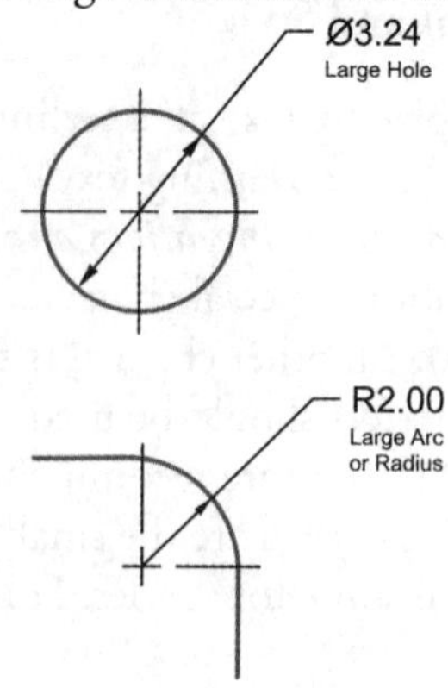

5.9 Dimensioning Large Holes and Arcs

Dimensioning Cylindrical Shapes

ASME Y14.5 specifies that cylinders and other outside diameters should be dimensioned in their profile (or side) view. The dimension should be preceded by the diameter symbol (see Figure 5.10).

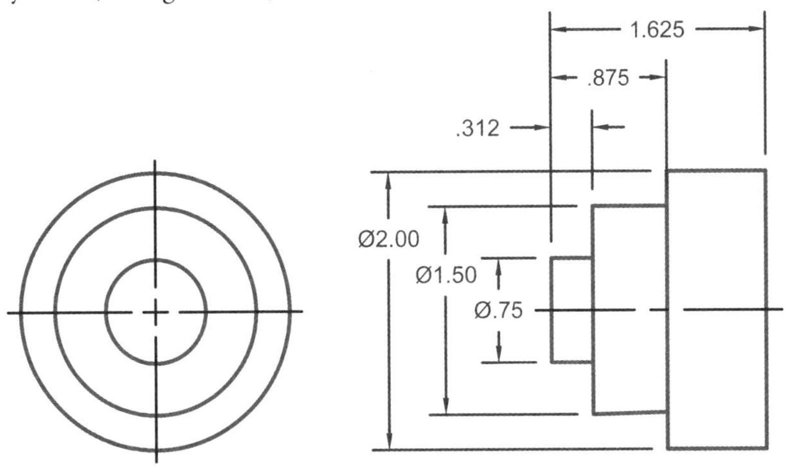

5.10 Dimensioning Cylindrical Shapes

Dimensioning Angles

An angle should also be dimensioned in its profile view, with the dimension value followed by the degree symbol (°) (see Figure 5.11). When greater accuracy for noting angles is desired, angles may be specified in degrees, minutes ('), and seconds ("). A minute equals 1/60th of one degree, and a second equals 1/60th of a minute (see Figure 5.12).

Ordinate Dimensioning

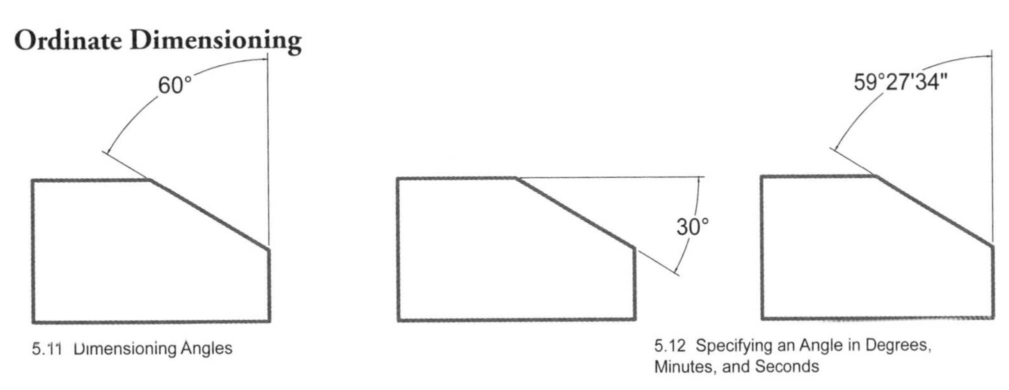

5.11 Dimensioning Angles

5.12 Specifying an Angle in Degrees, Minutes, and Seconds

In ordinate dimensioning a **0,0** (zero *X*, zero *Y*) datum point is defined on the object, and the location of the object's features are located along the X and Y-axcs as referenced from the 0,0 datum point.

In the example in Figure 5.13 the 0,0 datum is at the lower left corner of the object, and the dimensions shown are all relative to this point. A table, like the one in Figure 5.14, is often placed on the drawing to describe hole diameters when ordinate dimensioning is employed.

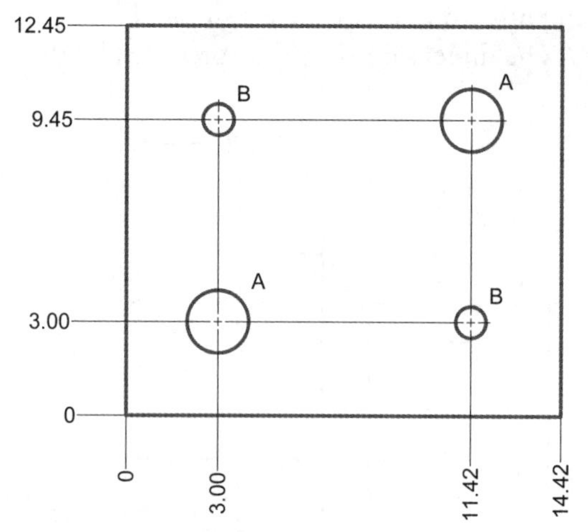

5.13 Ordinate Dimensioning

Hole Table	
Mark	Diameter
A	Ø2.00
B	Ø1.00

5.14 Hole Table

Ordinate dimensioning is useful for dimensioning parts that are to be manufactured by ***computer-aided manufacturing (CAM)*** machinery, such as a CAM drill press. The operator of the CAM drill press mounts the material to be drilled on the bed of the press and programs the location of the 0,0 point on the material. The operator then programs the location of holes along the *X*- and *Y*-coordinates. The drill bit moves along the *X*- and *Y*-axes relative to the 0,0 datum to drill holes in the material. (In some machines, the drill bit remains stationary, and the table on which the material is mounted moves to the bit instead.)

Notes for Drilling and Machining Operations

Notations for counterbored and countersunk holes are shown in Figure 5.15. Because the computer keyboard lacks the characters for the geometric symbols that represent a counterbore(⌴), countersink(⌵), or depth(↧), AutoCAD's Geometric Dimensioning and Tolerancing (GDT) font is often used to place these symbols.

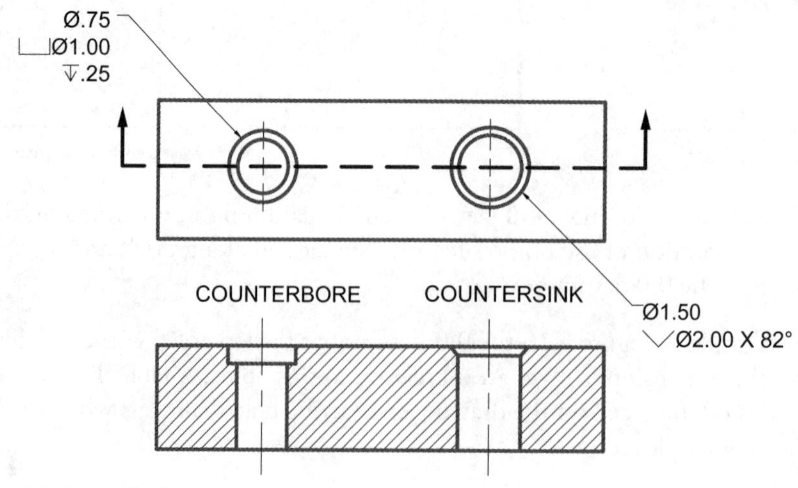

5.15 Notations for Counterbored and Countersunk Holes

In the GDT font, the lowercase alphabetical characters are replaced by the symbols and characters used in dimensioning and tolerancing notations. The lowercase alphabetical letter on the keyboard that corresponds to the dimensioning symbol in the GDT font is shown in Figure 5.16. For example, to insert the counterbore symbol into a notation for a hole, the drafter types a lowercase **v** into the note in place of the counterbore symbol and then highlights the v in the text window and changes it to the GDT font. The **v** is then replaced by the counterbore symbol in the note.

GDT SYMBOLS		
Keyboard Character (lowercase)	GDT Symbol	Interpretation
v	⌴	Counterbored Hole
w	⌵	Countersunk Hole
x	↧	Depth of Hole

5.16 Lowercase Letters Corresponding to Symbols in AutoCAD's GDT Font

JOB SKILL

When you are using the Text Editor to edit a hole note, you can access the GDT characters needed to represent the counterbore, countersink, and depth symbols by selecting the Symbol tool located on the Insert panel of the ribbon. When the drop-down list appears, select Other from the bottom of the list. When the Character Map opens, select the GDT font, pick a symbol from the map, and click the Select button. Once you have selected all the required characters, click the Copy button, and close the Character Map by clicking on the X in the upper right corner. Then, right-click the mouse and select the Paste option from the shortcut menu. The GDT symbols will be pasted into the text window associated with the hole note.

QUICK TIP

The symbols for degrees, plus/minus, and diameter may also be accessed by selecting the Symbol tool located on the Insert panel.

5.3 DO'S AND DON'TS OF MECHANICAL DIMENSIONING (ASME Y14.5)

Following the dimensioning guidelines listed here while creating a mechanical drawing will help ensure that the finished drawing is complete, orderly, and easy to interpret.

DO's

- ✓ Include every dimension and notation provided on the designer's sketch (or other form of input). If you have questions about the completeness of dimensions, consult the designer.
- ✓ Use the same precision (number of decimal places) identified by the designer when specifying dimension values.
- ✓ Use the datums identified on the designer's input when placing dimensions.
- ✓ Dimension to features in their profile (or most descriptive) view.
- ✓ Follow ASME Y14.5 spacing guidelines and dimension settings (see Figure 5.5).
- ✓ Place dimensions applying to two views between the views, but project extension lines from only one of the views.
- ✓ Describe a hole or cylinder (or other circular feature) with a diameter dimension (see Figures 5.8 and 5.9).
- ✓ Describe an arc with a radius dimension (see Figures 5.8 and 5.9).
- ✓ When dimensioning holes, provide the X- and Y-values of the center of the hole referenced from the datum features identified in the designer's input, and specify the diameter and depth of the hole.
- ✓ Dimension cylinders and other outside diameters in their profile (side) view and include the diameter symbol (see Figure 5.10).
- ✓ Include all notes required to manufacture the part (material, scale, etc.).
- ✓ Use a single-stroke Gothic (sans serif) text style for dimensions and notes. Minimum dimension text height should be .12" (3mm). All text should be fully legible when plotted or printed.
- ✓ When dimension values are in decimal inches, place decimal points in line with the bottom of the dimension text.
- ✓ When dimension values are in millimeters, place a zero before the decimal point when dimension values are less than 1mm (e.g., 0.7mm).

JOB SKILL

If the hole passes all the way through the part (referred to as a through hole) it is not necessary to specify a depth.

DON'Ts

- ✗ Don't over dimension. Features should be dimensioned only once. Do not include dimensions that were not on the designer's input without first consulting with the designer.
- ✗ Don't dimension to hidden lines or hidden features such as the centers of hidden holes.
- ✗ Don't cross dimension lines, or leaders, with extension lines (break the extension line).
- ✗ Don't run leaders through dimension lines if it can be avoided.
- ✗ Don't place dimensions on the object whenever possible.
- ✗ On drawings created with decimal inches, don't place a zero before the decimal point for a dimension value of less than 1" (suppress leading zeros).
- ✗ When dimension values are in millimeters, don't place a decimal point or a zero after a dimension that is a whole number (suppress trailing zeros).

5.4 ROLE OF DRAFTERS IN THE PREPARATION OF DIMENSIONED MECHANICAL DRAWINGS

The drafter's role in preparing a dimensioned drawing can be summarized as follows:

- In most cases, the drafter receives a design input—often a sketch—of the object to be drawn from a designer or engineer. The sketch will (or should) provide all the dimensions and other information necessary to fabricate the object. The drafter should ask for clarification from the designer if he or she feels that the sketch is missing a dimension or contains incorrect or unclear dimensions.
- The drafter determines which multiviews are necessary to describe the features of the object, as well as the sheet format and layout. The drafter then draws the views of the object.
- The drafter dimensions the views following the dimensions defined on the designer's sketch.
- The drafter is responsible for ensuring that the ASME Y14.5 (or other applicable) standard is followed with regard to the placement, spacing, and style of dimensions.

5.5 CHECKING DIMENSIONS ON THE FINISHED DRAWING

The finished dimensioned drawing should be compared carefully with the designer's input. To assure accuracy and completeness, the drafter should ask the following questions about the finished drawing:

- Does the drawing provide the multiviews necessary to describe the geometrical features of the object?

- Can each dimension and note included on the designer's input be accounted for on the final drawing?
- Have all applicable drafting and dimensioning standards been followed?
- Does the combination of the views, dimensions, and notes fully describe the object? Could this object be manufactured using only the views, dimensions, and notes provided on the drawing? (Novice drafters may find this difficult to answer, but as they gain experience with manufacturing processes and materials, it becomes easier.)

When the answers to all the questions posed are yes, the drawing is probably finished. However, in most offices, the final determination about whether a drawing is finished is made by an engineer, designer, or a ***checker***.

Checkers are usually very experienced designer-drafters with expertise in manufacturing, drafting techniques, and dimensioning conventions. Often, there is an *Approved* box in the drawing's title block for the checker's initials. When a checker initials this box, it indicates that the drawing has passed the checker's review. It also means that the drafter is no longer the only one responsible for the accuracy of the drawing.

An effective strategy for checking the sketch against the drawing is to use a yellow marker to highlight the dimensions on the sketch and the drawing, one by one, until all are accounted for.

5.6 DESIGN BASICS: HOW DESIGNERS CALCULATE DIMENSIONS

Mechanical engineers and designers are responsible for calculating the dimensions of an object. They determine the dimensional values by carefully considering the form, fit, and function of the object they are designing. For example, the material from which the part is to be manufactured, how the object fits with other parts in an assembly, and the role it plays in the overall design are all factors that may affect the size and location of the features of a part. Sometimes, the dimensions define the aesthetic rather than the functional qualities of the finished object. The aesthetic qualities of an object refer more to the object's appearance than to its function.

Because the designer's dimensions are carefully calculated, it is very important that they are faithfully reproduced by the drafter during the preparation of an engineering drawing.

To appreciate how crucial it is that the designer's dimensions be accurately portrayed in a mechanical drawing, drafters must have an understanding of a very important concept of mechanical design: ***tolerances***.

5.7 TOLERANCES

Manufacturing a machine part to an extreme degree of precision is difficult *and* expensive; therefore, designers must decide how much the size and location of a part's features can deviate from the dimensions specified on the drawing and still perform their design function. This allowable variation in the location, or size, of a feature is called the ***tolerance***. Once the acceptable tolerance for a feature has been determined, it is noted on the dimensioned drawing of the part.

QUICK TIP

To underscore the importance of tolerances in the dimensioning process, the *ASME Y14.5* standard states as one of its fundamental rules: "Each dimension shall have a tolerance except those dimensions specifically identified as reference, maximum, minimum, or stock (commercial stock) size."

On engineering drawings, the dimension to which the tolerances are applied is referred to as the ***nominal size*** of the feature. For example, a designer may decide the allowable diameter of a hole is 1.00", with a tolerance of ±.01". In this example, 1.00" in diameter is the nominal size of the hole. By applying a tolerance of ±.01" to the nominal size of the hole, we can calculate that the hole could range in diameter from .99" to 1.01" and still fall within the acceptable size specified by the designer. The tolerance is the allowable difference between the hole's minimum and maximum size limits, so in this case the tolerance for this hole would be .02" (1.01 minus .99 = .02).

The primary reason that designers calculate and specify tolerances on mechanical drawings is to control the size and location of the features of the part during the manufacturing process. After manufacture, the part is measured by quality control inspectors to verify that the features are within the allowable limits of size as defined by the tolerances on the drawing. Parts whose features measure within the allowable limits pass inspection. Parts whose features measure outside the limits are rejected.

Specifying tolerances on drawings has advantages for designers and manufacturers alike. The advantage for designers is that they can be confident that if the parts they design are manufactured within the tolerances specified on the drawing, their designs will fit and function as they intended. Manufacturers like toleranced parts because they can be confident that as long as the parts they make measure within the tolerances specified on the drawing, the client will purchase the parts.

Displaying Tolerances on Mechanical Drawings

Figures 5.17 and 5.18 show two examples of how tolerances may be specified for dimensions on a technical drawing. In Figure 5.17, the tolerance is shown beside the nominal dimension and noted with a plus/minus symbol (±). In Figure 5.18, the tolerance has been both added to and subtracted from the nominal size, and the allowable size limits are actually noted in the dimension.

Another method of specifying tolerance is to add notations in the field of the drawing or in the drawing's title block.

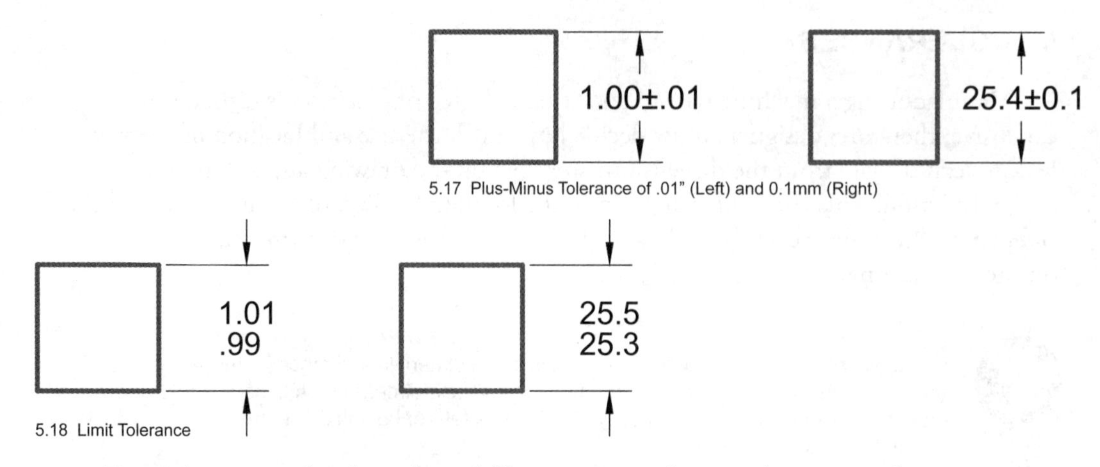

5.17 Plus-Minus Tolerance of .01" (Left) and 0.1mm (Right)

5.18 Limit Tolerance

Examples 1, 2 and 3 show three different ways to depict tolerances with notations and include an interpretation of the specified tolerance.

Example 1 **General tolerances** may be labeled in the title block or given as notes in the field of the drawing.

.X ± .05 Dimensions noted with one decimal place of precision on the drawing will have a tolerance of plus or minus .05".

.XX ± .02 Dimensions noted with two decimal places of precision on the drawing will have a tolerance of plus or minus .02".

.XXX ± .003 Dimensions noted with three decimal places of precision on the drawing will have a tolerance of plus or minus .003".

Example 2 **Decimal Dimensions to Be ±.005"** A general tolerance of .005" will apply to all dimensions labeled in decimal units.

Example 3 **Angular Tolerances ±1°** Dimensions on the drawing labeled as angles will have a tolerance of plus or minus 1°. For example, an angle labeled 30° on the drawing could measure between 29° and 31° on the manufactured part.

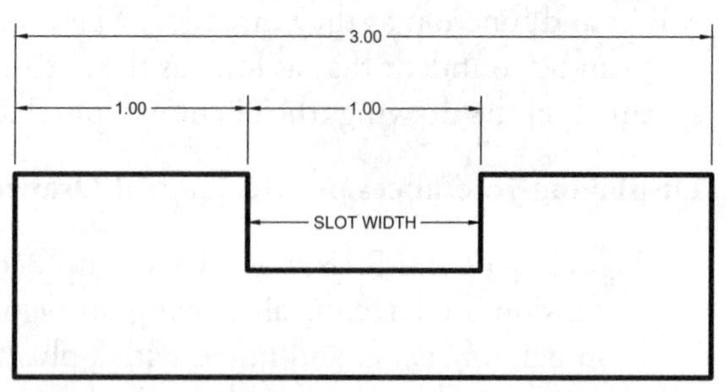

5.19 Part with Slot Dimensioned with a Continuous Dimension

Interpreting Tolerances on Technical Drawings

Figures 5.19 and 5.20 show examples of a machine part in which the dimensions *do not* include tolerances.

In Figure 5.19, the width of the slot is dimensioned with a ***continuous dimension***. A continuous dimension is referenced from the termination of the dimension that preceded it. The width of the slot in Figure 5.19 is 1.00".

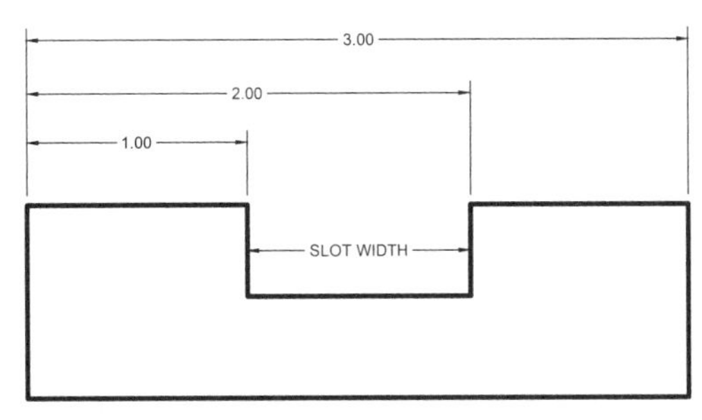

5.20 Part with Slot Dimensioned with a Baseline Dimension

In Figure 5.20, the width of the slot is dimensioned with ***baseline dimensions***. Baseline dimensions are referenced from a common datum. In Figure 5.20, the datum feature is the left edge of the part. The width of the slot can be determined by calculating the difference between locations of the right and left sides of the slot. In this case, the width of the slot equals 1.00 (2.00 minus 1.00).

Because no tolerances are specified for the dimensions in Figures 5.19 and 5.20, the width of the slot will be 1.00" whether it is defined by a single continuous dimension or by two baseline dimensions.

Figure 5.21 shows a view of a part that includes dimensions *and* tolerances. The overall length of this part is labeled 3.00" ± .01". This means that when the part is manufactured, its overall length must measure between 2.99" and 3.01" to pass a quality control inspection.

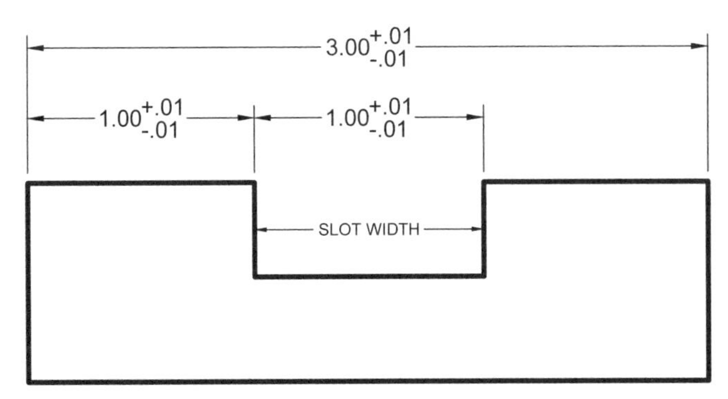

5.21 Part with Slot Dimensioned with a Toleranced *Continuous* Dimension

The left side of the slot in Figure 5.21 is located 1.00" ±.01" from the left edge of the part. This means that the location of the slot's left side must fall between .99" and 1.01" as measured from the left side of the part.

The width of the slot is dimensioned 1.00" ±.01", so as long as the width of the slot on the part measures between .99" and 1.01", it will pass inspection.

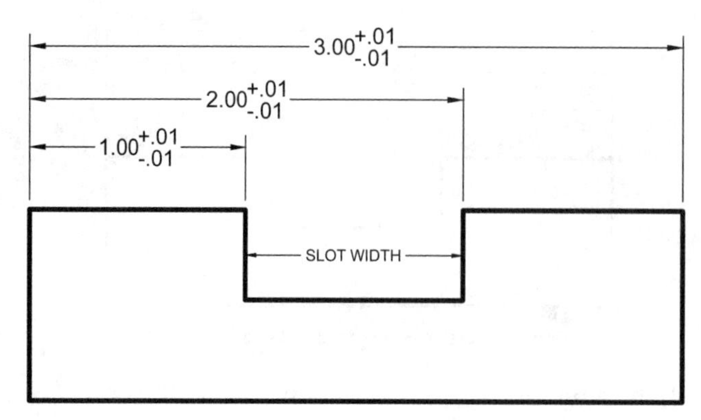

5.22 Part with Slot Size Defined by Toleranced *Baseline* Dimensions

Comparison of Continuous Dimensioning and Baseline Dimensioning

Figure 5.22 shows a part that is dimensioned with three baseline dimensions that include plus or minus tolerances. Here, the left edge of the part serves as the datum feature for each of the three dimensions.

The overall length of this part is labeled 3.00" ±.01", so when this part is manufactured, its overall length must measure between 2.99" and 3.01".

The width of the slot in Figure 5.22 is defined by two baseline dimensions instead of one continuous dimension, as in Figure 5.21. The location of the left side of the slot in Figure 5.22 must fall between .99" and 1.01" from the datum edge. The right side of the slot must be between 1.99" and 2.01" from the datum.

If the left edge of the slot is located .99" from the datum and the right edge is located 2.01" from the datum, the width of the slot will be 1.02". If the left and right sides of the slot are located 1.01" and 1.99", respectively, from the datum edge, the width of the slot will measure .98".

Using baseline dimensions to define the slot could result in its width measuring between .98" and 1.02". Dimensioning the slot with continuous dimensions as shown in Figure 5.21 resulted in its width measuring between .99" and 1.01".

As you can see from the comparison of the widths of the slots in Figures 5.21 and 5.22, the method used to dimension the slot (either baseline or continuous) affects its size, after tolerances are factored in. In this case, the width of the slot defined with baseline dimensions may vary by as much as .04" (.98" to 1.02"), whereas defining the width with continuous dimensions results in its varying by .02" (.99" to 1.01").

The point of this comparison is not that either dimensioning technique—continuous or baseline—is inherently better or worse than the other, but rather, that the method used to define the widths of the slot in Figures 5.21 and 5.22 affects the slot's size after tolerances are applied. The designer chooses the dimensioning technique that will ensure that the part's features are manufactured within size limits to allow the part to function in its intended application.

This comparison was intended to help you understand the importance of applying the same dimensioning method (continuous or baseline) to the creation of an engineering drawing as is shown on the design input. Mechanical drafters who do not follow the same dimensioning method as the designer run the risk of inadvertently changing the intended size or location of a feature.

Tolerancing Terminology

The following terms are used frequently by designers when discussing tolerances. As a drafter-in-training, you should become familiar with each term so that you can communicate effectively with designers and engineers.

- *Feature:* A geometric element that is added to the base part such as a slot, surface, or hole.
- *Nominal size:* A dimension used to describe the general size of the feature. Tolerances are applied to this dimension.

VIDEO There is a video tutorial available for this project inside the Chapter 5 folder of the book's Video Training downloads.

- *Tolerance:* The total permissible variation in a dimension value; the difference between the upper and lower size limits of a feature.
- *Limits:* The maximum and minimum sizes of a feature as defined by the toleranced dimension. For example, a hole dimensioned with a diameter of .50", with a tolerance of ±.02", has an upper limit of .52" and a lower limit of .48".
- *Allowance:* The minimum clearance or maximum interference between mating parts.
- *Datum:* A theoretically perfect element (an edge, plane, axis, point, or other geometric feature) from which dimensional information is referenced. Finished parts are inspected by measuring from the datum geometry identified on the drawing.
- *Actual size:* The measured size of a feature of a manufactured part. This size determines whether the part passes a quality control inspection.
- *Reference dimension:* A dimension without a tolerance that is provided only for information purposes. It is not used for manufacture or inspection of the part. A reference dimension is enclosed in parentheses.
- *Maximum material condition (MMC):* The condition of a feature when it contains the greatest amount of material. The MMC of an external feature, such as a shaft, is the upper limit of size. The MMC of an internal feature, such as a hole, is the lower limit of size.
- *Least material condition (LMC):* The condition of a feature when it contains the least amount of material. The LMC of an external feature is the lower limit. The LMC of an internal feature is the upper limit.

Interpreting Design Sketch 1

In the design sketch shown in Figure 5.23, the designer has provided the dimensions required to define the nominal sizes of the diameter (.98") and the length (2.00") of a cylinder. The designer has also specified a plus or minus tolerance of .01".

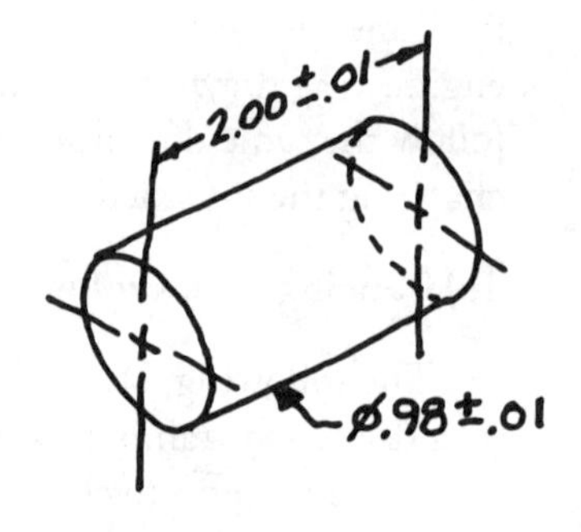

5.23 Design Sketch 1

Analyzing Design Sketch 1

Applying the tolerance to the cylinder's nominal dimensions results in a range for its diameter between a minimum of .97" (.98" minus .01") and a maximum of .99" (.98" plus .01"). Applying the same sort of calculation to the length of the cylinder results in a range between a minimum of 1.99" (2.00" minus .01") and a maximum of 2.01" (2.00" plus .01").

The designer has calculated the size of the cylinder so that it should be able to perform as intended as long as its features fall within the allowable size limits.

Calculating Maximum Material Condition (MMC) for Design Sketch 1

Applying the plus tolerance to the cylinder's nominal dimensions will define the MMC of the cylinder's diameter and length. According to this calculation, the MMC for the cylinder's diameter is .99" (.98" plus .01") and the MMC for the length of the cylinder is 2.01" (2.00" plus .01").

Calculating Least Material Condition (LMC) for Design Sketch 1

Applying the minus tolerance to the cylinder's nominal dimensions will define the cylinder's LMC. According to this calculation, the LMC of the cylinder's diameter is .97" (.98" minus .01"), and the LMC for the length of the cylinder is 1.99" (2.00" minus .01").

Interpreting Design Sketch 2

In the design sketch shown in Figure 5.24, the designer has provided the coordinate (X and Y) dimensions required to locate the center of the hole from the left side and bottom edge of the object. These sides will be used as *datum features* for referencing dimensions on the drawing and later for inspecting the finished part.

The designer has also noted a tolerance range of plus or minus one hundredth of an inch (± .01") that is to be applied to each nominal dimension.

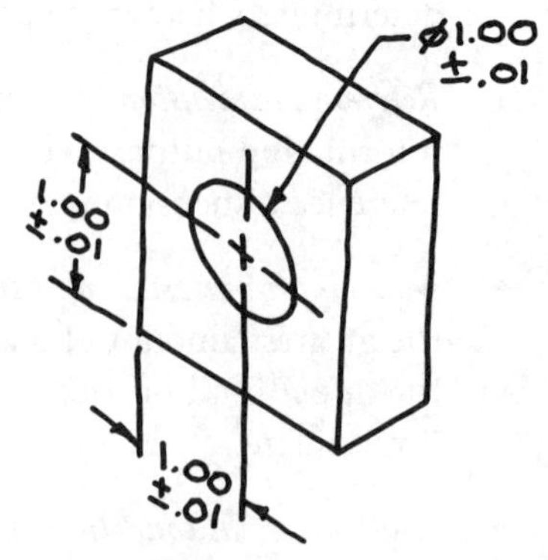

5.24 Design Sketch 2

Analyzing Design Sketch 2

When this part is manufactured, the location of the center of the hole, as well as the diameter of the hole, must comply with the conditions noted in Figure 5.25. Otherwise, the part will be out of tolerance and may be rejected during a quality assurance inspection check. When the part is inspected, the inspector will make measurements from the same datum features that are defined by the dimensions in the technical drawing.

Calculating Maximum Material Condition (MMC) for the Hole in Design Sketch 2

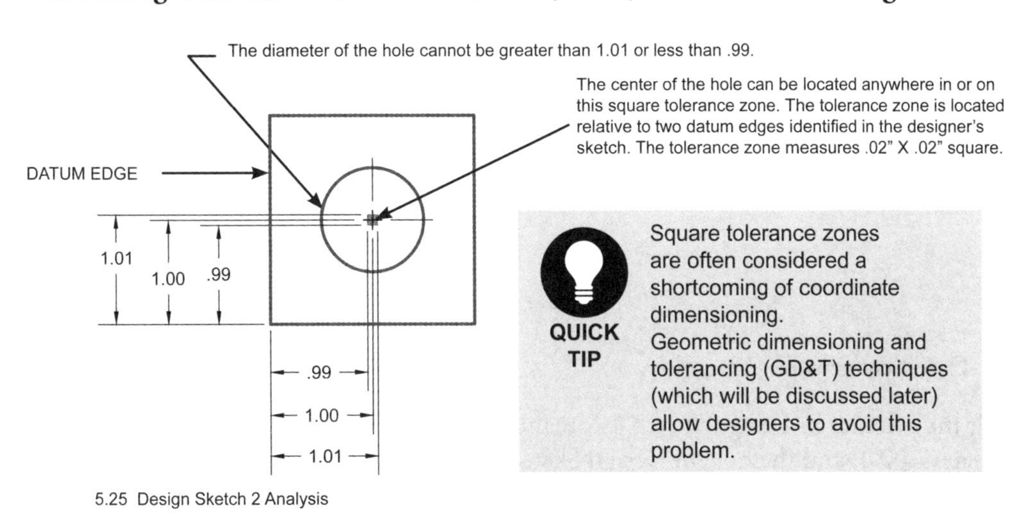

The diameter of the hole cannot be greater than 1.01 or less than .99.
The center of the hole can be located anywhere in or on this square tolerance zone. The tolerance zone is located relative to two datum edges identified in the designer's sketch. The tolerance zone measures .02" X .02" square.
DATUM EDGE
1.01
1.00
.99
.99
1.00
1.01
QUICK TIP
Square tolerance zones are often considered a shortcoming of coordinate dimensioning.
Geometric dimensioning and tolerancing (GD&T) techniques (which will be discussed later) allow designers to avoid this problem.

5.25 Design Sketch 2 Analysis

Subtracting the minus tolerance from the hole's nominal dimension will define its MMC. The nominal size of the hole is 1.00" in diameter, and the minus tolerance is .01", so in this case the MMC for the diameter of the hole is .99".

Calculating Least Material Condition (LMC) for the Hole in Design Sketch 2

Adding the plus tolerance to the hole's nominal dimension will define its LMC. The hole's nominal diameter is 1.00", and the plus tolerance is .01", so in this case the LMC for the hole's diameter is 1.01".

Calculating the Fit between the Parts in Design Sketches 1 and 2

Suppose the designer of the parts in Design Sketches 1 and 2 (see Figure 5.26) intended that the parts be assembled after they were manufactured. During the design of each part, the designer would have needed to analyze the possible limits of size of the cylinder in Part 1 and the hole in Part 2 to see if an interference could exist that would prevent the parts from being assembled.

There is a video tutorial available for this project inside the Chapter 5 folder of the book's Video Training downloads.
VIDEO

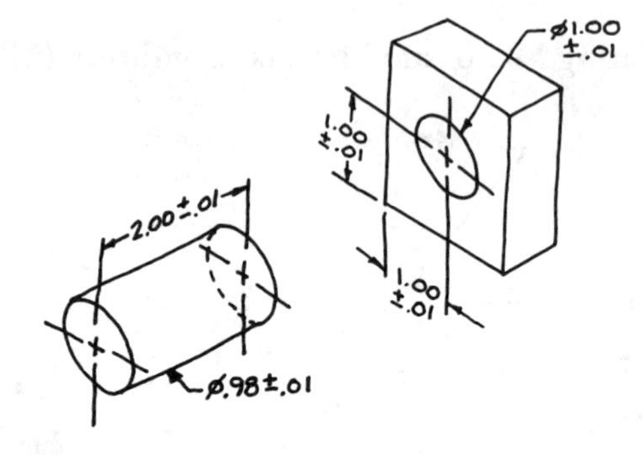

5.26 Assembling Parts 1 and 2

Best-Case Scenario for Assembly

If the cylinder in Design Sketch 1 is manufactured at its LMC, or smallest allowable diameter (.97"), and the hole in Design Sketch 2 is manufactured at its LMC, or largest allowable diameter (1.01"), the clearance between the two at LMC will be .04". With a clearance of .04", these parts can easily be assembled.

Worst-Case Scenario for Assembly

The worst-case scenario for assembly exists if the cylinder in Design Sketch 1 is manufactured at its MMC, or largest allowable diameter (.99"), and the hole in Design Sketch 2 is manufactured at its MMC, or smallest allowable diameter (.99"). At first it might seem that having the same diameter would cause interference when the parts are assembled, but this is not the case. The resulting fit would, however, be very tight, which would make assembling them more difficult.

If the designer desired a looser fit at MMC of both parts, the nominal diameter specified for the cylinder could be reduced to .97", or the nominal diameter of the hole could be enlarged to 1.01". Either would result in a clearance fit between the parts at MMC.

Reference Dimensions

Figure 5.27 shows an example of a machine part with an overall dimension of 3.00" labeled inside parentheses and a chain of three continuous dimensions, each labeled 1.00" ±.01". The dimension in parentheses is a *reference dimension.* A reference dimension is an untoleranced dimension that is provided only for informational purposes and is not used to manufacture or to inspect the part.

Because the reference dimension is not used to make the part in Figure 5.27, the overall length of the part will be a product of the cumulative effects of the three dimensions labeled 1.00" plus or minus their tolerances. Thus, the overall length of the part could range between 2.97" (.99" times 3) to 3.03" (1.01 times 3). This phenomenon, in which the sizes of toleranced features have a cumulative effect on the overall length, is called a *tolerance accumulation*.

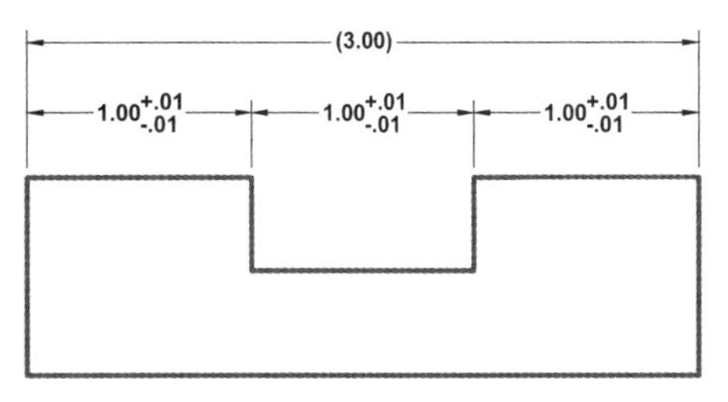

5.27 Reference Dimension

In Figure 5.27 if the overall length dimension had been labeled 3.00" ±.01" (instead of as a reference dimension), it would be impossible to reconcile its allowable limits (2.99" to 3.01") with the limits allowed by applying a tolerance to each of the dimensions labeled 1.00" (2.97" to 3.03").

In contrast, if the chain of 1.00" dimensions were broken by removing one of them, the overall dimension would no longer be a reference dimension and would need to have a tolerance.

QUICK TIP

If an unbroken dimension chain is specified on the designer's input, confirm with the designer that this is how it should be shown on the drawing.

Confirming the Tolerances of Manufactured Parts

It is important to confirm that after the part is manufactured, it falls within the allowable size limits defined by the dimensions and tolerances noted on the drawing. This step in the manufacturing cycle is performed during a ***quality control inspection***. Quality control (QC) inspectors use precise measuring (metrology) equipment to determine the actual size of the part. QC inspectors compare the actual size of the part with the dimensions noted on the technical drawing. Parts that measure within the allowable size limits will pass the QC inspection, whereas parts that measure outside the limits will be rejected.

QUICK TIP

For more information on geometric dimensioning and tolerancing, see Appendix D.

Tolerance Costs

Designers must consider cost when determining the tolerances for a feature, because as tolerances become tighter, the cost of manufacturing a part may increase. One reason is that as tolerance allowances become stricter, it may take longer to manufacture the part. Another reason is that due to tighter tolerances, fewer parts may pass a quality control inspection. The designer walks a fine line between the desired accuracy of the part and the cost of manufacturing the part within the budget constraints of the project.

Although tight tolerances usually add to the cost of a project, a type of tolerancing known as ***geometric dimensioning and tolerancing (GD&T)*** may actually lower the costs of producing a part. GD&T tolerances control the *form* (flatness, straightness, circularity, and cylindricity), *orientation* (perpendicularity, angularity, parallelism), or *position* of a part's features. By using GD&T, the odds that parts will pass a quality control inspection rise, and fewer rejected parts results in lower production costs. The *ASME Y14.5* standard covers the application of GD&T to technical drawings.

5.8 DIMENSIONING WITH AUTOCAD

The AutoCAD **Dimension** toolbar, containing all the commands necessary to add dimensions to a drawing, is shown in Figure 5.28. These commands are also located in the **Dimensions** panel of the **Annotate** tab of the ribbon. In Figure 5.28 each **dimension command** icon is labeled with its function. An explanation of each icon is presented on the following pages.

VIDEO

Video tutorials for the commands on the Dimension toolbar are located inside the Chapter 5 folder of the book's Video Training downloads. Other videos in this folder lead users through the creation of AutoCAD dimension styles based on the ASME 14.5 standard. These video downloads are available by redeeming the access code that comes with this book. Please see the inside front cover for further details.

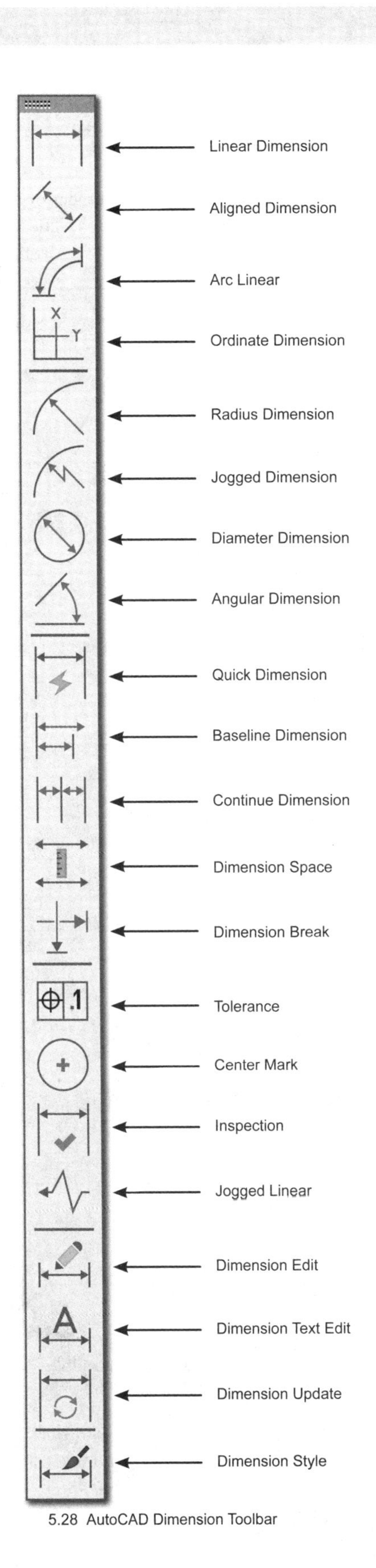

5.28 AutoCAD Dimension Toolbar

VIDEO

There are video tutorials available for the Dimensioning commands inside the Chapter 5 folder of the book's Video Training downloads.

JOB SKILL

Some, but not all, of the commands on the Dimension toolbar can also be found by picking on the down arrow next to the linear dimension symbol on the Annotation panel of the Home tab of the ribbon. Dimension commands are also available by left-clicking on the Dimension menu of the menu bar.

LINEAR DIMENSION Command			
	Ribbon	Annotate	The icon for the **Linear** dimension command is shown in Figure 5.29(a). This command displays the linear distance between two selected points. This option is used to dimension both vertical and horizontal features, as shown in Figure 5.29(b).
	Panel	Dimensions	
	Command Line	DimLinear	
5.29(a)	Alias	DLI	

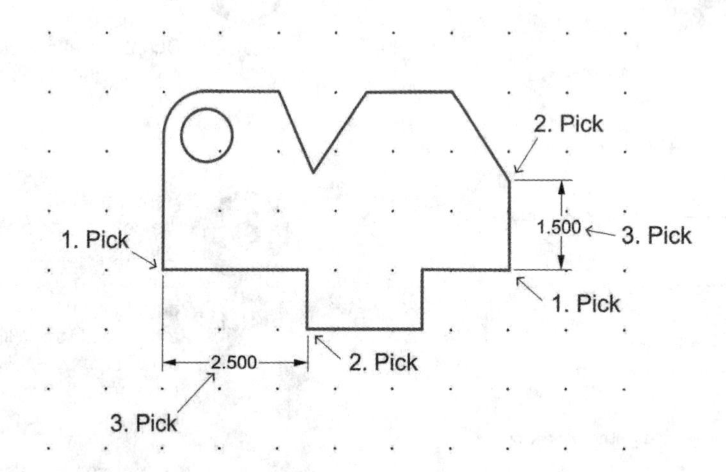

5.29(b) **Linear** Dimension Command

QUICK TIP

Whenever a dimension is placed on an AutoCAD drawing, AutoCAD automatically creates a new layer named Defpoints and places objects called definition points (defpoints) on this layer. The definition points reflect the points chosen by the user to define the dimension and are used by AutoCAD to calculate the value for the dimension. A special property of the Defpoints layer is that it does not print when the drawing is plotted, so don't be alarmed when you see the Defpoints layer appear in the list of layers shown in the Layer Properties Manager dialog box. Of course, since this layer does not print, it is not a good idea to set it *current* and create your drawing on it either.

ALIGNED DIMENSION Command			
	Ribbon	Annotate	The icon for the **Aligned** dimension command is shown in Figure 5.30(a). This command is used to display the length of an angled line, as shown in Figure 5.30(b).
	Panel	Dimensions	
	Command Line	DimAligned	
5.30(a)	Alias	DAL	

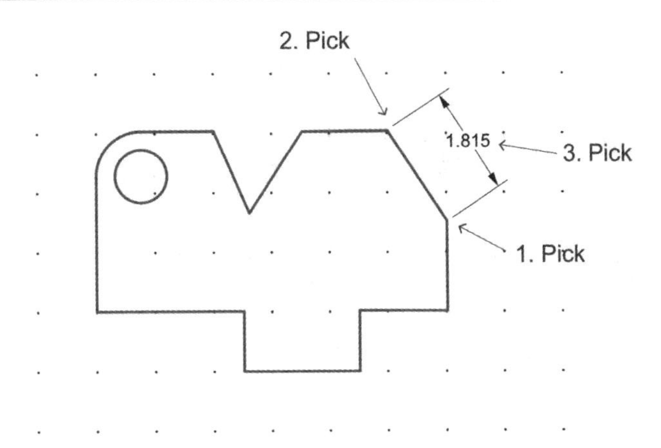

5.30(b) **Aligned** Dimension Command

ARC LENGTH Command			
	Ribbon	Annotate	The icon for the **Arc Length** command is shown in Figure 5.31(a). This command is used to denote the length dimension of an arc or a polyline arc segment as shown in Figure 5.31(b).
	Panel	Dimensions	
	Command Line	DimAligned	
5.31(a)	Alias	DAL	

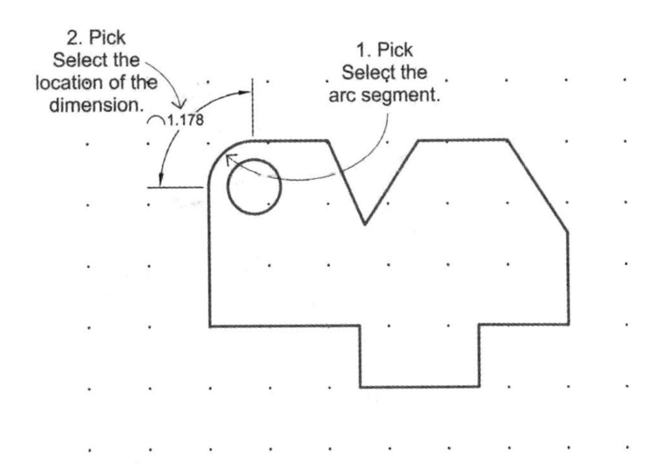

5.31(b) **Arc Length** Command

ORDINATE DIMENSION Command

	Ribbon	Annotate	The icon for the **Ordinate** dimension command is shown in Figure 5.32(a). This command is used to denote distances along *X*- and *Y*-axes relative to a defined origin point (usually labeled 0,0), as shown in Figure 5.32(b).
	Panel	Dimensions	
	Command Line	DimOrdinate	
5.32(a)	Alias	DOR	

In this example 0.000 X and 0.000 Y have been moved to the lower left corner of the object. To do this type UCS (User Coordinate System) and press enter. When prompted, type O and press enter. You will be prompted for the new origin point. Use object snaps to select the lower left corner.

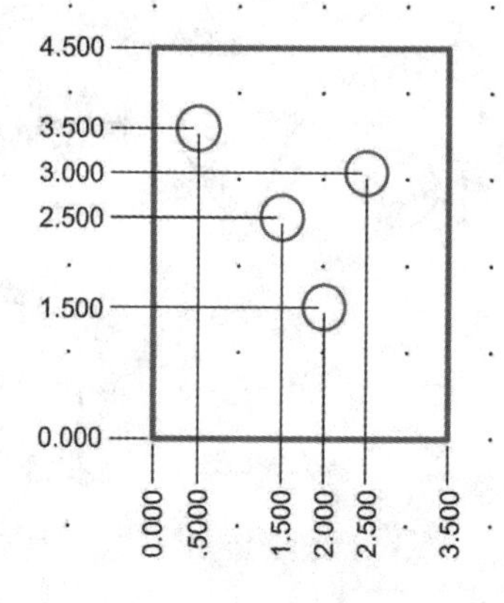

5.32(b) **Ordinate** Dimension Command

RADIUS DIMENSION Command

	Ribbon	Annotate	The icon for the **Radius** dimension command is shown in Figure 5.33(a). This command is used to denote the radius of an arc, as shown in Figure 5.33(b).
	Panel	Dimensions	
	Command Line	DimRadius	
5.33(a)	Alias	DRA	

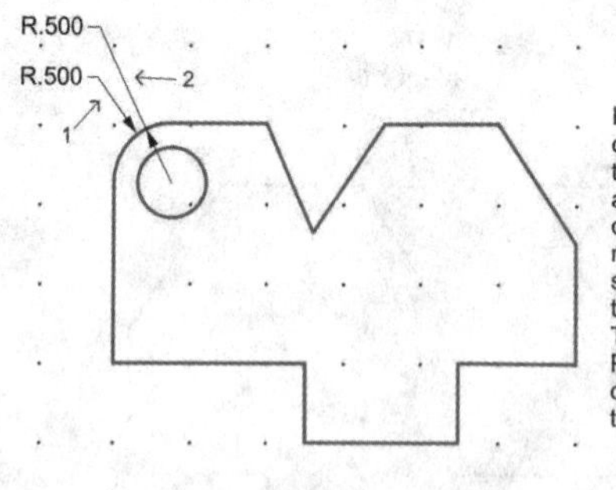

In this example two radius dimensions are given to show two methods to dimension an arc or circle. Number 1 is the default. To change it to style number 2, select dimension styles and Modify. Pick the Fit tab. Select the radio button Text (you may have to select Place text manually when dimensioning in Fine Tuning to get the desired results).

5.33(b) **Radius** Dimension Command

JOGGED DIMENSION Command			
	Ribbon	Annotate	The icon for the **Jogged** dimension command is shown in Figure 5.34(a). This command is used to create a jogged radius or diameter when dimensioning a large arc or circle whose center is outside the drawing area, as shown in Figure 5.34(b).
	Panel	Dimensions	
	Command Line	DimJogged	
5.34(a)	Alias	JOG	

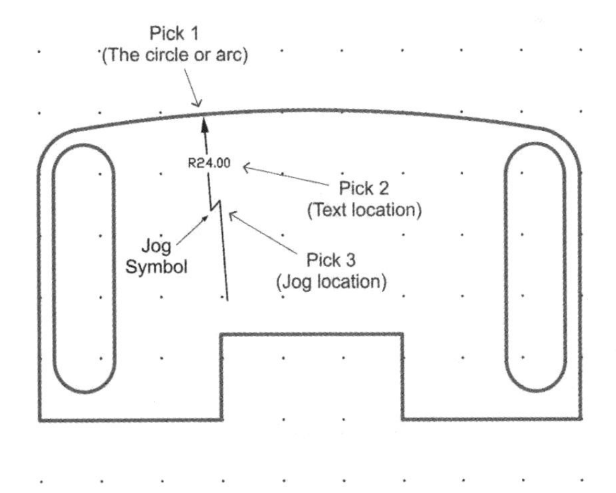

5.34(b) **Jogged** Dimension Command

DIAMETER DIMENSION Command			
	Ribbon	Annotate	The icon for the **Diameter** dimension command is shown in Figure 5.35(a). This command is used to denote the diameter of a circle, as shown in Figure 5.35(b).
	Panel	Dimensions	
	Command Line	DimDiameter	
5.35(a)	Alias	DDI	

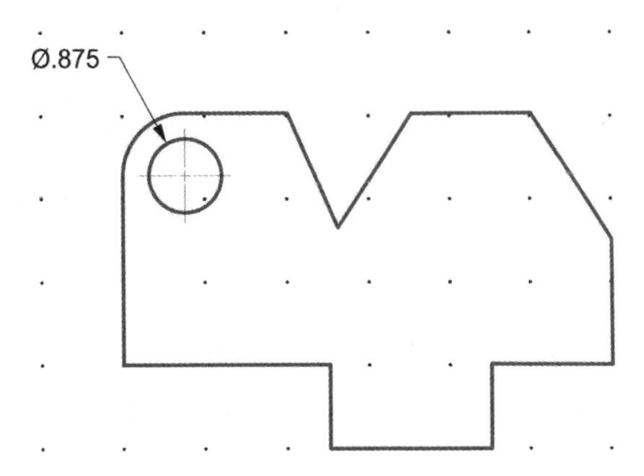

5.35(b) **Diameter** Dimension Command

ANGULAR DIMENSION Command			
	Ribbon	Annotate	The icon for the **Angular** dimension command is shown in Figure 5.36(a). This command is used to denote the angle between two features of an object, as shown in Figure 5.36(b).
	Panel	Dimensions	
	Command Line	DimAngular	
5.36(a)	Alias	DAN	

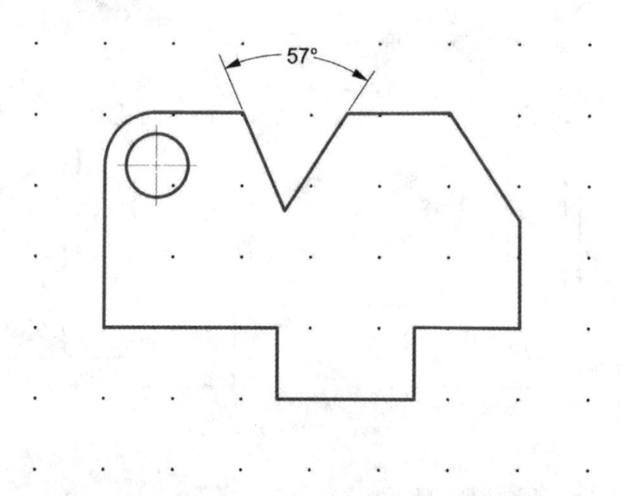

5.36(b) **Angular** Dimension Command

QUICK DIMENSION Command			
	Ribbon	Annotate	The icon for the **Quick** dimension command is shown in Figure 5.37(a). This command is used to create a group of dimensions quickly. You can either pick features on the object individually with the mouse or use a crossing window to select an area of the object to be dimensioned, as shown in Figure 5.37(b). Several options are available: **Continue, Staggered, Baseline, Ordinate, Radius,** and **Diameter.**
	Panel	Dimensions	
	Command Line	QDIM	
5.37(a)	Alias	QDIM	

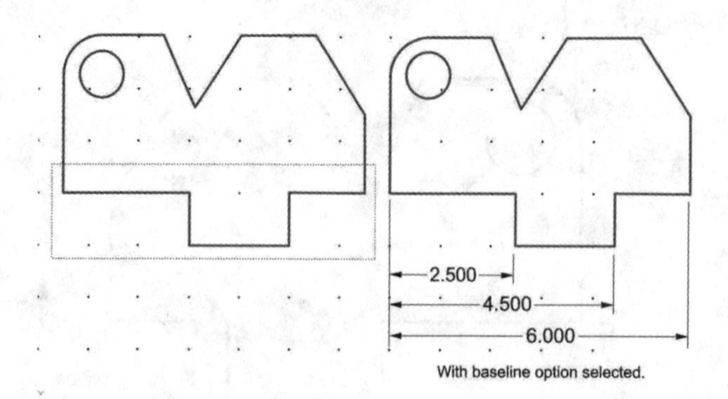

5.37(b) **Quick** Dimension Command

BASELINE DIMENSION Command

5.38(a)		
	Ribbon	Annotate
	Panel	Dimensions
	Command Line	DimBaseline
5.38(a)	Alias	DBA

The icon for the **Baseline** dimension command is shown in Figure 5.38(a). This command is used to create a series of dimensions measured from the same baseline (datum). The first extension line of the first dimension placed (using the Linear dimension command) becomes the default first extension line for dimensions placed after the Baseline option is selected, as shown in Figure 5.38(b).

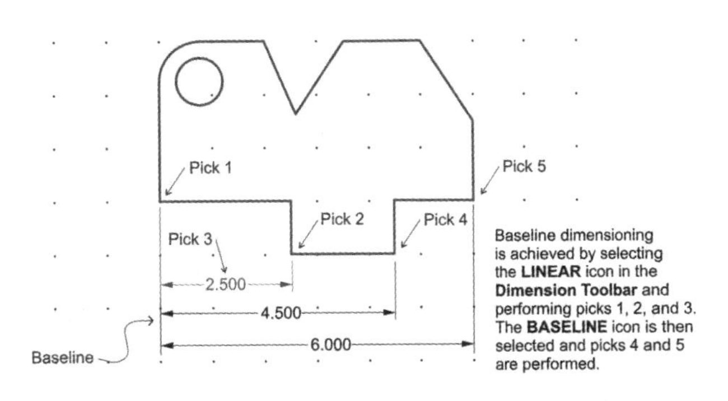

5.38(b) **Baseline** Dimension Command

CONTINUE DIMENSION Command

5.39(a)		
	Ribbon	Annotate
	Panel	Dimensions
	Command Line	DimContinue
5.39(a)	Alias	DCO

The icon for the **Continue** dimension command is shown in Figure 5.39(a). This command is used to create a string of continuous dimensions. The first dimension in the string is placed using the **Linear** dimension option. After the **Continue** option is selected, subsequent dimensions begin at the second extension line of the previously defined dimension, as shown in Figure 5.39(b). This dimensioning method is also called *chain dimensioning*.

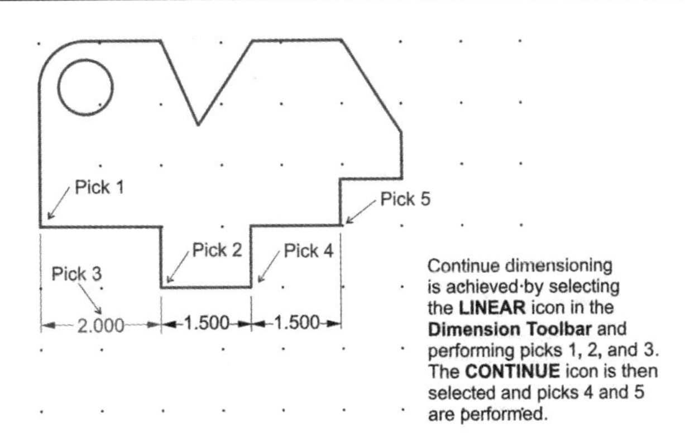

5.39(b) **Continue** Dimension Command

DIMENSION SPACE Command			
	Ribbon	Annotate	The icon for the **Dimension Space** command is shown in Figure 5.40(a). This command adjusts the space between parallel linear dimensions to match a defined distance, as shown in Figure 5.40(b).
	Panel	Dimensions	
	Command Line	DimSpace	
5.40(a)	Alias	DIMSPACE	

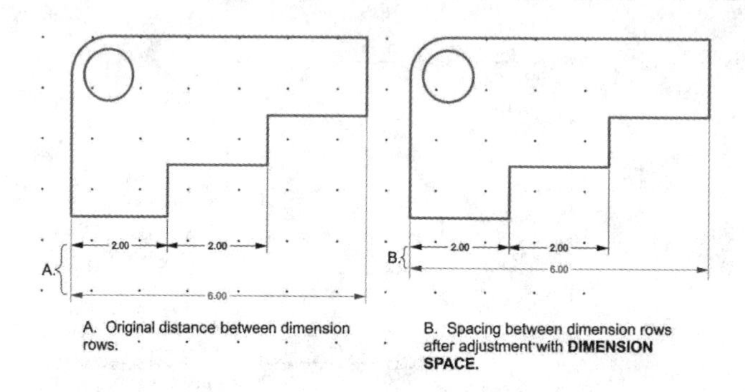
2.00
2.00
6.00
A.
B.
2.00
2.00
6.00
A. Original distance between dimension rows.
B. Spacing between dimension rows after adjustment with DIMENSION SPACE.

5.40(b) **Dimension Space** Command

DIMENSION BREAK Command			
	Ribbon	Annotate	The icon for the **Dimension Break** command is shown in Figure 5.41(a). This command is used to break dimension or extension lines where they overlap other lines, as shown in Figure 5.41(b).
	Panel	Dimensions	
	Command Line	DimBreak	
5.41(a)	Alias	DimBreak	

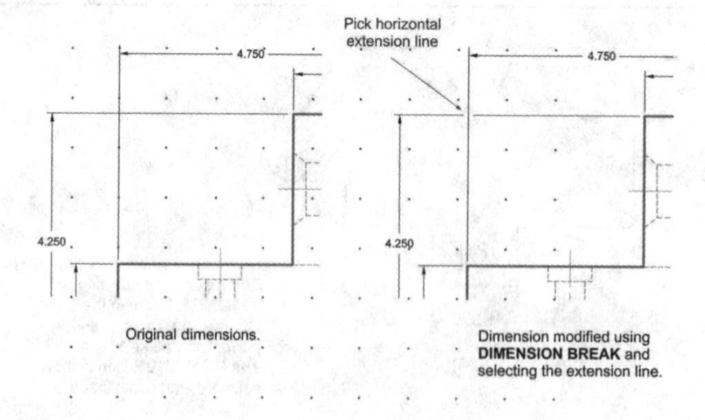
Pick horizontal extension line
4.750
4.250
4.750
4.250
Original dimensions.
Dimension modified using DIMENSION BREAK and selecting the extension line.

5.41(b) **Dimension Break** Command

TOLERANCE Command			
	Ribbon	Annotate	The icon for the **Tolerance** command is shown in Figure 5.42(a). This command is used to specify the symbols and values for geometric dimensioning and tolerancing, as shown in Figure 5.42(b).
	Panel	Dimensions	
	Command Line	Tolerance	
5.42(a)	Alias	TOL	

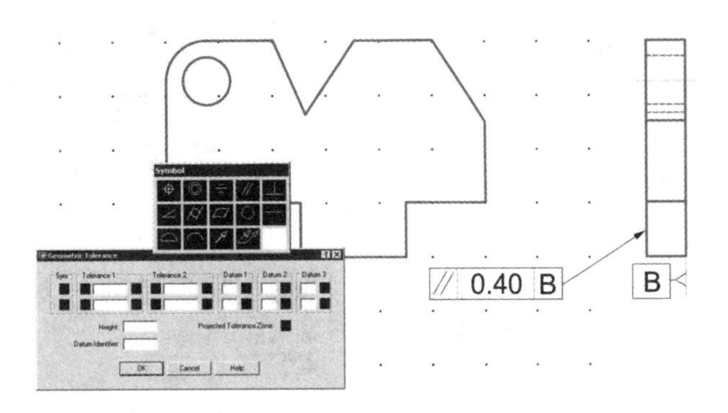

5.42(b) **Tolerance** Command

CENTER MARK Command			
	Ribbon	Annotate	The icon for the **Center Mark** command is shown in Figure 5.43(a). This command is used to create center marks or centerlines on circles and arcs, as shown in Figure 5.43(b). The option for centerlines or center marks can be found in the **Dimension Styles** dialog box in the **Symbols and Arrows** tab.
	Panel	Dimensions	
	Command Line	DimCenter	
5.43(a)	Alias	DCE	

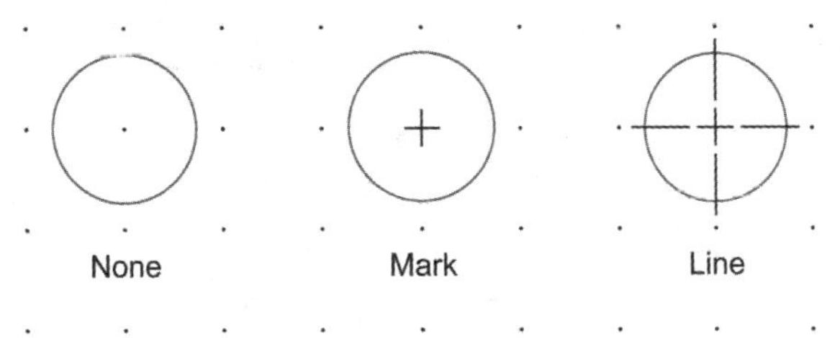

5.43(b) **Center Mark** Command

INSPECTION Command			
	Ribbon	Annotate	The icon for the **Inspection** command is shown in Figure 5.44(a). This command creates a dimension inside a frame that is used to provide inspection information about the feature, as shown in Figure 5.44(b).
	Panel	Dimensions	
	Command Line	DimInspect	
5.44(a)	Alias	DIMINSPECT	

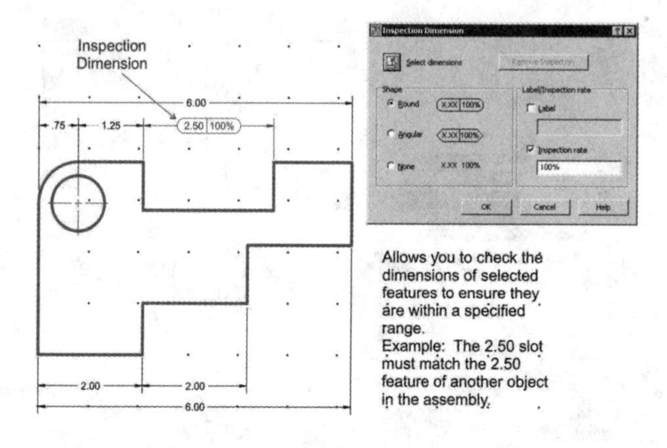

5.44(b) **Inspection** Command

JOGGED LINEAR Command			
	Ribbon	Annotate	The icon for the **Jogged Linear** command is shown in Figure 5.45(a). This command is used to create a jog in a linear dimension line when the feature is not drawn full size, as shown in Figure 5.45(b).
	Panel	Dimensions	
	Command Line	DimJogLine	
5.45(a)	Alias	DIMJOGLINE	

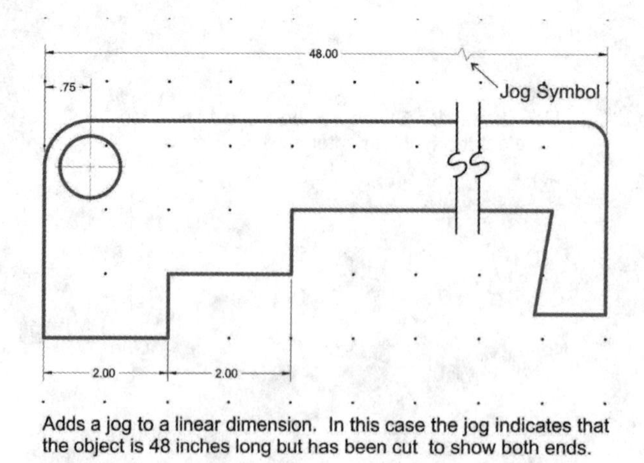

5.45(b) **Jogged Linear** Command

DIMENSION TEXT EDIT Command			
(icon)	Ribbon	Annotate	The icon for the **Dimension Text Edit** command is shown in Figure 5.46(a). This command is used to move and rotate dimension text, as shown in Figure 5.46(b).
	Panel	Dimensions	
	Command Line	DimTEdit	
5.46(a)	Alias	DIMTEDIT	

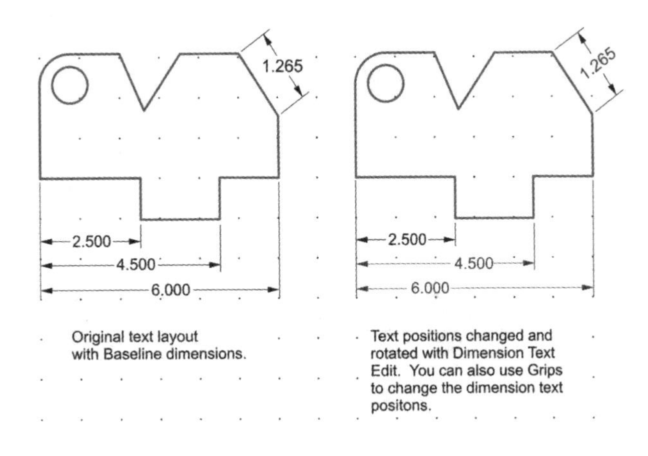

5.46(b) **Dimension Text Edit** Command

DIMENSION EDIT Command			
(icon)	Ribbon	Annotate	The icon for the **Dimension Edit** command is shown in Figure 5.47(a). This command is used to edit existing dimensions, as shown in Figure 5.47(b). Options in this command include **Home**, which changes rotated dimensions back to the default position; **New**, which changes dimension text with the **Multiline Text Editor; Rotate**, which rotates dimension text; and **Oblique**, which changes extension lines to oblique angles.
	Panel	Dimensions	
	Command Line	DimEdit	
5.47(a)	Alias	DED	

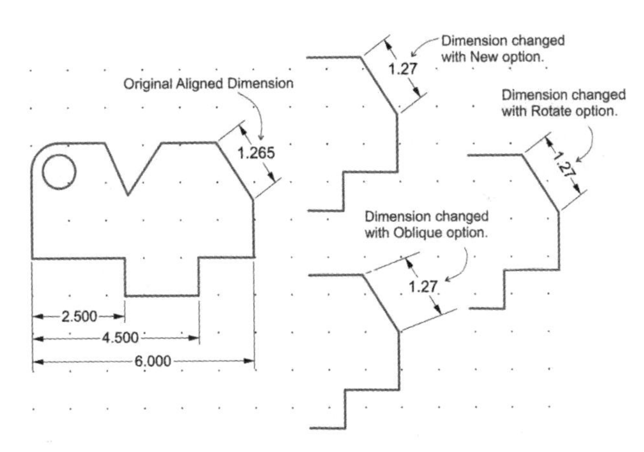

5.47(b) **Dimension Edit** Command

DIMENSION UPDATE Command

	Ribbon	Annotate	The icon for the **Dimension Update** command is shown in Figure 5.48(a). This command is used when selecting the individual dimensions that will be updated by dimension style overrides defined in the **Dimension Style Manager** dialog box (this is discussed later in this chapter). See Figure 5.48(b).
	Panel	Dimensions	
	Command Line	-DimStyle	
5.48(a)	Alias	-DIMSTYLE	

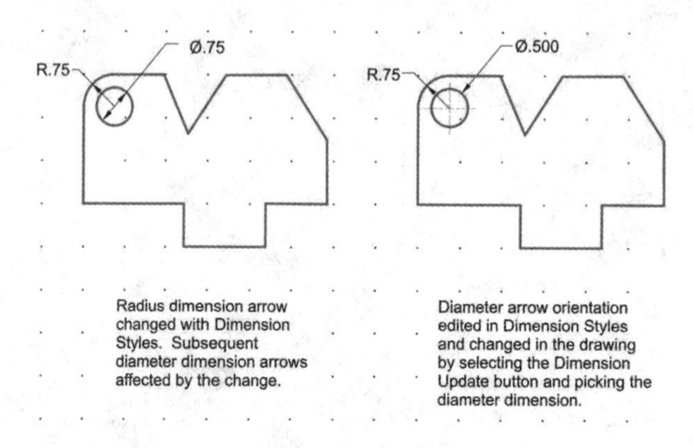

Ø.75
R.75
Ø.500
R.75
Radius dimension arrow changed with Dimension Styles. Subsequent diameter dimension arrows affected by the change.
Diameter arrow orientation edited in Dimension Styles and changed in the drawing by selecting the Dimension Update button and picking the diameter dimension.

5.48(b) **Dimension Update** Command

DIMENSION STYLE Command

	Ribbon	Annotate	A *dimension style* is a named set of values that define the appearance and format of dimensions, such as text height, precision, and arrowhead length. These values are assigned to the dimension style by entering them in the **Dimension Style Manager** dialog box. To open the **Dimension Style Manager** dialog box, click on the **Dimension Style** icon shown in Figure 5.49(a). The **Dimension Style Manager** dialog box is shown in Figure 5.49(b).
	Panel	Dimensions	
	Command Line	DimStyle	
5.49(a)	Alias	D	

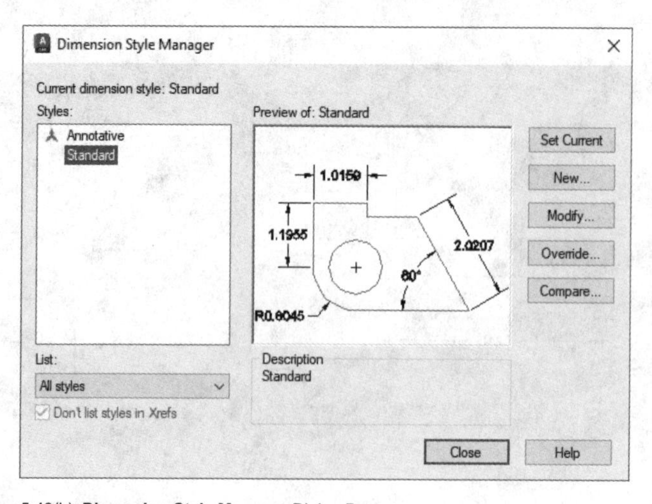

5.49(b) **Dimension Style Manager** Dialog Box

5.9 DEFINING DIMENSION SETTINGS WITH THE DIMENSION STYLE MANAGER

Before adding dimensions to an AutoCAD drawing, the drafter must set the values in the **Dimension Style Manager** that control the appearance and format of the dimensions. For example, entering the *ASME Y14.5* dimensioning standard's spacing, size, and formatting values (see Figure 5.5) in the **Dimension Style Manager** dialog box *before* dimensioning the drawing will ensure that when dimensions are added to the drawing they will automatically comply with the ASME standard.

The **Dimension Style Manager** dialog box, shown in Figure 5.49(b), can be opened either by selecting the **Dimension Style** icon located on the **Dimension** toolbar, by clicking on the arrow in the lower right corner of the Dimensions panel of the **Annotate** tab of the ribbon, or by typing **DIMSTYLE** and pressing **<Enter>**. By choosing from the buttons located on the right side of the **Dimension Style Manager** dialog box, shown in Figure 5.49(b), it is possible to create a new dimension style, modify or override the current dimension style, set a different dimension style current, or compare the settings of two dimension styles. Figure 5.50 provides detailed information about the functions of each of the buttons in this dialog box.

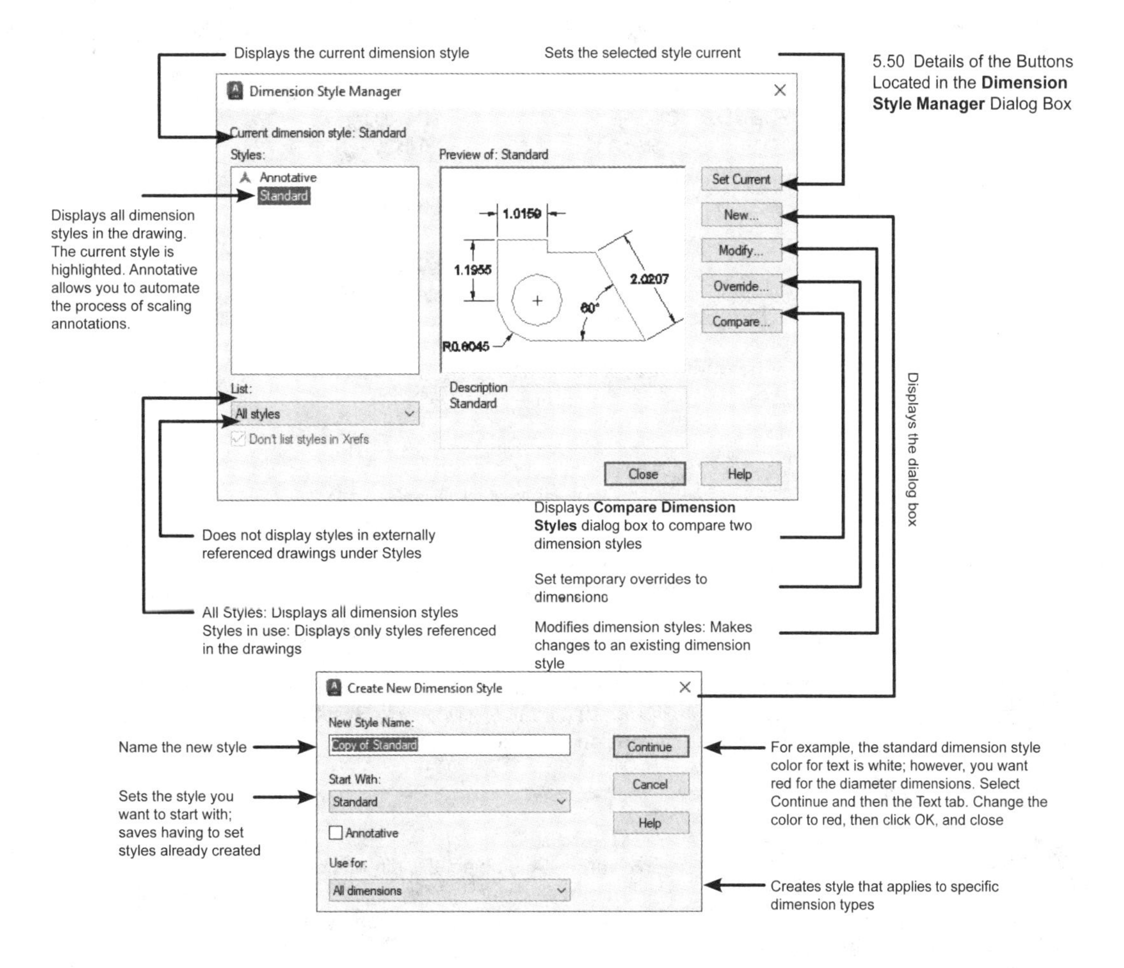

5.50 Details of the Buttons Located in the **Dimension Style Manager** Dialog Box

Selecting the Compare button on the **Dimension Style Manager** dialog box will open the **Compare Dimension Styles** dialog box. Selecting two different dimension styles in the **Compare** and **With** windows of this dialog box will display a comparison of the dimension style settings assigned to each of the dimension styles (see Figure 5.51).

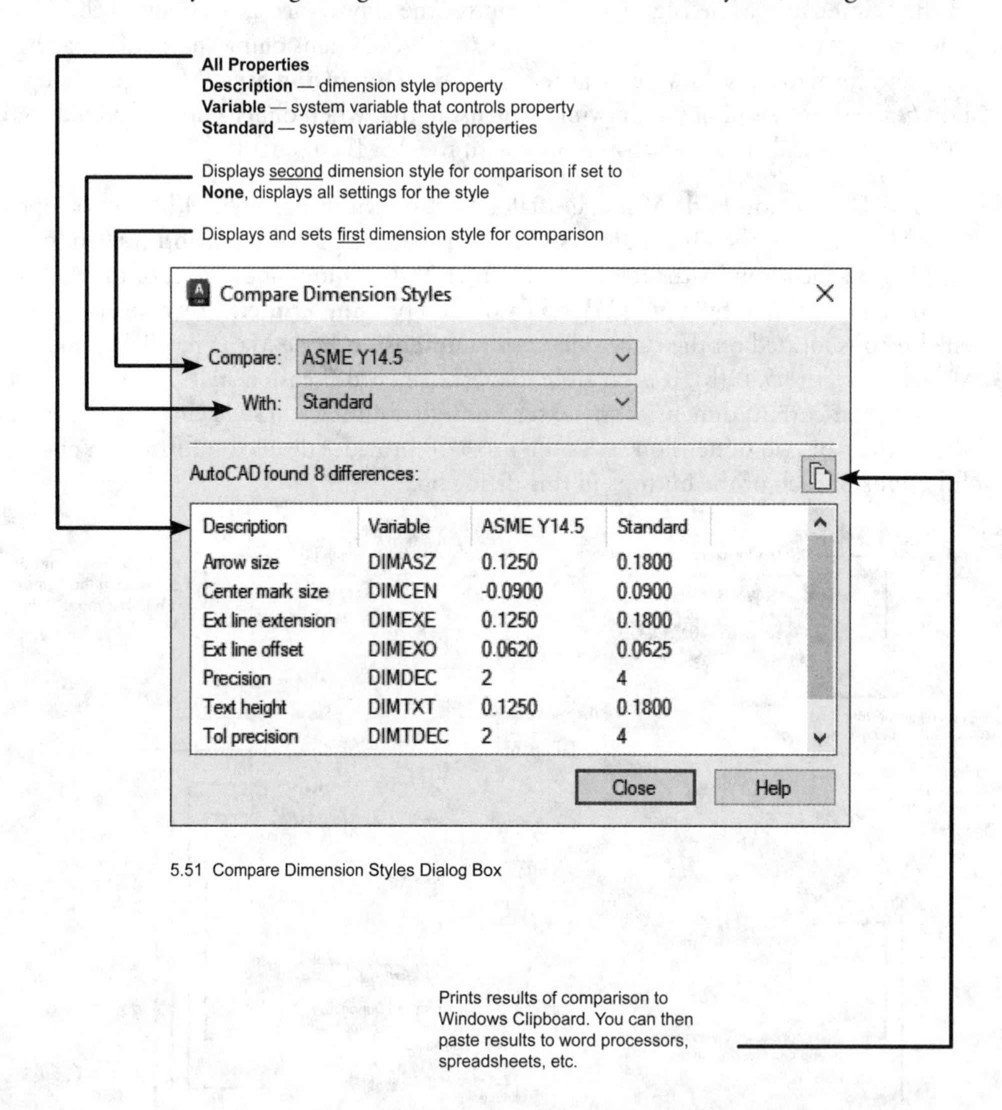

5.51 Compare Dimension Styles Dialog Box

A good way to check dimension system variables for a dimension in your drawing is to type or select List, then pick the dimension you wish to check. System variables and their properties will be listed for the dimension selected.

Another way to check a dimension's variables is select the dimension, right-click your mouse, and select Properties. The Properties palette will open, displaying the dimension settings for the dimension. Many of these settings can be edited by changing the values shown in the Properties palette.

STEP-BY-STEP

CREATING A NEW DIMENSION STYLE

Use the following steps to create a new dimension style:

Select the **Dimension Styles** icon, and when the **Dimension Style Manager** dialog box opens, select the **New** button.

Step 1. When the **Create New Dimension Style** box opens (see Figure 5.52), enter a name for the new dimension style in the **New Style Name** window. In Figure 5.52, **ASME Y14.5** has been entered as the name of the new style.

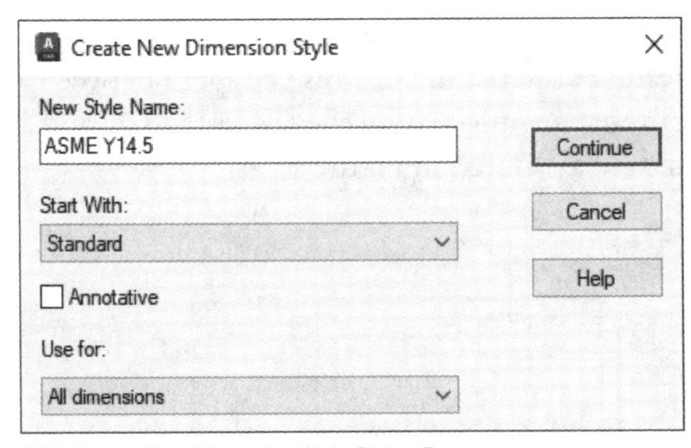

5.52 Create New Dimension Style Dialog Box

Selecting a style that already has some of the desired dimension settings in place can speed up the process of creating a new style. The *Start With* style defaults to the settings in AutoCAD's *Standard* dimension style unless a different text style is chosen from the drop-down list in this window.

Step 2. Locate the **Start With** window in Figure 5.52. AutoCAD uses the settings of the style shown in this window as a *template* for creating the new style.

Step 3. Select the **Continue** button, and when the **New Dimension Style** dialog box opens select from among the **Lines**, **Symbols and Arrows**, **Text**, **Fit**, **Primary Units**, **Alternate Units**, and **Tolerances** tabs (these tabs are explained in detail below) and enter values for dimension features such as text height, arrowhead style and size, and dimension spacing. Refer to Figures 5.54 through 5.61(b) to see the settings included on each of these tabs. When you have completed making these settings, click **OK**, and the **Dimension Style Manager** dialog box will reappear.

Step 4. Set the new dimension style current by selecting the **ASME Y14.5** style from the styles shown in the **Styles** pane of the **Dimension Style Manager** dialog box and click the **Set Current** button, shown in Figure 5.53. Pick the **Close** button to close the dialog box. When dimensions are added to the drawing, their settings will reflect the new dimension style.

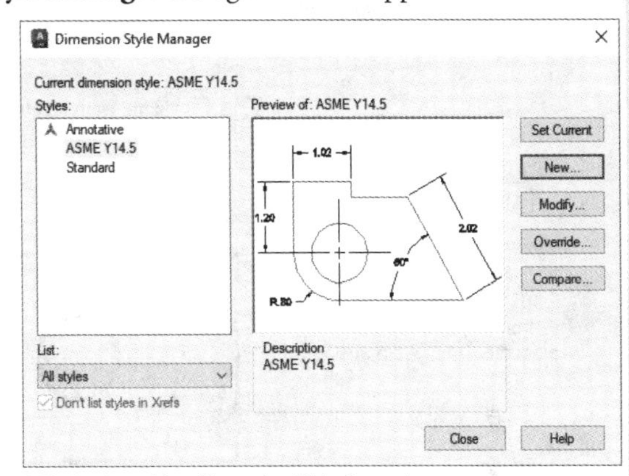

5.53 Setting a New Dimension Style Current

STEP–BY–STEP

Tabs of the New Dimension Style Dialog Box

Step 4 of the *Creating a New Dimension Style* section refers to selecting from among the **Lines, Symbols and Arrows, Text, Fit, Primary Units, Alternate Units,** and **Tolerances** tabs of the **New Dimension Style** dialog box to enter the values for the dimension features of the new dimension style. The settings for each of these tabs are described next.

Lines Tab

The **Lines** tab controls settings, such as the distance an extension line is offset from the object or extends past an arrowhead. This tab also controls the space between baseline dimensions. The settings of this tab that are affected by the suggested size and spacing guidelines of the *ASME Y14.5* standard are noted in Figure 5.54.

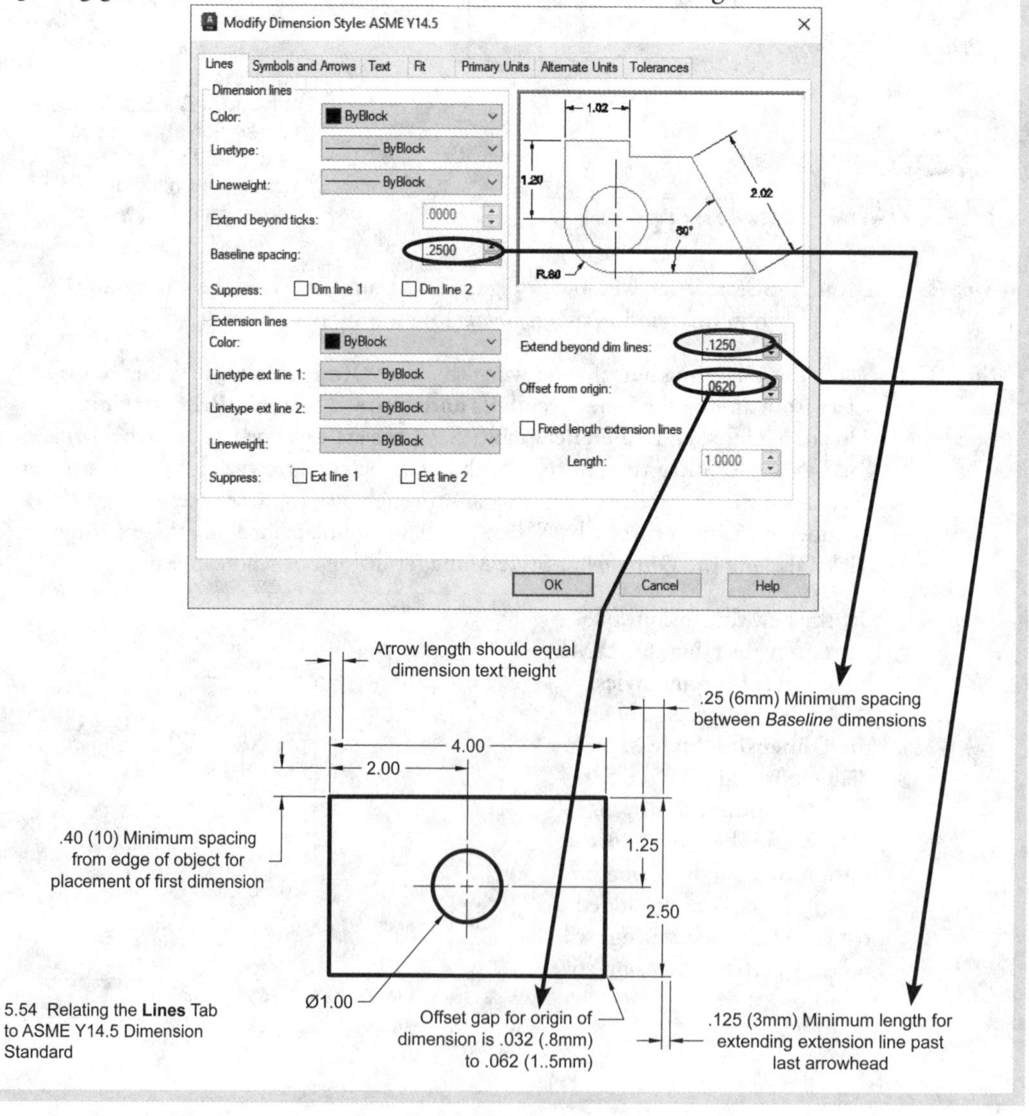

5.54 Relating the **Lines** Tab to ASME Y14.5 Dimension Standard

STEP–BY–STEP

Symbols and Arrows Tab

The **Symbols and Arrows** tab controls the size and type of arrowheads, including architectural tick marks, and the style of center marks used to dimension circles and arcs. The settings of this tab that are affected by the suggested size and spacing guidelines of the *ASME Y14.5* standard are noted in Figure 5.55.

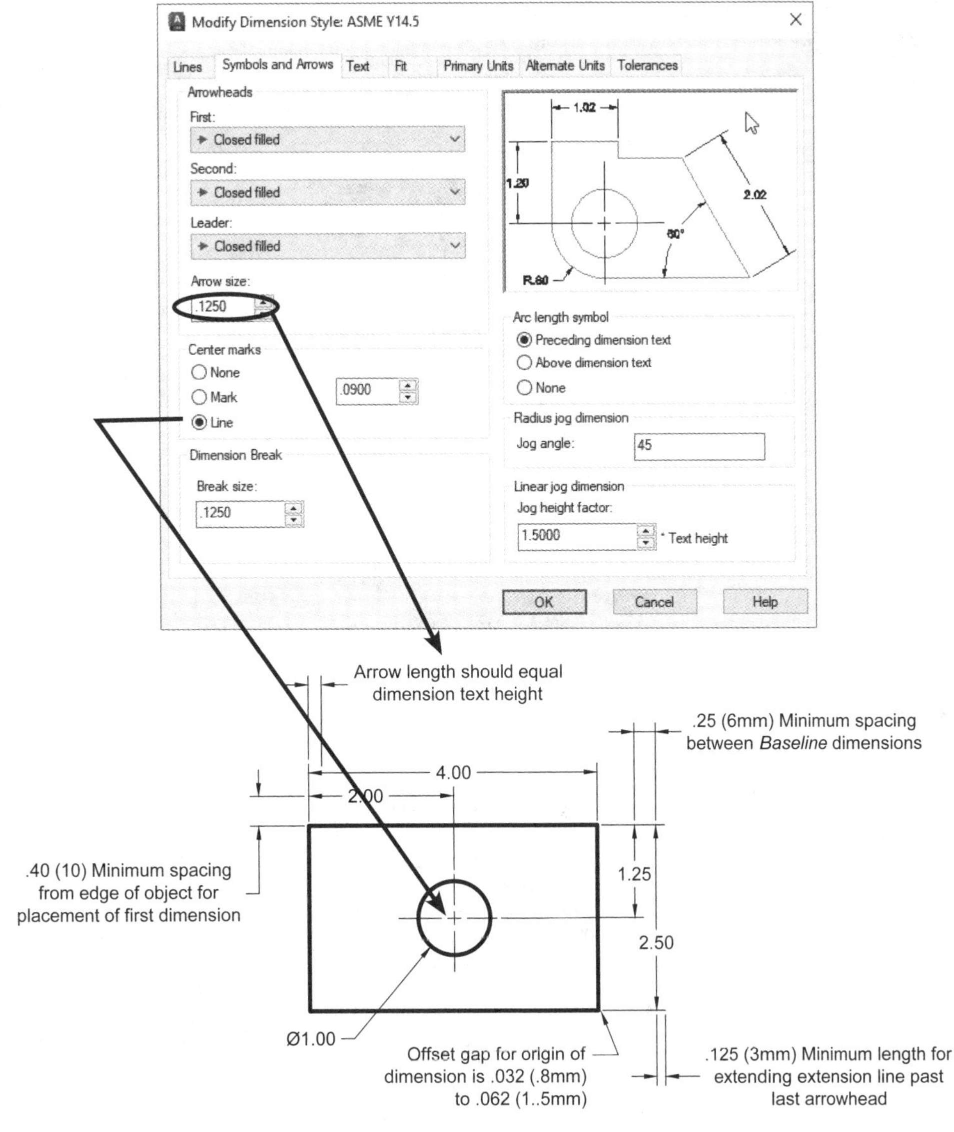

5.55 Relating the **Symbols and Arrows** Tab to ASME Y14.5 Dimension Standard

STEP–BY–STEP

Text Tab

The **Text** tab controls the text style, height, placement, and alignment of dimension text. The text style is based on the text properties defined in the **Text Style** dialog box. See Figure 5.56.

Text height for dimensions should not measure less than .12" (3mm) to comply with ASME standards for text height.

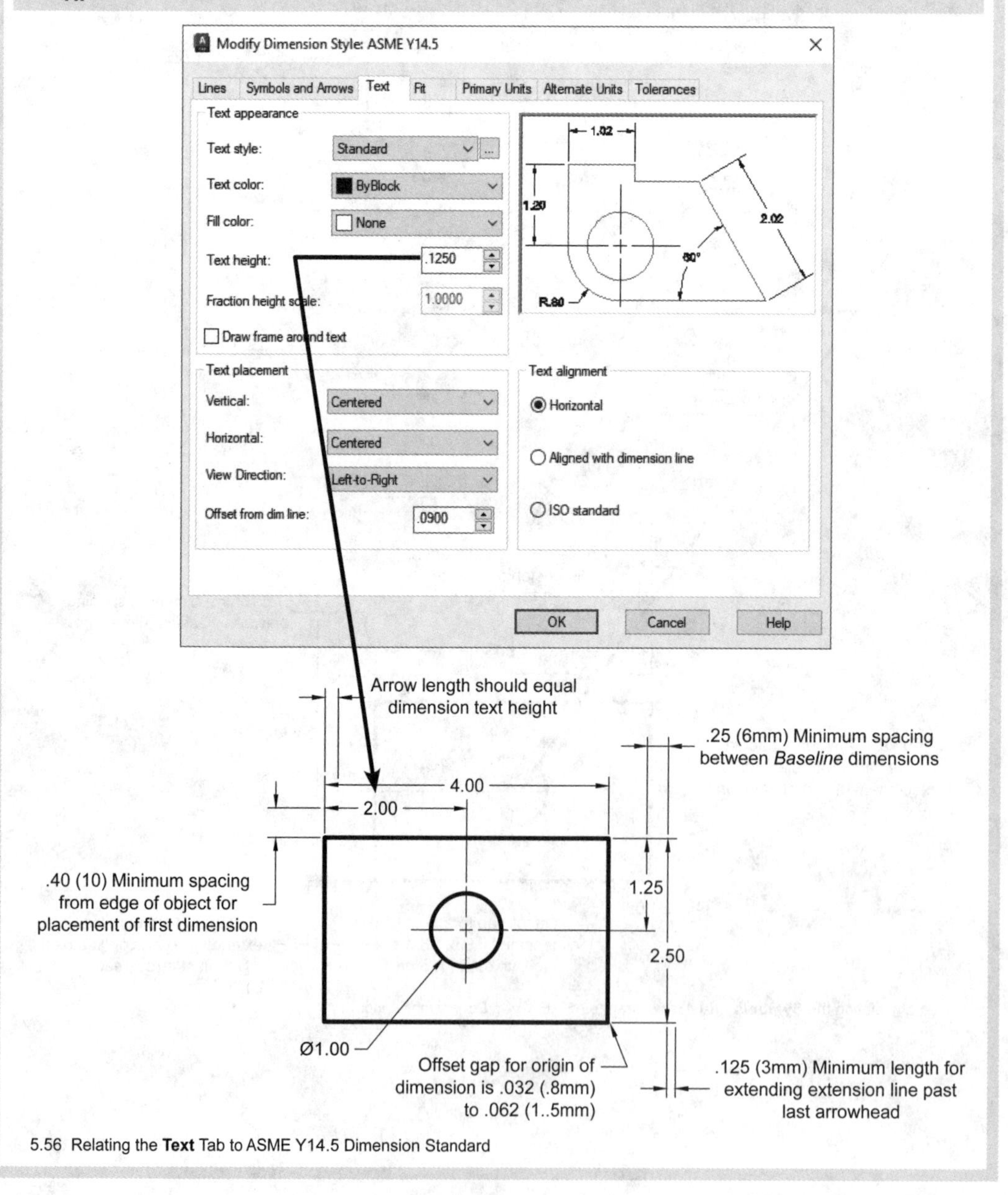

5.56 Relating the **Text** Tab to ASME Y14.5 Dimension Standard

STEP–BY–STEP

Fit Tab

The **Fit** tab controls the placement of text and the orientation of text and arrows on dimension leaders applied to diameters and radii. Different combinations of settings from this tab can be used to force text between arrows or to force arrows outside, for example. Study the examples in Figures 5.57(a) through 5.57(d) to see how different combinations result in different leader arrow placement and format.

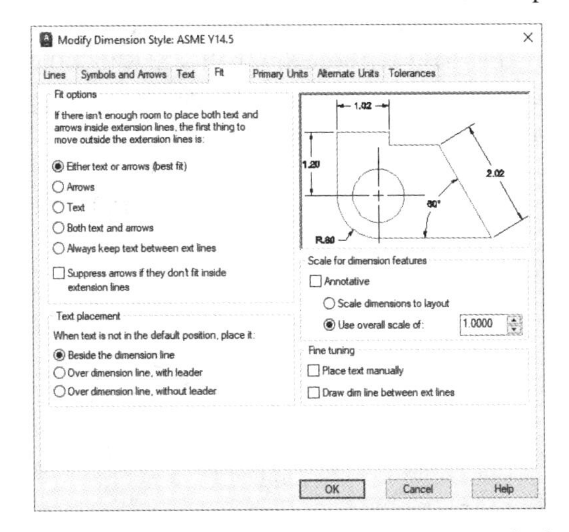

5.57(a) **Fit** Tab Settings for Controlling Arrow Orientation on Small Circles and Radii

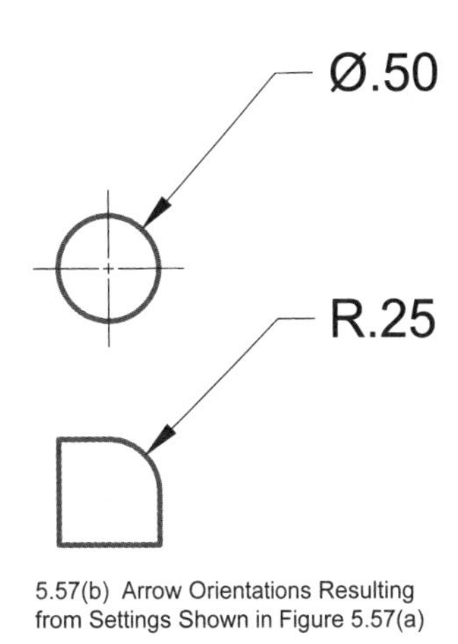

5.57(b) Arrow Orientations Resulting from Settings Shown in Figure 5.57(a)

QUICK TIP

Settings shown in Figure 5.57(a) might be used in a ***Small Radii*** dimension style and the settings shown in Figure 5.57(c) might be used in a ***Large Radii*** dimension style

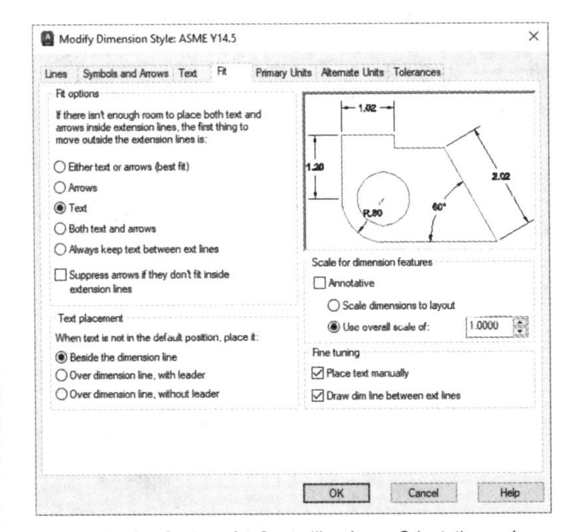

5.57(c) **Fit** Tab Settings for Controlling Arrow Orientation on Large Circles and Radii

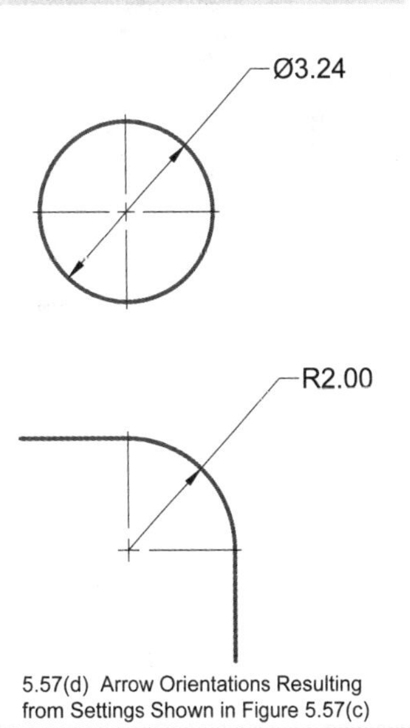

5.57(d) Arrow Orientations Resulting from Settings Shown in Figure 5.57(c)

STEP–BY–STEP

Primary Units Tab

The **Primary Units** tab controls the format of units (decimal, architectural, engineering, etc.) and the precision (the number of decimal places or fractional round-off) of dimensions. By checking the **Leading** box in the **Zero suppression** pane, the leading zero on decimal dimensions less than 1.00 unit in size will be suppressed. The setting in the **Scale factor** window located in the **Measurement scale** pane determines the numeric scale factor of the dimension display (see Figure 5.58). For example, if you add dimensions to a (model space) view that has been scaled to half-size (.5X), the dimensions will be displayed at half their full-scale value. By setting the **Scale factor** to 2 in the **Primary Units** tab, the numeric value for the dimensions will be multiplied by a factor of two and the view's full-size dimensions will be displayed. Likewise, if the view has been scaled to double-size (2X), setting the **Scale factor** to **.50** will display the dimension values at full size.

Alternate Units Tab

The **Alternate Units** tab allows dual dimensions to be shown side by side on the drawing; for example, decimal units shown alongside metric units. The alternate unit will be placed inside brackets. See Figure 5.59.

Tolerances Tab

The **Tolerances** tab allows tolerances to be incorporated into dimension text. Tolerances may be shown as **Limits** by applying the settings in Figures 5.60(a) and 5.60(b), or as **Symmetrical** (plus/minus) dimensions by applying the settings shown in Figures 5.61(a) and 5.61(b).

JOB SKILL

To change the dimension style of a dimension that has already been placed on the drawing, double-click on the dimension and when the Properties palette opens, left-click in the field next to Dim Style, and pick on the down arrow and select a different dimension style from the drop-down list.

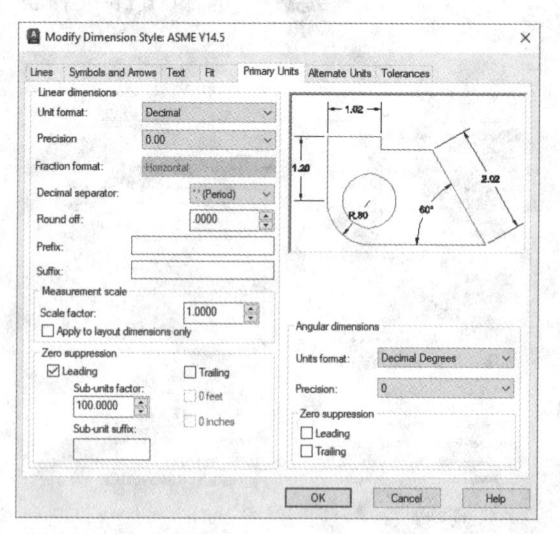

5.58 **Primary Units** Tab

STEP–BY–STEP

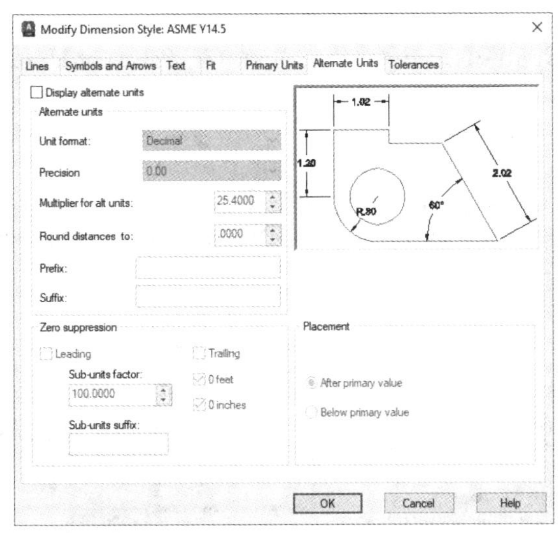

5.59 **Alternate Units** Tab

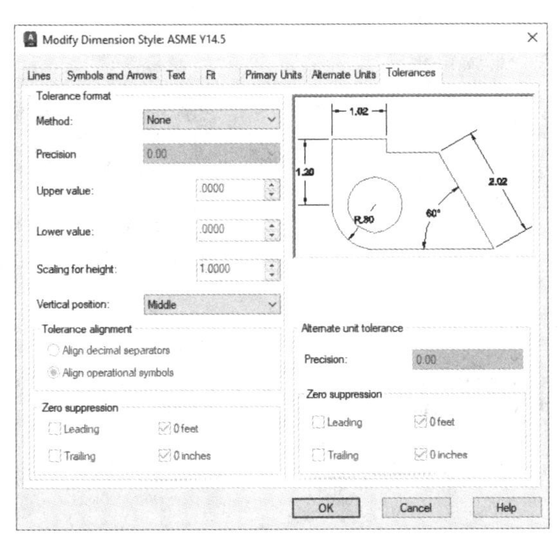

5.60(a) **Tolerances** Tab

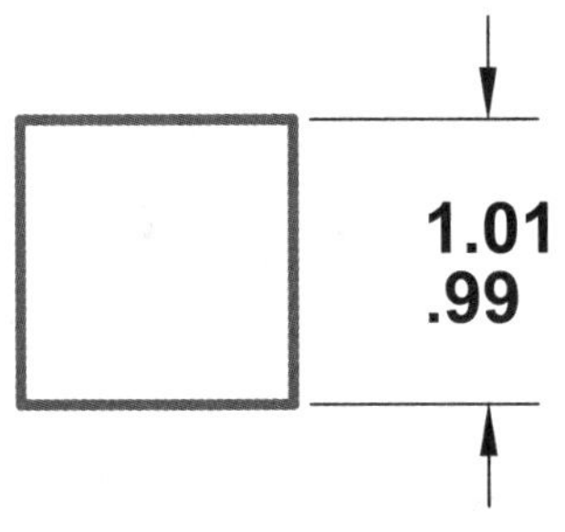

5.60(b) Example of Limits Dimensions

STEP-BY-STEP

Modifying a Dimension Style

By selecting the **Modify** button from the **Dimension Style Manager** dialog box, the **Modify Dimension Style** box will open. Like the **New Dimension Style** dialog box, this dialog box contains the **Lines**, **Symbols and Arrows**, **Text**, **Fit**, **Primary Units**, **Alternate Units**, and **Tolerances** tabs. The settings included in these tabs are exactly the same as the ones presented earlier in the *Tabs of the New Dimension Style Dialog Box* section. Refer to Figures 5.52 through 5.61(b) to see the settings included on each of these tabs. When the desired settings have been entered into the tabs, click the **OK** button in the **Modify Dimension Style** box and when the **Dimension Style Manager** dialog box appears, click **Close**.

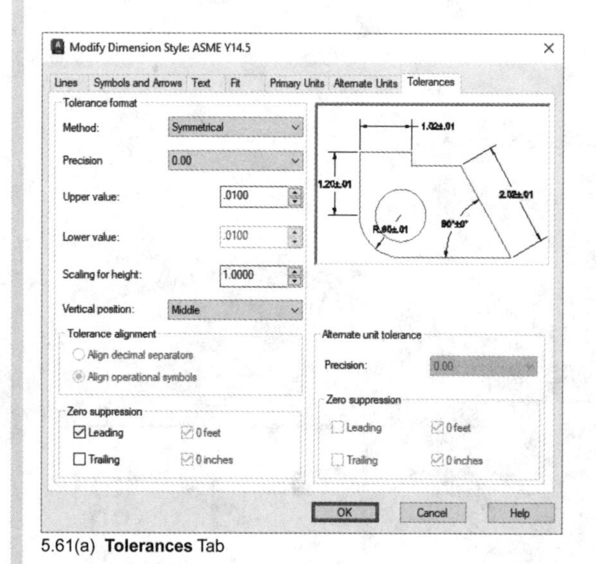

5.61(a) **Tolerances** Tab

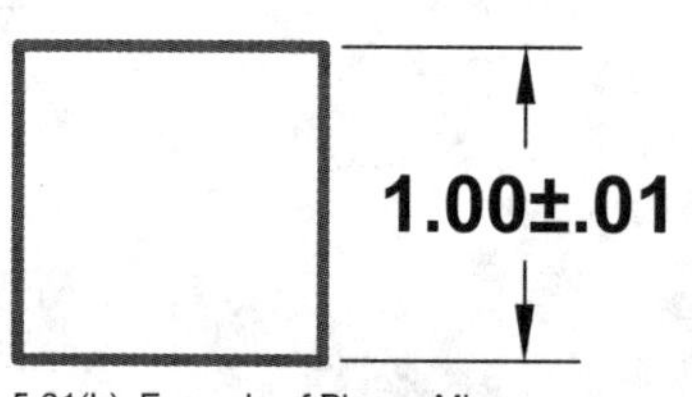

5.61(b) Example of Plus or Minus (Symmetrical) Dimensions

VIDEO

There are video tutorials available for the Dimensioning commands inside the Chapter 5 folder of the book's Video Training downloads.

QUICK TIP

Many experienced CAD users choose to create a new dimension style with custom settings rather than modify the settings of the Standard dimension style.

5.10 OVERRIDING A DIMENSION SETTING

Sometimes it is necessary to have a different dimension setting apply to only a few dimensions. In a case like this, a dimension style override can be performed.

To override a dimension style, open the **Dimension Style Manager** dialog box and select the **Override** button (see Figure 5.62). Next, select the appropriate tab(s) and assign the new setting(s), click **OK** then click **Close** to exit the **Dimension Style Manager**.

Updating a Dimension

Select the **Dimension Update** icon located on the **Dimension** toolbar or the **Dimensions** panel of the **Annotate** tab (see Figure 5.63), and select the dimension(s) to which you want to apply the overridden setting(s) and press **<Enter>**. The dimension's style will update to reflect the new setting(s).

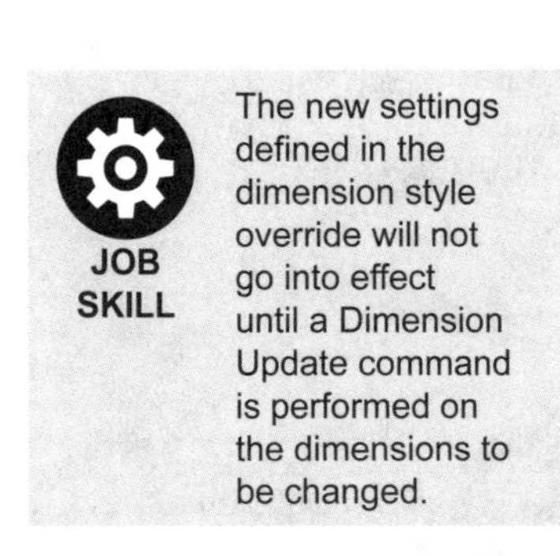

JOB
SKILL
The new settings defined in the dimension style override will not go into effect until a Dimension Update command is performed on the dimensions to be changed.

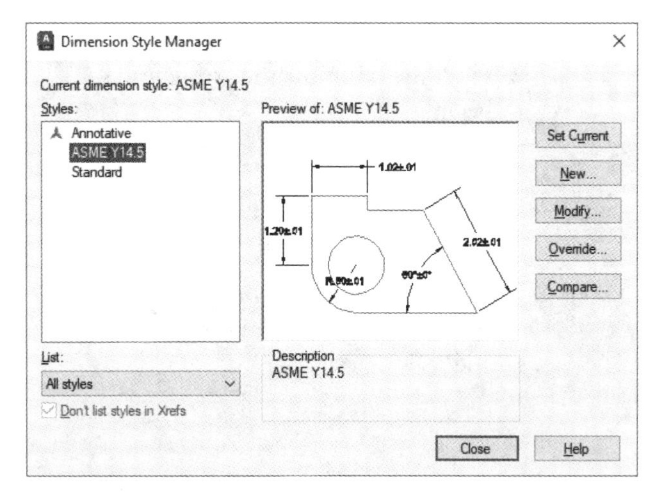

5.62 Override Dimension Style

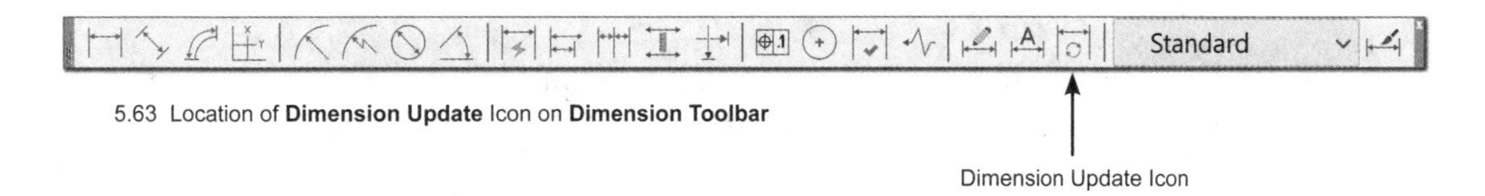

5.63 Location of **Dimension Update** Icon on **Dimension Toolbar**

5.11 ADDING A LEADER TO A DRAWING

A leader is an annotation created by drawing a line (or a spline) with an arrowhead at one end and text at the other end. Figure 5.64(a) shows a leader created for a mechanical engineering drawing including the three components of a leader: the arrowhead (which can be assigned a symbol other than an arrowhead), the leader line, and the landing line. Leaders are placed in drawings using either the **Quick Leader** or **Multileader** command.

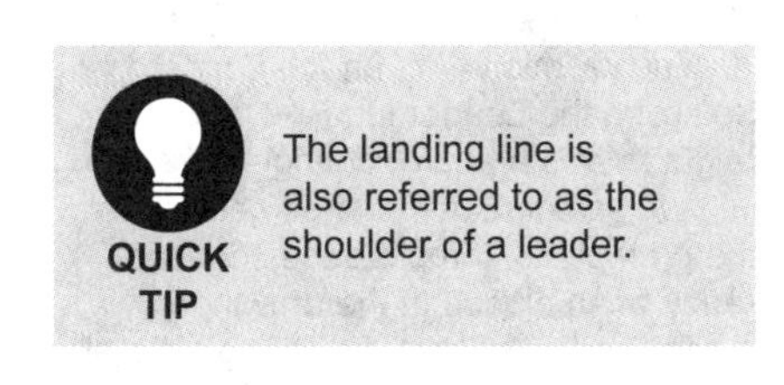

QUICK
TIP
The landing line is also referred to as the shoulder of a leader.

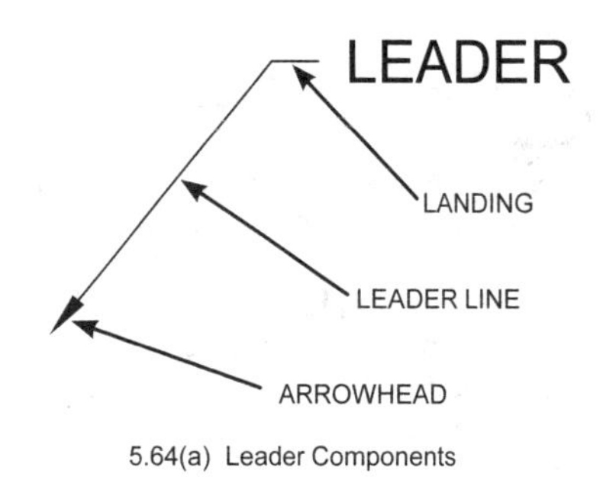

5.64(a) Leader Components

STEP-BY-STEP

QUICK LEADER COMMAND TUTORIAL (MECHANICAL STYLE LEADER)

Step 1. Type **QLEADER** on the command line and press **<Enter>**.

Step 2. At the *Specify first leader point:* prompt, select the point for the arrowhead to begin.

QUICK TIP

The alias for QLEADER is LE.

Step 3. At the *Specify next point:* prompt, select the point for the leader line to end.

Step 4. At the *Specify next point:* prompt, select the point where the landing line should end.

QUICK TIP

Turn Ortho *on* to draw a horizontal landing line.

Step 5. At the *Specify text width:* prompt, either press **<Enter>**, or type a width value and press **<Enter>**.

Step 6. At the *Enter first line of annotation text:* prompt, enter the text for the first line of the leader annotation in the Text box and press **<Enter>**.

Step 7. At the *Enter next line of annotation:* prompt, enter the text for the next line, or to end the command, press **<Enter>**.

JOB SKILL

To change the settings of a Quick Leader, begin the command, and when prompted for the first leader point, type S (for Settings) and <Enter>, this will open the Leader Settings dialog box, which contains three tabs. Selecting the Leader Line & Arrow tab allows you to change the leader line from straight to spline and/or change the arrowhead style (among other things). When you have selected the desired settings in the dialog box, select the OK button and follow the prompts to complete the command. To change the properties of a leader that has already been placed in the drawing, double-click on the leader and edit the fields in the Properties palette. To change the text of an existing leader, double-click on the text and make the changes inside the Text box.

STEP-BY-STEP

MULTILEADER COMMAND TUTORIAL

Step 1. Type **MLEADER** (or **MLD**) on the command line and press **<Enter>**, or select the **Multileader** tool from either the **Annotation** panel of the **Home** tab of the ribbon or from the **Leaders** panel of the **Annotate** tab. See Figure 5.64(b).

QUICK TIP The alias for Mleader is MLD.

Step 2. At the *Specify first leader point:* prompt, select the point for the arrowhead to begin.

Step 3. At the *Specify leader landing location:* prompt, select the point for the leader line to end.

Step 4. When the **Text** box opens, enter the desired text; to end the command, pick a point in the drawing window outside the **Text** box.

VIDEO There are video tutorials available for the Dimensioning commands inside the Chapter 5 folder of the book's Video Training downloads.

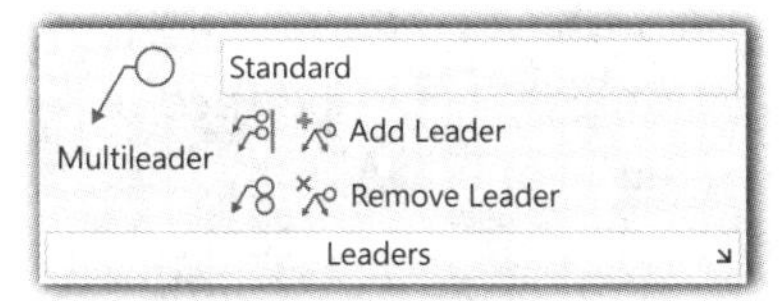

5.64(b) **Multileader** Tool Located on the **Leaders** Panel of the **Annotate** Tab

MODIFYING OR CREATING A MULTILEADER STYLE

The **Multileader** tool defaults to the settings in effect for the **Standard** multileader style found in the **Multileader Style Manager** dialog box. These settings include leader format (straight or spline), arrowhead type and size, and text height. However, it is possible to modify the **Standard** style settings or to create a new style(s) by changing the settings in this dialog box. Creating multiple styles allows drafters to utilize leaders with different style settings within the same drawing.

To modify an existing multileader style, or create a new style, follow these steps:

Step 1. Select the arrow next to the word **Leaders** on the **Leaders** panel of the **Annotate** tab of the ribbon.

Step 2. When the **Multileader Style Manager** dialog box opens, select the **Modify** button to modify the **Standard** style, or the **New** button to define a new style. Selecting either button opens a new dialog box containing three tabs. Make the desired changes in the windows of these tabs and select the **OK** button.

Step 3. If you are modifying the *current* style, select the **Close** button, and the settings will apply the next time the **Multileader** command is activated. If you created a new style in Step 2, left-click on the new style name in the **Styles** window then select the **Set Current** button and click the **Close** button. The new settings will apply the next time the **Multileader** tool is activated.

JOB SKILL Another way to change the settings of a multileader is to select the Multileader tool, type O (for Options), press <Enter> and choose a settings option from the pop-up list. For example, if you choose the Leader type option, you can choose either straight (for mechanical notes), or Spline (for architectural notes). After selecting the desired options, choose eXit options and follow the prompts to complete the Multileader command. To change the properties of a leader that has already been placed in the drawing, double-click on the leader and edit the fields in the Properties palette.

CHAPTER REVIEW

Dimensions communicate important information about the size and location of the features of an object. In the United States, an industry-wide standard for dimensioning mechanical drawings is published by the American Society of Mechanical Engineers. This standard is known as ASME Y-14.5. This standard guides the format, style, and placement of dimensions. Drafters who are employed by organizations that adopt the ASME standard are responsible for dimensioning objects in accordance with its guidelines.

Mechanical engineers and designers are responsible for calculating the dimensions to be included on a technical drawing, but drafters are responsible for creating drawings that accurately reflect the designer's intentions. In the mechanical engineering field, drafters must ensure that each dimension on the drawing reflects the exact dimensional value, including the exact degree of precision (the number of decimal places), as the designer's input. Drafters must also ensure that the dimensions on the drawing reference the same datum geometry that the designer specified in the input. If the drafter fails to portray the designer's input faithfully, the manufactured part may not fit or function as the designer intended. Also, parts that do not pass a quality control inspection may have to be scrapped, thus hampering the organization's ability to produce the product in the desired time frame, or worse, prevent the organization from profitably manufacturing the product.

Drafters must not only be familiar with dimensioning theory and standards, but they must also be able to manipulate the dimension tools and style settings of the CAD program they are using to create the drawing.

KEY WORDS

Aligned Text
American National Standards Institute (ANSI)
Baseline Dimensions Checker
Computer Aided Manufacturing (CAM)
Continuous Dimension
Datum
Dimension Standards
Geometric Dimensioning and Tolerancing (GDT)
International Standards Organization (ISO)
Nominal Size
Quality Control Inspection
Tolerances
Unidirectional Text

REVIEW

Short Answer

1. What is a tolerance?
2. What does ISO stand for?
3. What is a Datum?
4. Who is responsible for calculating tolerances on an engineering drawing?
5. Is the diameter of a hole a size or a location dimension?
6. In the United States, what is the name of the organization that publishes a dimensioning standard for engineering drawings?
7. What does the term CAM stand for?
8. What is a reference dimension?
9. What is meant by nominal size?
10. Which method of labeling text complies with the ASME standard: unidirectional or aligned?

Matching

Column A

a. Primary Units
b. Lines
c. Text
d. Fit
e. Symbols and Arrows

Column B

1. The Dimension Style Manager tab where dimension arrow size is defined
2. The Dimension Style Manager tab where dimension height is defined
3. The Dimension Style Manager tab where dimension precision is defined
4. The Dimension Style Manager tab where the orientation of arrowheads on leaders for circles and arcs is defined
5. The Dimension Style Manager tab where center mark style is defined

Step 1. Open the **Bracket** drawing you created in Chapter 4.

Step 2. Move the views away from each other to make room for the dimensions. Turn on Ortho to make sure they're still aligned with each other.

Step 3. Create a new layer named **Dimensions** and set it current.

5.65 Designer's Sketch of Bracket

Step 4. In the **Dimension Style Manager** dialog box:

a. Create two new dimension styles named **ASME Y14.5-Small Radii and ASME Y14.5-Large Radii** that contain the following dimension style settings. The only difference between these two styles will be in the Fit tab settings as shown in the table below. (Do not enable the settings of the **Alternate Units** or **Tolerances** tabs):

Dimension Style Settings for the Bracket:		
Dimension Style Tab	**Setting**	**Value**
Lines	Extend beyond dim lines	.125
	Offset from origin	.062
Symbols & Arrows	Arrow size	.125
	Center marks	Line
Text	Text height	.125
	Text style	Standard (font should be set to Arial)
Primary Units	Precision	0.00
	Zero suppression	Leading
Fit	See Figure 5.57(a) on page 287 and Figure 5.57(c) on page 287 for **Small** and **Large Radii** fit settings	

b. Set the new dimension style current.

Step 5. Add dimensions and notations to the views of the Bracket (see Figure 5.65).

Step 6. Follow the dimensioning rules presented earlier in this chapter.

Step 7. Follow your instructor's directions to print the drawing when you are finished placing the dimensions.

VIDEO

There are video tutorials available for the Dimensioning commands inside the *Chapter 5* folder of the book's *Video Training* downloads.

Step 1. Open the **Shaft Guide** drawing you created in Chapter 4.

Step 2. Move the views away from each other to make room for the dimensions. Turn on Ortho to make sure they're still aligned with each other.

Step 3. Create a new layer named **Dimensions** and set it current.

Step 4. In the **Dimension Style Manager** dialog box:

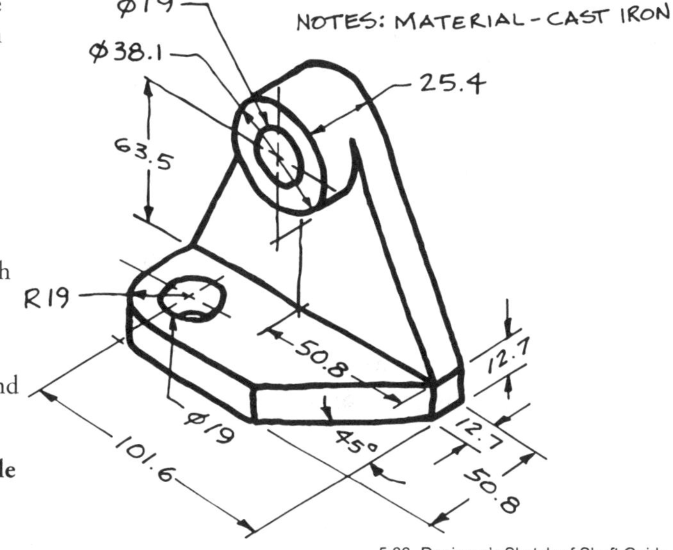

5.66 Designer's Sketch of Shaft Guide

a. Create two new dimension styles named **ASME Y14.5-Small Radii and ASME Y14.5-Large Radii** that contain the following dimension style settings. The only difference between these two styles will be in the Fit tab settings as shown in the table below. (Do not enable the settings of the **Alternate Units** or **Tolerances** tabs):

Dimension Style Settings for the Shaft Guide:		
Dimension Style Tab	**Setting**	**Value**
Lines	Extend beyond dim lines	3
	Offset from origin	1.5
Symbols & Arrows	Arrow size	3
	Center marks	Line
	Center mark size	2
Text	Text height	3
	Text style	Standard (font should be set to Arial)
	Offset from dim line	3
Primary Units	Precision	0.0
	Zero suppression	Trailing
Fit	See Figure 5.57(a) on page 287 and Figure 5.57(c) on page 287 for **Small** and **Large Radii** fit settings	

b. Set the new dimension style current.

Step 5. Add dimensions and notations to the views of the Shaft Guide (see Figure 5.66).

Step 6. Follow the dimensioning rules presented earlier in this chapter.

Step 7. Follow your instructor's directions to print the drawing when you are finished placing the dimensions.

QUICK TIP

Because the Shaft Guide is an SI drawing, the values of the dimension variables are in millimeters.

VIDEO

There are video tutorials available for the Dimensioning commands inside the *Chapter 5* folder of the book's *Video Training* downloads.

Step 1. Open the **Tool Holder** drawing you created in Chapter 4.

Step 2. Move the views away from each other to make room for the dimensions. Turn on Ortho to make sure they're still aligned with each other.

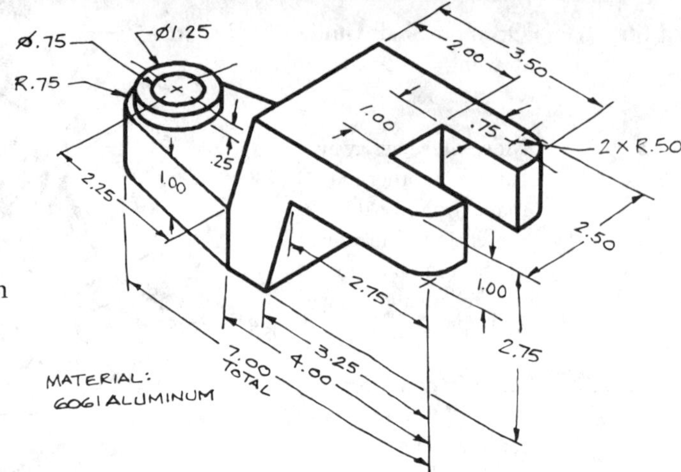

5.67 Designer's Sketch of Tool Holder

Step 3. Create a new layer named **Dimensions** and set it current.

Step 4. In the **Dimension Style Manager** dialog box:

a. Create two new dimension styles named **ASME Y14.5-Small Radii and ASME Y14.5-Large Radii** that contain the following dimension style settings. The only difference between these two styles will be in the Fit tab settings as shown in the table below. (Do not enable the settings of the **Alternate Units** or **Tolerances** tabs):

Dimension Style Settings for the Tool Holder:		
Dimension Style Tab	**Setting**	**Value**
Lines	Extend beyond dim lines	.125
	Offset from origin	.062
Symbols & Arrows	Arrow size	.125
	Center marks	Line
Text	Text height	.125
	Text style	Standard (font should be set to Arial)
Primary Units	Precision	0.00
	Zero suppression	Leading
Fit	See Figure 5.57(a) on page 287 and Figure 5.57(c) on page 287 for **Small** and **Large Radii** fit settings	

b. Set the new dimension style current.

Step 5. Add dimensions and notations to the views of the Tool Holder (see Figure 5.67).

Step 6. Follow the dimensioning rules presented earlier in this chapter.

Step 7. Follow your instructor's directions to print the drawing when you are finished placing the dimensions.

VIDEO

There are video tutorials available for the Dimensioning commands inside the *Chapter 5* folder of the book's *Video Training* downloads.

Step 1. Open the **Tool Slide** drawing you created in Chapter 4.

Step 2. Move the views away from each other to make room for the dimensions. Turn on Ortho to make sure they're still aligned with each other.

Step 3. Create a new layer named **Dimensions** and set it current.

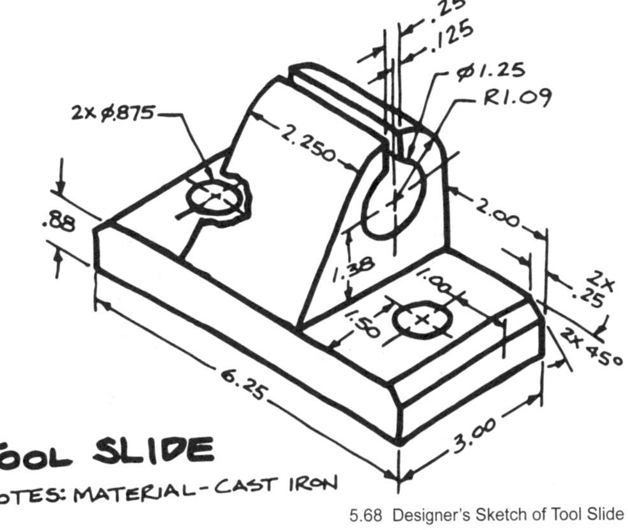

5.68 Designer's Sketch of Tool Slide

Step 4. In the **Dimension Style Manager** dialog box:

a. Create two new dimension styles named **ASME Y14.5-Small Radii and ASME Y14.5-Large Radii** that contain the following dimension style settings. The only difference between these two styles will be in the Fit tab settings as shown in the table below. (Do not enable the settings of the **Alternate Units** or **Tolerances** tabs):

Dimension Style Settings for the Tool Slide:		
Dimension Style Tab	**Setting**	**Value**
Lines	Extend beyond dim lines	.125
	Offset from origin	.062
Symbols & Arrows	Arrow size	.125
	Center marks	Line
Text	Text height	.125
	Text style	Standard (font should be set to Arial)
Primary Units	Precision	0.00
	Zero suppression	Leading
Fit	See Figure 5.57(a) on page 287 and Figure 5.57(c) on page 287 for **Small** and **Large Radii** fit settings	

b. Set the new dimension style current.

Step 5. Add dimensions and notations to the views of the Tool Slide (see Figure 5.68).

Step 6. Follow the dimensioning rules presented earlier in this chapter.

Step 7. Follow your instructor's directions to print the drawing when you are finished placing the dimensions.

VIDEO

There are video tutorials available for the Dimensioning commands inside the *Chapter 5* folder of the book's *Video Training* downloads.

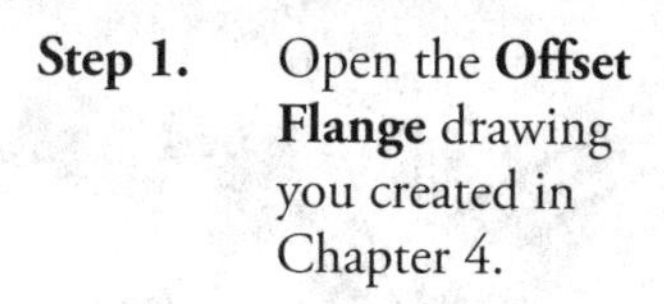

Step 1. Open the **Offset Flange** drawing you created in Chapter 4.

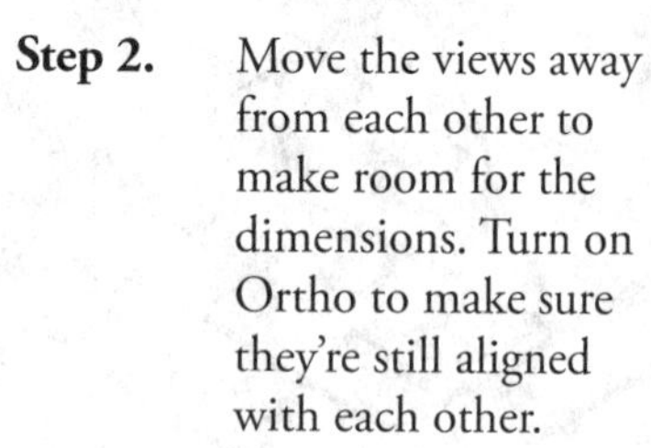

Step 2. Move the views away from each other to make room for the dimensions. Turn on Ortho to make sure they're still aligned with each other.

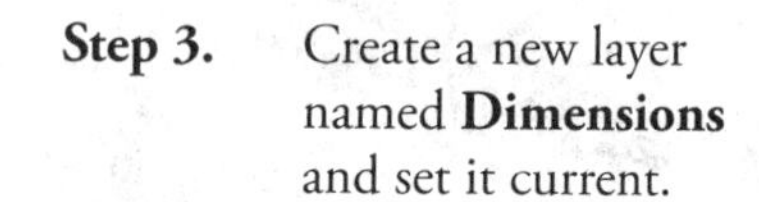

Step 3. Create a new layer named **Dimensions** and set it current.

Step 4. In the **Dimension Style Manager** dialog box:

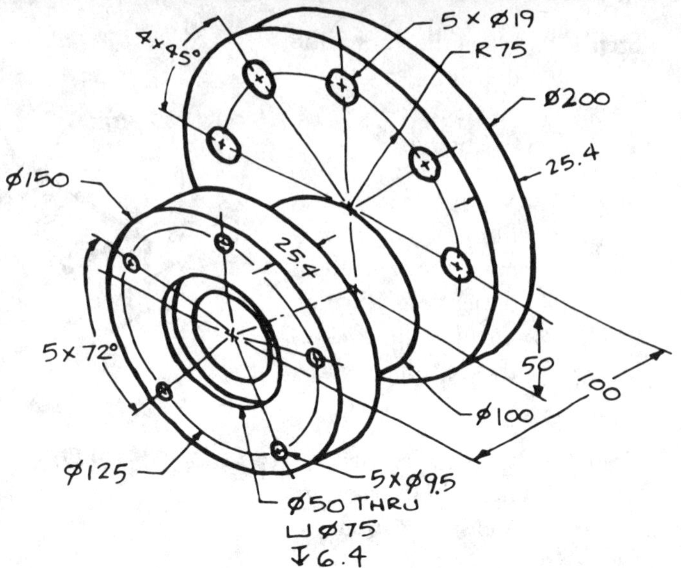

5.69 Designer's Sketch of Offset Flange

a. Create two new dimension styles named **ASME Y14.5-Small Radii and ASME Y14.5-Large Radii** that contain the following dimension style settings. The only difference between these two styles will be in the Fit tab settings as shown in the table below. (Do not enable the settings of the **Alternate Units** or **Tolerances** tabs):

Dimension Style Settings for the Offset Flange:		
Dimension Style Tab	**Setting**	**Value**
Lines	Extend beyond dim lines	3
	Offset from origin	1.5
Symbols & Arrows	Arrow size	3
	Center marks	Line
	Center mark size	2
Text	Text height	3
	Text style	Standard (font should be set to Arial)
	Offset from dim line	3
Primary Units	Precision	0.0
	Zero suppression	Trailing
Fit	See Figure 5.57(a) on page 287 and Figure 5.57(c) on page 287 for **Small** and **Large Radii** fit settings	

b. Set the new dimension style current.

Step 5. Add dimensions and notations to the views of the Offset Flange (see Figure 5.69).

Step 6. Follow the dimensioning rules presented earlier in this chapter.

Step 7. Follow your instructor's directions to print the drawing when you are finished placing the dimensions.

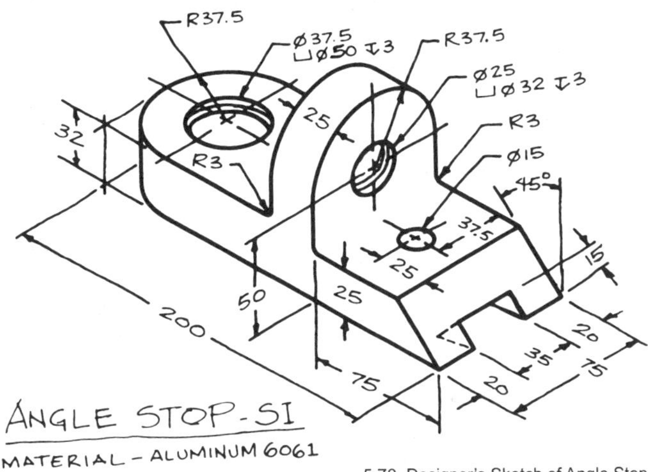

5.70 Designer's Sketch of Angle Stop

Step 1. Open the **Angle Stop** drawing you created in Chapter 4.

Step 2. Move the views away from each other to make room for the dimensions. Turn on Ortho to make sure they're still aligned with each other.

Step 3. Create a new layer named **Dimensions** and set it current.

Step 4. In the Dimension Style Manager dialog box:

a. Create two new dimension styles named **ASME Y14.5-Small Radii and ASME Y14.5-Large Radii** that contain the following dimension style settings. The only difference between these two styles will be in the Fit tab settings as shown in the table below. (Do not enable the settings of the **Alternate Units** or **Tolerances** tabs):

Dimension Style Settings for the Angle Stop:

Dimension Style Tab	Setting	Value
Lines	Extend beyond dim lines	3
	Offset from origin	1.5
Symbols & Arrows	Arrow size	3
	Center marks	Line
	Center mark size	2
Text	Text height	3
	Text style	Standard (font should be set to Arial)
	Offset from dim line	3
Primary Units	Precision	0.0
	Zero suppression	Trailing
Fit	See Figure 5.57(a) on page 287 and Figure 5.57(c) on page 287 for **Small** and **Large Radii** fit settings	

b. Set the new dimension style current.

Step 5. Add dimensions and notations to the views of the Angle Stop (see Figure 5.70).

Step 6. Follow the dimensioning rules presented earlier in this chapter.

Step 7. Follow your instructor's directions to print the drawing when you are finished placing the dimensions.

QUICK TIP

Because the Angle Stop is an SI drawing, the values of the dimension variables are in millimeters.

PROJECT

Step 1. Open the **Swivel Stop** drawing you created in Chapter 4.

Step 2. Move the views away from each other to make room for the dimensions. Turn on Ortho to make sure they're still aligned with each other.

Step 3. Create a new layer named **Dimensions** and set it current.

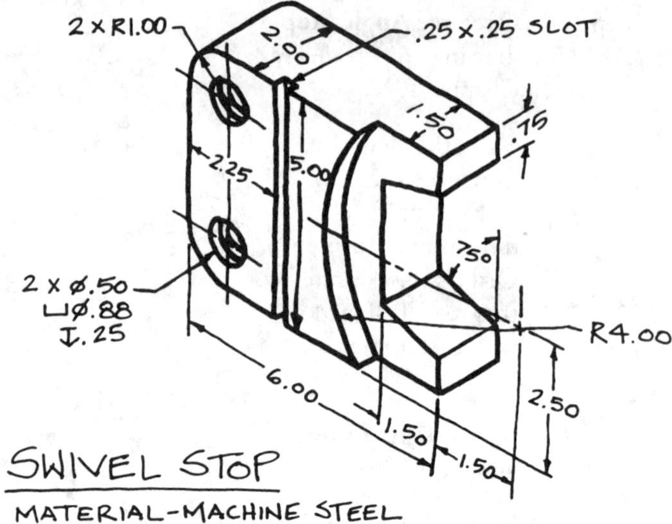

5.71 Designer's Sketch of Swivel Stop

Step 4. In the **Dimension Style Manager** dialog box:

a. Create two new dimension styles named **ASME Y14.5-Small Radii and ASME Y14.5-Large Radii** that contain the following dimension style settings. The only difference between these two styles will be in the Fit tab settings as shown in the table below. (Do not enable the settings of the **Alternate Units** or **Tolerances** tabs):

Dimension Style Settings for the Swivel Stop:		
Dimension Style Tab	**Setting**	**Value**
Lines	Extend beyond dim lines	.125
	Offset from origin	.062
Symbols & Arrows	Arrow size	.125
	Center marks	Line
Text	Text height	.125
	Text style	Standard (font should be set to Arial)
Primary Units	Precision	0.00
	Zero suppression	Leading
Fit	See Figure 5.57(a) on page 287 and Figure 5.57(c) on page 287 for **Small** and **Large Radii** fit settings	

b. Set the new dimension style current.

Step 5. Add dimensions and notations to the views of the Swivel Stop (see Figure 5.71).

Step 6. Follow the dimensioning rules presented earlier in this chapter.

Step 7. Follow your instructor's directions to print the drawing when you are finished placing the dimensions.

ALIGNMENT GUIDE
PROJECT 5.8

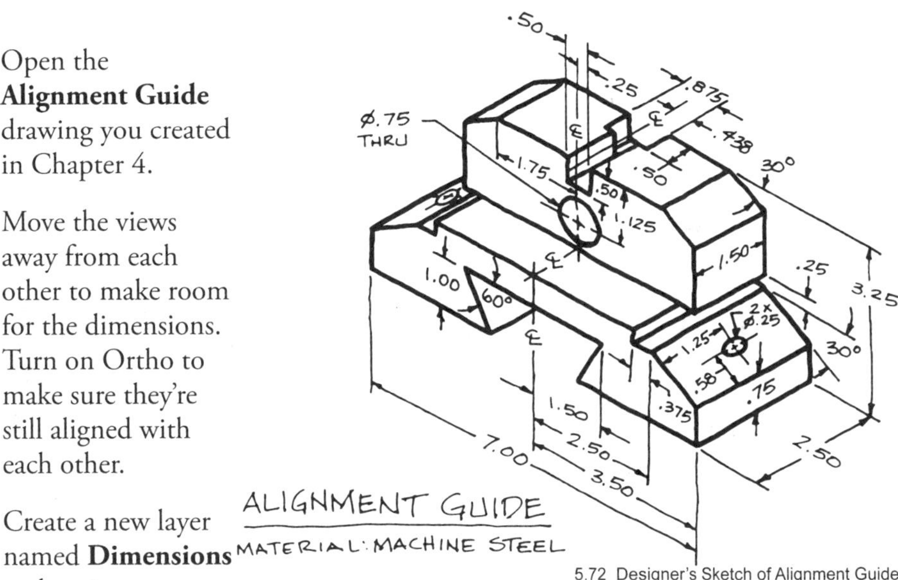

5.72 Designer's Sketch of Alignment Guide

Step 1. Open the **Alignment Guide** drawing you created in Chapter 4.

Step 2. Move the views away from each other to make room for the dimensions. Turn on Ortho to make sure they're still aligned with each other.

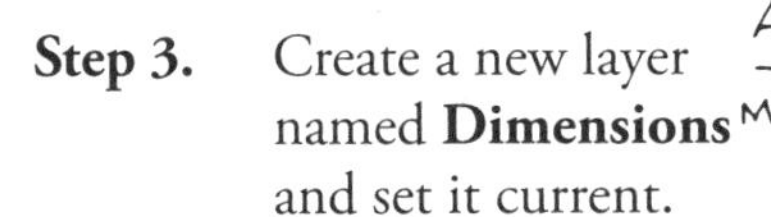

Step 3. Create a new layer named **Dimensions** and set it current.

Step 4. In the **Dimension Style Manager** dialog box:

a. Create two new dimension styles named **ASME Y14.5-Small Radii and ASME Y14.5-Large Radii** that contain the following dimension style settings. The only difference between these two styles will be in the Fit tab settings as shown in the table below. (Do not enable the settings of the **Alternate Units** or **Tolerances** tabs):

Dimension Style Settings for the Alignment Guide:		
Dimension Style Tab	**Setting**	**Value**
Lines	Extend beyond dim lines	.125
	Offset from origin	.062
Symbols & Arrows	Arrow size	.125
	Center marks	Line
Text	Text height	.125
	Text style	Standard (font should be set to Arial)
Primary Units	Precision	0.00
	Zero suppression	Leading
Fit	See Figure 5.57(a) on page 287 and Figure 5.57(c) on page 287 for **Small** and **Large Radii** fit settings	

b. Set the new dimension style current.

Step 5. Add dimensions and notations to the views of the Alignment Guide (see Figure 5.72).

Step 6. Follow the dimensioning rules presented earlier in this chapter.

Step 7. Follow your instructor's directions to print the drawing when you are finished placing the dimensions.

PROJECT

Step 1. Open the **Flange 1105** drawing you created in Chapter 4.

Step 2. Move the views away from each other to make room for the dimensions. Turn on Ortho to make sure they're still aligned with each other.

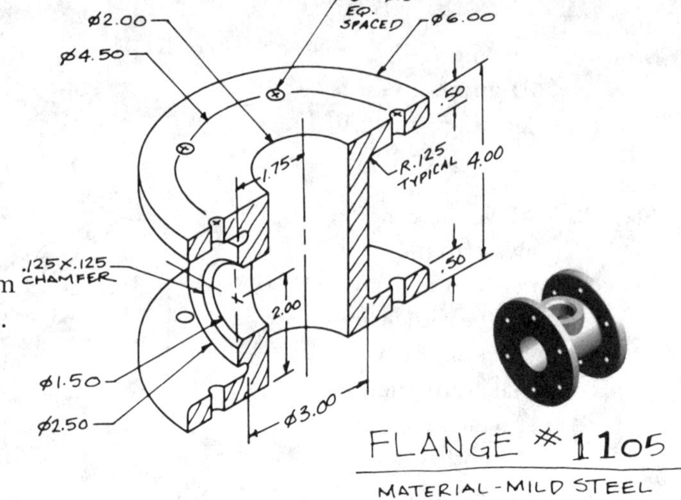

5.73 Designer's Sketch of Flange #1105

Step 3. Create a new layer named **Dimensions** and set it current.

Step 4. In the **Dimension Style Manager** dialog box:

a. Create two new dimension styles named **ASME Y14.5-Small Radii and ASME Y14.5-Large Radii** that contain the following dimension style settings. The only difference between these two styles will be in the Fit tab settings as shown in the table below. (Do not enable the settings of the **Alternate Units** or **Tolerances** tabs):

Dimension Style Settings for the Flange 1105:		
Dimension Style Tab	**Setting**	**Value**
Lines	Extend beyond dim lines	.125
	Offset from origin	.062
Symbols & Arrows	Arrow size	.125
	Center marks	Line
Text	Text height	.125
	Text style	Standard (font should be set to Arial)
Primary Units	Precision	0.00
	Zero suppression	Leading
Fit	See Figure 5.57(a) on page 287 and Figure 5.57(c) on page 287 for **Small** and **Large Radii** fit settings	

b. Set the new dimension style current.

Step 5. Add dimensions and notations to the views of the Flange 1105 (see Figure 5.73).

Step 6. Follow the dimensioning rules presented earlier in this chapter.

Step 7. Follow your instructor's directions to print the drawing when you are finished placing the dimensions.

CHAPTER

6

DIMENSIONING ARCHITECTURAL DRAWINGS

CHAPTER OBJECTIVES:

After studying the material in this chapter, you should be able to:

1. Explain how dimensions are determined on architectural drawings.
2. Describe standards that affect the creation of architectural drawings.
3. List the guidelines for adding dimensions to architectural drawings.
4. Use the commands on AutoCAD's **Dimension** toolbar.
5. Create and modify architectural dimension styles with AutoCAD's **Dimension Style Manager**.
6. Add dimensions and notes to a floor plan.

CHAPTER OVERVIEW

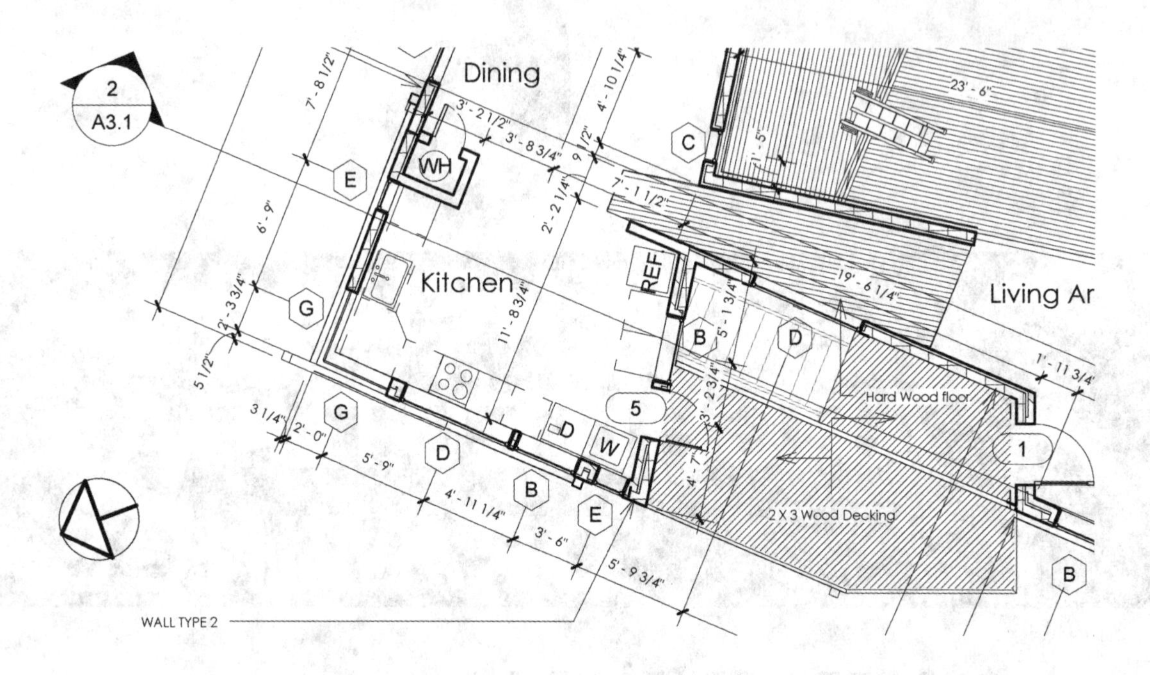

DIMENSIONING ARCHITECTURAL DRAWINGS:

In preparing a set of architectural plans, drafters add dimensions to floor plans, elevations, and construction details. Drafters are responsible for accurately transferring dimensions from the designer's input to the finished drawing. A mistake on an architectural drawing could lead to a costly revision on a construction site or, even worse, the failure of a structural system.

This chapter presents the theory and practice of dimensioning architectural drawings and the use of AutoCAD's dimensioning tools and settings for architectural design.

6.1 DIMENSIONING ARCHITECTURAL DRAWINGS

As with mechanical drawings, drawing and dimensioning standards apply to the creation of architectural drawings. Often, the standard is described in an in-house drafting manual that has been developed by the designers, architects, and drafters of the firm. This manual is used to guide placement and spacing of dimensions, text height for dimensions and notations, and naming conventions for the title block and layers.

Increasingly, however, national standards are being adopted by architectural design firms, especially those that bid on publicly funded projects such as schools and government buildings. At present, the ***United States National CAD Standard (NCS)*** is gaining acceptance by the building design and construction industry. The NCS is being developed by experts from the fields of architecture, engineering, and the construction industry to standardize building design data and improve communication among owners, designers, and construction professionals. The NCS defines standards for drafting conventions, CAD layer naming conventions (based on *AIA CAD Layer Guidelines*), dimensioning, drawing sheets, schedules, drawing sets, terms and abbreviations, graphic symbols, notations, and plotting.

Advances in CAD modeling and linking of digital information are driving another paradigm in design, construction, and building management called ***Building Information Modeling (BIM)***. Projects incorporating BIM technologies will move away from 2D drawings that do not have intelligence built into them toward a linked database that all users of the system can access. This standard, the *National Building Information Model Standard (NBIMS)*, is being developed by the National Institute for Building Sciences (NIBS). The NIBS website describes the mission of the NBIMS standard as an attempt to "improve the performance of facilities over their full life cycle by fostering common and open standards and an integrated life-cycle information model for the A/E/C & FM (Architectural/Engineering/Construction & Facilities Management) industry."

NOTE

To learn more about the United States National CAD Standard, see Appendix C. To learn more about the National Building Information Model Standard, visit the National Institute for Building Sciences website: www.nibs.org.

BIM covers all aspects of the project, from information management to the design, construction, and operation of the facility. BIM is not intended to compete with the NCS standard.

6.2 DETERMINING DIMENSIONS ON ARCHITECTURAL DRAWINGS

Architectural designers determine the dimensions and notes that will go into the design of a building and use this information to create a design input. Often, the design input comes to the drafter as a sketch showing placement of walls, doors, windows, and other features of the building. Drafters use the design input to draw the building's features and add the dimensions and notes necessary to build the project.

6.3 ARCHITECTURAL DRAFTING CONVENTIONS

Dimensioning conventions for architectural drawings differ from those for mechanical drawings. For example, the very small tolerances commonly noted on mechanical drawings are usually much less critical in architectural drawings—although sometimes a dimension may be marked clear, which indicates that after construction the actual dimension between finished surfaces may never be less than the clear dimension. Another difference is that either arrowheads or ***tick marks*** (short diagonal lines that NCS refers to as a *slashes*) can be used to show the termination of dimensions.

In the *drafting conventions* section of the NCS, three methods for dimensioning walls and partitions are presented: dimensioning from an exterior wall face to the face of stud walls or masonry units, dimensioning from an exterior wall face to the centerlines of walls (and from centerline to centerline of interior partitions), and dimensioning to the faces of *finished* walls (however, this requires the builder to know exactly what the final finish of walls will be during the layout of the wall). Windows and doors are dimensioned to the centers of their openings. The standard also states that dimension fractions should not be given in increments of less than 1/16".

In the detail of the floor plan shown in Figure 6.1, note how dimensions are given from the outside edges of framed exterior walls to the face of interior walls; doors and windows are dimensioned to their centers; and, unlike in mechanical drawings, unbroken chains of dimensions are the norm.

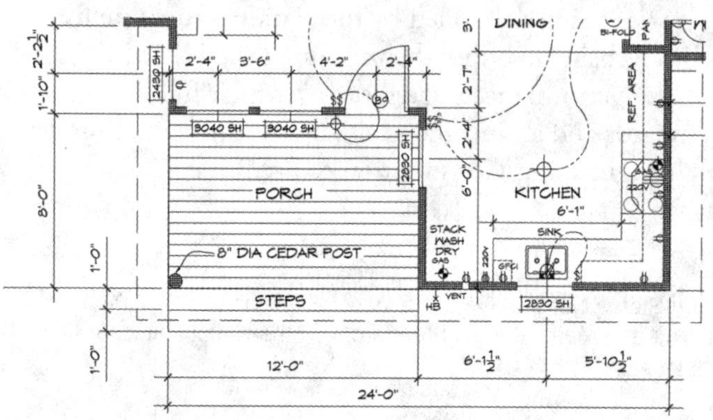

6.1 Detail from a Floor Plan

JOB SKILL

AutoCAD comes loaded with several architectural-style fonts that resemble hand lettered text. Many architects and designers like the traditional look that these fonts lend to their drawings (although these fonts may become less popular as architectural drawing conventions become more standardized). AutoCAD's architectural fonts include CityBlueprint, CountryBlueprint, and Stylus BT.

The NCS specifies that text used on architectural drawings should be ***sans serif*** (meaning without *serifs*). Serifs are the small flourishes found at the ends of the main strokes of characters in some font styles. Minimum text height should be not less than 3/32" (2.5 mm), and all notations should be capitalized. Italicized, underlined, or bold fonts should be avoided.

Dimensions and notations are also added to the elevation drawings of the project. These dimensions may include floor heights, overhangs of eaves, roof pitches, and building materials. It is helpful for architectural designers and drafters to have an understanding of construction processes, particularly framing, to determine the best placement of dimensions.

6.4 ALIGNMENT OF DIMENSION TEXT

Aligned text is used on architectural drawings. Aligned text faces the bottom and the right side of the drawing sheet. Unlike in mechanical drawings, the text is placed above the dimension line on architectural drawings. Examples of aligned text are shown on the dimensions in Figure 6.1.

6.5 ARCHITECTURAL DIMENSIONING GUIDELINES

The following is a review of architectural dimensioning guidelines:

- Dimensions are placed from the outside face of framing of exterior walls.
- The first line of dimensions locates interior walls and the centers of doors and windows.
- The second line of dimensions denotes distances between outside walls and interior walls.
- The third line of dimensions denotes the overall distance between outside walls.
- Interior dimensions usually locate interior walls and other features from edges of outside walls.
- Dimensions should be aligned with dimension lines and read from the bottom and right side of the sheet.
- Dimensions should be centered above dimension lines.
- The space between the outside edge of the building and the first dimension line (9/16" minimum) should be consistently applied throughout the drawing (see Figure 6.2).

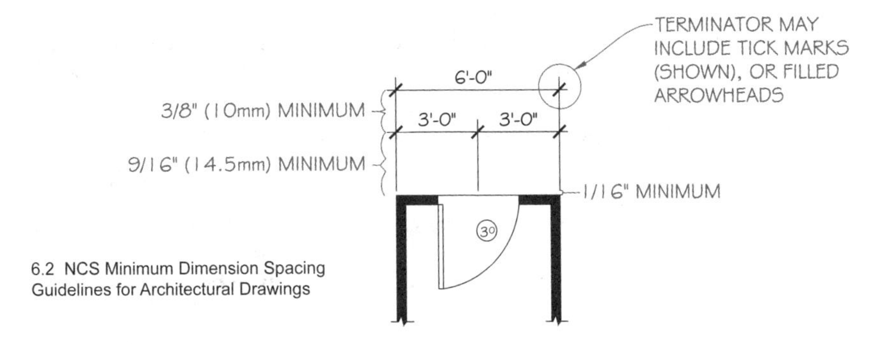

6.2 NCS Minimum Dimension Spacing Guidelines for Architectural Drawings

- The spacing between succeeding parallel dimension lines (3/8" minimum) should be consistent throughout the drawing (see Figure 6.2).
- Dimensions may terminate in tick marks or arrowheads. Tick marks should be parallel.
- Dimension text height should not be less than 3/32" on CAD drawings (1/8" on hand-drafted plans).
- Dimension font style should be sans serif, and all notations should be capitalized.
- Italicized, underlined, or bold fonts should be avoided.

6.6 ARCHITECTURAL DIMENSION SPACING

Minimum suggested spacing for architectural dimensions is shown in Figure 6.2. The spacing values shown apply to the plotted drawing. These spacings are entered into AutoCAD's Dimension Style Manager settings before applying to the drawing.

JOB SKILL

A 3/8" spacing would be equal to 18" if measured in model space in an AutoCAD drawing at a scale of 1/4" = 1'-0" and a 9/16" spacing would be equal to 27" (if 1/4" equals 12", then 1/16" equals 3").

6.7 ADDING A LEADER TO A DRAWING

Figure 6.3 shows an example of a leader that is appropriate for an architectural drawing. These leader styles can be created using either the **Quick Leader** or **Multileader** command.

QUICK TIP

Leaders on architectural drawings often employ a curving (spline) leader line.

VIDEO

There is a video tutorial available for this project inside the *Chapter 6* folder of the book's *Video Training* downloads.

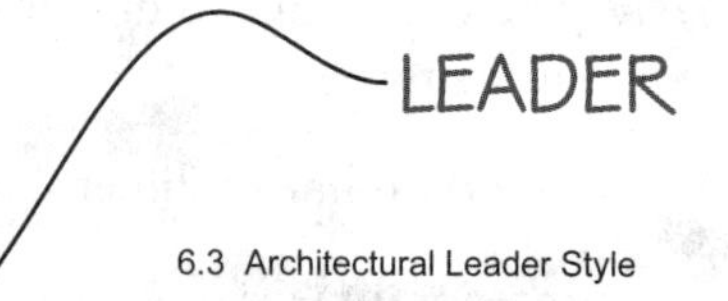

6.3 Architectural Leader Style

STEP-BY-STEP

QUICK LEADER COMMAND TUTORIAL

To create a Quick Leader with a curved (spline) leader line follow the following steps:

Step 1. Type **QLEADER** on the command line and press **<Enter>**.

Step 2. At the *Specify first leader point:* prompt, type **S** (for Settings) and press **<Enter>**.

Step 3. When the **Leader Settings** dialog box opens, select the **Leader Line & Arrow** tab and choose the **Spline** option for the **Leader Line** setting and click **OK**.

Step 4. At the *Specify first leader point:* prompt, select the point for the arrowhead to begin.

Step 5. At the *Specify next point:* prompt, select the point for the leader line to end.

Step 6. At the *Specify next point:* prompt, select the point where the landing line should end.

Step 7. At the *Specify text width:* prompt, either press **<Enter>** or type in a width value.

Step 8. At the *Enter first line of annotation text:* prompt, enter the text for the first line of the leader annotation and press **<Enter>**.

Step 9. At the *Enter next line of annotation:* prompt, enter the text for the next line or press **<Enter>**.

STEP-BY-STEP

MULTILEADER COMMAND TUTORIAL

To create a Multileader with a curved (spline) leader line follow the following steps:

Step 1. Type **MLEADER** (or **MLD**) on the command line and press **<Enter>**, or select the **Multileader** tool from either the **Annotation** panel of the **Home** tab of the ribbon or the **Leaders** panel of the **Annotate** tab. See Figure 5.64(b) on page 293.

Step 2. At the *Specify leader arrowhead location:* prompt type **O** (for Options) and click the **Leader Format** tab, the **Spline** option, then **eXit Options** and follow the prompts to complete the **multileader** command.

CHAPTER REVIEW

Architectural drafters are responsible for creating dimensioned drawings that are used by construction professionals to locate walls, windows, doors, and other features of the building project. Drafters are responsible for ensuring that the dimensions on the drawing accurately reflect the dimensions provided on the designer's input. Drafters must be familiar with architectural drawing and dimensioning standards and the appropriate dimension style settings for the preparation of CAD drawings.

KEY WORDS

Building Information Modeling (BIM)
Sans Serif
Tick Marks
United States National CAD Standard (NCS)

REVIEW

Short Answer

1. Who is responsible for determining dimensions on architectural drawing?
2. In the United States, what is the name of the organization that publishes a drawing standard for architectural drawings?
3. How are the locations of doors and windows defined with dimensions on architectural drawings?
4. What termination symbol is often used on architectural dimensions?
5. Which direction(s) does aligned text face?

Matching

Column A	Column B
a. Primary Units	1. The **Dimension Style Manager** tab where the size of tick marks (obliques) is defined
b. Lines	2. The **Dimension Style Manager** tab where the distance that dimension lines should extend beyond tick marks is defined
c. Text	3. The **Dimension Style Manager** tab where the format for dimension units is defined
d. Fit	4. The **Dimension Style Manager** tab where placing text above vertical dimension lines is defined
e. Symbols and Arrows	5. The **Dimension Style Manager** tab where the overall scale of dimension features is defined

In this project, you will open the **Guest Cottage** drawing that you created in Chapter 4 and apply the dimensions shown in Figure 6.4. Follow the rules of dimensioning outlined earlier in this chapter.

There is a video tutorial available for this project inside the *Chapter 6* folder of the book's *Video Training* downloads.

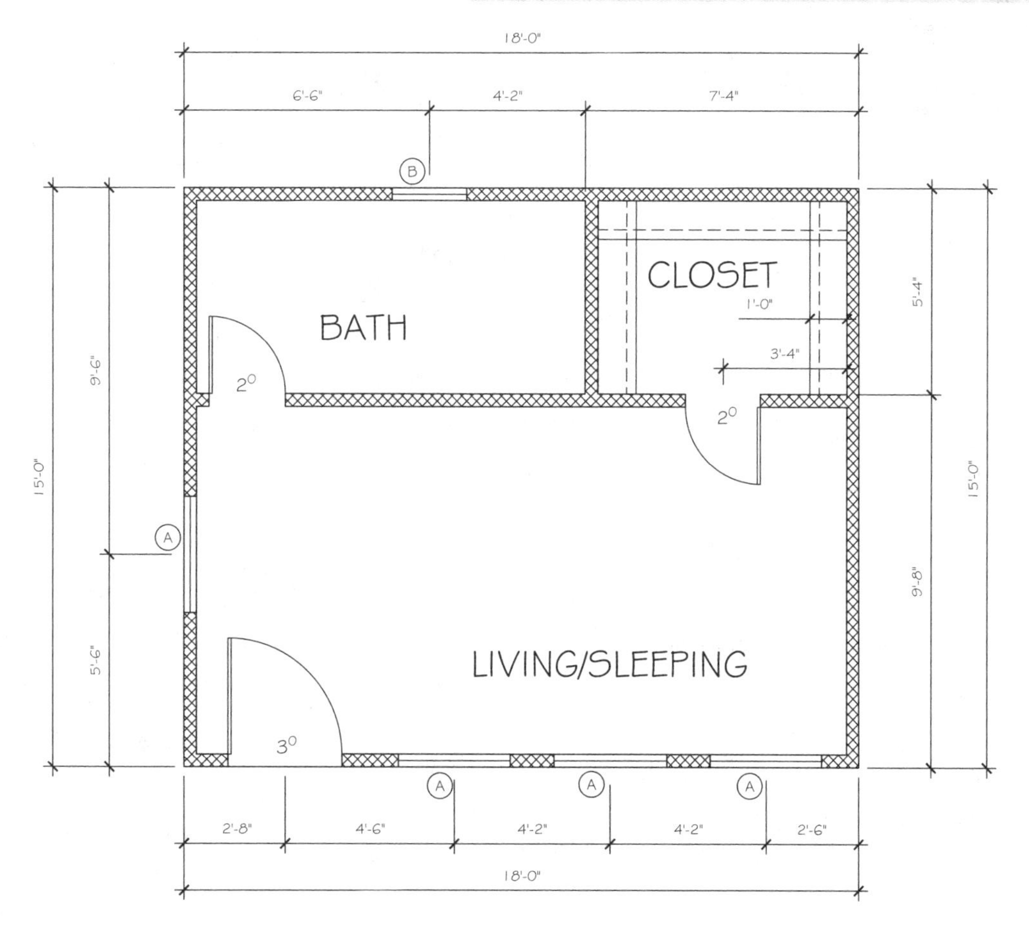

6.4 Designer's Sketch of the Guest Cottage

Step 1. Create a new dimension style named **ARCH48** and assign the dimension style settings shown in Figures 6.5(a)-6.5(e).

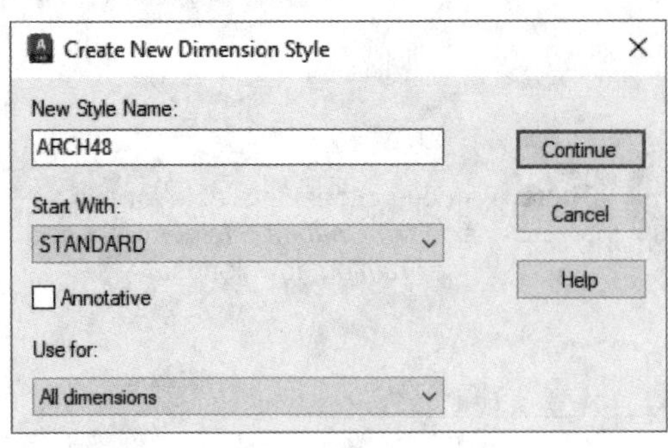

6.5(a) Create a new **Dimension Style** named ***ARCH48***

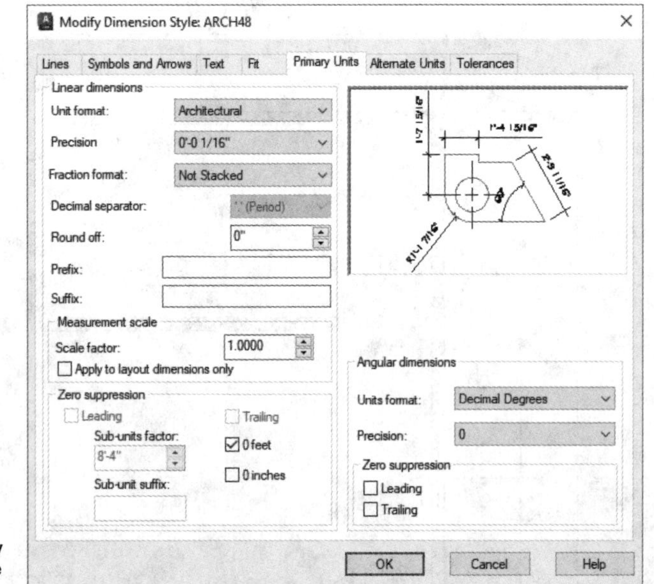

6.5(b) Dimension Settings for **Primary Units** Tab of ARCH48 Dimension Style

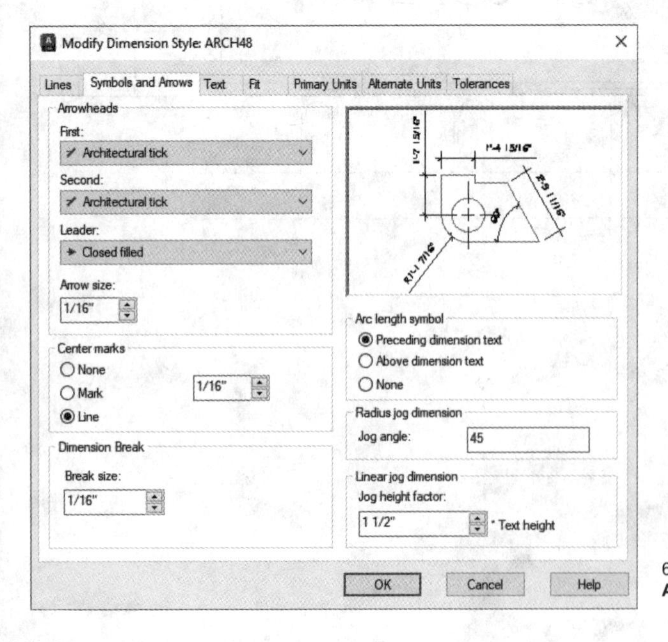

6.5(c) Dimension Settings for **Symbols and Arrows** Tab of ARCH48 Dimension Style

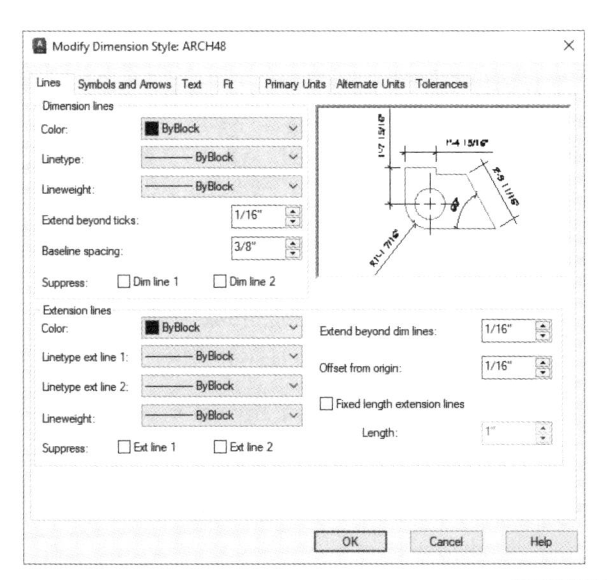

6.5(d) Dimension Settings for **Lines** Tab of ARCH48 Dimension Style

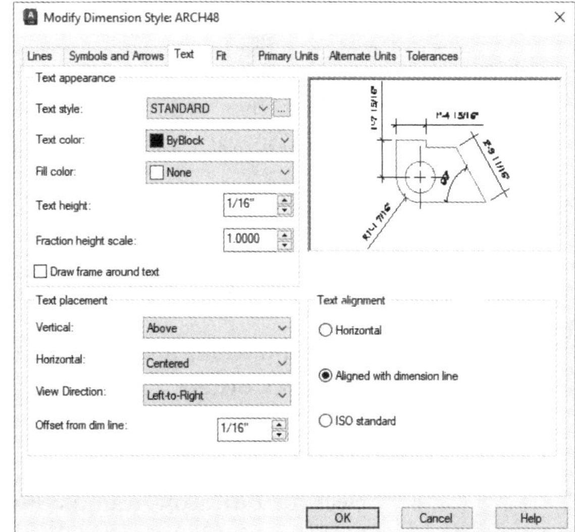

6.5(e) Dimension Settings for **Text** Tab of ARCH48 Dimension Style

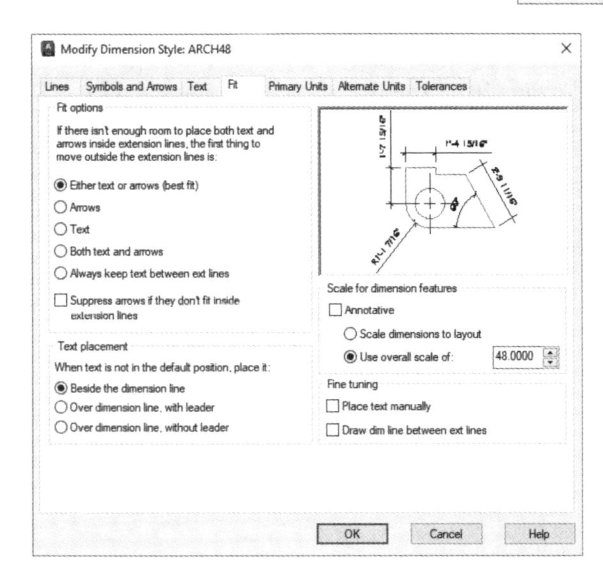

6.5(f) Dimension settings for **Fit** tab of ARCH48 dimension style

QUICK TIP

Why is the Dimension Style named ARCH48?
The name for this dimension style, ARCH48, was chosen because it is an *architectural* dimensioning style and because its dimension settings will be multiplied by *48* as shown in the *Scale for Dimension Features* pane of the Fit tab of the Dimension Settings dialog box.

48 is the *Drawing Scale Factor (DSF)* for a drawing plotted at 1/4"=1'-0". A quick way to calculate a Drawing Scale Factor of 1/4"=1'-0" is to invert the fraction and multiply by 12. In this case: (4/1) x 12 = *DSF* of 48 (there are forty eight 1/4" increments in 12").

Step 2. Set the new style current (refer to the *Creating a New Dimension Style* section of Chapter 5 if you need help creating a new dimension style).

Step 3. Create a **Dimensions** layer with continuous linetype and any color. Set this new layer current.

Step 4. Use the **Linear Dimension** tool to place the first dimension of the first row of dimensions.

a. Use the **Continue** dimension tool to place the rest of the dimensions that make up the first row.

b. The first rows of dimensions should be spaced a minimum of **2'-3"** from the outside walls of all four sides of the cottage.

c. The second rows of dimensions should be spaced **1'-6"** from the first rows.

d. Dimension spacing should be consistent on each side of the floor plan.

Step 5. Follow the directions in the plotting section of Chapter 4 to create a page setup named **Cottage**.

Step 6. To plot the page setup:

a. Select the **Plot** icon from the **Plot** panel of the **Output** tab of the ribbon (see Figure 4.120 on page 211).

b. When the **Plot** dialog box opens, select the down arrow in the **Page Setup** window and select **Cottage** from the list, shown in Figure 4.121 on page 211.

c. Click **OK** to send the print to the plotter or printer.

Step 7. Save the drawing file before closing AutoCAD.

CHAPTER

7

ISOMETRIC DRAWINGS

CHAPTER OBJECTIVES:

After studying the material in this chapter, you should be able to:

1. Define the term *isometric drawing*.
2. Correctly orient lines, ellipses, fillets, and rounds in isometric drawings.
3. Construct inclined planes in isometric drawings.
4. Construct isometric drawings with AutoCAD.

CHAPTER OVERVIEW

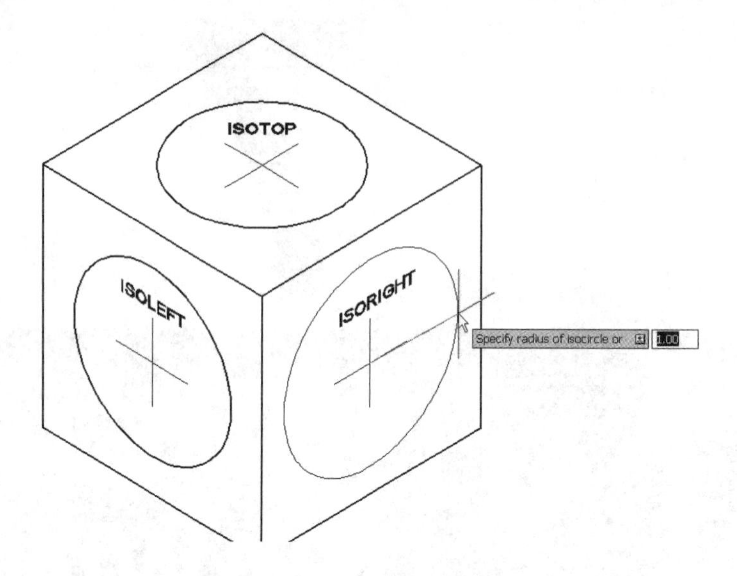

ISOMETRIC DRAWINGS:

An ***isometric drawing*** is a type of drawing known as a ***pictorial drawing***. In a pictorial drawing, an object appears to be three-dimensional, that is, it appears to have width, height, and depth. However, unlike a 3D model, a pictorial drawing is constructed using 2D drawing techniques. The type of drawing discussed in this chapter—isometric drawing—is generally used for drawing machine parts pictorially. Figure 7.1(a) shows an isometric drawing of a cube. In this drawing, the box appears to have width, depth, and height, but this drawing was constructed using only X- and Y- coordinates, so it is considered a 2D drawing.

In the architectural field, pictorial drawings are created using ***perspective drawing*** techniques. In a perspective drawing, lines appear to converge toward a vanishing point, whereas in an isometric drawing, the receding lines are drawn parallel and do not converge. Figure 7.1(a) shows an example of a perspective drawing. Study the differences in the two types of pictorial drawings—isometric and perspective—shown in Figures 7.1(a) and 7.1(b).

In many modern CAD applications, a drafter can construct a 3D model of an object and use it to generate a pictorial image instead of using isometric or perspective drawing techniques.

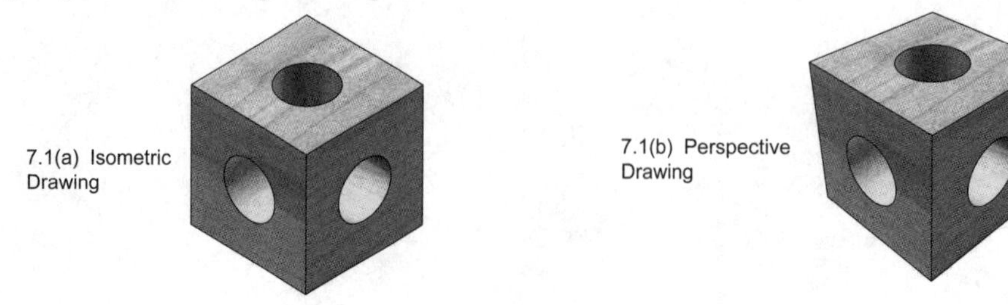

7.1(a) Isometric Drawing

7.1(b) Perspective Drawing

Even though 3D CAD models are used to generate pictorial images, designers and drafters often use the principles of isometric drawing to make freehand design sketches to facilitate communication quickly between other design team members or clients. The ASME standard covering isometric drawings is ASME Y.14.3 - 2012

7.1 ORIENTATION OF LINES IN ISOMETRIC DRAWINGS

In an isometric drawing, an object's horizontal lines are drawn at 30° angles relative to the horizon, and its vertical lines are drawn at a 90° angle relative to the horizon (in other words, the object's vertical lines are drawn vertically), as shown in Figure 7.2. When isometric lines meet at their ends to define a regular (not inclined) plane, the plane is classified in AutoCAD terminology as *isoright*, *isoleft*, or *isotop*, as shown in Figure 7.2.

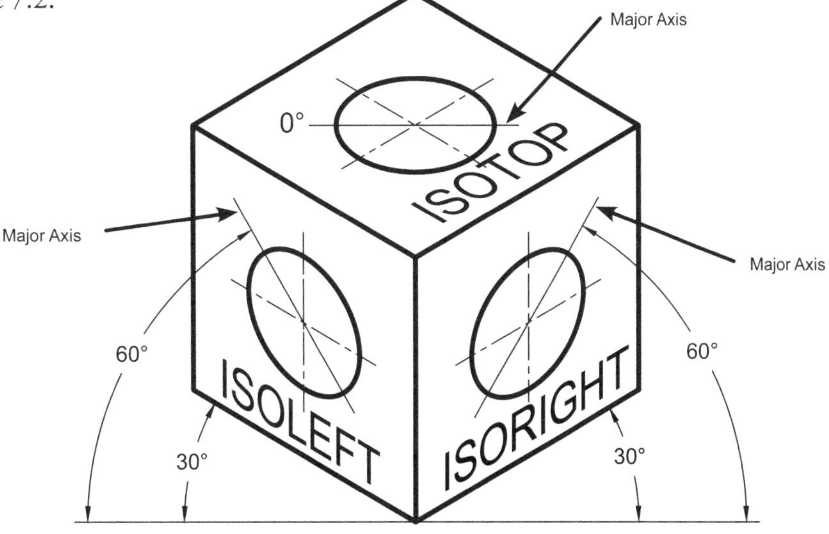

7.2 Orientation of Isometric Lines and Ellipses

The AutoCAD program uses the terms *isoright*, *isoleft*, and *isotop* to describe the isometric planes in Figure 7.2, but before AutoCAD came along, drafters referred to these planes as *right vertical*, *left vertical*, and *horizontal planes*, respectively.

7.2 ORIENTATION OF ELLIPSES IN ISOMETRIC DRAWINGS

Another characteristic of isometric drawings is that round shapes, such as circles and cylinders, appear as ellipses. For the drawing to look natural, however, an isometric ellipse must be aligned correctly along its *major axis*. Figure 7.2 shows the correct orientation of isometric ellipses on their respective planes. Note that although horizontal isometric lines are drawn at 30° angles, the major axes of ellipses located on isoright and isoleft planes are aligned along 60° angles. The major axis of an isometric ellipse on an isotop plane is aligned along a horizontal (0°) line. Study the differences in the orientation of the major axes of the isoright, isoleft, and isotop planes as shown in Figure 7.2.

STEP–BY–STEP

CONSTRUCTING AN ISOMETRIC DRAWING USING THE BOUNDING BOX TECHNIQUE

One of the easiest ways for beginners to construct an isometric drawing is to start by creating an isometric *bounding box*. A bounding box is a box drawn along isometric axes that can completely enclose the object. The bounding box is constructed using the overall width, depth, and height dimensions of the object. After drawing the bounding box, you can reference measurements from its corners to locate the object's features.

The steps in creating an isometric drawing of the object shown in Figure 7.3 using a bounding box are shown in Figures 7.4 through 7.6. These steps illustrate how to locate the object's features by measuring from the corners of the bounding box.

Step 1. Construct an isometric bounding box using height, width, and depth dimensions taken from Figure 7.3. The completed bounding box is shown in Figure 7.4.

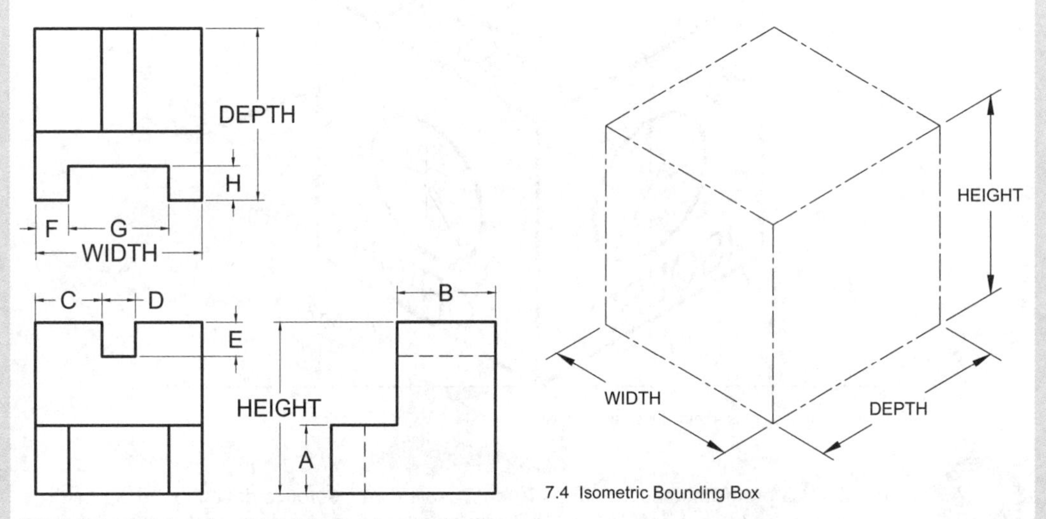

7.3 Multiview Drawing of an Object to be Drawing Isometrically

7.4 Isometric Bounding Box

Step 2. Transfer distances *A* and *B* from Figure 7.3 to the isometric drawing by measuring from the corners of the bounding box. Add lines as needed to construct the view shown in Figure 7.5.

Step 3. Transfer distances *C* through *H* from Figure 7.3 to locate the two slots. Trim and add lines as needed to complete the view as shown in Figure 7.6.

STEP-BY-STEP

7.5 Transferring Distances *A* and *B* to the Bounding Box

7.6 Transferring Distances *C* through *H* to the Bounding Box to Locate the Slots

STEP-BY-STEP

CONSTRUCTING INCLINED PLANES IN ISOMETRIC DRAWINGS

The angle of an inclined plane cannot be measured directly in an isometric drawing. You must instead locate the start and endpoints of the corners of the inclined plane by using measurements taken from a multiview drawing and connecting the points to define the angled plane. The bounding box technique discussed earlier is also helpful for locating the corners of inclined planes.

The object shown in Figure 7.7 contains an inclined plane. The steps in creating an isometric drawing of this object, including the construction of the inclined plane, are illustrated in Figures 7.8 through 7.10.

Step 1. Construct an isometric bounding box using height, width, and depth dimensions taken from Figure 7.7. The completed bounding box is shown in Figure 7.8.

Step 2. Transfer distances *A* and *B* from Figure 7.7 by measuring from the

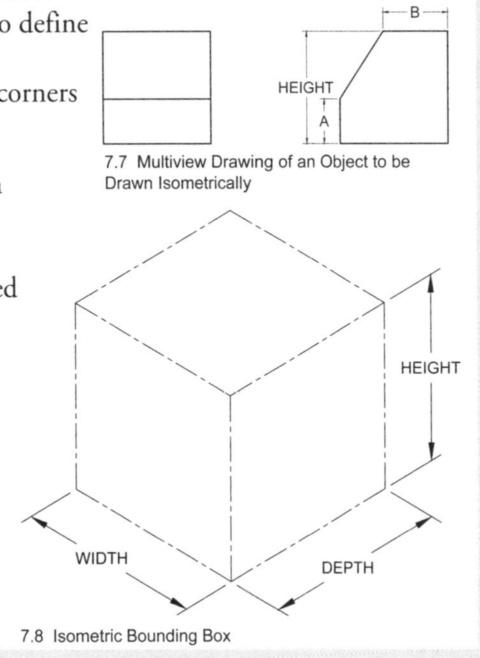

7.7 Multiview Drawing of an Object to be Drawn Isometrically

7.8 Isometric Bounding Box

STEP–BY–STEP

corners of the bounding box. Add lines as needed to construct the view shown in Figure 7.9.

Step 3. Connect the points that define the inclined plane to complete the view as shown in Figure 7.10.

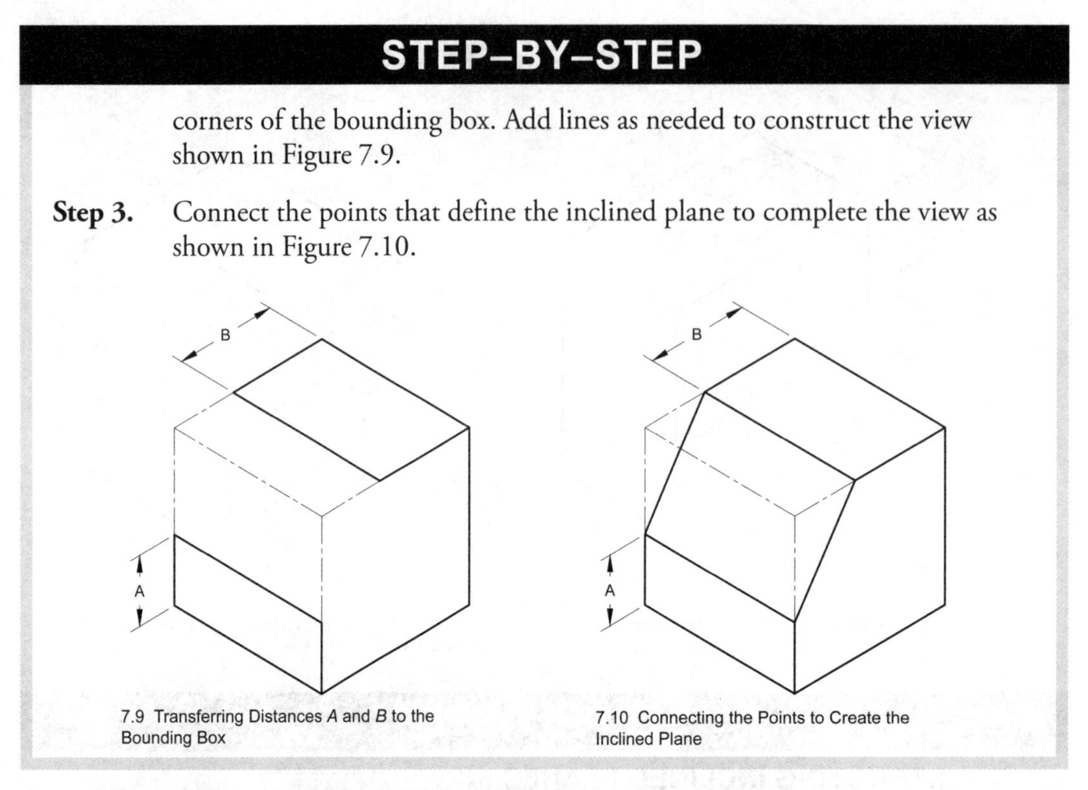

7.9 Transferring Distances *A* and *B* to the Bounding Box

7.10 Connecting the Points to Create the Inclined Plane

7.3 CREATING ISOMETRIC DRAWINGS WITH AUTOCAD

The first step in creating an isometric drawing is to set AutoCAD's snap type from rectangular to isometric. In AutoCAD, you simply left-click on the Isometric Drafting icon located on the Status Bar located along the lower edge of the graphic window. See Figure 7.11(a).

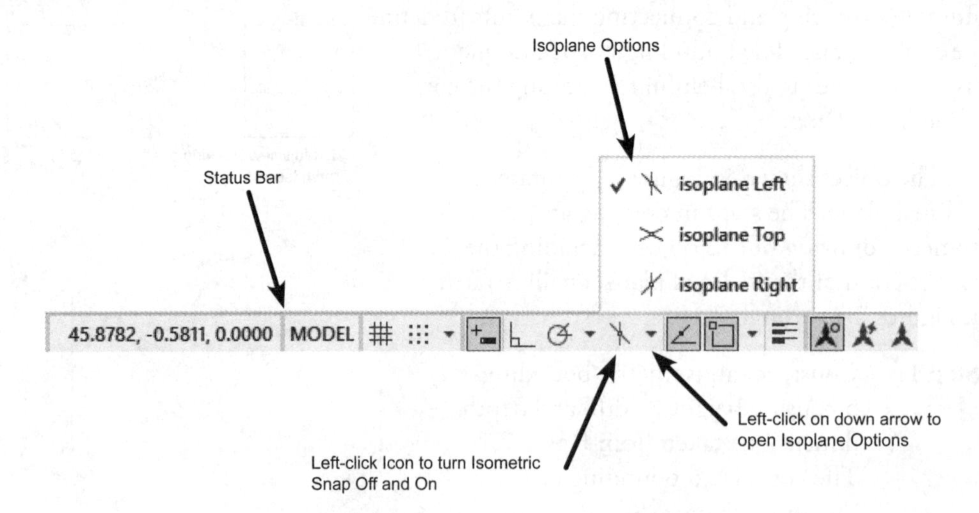

7.11(a) Changing the Isoplane from the Status Bar

Another method of turning on the isometric snap is by left-clicking on the **Tools** pull-down menu on the menu bar and selecting **Drafting Settings**. When the **Drafting Settings** dialog box opens, click the **Snap and Grid** tab and check the button next to **Isometric snap**, see Figure 7.11(b), and then click the **OK** button.

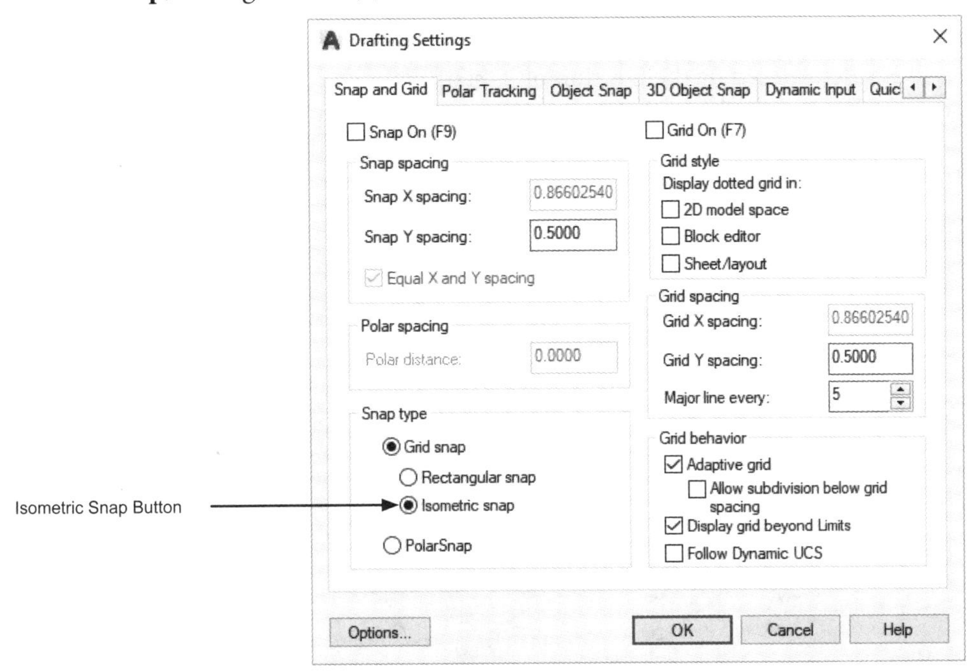

7.11(b) Setting the Isometric Snap Type

After changing the snap type from rectangular to isometric, you will notice that the crosshairs of the cursor are oriented at isometric angles. When **Ortho** is on, and the **LINE** command has been selected, lines are automatically limited to the isometric axes displayed by the angles of the lines in the cursor. To draw lines along different isometric axes than the ones displayed by the cursor, you must change the orientation of the cursor, either by pressing the **<F5>** key or by pressing the **<Ctrl>** and **<E>** keys simultaneously, or by picking on the down arrow to the right of the Isometric Snap icon located on the Status Bar; see Figure 7.11(a). Each time you press the **<F5>** key the orientation of the cursor's crosshairs, or *isoplane*, will change, cycling between the isotop, isoleft, and isoright isoplanes.

NOTE

The current isoplane setting will also be displayed on the command line each time you press **<F5>**.

JOB SKILL

When **Drafting Settings** are set to **Isometric snap**, and **Ortho** mode is turned on, lines drawn in an isometric drawing are automatically aligned along isometric axes. Toggling the **<F5>** key allows you to quickly change the isometric orientation between the isotop, isoleft, and isoright orientations. To draw lines that are not aligned to an isometric angle, you must turn **Ortho** off.

In an isometric drawing, horizontal isometric lines are drawn at 30° relative to the horizon. Creating an isometric drawing with AutoCAD, however, requires that the angle of isometric lines be converted to polar coordinates (where East equals 0°).

In Figure 7.12 the possible isometric angles for the lines of the object are noted on each corner (this example depicts the angles with **Dynamic Input** on). The angle of each line is determined by the position of its start point and its direction relative to East. Note that the angle of the inclined plane does not fall on an isometric angle and therefore would need to be drawn using the technique described in Figures 7.8 through 7.10.

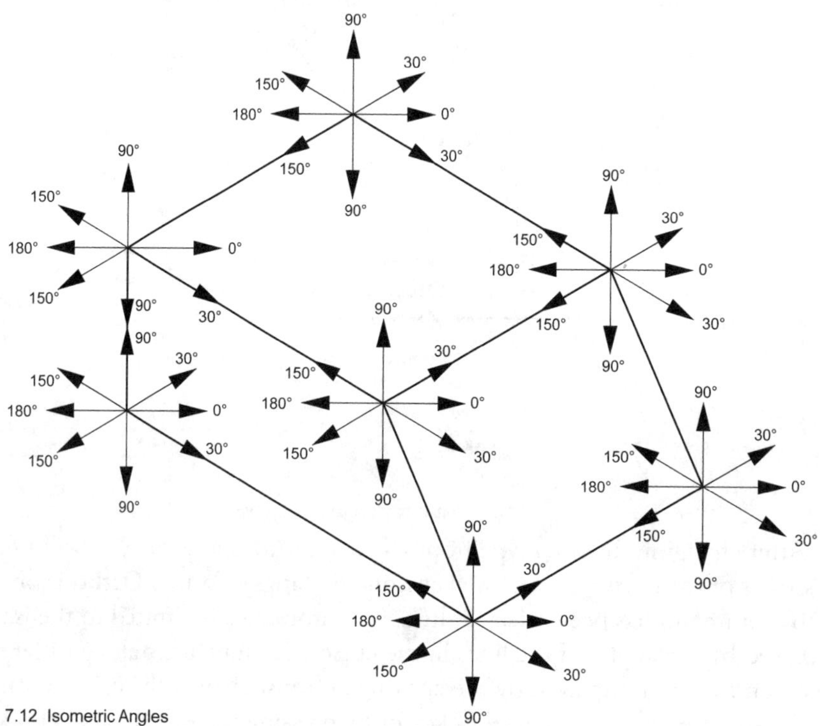

7.12 Isometric Angles

An example of the principal isometric axes for isometric lines using AutoCAD polar coordinates is shown in Figure 7.13. Some examples of polar coordinates for isometric lines (assuming **Dynamic Input** is on) are 6 **<Tab> 30, 4 <Tab> 90,** 7 **<Tab> 150,** and **6 <Tab> 180.**

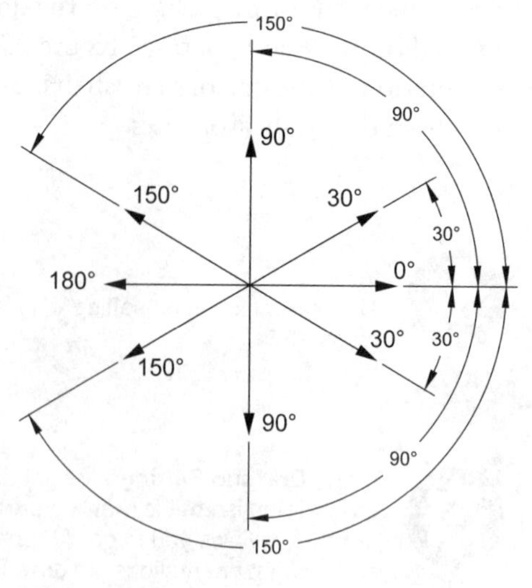

7.13 Isometric Axes

STEP–BY–STEP

DRAWING ISOMETRIC ELLIPSES WITH AUTOCAD

Figure 7.14 shows an isometric box with three isometric ellipses. Notice that each ellipse in Figure 7.14 is oriented differently from the others. This is because each ellipse was drawn on a different isoplane—isotop, isoright, or isoleft. AutoCAD refers to ellipses that are oriented on isometric planes as isocircles.

To place an isocircle in an isometric drawing, first determine the orientation of the isoplane on which the ellipse is to be drawn (isotop, isoright, or isoleft) by referring to Figure 7.14. Then, toggle to the appropriate isoplane by pressing **<F5>**. Next, select the **Ellipse** tool from the **Draw** toolbar or the **Draw** panel (**Axis End** option), type **I** for **Isocircle**, and press **<Enter>**. Next, specify the center of the ellipse and its radius and press **<Enter>** to complete the command.

QUICK TIP

The **Isocircle** option of the **ELLIPSE** command is available only when **Snap type** has been set to **Isometric snap** in either the **Snap and Grid** tab of the **Drawing Settings** dialog box or the **Isometric Drafting** icon located on the Status Bar is toggled *on*. See figures 7.11(a) and 7.11(b).

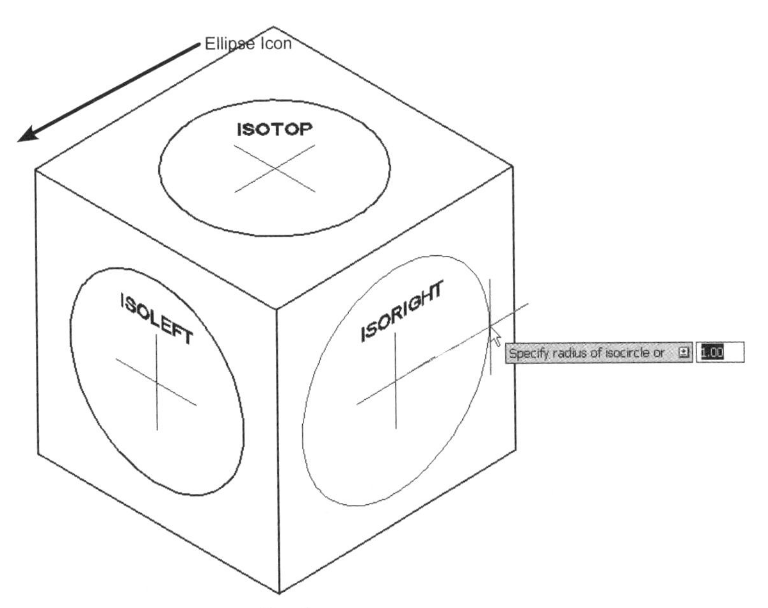

7.14 AutoCAD Isocircles and Isoplanes

The steps in constructing an isometric cylinder are shown in Figures 7.15(a)–7.15(c).

Step 1. Draw an ellipse (isocircle) at the desired diameter and orientation of the cylinder. Copy the ellipse along an isometric angle the desired length of the cylinder as shown in Figure 7.15(a).

Step 2. Draw lines from the quadrants of the front ellipse to the corresponding quadrants of the back ellipse, as shown in Figure 7.15(b).

Step 3. Trim the back ellipse to complete the cylinder as shown in Figure 7.15(c).

STEP–BY–STEP

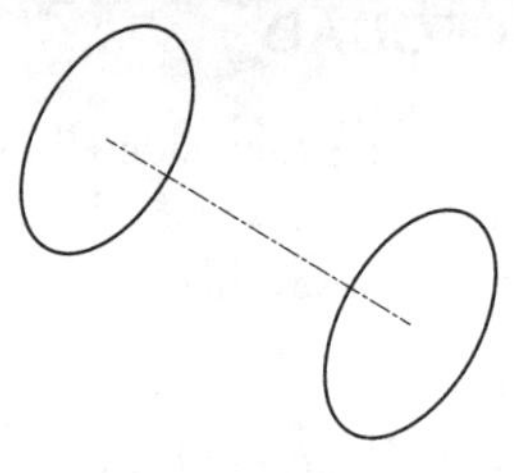

7.15(a) Copying an Ellipse

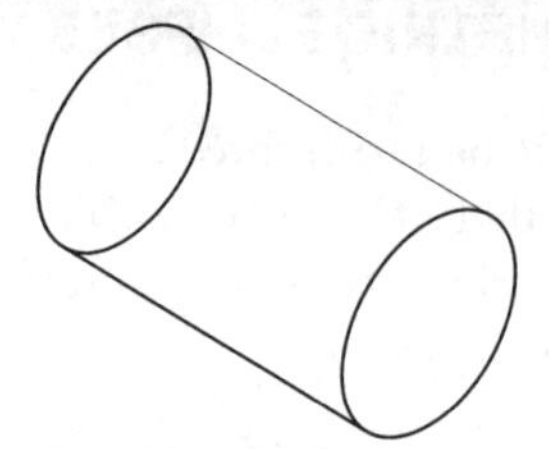

7.15(b) Drawing Lines from Quadrants of Ellipses

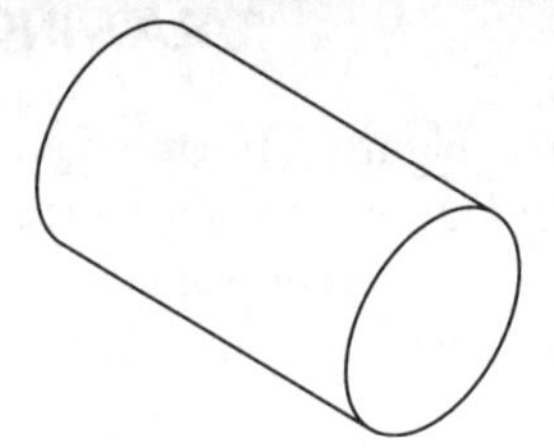

7.15(c) Completed Cylinder

Possible orientations for horizontal and vertical cylinders are shown in Figure 7.16.

Isocircles can be used to create *counterbored*, *countersunk*, and *through* holes in isometric drawings, as shown in Figure 7.17.

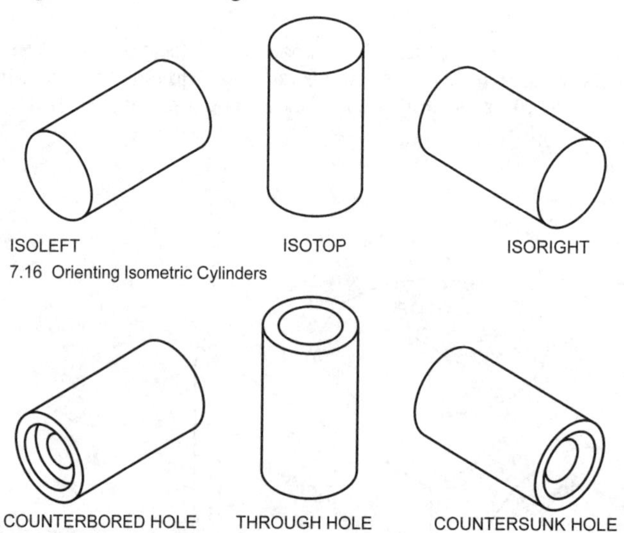

7.16 Orienting Isometric Cylinders

7.17 Orienting Machined Hole Features in Isometric Drawings

STEP-BY-STEP

CONSTRUCTING ISOMETRIC ARCS AND RADII (FILLETS AND ROUNDS)

In technical drawing, the terms **fillet** and **round** refer to a rounded inside and outside corner, respectively. On a multiview drawing, AutoCAD's **FILLET** command can be used to create fillets and rounds at an object's corners, but the **FILLET** command cannot be used in the creation of an isometric drawing because it creates a rounded, rather than an elliptical, corner. Therefore, drafters must use the steps illustrated in Figures 7.19 through 7.22 to add isometric fillets and rounds to the corners identified as *A*, *B*, *C*, and *D*, in Figure 7.18.

STEP-BY-STEP

Step 1. Copy edges **1** through **6** along isometric axes **1"** inside the edges of the part as shown in Figure 7.19.

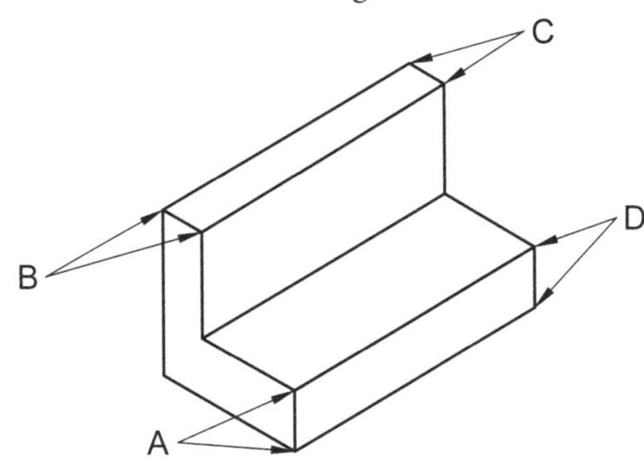

7.18 Add a 1" Radius to Corners *A*, *B*, *C*, and *D*

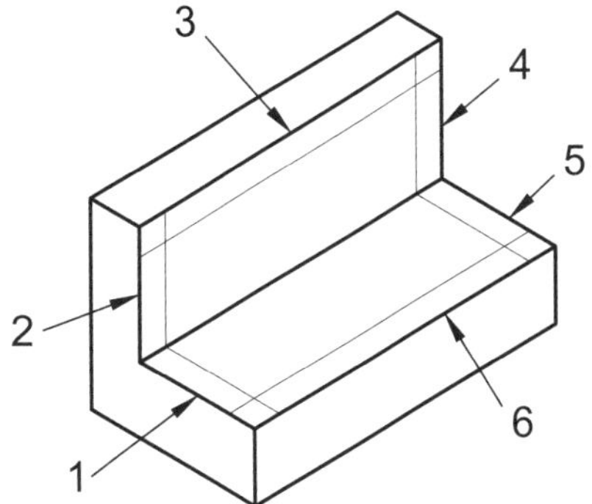

7.19 Copy Edges 1-6 at 1" along Isometric Axes

Step 2. Using the intersections of the lines copied in Step 1 for center points, draw four **2"**-diameter ellipses (isocircles) as shown in Figure 7.20.

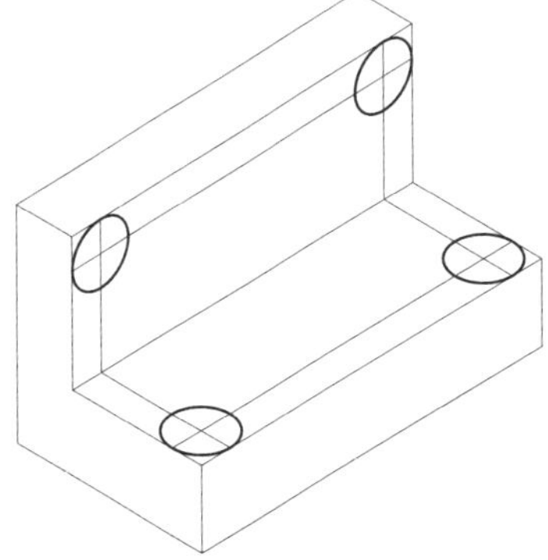

7.20 Add Isocircles at Intersections of Lines Copied in Step 1.

Step 3. Trim the ellipses to create a round for each corner, and copy the rounds along isometric axes to the corresponding corner as shown in Figure 7.21.

Step 4. Erase and edit construction lines as needed to complete the view as shown in Figure 7.22.

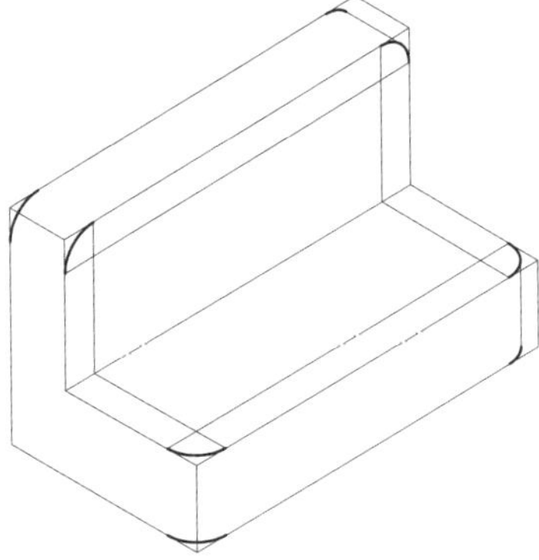

7.21 Trimming Ellipses to Create Rounded Corners

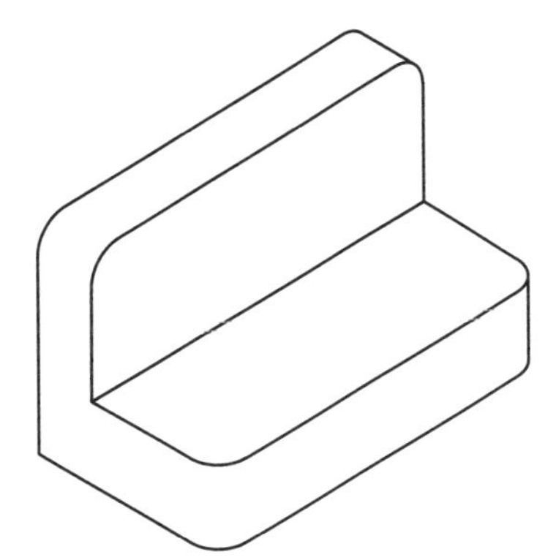

7.22 The Finished Isometric Drawing

Pictorial drawings appear to be three-dimensional but are constructed using only *X*- and *Y*-coordinates, so they are actually a form of two-dimensional drawing. To create isometric drawings in AutoCAD, drafters must master the concept of polar coordinates, be able to orient correctly the major axes of isometric ellipses, and be able to use the techniques needed to construct inclined planes.

Pictorial views created from 3D CAD models are replacing isometric and perspective drawings in many fields of technical graphics. However, the ability to create freehand pictorial sketches quickly to facilitate communication of design ideas remains an important job skill for designers and drafters and one that students seeking to become employable in this field should strive to develop.

The ASME standard covering isometric drawings is *ASME Y.14.3 - 2012 Orthographic and Pictorial Views.*

Fillet
Isometric Drawing
Perspective Drawing
Pictorial Drawing
Round

True or False

1. *True or False:* Angles of inclined planes can be measured directly in an isometric drawing.
2. *True or False:* Pressing **<F5>** or **<Ctrl>+E** are the two ways to change the orientation of the crosshairs in an isometric drawing created with AutoCAD.
3. *True or False:* When drawing an isometric ellipse, a drafter must select **Isoplane** after beginning AutoCAD's **ELLIPSE** command.
4. *True or False:* It is not important for drafters and designers to understand isometric drawing techniques.
5. *True or False:* In an isometric drawing the lines appear to recede toward a vanishing point.

REVIEW QUESTIONS

Multiple Choice

1. In which field of technical drawing are isometric techniques most often used to create a pictorial image?
   a. Archaeological
   b. Architectural
   c. Mechanical
   d. Civil
2. What is the angle of an object's horizontal lines in an isometric drawing?
   a. 60°
   b. 180°
   c. 30°
   d. None of the above
3. What is the angle of an object's vertical lines in an isometric drawing?
   a. 90°
   b. 45°
   c. 30°
   d. 60°
4. On which tab in the Drafting Settings dialog box is the Rectangular snap button located?
   a. Snap and Grid tab
   b. Polar Tracking tab
   c. Dynamic Input tab
   d. Object Snap tab
5. At what angle is the major axis of an ellipse drawn on an isoright plane?
   a. 90°
   b. 45°
   c. 30°
   d. 60°

PROJECT

Create an isometric drawing of the Tee Connector shown in Figure 7.23.

VIDEO There is a video tutorial available for this project inside the Chapter 7 folder of the book's Video Training downloads.

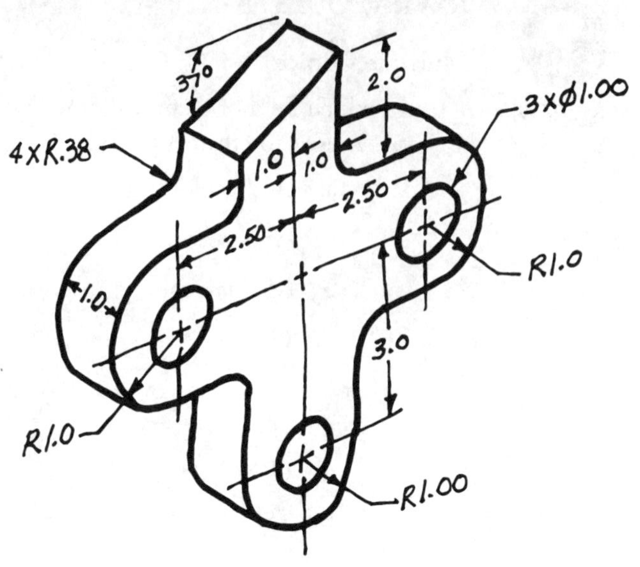

7.23 Designer's Sketch of the Tee Connector

Step 1. Getting Started

a. Download the **Imperial Prototype** drawing file located in the *Prototype Drawing Files* folder associated with this book.

b. Use **SAVE AS** to save the drawing to your **Home** directory, and rename the drawing **TEE CONNECTOR**.

c. Activate the **Model** tab in the lower-left corner of the graphics window (if it is not already highlighted) by left-clicking on the Model tab. **Note:** The views of Tee Connector will be drawn inside the magenta rectangle located in model space.

d. Set the units to **Decimal**.

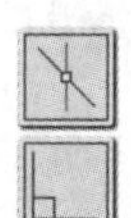

e. Set AutoCAD's snap type to **Isometric snap**—see figures 7.11(a) or 7.11(b)

f. Turn **Ortho** on (F8).

Step 2. Create an **Isometric** layer with the *Default* lineweight and set it current.

Step 3. Use the following steps to construct the Tee Connector.

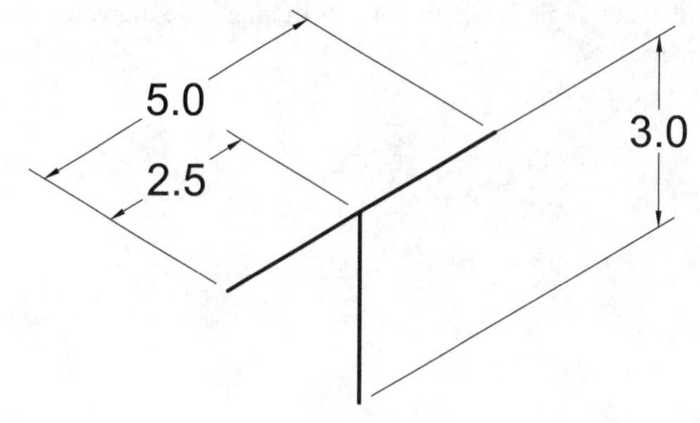

7.24 Isometric Construction Lines

Step 4. Concentric Isocircles: Draw two concentric isocircles at each of the three endpoints of the construction lines as shown in Figure 7.25.

a. Draw a **1" diameter** isocircle (ellipse command)

b. Draw a **2" diameter** isocircle

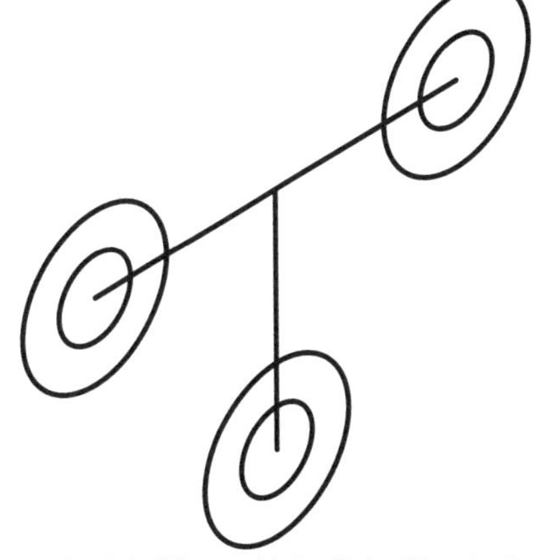

7.25 Isocircle Ellipses added to Ends of Construction Lines

QUICK TIP

You will probably need to add construction lines drawn from the center points of the isocircles (as shown in Figure 7.26) in order to locate snap points that are tangent to the ellipses.

Step 5. Draw tangency lines connecting the ellipses as shown in Figure 7.26.

Step 6. Trim the construction lines so that the front face of the Tee Connector resembles the view in Figure 7.27(b).

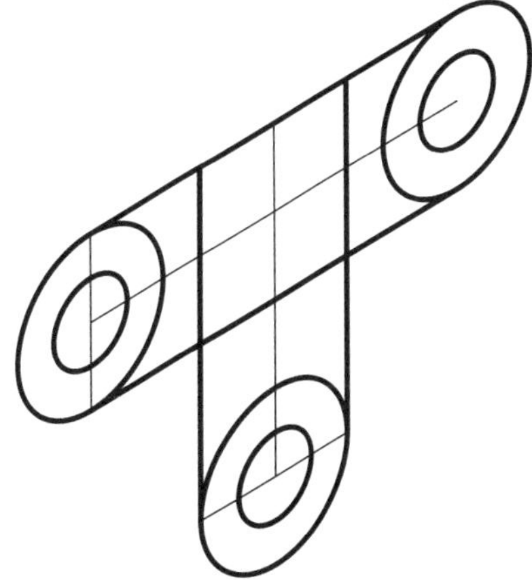

7.26 Adding Tangency Lines

Step 7. Draw angle

a. Since you cannot directly measure the **37°** angle shown in the designer's sketch in the isometric view, you will need to draw an *orthographic* front view of the top end of the Tee Connector and transfer distance **A**—see Figure 7.27(a)—to the isometric drawing

b. Draw the orthographic view using the dimensions shown in Figure 7.27(a).

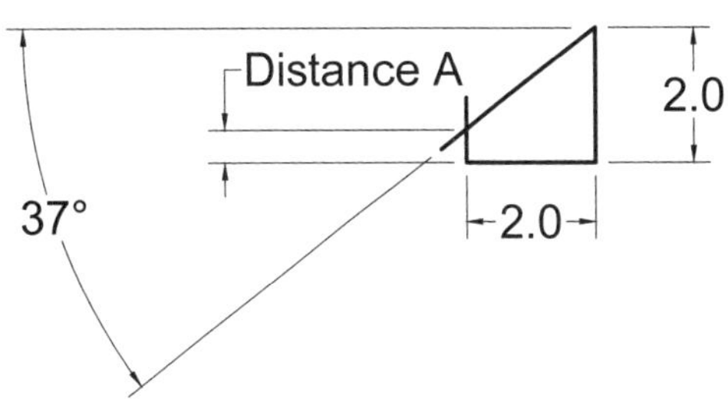

7.27(a) Orthographic View Drawn to Determine Distance A

NOTE

Before drawing this view, change the **Snap type** from **Isometric** to **Rectangular** by clicking on the **Isometric Drafting** icon located on AutoCAD's Status Bar; see Figure 7.11(a).

PROJECT

Step 8. Transfer distance **A** from the orthographic view—see Figure 7.27(a)—to the isometric view as shown in Figure 7.27(b).

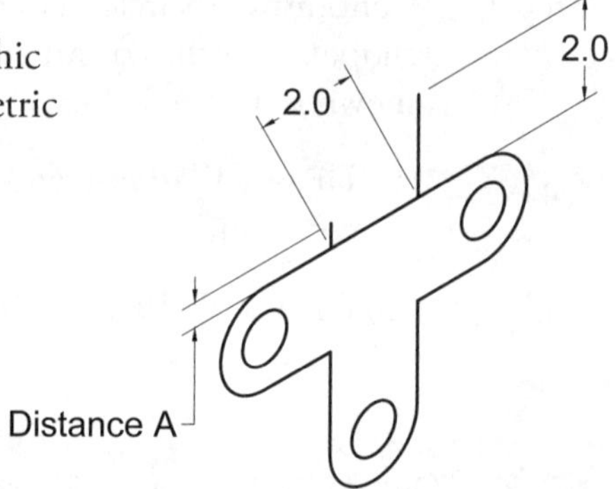

7.27(b) Transferring Distance A to the Isometric Drawing

Step 9. Set the **Snap type** back to **Isometric** (from **Rectangular** snap)

Step 10. Connect the endpoints of the two lines added to create the **37°** angle

Step 11. Trim any unnecessary lines.

The front plane of the Tee Connector should resemble the object shown in Figure 7.28

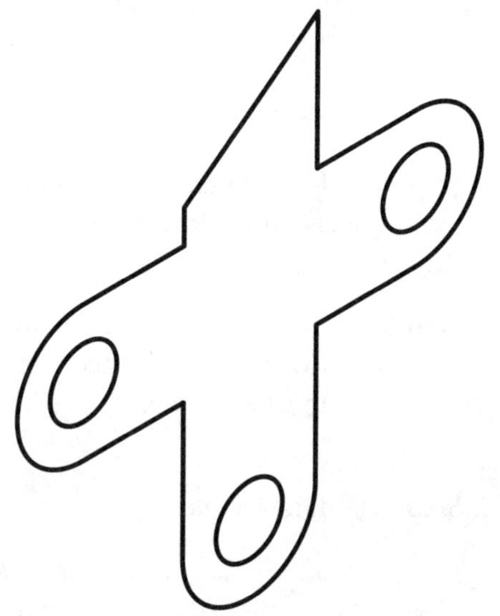

7.28 Front Plane of the Tee Connector

Step 12. With **Ortho** on, copy eight construction lines **.38"** away from each inside corner of the object as shown in Figure 7.29
(Note: Do not use the Offset command to accomplish this step).

Step 13. Draw four **Ø.76"** (R.38" X 2) isometric ellipses whose center points are located at the intersections of these lines as shown in Figure 7.29.

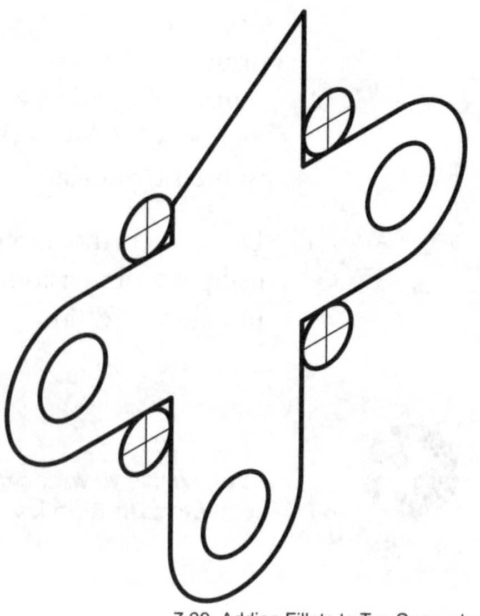

7.29 Adding Fillets to Tee Connector

Step 14. Trim the four **Ø.76"** isocircles created in Step 11 to create four **R.38"** fillets at the inside corners of the front plane of the Tee Connector as shown in Figure 7.30.

Step 15. Copy to give depth

a. Copy the Tee Connector's front face **1"** behind the front face and at an isometric angle of **150°** as shown in Figure 7.30

(N**ote: Ortho** should be on when placing this copy).

b. Add isometric lines connecting the front edges of the Tee Connector to the back edges as shown in Figure 7.31.

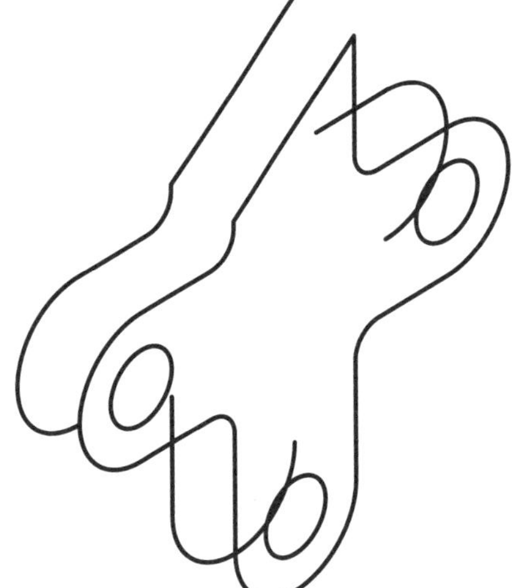

7.30 Copy the Front Face Back 1" at a 150° Angle

c. Trim any unnecessary lines to complete the construction.

Step 16. When finished drawing the Tee Connector, left-click on the Multiview tab in the lower-left corner of the graphics window to enter the Layout.

Step 17. Edit the text in the title block by double-clicking on it to enable the text editing option.

Step 18. Follow your instructor's directions to print the drawing.

Step 19. Be sure to save the drawing file before closing AutoCAD.

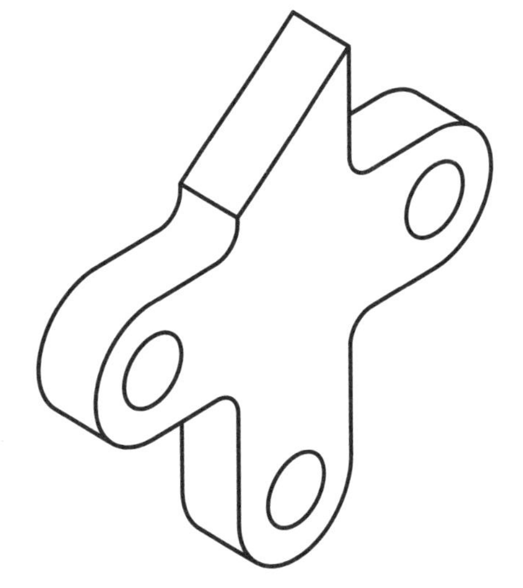

7.31 Completed Tee Connector Project

PROJECT

There is a video tutorial available for this project inside the Chapter 7 folder of the book's Video Training downloads.

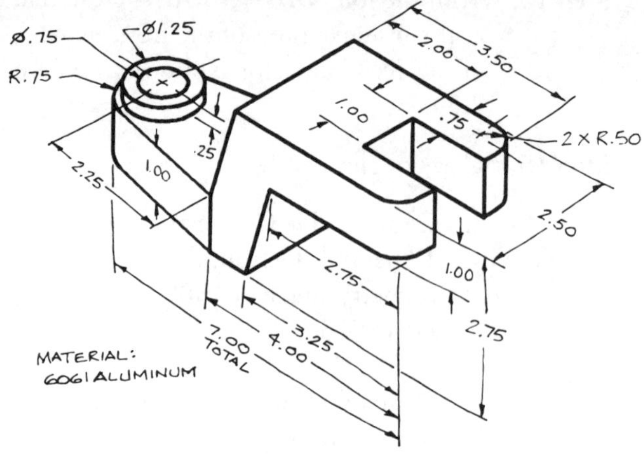

7.32 Designer's Sketch of the Tool Holder

Step 1. Getting Started

Step 2. Open the drawing titled **Tool Holder** that you created in Chapter 4.

Step 3. Use the following steps to create an isometric drawing from the sketch shown in Figure 7.32.

Step 4. Create an **Isometric** layer with the *Default* lineweight and set it current.

Step 5. Construct the isometric view in the space located to the right of the magenta rectangle (in model space).

Step 6. Move the finished isometric drawing inside the upper right corner of rectangle (to the right of the top view and above the right view).

Step 7. Set AutoCAD's **Snap type** to **Isometric snap**—refer to Figures 7.11(a) and 7.11(b).

Step 8. Turn **Ortho** on.

* When you are finished with this project, this drawing will include both the isometric drawing of the tool holder as well its dimensioned multiviews.

Step 9. Create a **7.00" X 2.50" X 2.75"** isometric bounding box that can contain the overall height, width, and depth of the Tool Holder as shown in Figure 7.33.

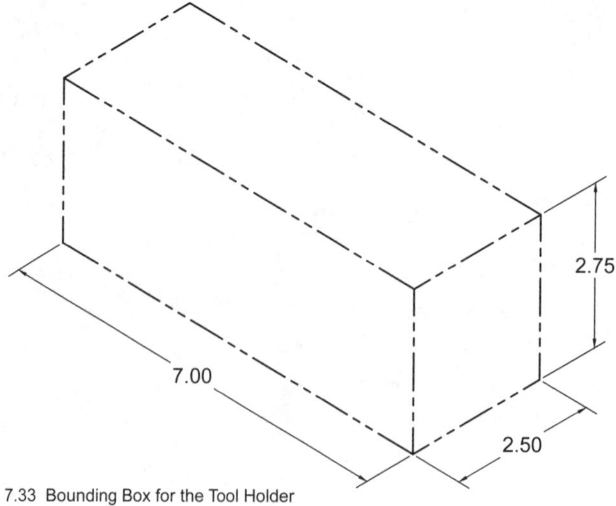

7.33 Bounding Box for the Tool Holder

Step 10. Begin Drawing

a. Using the dimensions shown in Figure 7.34, construct the basic geometry of the front of the Tool Holder within the bounding box.

b. Trim the ellipses shown in Figure 7.34 to create fillets at the front corners (see Figure 7.35).

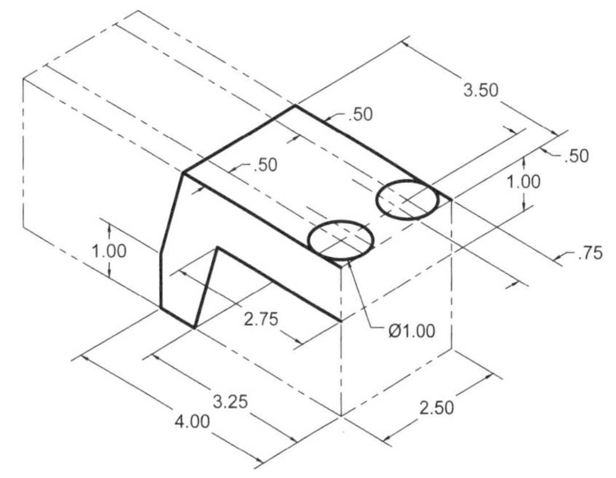

7.34 Constructing the Front of the Tool Holder

Step 11. Add the Slot

a. Add the **1.00" X 1.50"** slot using the dimensions shown in Figure 7.35.

b. Copy the slot (and the filleted corners created in Step 2) straight down **1.00"** along the Y-axis.

c. Add lines as needed to complete the front of the Tool Holder.

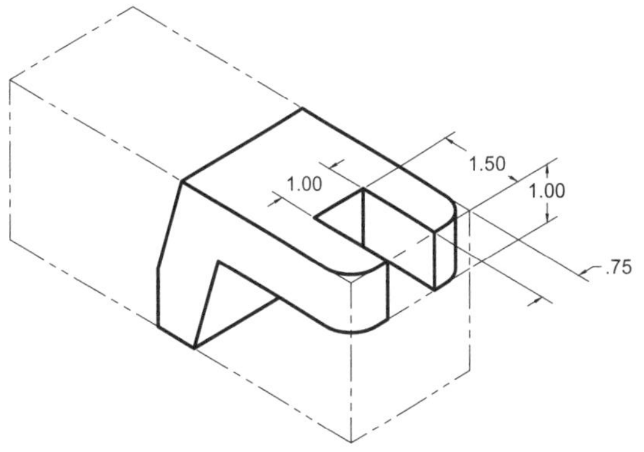

7.35 Completed Construction of the Front of the Tool Holder

Step 12. Draw Isocircles

a. Locate the two **Ø1.50"** ellipses at the back of the object.

b. Add tangency lines to the ellipses as shown in Figure 7.36.

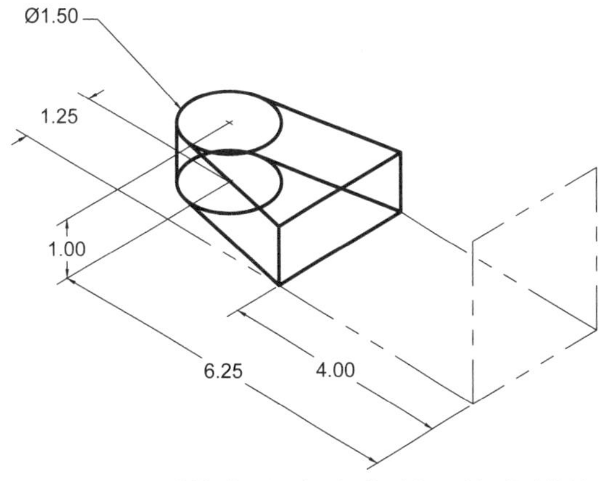

7.36 Constructing the Back Part of the Tool Holder

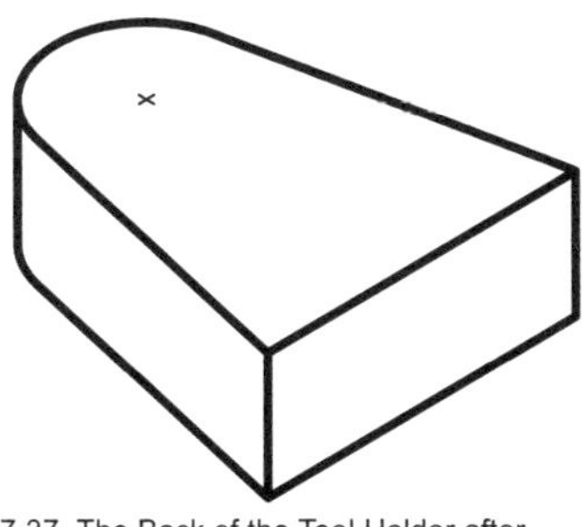

7.37 The Back of the Tool Holder after Trimming Ellipses

(Note: The front of the Tool Holder has not been shown in Figure 7.36 for clarity.) Trim the lines drawn in Step 4 so that the drawing resembles the one shown in Figure 7.37.

Step 13. Draw Boss

a. Draw two 1.25" diameter isocircles.

b. Space them .25" apart on the Y-axis to form the raised cylinder (also referred to as a boss) shown in Figure 7.38.

c. Add a .75" diameter ellipse at the top of the boss as shown in Figure 7.38.

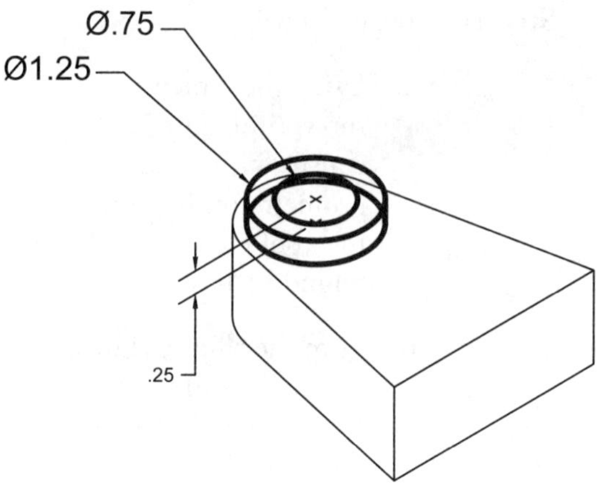

7.38 Constructing the Boss

d. Add vertical tangent lines to create the sides of the boss and trim other lines as needed to complete the boss as shown in Figure 7.39.

Step 14. Trim or erase any unnecessary lines to finish the isometric view of the Tool Holder.

Your drawing should resemble the one shown in Figure 7.40.

Step 15. Follow your instructor's directions to plot the drawing. Be sure to save the drawing file when you close AutoCAD.

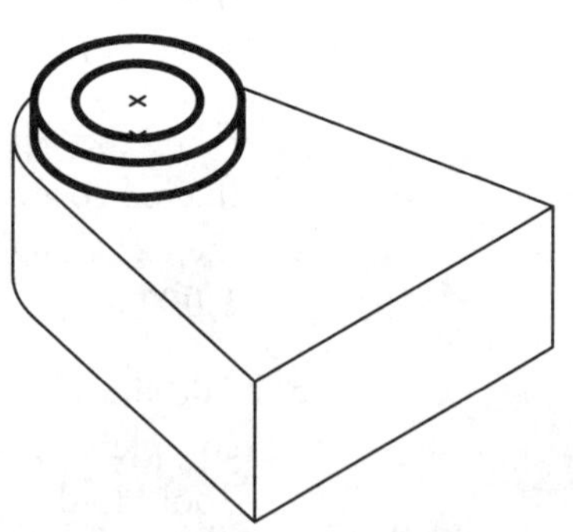

7.39 Completed Construction of the Back Part of the Tool Holder

7.40 Completed Isometric Drawing of the Tool Holder

OPTIONAL

Project 7.3 - Isometric Bracket

Open the drawing titled **Bracket** that you created in Chapter 4 and create an isometric drawing from the sketch shown in Figure 4.163 on page 236. Construct the isometric view in the space located to the right of the magenta rectangle (in model space), but move the finished isometric drawing inside the upper right corner of the rectangle (to the right of the top view and above the right view). Set AutoCAD's **Snap type** to **Isometric snap**—see figures 7.11(a) and 7.11(b)—and turn **Ortho** on. When you are finished with this project, this drawing will include both the isometric drawing of the Bracket as well as its multiviews.

OPTIONAL

Project 7.4 - Isometric Shaft Guide

Open the drawing titled **Shaft Guide** that you created in Chapter 4 and create an isometric drawing from the sketch shown in Figure 4.164 on page 237. Construct the isometric view in the space located to the right of the magenta rectangle (in model space), but move the finished isometric drawing inside the upper right corner of the rectangle (to the right of the top view and above the right view). Set AutoCAD's **Snap type** to **Isometric snap**—see Figures 7.11(a) and 7.11(b)—and turn **Ortho** on. When you are finished with this project, this drawing will include both the isometric drawing of the Shaft Guide as well its multiviews.

OPTIONAL

Project 7.5 - Isometric Tool Slide

Open the drawing titled **Tool Slide** that you created in Chapter 4 and create an isometric drawing from the sketch shown in Figure 4.166 on page 239. Construct the isometric view in the space located to the right of the magenta rectangle (in model space), but move the finished isometric drawing inside the upper right corner of the rectangle (to the right of the top view and above the right view). Set AutoCAD's **Snap type** to **Isometric snap**—see Figures 7.11(a) and 7.11(b)—and turn **Ortho** on. When you are finished with this project, this drawing will include both the isometric drawing of the Tool Slide as well its multiviews.

CHAPTER

8

SECTIONS

CHAPTER OBJECTIVES:

After studying the material in this chapter, you should be able to:

1. Define what section views are.
2. Describe how section views are used in technical drawings.
3. Provide the names and descriptions of the different types of sections and the terminology associated with section views.
4. Use AutoCAD to create section views, including properly placing cutting plane lines and hatch patterns.

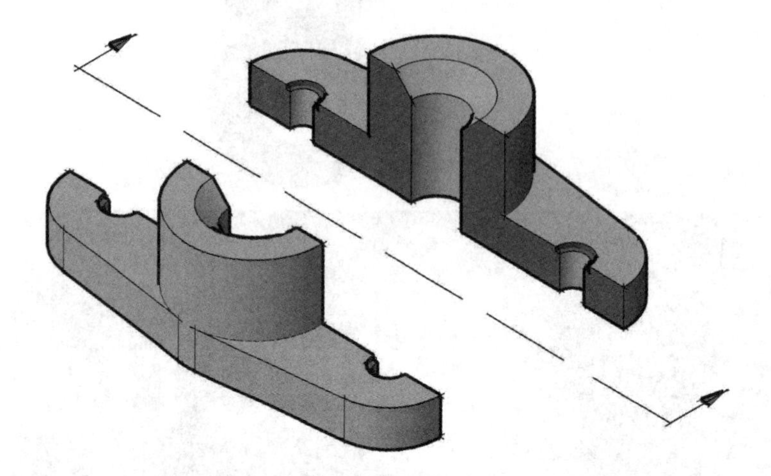

SECTIONS:

A ***section*** view is a type of drawing in which part of an object's exterior is removed to reveal its interior features. For example, in mechanical engineering drawings, sections are used to show interior features of machine parts that would not be clearly represented by hidden lines. In architectural drawings, sections are used to reveal the interior details of walls, roofs, and foundations.

In creating a section view, an imaginary ***cutting plane*** is used to slice through the object to reveal its interior features. In mechanical drawings, the sectioned areas are usually filled with diagonal ***section lines*** (also called *crosshatching*), which indicate where the cutting plane line passes through the part. In architectural drawings, the sectioned areas may be filled with hatch patterns that represent building materials such as concrete or insulation.

8.1 SECTIONS IN MECHANICAL DRAWINGS

In Figure 8.1, an object with five machined holes is shown as it would look if it were cut in half to show its inside detail. The heavy dashed line shown between the views is called a *cutting plane line*. The cutting plane line defines the line along which the object is cut. The arrows on the ends of the cutting plane line indicate the direction in which the viewer is observing the sectioned object. The diagonal lines drawn on the plane created by cutting the part in half are the *section lines*.

The principal reason for using section views in mechanical drawings is to define the complex interior features of an object by replacing hidden lines with visible lines in a view. This allows the drafter to dimension to interior features without dimensioning to hidden lines.

Figure 8.2 shows the front and top views of the object shown in Figure 8.1. The heavy dashed line running through the center of the top view is the cutting plane line. This line defines the line along which the object is cut. The upward-pointing arrows on the ends of the cutting plane line indicate the direction in which the viewer is observing the sectioned object. The diagonal lines shown on the front, or sectioned, view in Figure 8.1 indicate that the view is a section. The profile edges of the machined holes, which would have been represented by hidden lines in a regular multiview drawing, are represented by visible lines in the section view.

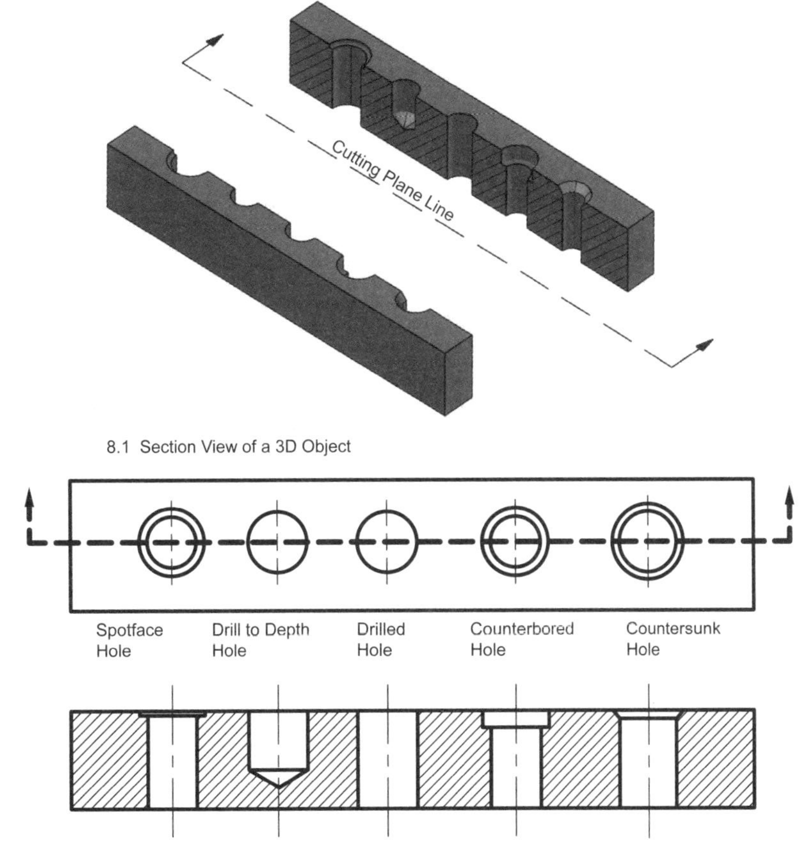

8.1 Section View of a 3D Object

8.2 Top View and Section Front View of an Object

8.2 SECTIONS IN ARCHITECTURAL DRAWINGS

In architectural drawings, sections are often included on foundation plans to show details through beams and footings, as shown in the designer's sketch in Figure 8.3. A CAD drafter might work from a sketch of this type to create a technical drawing.

Wall sections are prepared in architectural drawings to specify the composition of a wall, as shown in Figure 8.4. Often, a wall section of a building must be included in a set of architectural plans before a building permit will be granted for a project. Study Figure 8.4 and note the hatch patterns used to represent the insulation and concrete as well as the earth around the foundation.

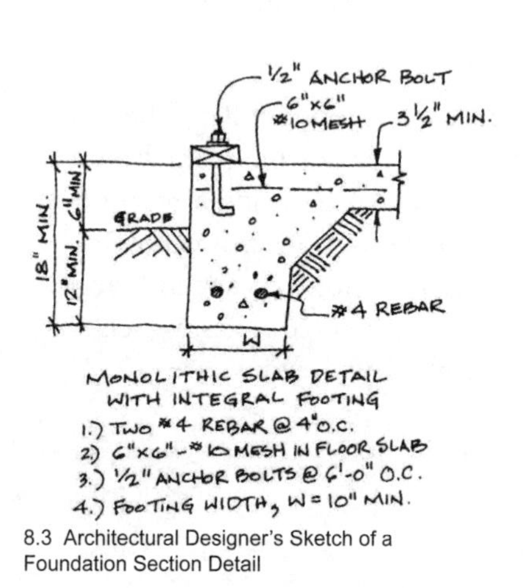

8.3 Architectural Designer's Sketch of a Foundation Section Detail

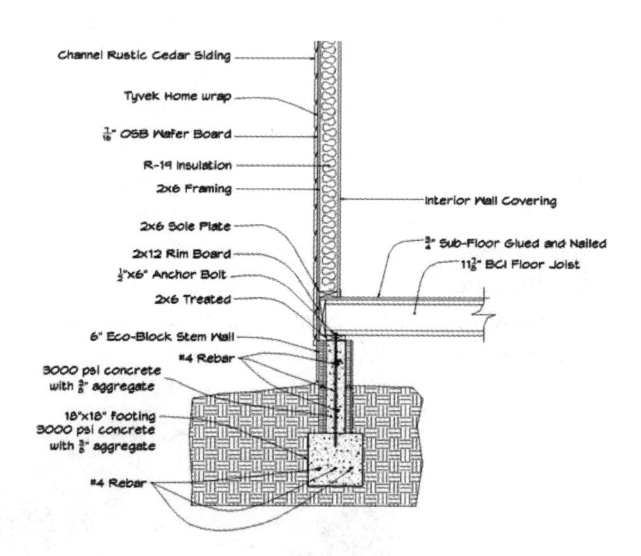

8.4 Architectural Wall Section

8.3 SECTIONS IN CIVIL DRAWINGS

In civil drawings, sections are commonly used to depict infrastructure. A few examples of civil sections include road designs, storm water ponds, utilities.

Road sections show details of the pavement structures like existing surface elevations, as well as proposed asphalt, sub-base depths, curbing, sidewalks and cut and fill earthworks. Figure 8.5 depicts a typical road section.

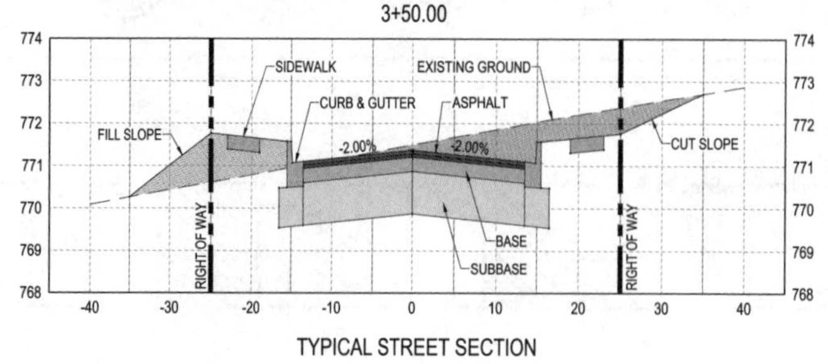

8.5 Civil Typical Road Section
Image courtesy of Jeffrey B. Muhammad

8.4 TYPES OF SECTIONS

Common section view types are *full, half, broken-out, revolved, removed,* and *offset.* Drafters decide on the type of section view to draw based on which one most clearly represents the necessary features.

Full Sections

A ***full section*** shows the object as if it has been cut in half. Figures 8.6(a) and 8.6(b) show the front and side views of an apple. The knife blade in the front view seen in Figure 8.6(a) represents the cutting plane line. In the side view seen in Figure 8.6(b), the apple appears as it would if it were sliced in half along the cutting plane line. This view represents a full section.

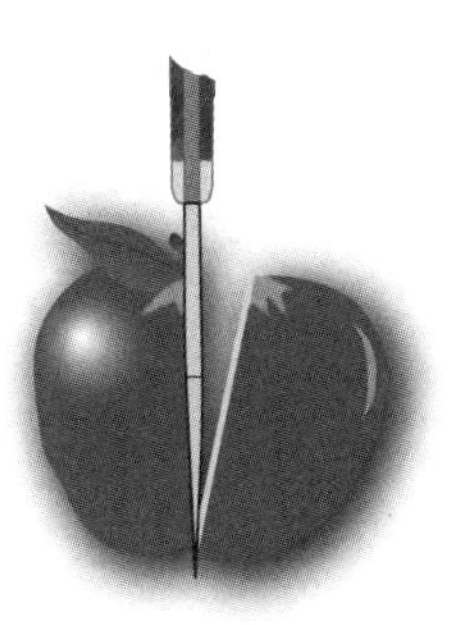

8.6(a) Knife Blade Represents the Cutting Plane Line in Front View

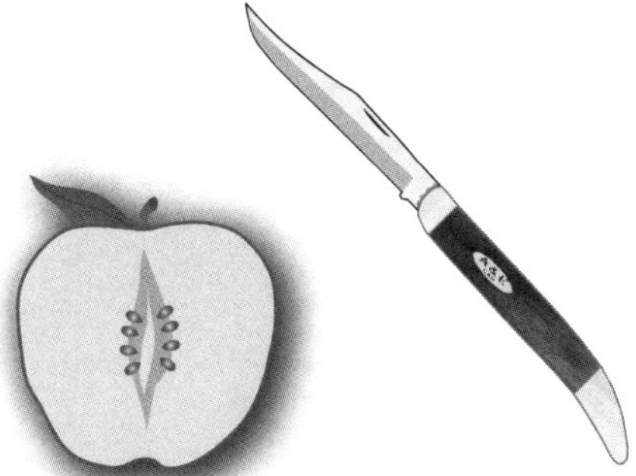

8.6(b) Side View Shown a Full Section

In Figures 8.7 and 8.8 the front and side views of a machine part are shown. The thick dashed lines with the arrows pointing to the left in the front view represent the cutting plane line (see Figure 8.7). The view shown in Figure 8.8 represents a full section. This view is drawn as it would appear if the part of the object in the front view that lies behind the cutting plane line had been removed. The diagonal lines shown in the side view are the section lines.

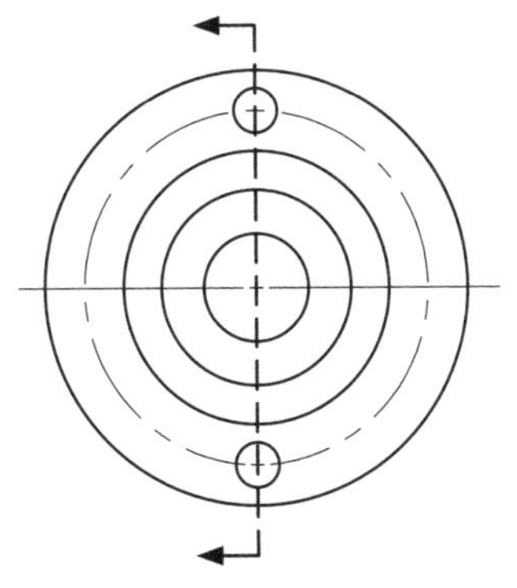

8.7 Front View with the Cutting Plane Line for a Full Section

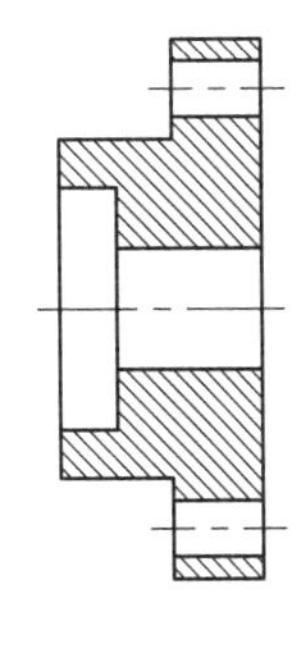

8.8 Side View Drawn as a Full Section

JOB SKILL

Cutting plane lines take precedence over centerlines. When a cutting plane line is used as the centerline, show the cutting plane line only. The ASME standard governing section views is ASME Y14.3 – 2012.

Half Sections

A ***half section*** shows the object as if one fourth of it has been removed. In Figures 8.9 and 8.10 the top and front views of an apple are shown. The knife blades in the top view (Figure 8.9) represent the cutting plane line. In the front view (Figure 8.10) the apple appears as it would if the part of the apple framed by the knife blades in the top view had been removed. This view represents a half section.

In Figures 8.11 and 8.12 the top and front views of a machine part are shown. The thick dashed line with the arrow pointing up in the top view represents the cutting plane line (see Figure 8.11). The view in Figure 8.12 represents a half section. This view is drawn as it would appear if the part of the object in the top view framed by the cutting plane line had been removed.

JOB SKILL

In a half section, it is not necessary to draw hidden lines in the unsectioned half of the view.

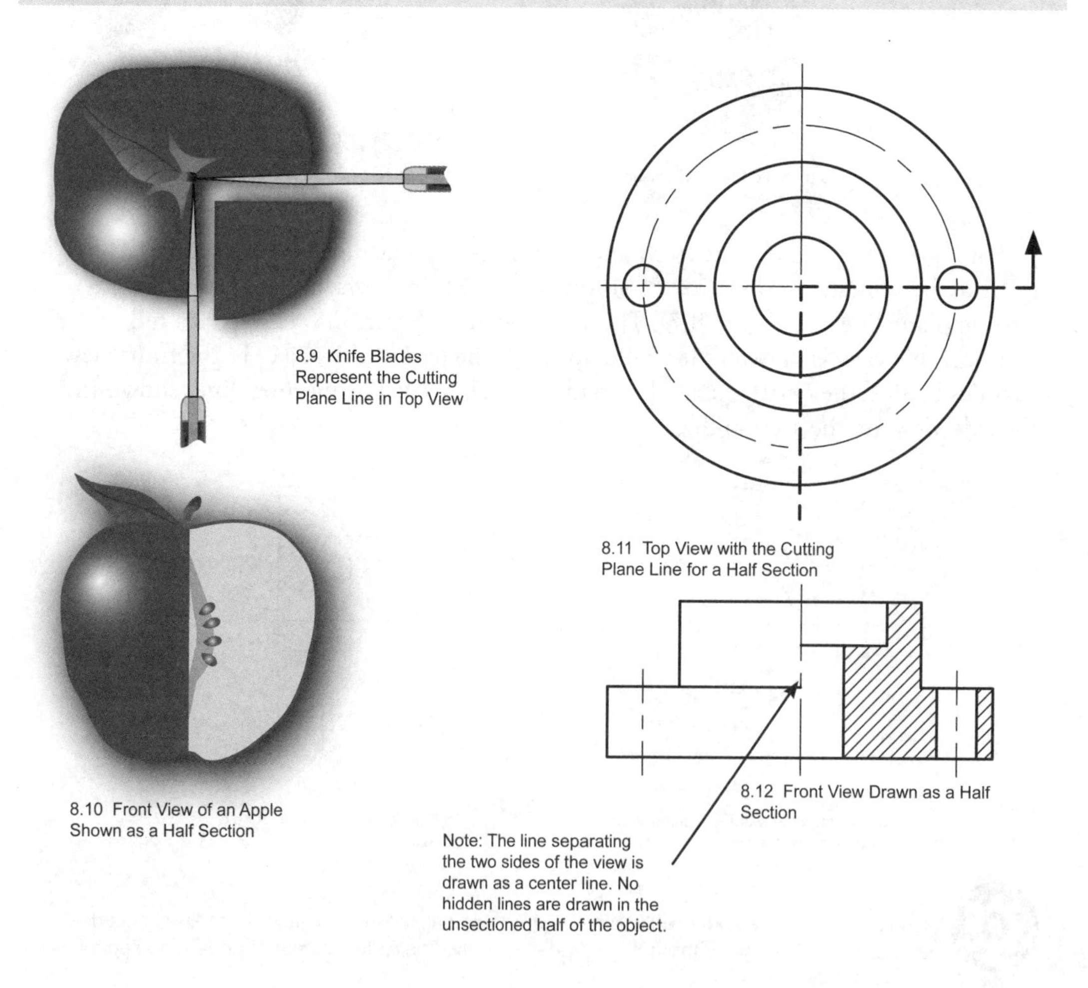

Note: The line separating the two sides of the view is drawn as a center line. No hidden lines are drawn in the unsectioned half of the object.

8.9 Knife Blades Represent the Cutting Plane Line in Top View

8.10 Front View of an Apple Shown as a Half Section

8.11 Top View with the Cutting Plane Line for a Half Section

8.12 Front View Drawn as a Half Section

In Figures 8.13 and 8.14 the front and side views of an apple are shown. The knife blades in the front view (Figure 8.13) represent the cutting plane line. In the side view (Figure 8.14), the apple appears as it would if the part of the apple inside the knife blades in the front view had been removed. This view represents a half section.

8.13 Knife Blades Represent the Cutting Plane Line in Front View

8.14 Side View of Apple Shown as a Half Section

In Figures 8.15 and 8.16 the front and side views of a machine part are shown. The thick dashed line with the arrow pointing toward the left in the front view represents the cutting plane line (see Figure 8.15). The view in Figure 8.16 represents a half section. This view is drawn as it would appear if the part of the object in the front view framed by the cutting plane line had been removed.

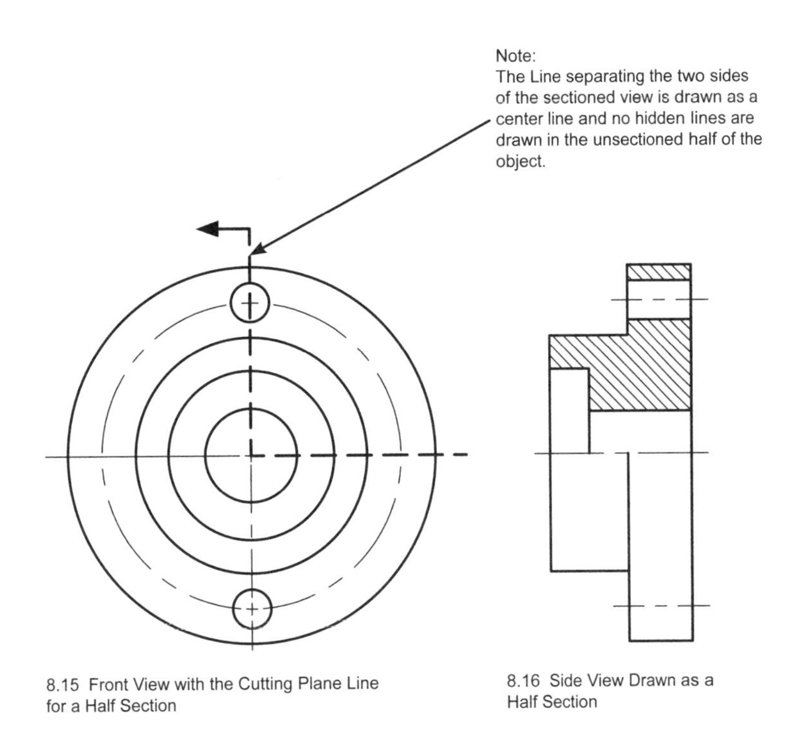
Note:
The Line separating the two sides of the sectioned view is drawn as a center line and no hidden lines are drawn in the unsectioned half of the object.

8.15 Front View with the Cutting Plane Line for a Half Section

8.16 Side View Drawn as a Half Section

Broken-Out Sections

A ***broken-out section*** is used when only a small portion of the object needs to be sectioned. Figure 8.17 shows the front and side views of an apple. In the side view, a piece of the apple has been broken out to show the worm on the inside.

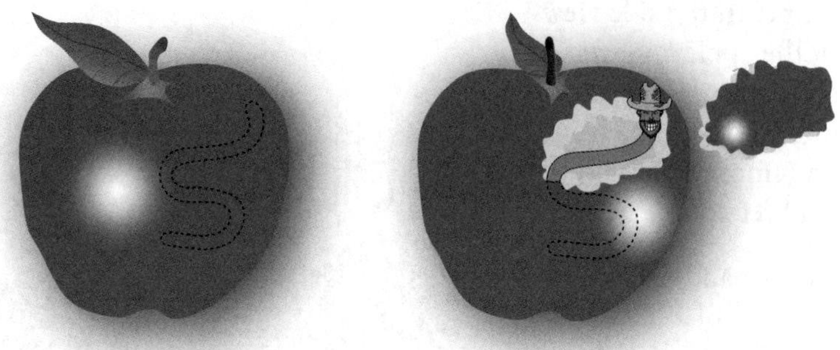

8.17 Front and Side Views of Apple with a Broken-Out Section Shown in the Side View

Figure 8.18 shows the front and top views of a machine part. The front view is drawn as it would appear if a small part of the object has been removed, or broken out, to reveal the desired interior feature. The broken line is drawn as a visible line, and diagonal section lines are added to the sectioned areas. The front view in Figure 8.18 represents a broken-out section.

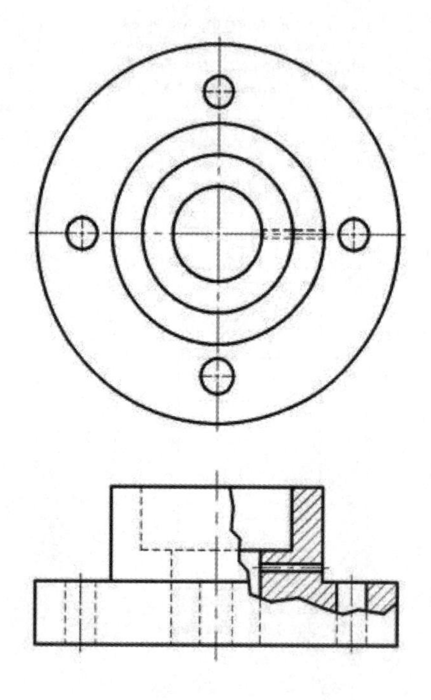

8.18 Front and Top Views with a Broken-Out Section in the Front View

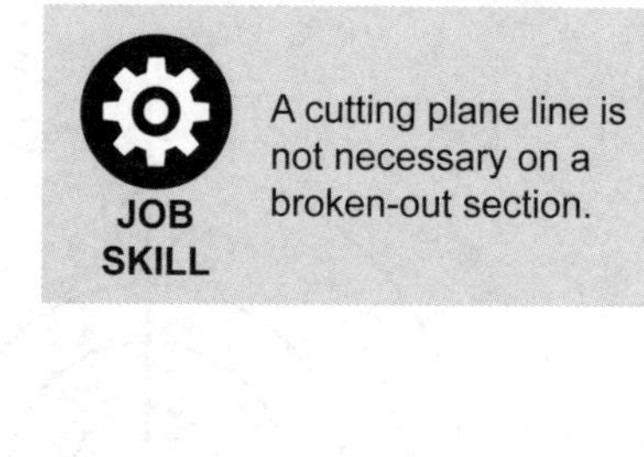

A cutting plane line is not necessary on a broken-out section.

Revolved Sections

In a ***revolved section*** the cross-sectional view of the object is drawn on the object as shown in Figures 8.19 and 8.20. Drafters can place dimensions directly on the revolved section.

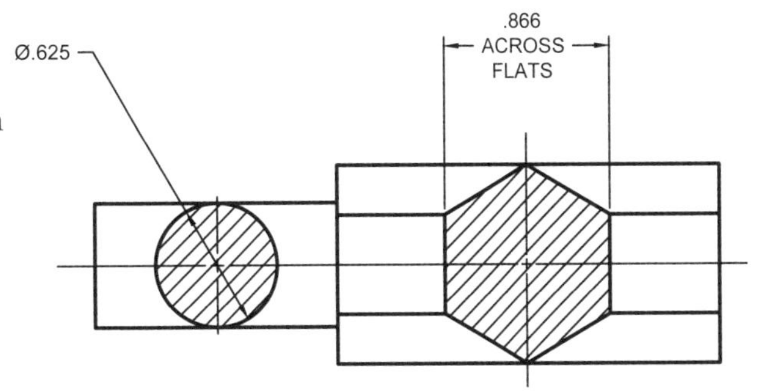

8.19 Revolved Section

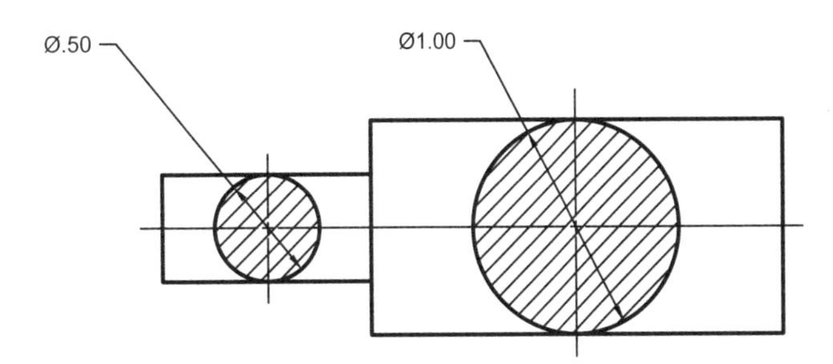

8.20 Revolved Section

Removed Sections

In a ***removed section*** the view created by the cutting plane line is not drawn in its normal "projected" position but is drawn somewhere else on the sheet. In Figure 8.21, two cutting plane lines are shown; one is labeled *A-A*, and the other is labeled *B-B*. Figure 8.22 shows the removed section of the object that results when the object is viewed through the cutting plane line labeled *A-A*. The section is labeled *SECTION A-A* to show its relationship to cutting plane *A-A*. Figure 8.23 shows the removed section that results when the object is viewed through the cutting plane labeled *B-B*. This view is labeled *SECTION B-B* to correspond to cutting plane *B-B*. Drafters must ensure that the cutting plane line and the resulting removed section view are labeled alike to avoid confusion when using this technique.

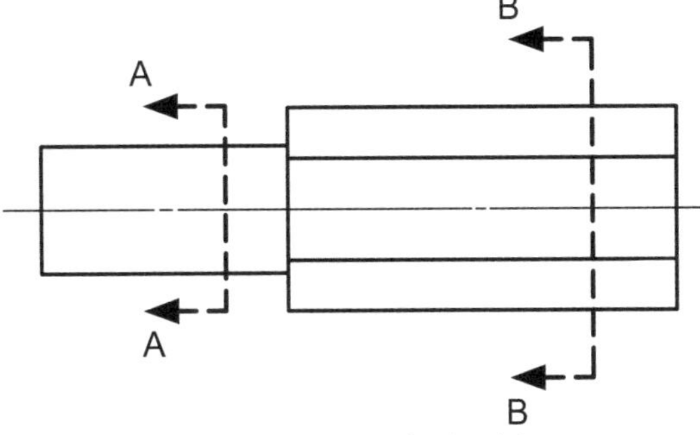

8.21 Cutting Plane Lines for Removed Sections A-A and B-B

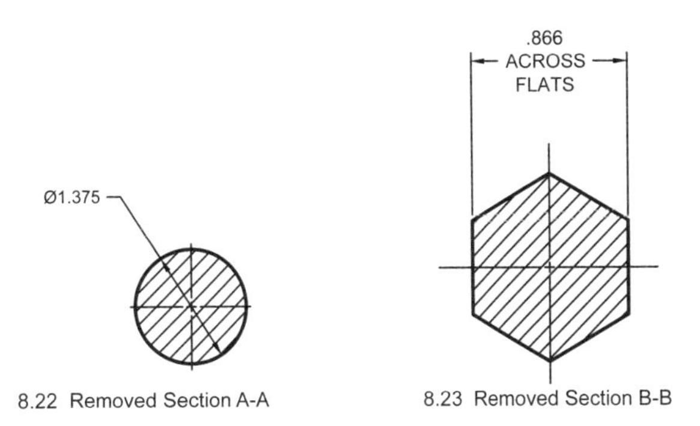

8.22 Removed Section A-A

8.23 Removed Section B-B

Offset Sections

Offset sections allow a drafter to create a section with features that would not lie on the path of a straight cutting plane line (see Figure 8.24). In an offset section, the cutting plane line is offset at 90° angles to allow it to pass through the features that the drafter would like to be shown in the resulting section view (see Figure 8.25).

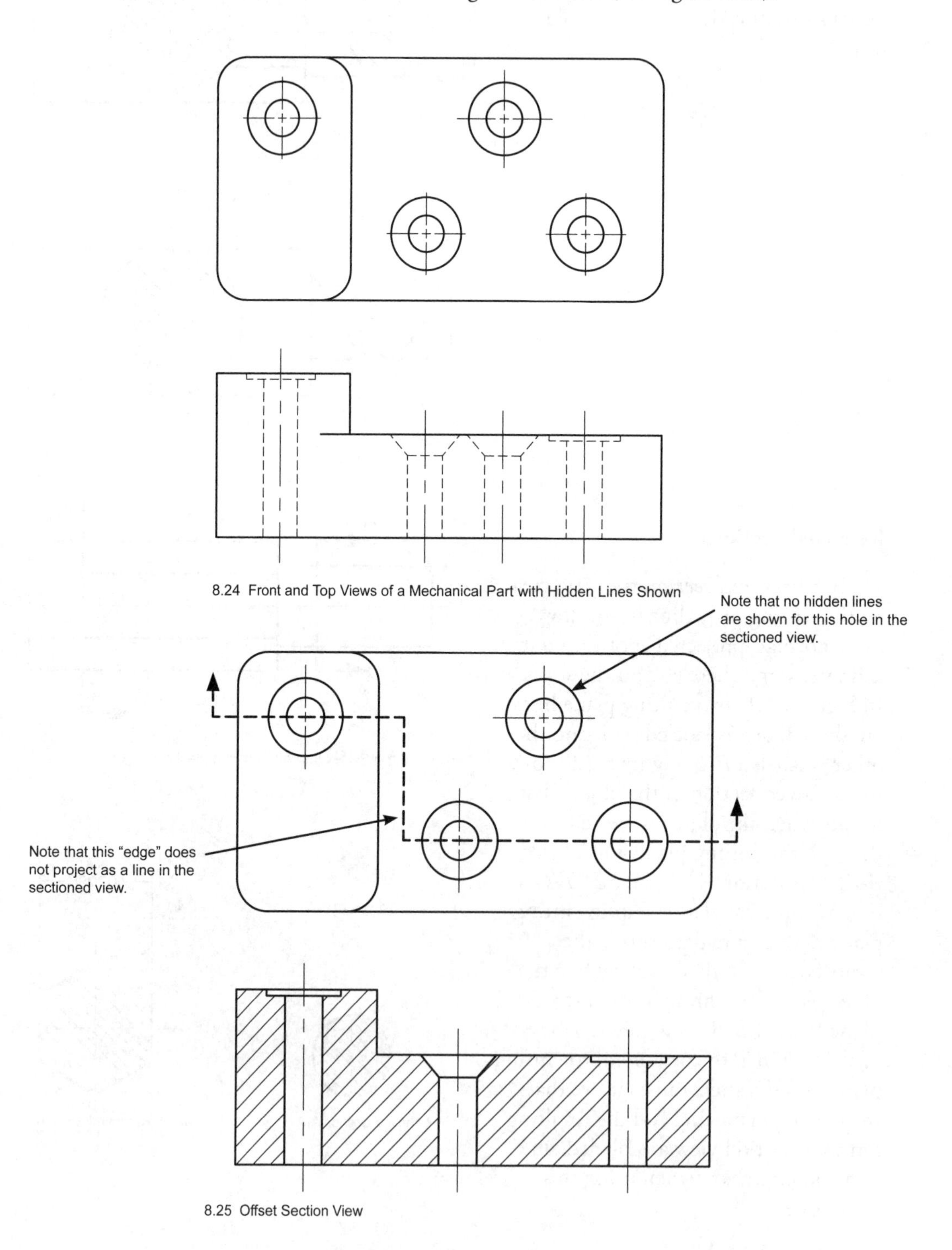
Note that no hidden lines are shown for this hole in the sectioned view.
Note that this "edge" does not project as a line in the sectioned view.

8.24 Front and Top Views of a Mechanical Part with Hidden Lines Shown

8.25 Offset Section View

STEP–BY–STEP

STEPS IN CREATING A SECTION VIEW

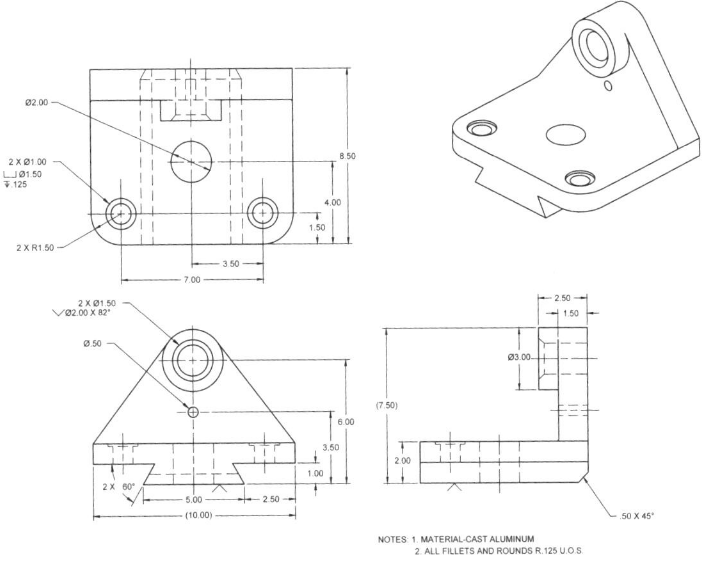

8.26 Multiview Drawing to Be Sectioned

The steps involved in creating a full section view of the object shown in Figure 8.26 are as follows:

Step 1. Determine the view to be sectioned and the location of the cutting plane line. In this case, the cutting plane line will be located on the front view (see Figure 8.27). The placement of the cutting plane line on the object's front view, along with the direction of the cutting plane line's arrows, indicates that the right-side view of the object will be drawn as a full section.

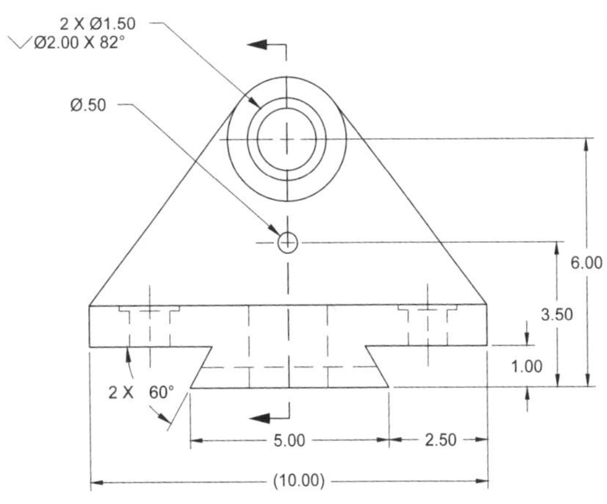

8.27 Placing a Cutting Plane Line on the Front View

QUICK TIP

The **Polyline** command can be used to make the large arrowheads needed at the ends of cutting plane lines. This is accomplished by drawing a short polyline (about .25" long) that has a starting width of 0 (zero) and an ending width of .10. See Figure 4.50(b) on page 150.

STEP-BY-STEP

Step 2. Study the right-side view (see Figure 8.28) and determine which of the object's hidden lines will become visible after it has been sectioned. Convert these hidden lines to visible lines as shown in Figure 8.29.

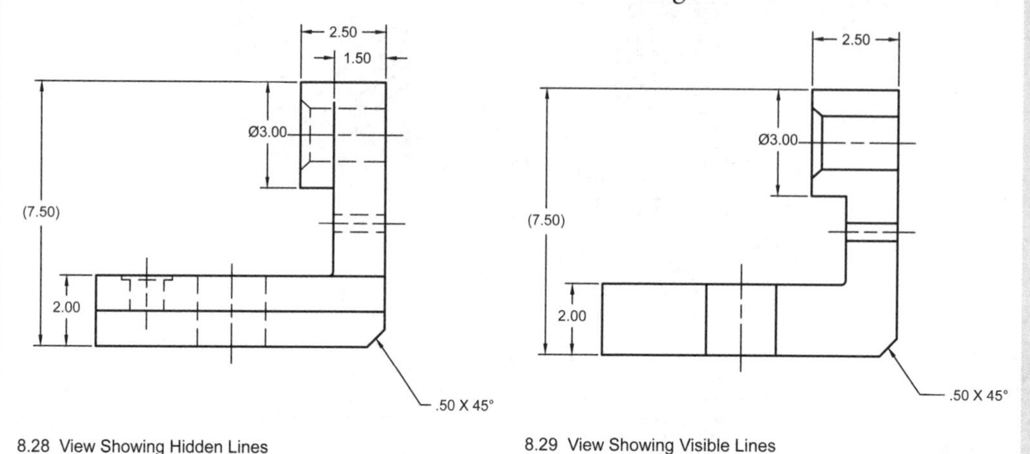

8.28 View Showing Hidden Lines

8.29 View Showing Visible Lines

Step 3. Place section lines inside the areas created by the cutting plane line's passing through the object. In an AutoCAD drawing, the **HATCH** command is used to place section lines or other hatch patterns into a sectioned area.

NOTE

The angle of the lines in the **ANSI31** pattern shown in Figure 8.30(b) will automatically be inserted at **45°** even though the default **Angle** of the pattern is displayed as **0°**. The angle settings for other predefined hatch patterns often follow this convention as well.

STEP-BY-STEP

Using the HATCH Command

Select the **HATCH** command icon shown in Figure 8.30(a) from the **Draw** toolbar or the **Draw** panel of the **Home** tab of the ribbon. The **Hatch Creation** tab shown in Figure 8.30(b) will open (replacing the **Home** tab). This tab contains six panels: **Boundaries**, **Pattern**, **Properties**, **Origin**, **Options**, and **Close**; however, the most important panels for beginners to master are the first three. These panels are explained next.

The Boundaries Panel

Left-clicking the **Pick Points** button in this panel—see Figure 8.30(b)—allows you to define the boundary area(s) to be hatched by picking points inside existing objects (such as circles, rectangles, polygons, and closed polylines) that form a closed boundary

STEP-BY-STEP

around the pick point. If the selected area is not a closed boundary, the **Hatch-Boundary Definition Error** dialog box will open and offer some assistance with solving the problem. Picking the **Select** button in the **Boundaries** panel, see Figure 8.30(b), rather than **Pick Points** allows you to define a hatch boundary by selecting the objects that form the closed boundary.

The Pattern Panel

This panel displays predefined hatch patterns that can be assigned when using the **HATCH** command. These patterns are represented in the panel by showing a preview of the pattern. You can scroll through the swatches by picking the up or down arrows on the right side of the panel. Left-click on the swatch you wish to assign as the current hatch pattern style.

The Properties Panel

Because this is a fundamentals text, we will focus on only two settings from this panel, the **Hatch Angle** setting and the **Hatch Pattern Scale** setting. The value assigned in the **Hatch Angle** window defines the angle for the hatch pattern relative to the *X*-axis of the current **UCS**. The default angle setting is **0°**.

The value assigned in the **Hatch Pattern Scale** window enlarges or reduces the size of a predefined (or custom) hatch pattern. The default scale setting is **1**. In metric and architectural drawings, this setting must often be increased to a very large value, or else the hatch pattern may appear so dense in the drawing that it may seem to be a solid fill.

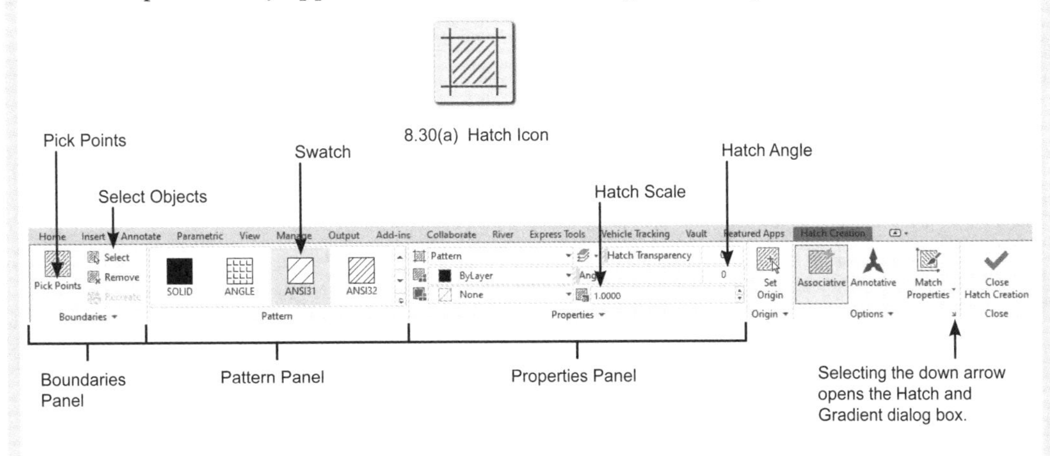

8.30(a) Hatch Icon

8.30(b) **Hatch Creation** Tab Settings

JOB SKILL

Two conditions that may affect the insertion of a hatch pattern in a drawing:
The objects that define the borders of the area to receive the pattern must create a closed boundary (closed boundaries do not have gaps in their boundary objects).
In older releases of AutoCAD, the entire hatch boundary must be visible inside the drawing window's display area. For this reason, the display window's scale must be adjusted (by zooming out) so that the entire hatch boundary is visible before beginning the **HATCH** command. **NOTE:** Zooming out in this fashion is not necessary in newer releases of AutoCAD.

STEP–BY–STEP

To place the hatch pattern in the section view shown in Figure 8.31, you would select the **Hatch** icon to access the **Hatch Creation** tab on the ribbon, then select **ANSI31** from the **Pattern** panel, and accept the default settings for **Hatch Angle** and **Hatch Pattern Scale** in the **Properties** panel. Next, you would select the **Pick Points** button from the **Boundaries** panel and pick inside the areas where the hatch pattern should be applied. When finished you would press **<Enter>**, and the hatch pattern would fill the areas defined by the pick points.

Figure 8.31 shows the side view with the ANSI31 hatch pattern applied. Note that no hidden lines are shown in this view.

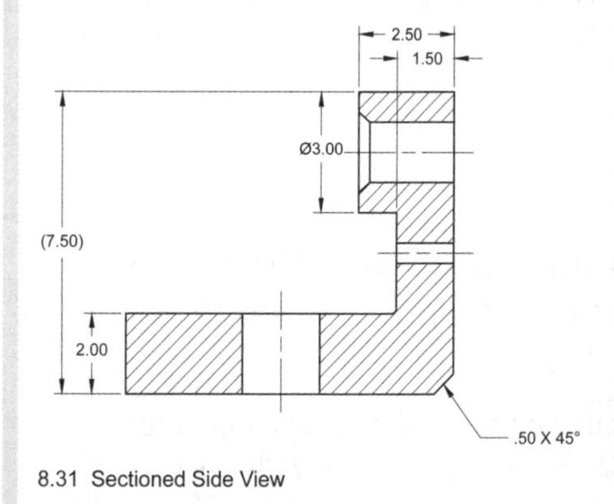

8.31 Sectioned Side View

JOB SKILL

The scale for hatching in metric units is as follows:

Decimal	Metric
.50	12.7
.75	19.5
1.00	25.4
1.25	31.75
1.50	38.1
1.75	44.45
2.00	50.8

Figure 8.32 shows the finished drawing with the right-side view replaced with a full section view. The full section provides a clear portrayal of the part's interior features.

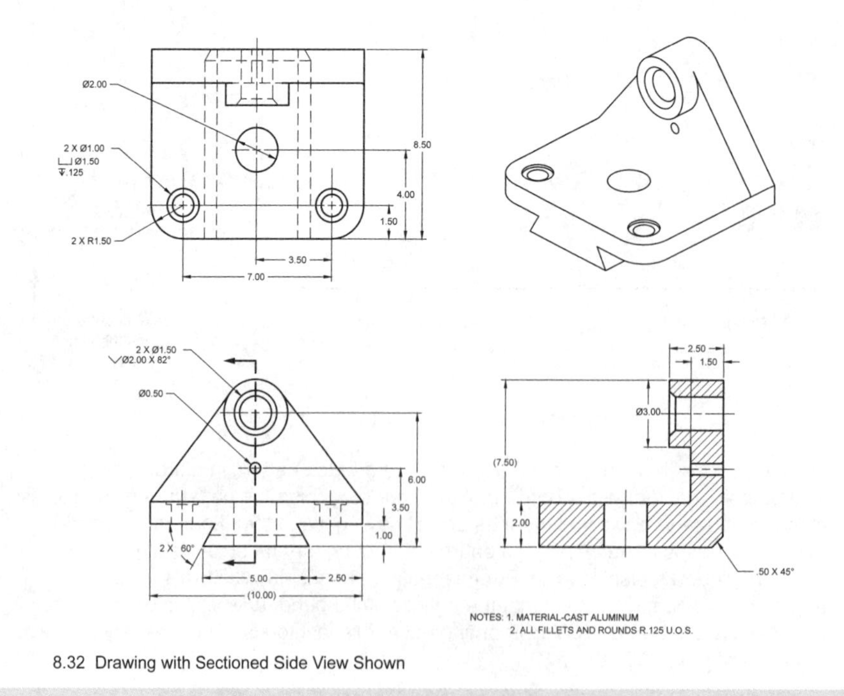

8.32 Drawing with Sectioned Side View Shown

Interpreting an object's complex interior features is often difficult, especially when the features are represented by hidden lines in a front, top, or side view. By creating a section view of the object, the drafter can clarify the interior details by replacing the hidden lines with visible lines.

Because dimensioning to hidden lines should be avoided in technical drawings, creating a section view also facilitates the placement of dimensions by making complex interior features visible.

To create and interpret section views, drafters and designers must be familiar with the different types of sections and the CAD techniques used to create them.

Before drawing a section view of an object, a drafter must imagine what the object will look like after it has been sliced along a cutting plane line. The ability to visualize section views is a skill that beginners can develop and improve with practice.

Broken-Out Section
Cutting Plane
Full Section
Half Section
Offset Section
Removed Section
Revolved Section
Section
Section Lines

Matching

Column A	**Column B**
a. Half	1. Section view that requires a label
b. Removed	2. Percentage of an object removed to create a half section
c. One fourth	3. The type of section on which a cutting plane line is not necessary
d. Revolved	4. Percentage of an object removed to create a full section
e. Offset	5. Section that cuts through features that do not lie on a straight line

Multiple Choice

1. What are the diagonal lines in a section view called?
   a. Section lines
   b. Cutting plane lines
   c. Horizontal plane lines
   d. Straight plane lines
2. In an offset section, the cutting plane line is offset at what angle?
   a. 45°
   b. 35°
   c. 90°
   d. 25°
3. Where is a removed section placed on a drawing?
   a. On another sheet
   b. Somewhere else on the sheet
   c. Deleted from the sheet
   d. On another tab
4. What do the arrows on a cutting plane line indicate?
   a. Direction in which section is viewed
   b. Placement of centerlines
   c. The top of the sheet
   d. None of the above
5. What is the name of the HATCH setting that controls the density of the hatch pattern?
   a. Volume
   b. Scale
   c. Hue
   d. All the above

In this project, you will open the **Tool Holder** drawing you created in Chapter 4 and convert its front view to a full section.

Studying Figure 8.33 will help you visualize the tool holder after it has been cut along the cutting plane line. Figure 8.34 shows the location of the cutting plane line in the top view.

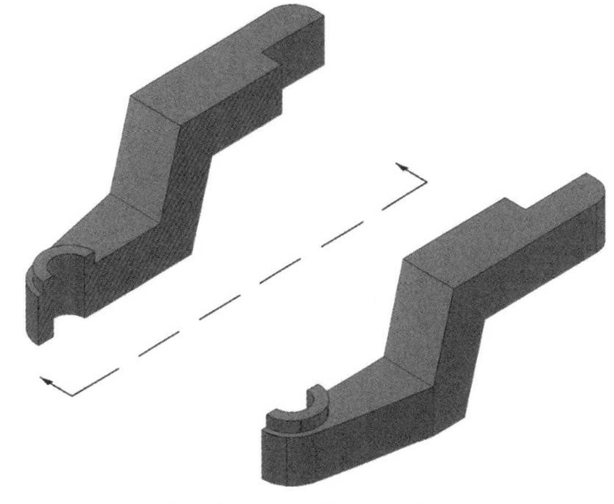

8.33 3D Sketch of the Tool Holder Shown as a Full Section

Step 1. Open the **Tool Holder** drawing you created in Chapter 4.

Step 2. Make two new layers:

a. **Cutting Plane**

b. Set the linetype to **DashedX2**

- Set the lineweight to **.60mm**.

c. **Hatch Pattern**.

- Set the linetype for the **Hatch Pattern** layer to **Continuous**
- Set the lineweight to **Default**.

Step 3. Use the **ANSI31** hatch pattern to draw the section lines.

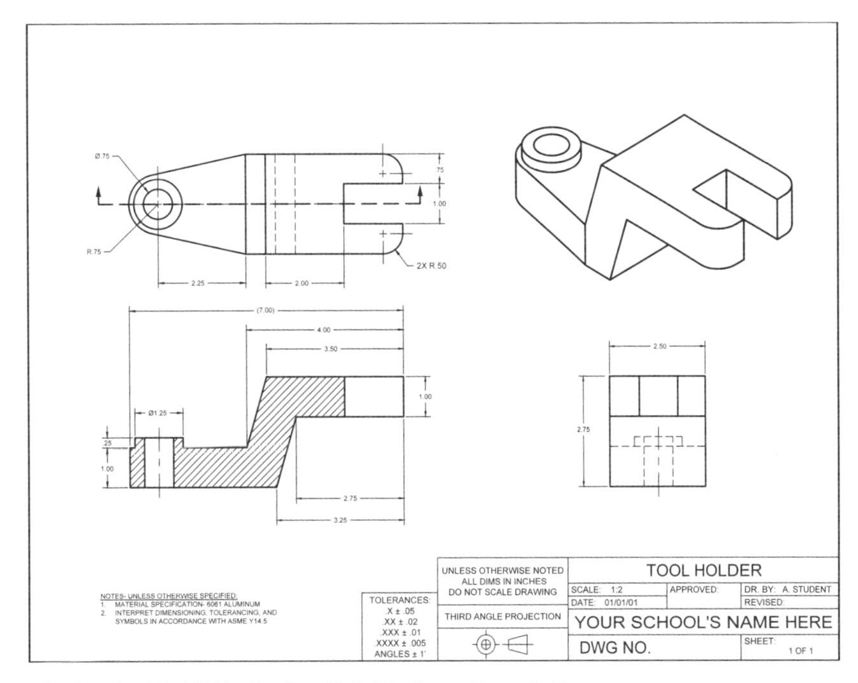

8.34 Completed Tool Holder Drawing with Full Section and Isometric View

VIDEO There is a video tutorial available for this project inside the Chapter 8 folder of the book's Video Training downloads.

Step 1. Getting Started

a. Download the **Flange Bearing Prototype** drawing file located in the *Prototype Drawing Files* folder associated with this book.

- This prototype contains the layers, text styles and drawing settings needed to begin the Flange Bearing project.

b. Use **SAVE AS** to save the drawing to your Home directory and rename the drawing **FLANGE BEARING**.

c. Activate the **Model** tab in the lower-left corner of the graphics window (if it is not already highlighted) by left-clicking on the Model tab.

Note: The views of Flange Bearing will be drawn inside the magenta rectangle located in model space. The prototype file provides the layers you need for this project.

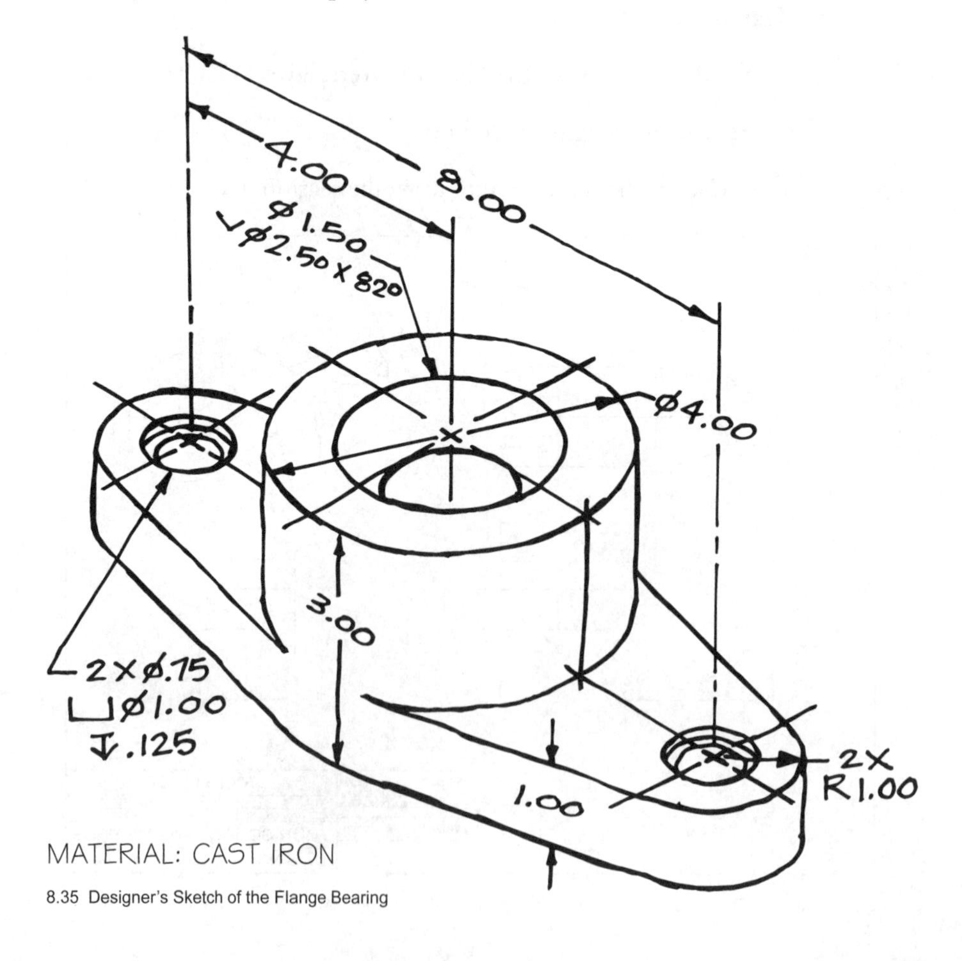

8.35 Designer's Sketch of the Flange Bearing

Step 2. Refer to the designer's sketch in Figure 8.35 and draw the front and top views of the flange bearing

a. Draw the top view as a full section (see Figure 8.36 for help visualizing section view).

NOTE

The front view should be the view that depicts the holes in the flange bearing as circles, and the cutting plane line should run horizontally through this view with its arrows pointing down.

Step 3. In the **Dimension Style Manager** dialog box, create a new dimension style named **ASME Y14.5** that contains the following dimension style settings, and set the new style current:

Dimension Style Settings for the Flange Bearing	
Text height:	.125
Arrow size:	.125
Center marks:	Line
Extend beyond dim lines:	.125
Precision:	Varies - match precision of dimensions on sketch
Zero suppression:	Leading
Offset from origin:	.062

Step 4. Add dimensions to the views, including any necessary notes. The notations and construction of the geometry for the countersunk and counterbored holes shown in Figure 8.35 are explained on the following pages.

Step 5. When you are finished, follow your instructor's directions to print the drawing.

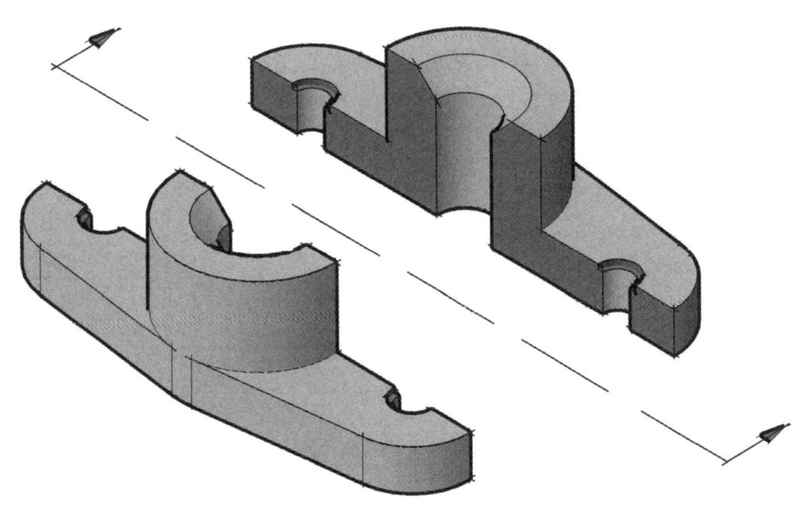

8.36 3D Flange Bearing

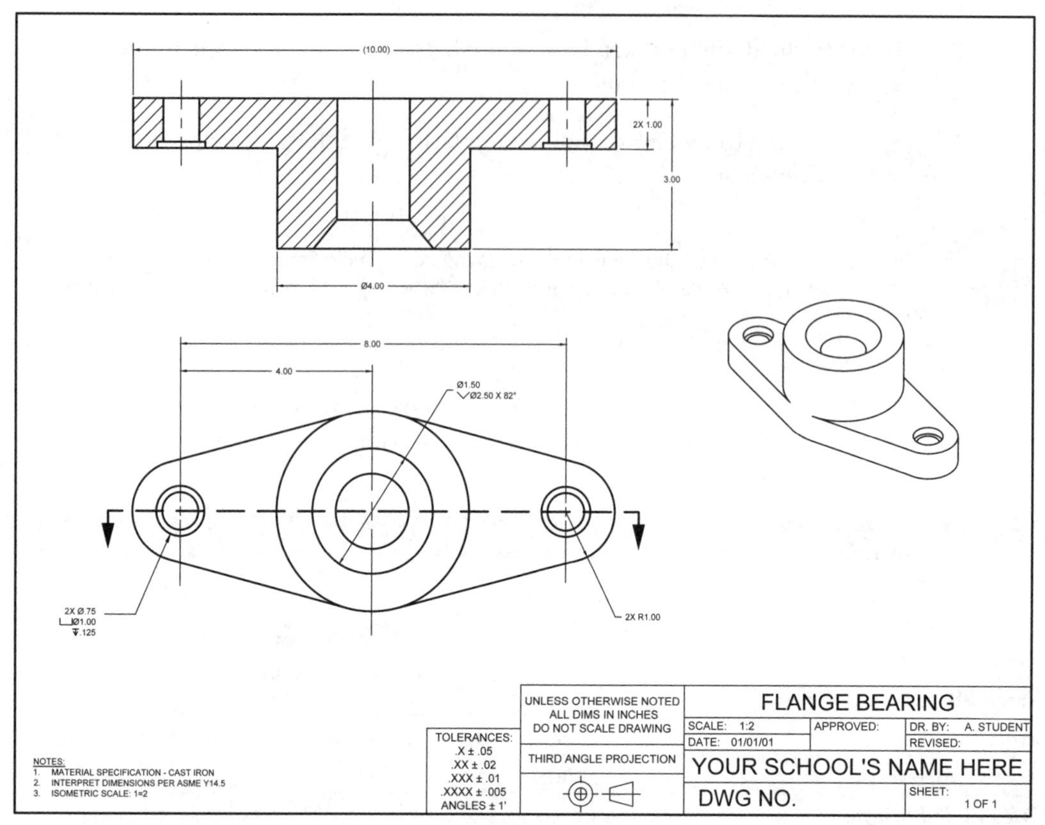

8.37 Completed Flange Bearing Drawing with Full Section and Isometric View

Interpreting the Counterbored Hole Note

Before you can draw a counterbored hole, you must first be able to interpret the annotation and symbology used in notating the hole. An explanation of the note attached to the leader pointing to the counterbored hole in Figure 8.38 follows.

Line 1: A **.75"** diameter hole passes all the way through the part.

Lines 2 and 3: A **1.00"** diameter, flat-bottomed hole (a counterbore) is to be bored into the part to a depth of **.125"**. The symbols for notating *counterbore* and *depth* in technical drawings are ⌴ and ↧ respectively.

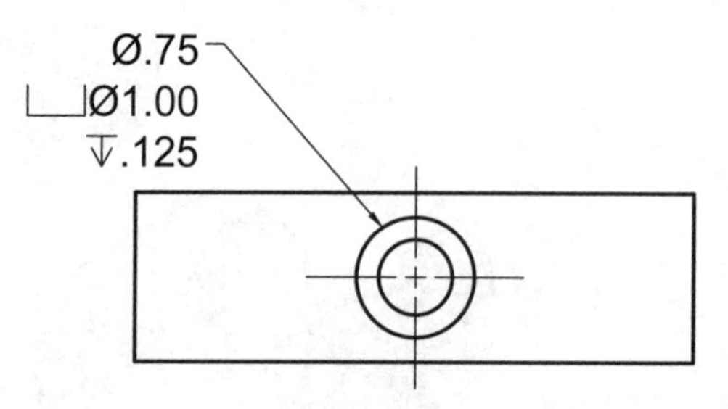

8.38 Counterbored Hole Specification

To enter the symbol for *diameter*, click on the **Symbol** icon located on the **Insert** panel of the **Text Editor** tab of the ribbon, and select **Diameter** from the drop-down list (or type %%c). See Figure 8.42(b).

STEP-BY-STEP

Constructing a Counterbored Hole

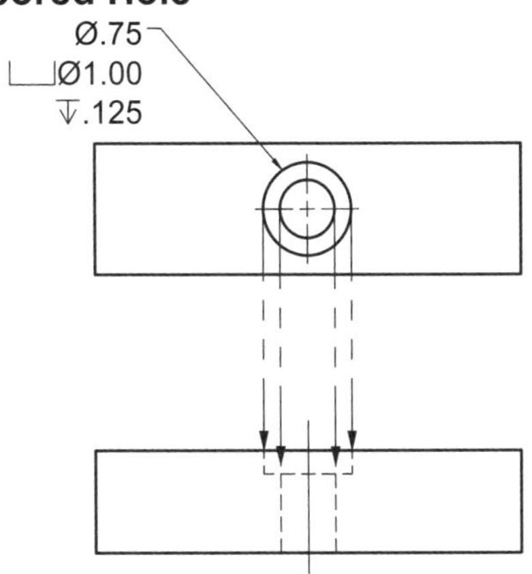

8.39 Multiview Drawing of Counterbored Hole

Step 1. Begin the construction by drawing two concentric circles in the top view, **.75"** diameter and **1.00"** diameter, respectively.

Step 2. Project lines from the quadrants of these circles to the front view to construct the hidden lines for the through hole and **.125** deep counterbore as shown in Figure 8.39.

STEP-BY-STEP

Adding Annotation to a Counterbored Hole

The following steps describe how to annotate a counterbored hole. The desired notation is shown in Figure 8.40.

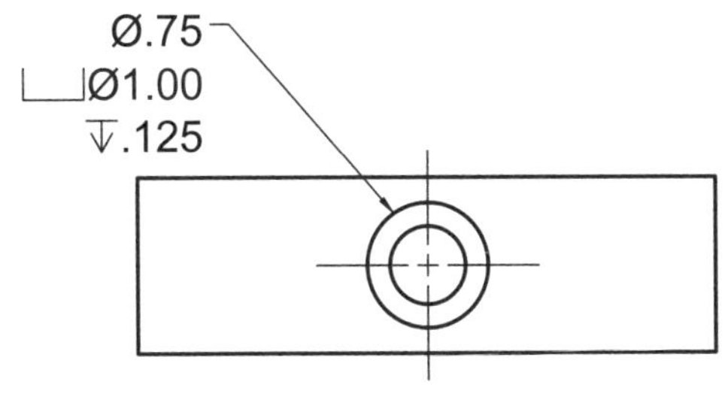

8.40 Counterbored Hole Note

8.41 **Diameter** Dimension Icon

Step 1. Select the **Diameter** dimension icon from the **Dimension** toolbar (see Figure 8.41) and place a diameter dimension on the larger circle.

Step 2. Select the diameter dimension created in Step 1, type **ED** (to open the **Text Editor**), and press **<Enter>**. Make edits inside the **Text Editor** box to create the three lines of text shown in Figure 8.42(a).

Step 3. Highlight the v and pick on the down arrow in the **Font** window located on the **Formatting** panel and change the font to **GDT** (geometric dimensioning and tolerancing). See Figure 8.43. When the font is changed, the **v** will be replaced by the **counterbore** symbol (⌴). Repeat this step for the **x**, and it will be replaced by the depth symbol (↧).

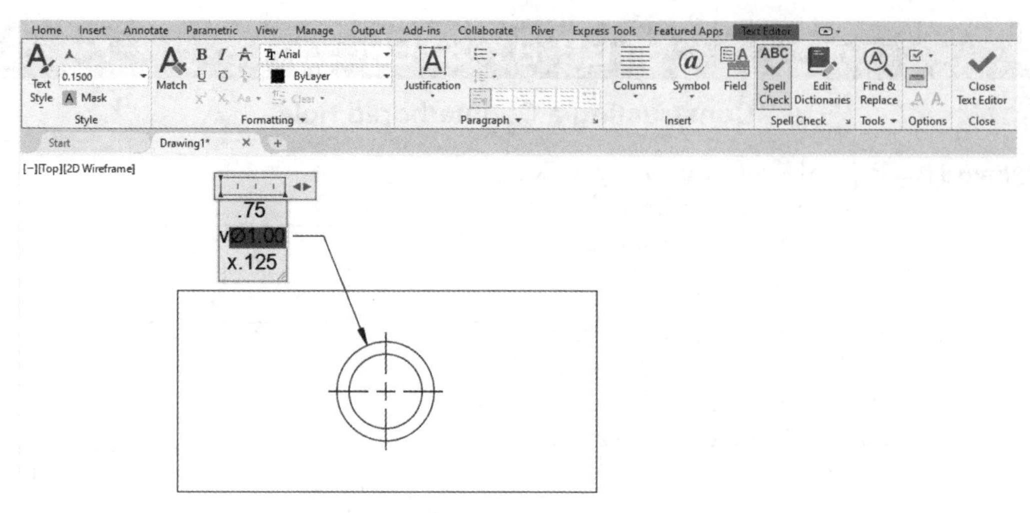

8.42(a) Entering Text for the Counterbored Hole

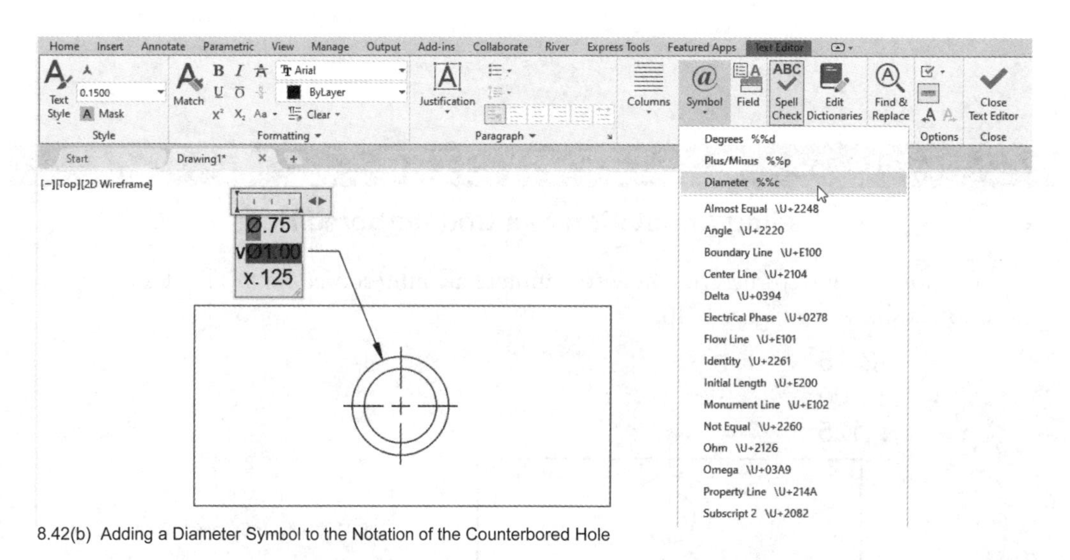

8.42(b) Adding a Diameter Symbol to the Notation of the Counterbored Hole

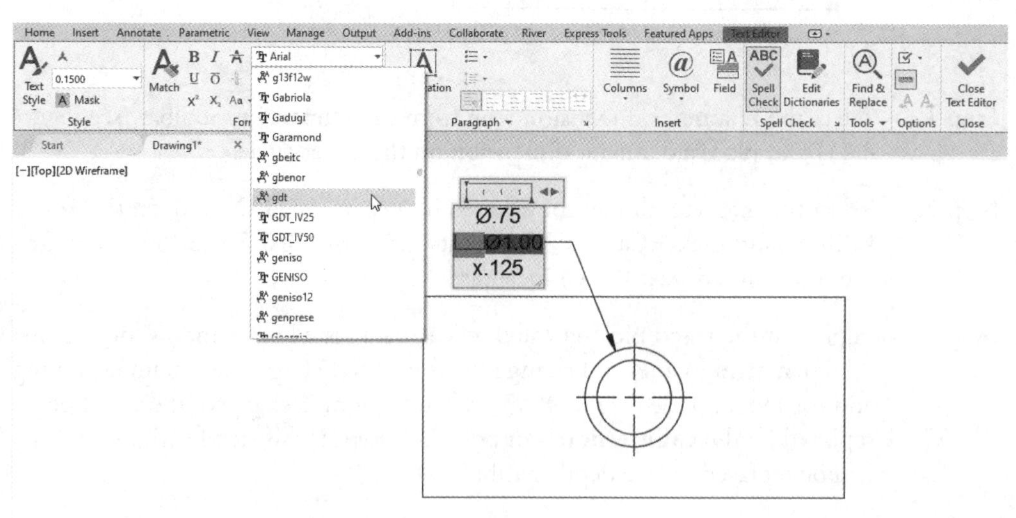

8.43 Converting an ***Arial*** Font Character to ***GDT*** Font in the Notation of the Counterbored Hole

Interpreting a Countersunk Hole Note

Before you can draw a countersunk hole, you must first be able to interpret the annotation and symbology used in notating the hole.

An explanation of the note attached to the leader pointing to the countersunk hole in Figure 8.44 follows.

Line 1: A **1.50"** diameter hole passes all the way through the part.

Line 2: A **2.50"** diameter countersunk hole is to be drilled into the part. The angle between the sloping sides of the countersink is **82°**. The symbol for notating a countersink on a technical drawing is ⌵.

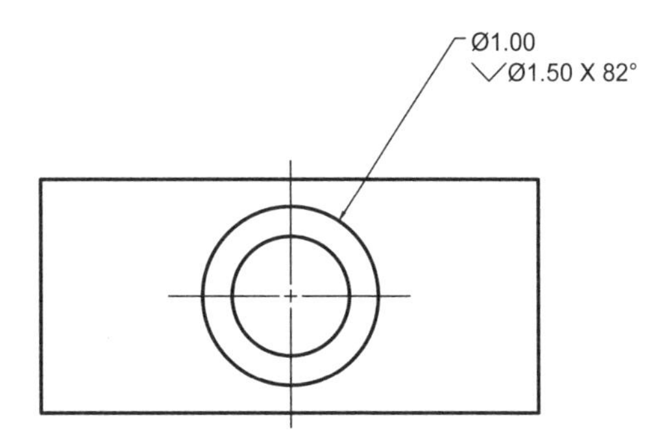

8.44 Countersunk Hole Specification

Steps in Constructing a Countersunk Hole

Step 1. Begin the construction by drawing two concentric circles in the front view, **1.50"** diameter and **2.50"** diameter, respectively. Project construction lines from the quadrants of these circles to the top view as shown in Figure 8.45.

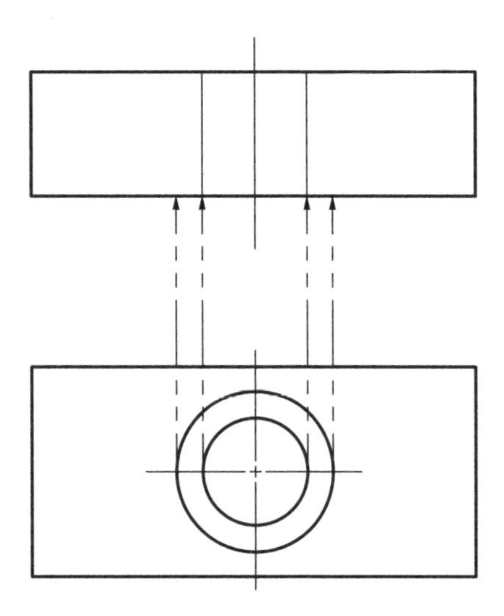

8.45 Constructing a Countersunk Hole

Step 2. From the lines projected from the circle quadrants in the front view, construct the **82°** angled sides of the countersunk hole and the sides of the through hole in the top view (see Figure 8.46). For help constructing the angles, see Figures 8.47 through 8.49.

Use the protractor in Figure 8.47 to determine countersink angles. Once you have determined the beginning points of the countersunk diameter, draw lines by using either polar coordinates or polar tracking with the required angles added in as illustrated in Figures 8.48 and 8.49. Do not be concerned with the length of the lines; concentrate on the angle. If the lines are too short, use **EXTEND** to extend them to the vertical (or horizontal) lines. If they overlap, use **TRIM** to trim back to the vertical (or horizontal) lines.

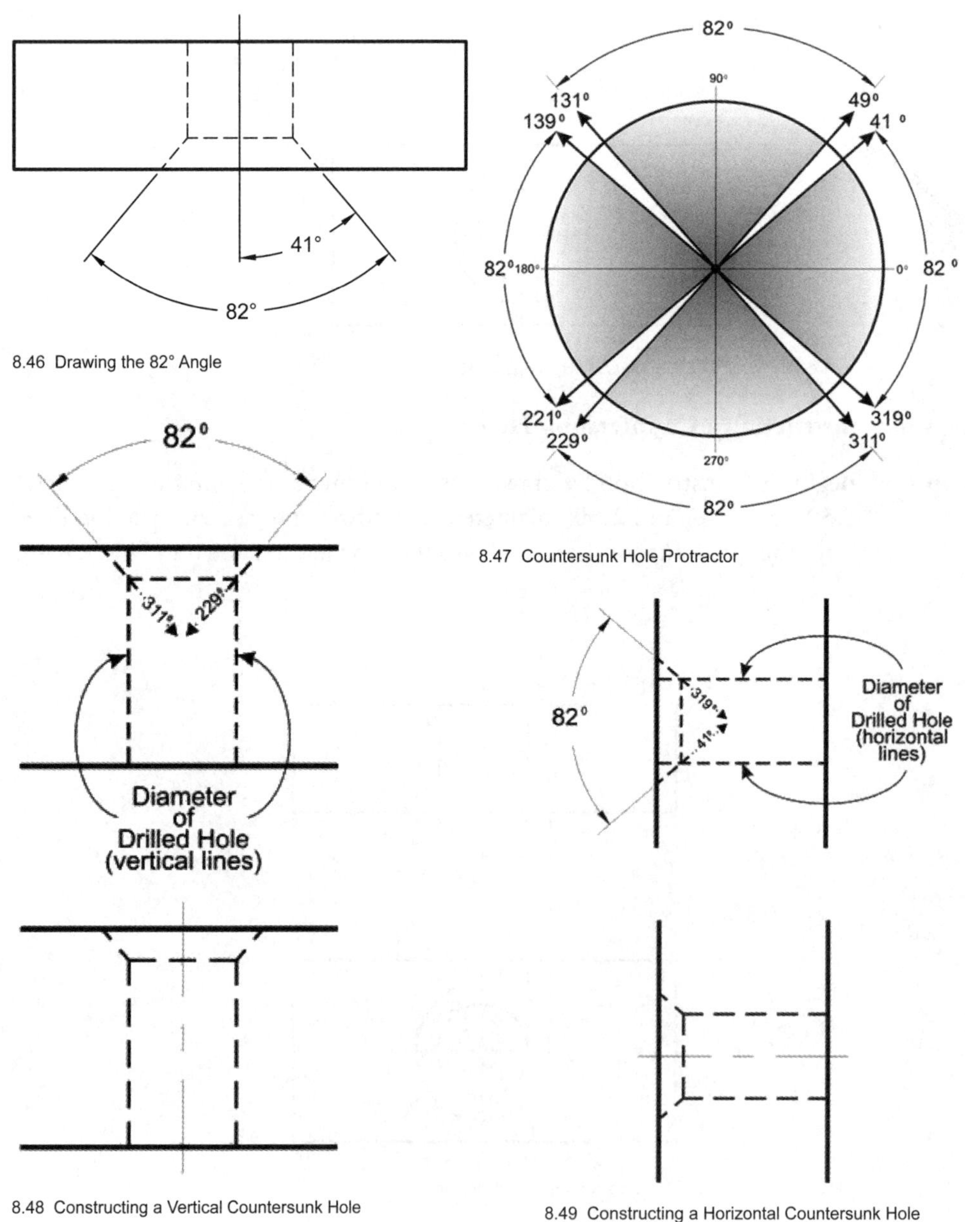

8.46 Drawing the 82° Angle

8.47 Countersunk Hole Protractor

8.48 Constructing a Vertical Countersunk Hole

8.49 Constructing a Horizontal Countersunk Hole

The completed front and top views of the countersunk hole are shown in Figure 8.50.

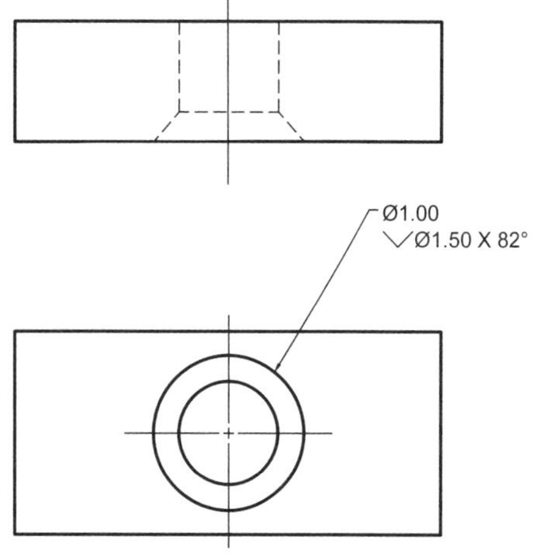

8.50 Completed Multiviews of the Countersunk Hole

STEP-BY-STEP

Adding Annotation to a Countersunk Hole

The steps required to annotate a countersunk hole are as follows. The desired notation is shown in Figure 8.51.

Step 1. Select the **Diameter** dimension icon from the **Dimension** toolbar (see Figure 8.52) and place a diameter dimension on the larger circle.

Step 2. The diameter dimension created in Step 1, type **ED** (to open the **Text Editor**), and press **<Enter>**. Make edits inside the **Text Editor** box to create the two lines of text shown in Figure 8.53.

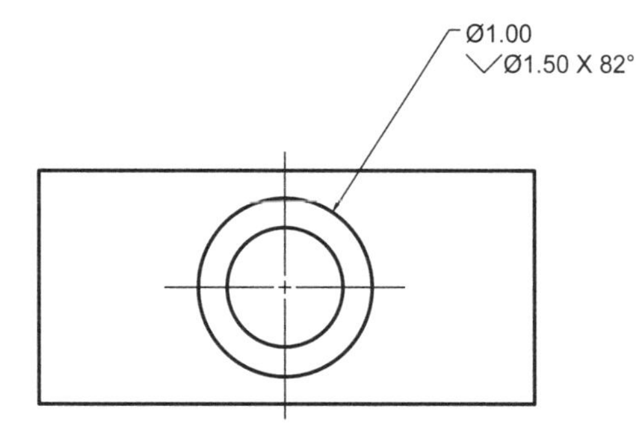

8.51 Countersunk Hole Specification

8.52 **Diameter** Dimension Icon

QUICK TIP

To directly enter the symbols for *diameter* or *degree* from the **Text Editor** tab of the ribbon, click on the **Symbol** icon located on the **Insert** panel of the **Text Editor** tab, and select **Diameter** or **Degrees** from the drop down list. See Figure 8.42(b).
You can enter these symbols from the keyboard in AutoCAD's **Text Editor** by typing **%%C** to produce a diameter symbol, or typing **%%d** to produce a degree symbol.

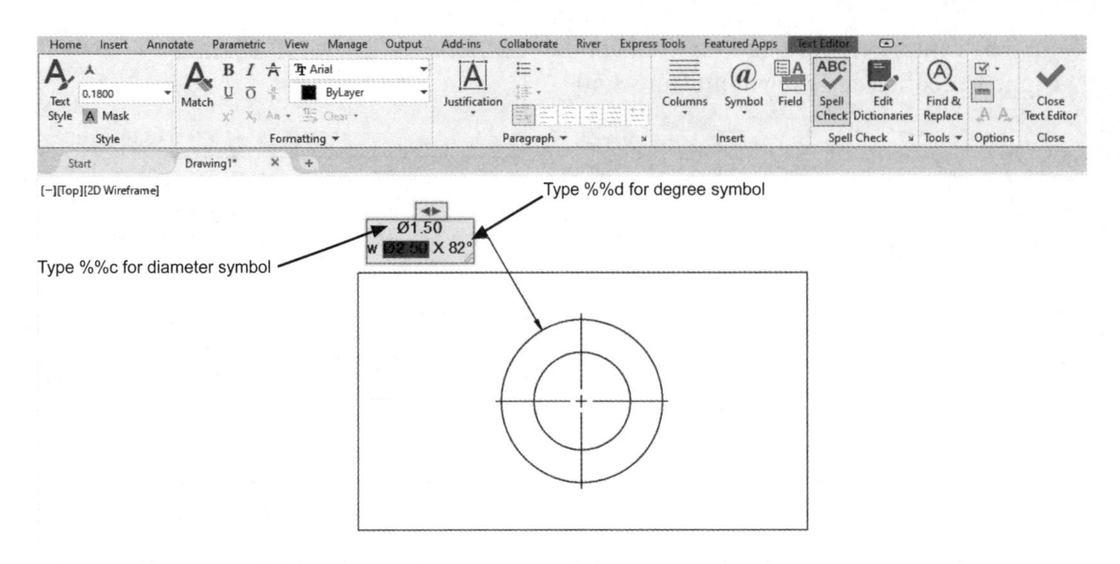

8.53 Entering Text for the Countersunk Hole

Step 3. Highlight the **w** and pick on the down arrow in the **Font** window located on the **Formatting** panel and change the font to **GDT** (see Figure 8.54). When the font is changed, the **w** will be replaced by the countersink symbol (⌵).

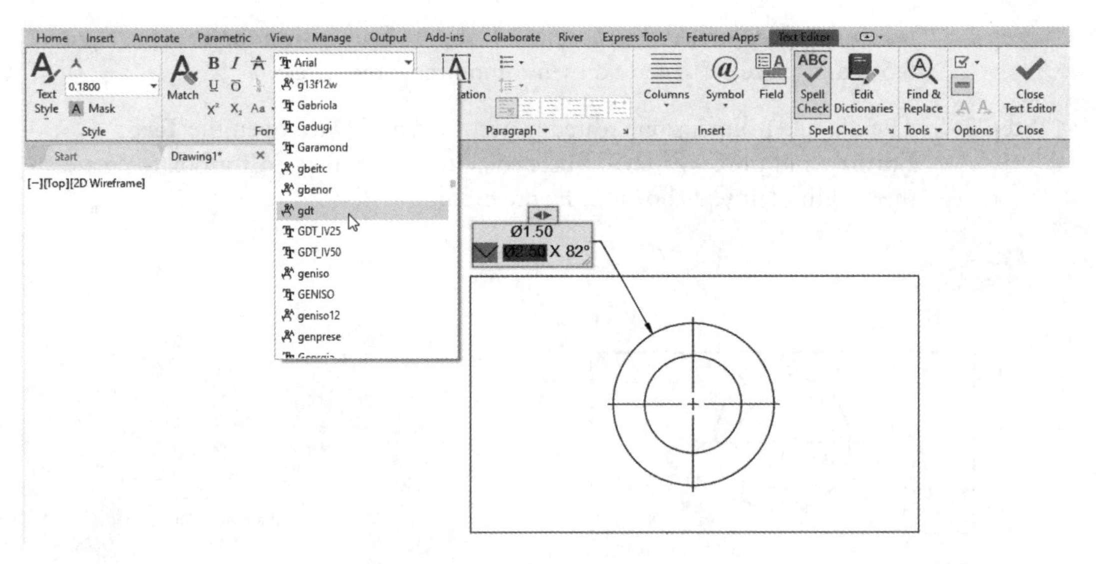

8.54 Converting an ***Arial*** Font Character to ***GDT*** Font in the Notation of the Countersunk Hole

When finished dimensioning the Flange Bearing, left-click on the **Layout1** tab in the lower-left corner of the graphics window and edit the text in the title block by double-clicking on it to enable the text edit option. Save the file and plot the drawing as instructed by your instructor.

CHAPTER

9

BLOCKS

CHAPTER OBJECTIVES:

After studying the material in this chapter, you should be able to:

1. Describe what blocks are and how they are used in technical drawings created with AutoCAD.
2. Create, insert, and edit blocks with AutoCAD software.
3. Create a block library of architectural symbols and use them to produce a floor plan.

CHAPTER OVERVIEW

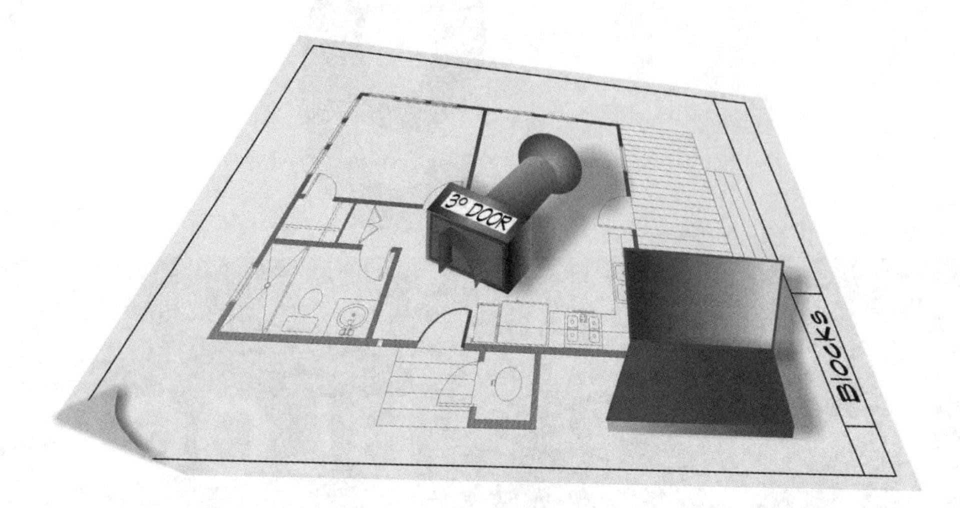

BLOCKS:

A ***block*** is an AutoCAD term that refers to a predrawn object stored in an AutoCAD drawing file that can be placed or inserted into the drawing whenever it is needed. For example, an architectural firm may create a block of door and window symbols. Later, when a door or window symbol is needed in a floor plan, the firm's drafters can select the symbol from the ***block library*** and insert it into the drawing rather than drawing each door and window from scratch. Thus, drafters save a huge amount of time and lower the cost of producing a drawing. Another advantage of using blocks is that when inserting a block, the drafter can change its scale, proportion, and rotation without redrawing the object.

Sometimes, architectural and engineering firms purchase premade block libraries from vendors, or in some cases, download them directly from vendor's websites.

9.1 CONSIDERATIONS FOR CREATING BLOCKS

Prior to creating a block it is important to consider the layer on which the entities of the block are drawn. This is important because objects that are on layer 0 when the block is created will assume the properties (color, linetype, etc.) of the layer that is current when the block is inserted into the drawing. Conversely, if the entities of a block are on a layer other than 0 when the block is created, the block will retain the characteristics (color, linetype, etc.) of the original layer, even if the block is inserted on a layer with different layer properties.

Another important consideration in block creation is the definition of the block's ***base point***. The base point is the point on the block that aligns to the location of the ***insertion point*** that AutoCAD prompts you to define when you insert the block into a drawing. For example, if you were making a block of the door shown in Figure 9.1, the corner of the door where the hinge would meet the door's opening is a logical base point. With this corner defined as the door's base point, the hinged corner of the door can be placed precisely into a floor plan by snapping to the corner of the door opening when you are prompted to define the block's insertion point.

Figure 9.2 shows the drawing for a ceiling fan. In a block of the fan, a logical base point is its center. Thus, when prompted to define the base point of the fan, you would select the center of the circle around which the blades are arrayed. Later, when the fan is inserted into a drawing, the fan's center point will align to the pick point, or coordinates, that you define when prompted by AutoCAD to define the block's insertion point.

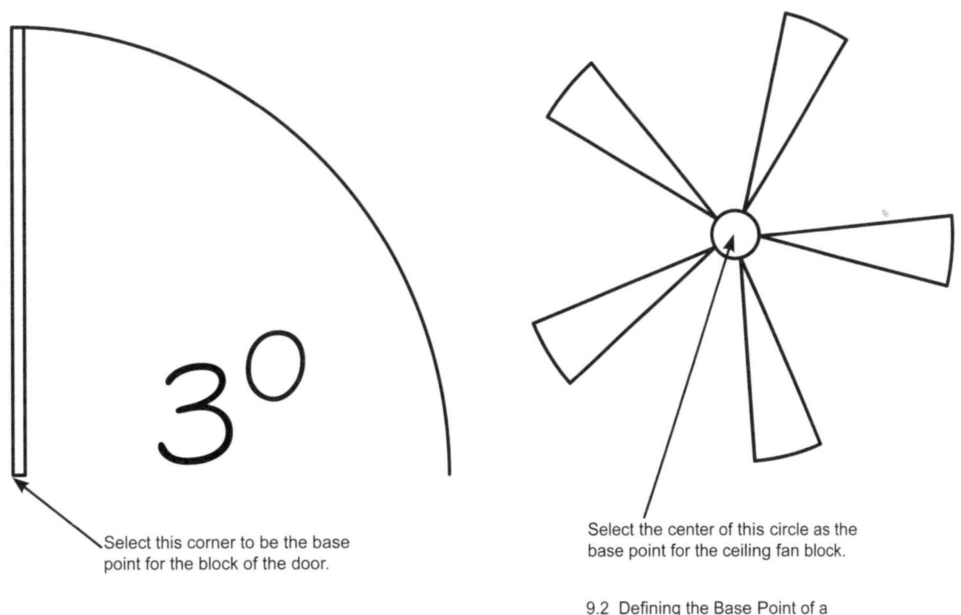
3O
Select this corner to be the base point for the block of the door.
Select the center of this circle as the base point for the ceiling fan block.

9.1 Defining the Base Point of a Block of a Door.

9.2 Defining the Base Point of a Block of a Ceiling Fan.

STEP-BY-STEP

CREATING BLOCKS

Step 1. Draw the object to be blocked.

NOTE

The layer on which the entities of a block are drawn is very important. For example, objects that are on layer **0** when the block is created will assume the color, linetype, and lineweight of the layer that is current when the block is inserted. Blocks created on a layer other than layer 0 will retain the characteristics of that layer, even when inserted on a different layer.

Step 2. Select the **Make Block** icon from the **Draw** toolbar or the **Block** panel of the **Home** tab of the ribbon (see Figure 9.3). The **Block Definition** dialog box shown in Figure 9.4 will open.

9.3 **Block** Icon

Step 3. Enter the block name in the **Name:** text box.

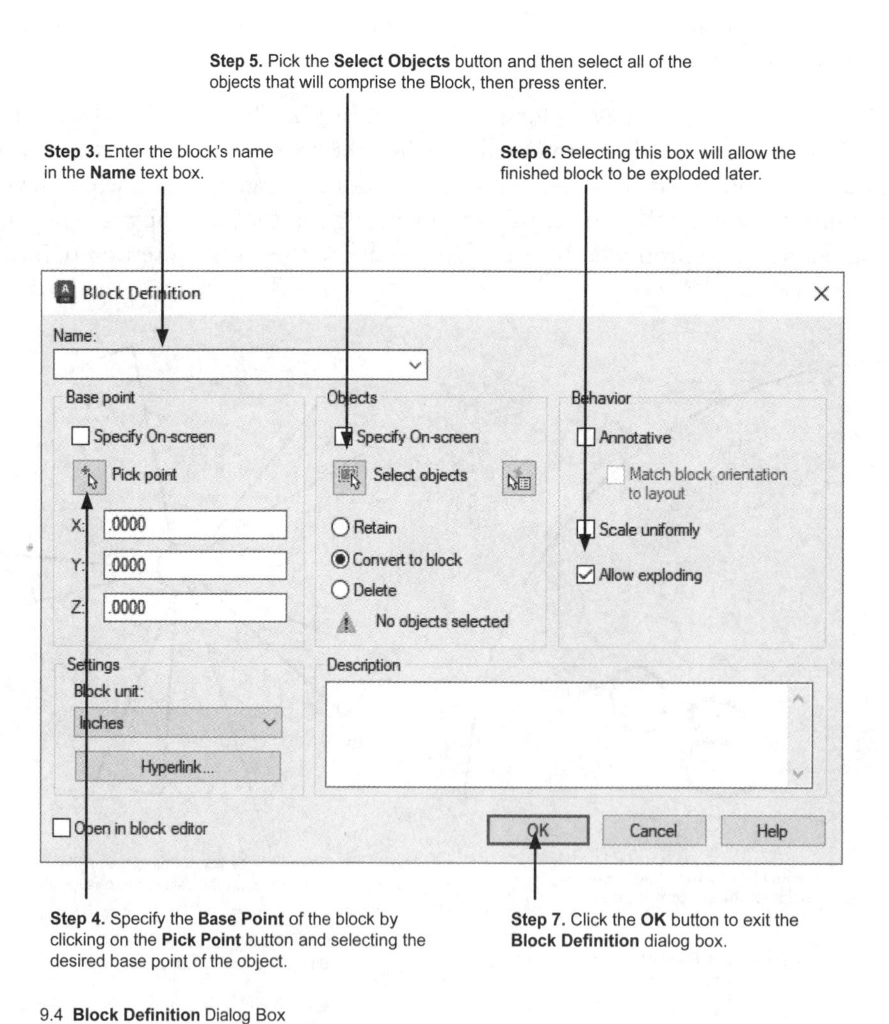
Step 5. Pick the Select Objects button and then select all of the objects that will comprise the Block, then press enter.
Step 3. Enter the block's name in the Name text box.
Step 6. Selecting this box will allow the finished block to be exploded later.
Block Definition
Name:
Base point
Specify On-screen
Pick point
X: .0000
Y: .0000
Z: .0000
Objects
Specify On-screen
Select objects
Retain
Convert to block
Delete
No objects selected
Behavior
Annotative
Match block orientation to layout
Scale uniformly
Allow exploding
Settings
Block unit:
Inches
Hyperlink...
Description
Open in block editor
OK
Cancel
Help
Step 4. Specify the Base Point of the block by clicking on the Pick Point button and selecting the desired base point of the object.
Step 7. Click the OK button to exit the Block Definition dialog box.

9.4 **Block Definition** Dialog Box

STEP-BY-STEP

Step 4. Specify the base point of the block by picking on the **Pick point** button and selecting the desired base point located on the object. Select a point that will make a "logical" base point for inserting the block into the drawing.

Step 5. Pick the **Select objects** button and select all the objects that will comprise the block. Press **<Enter>** when you have finished selecting objects.

Step 6. Checking the **Allow exploding** box allows the finished block to be exploded later if necessary.

Step 7. Click **OK** to exit the **Block Definition** dialog box.

STEP-BY-STEP

INSERTING BLOCKS INTO A DRAWING

Blocks are placed into drawings by using the **Insert Block** command (see Figure **9.5**). The steps involved in inserting a block into a drawing are as follows:

Step 1. Select the **Insert Block** icon from either the **Home** or **Insert** tabs of the ribbon (see Figure 9.5).

9.5 **Insert** Icon

Step 2. A drop-down list appears; you will see a visual preview of all the blocks that exist in the drawing's **Block Library**. You may "drag and drop" the blocks directly into your drawing from this drop-down list (see Figure 9.6).

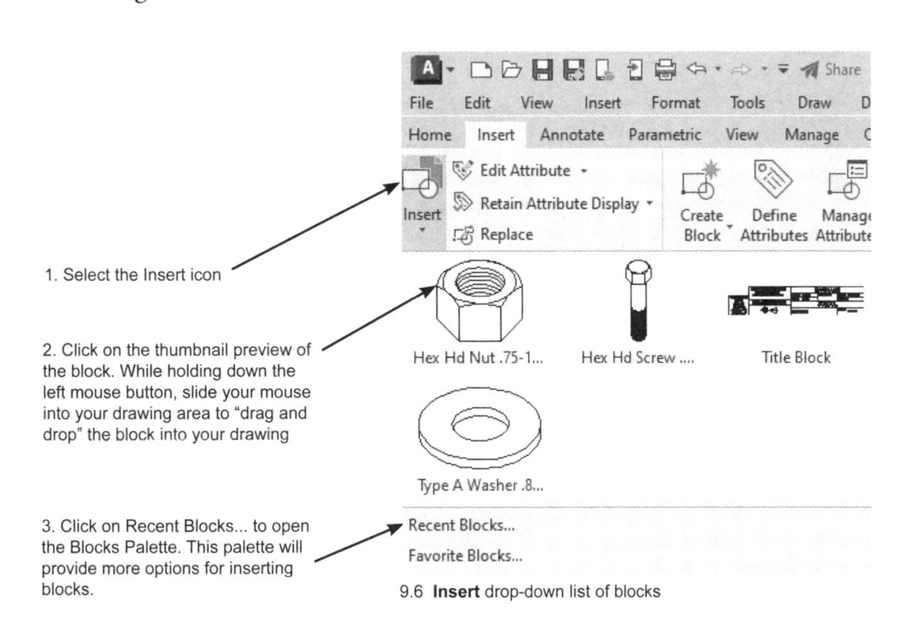
Share
File
Edit
View
Insert
Format
Tools
Draw
Home
Insert
Annotate
Parametric
View
Manage
Edit Attribute
Retain Attribute Display
Replace
Insert
Create Block
Define Attributes
1. Select the Insert icon
2. Click on the thumbnail preview of the block. While holding down the left mouse button, slide your mouse into your drawing area to "drag and drop" the block into your drawing
Hex Hd Nut .75-1...
Hex Hd Screw ....
Title Block
Type A Washer .8...
3. Click on Recent Blocks... to open the Blocks Palette. This palette will provide more options for inserting blocks.
Recent Blocks...
Favorite Blocks...

9.6 **Insert** drop-down list of blocks

STEP–BY–STEP

Step 3. When the block palette appears, you have the option to specify the exact insertion coordinate, rotate and scale on placement. (see Figure 9.7).

Step 4. Either drag and drop the block from the palette into your drawing, or left click on the thumbnail in order to insert the block in a specific location by the base point. You will also be prompted to define the block's scale and rotation if either of these boxes was checked in the palette.

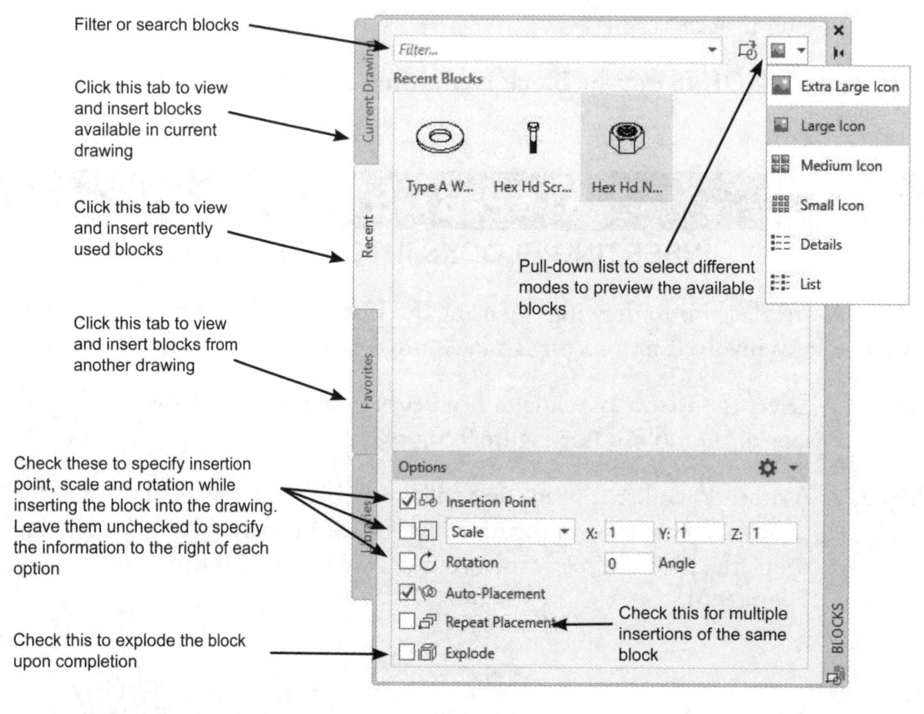
Filter or search blocks
Click this tab to view and insert blocks available in current drawing
Click this tab to view and insert recently used blocks
Click this tab to view and insert blocks from another drawing
Check these to specify insertion point, scale and rotation while inserting the block into the drawing. Leave them unchecked to specify the information to the right of each option
Check this to explode the block upon completion
Filter...
Recent Blocks
Type A W...
Hex Hd Scr...
Hex Hd N...
Current Drawing
Recent
Favorites
Libraries
Extra Large Icon
Large Icon
Medium Icon
Small Icon
Details
List
Pull-down list to select different modes to preview the available blocks
Options
Insertion Point
Scale
X: 1
Y: 1
Z: 1
Rotation
0
Angle
Auto-Placement
Repeat Placement
Explode
Check this for multiple insertions of the same block
BLOCKS

9.7 **Block** Insertion Palette

In most cases, exact placement of the block is important—at the endpoint of a line, for example. For this reason, use an appropriate **OSNAP** setting to facilitate accurate block placement whenever possible.

By left-clicking on the ribbon's **Insert** tab and then left-clicking on the **Insert** button, a palette displaying the drawing's blocks appears. Left-clicking on the image of a block in the palette initiates the steps involved in inserting the block into the drawing beginning with the *Specify insertion point prompt*. Follow the prompts on the command line to complete the command.

You can insert an entire existing AutoCAD drawing into your current drawing by selecting the **Browse...** button from the **Insert** dialog box, browsing to the AutoCAD drawing file you wish to insert, selecting its file name, and clicking **OK**. The **0,0** point on the inserted drawing will default as its base point. The inserted drawing will behave like a block (it will be inserted as one entity), and its layers, linetypes, and blocks will become part of the current drawing.

STEP–BY–STEP

EDITING BLOCKS WITH THE BLOCK EDITOR COMMAND

The easiest way to make changes to a block is with the **Block Editor** command. The steps for using this command follow.

Step 1. Open the **Block Editor** by selecting the **Block Editor** tool located on the **Block Definition** panel of the **Insert** tab of the ribbon or by double-clicking on a block that has already been inserted into the drawing or by typing **BEDIT** and pressing **<Enter>**.

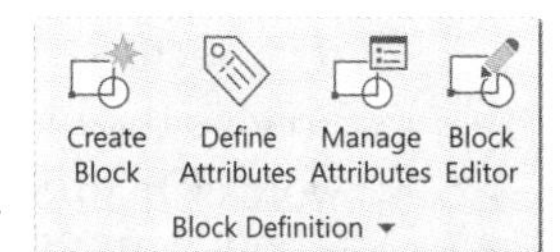

9.8 Block Definition panel in the Insert ribbon

Step 2. When the **Edit Block Definition** dialog box opens, as in Figure 9.8(a), select the name of the block you wish to edit from the list on the left side of the box and click **OK**.

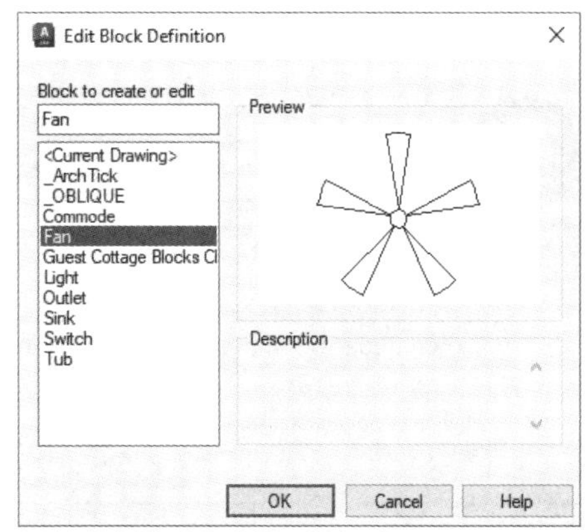

9.8(a) **Edit Block Definition** Dialog Box

Step 3. At this point two things will occur:

a. A window with a gray background will open that displays the block you selected for editing in Step 2. See Figure 9.9(a).

b. The **Home** tab of the ribbon will be replaced by the **Block Editor** tab.

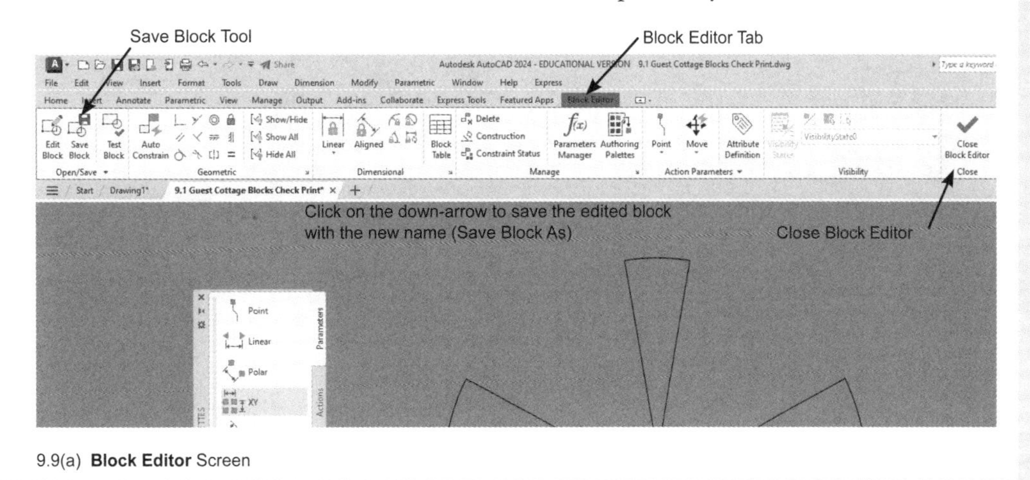

9.9(a) **Block Editor** Screen

STEP–BY–STEP

Step 4. Use the tools available on the Draw and Modify toolbars and on the Block Editor tab to make the desired changes to the block.

Step 5. When you have finished editing the block, choose the Save Block tool from the Open/Save panel of the Block Editor tab of the ribbon. See Figure 9.9(a). To give the block a different name, pick on the down arrow next to the words Open/Save and select the Save Block As option. See Figure 9.9(a). To end the editing operation, choose the Close Block Editor tool from the Close panel.

QUICK TIP

Blocks with the same name as the redefined block that have been inserted into the drawing prior to editing will also be updated to reflect the redefined properties. This situation can be avoided by choosing the **Save Block As** tool from the **Open/Save** panel and giving the edited block a different name.

QUICK TIP

Another way to edit a block that has already been inserted into the drawing is to explode the block by selecting it and clicking the **Explode** tool located on the **Modify** toolbar. After a block has been exploded, its entities revert to individual objects that no longer constitute a block. It should be noted that block entities that were on layer 0 when the block was created will remain on layer 0 after the block is exploded. Block entities that were on layers other than layer 0 when the block was created will revert to their *original* layers after the block is exploded. After the entities have been edited, you can use the **Make Block** command to convert the objects back to a block if desired.

BLOCK COUNT PALETTE			
Count	Ribbon	View	AutoCAD offers a quick and easy way to count how many blocks are in use in the drawing. The Count command can be found on the Palettes panel in the View ribbon. Simply click on the Count icon (see image to the right) and a palette will appear with a count of every block in use in the active drawing.
	Panel	Palettes	
	Command Line	bcount	
9.10	Alias	-	

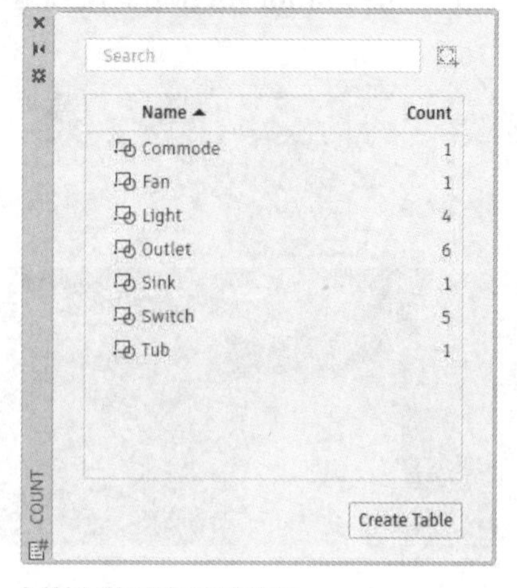

9.10(a) **Block Count Palette**

NEW

SMART BLOCK PLACEMENT
AutoCAD 2024 now offers suggestions on placement of blocks based on where you've previously placed the block in the drawing. AutoCAD learns how existing blocks have been placed in the drawing and infers the next placement of the same block.

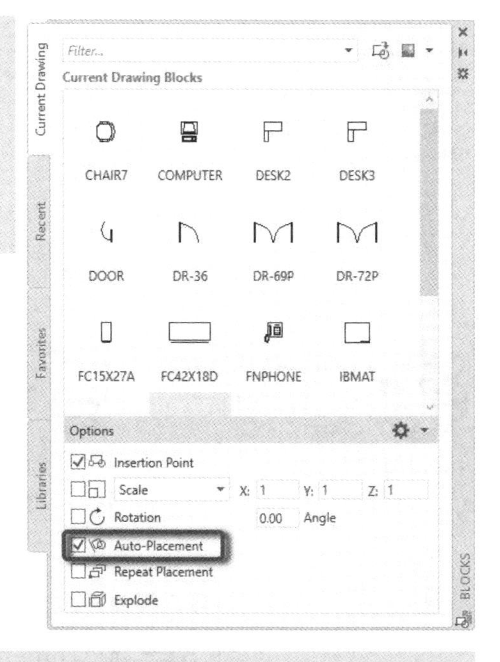

9.10(b) **Smart Block Placement**

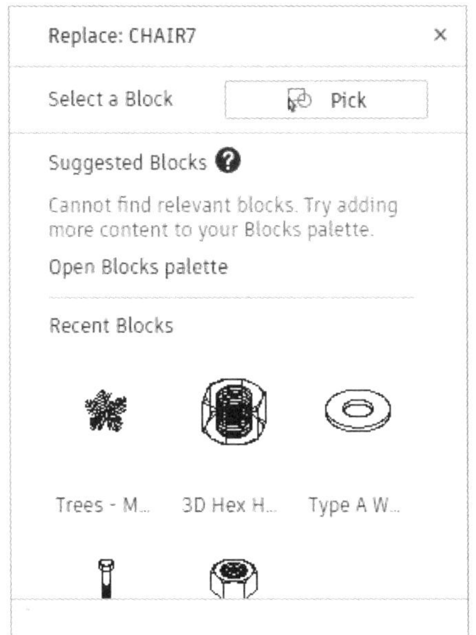

NEW

SMART BLOCK REPLACEMENT
AutoCAD 2024 now has the ability to replace specified block references by selecting from a palette of suggested blocks.

9.10(c) **Smart Block Replacement Window**

CHAPTER REVIEW

Creating and using blocks is another method employed by AutoCAD users to work more quickly and efficiently. This chapter presented the basics of creation, usage, and editing of blocks. Other, more powerful, methods of using blocks, including creating *attributed blocks* (an attribute is a label or tag that attaches data to a block) and *dynamic blocks* (blocks that can be edited in place rather than redefining them), are usually covered in more advanced AutoCAD texts.

Later in this text you will learn how to use an AutoCAD feature called *DesignCenter* to insert blocks from other drawings into the current drawing.

Base Point
Block
Block Library
Insertion Point

Multiple Choice

1. What is a block?
    a. One set of a cube
    b. Predrawn object
    c. Group of drawing files
    d. None of the above
2. What AutoCAD command is used to place a block into a drawing?
    a. INSERT
    b. ADD
    c. COPY & PASTE
    d. All the above
3. What AutoCAD command is used to turn an object into a block?
    a. GROUP
    b. BLOCK
    c. Both A and B
    d. Insert Block
4. A label or tag that attaches data to a block is called what?
    a. Dynamic block
    b. Attributed block
    c. String block
    d. Wooden block
5. What command is used to edit a block?
    a. BLOCKX
    b. BEDIT
    c. BEXPLODE
    d. BCLOSE

In this project you will add plumbing and electrical symbols to the floor plan of the guest cottage you created in Chapter 4. This project will require you to create a block for each plumbing and electrical symbol. These blocks will be inserted into the floor plan in the locations shown in the designer's sketch in Figure 9.11.

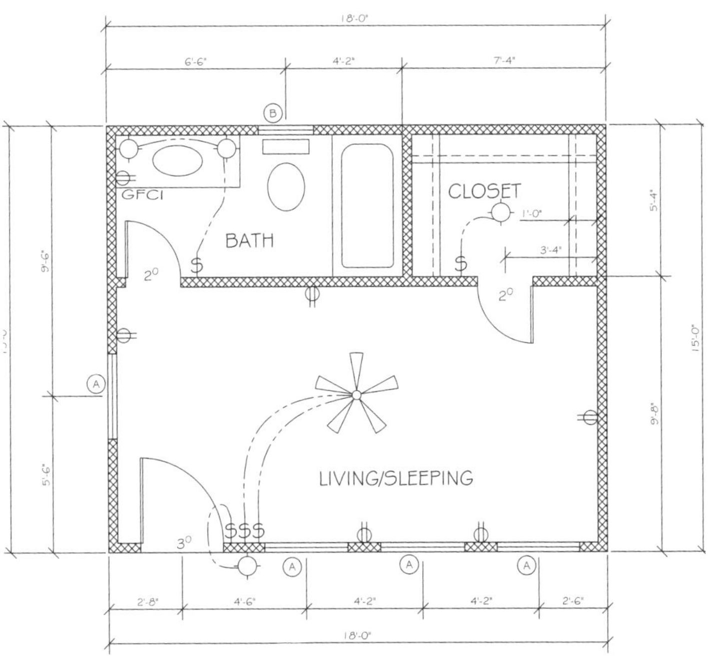

VIDEO There is a video tutorial available for this project inside the *Chapter 9* folder of the book's *Video Training* downloads.

9.11 Designer's Sketch of the Guest Cottage Showing Plumbing, Electrical, and Wiring Symbols

Step 1. Getting Started

a. Open the **Guest Cottage** drawing file you drew in Chapter 4 and dimensioned in Chapter 6.

b. Create the following layers: **Plumbing**, **Electric**, and **Wiring**.

- Assign a color to each layer.
- Set the Wiring layer's linetype to **Phantom**.

Follow the steps detailed earlier in this chapter to make blocks for the plumbing and electrical symbols. You will make five blocks in all. Insert the blocks into the **Guest Cottage** drawing in the locations shown in the designer's sketch in Figure 9.11.

QUICK TIP

Draw and create the blocks on layer **0**. Insert the plumbing symbols on the **Plumbing** layer and the electrical symbols on the **Electric** layer.

Step 2. Draw the Tub

a. Draw a **2'-6" X 5'-0"** rectangle.

b. Offset it **4"** to the inside.

c. Use the **FILLET** command to fillet the corners of the inside rectangle.

- Fillet the top corners **R6"**
- Fillet the bottom corners **R2"** as shown in Figure 9.12.

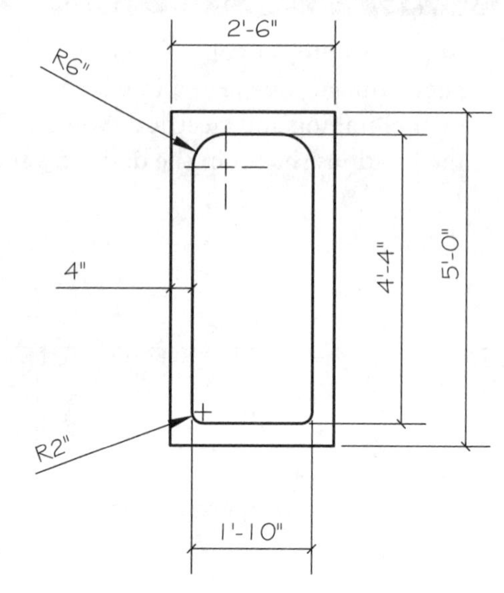

9.12 Size Dimensions for Tub

Step 3. Make a Tub Block

a. Select the tub's upper right corner as the base point.

b. Set the **Plumbing** layer current and insert the block of the tub into the corner of the bathroom as shown in Figure 9.11.

Step 4. Draw the Lavatory

a. Draw a **4'-6" X 1'-10"** rectangle for the sink top.

b. Select the **ELLIPSE** command.

c. When prompted to *Specify axis end-point:*, pick a point and draw a horizontal line **1'-8"** long.

d. When prompted for the length of the other axis, type **7" to create the bowl** (1'-2" divided by 2).

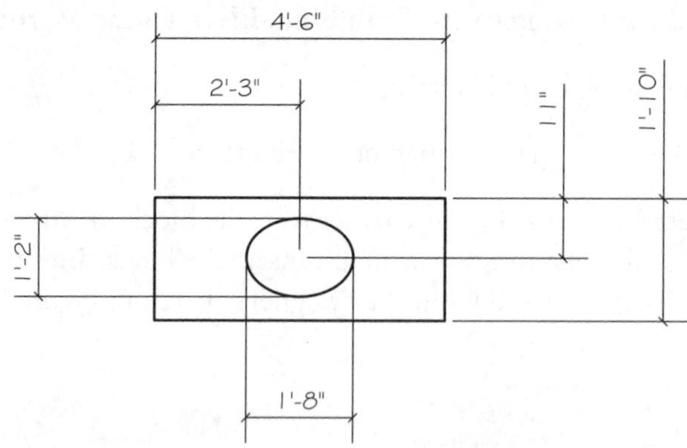

9.13 Size Dimensions for Lavatory

e. Move the center of the bowl to the location on the rectangle shown in Figure 9.13.

Step 5. Make a lavatory block

a. Select the lavatory's upper left corner as the base point.

b. Insert the block of the lavatory into the corner of the bathroom as shown in Figure 9.11.

Step 6. Draw the toilet (also called a *commode* or *water closet*)

a. Draw a **1'-8" X 6"** rectangle for the toilet tank.

b. Draw a vertical line from the midpoint of the top edge of the rectangle to a point **1'-8"** below the first point.

c. Select the **ELLIPSE** command

- When prompted to *Specify axis end-point:*, pick a point and draw a horizontal line **1'-4"** long.
- When prompted for the length of the other axis, type **10"** (1'-8" divided by 2) to create the toilet bowl.

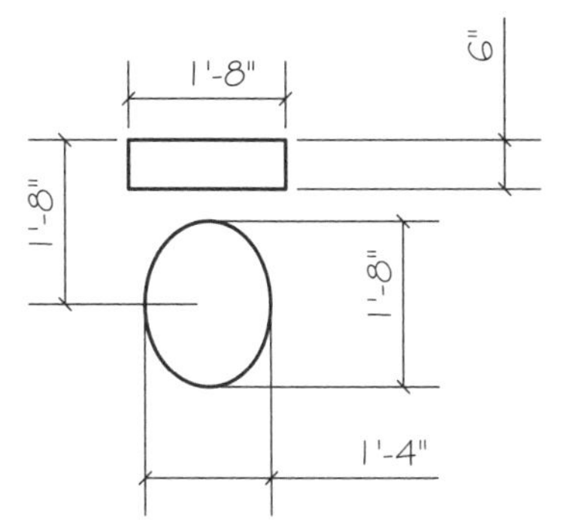

9.14 Size Dimensions for Toilet

d. Move the ellipse by its center point to the endpoint of the **1'-8"** line drawn earlier and erase the line (see Figure 9.14).

Step 7. Make a toilet block

a. Select the midpoint of the toilet tank's top edge as the block's base point.

b. Insert the block of the toilet into the **Guest Cottage** floor plan by centering it between the right edge of the lavatory and the left edge of the tub as shown in Figure 9.11.

c. Move the toilet **2"** away from the inside edge of the wall.

Step 8. Draw the ceiling fan

a. Draw the fan blade and circle shown in Figure 9.15.

b. Construct the fan inside a circle **3'** in diameter.

c. Use **Polar Array**, selecting the fan blade as the object to array and the center of the **4"** circle as the center point of the array.

- In the **Array** dialog box, enter **5** as the total number of items and **360** as the angle to fill.
- Check the box for **Rotate items as copied**.
- The finished fan should look like the one shown in Figure 9.16.

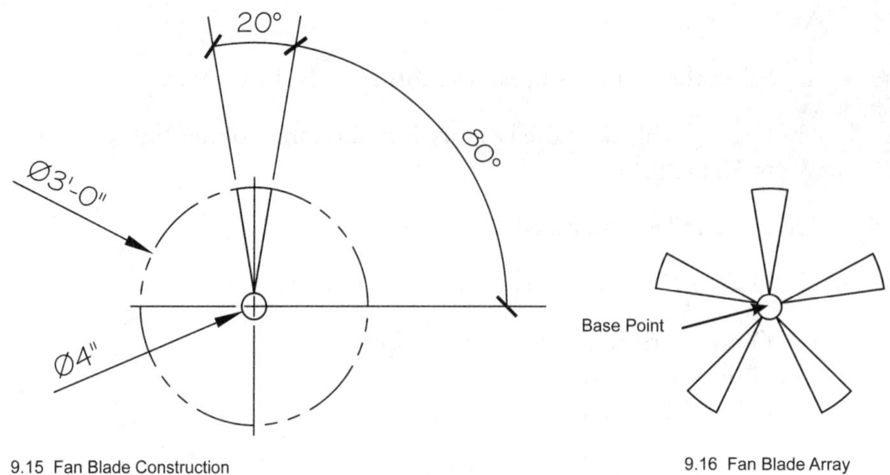

9.15 Fan Blade Construction

9.16 Fan Blade Array

Step 9. Make a fan block

a. Choose the center of the array as the base point.

b. Set the **Electric** layer current, and insert the fan in the location shown in Figure 9.11.

Step 10. Draw the light fixture (also called a *luminaire*)

a. Draw a **6"** diameter circle.

b. Draw a 2" line from each of the circle's quadrants as shown in Figure 9.17.

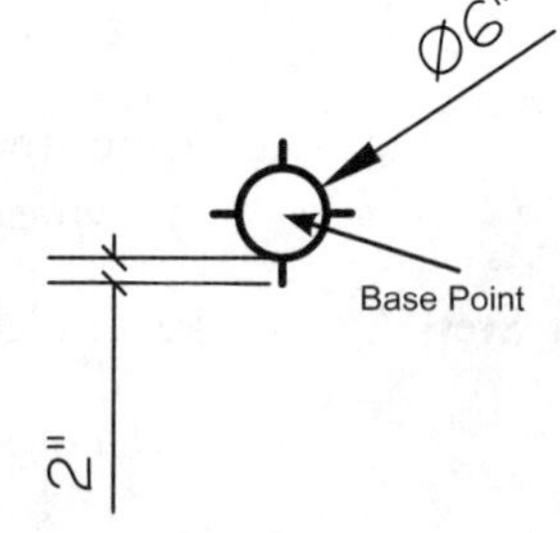

9.17 Size Dimensions for Light Fixture

Step 11. Make a light fixture block

a. Select the center of the circle as the base point.

b. Insert two light fixture blocks above the lavatory as shown on the sketch of the floor plan in Figure 9.11.

Step 12. Switches and outlets

a. Blocks for the switches and outlets are included in the prototype drawing for the **Guest Cottage** drawing.

b. Use the **INSERT** command to place the electrical outlets (see Figure 9.18) and the switches (see Figure 9.19) from the list of blocks.

c. Refer to the floor plan sketch in Figure 9.11 for placement of these symbols.

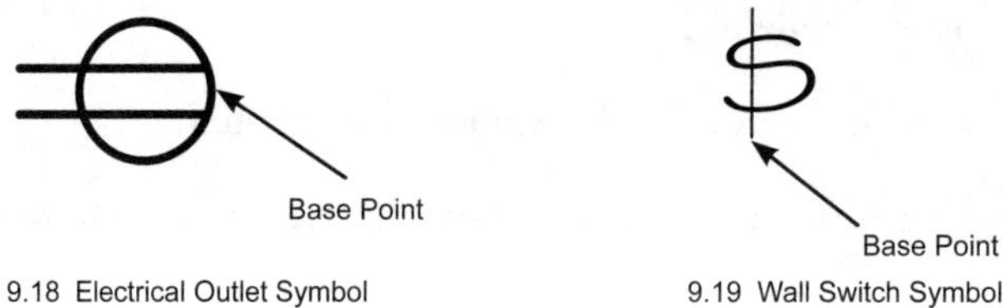

9.18 Electrical Outlet Symbol

9.19 Wall Switch Symbol

Step 13. Label the drawing

a. Set the **Labels** layer current

- Label the outlet in the bathroom **GFCI** using **3"** text, as shown in Figure 9.11.
- **GFCI** indicates that the type of outlet is a *ground fault circuit interrupt.*
- GFCI outlets are required near plumbing fixtures where an electric appliance, like a hair dryer, might come in contact with water and pose a risk of electric shock.

Step 14. Wiring

a. Set the **Wiring** layer current.

b. Use the **Spline** command—refer to figure 4.56(b) on page 156—to create the switch lines connecting the switches to the electrical fixtures as shown in Figure 9.20.

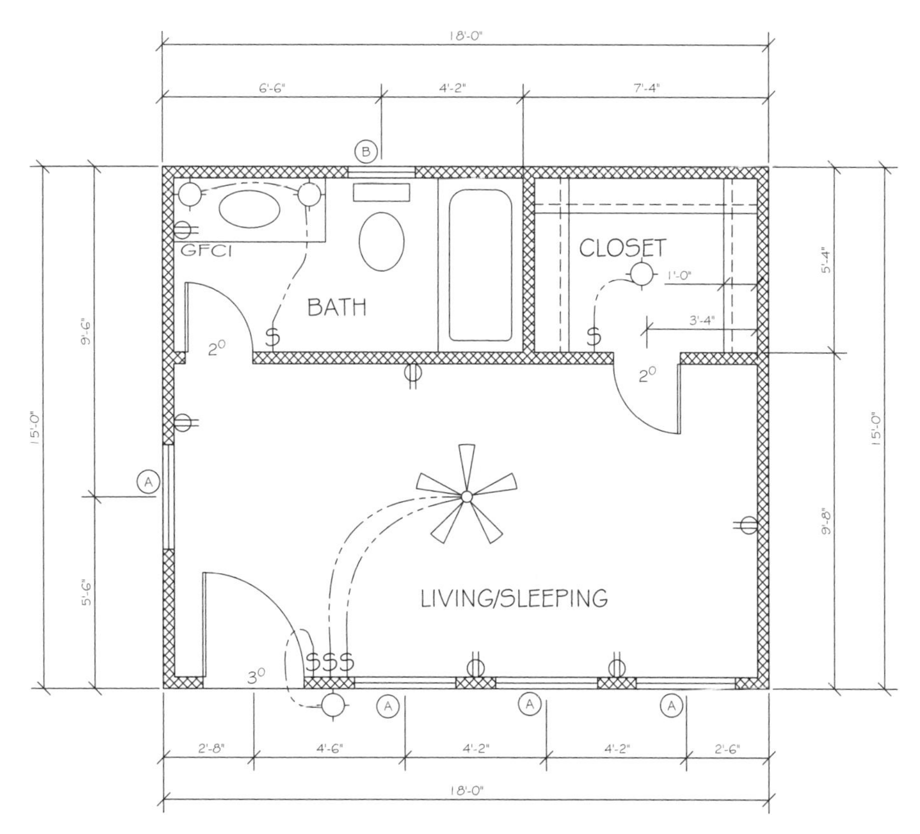

9.20 Switch Lines Added to Lights and Ceiling Fan

Step 15. You may need to set the **LTSCALE** (linetype scale) to a different (larger or smaller) value for dashes to appear in the phantom lines.

a. **LTSCALE** controls the size of dashes in noncontinuous lines such as hidden, center, and phantom lines.

b. To change the **LTSCALE**, type **LTS** and press **<Enter>**, enter a new value (the default value is **1**), and press **<Enter>**.

c. To change the **LTSCALE** of only one line, select the line, right-click, and select **Properties** from the menu; this will open the **Properties** palette.

d. In the field next to **Linetype Scale**, set the value to a larger number.

This step completes the Guest Cottage project.

Step 16. Print the guest cottage

a. Left-click on the **Layout1** tab in the lower-left corner of the graphics window.

b. Edit the text in the title block by double-clicking on it to enable the text editing option.

c. Follow your instructor's directions to print the drawing.

d. Be sure to save the drawing file before closing AutoCAD.

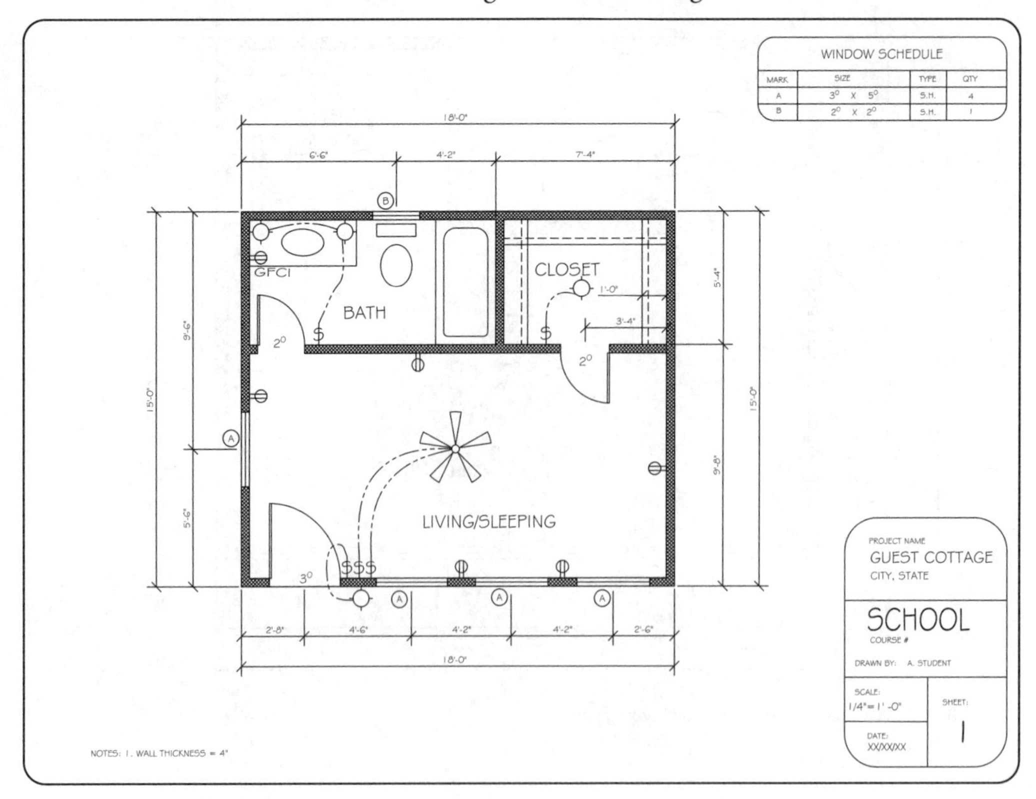

9.21 Completed Guest Cottage With Dimensions, Notes and Blocks

Draw the schematic of the FM tuner shown in Figure 9.22. The symbols for the electronic components needed in this drawing are already drawn, but you will need to create a block for each symbol.

There is a video tutorial available for this project inside the Chapter 9 folder of the book's Video Training downloads.

Step 1. Getting Started

a. Download the **Electronic Schematic Template** drawing file located in the *Prototype Drawing Files* folder associated with this book.

b. Use **SAVE AS** to save the drawing to your **Home** directory.

- Rename the drawing **FM TUNER**.

c. Activate the **Model** tab in the lower-left corner of the graphics window (if it is not already highlighted) by left-clicking on the Model tab.

- **Note:** The FM Tuner will be drawn inside the magenta rectangle located in model space.

Step 2. Create two new layers named **Circuit** and **Symbols.**

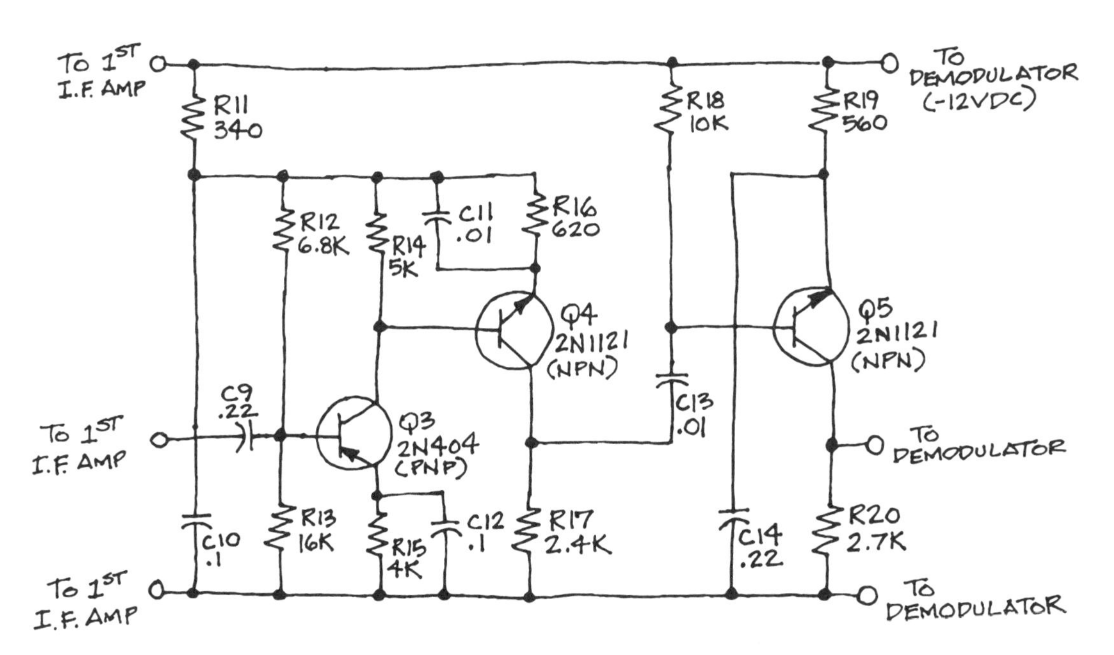

9.22 Designer's Sketch of the FM Tuner Schematic

PROJECT

Step 3. Blocks

a. Make a block for each of the electronic symbols included on the prototype drawing.

- There will be six blocks in all.

b. Use the block names shown in Figure 9.23.

c. Select the point displayed in red on each symbol as the block's base point.

Choose red *point* included on each symbol as *Base Point* for each Block

Pin Capacitor Dot Resistor npn Transistor pnp Transistor

9.23 Electronic Component Symbols Used in FM Tuner

QUICK TIP

You will need to have the **Node** object snap setting running to snap to the red *point* included on each component symbol, but **do not** include these points when selecting the *objects* to be included in the block.

Step 4. When you have finished making the blocks, turn the **Nodes** layer off and erase the symbols remaining on the screen.

Step 5. Set the Circuit layer current.

Step 6. Use **OFFSET** or **Rectangular Array** to draw a grid like the one shown in Figure 9.24.

a. Use **1.125"** spacing between lines.

b. Begin the lower left corner of the grid at absolute coordinates 2.00, 2.75.

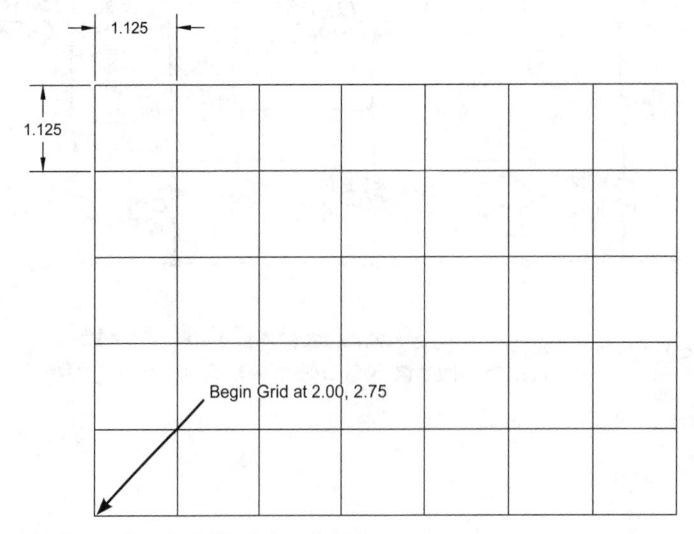

9.24 Construction of a Grid for Schematic Diagram

Step 7. Set the **Symbols** layer current.

Step 8. Insert the blocks representing the components on the grid lines as shown in Figure 9.25.

a. Visually center the symbols between the lines.

b. You may need to move some lines at this point to reflect the sketch of the circuit accurately.

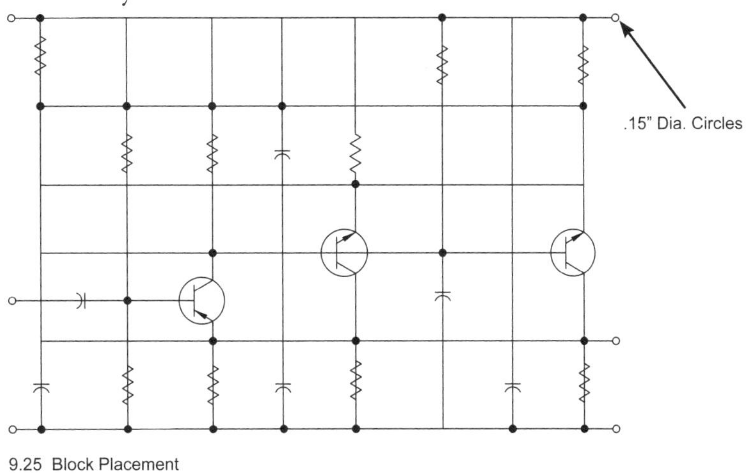

9.25 Block Placement

Step 9. Trim

a. Trim the lines that run through the blocks of the electronic symbols.

b. Trim or delete other layout lines as needed to create the circuit configuration shown in Figure 9.26.

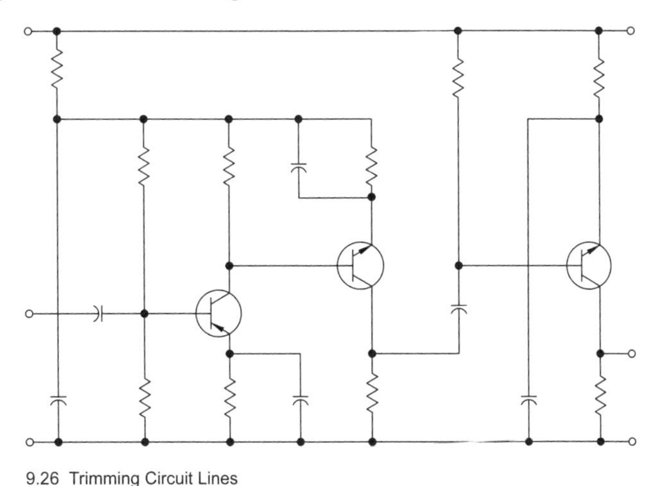

9.26 Trimming Circuit Lines

QUICK TIP

Text height should be **.125"** for both labels and notes.

Step 10. Labels

a. Refer to the sketch of the schematic shown in Figure 9.27 to add labels next to the components.

b. Place labels on the **Text** layer.

c. Labels are centered to the right of each component.

d. Labels are **.125"** text height.

e. When finished with this step, you can move the schematic to better center it on the sheet if necessary.

Step 11. Notes

a. Place Notes 1 and 2 (Figure 9.22) on the **Text** layer.

b. Notes are located in the area to the left side of the title block.

c. Notes are **.125"** text height. When finished with this step, you can move the schematic to better center it on the sheet if necessary.

Step 12. Print the FM Tuner

a. Left-click on the **Layout1** tab in the lower-left corner of the graphics window.

b. Edit the text in the title block by double-clicking on it to enable the text editing option (in Paper Space).

c. Follow your instructor's directions to print the drawing.

d. Be sure to save the drawing file before closing AutoCAD.

When finished drawing the FM Tuner, left-click on the **Layout1** tab in the lower-left corner of the graphics window and edit the text in the title block by double-clicking on it to enable the text editing option. Follow your instructor's directions to print the drawing. Be sure to save the drawing file before closing AutoCAD.

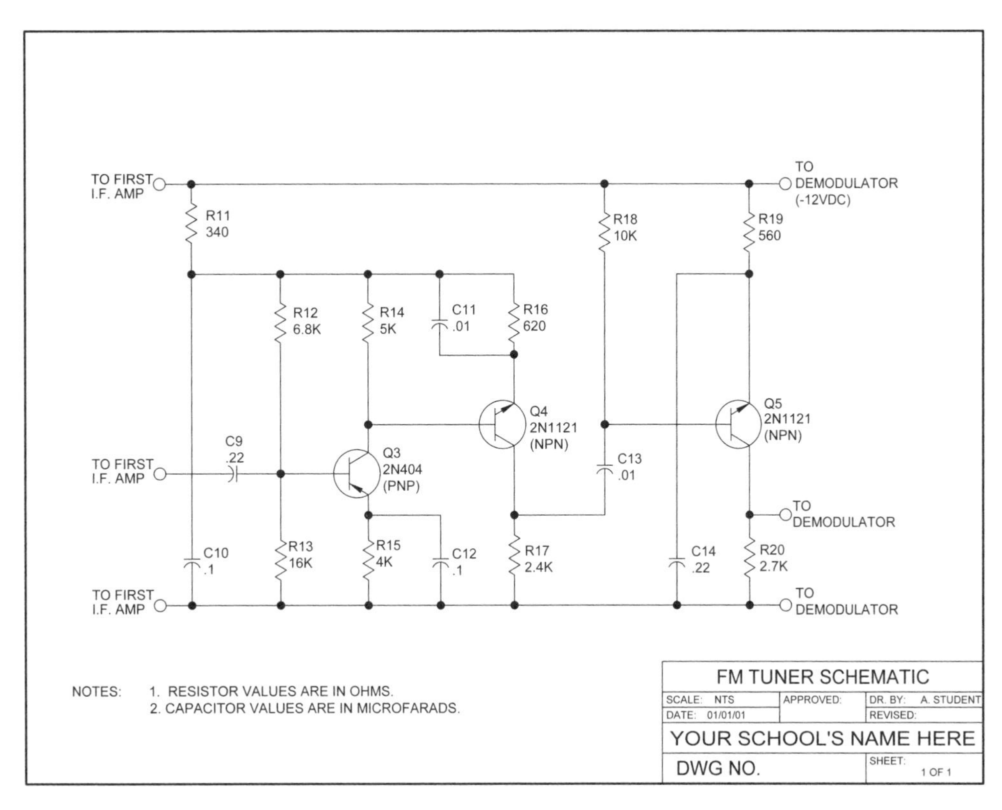

9.27 Completed FM Tuner Schematic Diagram

CHAPTER

10

3D MODELING BASICS

CHAPTER OBJECTIVES:

After studying the material in this chapter, you should be able to:

1. Define coordinates along X-, Y-, and Z-axes.
2. Change the viewpoint to reveal the Z-axis of the user coordinate system in an AutoCAD drawing.
3. Use the PRESSPULL, EXTRUDE, and REGION commands to convert 2D entities into 3D objects.
4. Use the 3DROTATE command to rotate objects as needed in the creation of 3D models.
5. Employ the SUBTRACT and UNION commands to create complex 3D objects with AutoCAD.
6. Represent 3D models in wireframe or shaded form.

CHAPTER OVERVIEW

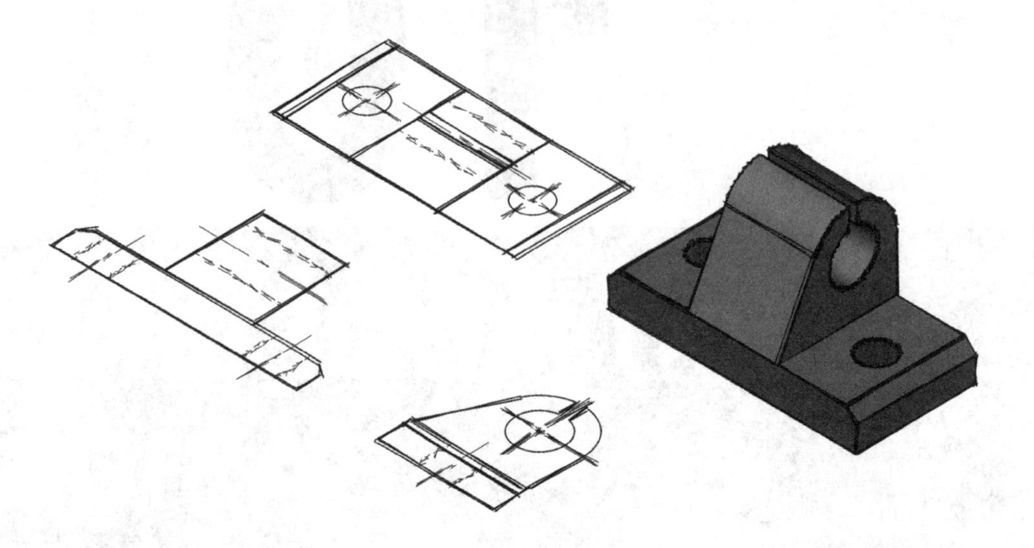

3D MODELING BASICS:

For years, drafters and designers in the mechanical engineering field have used CAD programs such as Inventor, SOLIDWORKS, and Creo to produce 2D working drawings from 3D models. With the release of CAD software products such as Civil 3D and Revit, this technique is rapidly becoming the norm for civil and architectural drafters and designers as well. Working with 3D models gives designers a much more powerful way to conceptualize, edit, and analyze their designs.

Most 3D CAD software programs use a technique known as ***parametric modeling*** to create 3D geometry. In parametric models the geometry of the model is driven by the dimensions associated with the geometry. This allows designers to modify the features of the model by simply editing the dimensions. When a parametric dimension is changed, the 3D model updates to reflect the new dimension value.

JOB SKILL

Finite element analysis (FEA) software is often used in conjunction with CAD modeling software to perform advanced design analysis on 3D models. FEA allows engineers and designers to calculate such properties as an object's mass, center of gravity, strength, distribution of stresses, and bending moments.

10.1 2D VERSUS 3D

Two-dimensional objects drawn with AutoCAD are described by specifying their *X*- and *Y*-coordinates. In a 2D drawing, the coordinate value of points located on the *Z*-axis is zero. Figure 10.1 shows a 2D drawing of an object created with AutoCAD. In the lower left corner of the screen, the user coordinate system (UCS) icon is visible. As you learned in Chapter 3, the UCS icon orients the drafter to the location of the 0,0 coordinates and the direction of the positive *X*- and *Y*-axes.

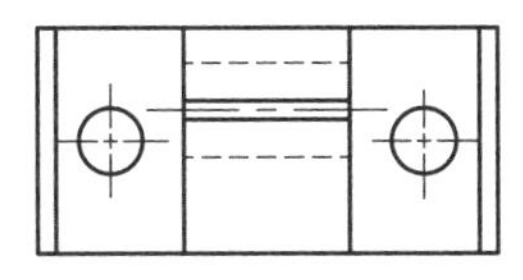

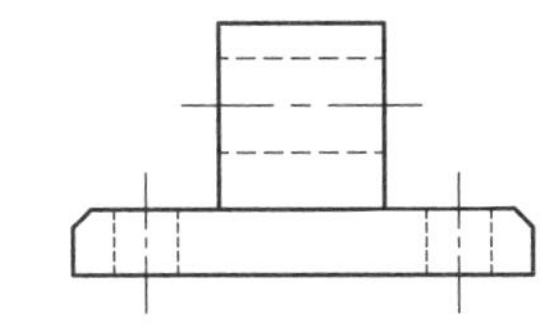

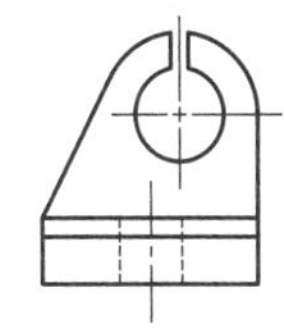

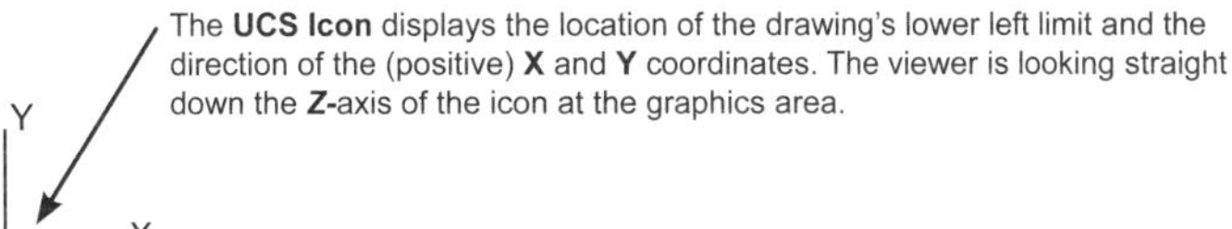

The **UCS Icon** displays the location of the drawing's lower left limit and the direction of the (positive) **X** and **Y** coordinates. The viewer is looking straight down the **Z**-axis of the icon at the graphics area.
Y
X

10.1 2D Views of an Object

In 3D drawing, objects are defined with coordinates that have *X*-, *Y*-, *and Z*- values. For example, the coordinates of the endpoint of a 3D line may be described as **6**, **2**, **4**, with **6** representing the *X*-value, **2** representing the *Y*-value, and **4** representing the *Z*-value. However, this line does not appear to be 3D if the graphics area is viewed from directly overhead, as in Figure 10.1. To view the *Z*-axis, it is necessary to change the point of view from which the 3D line is viewed.

NOTE

Figure 10.1 shows the views of the tool slide as they would appear if the **Top** icon were selected from the **View** toolbar. With the view set to **Top**, the point of view is perpendicular to the *X-Y* plane. In other words, the viewer is looking straight down the Z-axis of the UCS icon to view objects on the *X-Y* plane.

10.2 CHANGING THE POINT OF VIEW OF AN AUTOCAD DRAWING

The AutoCAD user's point of view can be changed to show the *Z*-axis (as shown in Figure 10.3) by opening AutoCAD's **View** toolbar and by selecting the SE Isometric view rather than the Top view (see Figure 10.2). To open the **View** toolbar, left-click on the **Tools** tab located on the menu bar and then left-click on **Toolbars**, then left-click on AutoCAD, and finally, left-click on **View**. When the View toolbar opens, left-click on **SE Isometric** view icon.

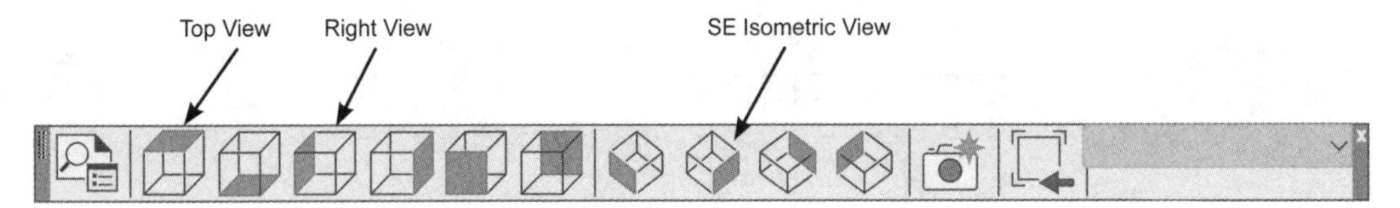

10.2 View Toolbar

The AutoCAD drawing window will adjust to resemble the example shown in Figure 10.3. In this figure 2D multiviews of the tool slide (drawn in Chapter 4) appear as they would if viewed from an elevated vantage point located on the southeast side of the drawing window. In fact, the SE in the **SE Isometric** icon on the toolbar refers to this *southeast* point of view. Likewise, *SW* refers to southwest, *NE* refers to northeast, and *NW* refers to a viewpoint located northwest of the viewed object.

More details about changing the point of view using the tools located on the ribbon will be presented later in this chapter.

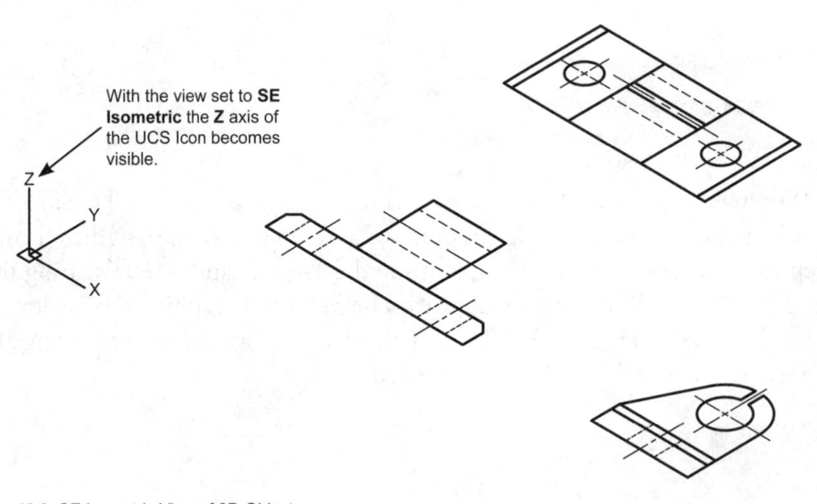

10.3 SE Isometric View of 2D Objects

In Figure 10.4, the menu bar is visible, and the **Draw**, **Modify**, **Dimension**, and **Osnap** toolbars have been docked along the left side of the drawing window. It should be noted, however, that the choice of whether or not to display the menu bar and toolbars while in the **3D Modeling** workspace is left up to the user's discretion. The authors of *Technical Drawing 101 with AutoCAD* believe that having these tools available is useful for new users of AutoCAD (refer to Figure 4.4 on page 118 for help opening the menu bar and the toolbars).

10.3 AUTOCAD'S 3D MODELING ENVIRONMENT

AutoCAD's 3D environment has its own unique workspace named **3D Modeling**. When the **3D Modeling** or **3D Basics** workspace is selected, many of the tools found on the ribbon panels are replaced with specialized tools that are used in the creation, editing, and viewing of 3D models. In Figure 10.4 the workspace is set to **3D Modeling** and the view is set to **SE Isometric**. The **SE Isometric** tool is located on the **Views** panel of the **View** tab of the ribbon, as shown in Figure 10.4 (although the **View** tools can also be accessed by opening the **View** toolbar as discussed earlier in this chapter).

As you can see in Figure 10.4, the **3D Modeling** workspace looks different from AutoCAD's 2D workspace. For example, with the view set to **SE Isometric**, the *Z*-axis of the UCS icon is visible, the cursor displays the *X*-, *Y*-, and *Z*-axes, and the ViewCube (located in the upper right corner of the drawing window) resembles a 3D cube with the words **Top**, **Front**, and **Right** written on its surfaces. By clicking on a surface of the ViewCube, the view displayed in the drawing window will be changed to match the selected surface; for example, clicking on the ViewCube surface labeled **Top** will display objects as viewed from directly above. Clicking on a *corner* of the ViewCube (instead of a flat surface) will change the display to the corresponding isometric view. Right-clicking on the ViewCube and making a choice from the shortcut menu changes the display to show objects from either a perspective or a parallel (no perspective) point of view.

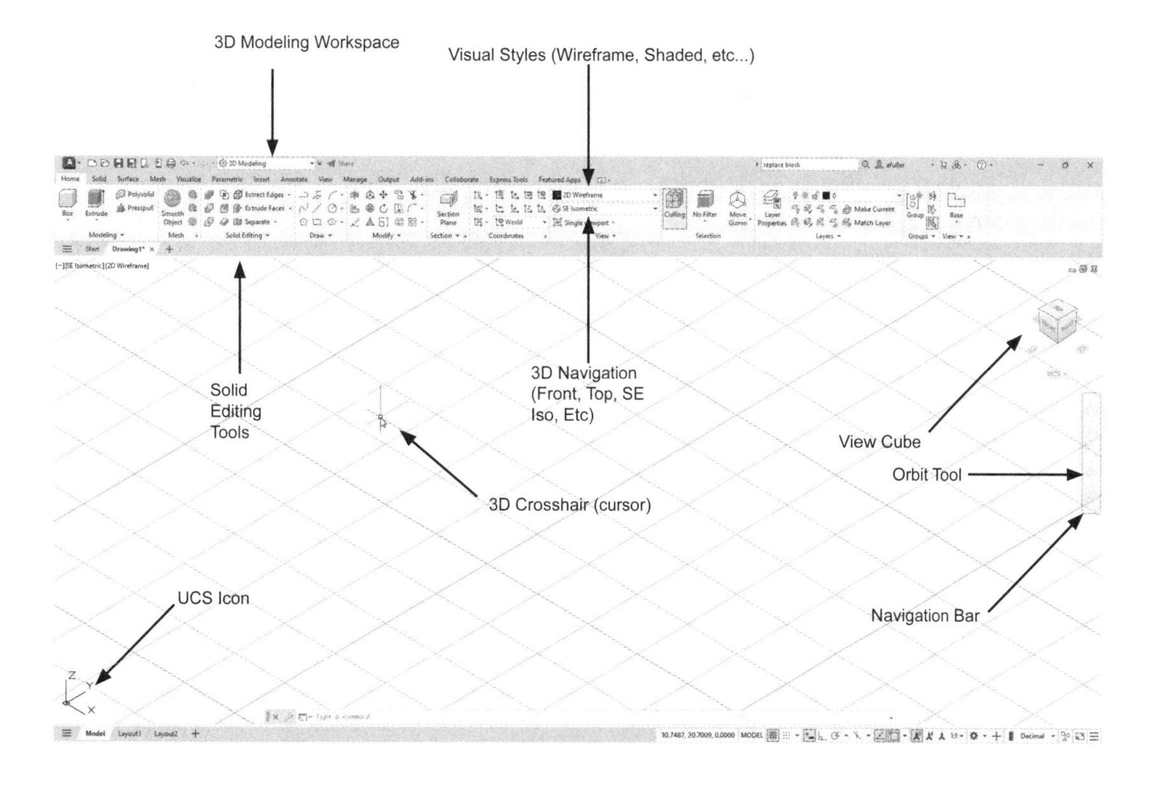

10.4 3D Modeling Workspace Environment

STEP-BY-STEP

BEGINNING A NEW 3D DRAWING USING A 3D TEMPLATE FILE

To begin a new drawing using a template file that is preset to the 3D drawing environment, follow these steps:

Step 1. Select **File**, pick **New**, select the **acad3D.dwt** template file from the **Select Template** dialog box, and click **Open**.

Step 2. When the new drawing opens, select the **3D Modeling** workspace from the **Workspace Switching** drop-down menu located in the upper left corner of the interface.

Step 3. Next, pick on the down arrow next to the **Unsaved View** button located on the **View** panel and select **SE Isometric** from the drop-down menu (see Figure 10.4). Another way to access AutoCAD's views is to select the **Visualize** tab and pick on the **Views** button and select a view from the drop-down menu.

Step 4. Select the **Home** tab of the ribbon. The **Home** tab contains most of the tools necessary to create and edit 3D models.

STEP-BY-STEP

Converting the 2D Drawing Environment to 3D

To convert the drawing environment of an existing 2D drawing to a 3D modeling environment, follow these steps:

Step 1. Open the drawing and select **3D Modeling** in the **Workspace Switching** drop-down menu, as shown in Figure 10.4.

Step 2. Next, pick on the down arrow next to the **Unsaved View** button located on the **View** panel and select **SE Isometric** from the drop-down menu (see Figure 10.4).

Step 3. Open the **Drafting Settings** dialog box. Under the **Grid Behavior** settings, check the box next to **Display Grid Beyond Limits** (refer to Figure **4.96 on page 192**) and turn the grid on.

Step 4. Select the **Home** tab of the ribbon. The **Home** tab contains most of the tools necessary to create and edit 3D models.

When Steps 1 through 4 have been completed, the drawing's environment will resemble the screen layout shown in Figure 10.42.

10.4 3D MODELING TOOLS

Most of the tools necessary to create 3D solid models can be found on the **Modeling**, **Solid Editing**, and **Modify** panels of the ribbon's **Home** tab (when the workspace is set to **3D Modeling** as in Figure 10.4) or on the **Modeling** toolbar. In Figures 10.5(a) and 10.5(b) the icons for six essential 3D **modeling commands** are identified: **EXTRUDE**, **PRESSPULL**, **REVOLVE**, **UNION**, **SUBTRACT**, and **3DROTATE.**

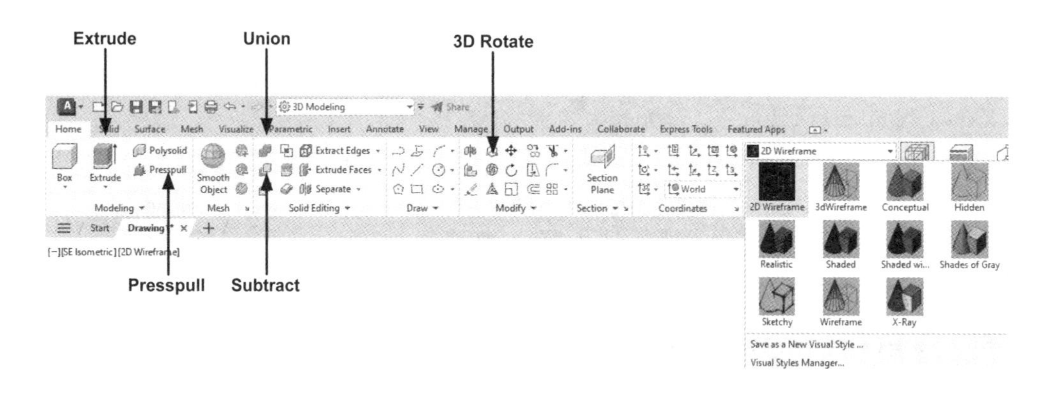

10.5(a) Modeling Tools Located on the Home Tab of the 3D Modeling Workspace

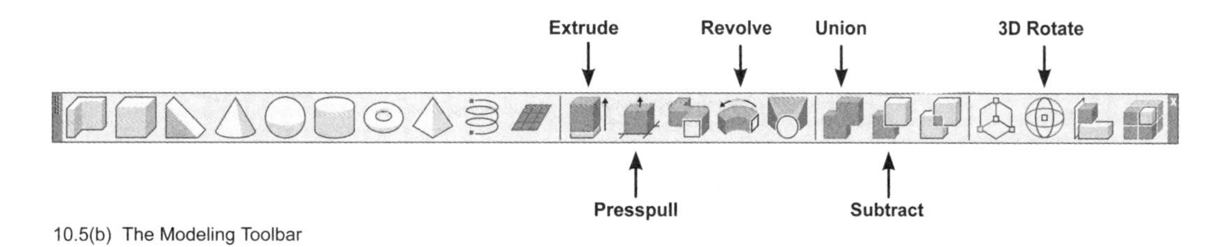

10.5(b) The Modeling Toolbar

Extruding 2D Entities to Create 3D Objects

In most cases, 2D objects drawn on the *XY*-plane can be converted to 3D objects by extruding them along their *Z*-axis using AutoCAD's **EXTRUDE** or **PRESSPULL** commands; see Figures 10.5(a), 10.5(b), 10.6(a) and 10.6(b). Objects that can be extruded into 3D models include lines, arcs, circles, rectangles, ellipses, polygons, closed polylines, and profiles that have been *regioned* with the **REGION** command, see Figure 10.6(c). The steps involved in using the **REGION** command are presented later in this chapter.

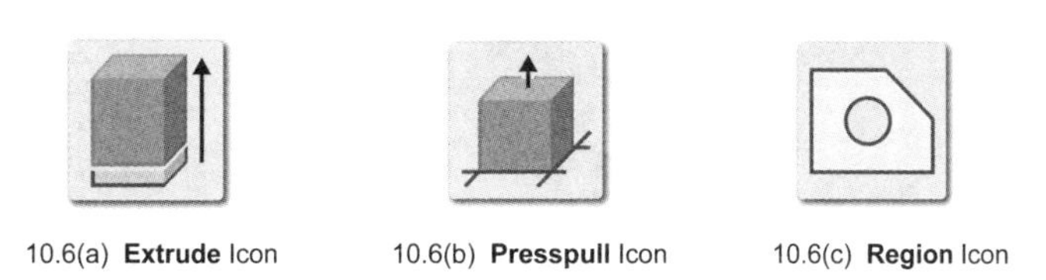

10.6(a) **Extrude** Icon 10.6(b) **Presspull** Icon 10.6(c) **Region** Icon

Using the EXTRUDE Command

Select the **EXTRUDE** command and at the *Select objects:* prompt, pick the 2D object(s) to be extruded and press **<Enter>**. At the *Specify height of the extrusion (path):* prompt, type the desired value for the extrusion height as measured along the *Z*-axis (the height may be given as either a positive or a negative value depending on which direction is desired for the extrusion) and press **<Enter>**. The 2D object will be extruded along its *Z*-axis, becoming a 3D model. Note: If the object being extruded is not a closed object (a line or an arc, for example) it will extrude as a 3D *surface* model. Closed boundaries like circles, rectangles, and closed polylines will extrude as 3D *solid* models. If an object's profile is not a closed boundary but you wish to extrude it as a 3D solid model, you may need to ***region*** the area first. The **REGION** command converts objects that enclose an area into a 2D *region*.

QUICK TIP

At the *Specify height of the extrusion:* prompt, you can move the cursor in the desired direction of the extrusion along the *Z*-axis, then enter an extrusion height and press **<Enter>**, and the object will be extruded in the direction defined by the cursor.

Using the REGION Command

The **REGION** command is located on both the **Draw** panel of the Home tab and the **Draw** toolbar. Select the **REGION** icon (see Figure 10.7), or type **REGION** and press **<Enter>**, and at the *Select objects:* prompt, select all the objects that enclose the area to be regioned and press **<Enter>**. A note will appear on AutoCAD's command line indicating whether or not a region was created. A region can be extruded into a 3D solid model using the **EXTRUDE** or **PRESSPULL** commands.

Using the PRESSPULL Command

Select the **PRESSPULL** command and at the *Select object or bounded area:* prompt, left-click on the object, or left-click inside the boundary of the closed object, to be extruded. When the object or boundary has been selected, move the cursor in the desired direction of the extrusion along the *Z*-axis and at the *Specify extrusion height:* prompt, type in the extrusion height as measured along the *Z*-axis (the height may be given as either a positive or a negative value depending on which direction is desired for the extrusion) and press **<Enter>**. The 2D shape will be extruded along its *Z*-axis, becoming a 3D object. Boundary shapes that are closed will extrude as 3D solid models and objects that are not closed will extrude as 3D surface models.

Unioning 3D Objects

10.7 Union Icon

Two or more 3D solid objects can be combined to form one object using the **UNION** command. The **UNION** command can be selected from either the **Solid Editing** panel of the **Home** tab of the ribbon or the **Modeling** toolbar. See Figures 10.5(a), 10.5(b), and 10.8.

NOTE

Unioning the solid objects combines their total volumes into one object, and they can no longer be edited as separate solids. An example of the union of the volumes of two solid cylinders is shown in Figure 10.9.

Using the UNION Command

Select the **UNION** command and at the *Select objects:* prompt, select the 3D solid objects you would like to combine. When you have selected all the objects to be unioned, press **<Enter>**.

10.8 Object Resulting from the Union of Two 3D Cylinders with Different Diameters

Subtracting 3D Objects

When solid objects overlap, the **SUBTRACT** command can be used to remove the shared volume of one solid from the volume of the other solids. For example, subtracting a cylinder with a small diameter from a cylinder with a larger diameter forms a hole through the larger cylinder. The **SUBTRACT** command can be selected from either the **Solid Editing** panel of the **Home** tab of the ribbon or the **Modeling** toolbar. See Figures 10.5(a), 10.5(b), and 10.9.

Using the SUBTRACT Command

10.9 Subtract Icon

Select the **SUBTRACT** command. At the *Select solids or regions to subtract from:* prompt, pick the principal object and press **<Enter>**. At the *Select solids or regions to subtract:* prompt, pick the object whose mass you would like to subtract from the mass of the principal object and press **<Enter>**. An example of the subtraction of a small solid cylinder from a larger solid cylinder is shown in Figures 10.11 and 10.12.

10.10 Two Intersecting Solid Cylinders (with Different Diameters) before the **SUBTRACT** Command

10.11 Object Resulting from the Subtraction of a Solid Cylinder with a Smaller Diameter from a Solid Cylinder of Greater Diameter

10.5 ROTATING 3D OBJECTS

The **3DROTATE** command is used to rotate 3D objects around the *X*-, *Y*-, or *Z*-axis. The **3DROTATE** command can be selected from the **Modify** panel of the **Home** tab of the ribbon or the **Modeling** toolbar. See Figures 10.5(a), 10.5(b), and 10.12. When using the **3DROTATE** command, you are prompted to *Select objects:*, *Specify base point:*, *Pick a rotation axis:*, and *Specify angle start point or type an angle:*. The selected object(s) rotates around the base point, and the axis of rotation passes through the base point. The angle of the rotation is specified in degrees. Entering a positive value for the rotation angle will result in the counterclockwise rotation of the object, and entering a negative value for the angle will result in the clockwise rotation of the object.

10.12 Rotate 3D Icon

The step-by-step instructions for using the **3DROTATE** command are presented in Project 10.1 at the end of this chapter.

NOTE

10.6 VIEWING 3D OBJECTS

In addition to the viewpoints available on the **View** toolbar, the **Orbit** tool is also helpful when viewing 3D objects. Using this tool allows you to "orbit" around a 3D object and view it from any angle. The **Orbit** tool can be selected from either the **Navigation Bar** located on the right side of the graphics window (see Figure 10.4) or the **Orbit** toolbar shown in Figure 10.13. The **Orbit** tool is constrained to a horizontal and vertical orbit only. The **FREE ORBIT** tool shown in Figure 10.13 rotates the view in 3D space with no constraint on the roll. **Note:** If the **Navigation Bar** is not visible, select the **View** tab and left-click on the **Navigation Bar** button.

Using the Free Orbit Tool

Select **FREE ORBIT** from the **Orbit** toolbar, and while holding down the left-click button of the mouse, move the mouse up and down or side to side. The viewpoint will change dynamically as the mouse is moved. To end the command, right-click and select **Exit**. Reset the object to a **SE Isometric** view by picking on the down arrow next to the **Unsaved View** button located on the **View** panel and select **SE Isometric** from the drop-down menu.

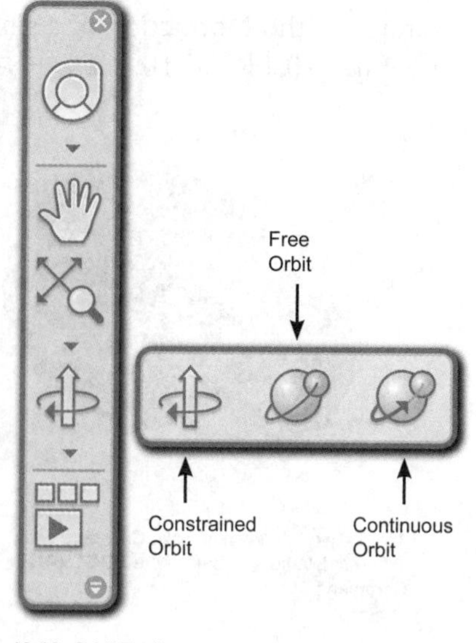

10.13 Orbit Toolbar

10.7 REPRESENTING 3D OBJECTS AS SHADED OR WIREFRAME MODELS

The **Visual Styles** tools control whether a solid is shown as a shaded, unshaded, or wireframe image in the drawing window. These tools can be accessed by picking on the down arrow located next to the word **2D Wireframe** located on the **View** panel of the **Home** tab of the ribbon [see Figure 10.14(a)] or by selecting them from the **Visual Styles** toolbar. See Figure 10.14(b).

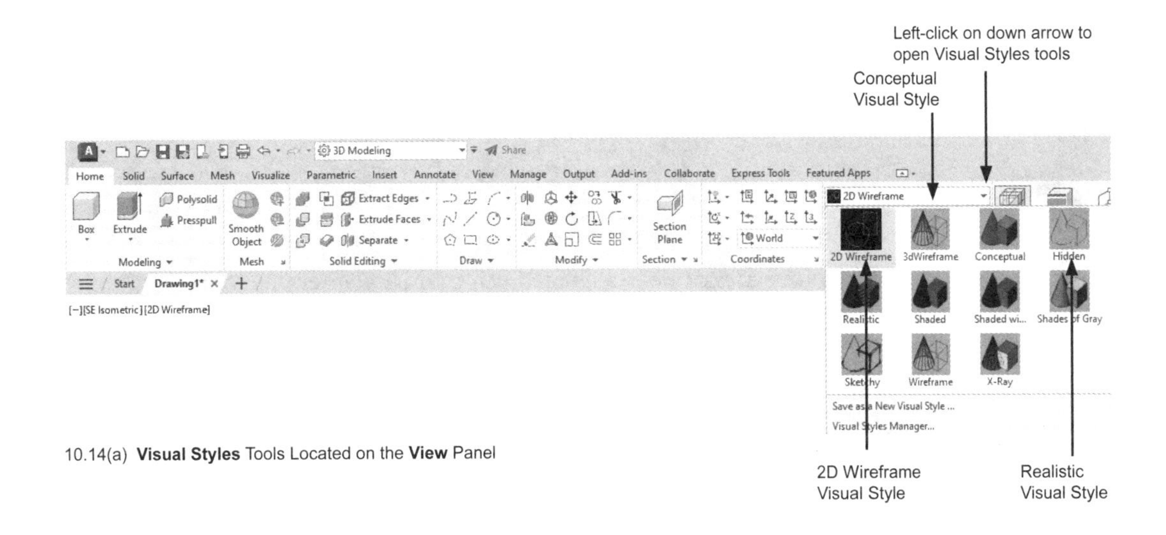

10.14(a) **Visual Styles** Tools Located on the **View** Panel

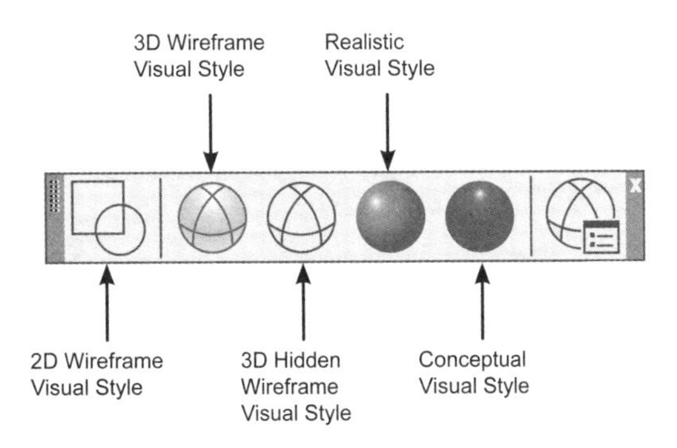

10.14(b) **Visual Styles** Toolbar Icons

NOTE

Selecting the **Conceptual** or **Realistic** visual styles results in shaded models. Shading works best on objects that have been assigned a color other than black or white. Selecting the **2D Wireframe**, **3D Wireframe**, or **3D Hidden Visual Style** icons will result in wireframe images of 3D objects. See Figures 10.14(a) and 10.14(b).

CHAPTER REVIEW

Compared with 2D CAD drafting, creating 3D models gives designers a more powerful and dynamic means to conceptualize, edit, and analyze their designs. The evolution of 3D CAD tools will allow many of the design functions currently performed by engineers to be performed by designers and designer/drafters. This will create career advancement opportunities for drafters who adapt to these changes in CAD technology and master the design capabilities of these programs.

KEY WORDS

Finite Element Analysis (FEA)
Modeling Commands
Parametric Modeling

REVIEW

Short Answer

1. Define the term parametric modeling.
2. What properties can be calculated using finite element analysis software?
3. What is the name of the AutoCAD command that combines the volumes of two or more 3D objects into one object?
4. Name the AutoCAD toolbar on which the **SUBTRACT** command is located.
5. When using the EXTRUDE command to create a 3D object, the height of the extrusion is defined along which axis (X, Y, or Z)?

VIDEO

There is a video tutorial available for this project inside the Chapter 10 folder of the book's Video Training downloads.

Create a 3D solid model of the Tool Slide shown in the designer's sketch in Figure 10.15.

10.15 Designer's Sketch of the Tool Slide

Step 1. Getting Started

a. Download the **Tool Slide 3D Exercise** drawing file located in the *Prototype Drawing Files* folder associated with this book.

b. Use **SAVE AS** to save the drawing to your **Home** directory.

c. Rename the drawing **TOOL SLIDE 3D**.

d. Activate the **Model** tab in the lower-left corner of the graphics window (if it is not already highlighted) by left-clicking on the tab.

e. The front, top, and side views of the Tool Slide should appear in model space as shown in Figure 10.16. If the views do not appear correctly, follow the four steps outlined in the "Converting the 2D Drawing Environment to 3D" presented earlier in Section 10.3 of this chapter.

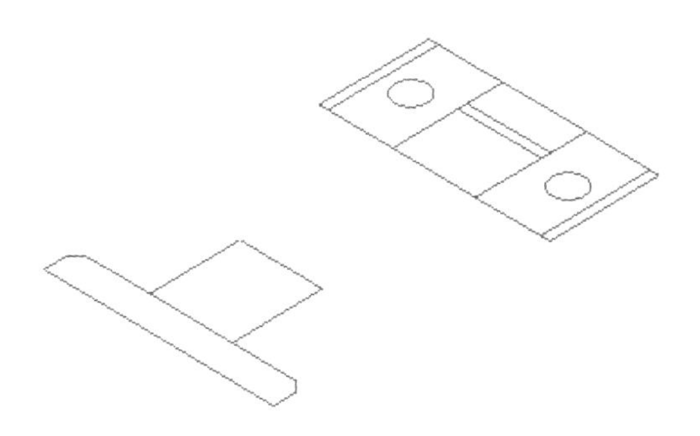

10.16 Beginning the Tool Slide 3D Drawing

f. Then, follow the steps below to create a 3D solid model of the Tool Slide.

Step 2. Make a copy of the front and side views of the Tool Slide and place the copies to the right of the original views.

Step 3. Erase and trim the lines of the copied views as necessary to create the two profiles shown in Figure 10.17.

- These profiles will be extruded in Step 2 to create the building blocks of the 3D model.

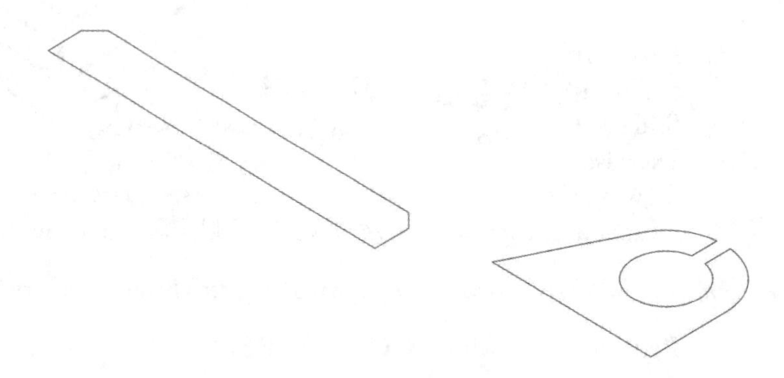

10.17 Profiles of the Front and Side Views of the Tool Slide

Step 4. Select the **PRESSPULL** command and left-click inside the profile shown on the left in Figure 10.17.

a. At the *Select object or bounded area:* prompt, move the cursor *up* (in the *positive Z*-axis) to define the direction for the extrusion along the *Z*-axis.

b. Type **3** for the extrusion height and press **<Enter>**.

Step 5. Select the **PRESSPULL** command again

a. Left-click inside the profile shown on the right in Figure 10.17 and move the cursor up and type **2.250** for the extrusion height and press **<Enter>**.

10.18 **REGION** Command Icon

QUICK TIP

If the **REGION** command is successful, a note will appear on the command line indicating that one or more regions were created. If the note says instead *0 Regions created*, then there is a problem with the objects that were selected for regioning. The most common reason that a region fails to be created is that the selected objects do not form a closed shape or loop (there are breaks in the objects that make up the shape). To correct the problem, edit the objects to close any breaks or gaps in the shape.

b. Note: If you prefer to use the **EXTRUDE** command instead of the **PRESSPULL** command to create the solid models shown in Figure 10.19, it is necessary to first use the **REGION** tool (see Figure 10.18) to create a separate region for each of the two profiles created in Step 1.

- Do this by selecting the REGION icon and when prompted by the command to *Select objects:*, select all the lines making up the boundary shown on the left in Figure 10.17 and press **<Enter>**.
- Next, select the EXTRUDE icon and select the region and press **<Enter>**. At the *Specify height of the extrusion (path):* prompt, type **3** for the extrusion height and press **<Enter>**.
- Repeat these steps to extrude the region shown on the right in Figure 10.17 except enter **2.250** for the height of the extrusion.

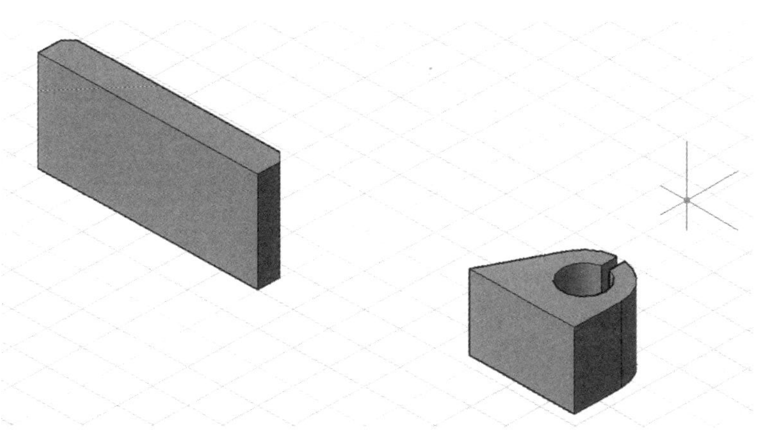

10.19 3D Models Created with the PRESSPULL Command

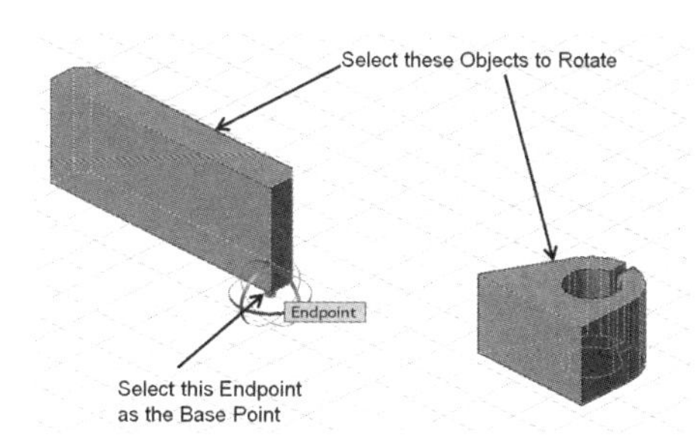

10.20 Specifying the Base Point for the **3D Rotate** Command

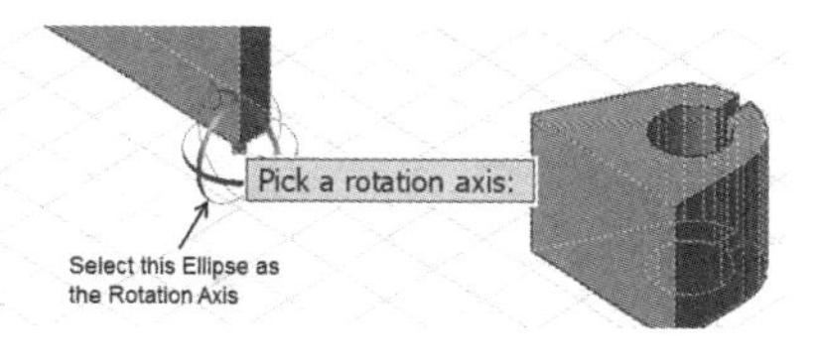

10.21 Selecting the Rotation Axis of the 3D Rotate Command

PROJECT

Step 6. Rotate the 3D objects created in Step 4.

a. Select the **3D Rotate** icon; refer to Figures 10.5(a) and 10.5(b).

b. When prompted to *Select objects:*, pick both of the 3D objects created in Step 4 and press **<Enter>**.

c. When prompted to *Specify base point:* select the endpoint shown in Figure 10.20.

d. When prompted to *Pick a rotation angle:*, pick the axis (ellipse) shown in Figure 10.21.

e. At the *Specify angle start point:* prompt, type **90** and press **<Enter>**. The object will rotate 90° around the base point in the positive direction of the rotation axis.

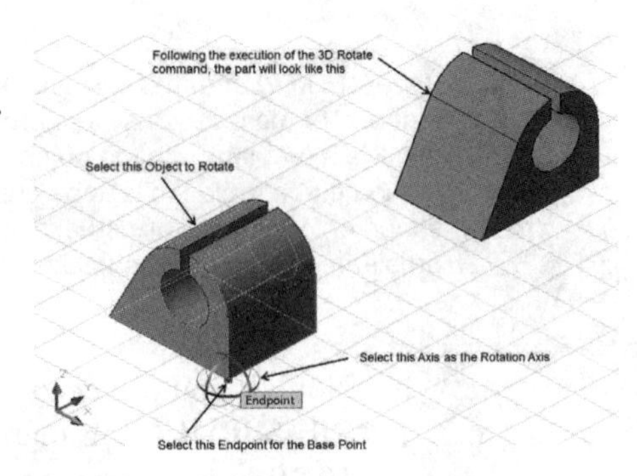

10.22 Rotating the Object 90° around the Z-Axis

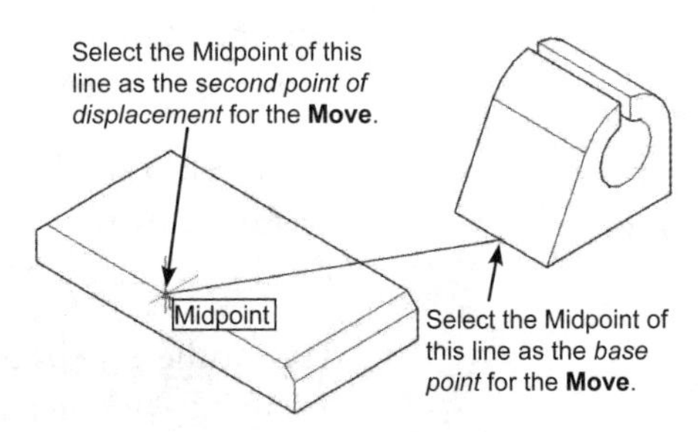

10.23 Positioning the 3D Objects with the **MOVE** Command

Step 7. Use the **3DROTATE** command to rotate the object shown in Figure 10.22 once more.

a. This time, the rotation will be **90°** around the *Z*-axis.

Step 8. Use the **MOVE** command to center the smaller 3D object on top of the larger 3D object as shown in Figure 10.23.

Step 9. Use the **UNION** command—refer to Figures 10.5(a) and 10.5(b)—to combine the two solids into one solid. Copy the two circles from the top view of the **Tool Slide** drawing to the 3D model.

a. Select the midpoint of the top view as the *base point* and the midpoint of the 3D model as the *Second point of displacement* as shown in Figure 10.24.

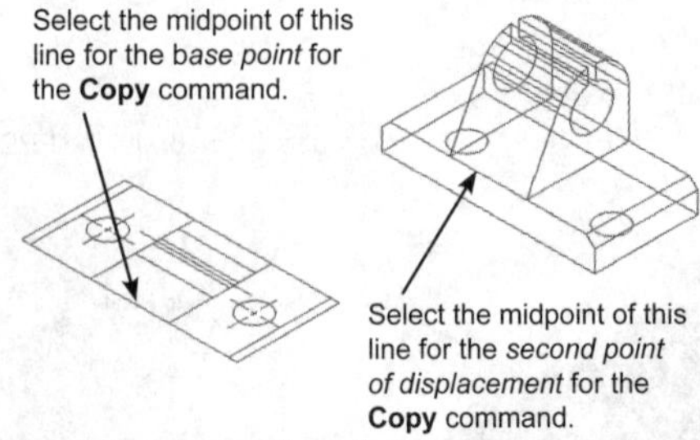

10.24 Copying the Circles from the Top View to the 3D Model

Step 10. Extrude, or Presspull, the two circles copied in Step 9 to a height of **1.00"** to form two cylinders representing the holes in the object as shown in Figure 10.25.

Step 11. If the extrude command was used, now use the **SUBTRACT** command—refer to Figures 10.5(a) and 10.5(b)—to subtract the two cylinders (created in Step 10) from the larger solid object (created in Step 9) to create the holes in the solid. (If Presspull was used, the subtract command isn't needed.)

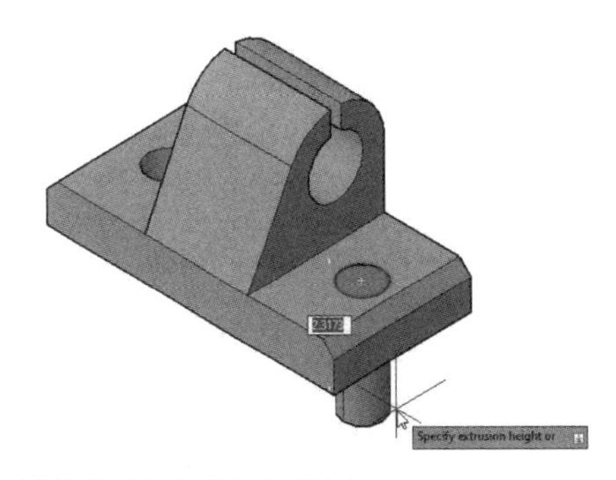

10.25 Model with Extruded Circles

- This step completes the construction of the 3D model of the tool slide.

Step 12. Select the **Conceptual Visual Style** icon [refer to Figures 10.14(a) and 10.14(b)] to add shading to the tool slide.

- The finished model should resemble the one in Figure 10.26 after shading.

Step 13. When finished modeling the Tool Slide, left-click on the ***Layout1*** tab in the lower-left corner of the graphics window.

a. Edit the text in the title block by double-clicking on it to enable the text editing option.

Step 14. Follow your instructor's directions to print the drawing.

Step 15. Be sure to save the drawing file before closing AutoCAD.

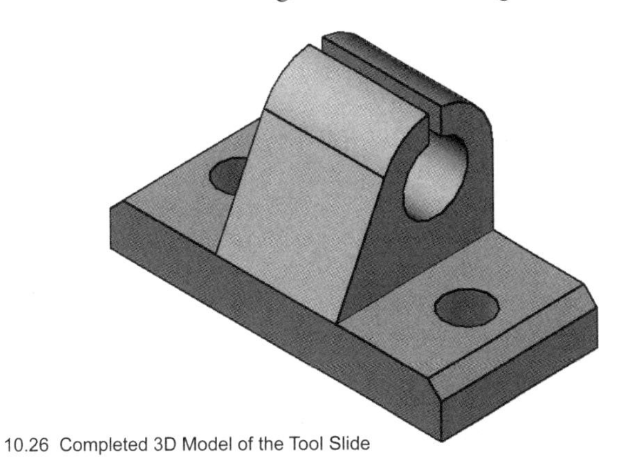

10.26 Completed 3D Model of the Tool Slide

PROJECT

VIDEO

There is a video tutorial available for this project inside the Chapter 10 folder of the book's Video Training downloads.

Create a 3D solid model of the Bracket shown in Figure 10.27.

Step 1. Getting Started

a. Download the **Bracket 3D Exercise** drawing file located in the *Prototype Drawing Files* folder associated with this book.

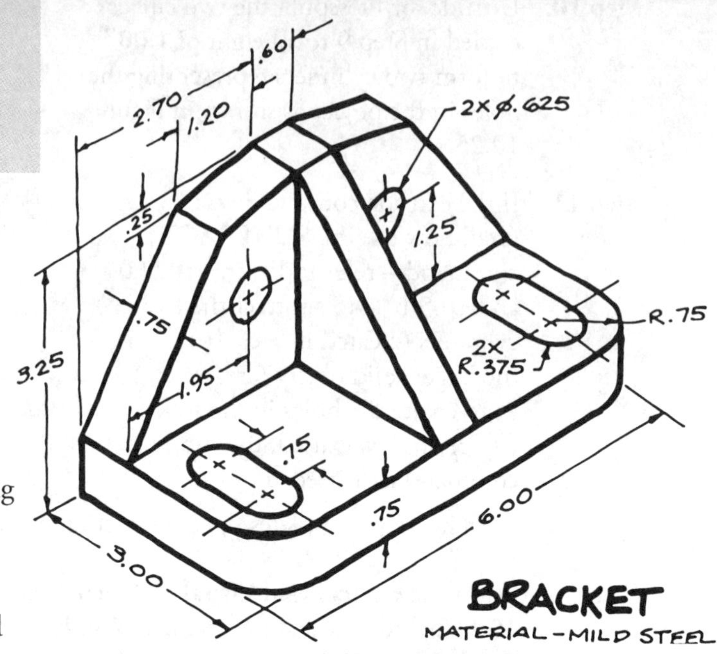

10.27 Designer's Sketch of the Bracket Project

b. Use **SAVE AS** to save the drawing to your **Home** directory, and rename the drawing **BRACKET 3D**.

c. Activate the **Model** tab in the lower-left corner of the graphics window (if it is not already highlighted) by left-clicking on the tab.

- **Note:** The front, top, and side views of the Bracket should be visible in model space.

Step 2. Make copies of the front, top, and side views of the Bracket and edit the copies to resemble the 2D profiles shown in Figure 10.28

- Note: To position the views as shown in Figure 10.28, set the **View** to **SE Isometric**). These profiles will be extruded in Step 3 to create the building blocks of the 3D model.

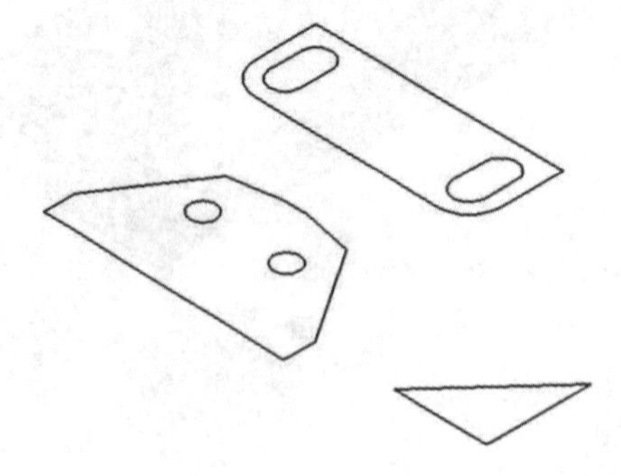

10.28 Profiles of Bracket Prior to Extruding

Step 3. Use the **PRESSPULL** or **REGION/EXTRUDE** methods explained in Step 2 and Step 3 of the **Tool Slide 3D Exercise** to extrude the regions to the desired heights (see Figure 10.29).

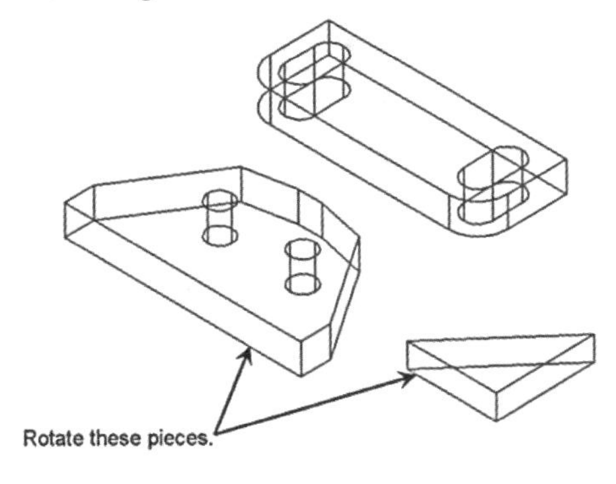

10.29 Extruded Profiles

Step 4. Use the **3DROTATE**, **UNION**, **SUBTRACT**, and **MOVE** commands as necessary to construct the 3D model (see Figure 10.30).

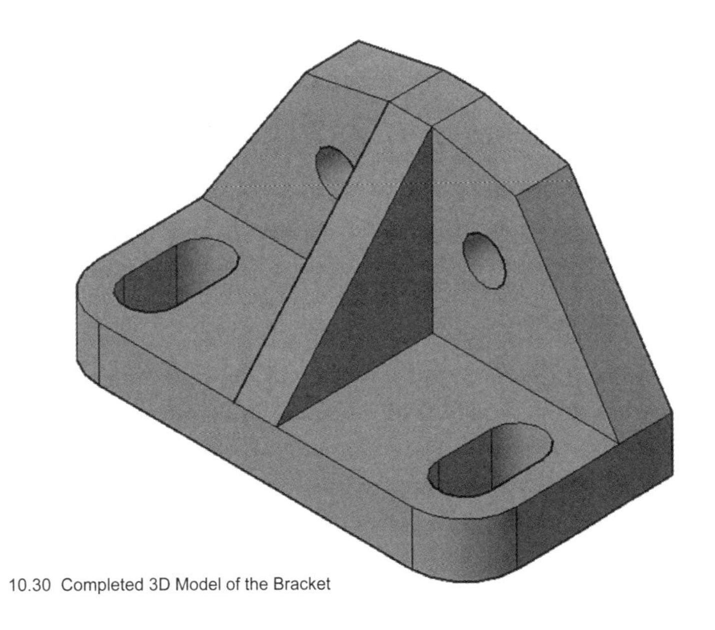

10.30 Completed 3D Model of the Bracket

Step 5. When finished modeling the Bracket, left-click on the ***Layout1*** tab in the lower-left corner of the graphics window.

Step 6. Edit the text in the title block by double-clicking on it to enable the text editing option.

Step 7. Follow your instructor's directions to print the drawing.

Step 8. Be sure to save the drawing file before closing AutoCAD.

VIDEO

There is a video tutorial available for this project inside the Chapter 10 folder of the book's Video Training downloads.

Create a 3D model of the Tool Holder shown in Figure 10.31.

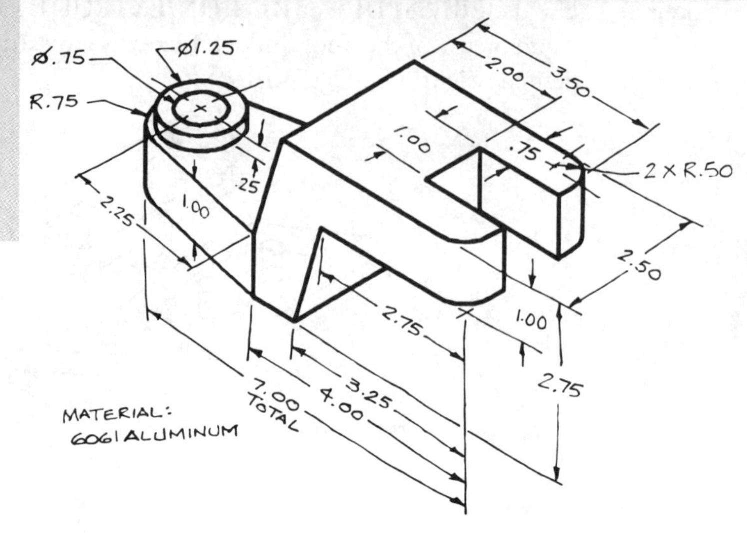

10.31 Designer's Sketch of the Tool Holder Project

Step 1. Getting Started

a. Download the **Tool Holder 3D Exercise** drawing located inside the *Prototype Drawing Files* folder associated with this book.

b. **SAVE AS** to save the drawing to your **Home** directory.

c. Rename the drawing **TOOL HOLDER 3D.**

d. Activate the **Model** tab in the lower-left corner of the graphics window (if it is not already highlighted) by left-clicking on the tab.

- **Note:** The front, top, and side views of the 3D Tool Holder should be visible in model space.

Step 2. Make copies of the front and top views of the Tool Holder and edit the copies to resemble the 2D profiles shown in Figure 10.32. These profiles will be extruded in Step 3 to create the building blocks of the 3D model.

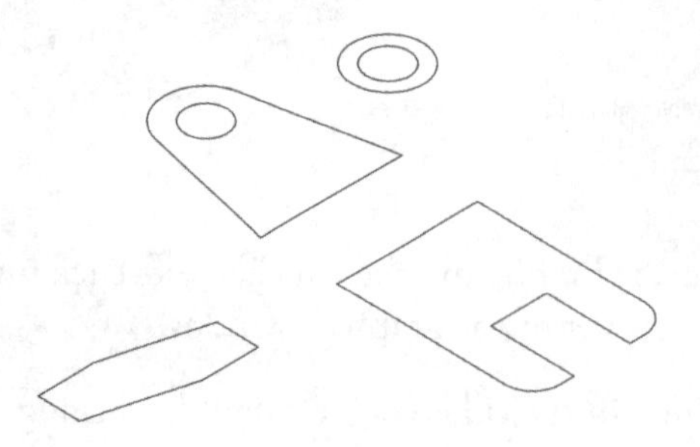

10.32 Profiles of Tool Holder Prior to Extruding

Step 3. Use the **PRESSPULL** or **REGION/EXTRUDE** methods explained in Step 2 and Step 3 of the **Tool Slide 3D Exercise** to extrude the regions to the desired heights, see Figure 10.33(a). Rotate the middle section 90 degrees and move the small cylinder to the center of the arc as shown in Figure 10.33(b).

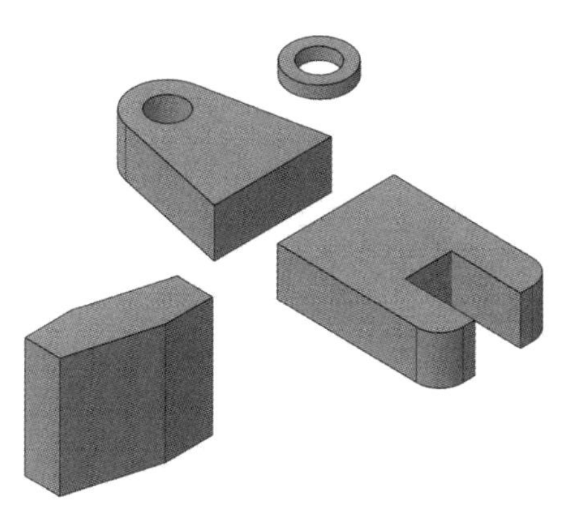

10.33(a) Extruded Profiles

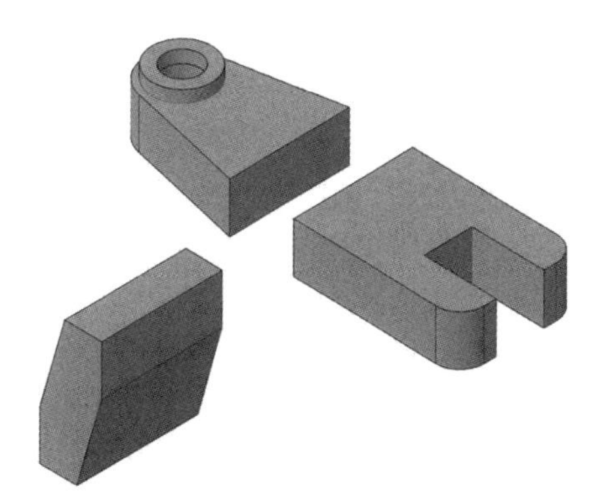

10.33(b) Extruded Profiles with Rotated Middle and Cylinder in place

Step 4. Use the **3DROTATE**, **UNION**, **SUBTRACT**, and **MOVE** commands as necessary to construct the 3D model. See Figures 10.34(a) and 10.34(b).

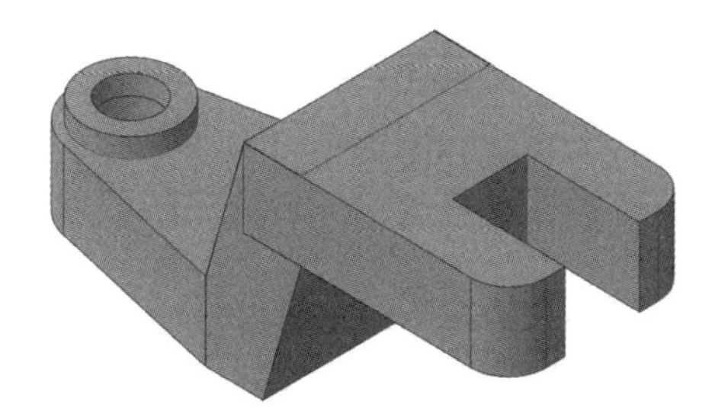

10.34(a) Assembled Tool Holder, Before Union

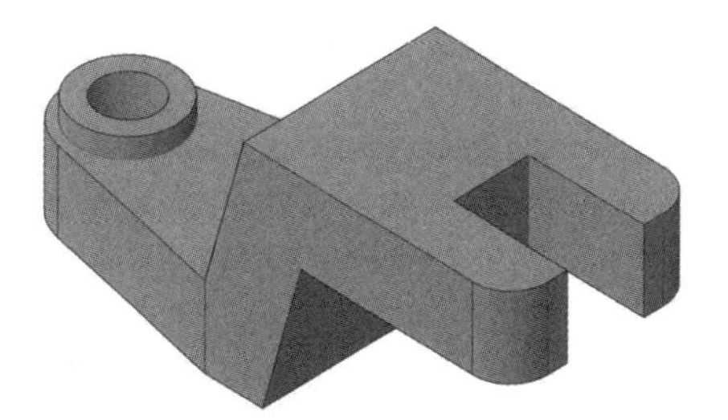

10.34(b) Assembled Tool Holder, After Union

Step 5. When finished modeling the Tool Holder, left-click on the ***Layout1*** tab in the lower-left corner of the graphics window.

a. Edit the text in the title block by double-clicking on it to enable the text editing option.

Step 6. Follow your instructor's directions to print the drawing.

Step 7. Be sure to save the drawing file before closing AutoCAD.

There is a video tutorial available for this project inside the Chapter 10 folder of the book's Video Training downloads.

VIDEO

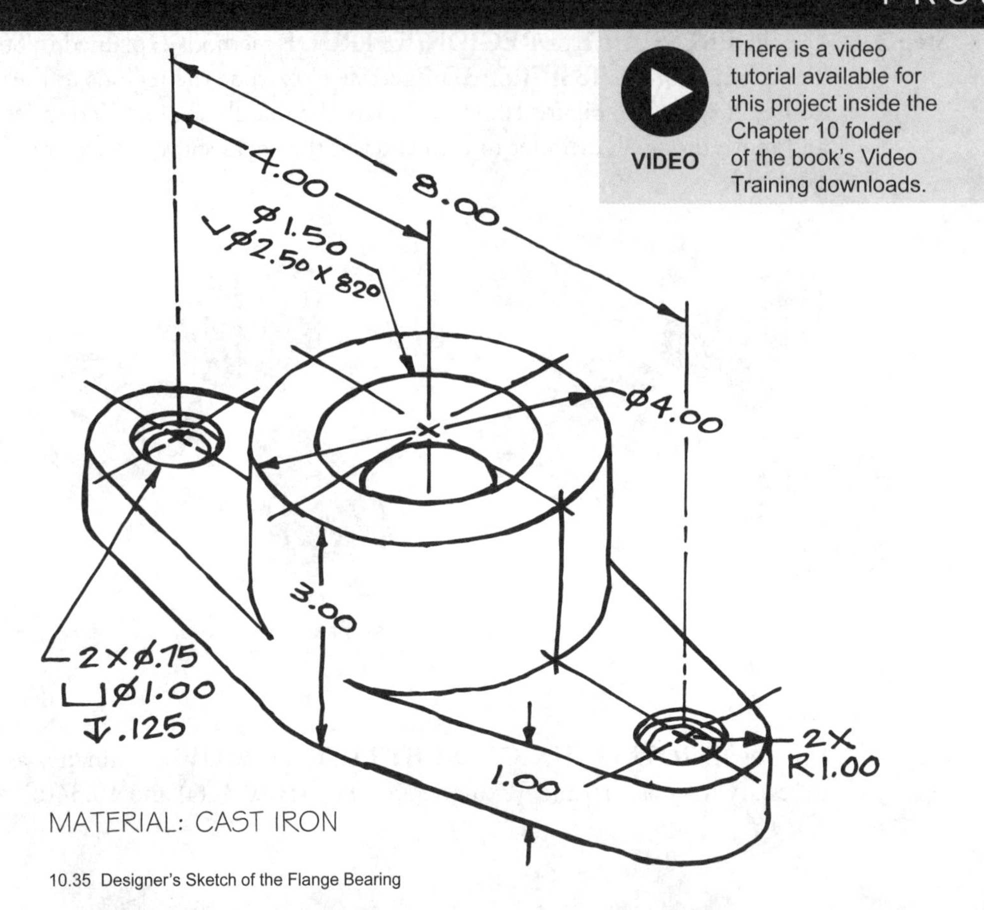

10.35 Designer's Sketch of the Flange Bearing

Create a 3D solid model of the Flange Bearing shown in Figure 10.35.

Step 1. Getting Started

a. Download the Flange Bearing 3D Exercise drawing file located in the *Prototype Drawing Files* folder associated with this book.

b. Use **SAVE AS** to save the drawing to your **Home** directory, and rename the drawing **FLANGE BEARING 3D**.

c. Activate the ***Model*** tab in the lower-left corner of the graphics window (if it is not already highlighted) by left-clicking on the tab.

- The top and section views of the Flange Bearing drawn in Chapter 8 should appear in model space.

- Follow the steps below to create a 3D solid model of the Flange Bearing.

Step 2. Make a copies of the top view of the Flange Bearing and of the profiles of the countersunk and counter-bored holes found in the section view and move the copies to the side of the original views.

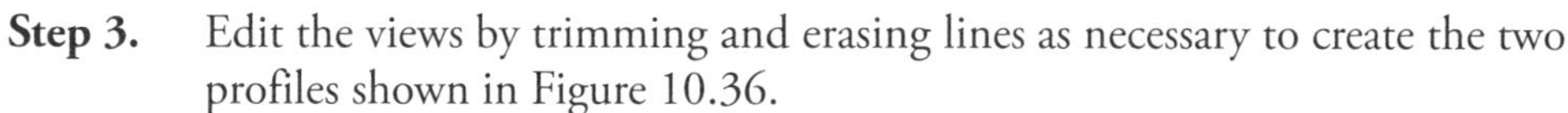

Step 3. Edit the views by trimming and erasing lines as necessary to create the two profiles shown in Figure 10.36.

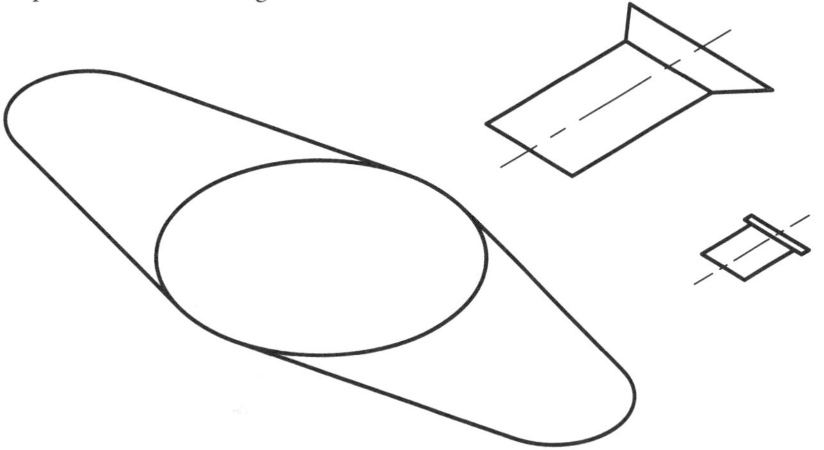

10.36 Profiles created from Front and Top Views of Flange Bearing

Step 4. Presspull Sleeve and Base

a. Select the **PRESSPULL** command

- At the *Select object or bounded area:* prompt, left-click inside the circular profile shown in Figure 10.37(a).
- Then, at the *Specify extrusion height:* prompt, move the cursor up (in the positive Z-axis) to define the direction for the extrusion along the Z-axis and type **3** for the extrusion height and press **<Enter>**.
- The resulting cylindrical feature is referred to as the *sleeve* of the Flange Bearing.

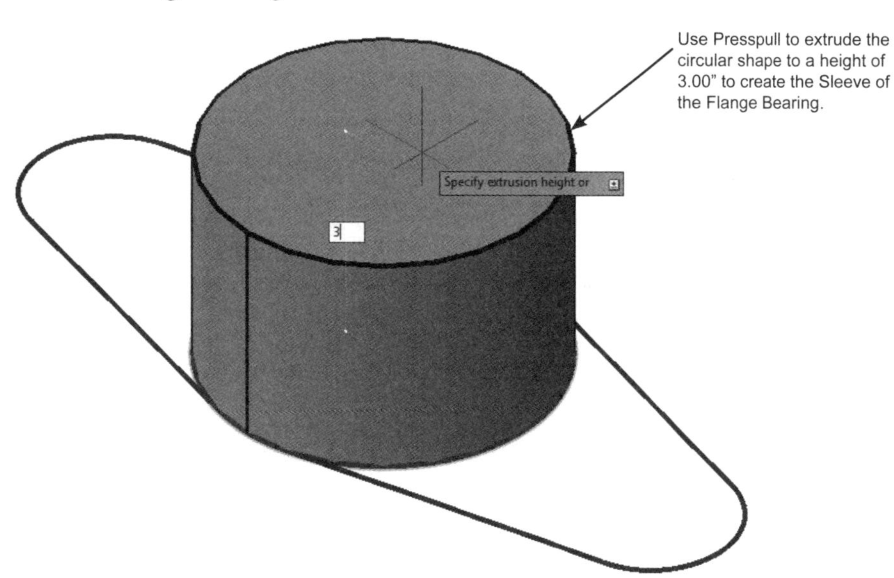
Use Presspull to extrude the circular shape to a height of 3.00" to create the Sleeve of the Flange Bearing.
Specify extrusion height or
3

10.37(a) Extruding the Profile of the Sleeve

b. Select the **PRESSPULL** command

- While holding down the shift key, left-click inside the two shapes located on either side of the sleeve as shown in Figure 10.37(b) and press **<Enter>**.
- Move the cursor up along the Z axis and type **1** for the extrusion height and press **<Enter>**. The resulting features comprise the *base* of the Flange Bearing.

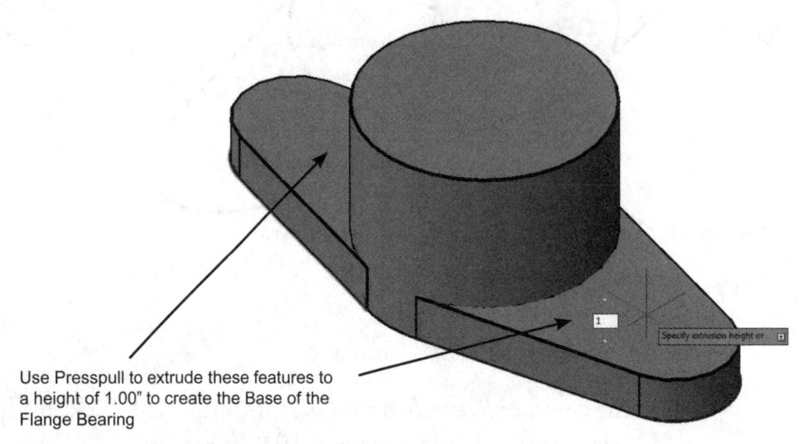
Use Presspull to extrude these features to a height of 1.00" to create the Base of the Flange Bearing
1
Specify extrusion height or

10.37(b) Extruding the Base of the Flange Bearing

Step 5. Adding the Countersunk Hole

a. Edit the profile of the countersunk hole shown in Figure 10.38(a) to match the shape in Figure 10.38(b).

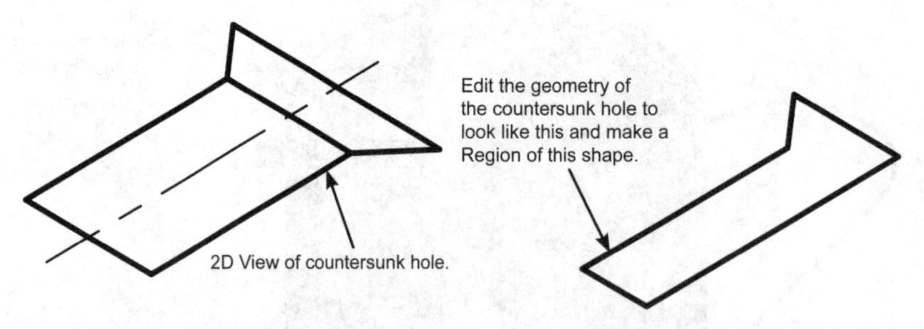
2D View of countersunk hole.
Edit the geometry of the countersunk hole to look like this and make a Region of this shape.

10.38(a) 2D Profile of Countersunk Hole

10.38(b) 2D Profile of Countersunk Hole after edits but before REGION command

b. Use the **REGION** command (see directions on page 401) to convert this shape into a closed loop as shown in Figure 10.38(c).

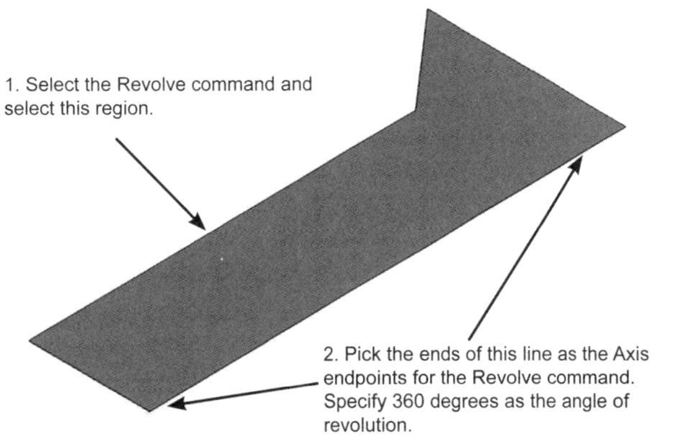

1. Select the Revolve command and select this region.
2. Pick the ends of this line as the Axis endpoints for the Revolve command. Specify 360 degrees as the angle of revolution.

10.38(c) 2D Profile of Countersunk Hole after Regioning.

10.38(d) 3D Revolve Command

c. Select the **REVOLVE** command, see figure 10.38(d).

- At the *Select object to revolve:* prompt, left-click inside the region and press **<Enter>**.
- At the *Specify axis start point or define axis:* prompt, select the end points of the line shown in Figure 10.38(c), and at the *Specify angle of revolution:* prompt, type **360** and press **<Enter>**.
- The resulting geometric shape should look like the example in Figure 10.38(e).

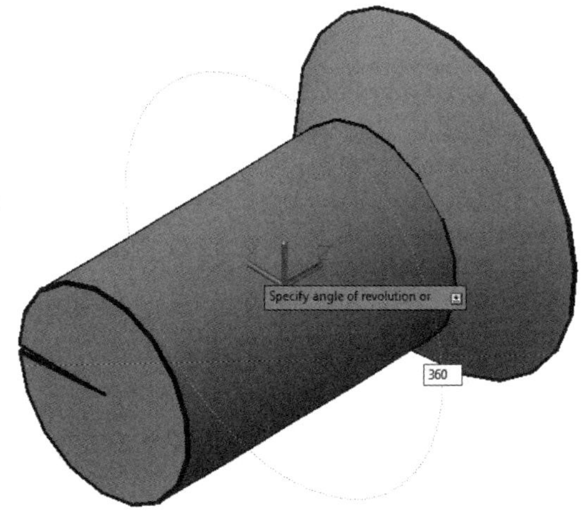

10.38(e) 3D Geometric Shape of Countersunk Hole

d. Use the **3D Rotate** command to rotate the geometric shape of the countersunk hole shown in Figure 10.39(a) **90** degrees to match the orientation of the shape shown in Figure 10.39(b).

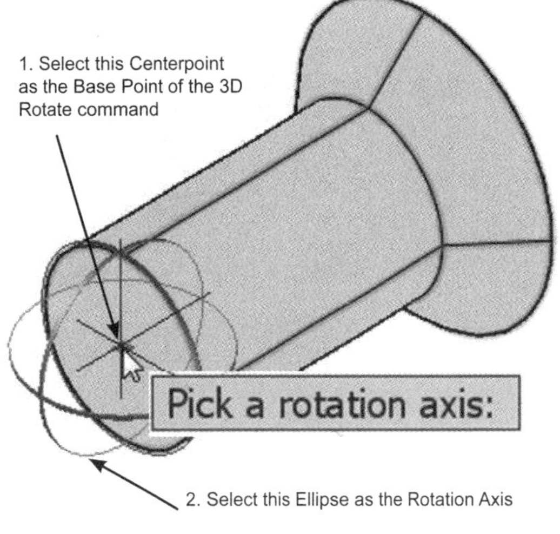

10.39(a) Rotating the Countersunk Hole Shape

e. Move the rotated shape to the position on the Flange Bearing model shown in Figure 10.39(b).

- Note: Select the center points shown in the figure as the base point and destination points for this move.

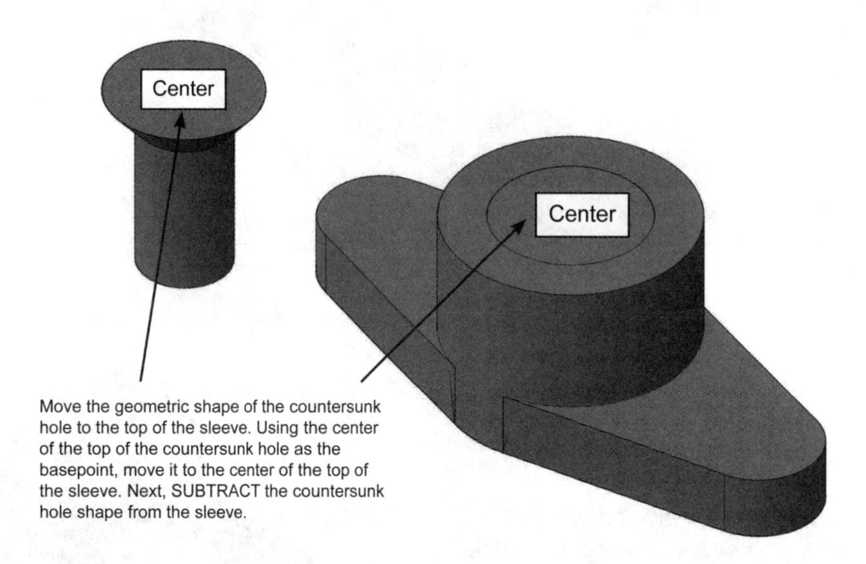

Center
Center
Move the geometric shape of the countersunk hole to the top of the sleeve. Using the center of the top of the countersunk hole as the basepoint, move it to the center of the top of the sleeve. Next, SUBTRACT the countersunk hole shape from the sleeve.

10.39(b) Moving the Countersunk Hole's geometric shape to the Flange Bearing's cylinder

f. Then, use the **SUBTRACT** command to subtract the countersunk shape from the Flange Bearing model to create the shape shown in Figure 10.40.

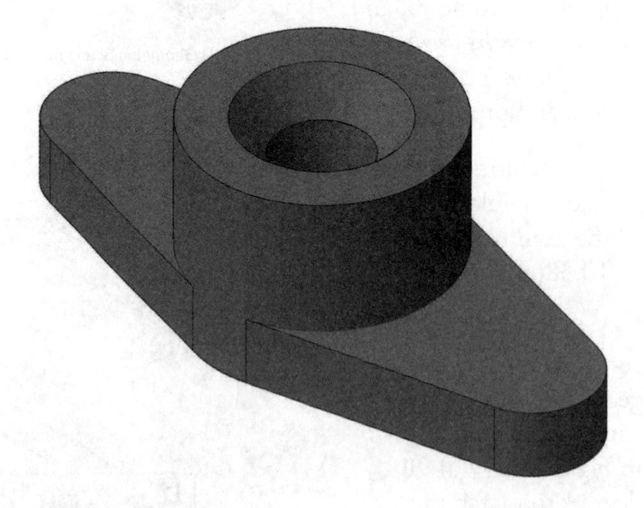

10.40 Flange Bearing model after countersunk hole is subtracted

Step 6. Adding the Counterbored Hole

a. Edit the profile of the counterbored hole shown in Figure 10.41(a) to match the shape in Figure 10.41(b).

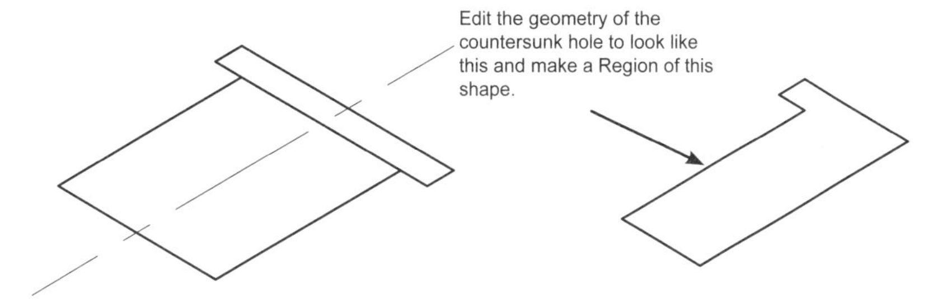

Edit the geometry of the countersunk hole to look like this and make a Region of this shape.

10.41(a) 2D Profile of Counter-bored Hole

10.41(b) 2D Profile of Counter-bored Hole after edits but before REGION command

b. Use the **REGION** command (see directions on page 401) to convert this shape into a closed loop as shown in Figure 10.41(c).

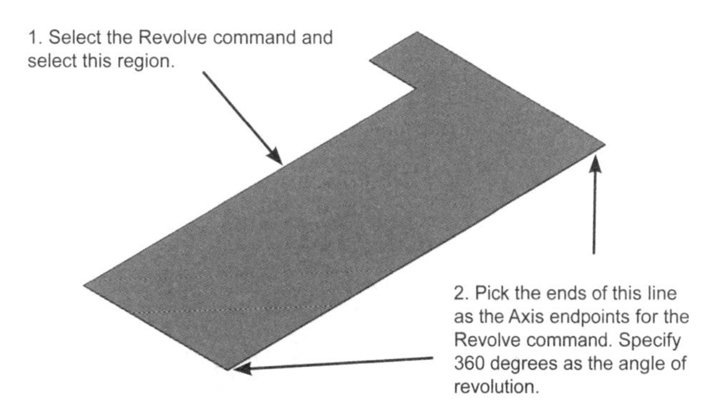

10.41(c) 2D Profile of Counter-bored Hole after Regioning.

10.41(d) 3D Revolve Command

c. Select the **REVOLVE** command, see figure 10.41(d).

- At the *Select object to revolve:* prompt, left-click inside the region and press **<Enter>**.
- At the *Specify axis start point or define axis:* prompt, select the end points of the line shown in Figure 10.41(c), and at the *Specify angle of revolution:* prompt, type **360** and press **<Enter>**.
- The resulting geometric shape should look like the example in Figure 10.41(e).

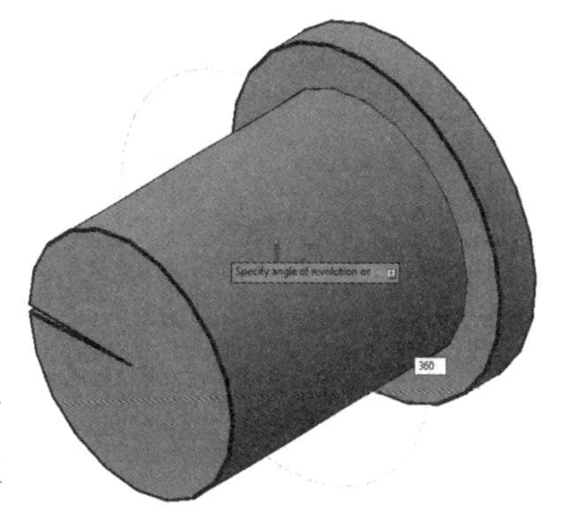

10.41(e) 3D Geometric Shape of Counterbored Hole

PROJECT

d. Use the **3D Rotate** tool to rotate the geometric shape of the counter-bored hole shown in Figure 10.42(a) **90** degrees to match the orientation of the shape shown in Figure 10.42(b).

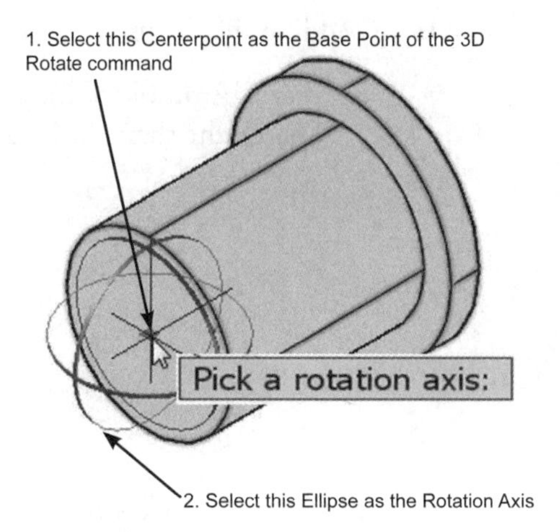
1. Select this Centerpoint as the Base Point of the 3D Rotate command
Pick a rotation axis:
2. Select this Ellipse as the Rotation Axis

10.42(a) Rotating the Counterbored Hole Shape

e. Copy the rotated shape to the positions on the Flange Bearing model shown in Figure 10.42(b).

- Note: Select the center points shown in the figure as the base point and destination points for this move.

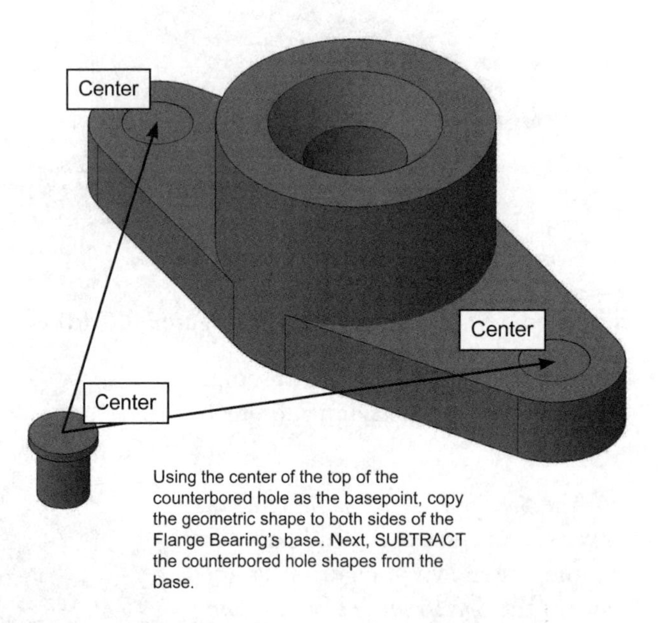
Center
Center
Center
Using the center of the top of the counterbored hole as the basepoint, copy the geometric shape to both sides of the Flange Bearing's base. Next, SUBTRACT the counterbored hole shapes from the base.

10.42(b) Copying the Counterbored Hole's geometric shape to the Flange Bearing

f. Use the **SUBTRACT** command to subtract the geometric shapes of the counter-bored holes from the Flange Bearing model to create the shape shown in Figure 10.43.

g. Select the **Conceptual Visual Style** icon [refer to Figures 10.14(a) and 10.14(b)] to add shading to the Flange Bearing.

- The finished model should resemble the one in Figure 10.43 after shading.

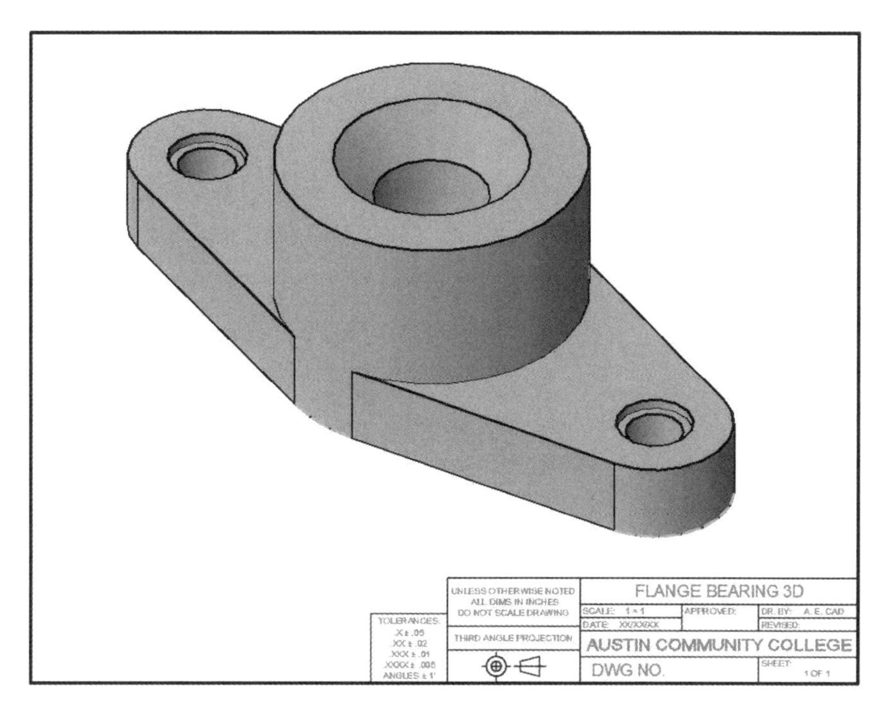

10.43 Completed 3D Model of the Flange Bearing

Step 7. When finished modeling the Flange Bearing, left-click on the ***3D View*** tab in the lower-left corner of the graphics window

a. Edit the text in the title block by double-clicking on it to enable the text editing option.

Step 8. Follow your instructor's directions to print the drawing.

Step 9. Be sure to save the drawing file before closing AutoCAD.

[illegible]

[illegible]

[illegible]

[illegible]

[illegible]

[illegible]

CAPSTONE PROJECT 1

MECHANICAL WORKING DRAWINGS

PROJECT OBJECTIVES:

After studying the material in this project, you should be able to:

1. Describe what mechanical working drawings are and how they are produced.
2. Use AutoCAD to create an exploded, isometric assembly view of a mechanism including balloons, part numbers, and a parts list.
3. Use AutoCAD to create detail drawings of mechanical parts, including all the necessary multiviews, dimensions, and notations required to manufacture each part.
4. Represent and specify fasteners and other hardware in a mechanical working drawing.

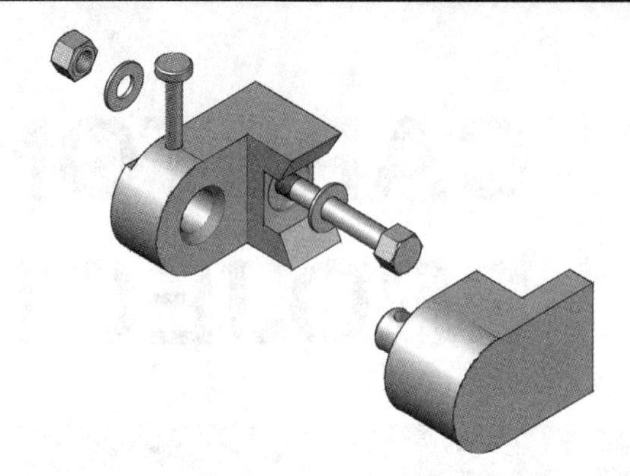

MECHANICAL WORKING DRAWINGS:

In the mechanical engineering field, drafters are often required to create complex sets of ***mechanical working drawings*** for entire mechanical assemblies. To better understand the creation of these drawings, it is helpful to step back and take a wider view of the mechanical design process itself.

For most mechanical designers, the first phase of a design project involves clearly defining the design problem and specifying the criteria that the finished design must meet to be considered a success. As stated earlier in Chapter 1, designers often refer to this phase in the design process as *problem identification*. For example, before beginning a design for a machine part, a mechanical designer must have a clear understanding of the following: the function the part serves; the ability of the part to work in conjunction with other parts; an idea of the shape, size, strength, and material of the finished part; any safety and reliability concerns the part may present; and an estimated budget for the project.

After the design problem is clearly defined, the *preliminary design* phase begins. This stage is also referred to by designers as the *ideation*, or *brainstorming*, phase of the process. During this phase, multiple solutions to the design problem are generated. The documentation for these preliminary designs may be in the form of freehand sketches or 2D and 3D CAD models.

In the next phase, the preliminary designs are analyzed and evaluated by the design team to decide which one best meets the design criteria defined in the first phase. During this process, the design may be further refined, and the best features of some of the rejected designs may be incorporated into the final design solution. The analysis of the designs may involve computer modeling or the preparation of actual prototypes of the part, which are subjected to performance testing. It is not unusual for a design to go through many revisions, or *iterations*, during this phase.

After the team decides on the best solution, the designer begins preparing design inputs that more clearly define the details of the project. Design inputs may include freehand sketches or CAD models that provide dimensional information and detailed notes about the project.

When the design inputs are finished, they are given to the drafter(s) responsible for preparing the working drawings for the project. In this phase, drafters usually work closely with designers, checkers, engineers, and other drafters to create the set of drawings. Drafters must follow any applicable drawing standards (such as ASME or ISO) during this phase.

As the drafter finishes each sheet in the set of drawings, the sheet is checked carefully for mistakes by designers or checkers. If mistakes are found, or if the design needs to be revised, the drafter makes the necessary corrections or revisions to the drawings. This process is repeated until the drawings are complete. Revisions are such an integral part of the design process that they are usually noted in a *revision history block* located on the upper right corner of the sheet.

JOB SKILL

The layout of title blocks for the first sheet and continuing sheets, including the information that goes into the fields of the title block, is covered in the *ASME Y14.1-2012* standard.

The finished set of working drawings represents the master plan for the project. All the information required to manufacture and assemble the project should be included in these drawings.

CP1.1 PREPARING MECHANICAL WORKING DRAWINGS

Working drawings usually include assembly and detail drawings. ***Assembly drawings*** show how the separate parts of the assembly are related to each other, for example, how mating parts fit together. ***Detail drawings*** provide all the information required to manufacture or purchase each part in the assembly, including the necessary views, dimensions, notations, and specifications.

Assembly Drawings

The assembly drawing usually acts as the "cover" sheet in a set of working drawings and is numbered as the first sheet in the set of plans. For example, if a set of working drawings contains a total of 20 sheets, the drafter will put the label *SHEET 1 OF 20* in the assembly drawing's title block. The details for each part in the assembly are drawn on subsequent sheets, and these sheets are numbered sequentially, for example, *SHEET 2 OF 20*, *SHEET 3 OF 20*, and so on, through *SHEET 20 OF 20*.

JOB SKILL

The ASME standards governing the size and format of drawing sheets, including borders and title blocks, are *ASME Y14.1-2012* and *ASME Y14.1M-2012* (metric). The ASME standard governing parts list is *ASME Y14.34 - 2013* and the standard governing revision history blocks is *ASME Y14.35 - 2014*.

Assembly drawings are often drawn pictorially with the parts pulled apart, or "exploded," to show how the device is assembled. Figure CP1.1 shows an example of an exploded isometric drawing that is used to show how the parts in the assembly fit together.

The assembly drawing often includes a ***parts list*** (sometimes referred to as a *bill of materials* or *BOM*) itemizing all the parts in the assembly. In Figure CP1.1 the parts list is located above the title block, and a revision history block is located in the upper right corner of this drawing.

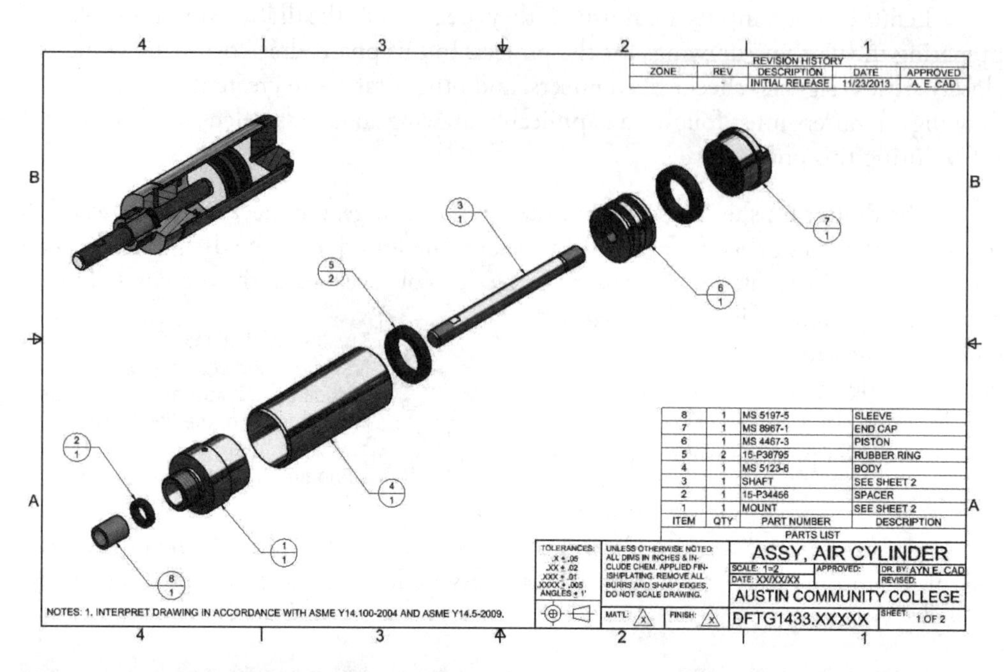

CP1.1 Exploded Assembly Drawing

STEP-BY-STEP

CREATING AN EXPLODED ASSEMBLY DRAWING

Step 1. Study the individual parts of the assembly and determine how they fit together. Make an isometric planning sketch of the assembly similar to the one in Figure CP1.2. Sketch the parts of the assembly as if they were pulled apart along isometric axes.

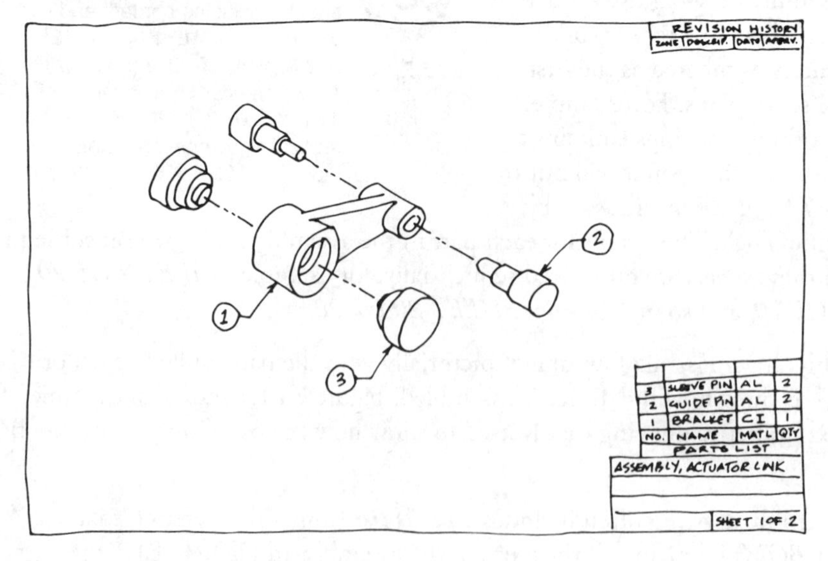

CP1.2 Planning Sketch for Exploded Assembly

STEP–BY–STEP

Assign item numbers to the separate parts and enclose each number in a circle with a leader attached. Item numbers are referred to in the ASME Y14.34 standard as *find* numbers. In an assembly drawing, this combination of an item (or find) number, circle, and leader is called a *balloon.*

Refer to the item numbers assigned to the sketch, and plan a parts list. The categories that the parts list must include are the item (or find) number, part number, description, and the quantity required. The parts list may also include categories for material, *Commercial and Governmental Entity* Code (CAGE Code), and other information about the assembly.

Step 2. Begin a new AutoCAD drawing using an appropriate sheet size, border, and title block. Refer to the planning sketch made in Step 1, and construct an isometric view for each part in the assembly. Whenever possible, orient the isometric view to show as much detail about the assembly as possible. Align the parts as they would appear if they had been pulled apart from their normal assembled arrangement, as shown in Figure CP1.3.

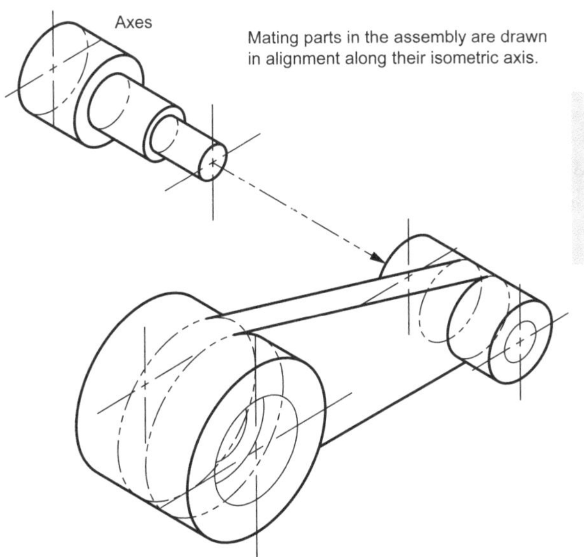

Axes
Mating parts in the assembly are drawn in alignment along their isometric axis.

CP1.3 Aligning Mating Parts along their Isometric Axes

NOTE Duplicate parts in an assembly receive the same item (or find) number.

Step 3. Add balloons and item numbers to each part. Add phantom lines along isometric axes to show the relationship between mating parts. See Figure CP1.4.

Step 4. Add a parts list to the drawing. An example of a parts list is shown in Figure CP1.5. Parts lists are often placed above the title block in assembly drawings but may be placed in other areas of the drawing with the exception of the upper right corner which is reserved for the Revision History Block (see Figure CP1.6).

STEP–BY–STEP

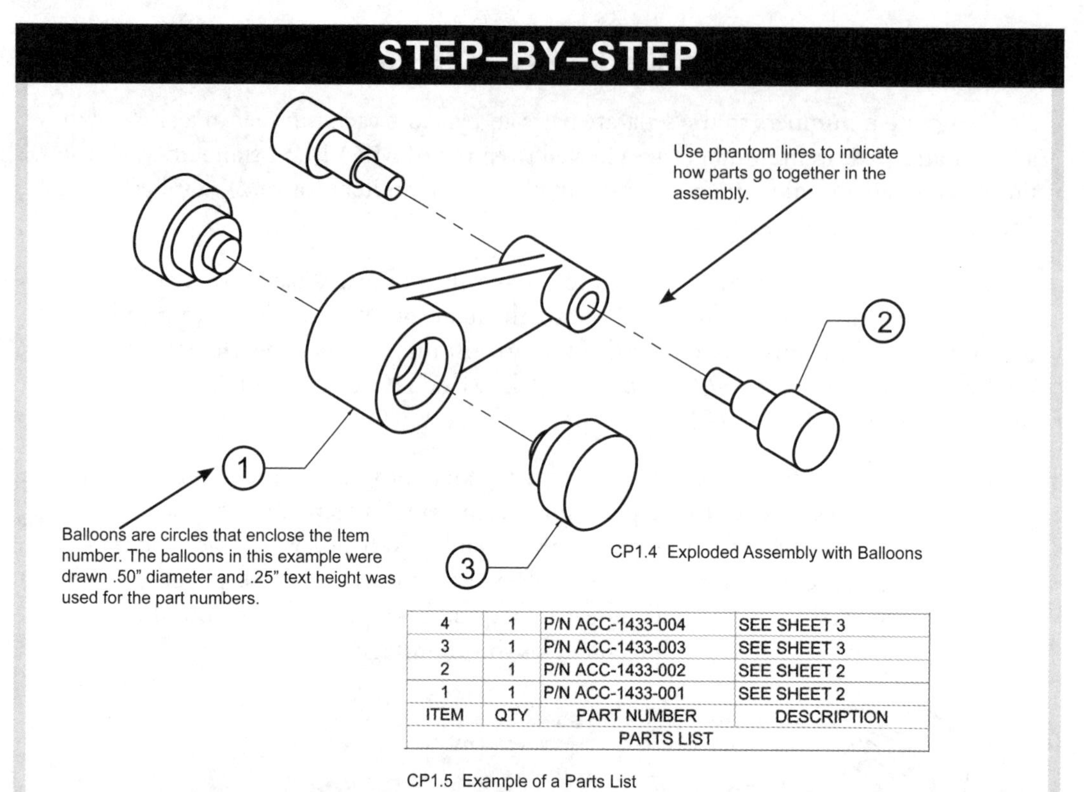

CP1.4 Exploded Assembly with Balloons

4	1	P/N ACC-1433-004	SEE SHEET 3
3	1	P/N ACC-1433-003	SEE SHEET 3
2	1	P/N ACC-1433-002	SEE SHEET 2
1	1	P/N ACC-1433-001	SEE SHEET 2
ITEM	QTY	PART NUMBER	DESCRIPTION
PARTS LIST			

CP1.5 Example of a Parts List

Step 5. Complete the information in the title block including the name, scale, and sheet number of the assembly. Compare the planning sketch shown in Figure CP1.2 with the completed CAD drawing shown in Figure CP1.6.

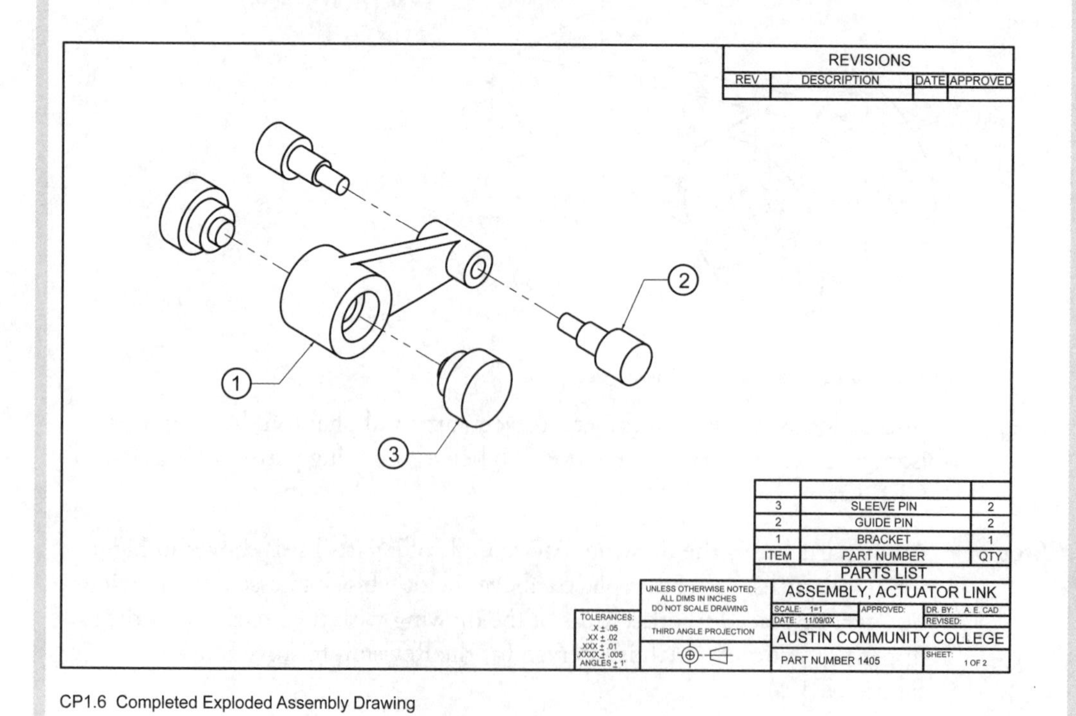

CP1.6 Completed Exploded Assembly Drawing

Detail Drawings

An example of a detail drawing is shown in Figure CP1.7. To create this drawing, the drafter works from design inputs provided by an engineer or designer. Often, the design input is in the form of a rough sketch; however, in the modern practice of technical drawing, the drafter may receive this input in the form of a 3D CAD drawing file created by the designer.

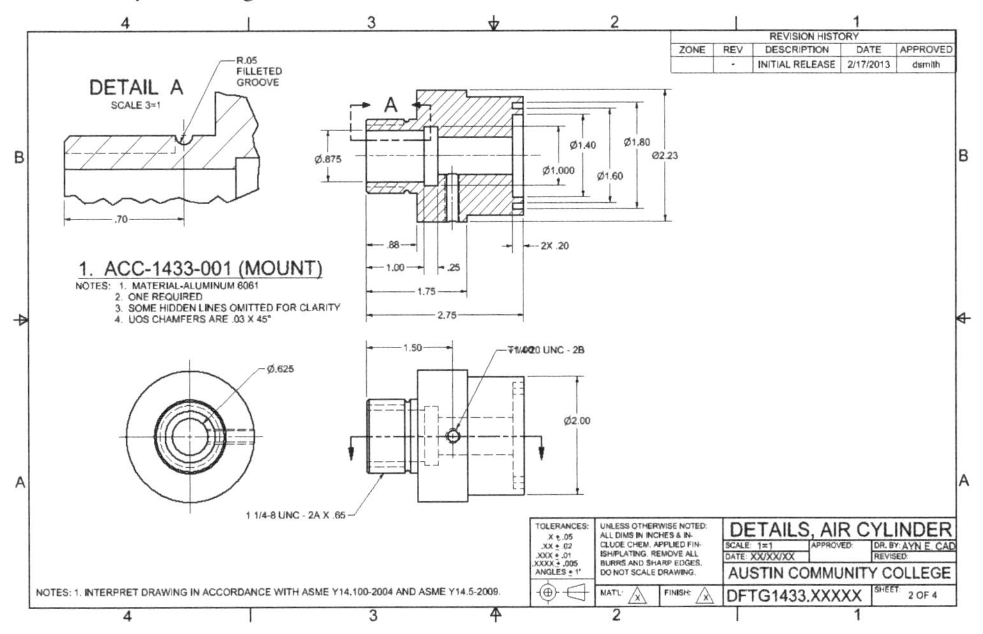

CP1.7 Example Detail Drawing

When the individual part is "detailed," all the views, dimensions, and notes necessary to manufacture the part are included. Because the drafter's expertise is in creating technical drawings, decisions regarding the necessary views and correct dimensioning of the object are often left up to the drafter's judgment. The drafter may also be responsible for determining the sheet size and making sure that the drawing complies with any appropriate drafting or dimensioning standards (such as *ASME Y14.5*).

JOB SKILL

When creating a detail drawing of an object, the drafter must be careful not to change the design intent of the designer by incorrectly noting dimension values, changing the precision of dimensions, or referencing different datum features than the ones shown on the designer's sketch.

STEP–BY–STEP

CREATING A DETAIL DRAWING

Step 1. Make a planning sketch of the multiviews of the parts to be detailed similar to the one shown in Figure CP1.8. This sketch may also include dimensions and notations.

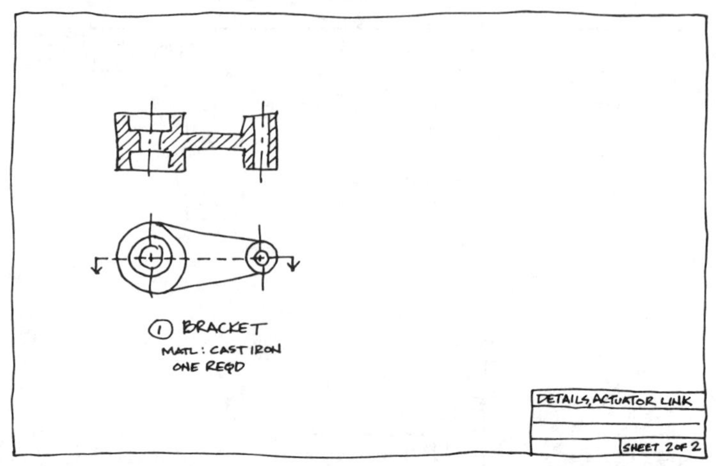

CP1.8 Planning Sketch for a Detail Sheet

If multiple parts are drawn on the same sheet, space the views of separate parts so there is no confusion concerning which view goes with which part.

Using the same part names and numbers defined in the assembly drawing, label each part. Include in the drawing the part's material, quantity (the number required for the assembly), and any other notes required for manufacture.

Step 2. Begin a new AutoCAD drawing using the same sheet size, border, and title block used in the assembly drawing. Refer to the planning sketch and draw the necessary views for each part. Consider drawing section views of parts with complex interior details.

Include all the dimensions necessary to manufacture the part. When dimensioning, follow guidelines established by the *ASME Y14.5* standard (refer to Chapter 5).

Add part numbers and part names. These are often placed using **.25"** text height for emphasis.

Include any notes required for manufacture and the total number of each part required for the assembly. These notes are often placed beneath the part name and number using the same text height as the dimension text.

Step 3. Complete the information in the title block including the name, scale, and sheet number of the detail. Compare the planning sketch shown in Figure CP1.8 with the completed CAD drawing shown in Figure CP1.9.

STEP–BY–STEP

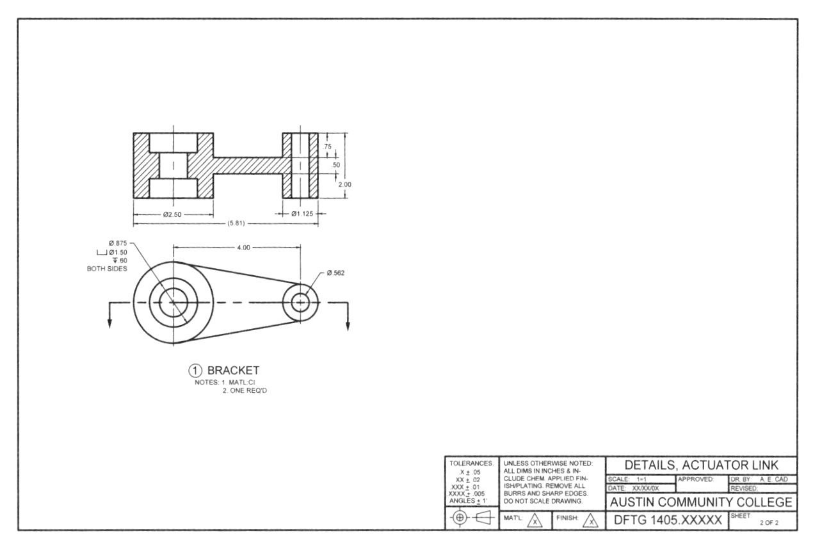

CP1.9 Completed Detail Drawing

PROJECT REVIEW

To be successful in the mechanical design field, drafters and CAD operators must be able to create working drawings consisting of both assembly and detail drawings. Assembly drawings illustrate how the parts in the assembly fit together and should include part numbers, balloons, and a parts list. Detail drawings should include all the views, dimensions, and notations required to manufacture each part of the assembly. Drafters and designers should also be familiar with specifying and representing threaded holes, shafts, and fasteners. It is important that the drafter represent every component of the assembly exactly as specified in the designer's inputs to ensure that the finished project is manufactured and assembled as intended.

Mechanical drafters must also be familiar with the drafting and dimensioning standards that control the format, dimensioning, and tolerancing of engineering working drawings.

KEY WORDS

Assembly Drawing
Mechanical Working Drawings
Detail Drawings
Parts List

REVIEW

Short Answer

1. Define what is meant by the term *assembly drawing*.
2. Define what is meant by the term *detail drawing*.
3. Explain the sheet numbering system used in mechanical working drawings.
4. Name four column headings that might be included on a parts list.
5. Explain how planning sketches are used in the creation of mechanical working drawings.

Threads and Fasteners in Mechanical Working Drawings

Mechanical drawings often include threaded holes, shafts, and fasteners (hex nuts, bolts, screws, etc.). The threads on these features are specified with a thread note detailing the size and type of thread.

The Toe Stop Assembly project calls for a hex cap screw and a mating hex nut (see Figure CP1.12). The thread specification for the screw is .75-10UNC-2A. This thread note is interpreted in the following way: the major diameter of the thread is **.75** inches, the number of threads per inch is **10**, the thread series is **Unified National Coarse (UNC)**, the thread class is **2**, and the **A** indicates it is an external thread. The thread specification for the mating hex nut is exactly the same, except the **A** is replaced with a **B** to indicate an internal thread. Figure CP1.10 shows how to interpret the thread specification for the hex cap screw.

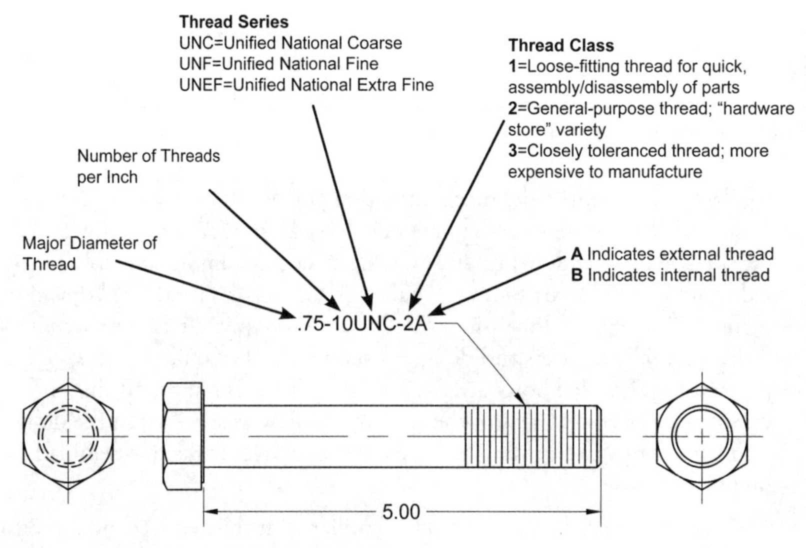

CP1.10 Interpreting a Thread Note for a Hex Cap Screw

The ASME standard governing the specification and dimensioning screw threads is ASME Y14.6-2001 (Reaffirmed 2007).

Representing External Screw Threads on Mechanical Drawings

There are three methods of representing external screw threads on a drawing: *schematic*, *detailed*, and *simplified*. The schematic method shown in Figure CP1.11 is widely used because it can be drawn quickly using AutoCAD tools such as **OFFSET** and **ARRAY**. The detailed method shown in Figure CP1.12 is used less frequently because it is more time consuming to construct. The simplified method shown in Figure CP1.13 is also commonly used to represent screw threads on mechanical drawings.

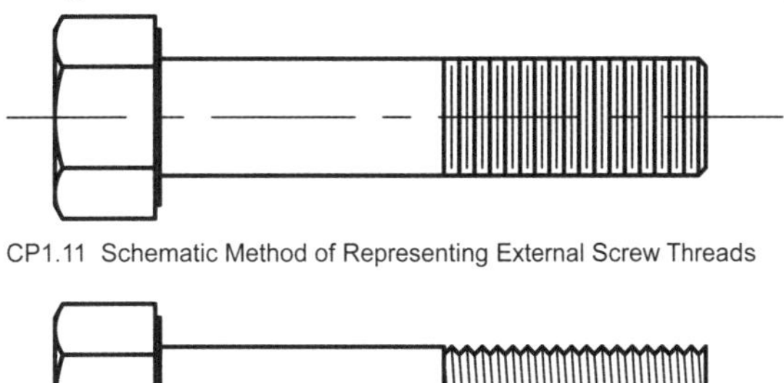

CP1.11 Schematic Method of Representing External Screw Threads

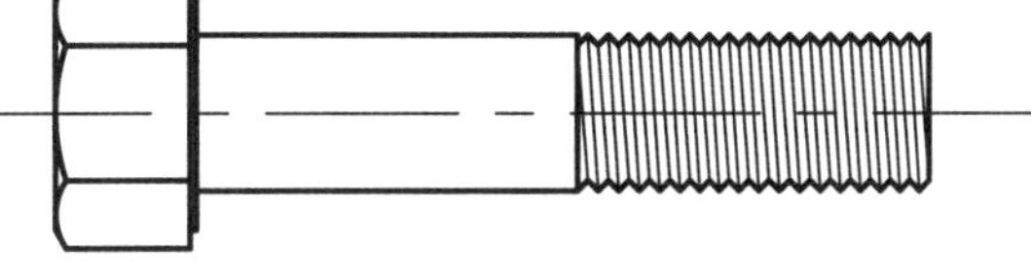

CP1.12 Detailed Method of Representing External Screw Threads

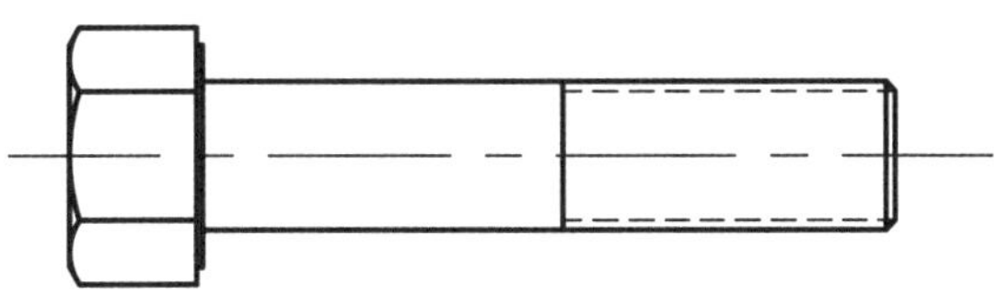

CP1.13 Simplified Method of Representing External Screw Threads

Representing Internal Screw Threads on Mechanical Drawings

The end view of a threaded hole is shown in Figure CP1.14(a). In this view the major diameter of the thread is represented by a circle drawn with a hidden line. The circle drawn with a visible line represents the minor diameter of the thread. The threaded hole is represented in the side view by four hidden lines, as shown in Figure CP1.14(b). These hidden lines are projected from the quadrants of the circles in the front view.

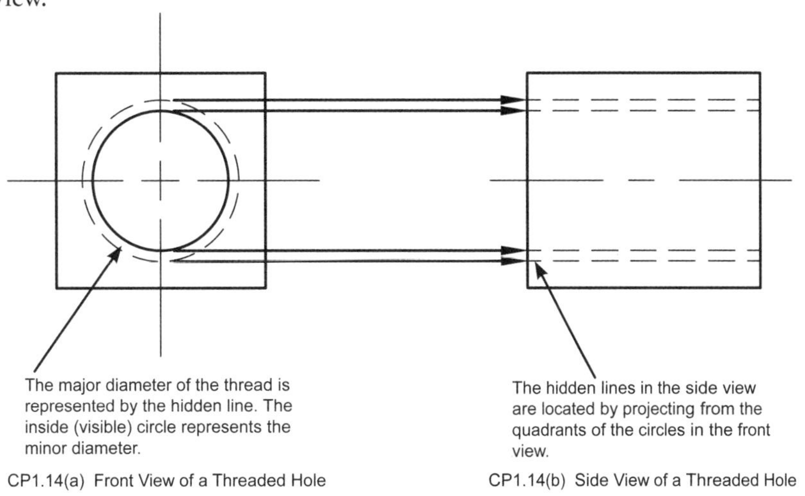

The major diameter of the thread is represented by the hidden line. The inside (visible) circle represents the minor diameter.
The hidden lines in the side view are located by projecting from the quadrants of the circles in the front view.

CP1.14(a) Front View of a Threaded Hole

CP1.14(b) Side View of a Threaded Hole

Create mechanical working drawings for the Toe Stop Assembly detailed in Figures CP1.15, CP1.16, and CP1.17. Draw the assembly as an exploded isometric view, or as an exploded 3D model, including balloons, part numbers, and a parts list. On the detail sheets, fully dimension each part and provide specifications for the standard hardware (hex screw, hex nut, and washer).

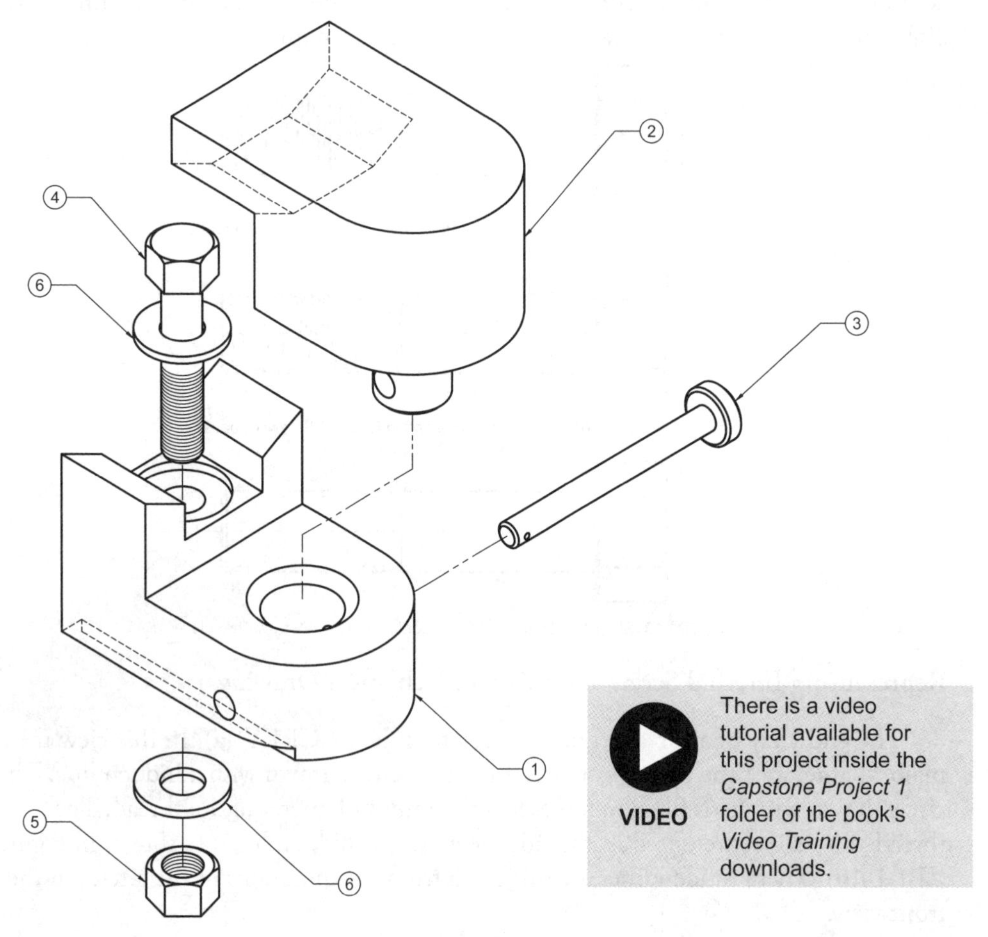

There is a video tutorial available for this project inside the *Capstone Project 1* folder of the book's *Video Training* downloads.

VIDEO

CP1.15 Toe Stop Assembly

PART NO.	PART DESCRIPTION	MATERIAL	QTY
6	TYPE A WASHER	SEE SHEET 3	2
5	HEX NUT	SEE SHEET 3	1
4	HEX CAP SCREW	SEE SHEET 3	1
3	CLEAT PIN	MILD STEEL	1
2	CLEAT	MILD STEEL	1
1	TOE STOP BASE	MILD STEEL	1
PARTS LIST			

CP1.16 Toe Stop Assembly Parts List

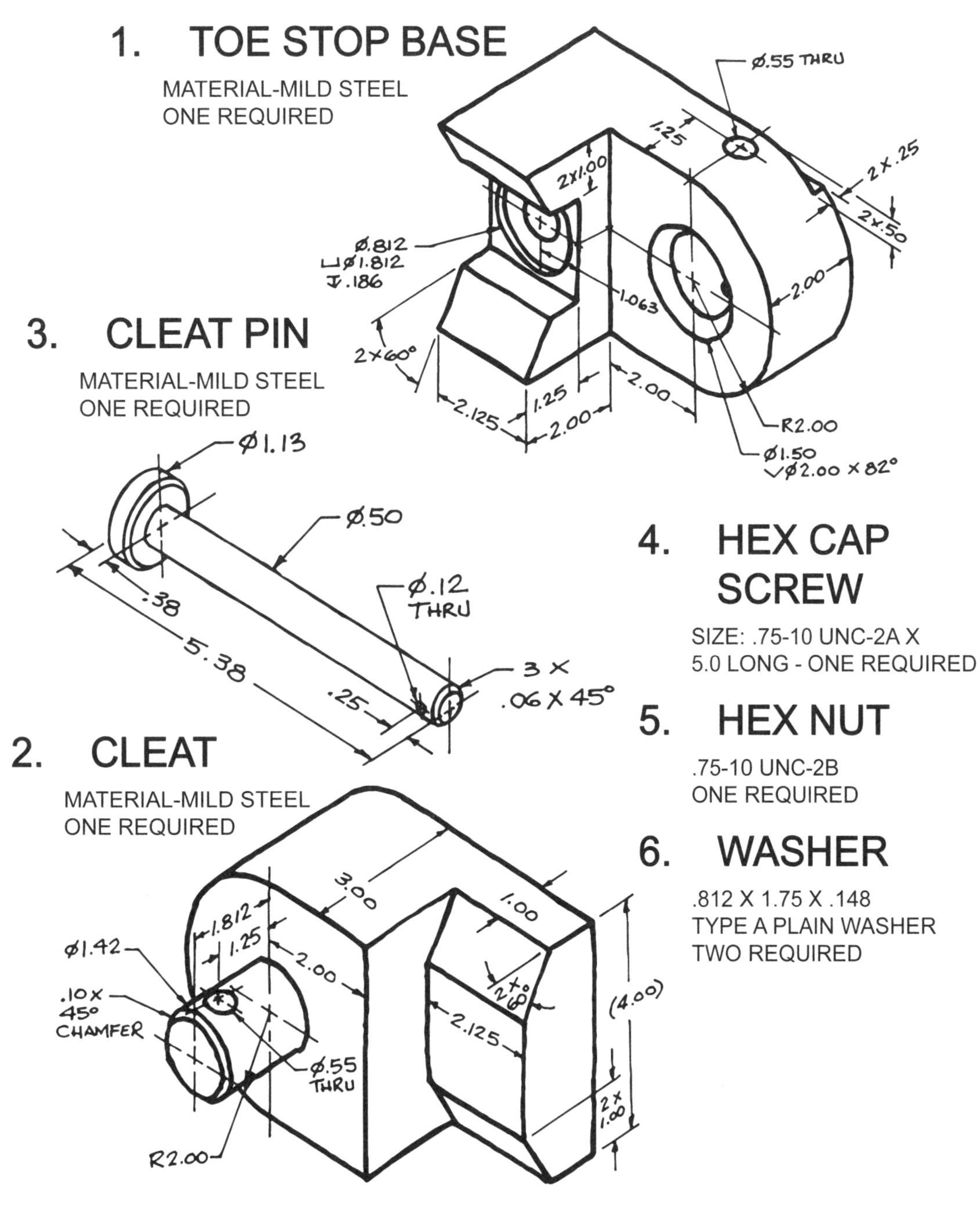

CP1.17 Designer's Sketch of Toe Stop Assembly Parts

NOTE

The specification for the washer called out in Figures CP1.17 and CP1.19 indicates the washer's inside diameter (.812), outside diameter (1.75), and thickness (.148). The thread specifications provided for the hex cap screw and hex nut are explained in the next section.

Follow the directions below to draw the multiviews and an isometric, exploded assembly of the Toe Stop Base, Cleat, and Cleat Pin. Refer to the planning sketches in Figures CP1.18, CP1.19, and CP1.20 for suggested sheet layouts.

- Note: The views will be drawn inside the magenta rectangles labeled *ASSEMBLY*, *DETAILS 1*, and *DETAILS 2* that are located in model space.

Step 1. Getting Started

- Download the **CP 1.1 Mechanical Working Drawing Prototype** drawing file located in the *Prototype Drawing Files* folder associated with this book.
- Use **SAVE AS** to save the drawing to your **Home** directory, and rename the new drawing **TOE STOP ASSEMBLY**.
- Activate the **Model** tab in the lower-left corner of the graphics window (if it is not already highlighted) by left-clicking on it.

Step 2. In the **Dimension Style Manager** dialog box, create two new dimension styles named **ASME Small Radii** and **ASME Large Radii** that contain the following dimension style settings:

Dimension Style Settings for the Toe Stop Assembly:	
Text Height	.15
Arrow Size:	.15
Center Marks:	Line
Extend Beyond Dim Lines:	.125
Precision:	Varies-match precision of dimensions on sketch
Zero Suppression:	Leading
Offset from Origin:	.062

a. For the **ASME Small Radii** style, set the **Fit** tab settings shown in Figure 5.57(a) on page 287.

- Use this dimension style when dimensioning small circles and arcs.

b. For the **ASME Large Radii** style settings refer to 5.57(c) on page 287.

- Use this dimension style when dimensioning large circles and arcs.

Step 3. Planning the Sheets

a. This project requires three D-size (34" X 22") sheets.

- **Note:** The D-size sheets can be printed to 17" X 11" or 11" X 8.5" sheets.

Step 4. Detail Sheet 1

a. Sheets 2 and 3 will contain the details (multiviews, dimensions, and notations) for each part.

b. Begin the project by drawing the first sheet of details.

- Figure CP1.18 shows a planning sketch created for Sheet 2.
- This sheet will be numbered Sheet 2 of 3.

c. This sheet will include the details of the Toe Stop Base and the Cleat Pin (Note: In most offices only one part is drawn per sheet).

d. The Toe Stop Base is considered the principal part in this assembly because all the other parts are aligned to or mate with it.

- Because it is the principal part, it will be identified as part number 1 in the parts list.

e. Begin this sheet by determining which view of the Toe Stop Base will be drawn as the front, or principal, view and then determine the other multiviews necessary to describe the part.

f. For the Toe Stop Base, a full section should be included to show the interior detail of the object.

g. Add the necessary dimensions and notations and follow the *ASME Y14.5* dimensioning guidelines described in Chapter 5.

- Note: This sheet should be drawn inside the magenta rectangle in model space labeled *DETAILS 1*.

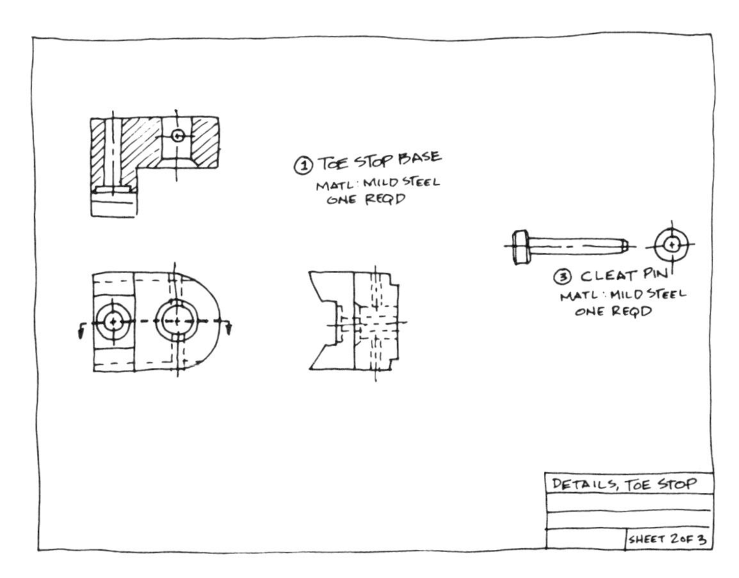

CP1.18 Sheet 2 of the Planning Sketch

PROJECT

Step 5. Detail Sheet 2

a. On this sheet you will draw the details of the Cleat and include the specifications for the hex cap screw, hex nut, and washer.

- A planning sketch for this sheet showing the suggested views is shown in Figure CP1.19.
- **Note:** This sheet should be drawn inside the magenta rectangle in model space labeled *DETAILS 2.*

b. Begin this sheet by determining which view of the Cleat will be drawn as the front, or principal, view and then determine the other required views.

c. Add the dimensions and notations required to manufacture the part.

d. This sheet is numbered sheet 3 of 3.

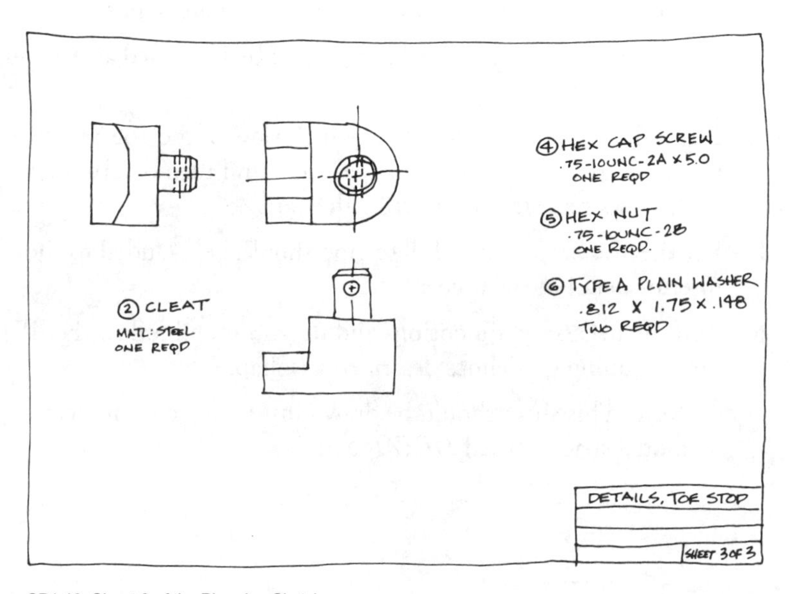

CP1.19 Sheet 3 of the Planning Sketch

Note in Figure CP1.19 that instead of creating detail drawings for the hex cap screw, hex nut, and washer only their specifications for size and quantity are provided. This is because these parts are standard hardware that can be purchased from an outside vendor. However, it is necessary to note on the drawing the part name, part number, specification, and quantity for each part so that the correct hardware can be ordered.

QUICK TIP

Exploded assemblies are often drawn *after the detail sheets are drawn* because drafters may need to transfer measurements taken from the multiviews on the detail drawings in order to construct isometric views of features such as inclined planes and the depths of countersunk holes.

Step 6. Exploded Assembly - Isometric Drawing Method

a. Sheet 1 will contain an exploded assembly (including a parts list, balloons, and part numbers).

- The exploded assembly including the parts list will be numbered Sheet 1 of 3.
- A planning sketch for this sheet is shown in Figure CP1.20.
- In this sketch the parts are sketched isometrically and positioned as they would look if the assembly was pulled apart.
- Even though this is Sheet 1 of the project, it is best to draw this sheet last because you will need to transfer the depth of the countersunk hole from the multiview drawn on Sheet 2 in order to construct its isometric view.
- The isometric assembly view should be drawn inside the magenta rectangle in model space labeled *ASSEMBLY*.
- Balloons, part numbers, and the parts list are added in paper space so click on the *Assembly* tab in the lower-left corner of the graphics window before adding or editing these items.

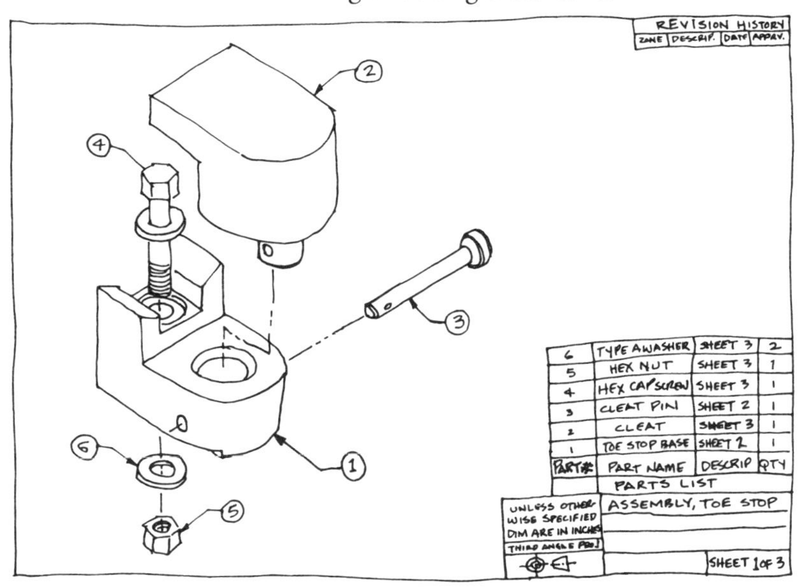

CP1.20 Planning Sketch for Sheet 1

STEP-BY-STEP

Drawing an Isometric Counterbored Hole

In creating the exploded assembly drawing for the Toe Stop Base, you will need to construct an isometric view of a countersunk hole and a counterbored hole. Figures CP1.21 through CP1.25 illustrate the steps in constructing an isometric counterbored hole.

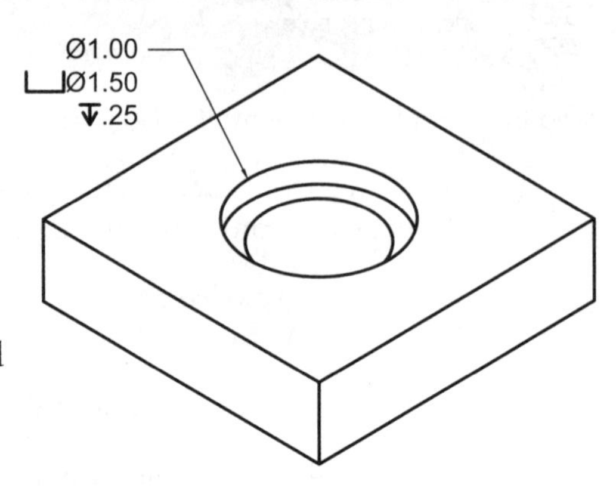

CP1.21 Isometric View of a Counterbored Hole

The specifications for this counterbored hole are shown in Figure CP1.21. In this example a **1.00"** diameter hole goes all the way through the object. Centered on this hole is a **1.50"** diameter counterbore that has a depth of **.25"**.

Step 1. Draw a **1.50"** diameter isometric ellipse, and copy it directly below the first ellipse at a distance of **.25"** as shown in Figure CP1.22.

Step 2. Trim the bottom ellipse to the edge of the top ellipse as shown in Figure CP1.23.

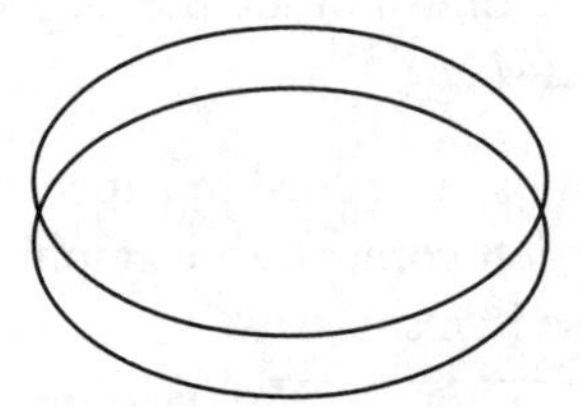
CP1.22 Isometric Ellipses

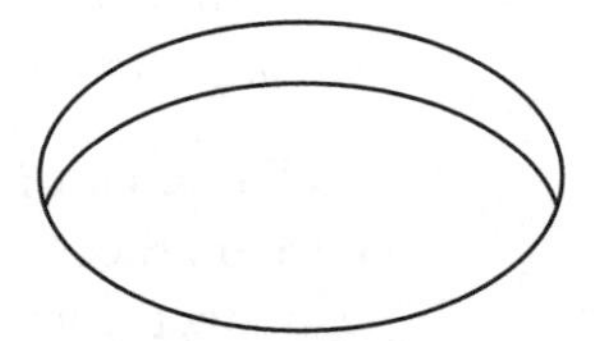
CP1.23 Ellipses after Trimming

Step 3. Draw a **1.00"** diameter ellipse at the center of the lower ellipse as shown in Figure CP1.24.

Step 4. Trim the part of the ellipse drawn in Step 3 that extends below the first ellipse as shown in Figure CP1.25. This completes the construction of the counterbored hole.

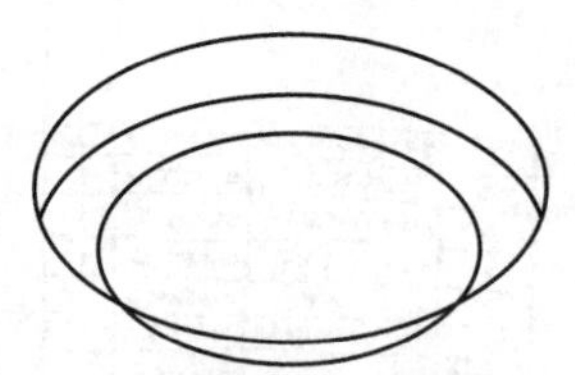
CP1.24 Adding the 1.00" Ellipse

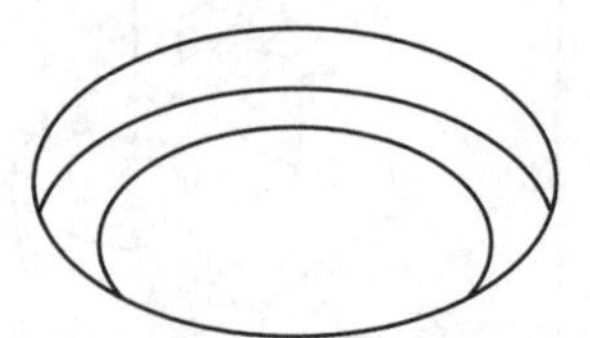
CP1.25 Finished Isometric Counterbored Hole

NOTE

When constructing the isometric counterbored hole for the Toe Stop Base, substitute the dimensions noted on the designer's sketch in Figure CP1.17 for those used in these examples.

STEP-BY-STEP

Drawing an Isometric Countersunk Hole

Figures CP1.26 through CP1.29 illustrate the steps in constructing an isometric countersunk hole.

The specifications for this countersunk hole are shown in Figure CP1.26. In this example a **1.00"** diameter hole passes all the way through the object. Centered on this hole is a **1.50"** diameter countersunk hole. The angle between the sides of the countersunk hole is **82°**.

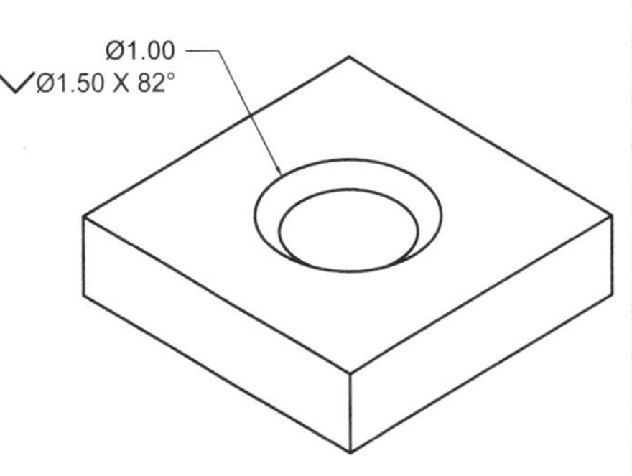

CP1.26 Isometric Countersunk

Step 1. Draw two concentric isometric ellipses, one with a diameter of **1.00"**, and the second with a diameter of **1.50"** as shown in Figure CP1.27.

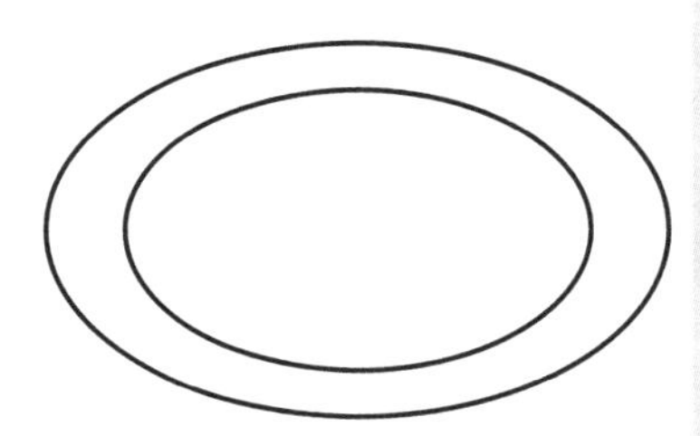

CP1.27 Isometric Ellipses

Step 2. Determine the value for **D** by measuring from the multiview of the countersunk hole's side view as shown in Figure CP1.28(a), and move the **1.00"** diameter ellipse straight down at distance **D** as shown in Figure CP1.28(b).

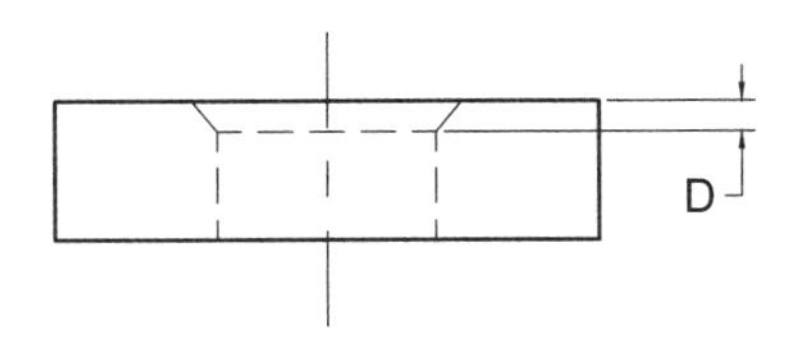

CP1.28(a) Finding the Depth (D) of the Countersunk Hole

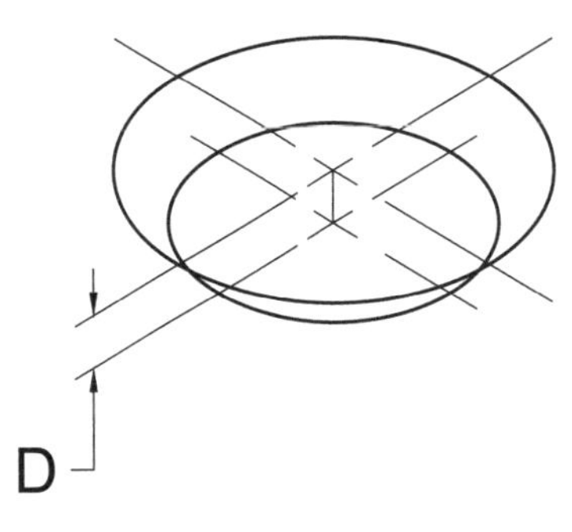

CP1.28(b) Transferring the depth (D) into the construction of the isometric view of hole

Step 3. Trim the smaller ellipse where it extends beyond the larger ellipse to complete the construction of the countersunk hole (see Figure CP1.29).

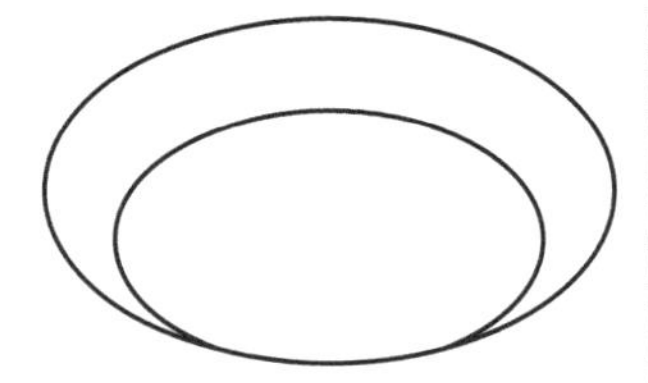

CP1.29 Completed Countersunk Hole

NOTE

When drawing the countersunk hole in the Toe Stop's exploded assembly, measure the depth of the countersink in either the top or side views of the Toe Stop Base's multiviews and transfer this distance (D) to the isometric view as shown in Figures CP1.28(a) and CP1.28(b).

Step 4. Constructing the Slots in the Isometric View of the Toe Stop Base

a. On each side of the bottom of the Toe Stop Base is a slot that is **.50"** wide by **.25"** tall. The construction for the isometric drawing of these slots is shown in Figure CP1.30(a). Figure CP1.30(b) shows the Toe Stop Base after the construction lines for the slots have been trimmed.

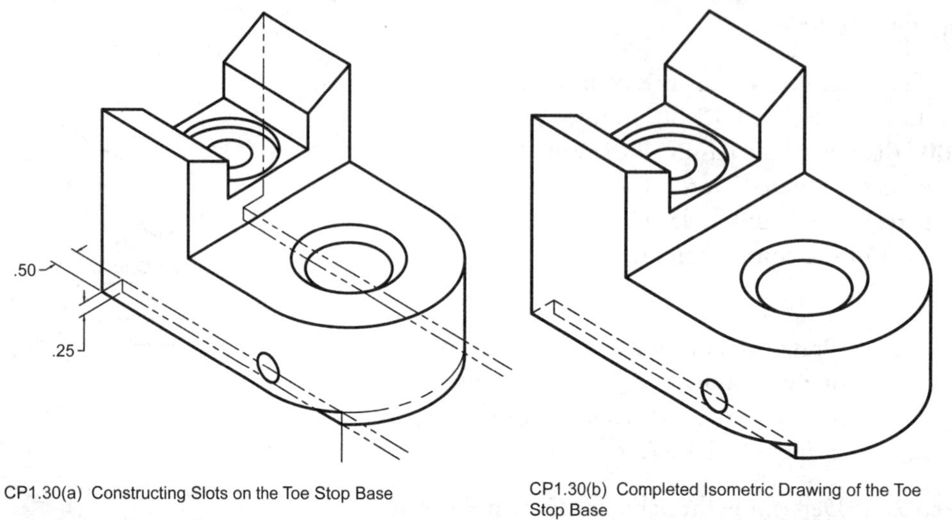

CP1.30(a) Constructing Slots on the Toe Stop Base

CP1.30(b) Completed Isometric Drawing of the Toe Stop Base

NOTE

Hidden lines showing the profile of the slot in the front side of the Toe Stop Base have been shown for clarity in Figure CP1.30(b).

Step 5. Completing the Title Block and Parts List and Plotting the Sheets

a. When finished drawing the assembly and multiviews of the parts, left-click on the **Layout** tabs named **Assembly**, **Details 1** and **Details 2** in the lower-left corner of the graphics window.

b. Edit the text in the title block of each one by double-clicking on it to enable the text edit option.

c. On the *Assembly* Layout edit the parts list and add balloons as needed.

d. Follow your instructor's directions to plot the sheets.

- If your plotter or printer allows, each sheet can be plotted on a 17" X 11" sheet using a Viewport Scale of **1=2** (half size).

e. Sheets should be plotted using the *monochrome* plot style setting.

QUICK TIP

Follow the steps presented in Chapter 4 in **"4.18 Creating a Page Setup for Plotting"** to create a **Page Setup** for plotting each sheet of this project.

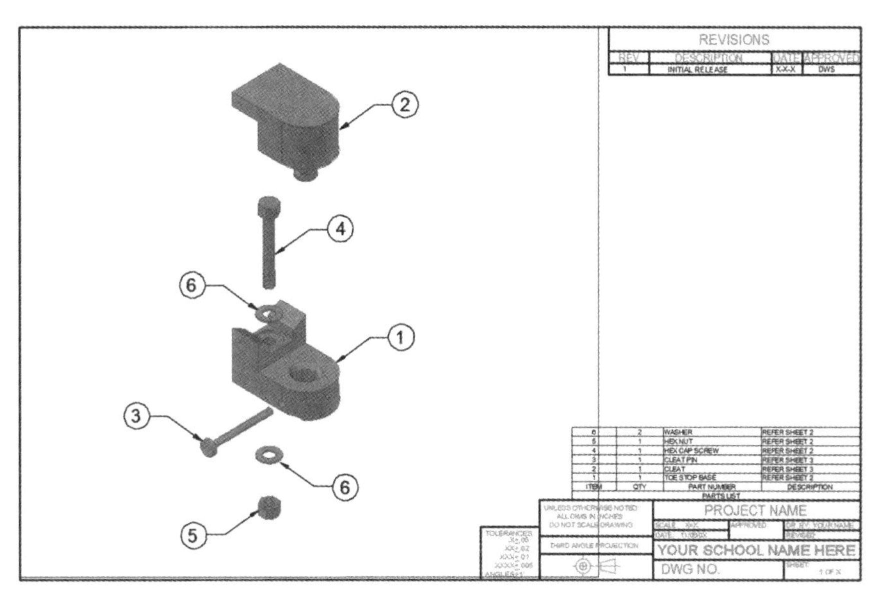

CP1.31 Toe Stop 3D Assembly Example

OPTIONAL: Exploded Assembly - 3D Modeling Method

Instead of drawing an exploded isometric view of the assembly, use the 3D modeling techniques presented in Chapter 10 to construct an exploded 3D model of the Toe Stop assembly. The completed project will resemble the example in Figure CP1.31.

Step 6. Getting Started

a. Download the **Toe Stop Assy 3D Prototype** drawing file located inside the *Prototype Drawing Files* folder associated with this book.

b. Use **SAVE AS** to save the drawing to your **Home** directory.

c. Rename the drawing **TOE STOP 3D ASSY**.

d. Activate the **Model** tab in the lower-left corner of the graphics window (if it is not already highlighted) by left-clicking on the tab.

Step 7. Using the **Insert** command, insert the **TOE STOP ASSEMBLY** drawing you created in earlier in this project into the **TOE STOP 3D ASSEMBLY** drawing.

a. Use 0,0 as the insertion point.

- **Note:** After inserting the drawing, you may need to use the **Explode** command to explode the views before you can edit them.

Step 8. Change AutoCAD's Workspace setting to 3D Modeling and edit the 2D views as necessary to create the profiles needed to model the parts and use the **PRESSPULL**, **EXTRUDE**, **REGION**, **REVOLVE**, **ROTATE 3D**, **UNION**, and **SUBTRACT** commands to create 3D models of the Toe Stop Base, Cleat, and Cleat Pin. Refer to Figures CP1.32, CP1.33, and CP1.34 to assist you with creating the 3D models from the 2D drawings.

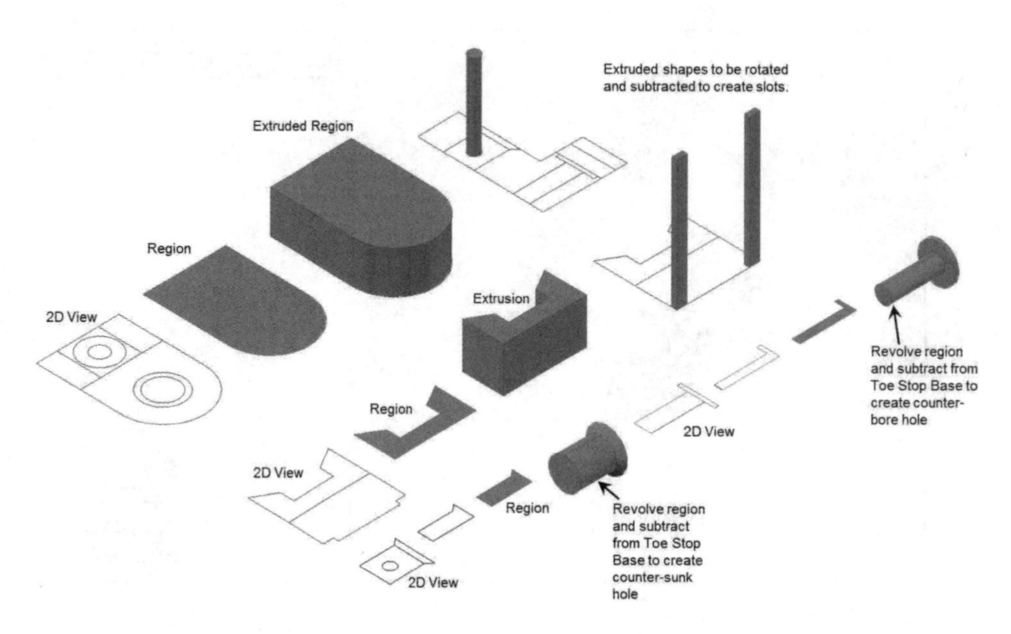

CP1.32 Creating 3D Models of the Toe Stop Base from the 2D Views

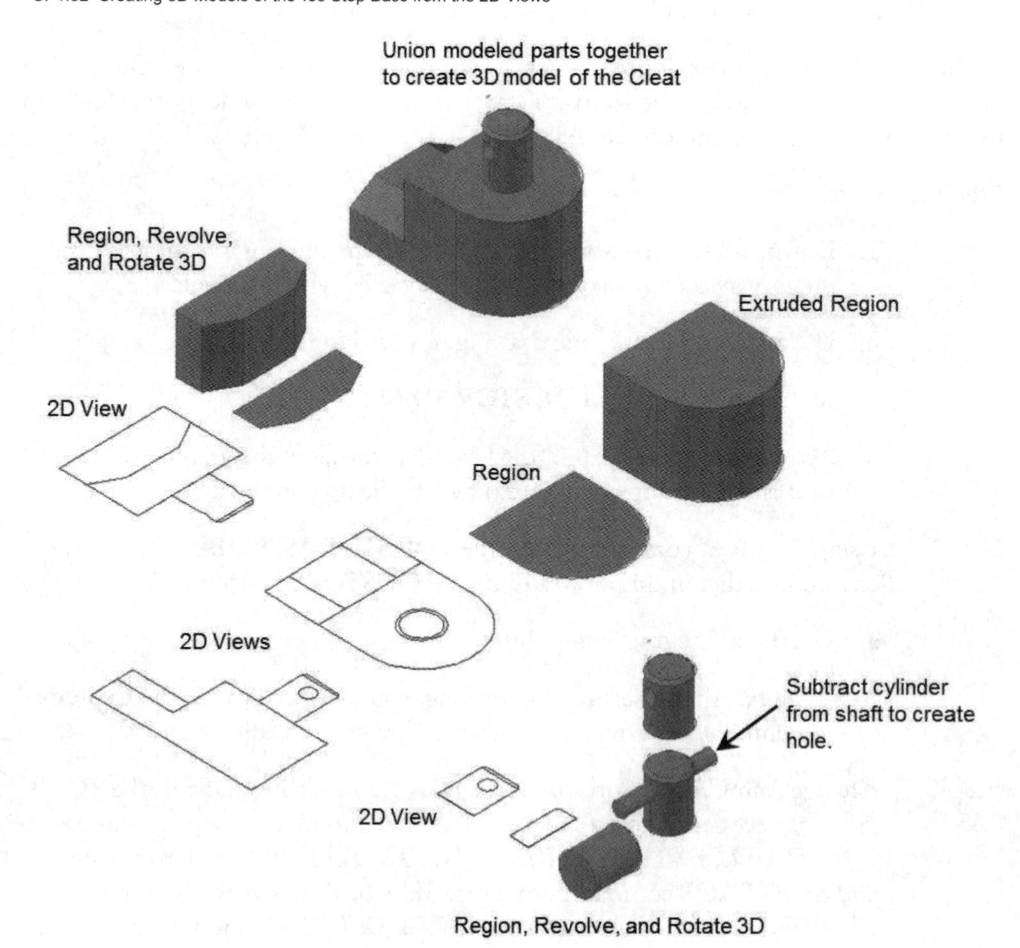

CP1.33 Creating 3D Models of the Cleat from the 2D Views

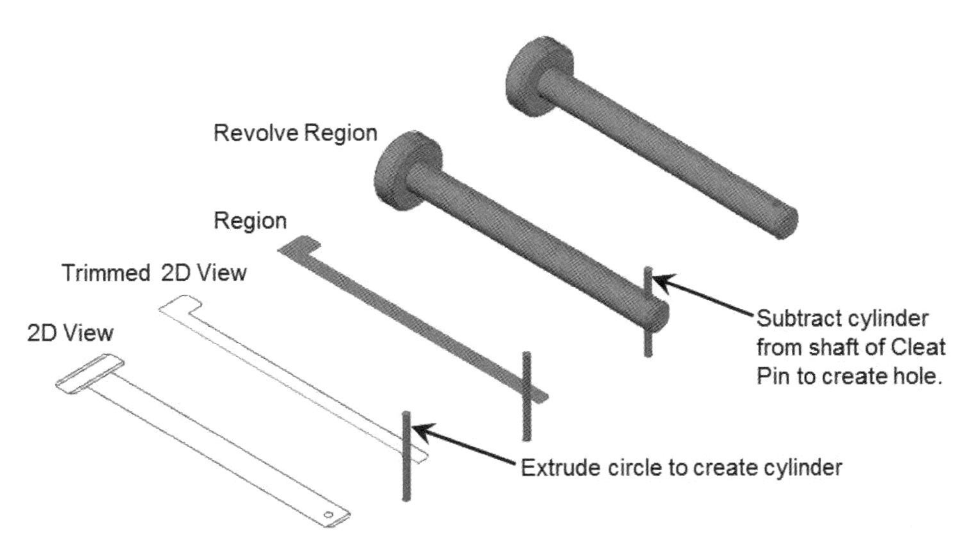

CP1.34 Creating 3D Models of the Cleat Pin from the 2D Views

Step 9. When finished creating the 3D models, align them as shown in Figure CP1.31 to create an exploded assembly view.

a. Insert the pre-made blocks of the 3D Hex Head Screw, 3D Hex Nut, and 3D washer that are included in the prototype drawing and align them as shown in Figure CP1.31.

Step 10. Toe Stop Assembly Using Base View

a. Click on the tab at the lower left corner of the graphics window that is labeled Assembly (see Figure CP1.35).

b. Click the Layout tab located on the ribbon.

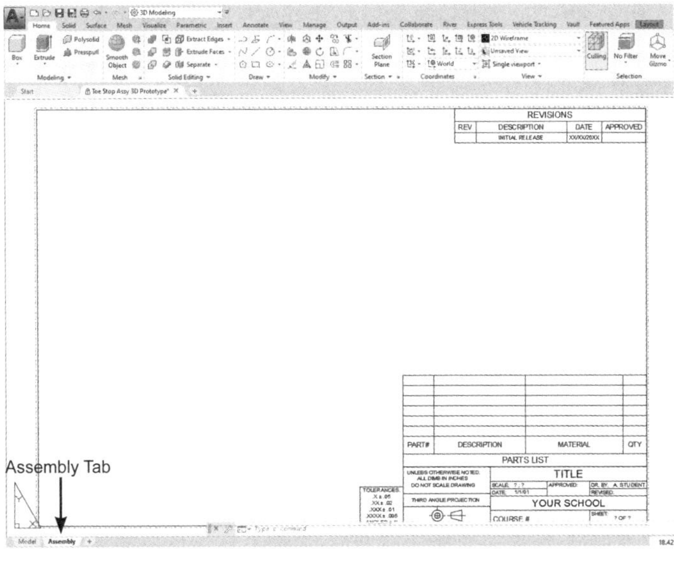

CP1.35 Assembly Layout

c. Pick the Base icon

- When the drop down menu opens select the From Model Space option (see Figure CP1.36).

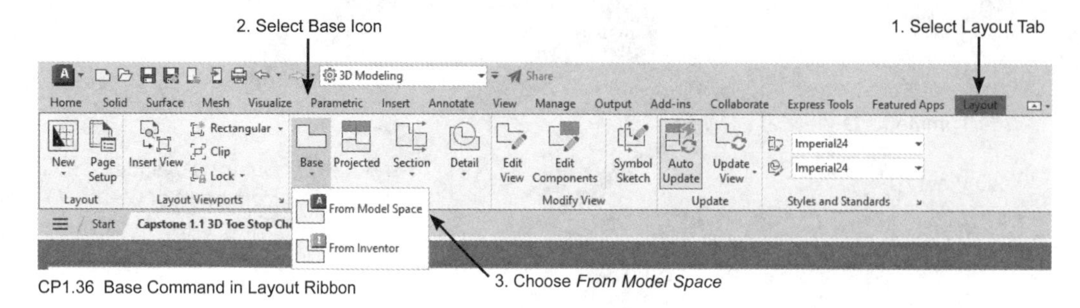

CP1.36 Base Command in Layout Ribbon

d. When prompted to Specify Location of base view pick a point near the center of the open area located to the left of the parts list.

e. Click on the Orientation option from the menu located next to the view and select the NE ISO option (North East Isometric).

f. Click on the Hidden lines option from the menu and select the Shaded with visible lines option.

g. Click on the Scale option from the menu and type 1/3 or .33 and press <Enter> (this sets the scale of the view to 1/3 of full size).

h. To end the Base View, click eXit from the menu and press <Enter>. Note: the Move command can be used to move the view if necessary.

Step 11. Parts List

a. Complete the parts list and title block by editing the text in the fields of the parts list and title block.

Step 12. Add Balloons

a. Using the Multileader tool (select the Balloon style), add balloons to the parts in paper space as shown in Figure CP1.37.

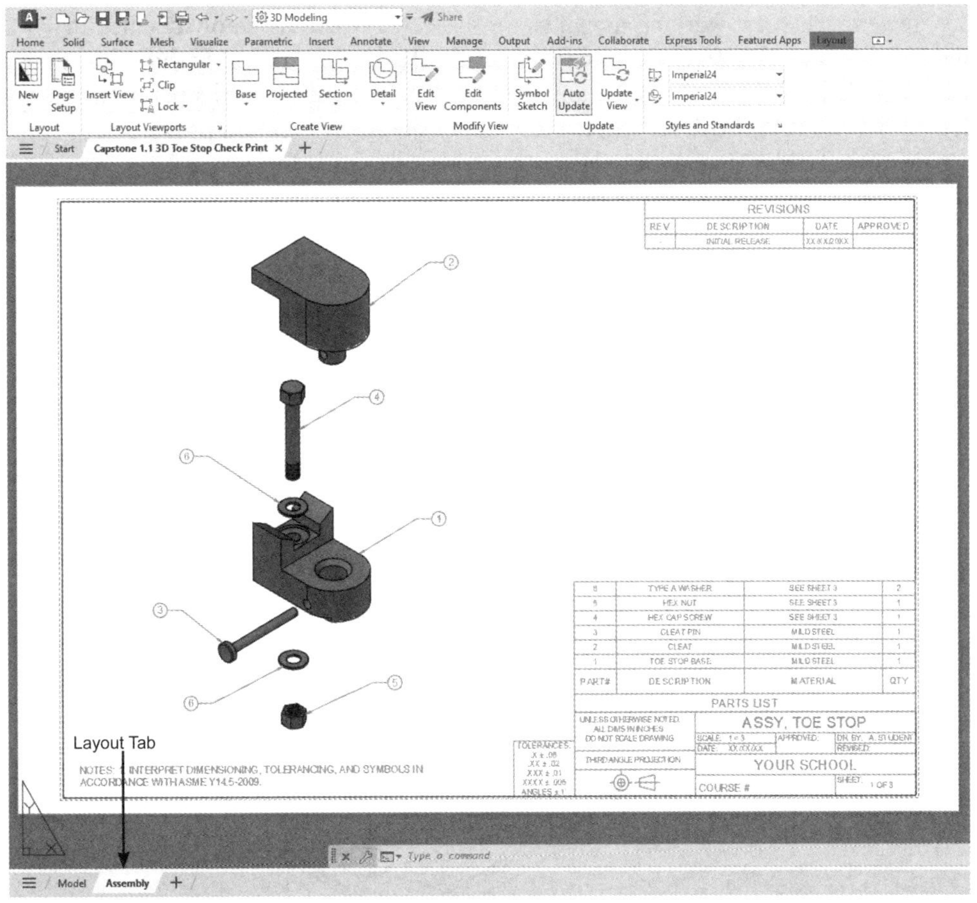

CP1.37 Formatting the 3D Assembly Model in the Layout Tab

Step 13. Plot the Drawing

Follow your instructor's directions to print the sheet.

a. Select the **Plot** command and select layout at the *What to plot* prompt in the Plot dialog box.

b. If your plotter or printer allows, each sheet can be plotted on a 17" X 11" sheet using a Viewport Scale of **1=2**.

c. Sheets should be plotted using the *monochrome* plot style setting.

Create mechanical working drawings for the Test Fixture Assembly detailed in Figures CP1.38, CP1.39, CP1.40, CP1.41 and CP1.42. The assembly should be drawn as an exploded isometric view including balloons, part numbers, and a parts list (see Figure CP1.39). On the detail sheets, fully dimension each part and provide specifications for the standard hardware (hex screw, hex nut, and washer).

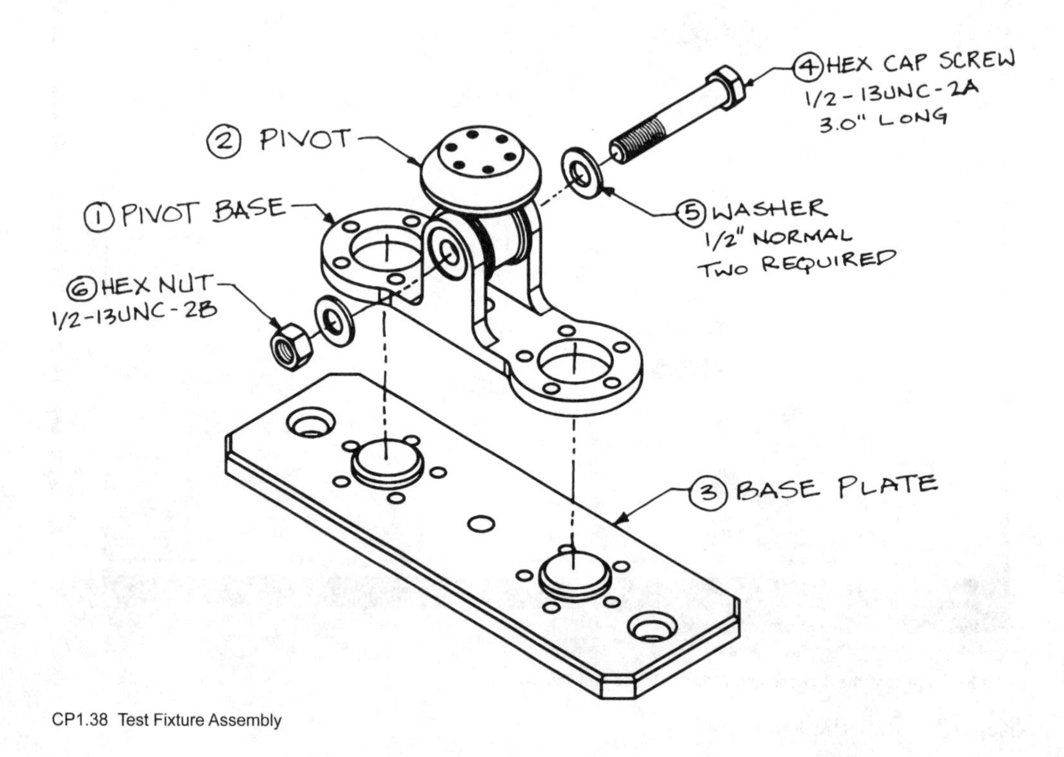

CP1.38 Test Fixture Assembly

6	HEX NUT	1	SEE SHEET 3
5	WASHER	2	SEE SHEET 3
4	HEX CAP SCREW	1	SEE SHEET 3
3	BASE PLATE	1	SEE SHEET 4
2	PIVOT	1	SEE SHEET 3
1	PIVOT BASE	1	SEE SHEET 2
FIND NO.	PART NUMBER	ITEM QTY	DESCRIPTION
PARTS LIST			

CP1.39 Test Fixture Assembly Parts List

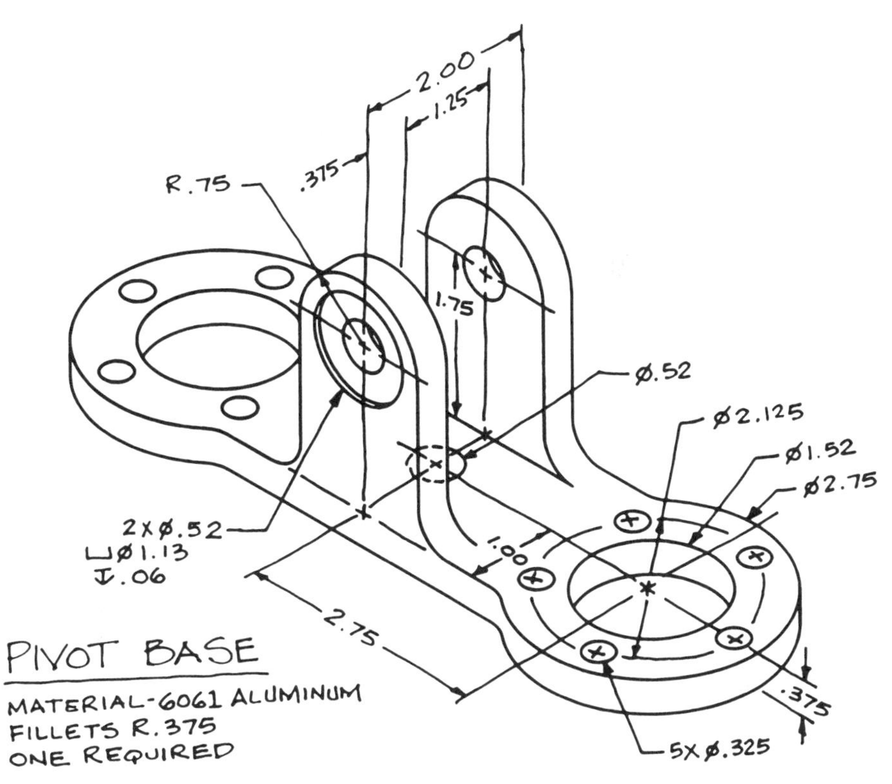

CP1.40 Designer's Sketch of the Pivot Base

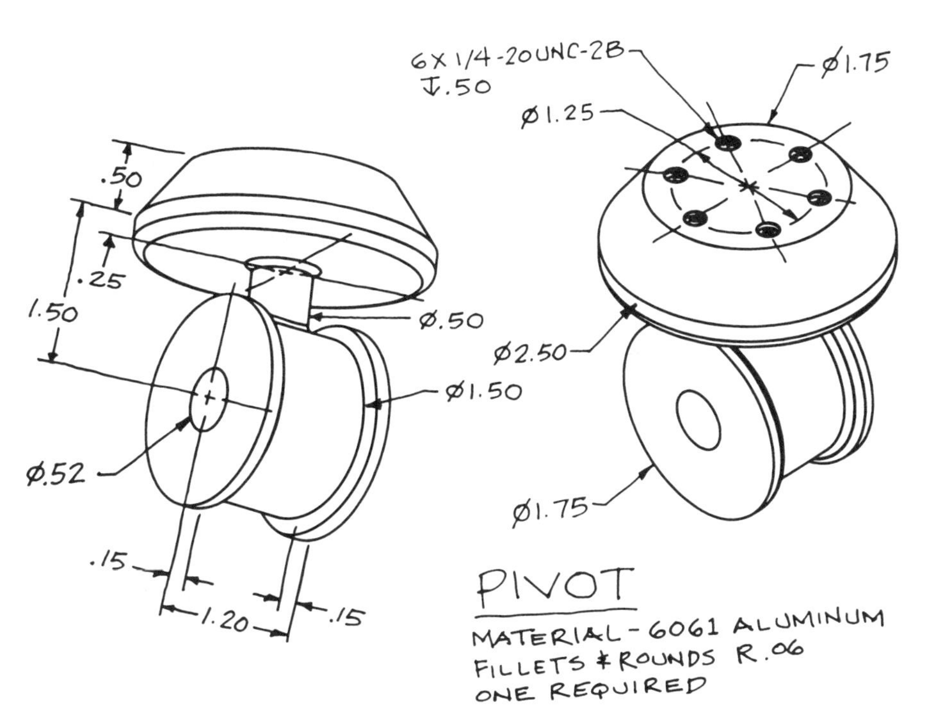

CP1.41 Designer's Sketch of the Pivot

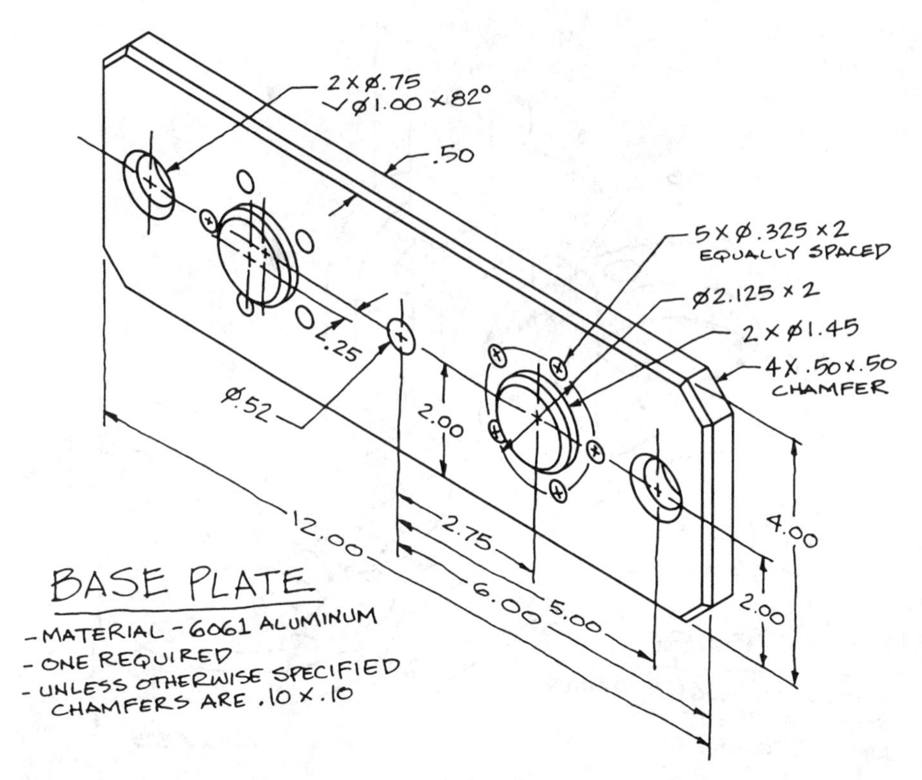

CP1.42 Designer's Sketch of the Base Plate

Step 1. Getting Started

a. Download the **CP 1.2 Mechanical Working Drawing Prototype** drawing file located in the *Prototype Drawing Files* folder associated with this book.

b. Use **SAVE AS** to save the drawing to your **Home** directory.

c. Rename the new drawing **TEST FIXTURE ASSEMBLY**.

d. Activate the **Model** tab in the lower-left corner of the graphics window (if it is not already highlighted) by left-clicking on it.

Step 2. Draw and dimension the necessary views of the Pivot, Pivot Base, and Base Plate.

a. Create a full section view of the Base Plate.

b. Draw and dimension the views inside the magenta rectangles labeled *DETAILS 1* and *DETAILS 2*.

c. Create two new dimension styles using the same settings defined in Project CP1.1 (Toe Stop Assembly Project).

Step 3. Draw an exploded isometric assembly showing all the parts in the assembly.

a. This exploded assembly should be drawn inside the magenta rectangle in model space labeled *ASSEMBLY.*

b. Insert the pre-made blocks representing the isometric views of the hex cap screw, hex nut, and washer contained in the prototype drawing.

c. Balloons, find numbers, and the parts list are added in paper space.

- Click on the *Assembly* tab in the lower-left corner of the graphics window to enter paper space before adding or editing these items.

Step 4. Completing the Title Block and Parts List and Plotting the Sheets

a. When finished drawing the assembly and multiviews of the parts, left-click on the **Layout** tabs (**Assembly**, **Details 1** and **Details 2**) in the lower-left corner of the graphics window.

b. Edit the text in the title block by double-clicking on it to enable the text edit option.

c. Click on the *Assembly* Layout tab.

- Edit the parts list, and add balloons as needed.

Step 5. Follow your instructor's directions to plot the sheets.

a. If your plotter or printer allows, each sheet can be plotted on a 17" X 11" sheet using a Viewport Scale of **1=2**.

b. Sheets should be plotted using the *monochrome* plot style setting.

QUICK TIP

Follow the steps presented in Chapter 4 in **"4.18 Creating a Page Setup for Plotting"** to create a **Page Setup** for plotting each sheet of this project.

CAPSTONE PROJECT 2

ARCHITECTURAL WORKING DRAWINGS

PROJECT OBJECTIVES:

After studying the material in this chapter, you should be able to:

1. Describe architectural working drawings and their importance to the field of architecture.
2. Describe how floor plans and elevation drawings are planned and prepared.
3. Use AutoCAD to create a floor plan for a small house.
4. Use AutoCAD's **DesignCenter** to place blocks of electrical and plumbing symbols into the floor plan.
5. Use AutoCAD to create elevation drawings for a small house.

PROJECT OVERVIEW

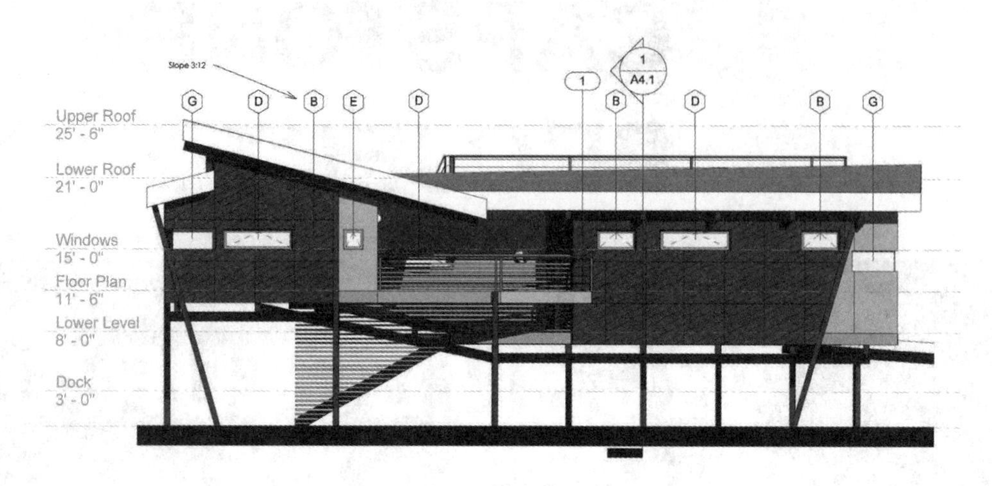

ARCHITECTURAL WORKING DRAWINGS:

In many architectural offices, drafters work with architects and designers to prepare the drawings used in the construction of residential and commercial buildings. These drawings, which may include floor plans, elevations, foundations, wall sections, and roof framing plans, are called **construction documents (CDs)**. Often, the separate sheets for a full set of plans are created on different CAD layers within the same CAD drawing file so that the drafter can selectively view and print the layers as needed.

CP2.1 FLOOR PLANS

Floor plans, like the one shown in Figure CP2.1, provide home builders and contractors with the necessary information to lay out the building, including the locations of features such as walls, doors, electrical components (switches, lamps, etc.) and plumbing fixtures (tubs, toilets, sinks, etc.). Floor plans should include all the dimensions and notations required by the workers on the job site. Doors and windows are dimensioned to their centers, and continuous (also known as chain) dimensioning is typically employed on floor plans. Dimensions are labeled above the dimension line, and tick marks replace arrowheads.

Architectural firms create or purchase block libraries of doors, windows, electrical, plumbing, and other symbols frequently used on floor plans.

The efficient use of blocks and layering techniques by drafters can increase productivity and lower the cost of creating a set of plans.

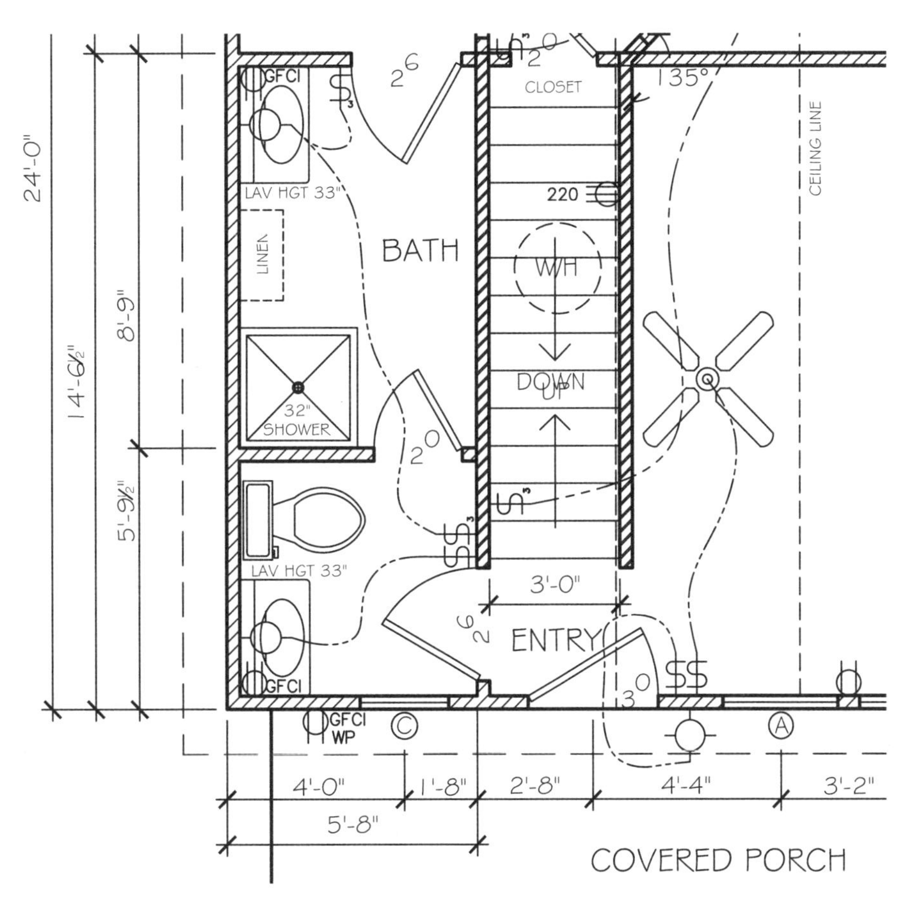

CP2.1 Detail from a Floor Plan Including Dimensions, and Electrical and Plumbing Symbols

CP2.2 ELECTRICAL PLANS

An **electrical plan** provides electrical contractors information about the type, location, and installation of electrical components (switches, lamps, ceiling fans, electrical outlets, cable TV jacks, etc.) used in the project. All the information needed by the electrical contractor to wire the building should be provided by this plan.

Drafters can create a block library of electrical symbols to speed the process by which electrical plans are created. The electrical components and wiring are usually drawn on a separate layer that is superimposed on the floor plan layer(s). Figure CP2.2 shows a detail from an electrical plan. A legend is included on the electrical plan to help workers identify all the components on the floor plan. Figure CP2.3 shows an example of an electrical legend.

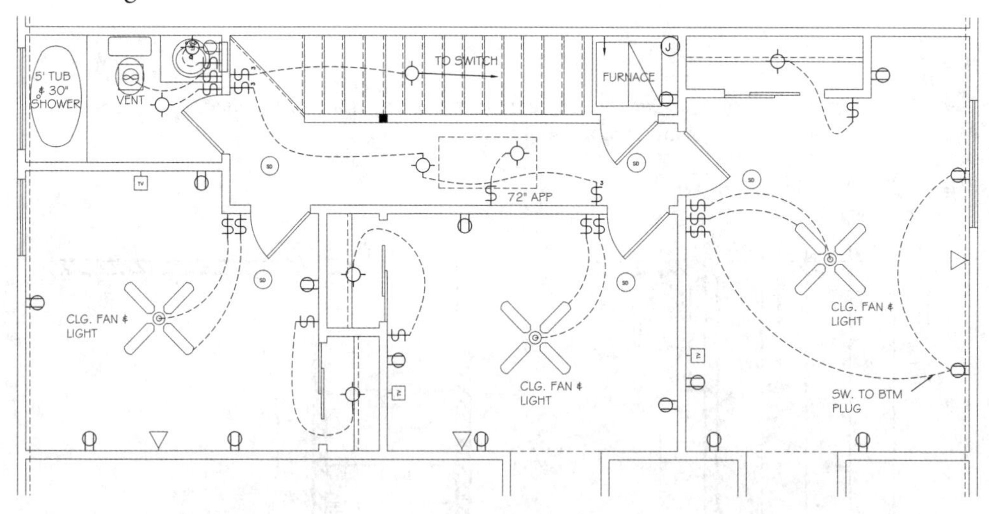

CP2.2 Electrical Plan Detail

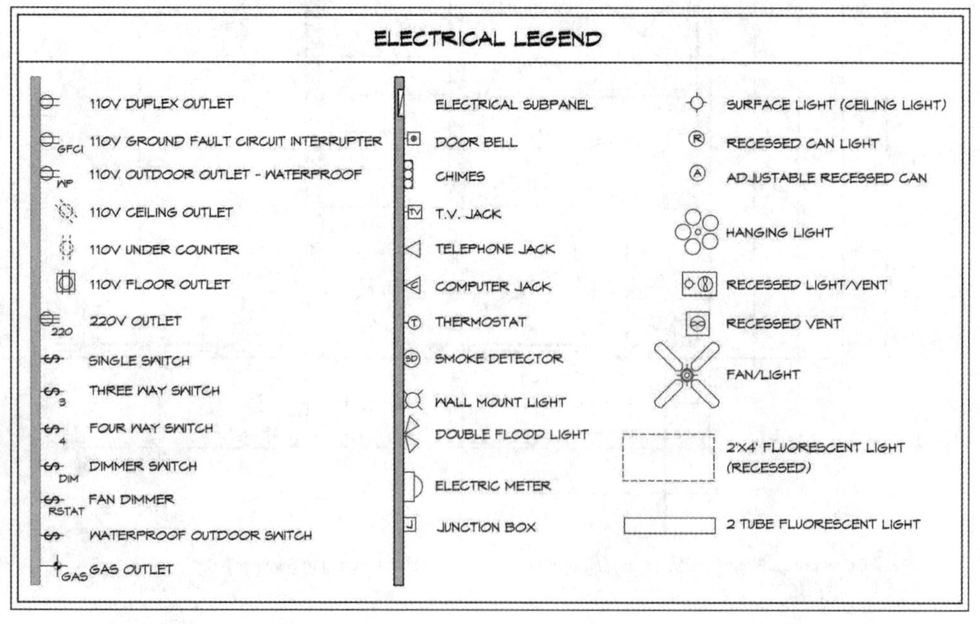

CP2.3 Electrical Legend

CP2.3 ELEVATIONS

Elevation drawings provide information about the exterior details of a building. This information may include roof pitch, exterior materials and finishes, overall heights of features, and window and door styles, as shown in Figure CP2.4. All the dimensions and notations required by workers on the job site should be included on this sheet.

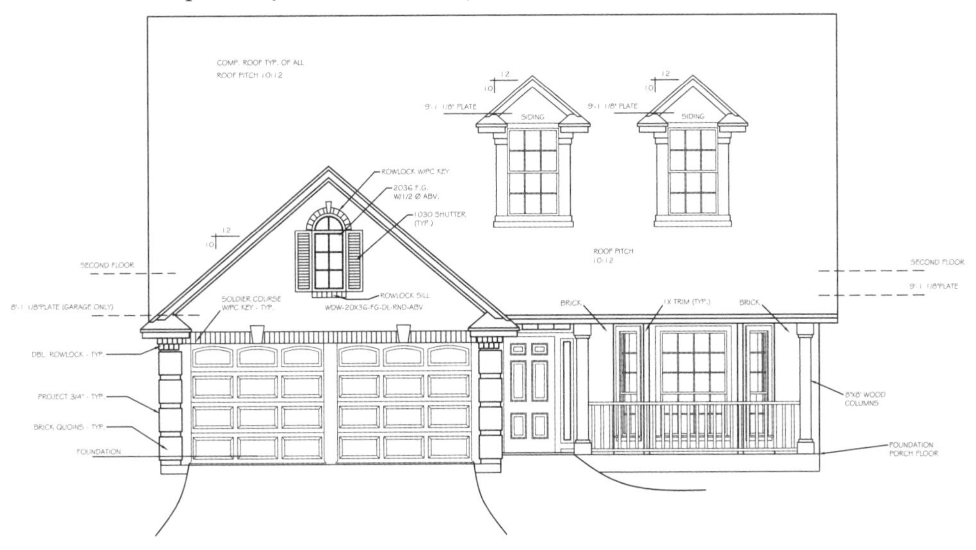

CP2.4 Front Elevation of a House

Creating Elevations Using Multiview Drawing Techniques

In Figure CP2.5 lines and arrows have been drawn between the views to show how the location and size of features on the house's exterior can be projected from one view to another using multiview drawing techniques. In fact, sometimes it is not possible to complete the construction of one elevation view without constructing an adjacent elevation and projecting information from the new elevation back to the original view.

For example, in Figure CP2.5 it would not be possible to locate the top edge of the roof plane in either of the side elevations without first drawing the front elevation and projecting the roof peak to the side elevations.

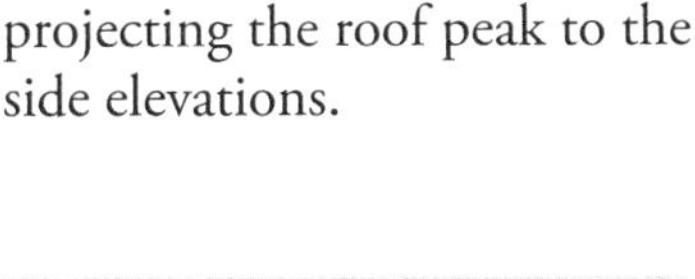

NOTE Although Figure CP2.5 shows the top view of the house, this view is not included on the elevations sheet because the top view reflects the *roof plan* of the building, which is typically drawn on a separate sheet.

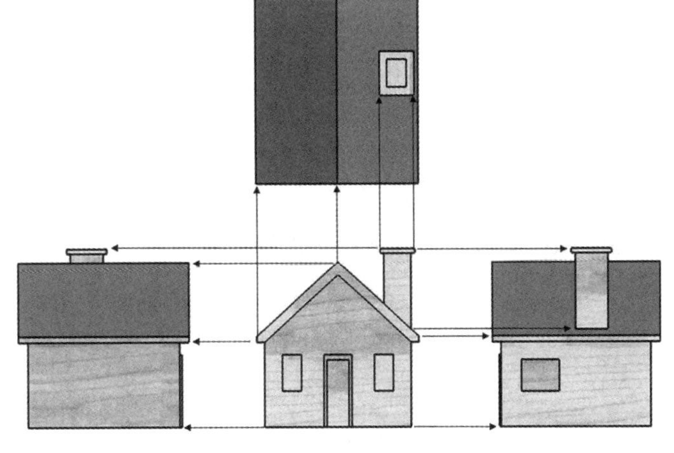

CP2.5 Projecting Points and Planes between Views of a House Using Multiview Drawing Techniques

Architectural Wall Sections

In architectural drawings, *wall sections* are often included to specify the composition of a wall as shown in Figure CP2.6. Drafters often refer to an exterior wall section when determining roof angles, overhangs of rafters, and heights of walls and ceilings in the elevation view.

Sections are also included on *foundation* plans to show interior details of the composition of foundation beams, slabs, and footings.

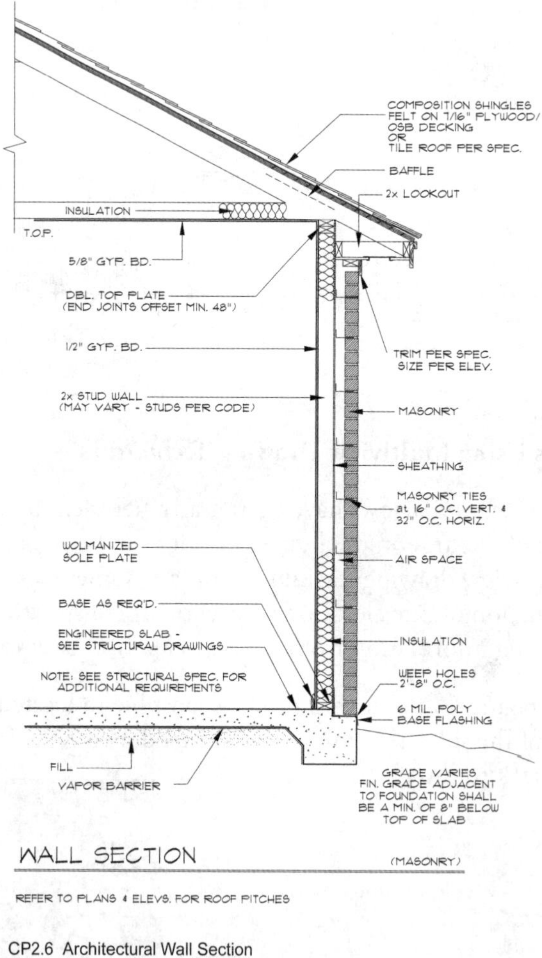

CP2.6 Architectural Wall Section
Image courtesy of Kipp Flores Architects

Roof Profiles on Architectural Elevations

The angle of a roof is called its **roof pitch**. Pitch is specified as a ratio of the vertical *rise* of the roof (measured in inches) to the horizontal **run** of the roof (measured in inches). Using this notation, a roof with a "four-twelve" pitch (labeled as **4/12** on the drawing) would rise 4" for every 12" of horizontal run. A roof with a **12/12** pitch would rise 12" for every 12" of run. A **12/12** pitch would result in a roof angle of 45°.

A roof pitch symbol is created by drawing a horizontal line that is crossed near one end by a vertical line like the ones shown along the roof profiles in Figure CP2.7. The rise is labeled next to the vertical line, and the run (usually 12") is noted above the horizontal line.

In the roof profile shown in Figure CP2.7, for every 12" the roof runs along its horizontal axis, it rises 10". A drafter would label this pitch specification as **10/12** on the drawing.

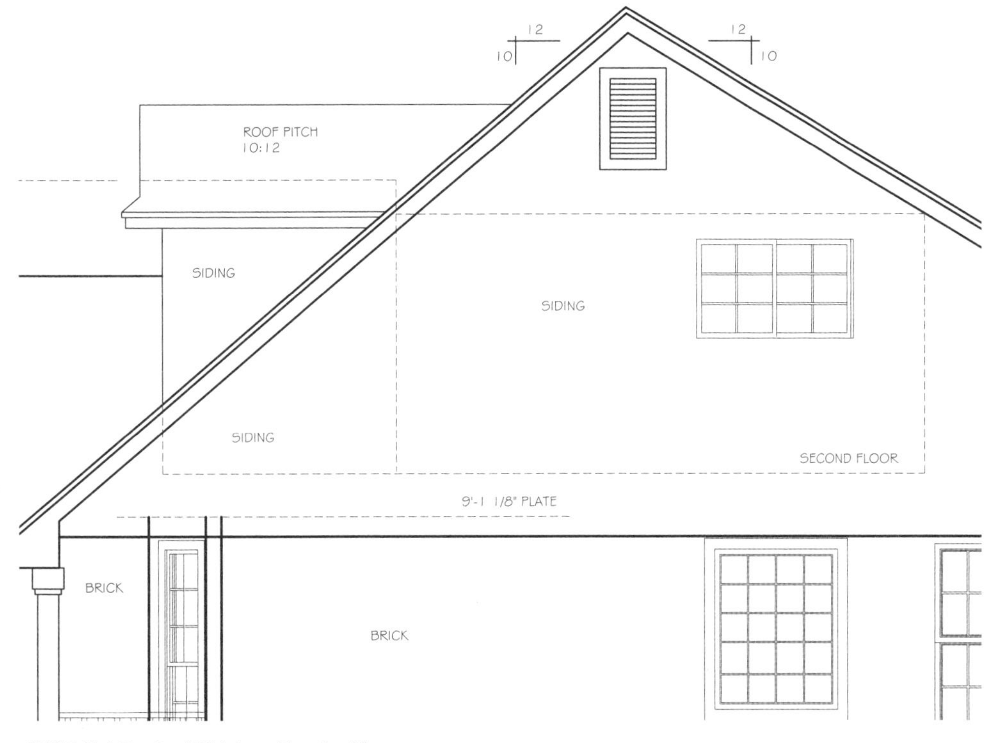

CP2.7 Notating Roof Pitch in an Elevation View

Using the Floor Plan to Locate Features on Elevations

When elevation drawings are created, the physical location of features on the floor plan, such as doors and windows, can be used to locate these same features in the elevation drawing.

Figure CP2.8 shows the floor plan of the guest cottage you drew earlier in this course. Also shown in this figure are the front, back, and side elevations of the guest cottage. To speed the creation of the elevation views, information about the size and location of the windows, outside walls, and the front door was projected from the floor plan to the elevation views.

Locate the 45° miter lines in the corners of Figure CP2.8 and note how information is projected among the front, back, and side views through these miter lines.

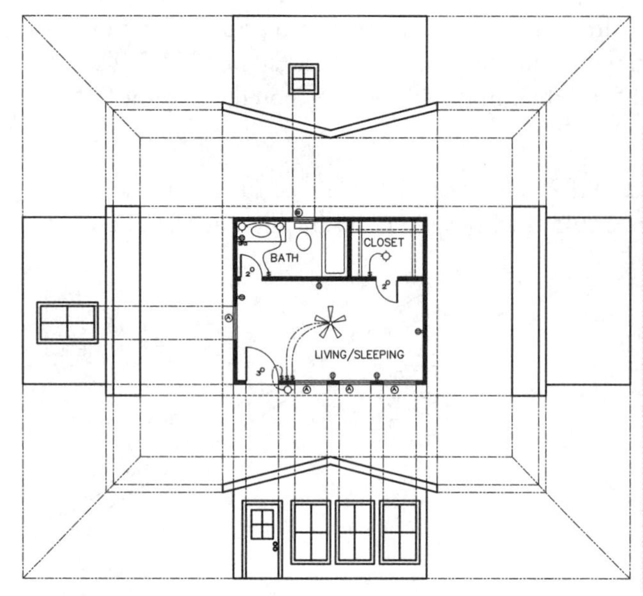

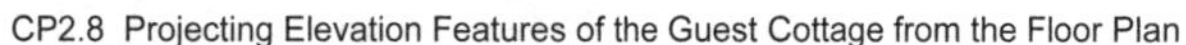

CP2.8 Projecting Elevation Features of the Guest Cottage from the Floor Plan

JOB SKILL

The multiview projection technique shown in Figure CP2.8 is often used by architectural drafters to create elevation views. This technique involves the creation of *named* views in AutoCAD. Named views allow drafters to quickly rotate the side and back views to the upright position for easier editing. Creation of named views is often covered in intermediate CAD courses.

PROJECT REVIEW

In the early stages of an architectural project, designers and clients work together to produce a design that meets the client's needs and budget. During the design stage, the designer may communicate with the client through sketches, rendered CAD models, or scale models built of cardboard and foam-core.

When the client is satisfied with the initial design, drafters work under the supervision of the designer to create a set of construction documents (CDs) containing all the information necessary to build the project. CDs are the centerpiece of every construction project, and almost all who have a role in the construction of the project rely on architectural working drawings to accomplish their jobs, from lenders who review CDs to determine the levels of funding for the project, to contractors who use CDs during the bidding and construction phases of the project.

Drafters must understand how to apply CAD techniques, such as the use of block libraries, to produce CDs quickly without sacrificing detail and accuracy.

KEY WORDS

Construction Documents (CDs)
Electrical Plans
Elevation Drawings
Floor Plans
Roof Pitch

REVIEW

Short Answer

1. What are construction documents?
2. Name four block libraries that might be used by architectural firms to produce drawings.
3. What is meant by the term pitch when used in the context of a roof?
4. Describe what is meant by a **3/12** notation on a roof plan.
5. Name three types of architectural drawings that may be drawn as section views.

CABIN PROJECT
CAPSTONE PROJECT 2.1

In this project you will create the floor plan and elevations for a small cabin. The finished sheets will resemble the ones shown in Figures CP2.9 and CP2.10 (detailed views are provided later).

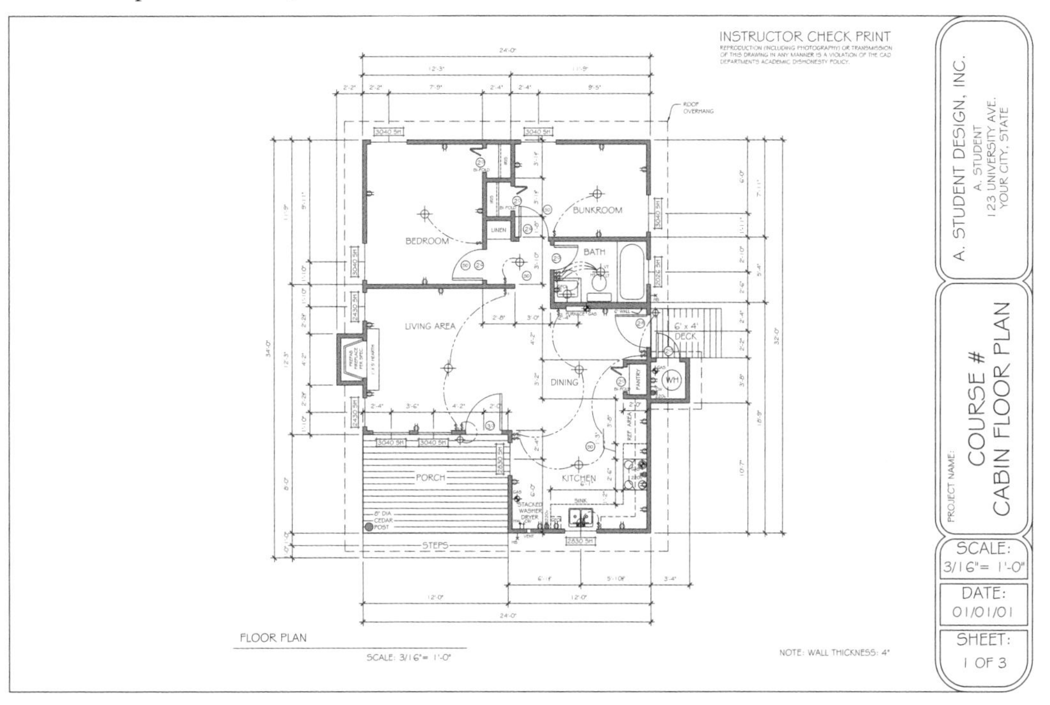

CP2.9 Cabin Floor Plan

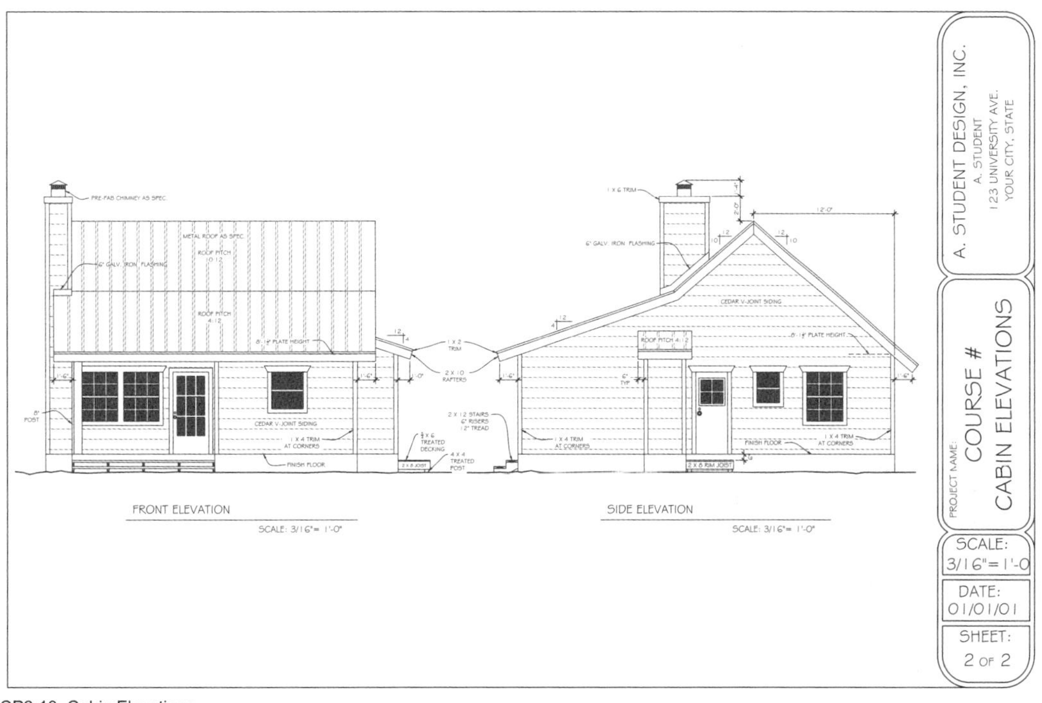

CP2.10 Cabin Elevations

PROJECT

Step 1. Getting Started

a. Download the **CP 2.1 Arch Wrkg Dwg Prototype** drawing file located in the *Prototype Drawing Files* folder associated with this book.

VIDEO

There is a video tutorial available for this project inside the Capstone Project 2 folder of the book's Video Training downloads.

b. Use **SAVE AS** to save the drawing to your **Home** directory.

c. Rename the drawing **CABIN PROJECT**.

d. Activate the **Model** tab in the lower-left corner of the graphics window (if it is not already highlighted) by left-clicking on the tab.

e. Ensure that the Drawing Units are set to Architectural.

Step 2. Create the following layers:

LAYER NAME	COLOR	LINETYPE	LINEWEIGHT
Appliances	Magenta	Continuous	Default
Cabinets- Base	Cyan	Continuous	Default
Cabinets- Wall	Cyan	Hidden	Default
Closet Rod	Magenta	Hidden	Default
Closet Shelf	Magenta	Continuous	Default
Dimensions	Green	Continuous	Default
Doors	Blue	Continuous	Default
Electric Plan	Green	Continuous	Default
Elevations	Red	Continuous	Default
Fireplace	Cyan	Continuous	Default
Notes	Cyan	Continuous	Default
Plumbing Plan	Magenta	Continuous	Default
Porch	Blue	Continuous	Default
Roof Overhang	Magenta	Dashedx2	Default
Room Labels	Cyan	Continuous	Default
Walls	Red	Continuous	Default
Windows	Blue	Continuous	Default
Wiring	Green	Phantom	Default

NOTE

ENTERING ARCHITECTURAL UNITS OF MEASUREMENT

When architectural units are assigned to the drawing, lengths and distances will default to inches unless you enter a foot mark (') symbol. For example, to draw a line 24' 6" in length, type **24'-6** (you do not need to include an inch mark after the **6** because AutoCAD defaults to inches).

To enter lengths containing fractions, type a dash between the inch value and the fractional value - for example, type **15'9-1/2**.

Step 3. Draw the exterior walls

a. Set the **Walls** layer current.

b. Draw the perimeter of the cabin using the dimensions shown on the architect's sketch in Figure CP2.11.

- **Note:** Draw the floor plan inside the magenta rectangle located in model space named *FLOOR PLAN*.

c. Draw the perimeter of the floor plan using the POLYLINE command.

d. Use the OFFSET command to offset the completed walls **4"** to the inside of the perimeter to create exterior walls that are 4 inches wide.

- **Note:** Walls drawn with the POLYLINE command behave as one entity and are easier to OFFSET when creating the 4" wall thickness.

e. Explode the walls so they can be edited further.

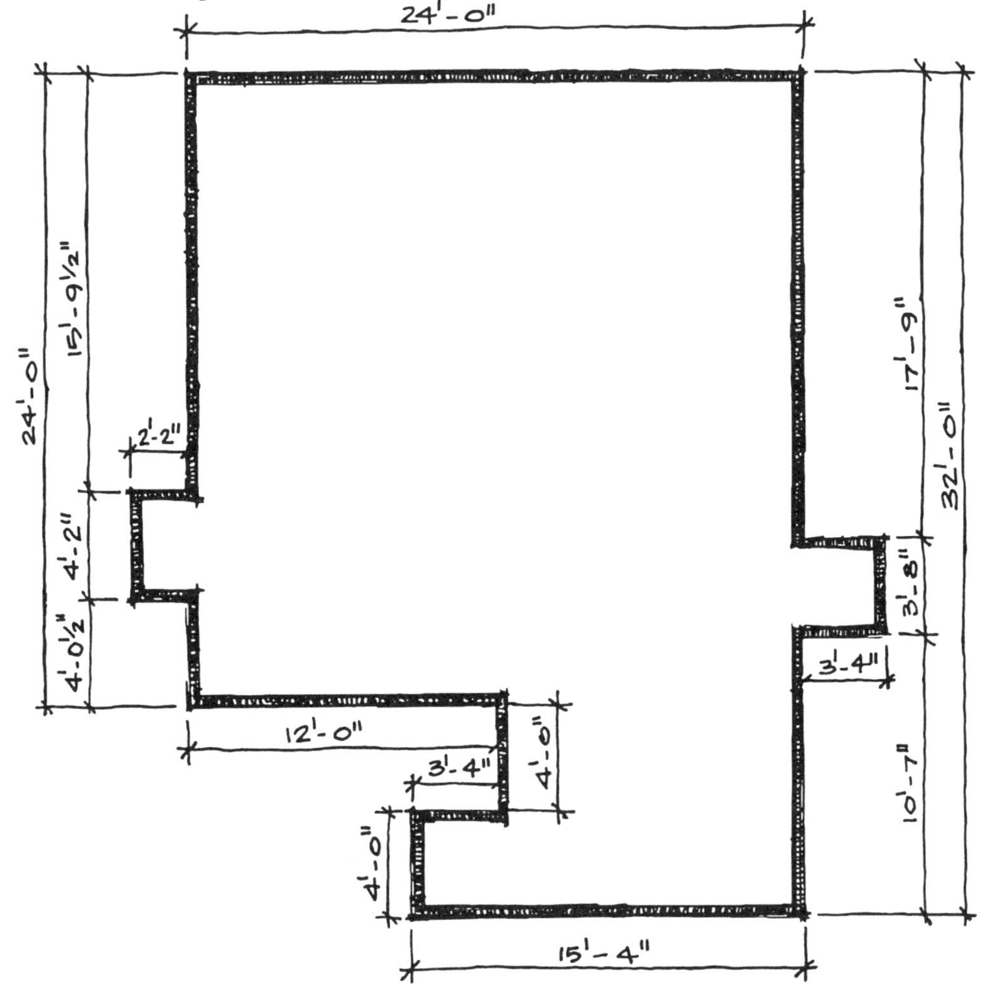

CP2.11 Architect's Sketch Showing Dimensions for Perimeter Walls

Step 4. Draw the interior walls

a. Use the dimensions shown in Figure CP2.12.

b. You can locate these walls by offsetting their edges from the edges of the walls drawn in Step 3.

c. Draw all interior walls **4"** wide except for the 6" wide wall noted in the figure.

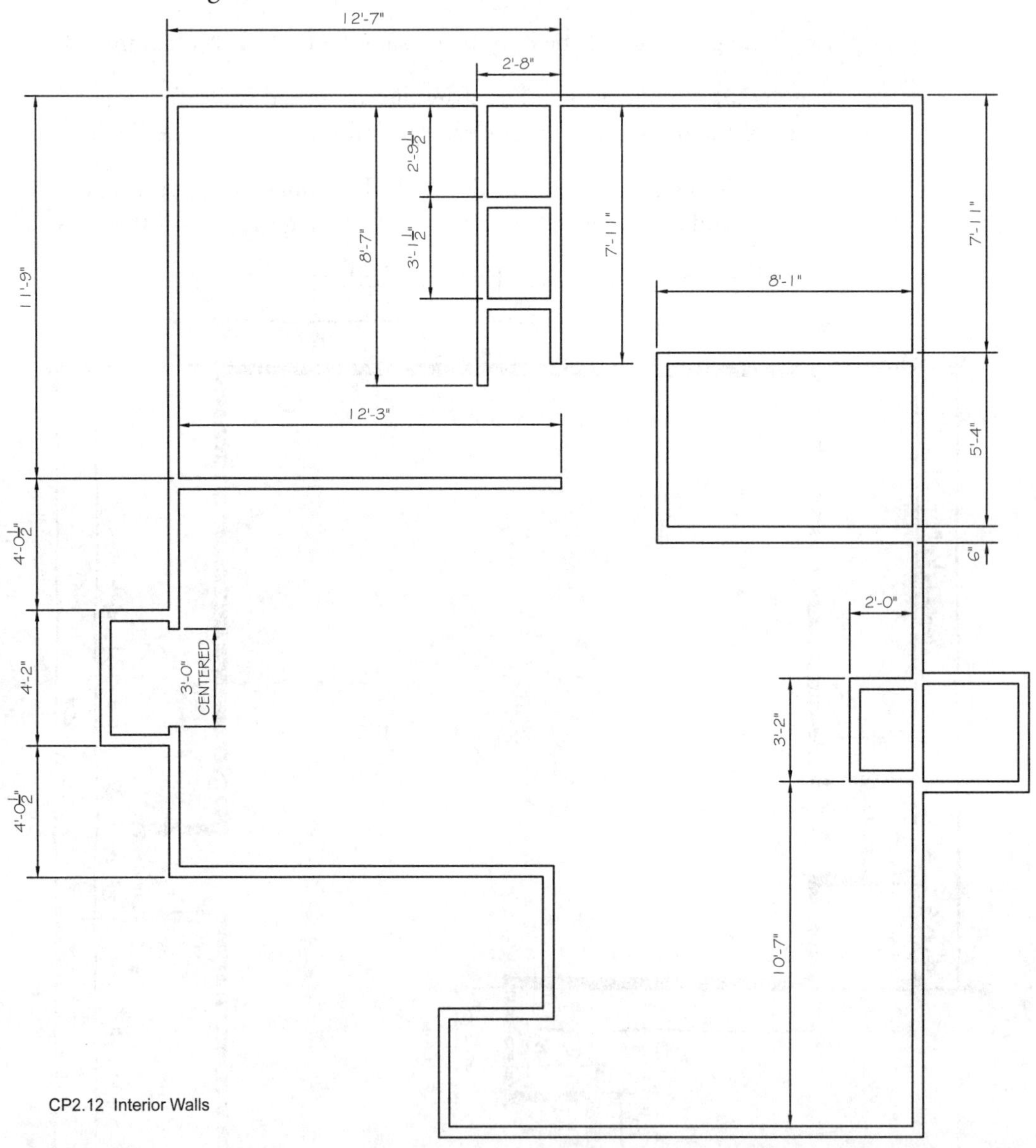

CP2.12 Interior Walls

NOTE

Do not add dimensions to the floor plan until you are instructed to do so in Step 18.

Step 5. Create blocks

a. In this step you will create a block library of architectural symbols that will be inserted into the floor plan.

b. To create the block library, download the **CP 2.1 Cabin Symbols** drawing file located in the *Prototype Drawing Files* folder associated with this book.

c. Save it to your **Home** folder.

d. Make a separate block for each of the symbols shown in Figure CP2.13 (refer to Chapter 9 on block creation and editing if necessary).

- Make the geometry for the blocks on layer **0.**

e. Assign base points that you think will facilitate the placement of the blocks into the drawing.

f. When you have finished creating all the blocks, save and close the **CP 2.1 Cabin Symbols** drawing (to your **Home** folder).

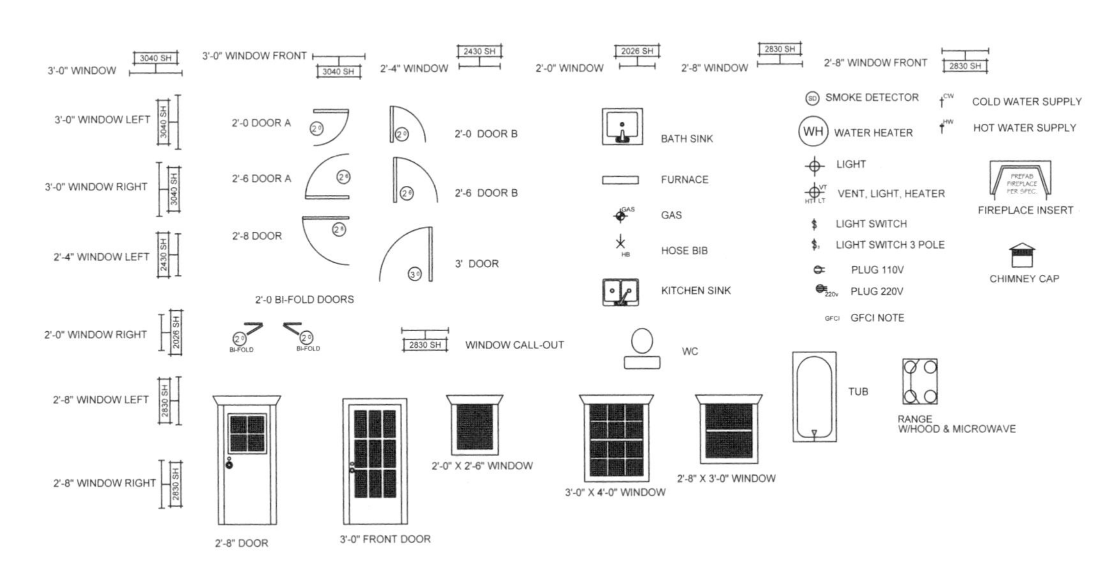

CP2.13 Architectural Symbols to be Blocked in Step 3(a)

The next section explains how to insert the **Cabin Symbols** drawing's blocks into the **Cabin Project** drawing.

Step 6. Using **DesignCenter** to place blocks

Design Center is an AutoCAD feature that allows drafters to insert the blocks, layers, linetypes, dimension styles, and text styles created for one drawing file into a different drawing file.

The following steps explain how to use **DesignCenter** to insert the blocks associated with the **Cabin Symbols** [that were created in Step 3(a)] into the **Cabin Project** drawing file.

a. Pick the **DesignCenter** tool located on the **Palettes** panel of the **View** tab of the ribbon (see Figure CP2.14) or type **ADCENTER** or **DC** and press <**Enter**>.

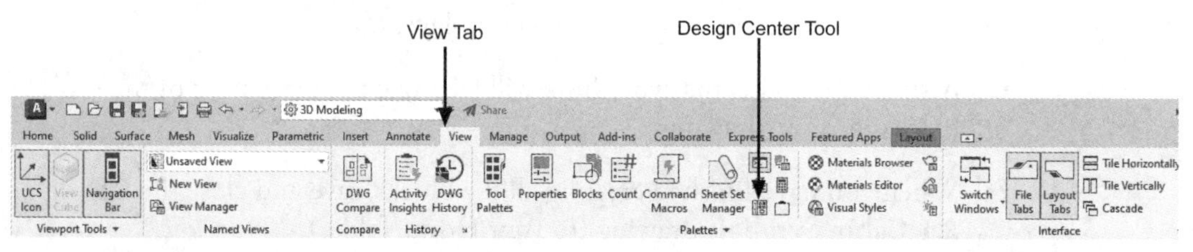

CP2.14 Location of the **DesignCenter** Icon on the **Ribbon Toolbar**

b. Selecting this tool opens the **DesignCenter** window (see Figure CP2.15). Note the location of the **Tree** pane and the **Contents Area** pane in the **DesignCenter** window shown in Figure CP2.15.

c. Move the cursor into the **Tree** pane and scroll through the file tree until you have located the **Cabin Symbols** drawing that you saved in your **Home** folder in Step 3(a).

d. Double-click on the **Cabin Symbols.dwg** file name.

e. Click on the word **Blocks** located in the tree below the **Cabin Symbols** file name (see Figure CP2.15).

- The blocks associated with the **Cabin Symbols** drawing will appear in the **Contents Area** pane (see Figure CP2.15).

After creating the blocks inside the Cabin Symbols drawing, navigate to the Cabin Symbols drawing file in the Tree Pane

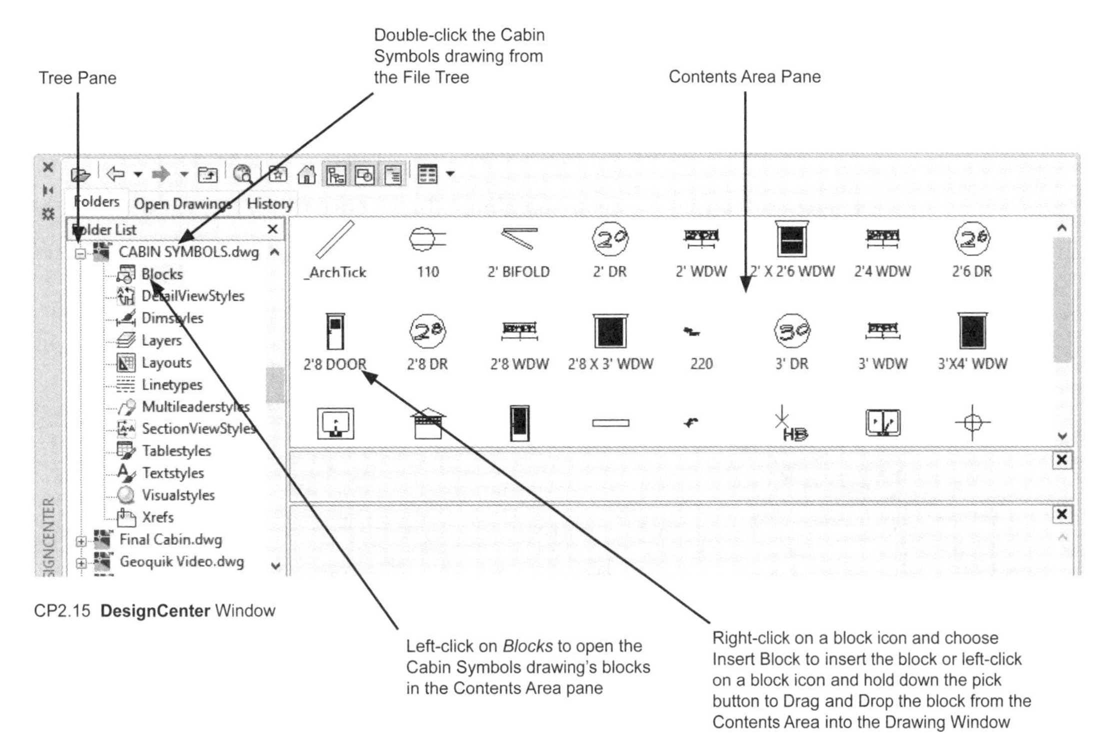

CP2.15 **DesignCenter** Window

f. Insert the blocks shown in the **Contents Area** pane into the **Cabin Project** floor plan by "dragging and dropping" them into the drawing.

- You do this by clicking on a block icon and holding down the pick button of the mouse and dragging the block into the drawing window.
- You can also insert a block by selecting its icon in the **Contents Area** pane, right-clicking, choosing **Insert Block** from the menu, and inserting the block using the techniques presented in Chapter 9 (see Figure 9.6).

Step 7. Insert window blocks

a. Set **Windows** as the current layer.

b. To locate the centers of the windows, offset the perimeter wall lines using the dimensions shown in Figure CP2.16.

c. Use **DesignCenter** to access the block library created for the **Cabin Symbols** drawing and follow Steps 3(b) through 3(e) to insert the window blocks into the walls of the floor plan.

d. Trim the walls to the edges of the windows as shown in Figure CP2.16.

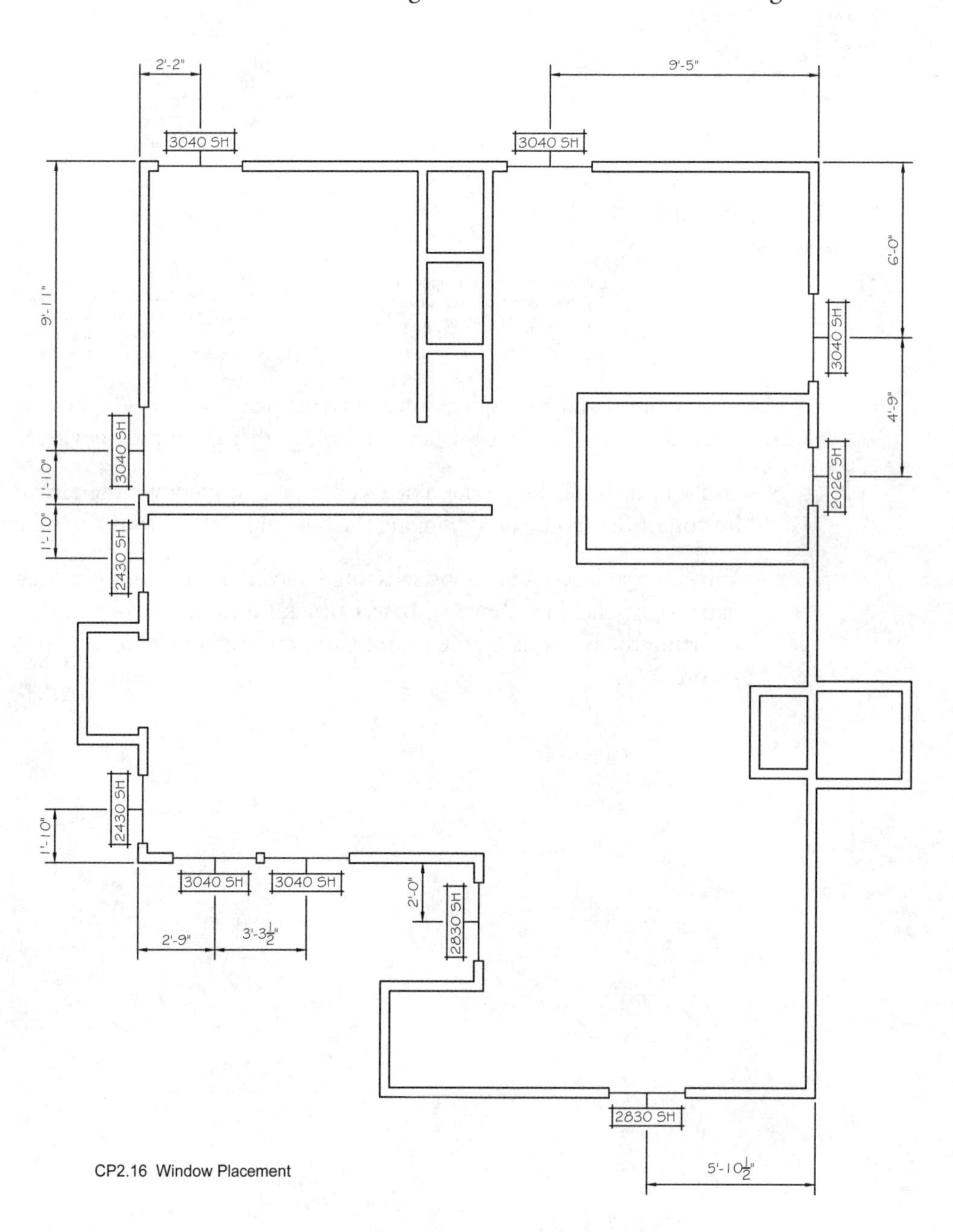

CP2.16 Window Placement

Step 8. Insert door blocks

a. Set the **Doors** layer current.

b. Offset the wall lines to locate the centers and edges of doors.

- For example, the opening for a door marked 2^6 will be 2'-6" wide.

c. Allow a minimum length of **4"** for door jambs located on the hinged side of the bedroom, bunk room, and bath doors.

d. The pantry door and the bi-fold closet doors and the water heater should be centered on their respective enclosures (see Figure CP2.17).

e. Insert the desired door block from the **DesignCenter** location and trim to the edges of the door block as shown in Figure CP2.17.

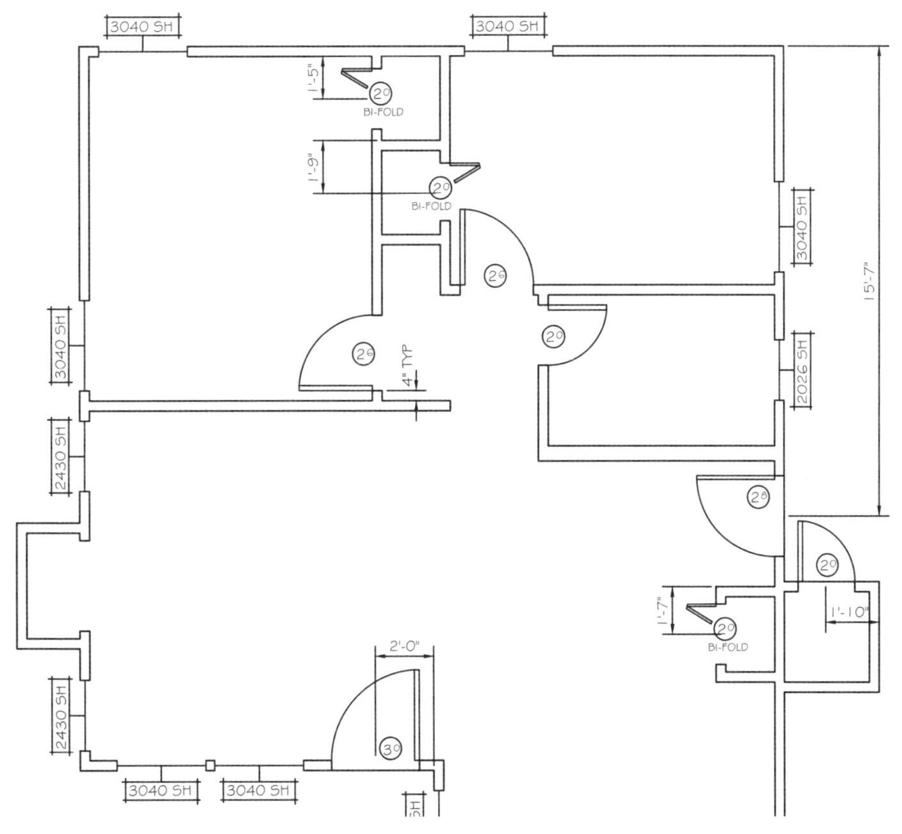

CP2.17 Door Placement

JOB SKILL

The side of the door frame where the hinges are attached is referred to as the *hinge* jamb. The side of the door frame where the latching mechanism attaches is referred to as the *strike* jamb.

Step 9. Insert the fireplace

a. Set the Fireplace layer current.

b. Insert the fireplace created from the Cabin Symbol drawing file in Step 5.

c. The fireplace is located in the floor plan as shown in Figure CP2.18.

d. The *hearth* shown in Figure CP2.18 is a raised stone or brick ledge rising from the floor and running across the front of the fireplace.

- The hearth is drawn **1'** wide by **5'** long and should be centered on the fireplace.

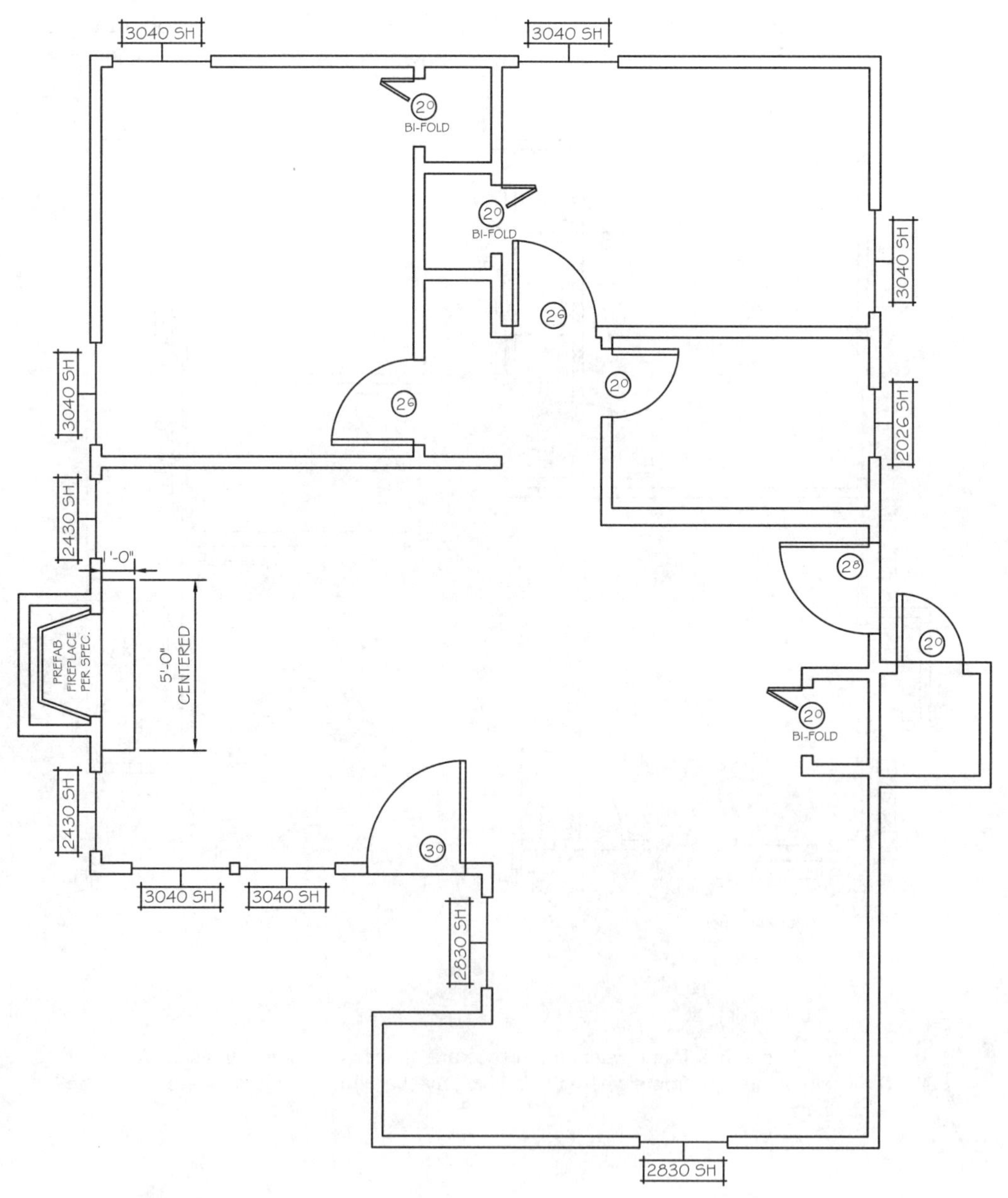

CP2.18 Adding the Fireplace and Hearth

Step 10. Kitchen cabinets and closet shelves

a. Set the **Cabinets- Base** layer current

- The lower kitchen cabinets are **24"** deep.

b. Set the **Cabinets-Wall** layer current.

- The upper wall cabinets (dashed) are **12"** deep.
- Upper wall cabinets are offset in 4" from the window.

c. Set the **Closet Shelf** layer current.

- Offset 1'-6" from the back wall of the linen closet and pantry
- Offset 12" from the back wall of the bedroom closets

d. Set the **Closet Rod** layer current.

- Offset 10" from the back wall of the bedroom closets

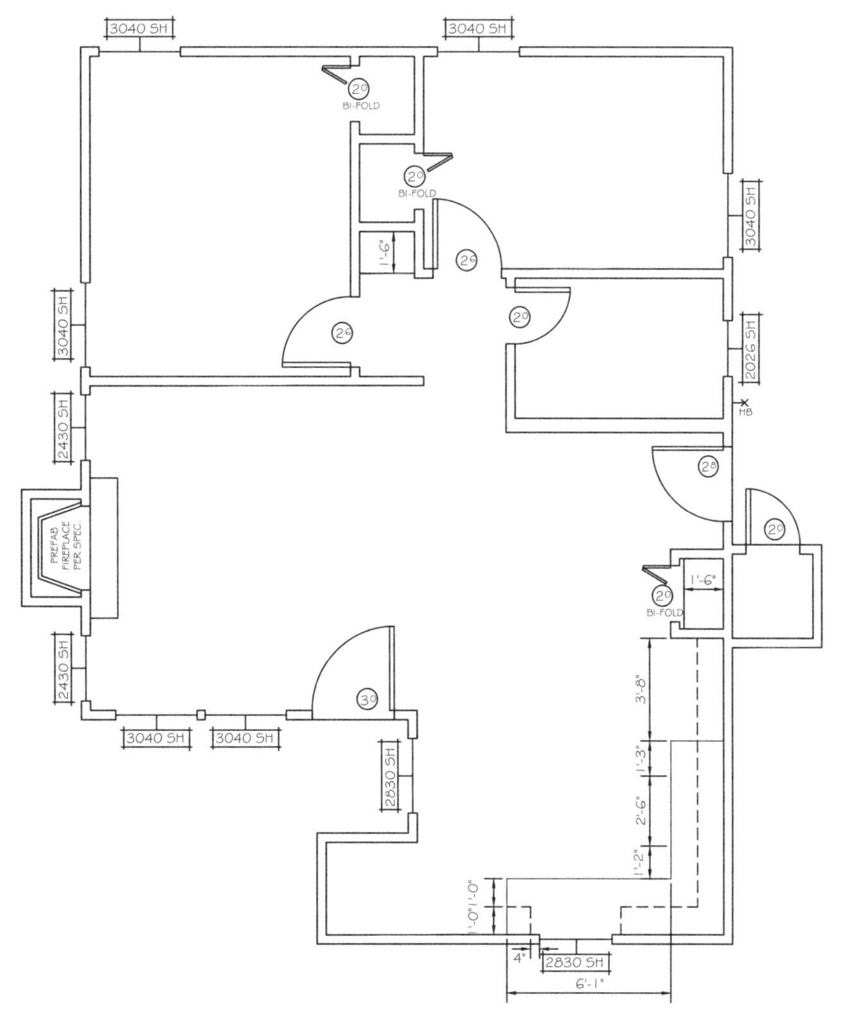

CP2.19 Location of Cabin's Plumbing Fixtures and Kitchen Cabinets

Step 11. Add Plumbing Blocks

a. Open **DesignCenter** and insert the following blocks by selecting them from the **Cabin Symbols** drawing file as shown in Figure CP2.20.

b. Set the **Appliances** layer current.

- Place stove on *Appliances* layer.

c. Set the **Plumbing Plan** layer current.

- Tub
- Toilet (WC)
- Bath sink
- Kitchen sink
- Furnace
- Water heater
- Hose bib (outdoor water faucets)
- Gas Outlets

d. Draw a **4" X 4"** square dryer vent in the wall and as shown in Figure CP2.20.

Step 12. Add the electrical plan

a. Set the **Electric Plan** layer current.

b. Open **DesignCenter** and insert the blocks of the electric symbols from the **Cabin Symbols** drawing file into the floor plan as shown in Figure CP2.20.

c. Set **Wiring** as the current layer.

d. Using the **Spline** command, draw the *switch legs* from the switches to the overhead and wall-mounted lights as shown in Figure CP2.20.

e. Set the **LTSCALE** to 10 to see the dashes in the phantom lines.

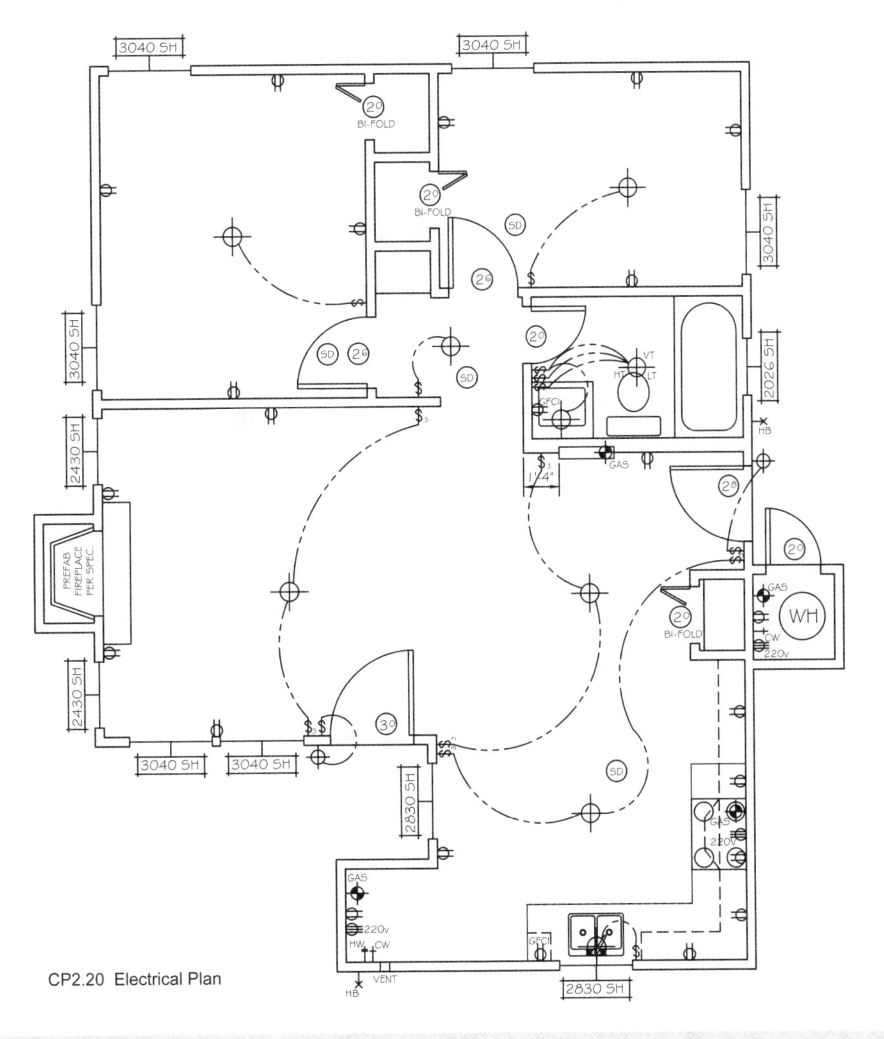

CP2.20 Electrical Plan

JOB SKILL

In an electrical plan, switch legs represent the electrical circuit that connects switches to lights (or other electric fixtures).
The switch legs are often drawn with phantom lines.

Step 13. Adding labels

a. Change the **Standard** text style's font to **Stylus BT.**

b. Add the labels shown in Figure CP2.21.

c. Set the **Room Labels** layer current.

- Room Labels are set to 6" text height.

d. Set the **Notes** layer current.

- Detail notes, or *callouts* are set to **4"** text height.
- Very small text is set to **3"** text height.

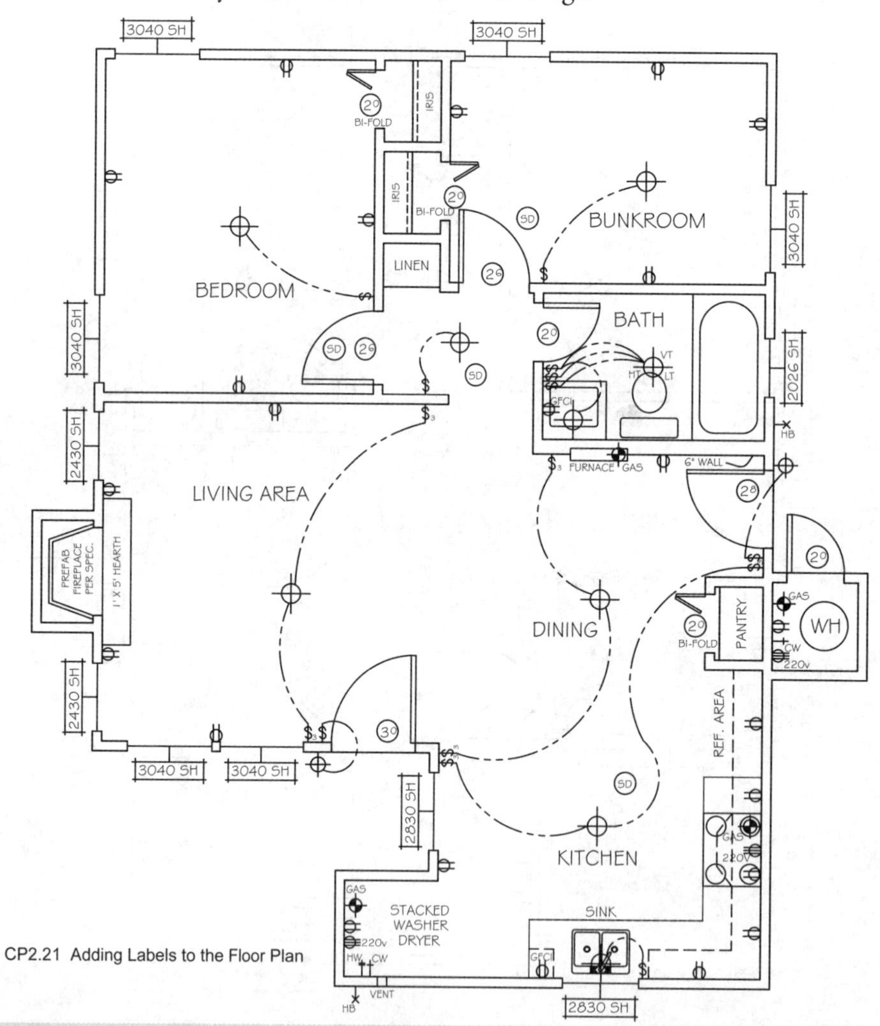

CP2.21 Adding Labels to the Floor Plan

QUICK TIP

The notes shown in the closets in Figure CP2.21 indicate that there is 1 Rod and 1 Shelf (1R1S). Refer to Figure 4.158 on page 233 for more information.

Step 14. Draw the porch

a. Set the **Porch** layer current.

b. Draw the porch and deck as shown in Figure CP2.22(b).

c. Draw the side deck **6'** long and **4'** wide.

- Use **6"** wide boards for the porch and deck flooring (also referred to as decking).

QUICK TIP

Decking can be added using the "User Defined" hatch pattern, setting the scale to 6" in the contextual Hatch ribbon as shown in figure CP2.22(a)

CP2.22(a) User Defined Hatch Pattern

- The treads of the porch steps should be drawn **12"** wide.

d. Add an **8"** diameter cedar post to the front left corner of the porch roof. Place it 2" from the edges of the porch.

- This post supports the roof above the front porch.

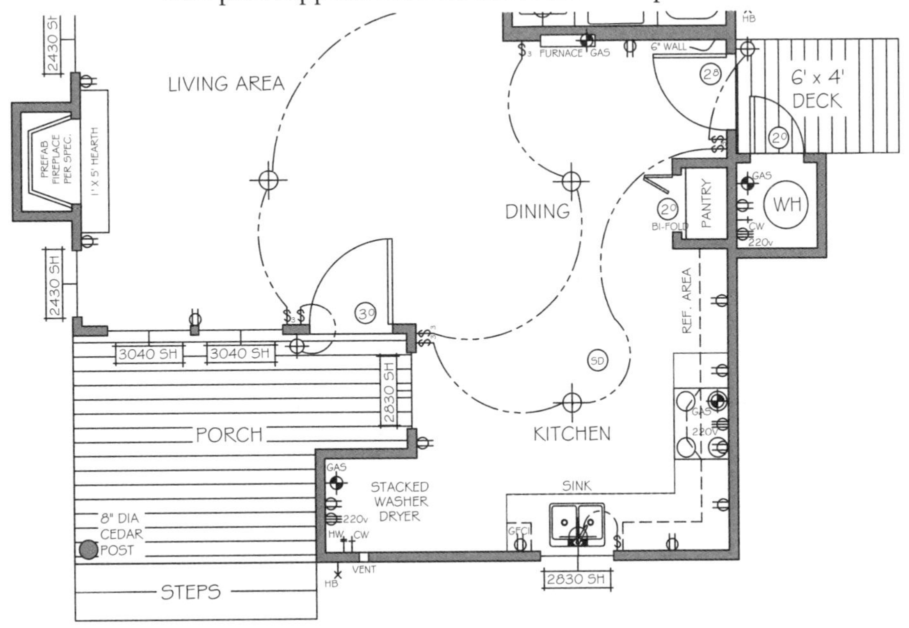

CP2.22(b) Floor Plan with Porch, Deck, and Hatched Walls

Step 15. Hatch the Walls

a. Set **Wall Hatch** layer current (this layer was provided in your prototype).

b. Hatch the walls and the cedar post with the **Solid** hatch pattern.

- The wall hatch layer provided in this template is set to plot in grayscale, instead of monochrome.
  ****Do not change the color of this layer.***

Step 16. Change order

After a meeting between the project architect and the Cabin Project's client, the client has decided to make changes to the layout of the cabin's entry and kitchen area.

The new layout is shown in Figure CP2.23.

Making these changes at this point in the design process will increase the fees the architectural firm will charge the client for this project. These changes may also affect the final construction cost of the cabin.

If the client instructs the architect to proceed with the design changes, the architect will issue an *Architectural Change Order* to the drafting department requesting the changes be made to the drawings.

The change order will note the specific changes that need to be made to the floor plan. By following a formal process in making these changes, the architectural firm creates a record of what changes were made, when they were made, and by whom.

a. Use the Stretch command to edit your floor plan to reflect all the changes to the windows in the living area and kitchen, which are highlighted with revision clouds, in Figure CP2.23. Reapply the hatch pattern to the walls as needed when the changes are complete (don't add the dimensions).

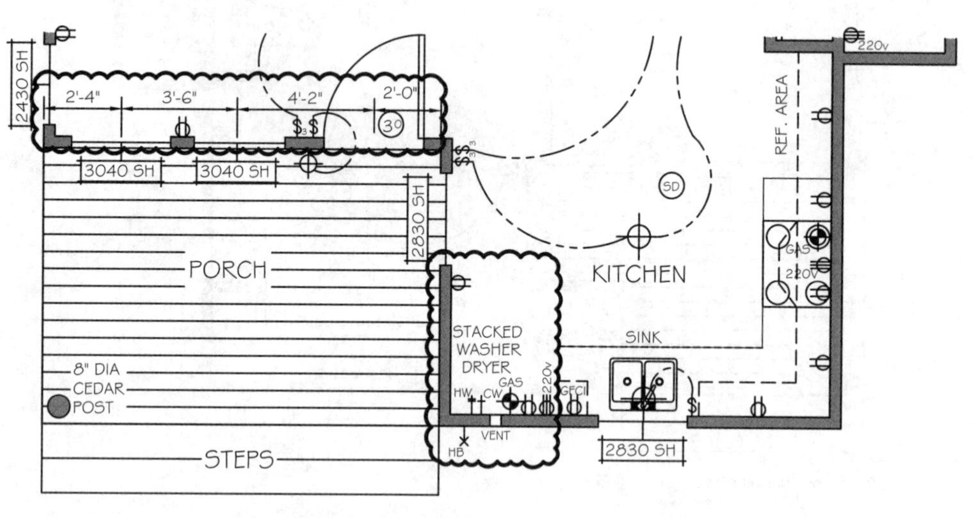

CP2.23 Changes to Floor Plan resulting from the Architectural Change Order

Step 17. Roof overhang

Add dashed lines around the outside of the floor plan's exterior walls. The dashed lines represent the edges of the *roof overhang* as shown in Figure CP2.24. Drafters can determine the placement of the dashed lines that represent the roof overhang by referring to the elevation sketches of the cabin and noting the roof overhang distances.

a. Set the Roof Overhang layer current.

b. The roof overhang around the main living area of the cabin is **18"**.

c. The roof overhang around the water heater closet is **6"** on the top and bottom edges and **12"** along the right edge. Refer to Figure CP2.24.

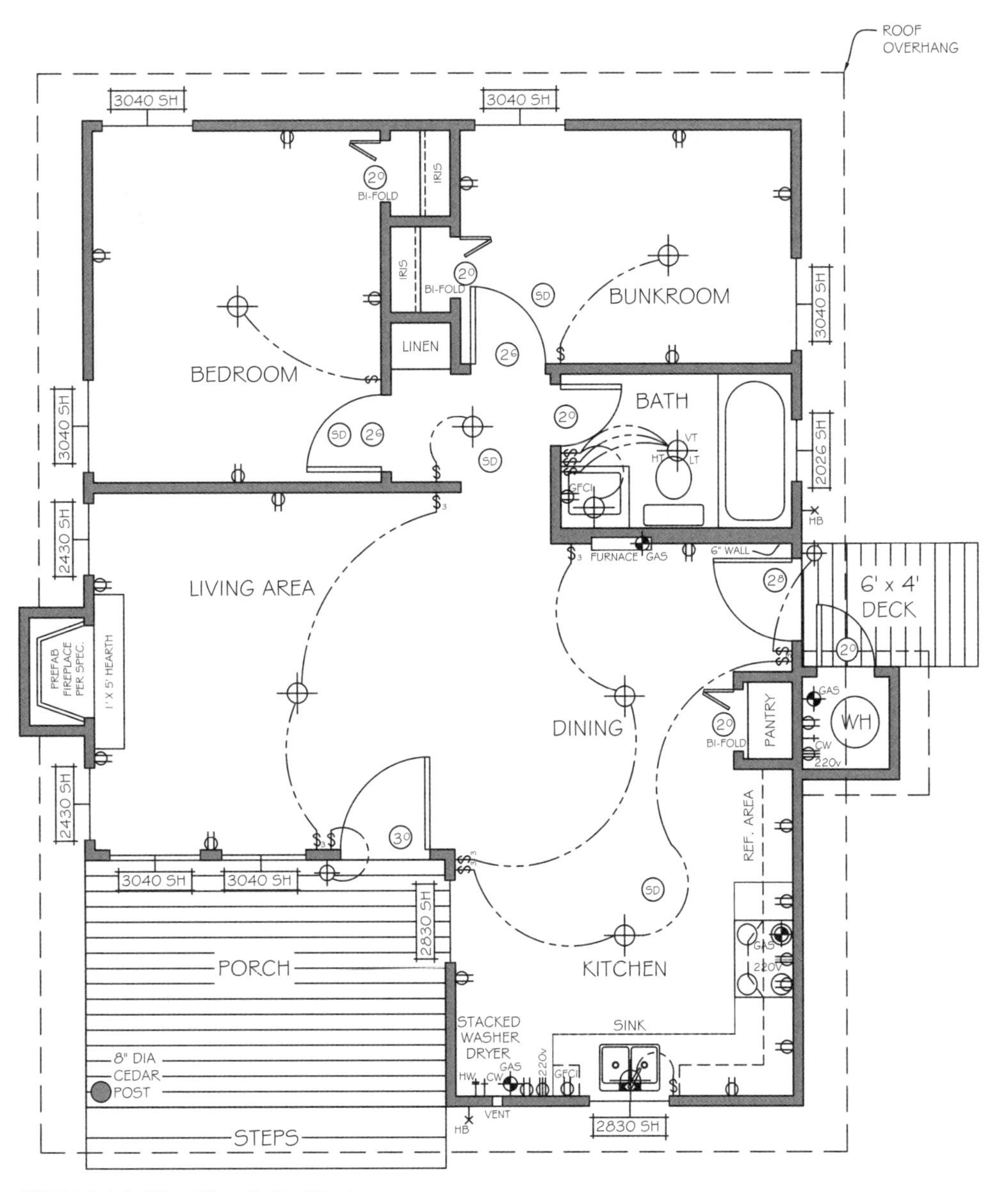

CP2.24 Detail of Floor Plan with Roof Overhang

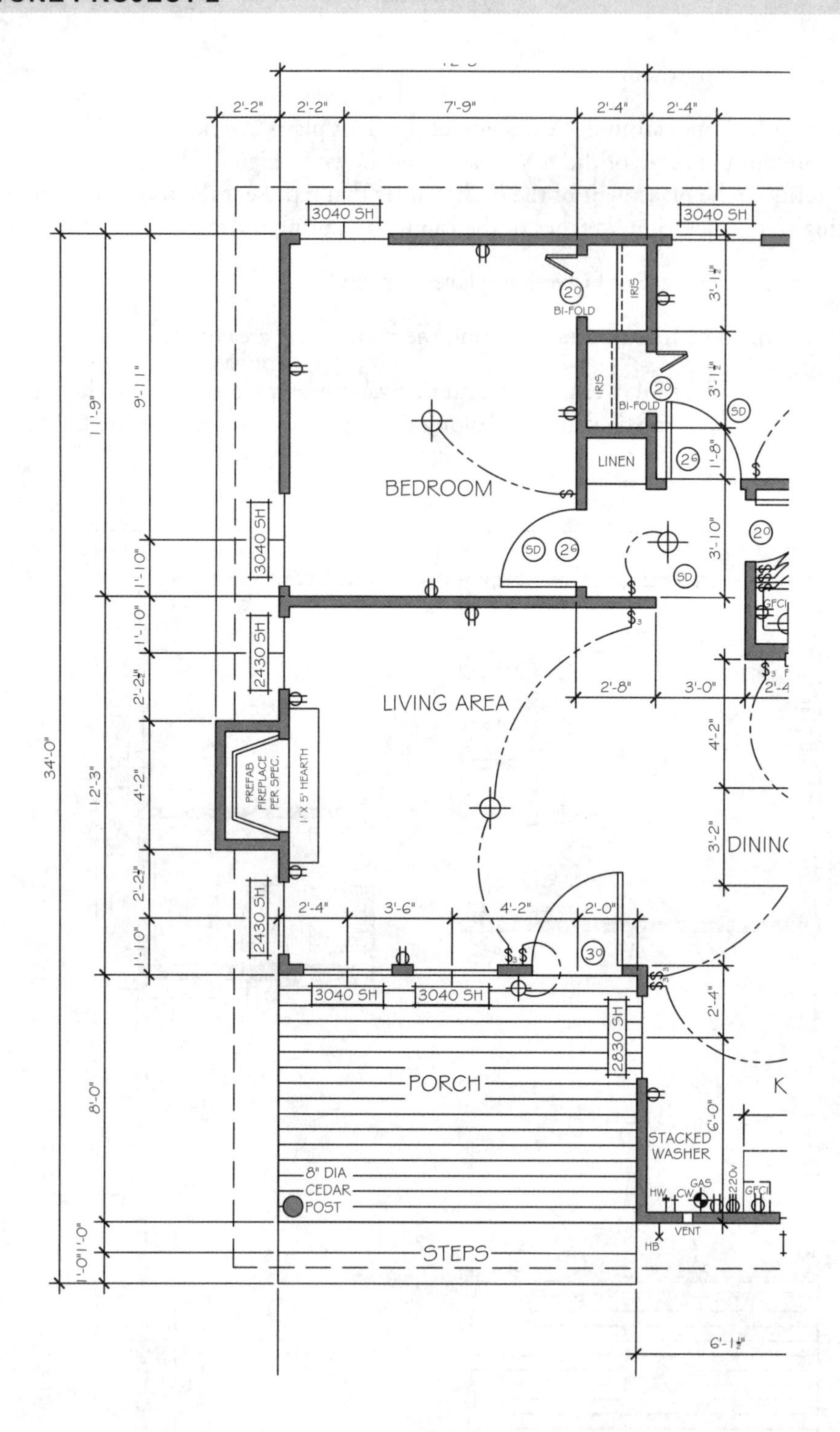

CP2.25(a) Dimensioning the Floor Plan (Left Half Shown)

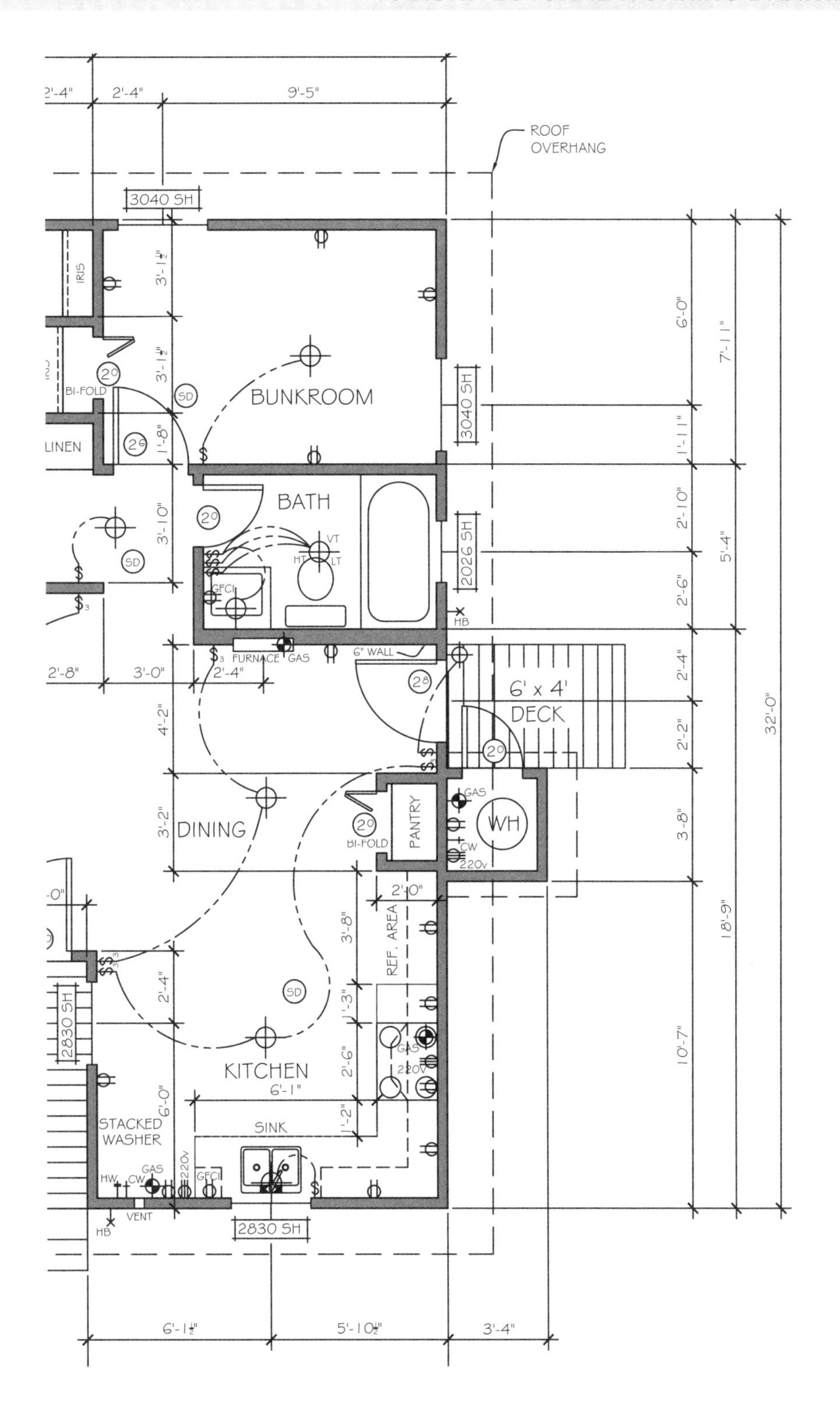

CP2.25(b) Dimensioning the Floor Plan (Right Half Shown)

Step 18. Dimensioning the cabin

a. Create a new dimension style named **Arch 64** that has the dimension style settings shown below.

- **Note:** The values entered in the **Dimension Style** tabs may seem small — for example, setting the text height to 3/32"— but because the value for *Scale For Dimension Features* in the **Fit** tab is set to *Use overall scale of: 64*, all of the dimension settings will be multiplied by 64.

Dimension Style Settings for Arch 64		
Dimension Style Tab	**Setting**	**Value**
Primary Units *(Linear dimensions)*	Unit format	Architectural
	Precision	1/16"
	Fraction Format	Horizontal
	Zero suppression	0 Feet
Fit	Text placement	Over dimension line, with leader
	Use overall scale of	64
Text	Text style	Standard (font should be set to Stylus BT)
	Text height	1/16"
	Fraction height scale	.50
	Text placement- vertical	Above
	Offset from dim line	1/16"
	Text alignment	Aligned with dimension line
Symbols & Arrows	Arrowheads (first and second)	Architectural tick
	Arrow size	1/16"
Lines	Extend beyond ticks	1/16"
	Extend beyond dim lines	1/16"
	Offset from origin	1/16"

b. When the **Arch 64** dimension style has been created, set it as the *current* dimension style.

c. Set the **Dimensions** layer current.

d. Dimension the floor plan as shown in Figures CP2.25(a) and CP2.25(b).

After Step 18 is completed, the floor plan sheet of the cabin is finished (except for the title block).

JOB SKILL

Why ARCH 64?
64 is the *Drawing Scale Factor (DSF)* for a drawing plotted at 3/16"=1'-0". A quick way to calculate a Drawing Scale Factor of 3/16"=1'-0" is to invert the fraction and multiply by 12. In this case: (16/3) x 12 = *DSF* of 64 (there are sixty four 3/16" increments in 12").
Using the formula above, the *DSF* for a drawing plotted at 1/4"=1'-0" would be 48 (there are forty eight 1/4" increments in 12").
The *DSF* for a drawing plotted at 1/8"=1'-0" would be 96 (there are ninety six 1/8" increments in 12").

PROJECT REVIEW

Floor Plan Checklist

When you've completed the Floor Plan, review your drawing:

- Make sure no lines cross through dimensions or text
- Make sure you can read all text and dimensions clearly
- Make sure all layers are set to the *Default* lineweight
- Make sure you have all necessary notes (Note: 1. Wall thickness = 4")
- Make sure your Viewport Scale is set to ***3/16"=1'-0"***
- 61 linear or continuous dimensions (40 outside of floorplan and 21 inside).
- Twenty-one 120VAC outlet (not GFCI type)
- Two 120VAC outlets labeled GFCI
- Three 220VAC outlets
- Four Gas outlets
- Six 3-way switches
- Two hose bibs
- Four smoke detectors
- One dryer vent
- 1R1S is the abbreviation for 1 Rod 1 Shelf
- In the bath, the overhead fixture's labels are HT (Heater), VT (Vent) and LT (Light)
- The dimension spacing for dimensions placed outside the floor plan are:
    - 27" for first row of dimensions and
    - 18" for additional rows.
    - Note: The 27" inches is measured from edge of farthest feature. For example, on the top of the floor plan, place the first row of dimensions 27" above the roof overhang.

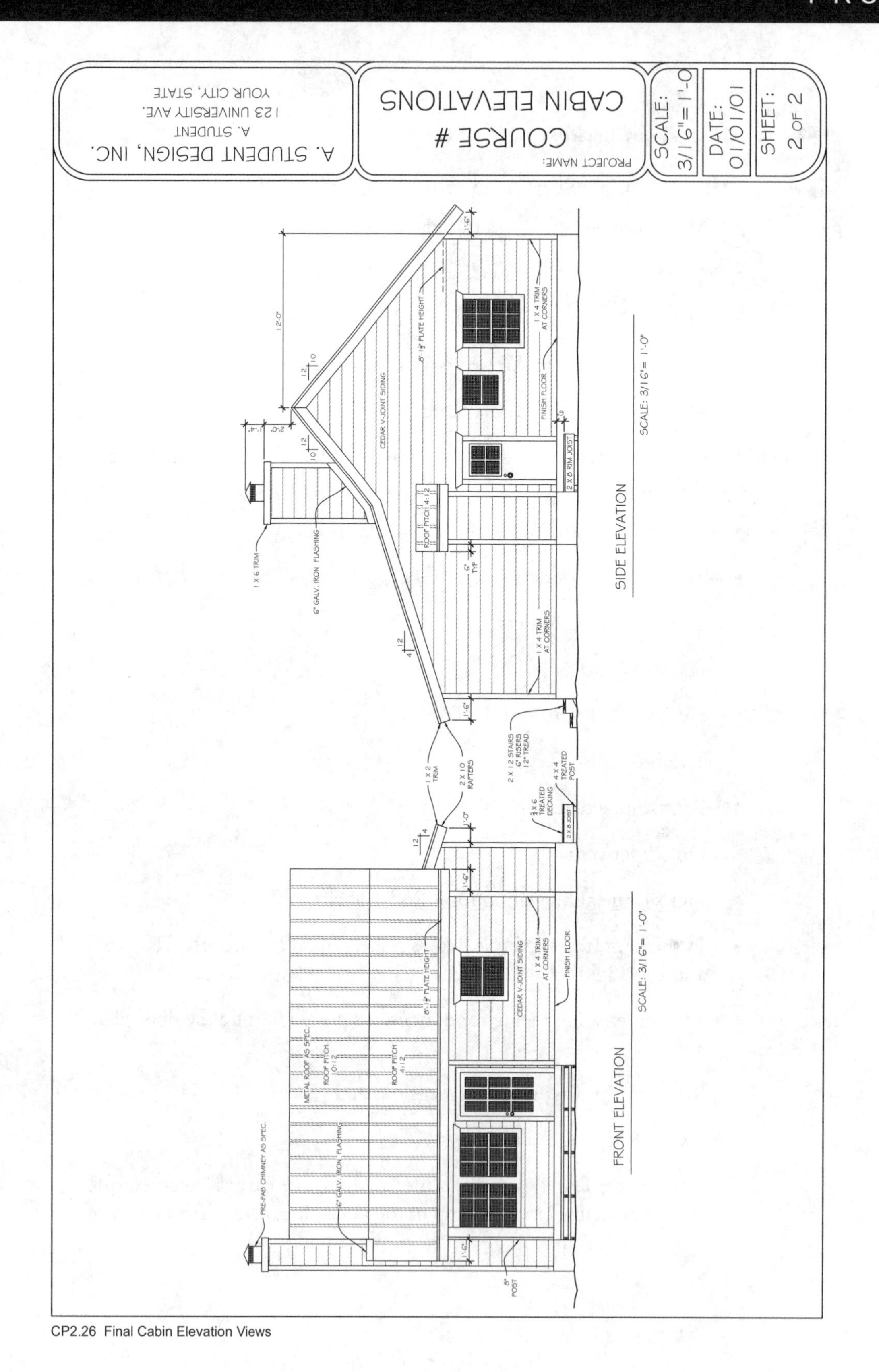

CP2.26 Final Cabin Elevation Views

You will draw the front and right elevations of the cabin (as shown in Figure CP2.26) inside the magenta rectangle located in model space named *ELEVATIONS*. Architectural elevations are constructed using the principles of multiview drawing which allows drafters to project geometric features from one view to the next as they construct the elevation views.

Relating the Features on the Floor Plan to the Cabin's Elevation Views

When elevation drawings of the cabin are created, the dimensions shown on the floor plan are used to locate exterior features such as doors, windows, porches, and chimneys.

The dashed lines in Figure CP2.27 illustrate the relationship between features on the floor plan—like doors and windows—and their locations on the front and side elevations. In Figure CP2.27, a 45° miter line in the lower right corner is used to relate information between the front and side elevations in the same manner that the features of a machine part can be projected through a miter line.

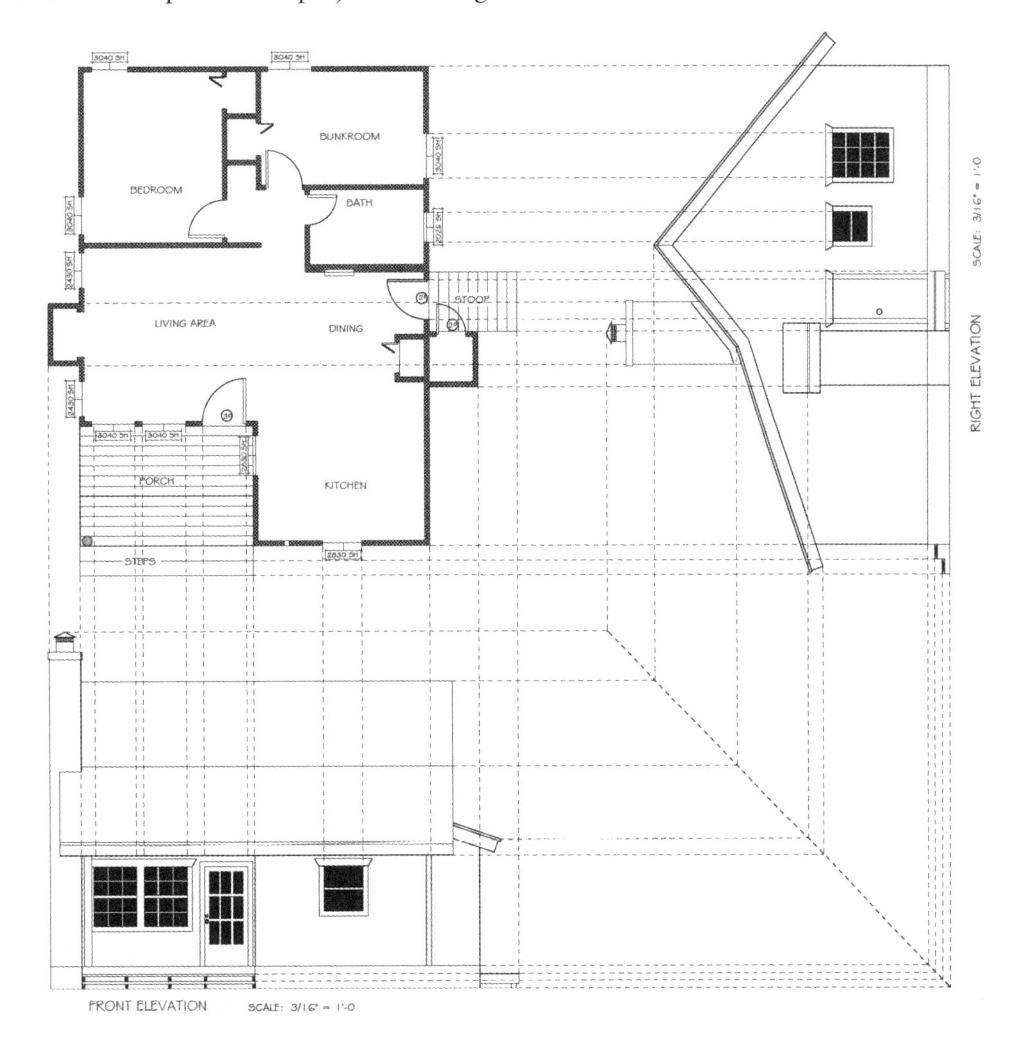

CP2.27 Relating the Cabin's Elevations to the Floor Plan

Wall Framing 101

An understanding of the basics of framing is very helpful to drafters in constructing elevation drawings. The method used to frame the wall will determine the heights of ceilings, windows, and other exterior features.

The example in Figure CP2.28 shows the front and end views of a length of wall framing. In this example, the studs are placed 16" on center. Figure CP2.28 also shows the framing for the rough openings for a window and a door. The rough opening is sized by the framers to accommodate the size of the window or door specified on the plan. Generally, rough openings will be about 2 1/2" wider and taller than the window or door specified on the floor plan.

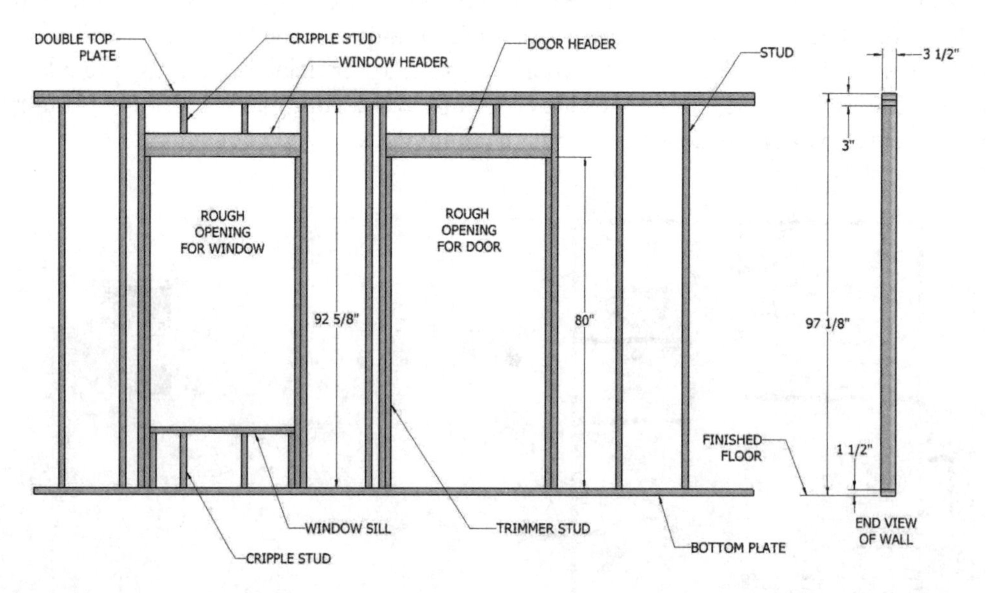

CP2.28 Wall Framing Example

VIDEO There is a video tutorial available for this project inside the Capstone Project 2 folder of the book's Video Training downloads.

NOTE Draw the elevations of the cabin inside the magenta rectangle located in model space named *ELEVATIONS.*

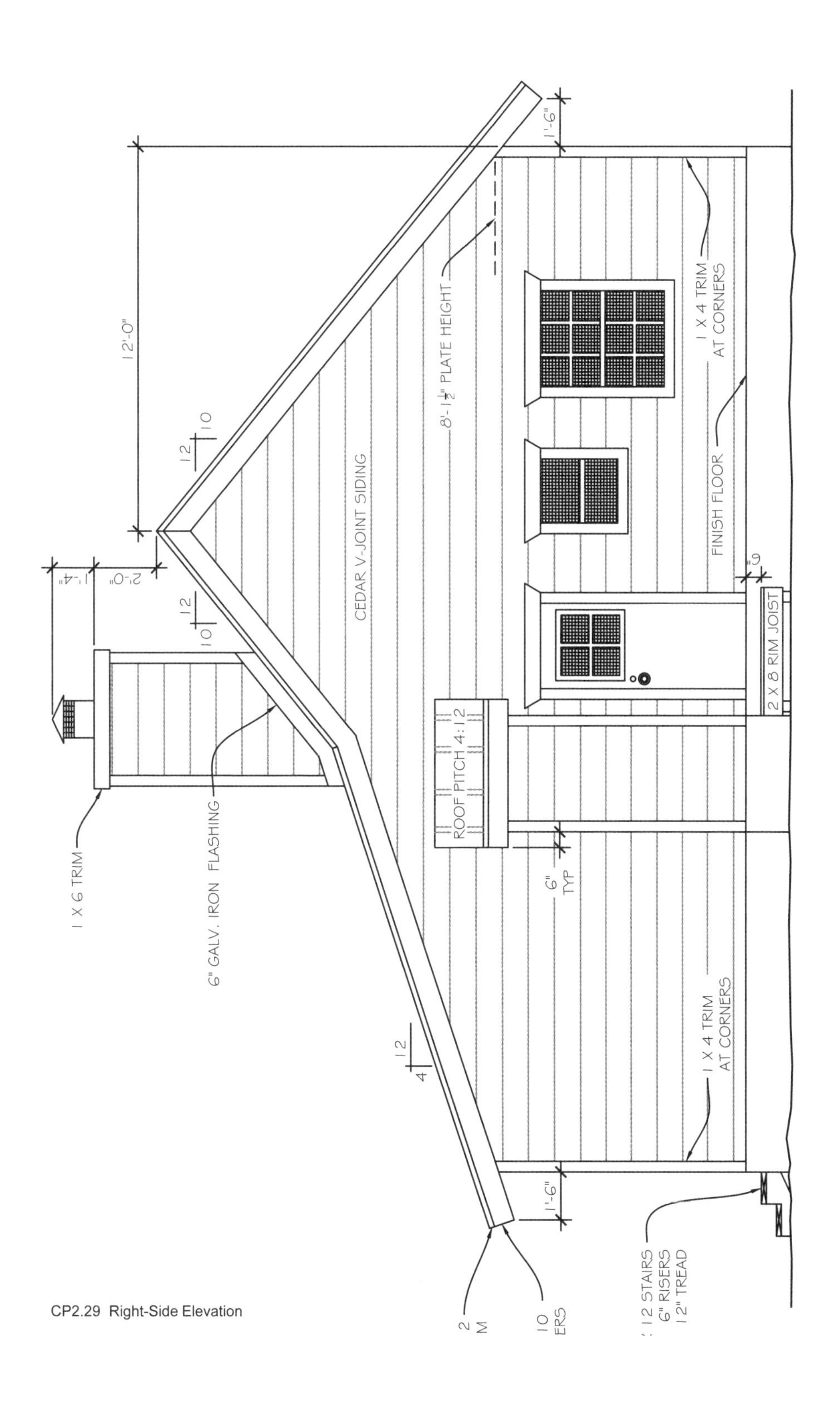

CP2.29 Right-Side Elevation

CONSTRUCTING THE RIGHT-SIDE ELEVATION

Follow the next steps to draw the elevations of the cabin in model space. The first view to be drawn is the right-side elevation shown in Figure CP2.29. This view was chosen because it shows the profile of the roof's pitch. When this view is complete, construction lines will be projected from its features to assist in the construction of the front view of the cabin.

Step 1. Set the **Elevations** layer current

Step 2. Use the dimensions shown in Figure CP2.30 to construct the cabin's right exterior wall.

Step 3. Add the Plate Height dashed line, dimension and note as shown. (See note below.)

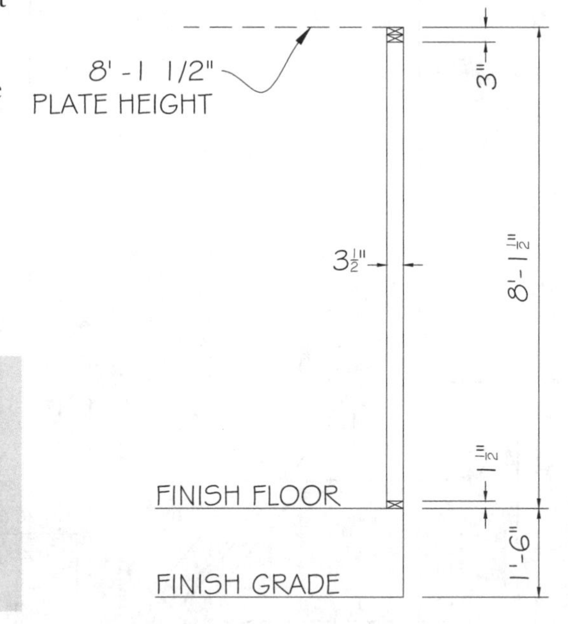

CP2.30 Exterior Wall Dimensions

NOTE

The line labeled *FINISH GRADE* in Figure CP2.30 is the point where the foundation of the cabin meets the ground line.

The line labeled 8'-1 1/2" PLATE HEIGHT represents the height where the exterior walls meet the roof rafters.

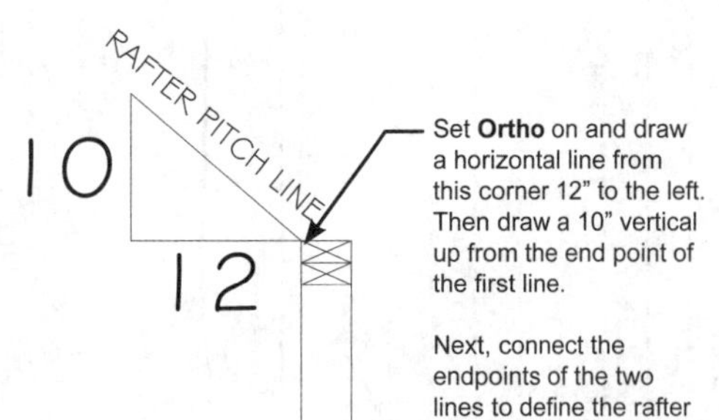

CP2.31 Drawing a 10/12 Rafter Pitch

Step 4. To draw the rafters, follow the directions shown in Figure CP2.31 to define the **10/12** rafter pitch that begins at the upper left corner of the exterior wall constructed in the previous step.

Step 5. Offset the rafter pitch line **10"** and **8"** as shown in Figure CP2.32.

- The first offset line represents the rafter's width of 10".
  - The second offset line is used to represent a 2"-wide piece of trim attached along the rafter's top edge.

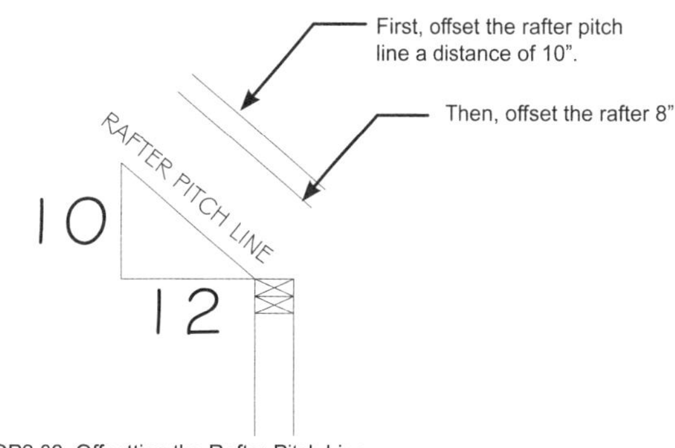
First, offset the rafter pitch line a distance of 10".
Then, offset the rafter 8"
RAFTER PITCH LINE
10
12

CP2.32 Offsetting the Rafter Pitch Line

Step 6. Creating the rafter

a. The rafter overhangs the outside edge of the wall **1'-6"**, as shown in Figure CP2.33.

b. Offset the outside wall line **1'-6"** to the right and extend the bottom edge of the rafter pitch line to the offset line.

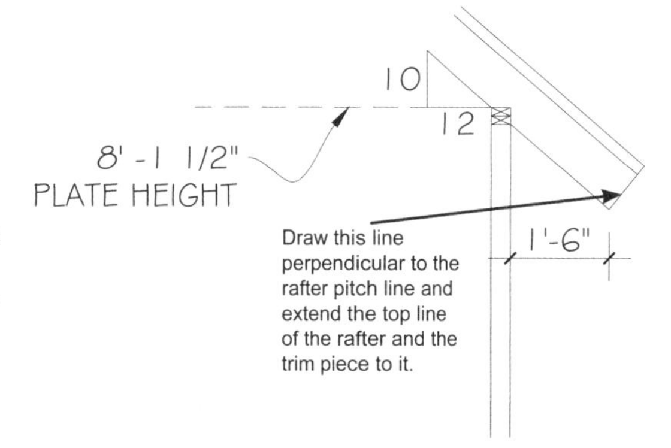
10
12
8' - 1 1/2"
PLATE HEIGHT
Draw this line perpendicular to the rafter pitch line and extend the top line of the rafter and the trim piece to it.
1'-6"

CP2.33 Detail of the Rafter Overhang

c. From the end of the rafter pitch line, draw a line perpendicular to the **10/12** pitch and extend the lines that were offset in the previous step to the perpendicular line to define the end of the rafter.

d. Offset the outside right edge of the exterior wall **12'** to the left.

e. Turn **Ortho** on, and use *grips edit* to lengthen the line so that the lines representing the rafter and the finished floor and grade can be extended to it.

f. Extend the rafter and the finished floor and grade lines to the offset line as shown in Figure CP2.34.

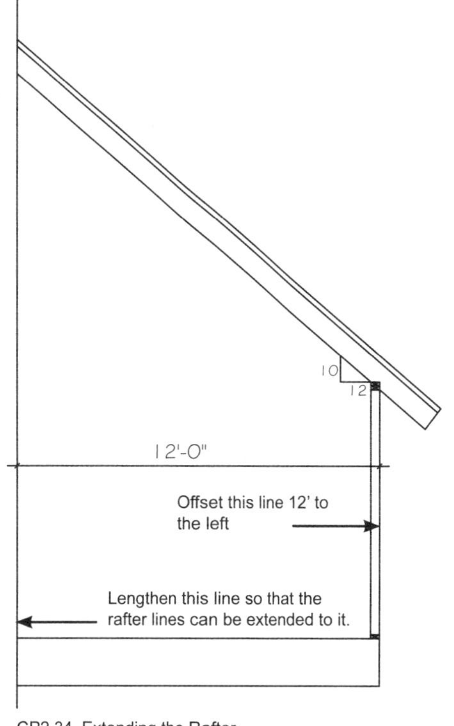
10
12
12'-0"
Offset this line 12' to the left
Lengthen this line so that the rafter lines can be extended to it.

CP2.34 Extending the Rafter

g. Mirror the rafter, foundation, and right outside wall as shown in Figure CP2.35. Select the offset line created in Step 5(e) as the mirror line.

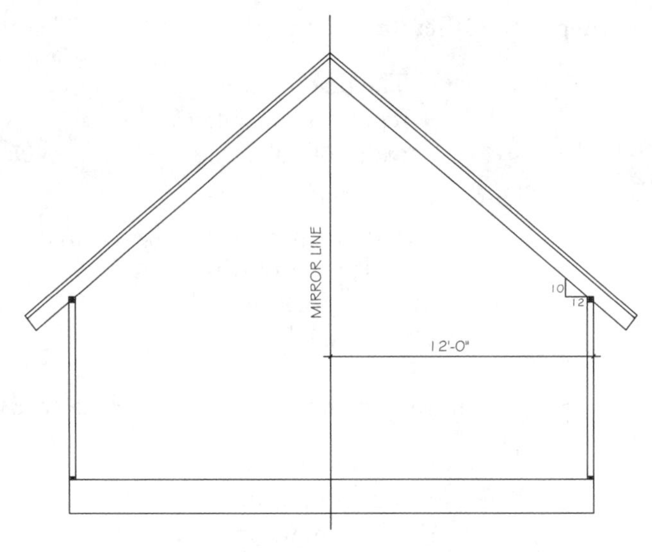

CP2.35 Mirroring the Roof Line

Step 7. Left side wall and roof profile

a. Copy the side wall created with the **MIRROR** command in the previous step, **8'** toward the left (**Ortho** should be on) as shown in Figure CP2.36.

b. Construct the rafter as you did in the previous steps.

- This time use a **4/12** pitch.

c. Extend the rafter lines until they intersect the **10/12** pitch roof section as shown in Figure CP2.36.

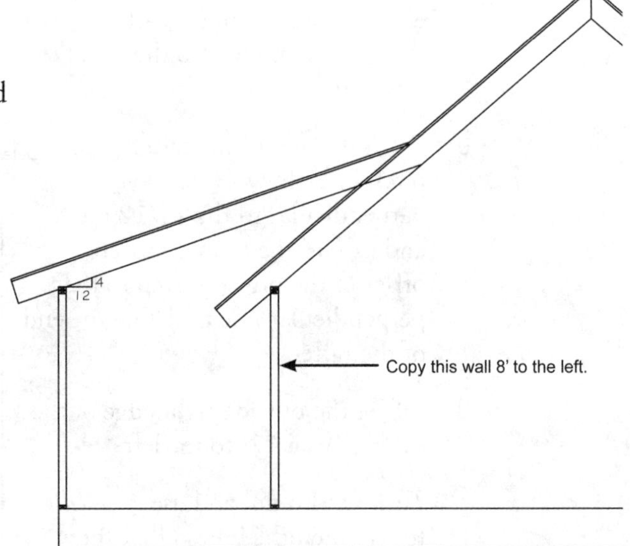

CP2.36 Constructing the Left Side Wall and Roof

NOTE

You will not be able to add the roof of the water heater closet to the Right Elevation until it is constructed in the Front Elevation. Once it is drawn in the Front Elevation, project the roof to the Right Elevation as shown in Figure CP2.39(c).

Step 8. Use **TRIM**, **EXTEND**, and **ERASE** to complete the profile of the right elevation as shown in Figure CP2.37.

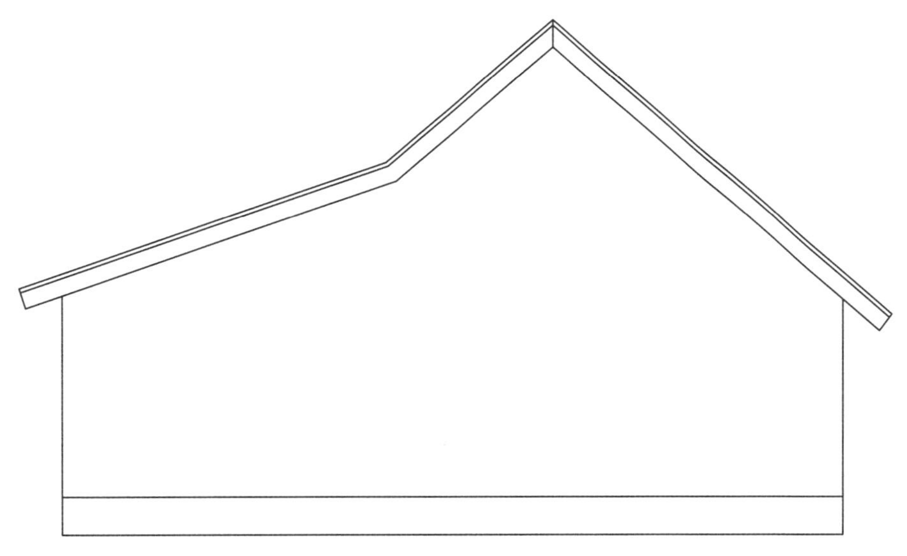

CP2.37 Right Elevation after Removal of Construction Lines (Shown)

Step 9. Insert the windows

Window dimensions for the **3°4°** window are shown in Figure CP2.38. The dimensions for the wood trim around the windows will be the same for other windows and exterior doors.

a. To add windows and doors, open **DesignCenter** and insert blocks of the exterior door and windows from the **Cabin Symbols** drawing into the elevation view.

b. Use a **6"** wide trim piece along the top edges of doors and windows, and draw the angled cuts at **15°** from vertical – other vertical trim is drawn **4"** wide.

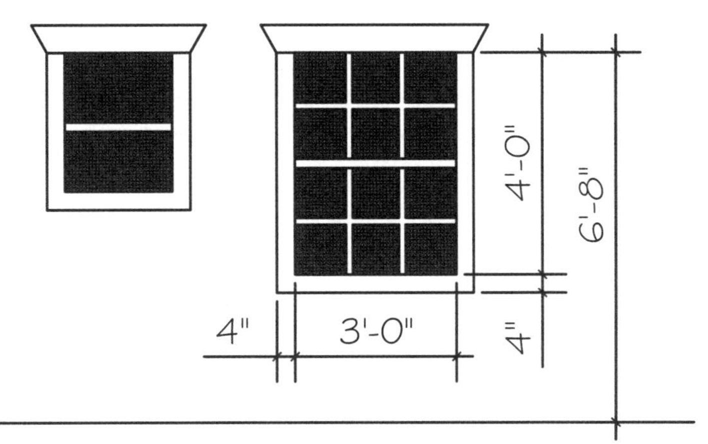

CP2.38 Window Dimensions

Step 10. Draw the porch and deck elevations

a. Refer to Figures CP2.39(a) and CP2.39(b) to construct the steps in the Front and Right side elevations.

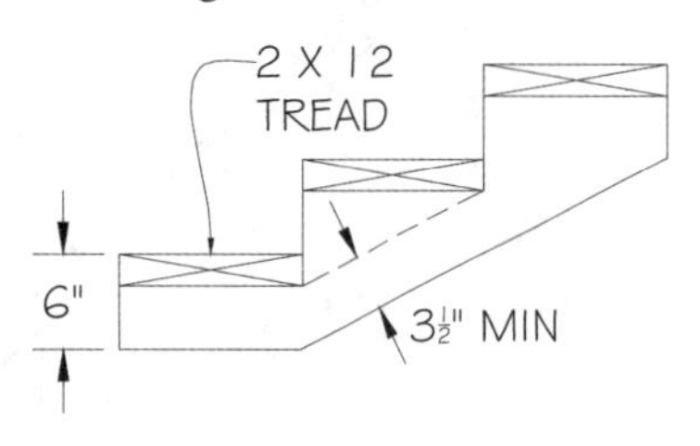

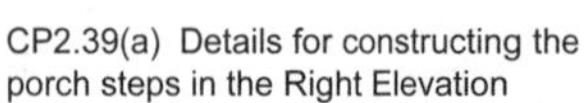
CP2.39(a) Details for constructing the porch steps in the Right Elevation

CP2.39(b) Details for constructing the 4' x 6' side deck

Step 11. Create the chimney and water heater closet

a. Using dimensions from the floor plan, add the chimney and locate the edges of the water heater closet in the right elevation (refer to Figure CP2.27).

CONSTRUCTING THE FRONT ELEVATION

Step 1. Projecting elevation features

Information from features in the right elevation such as the roof height and location and height of the porch steps and chimney can be projected to the front elevation. The dashed lines in Figure CP2.39(c) show where geometry is projected between the views.

a. Construct the side wall and the **4/12** roof pitch for the water heater closet in the front elevation using the same techniques used during the construction of the right elevation.

- Notice how the water heater closet's roof extends to the cabin's right side wall in the front elevation.
- It will be necessary to project the roof geometry of the water heater closet from the front elevation back to the right elevation to complete the construction of the water heater closet in the right elevation.

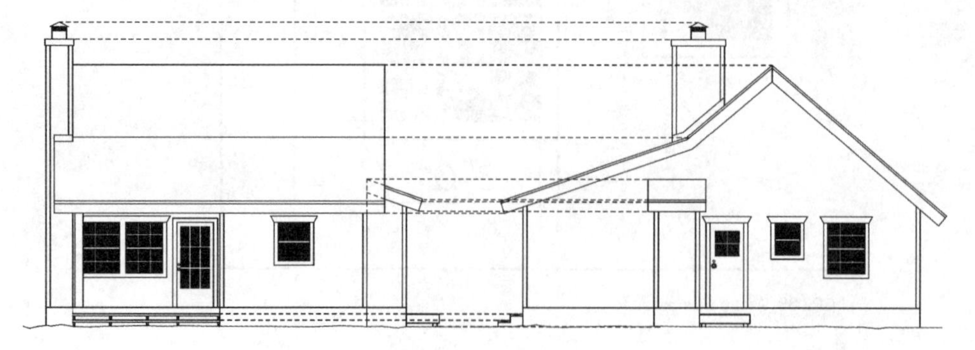

CP2.39(c) Projecting Features between the Right Elevation and the Front Elevation

Step 2. To complete the front elevation, open **DesignCenter** and insert blocks for the front door and windows from the **Cabin Symbols** drawing into the elevation view.

Figure CP2.40 shows the correct spacing for the front door and windows. The dimensions reflect the distances between centers. These dimensions are for reference only and should not be shown on the elevation view.

Step 3. Add leaders and notes to the elevation views as shown in Figures CP2.29 and CP2.43.

a. Set the Notes layer current.

b. Select the **Spline with Arrow** multileader style provided in the template and follow the prompts to place the leaders as shown in Figures CP2.29 and CP2.43.

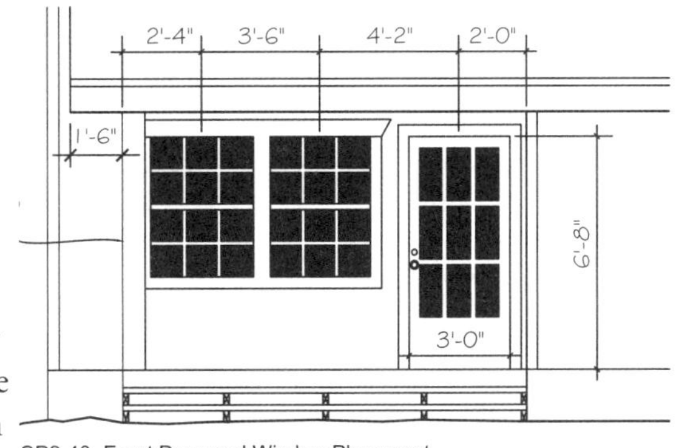

CP2.40 Front Door and Window Placement

c. Use Single Line Text with a text height of 4" for all elevation notes.

Step 4. Use either the Spline or Polyline command to draw the rough ground surface in front of each elevation.

Step 5. Hatch the siding in the elevation views

a. Set the Hatch layer current.

b. Refer to CP2.41 to create a User defined hatch pattern to create the Cedar V-Joint siding on the cabin.

c. The hatch pattern will go around text and multileaders.

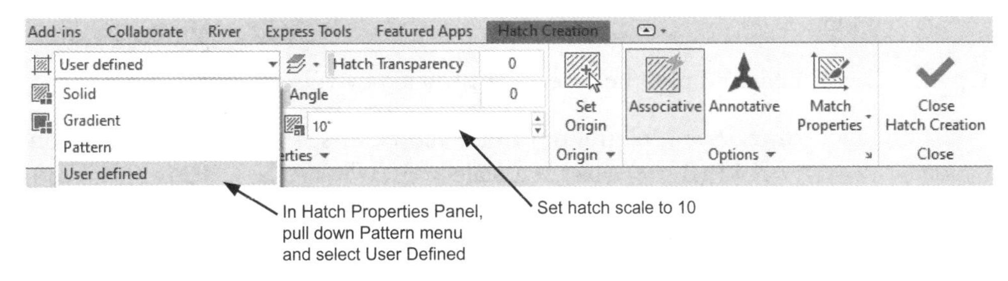

CP2.41 User defined hatch pattern for siding

Step 6. Add pattern for metal roof in the front view

a. On the right side of the roof in the front view, offset the vertical line that projects from the top of the roof to the bottom of the roof at a distance of 7" to the left (don't include the 1 x 2 trim or 2 x 10 rafter).

b. Offset the new line 2" to the left.

c. Refer to Figure CP2.42(a) to array the two new roof lines using the Rectangular Array command.

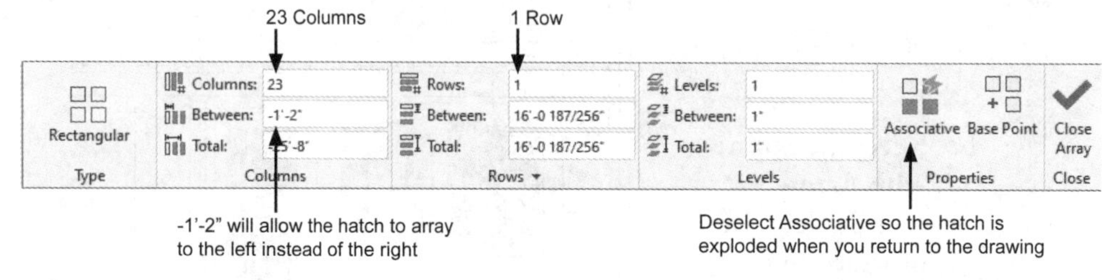

CP2.42(a) Polar Array for Metal Roof

d. Trim arrayed roof lines as-needed to form to the shape of the roof.

e. Trim arrayed roof lines as needed around Single-line Text.

f. Refer to Figures CP2.42(b) and CP2.42(c) to mask a border around leaders and Multi-line Text.

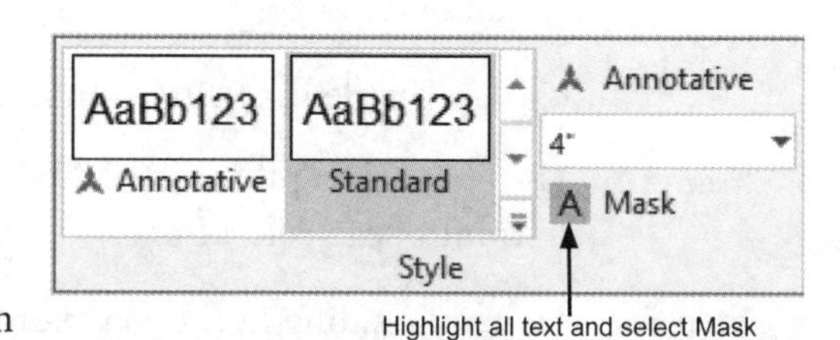

CP2.42(b) Style Panel of Text Editor

Step 7. When you are finished drawing the elevation views and floor plan in model space, left-click on the Layout Tabs labeled *Floor Plan* and *Elevations* in the lower-left corner of the graphics window and edit the text in the title blocks by double-clicking on the text to enable the text edit option.

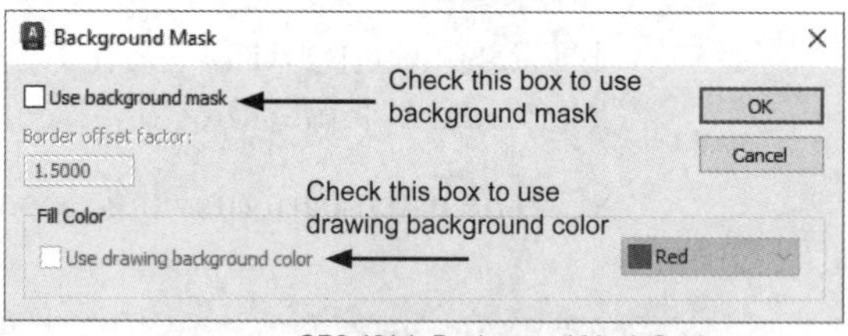

CP2.42(c) Background Mask Settings

Step 8. Save the file and plot the sheets as instructed by your teacher.

- **Note:** If your plotter or printer allows, these sheets can be plotted on a 17 X 11 (B size) sheet at a scale of 3/16"= 1'-0".
- The *plot style table* should be set to monochrome.

QUICK TIP

Follow the steps presented in Chapter 4 **"4.18 Creating a Page Setup for Plotting"** to create a **Page Setup** for plotting each sheet of this project. The plotted sheets will resemble the examples shown in Figures CP2.9 and CP2.10.

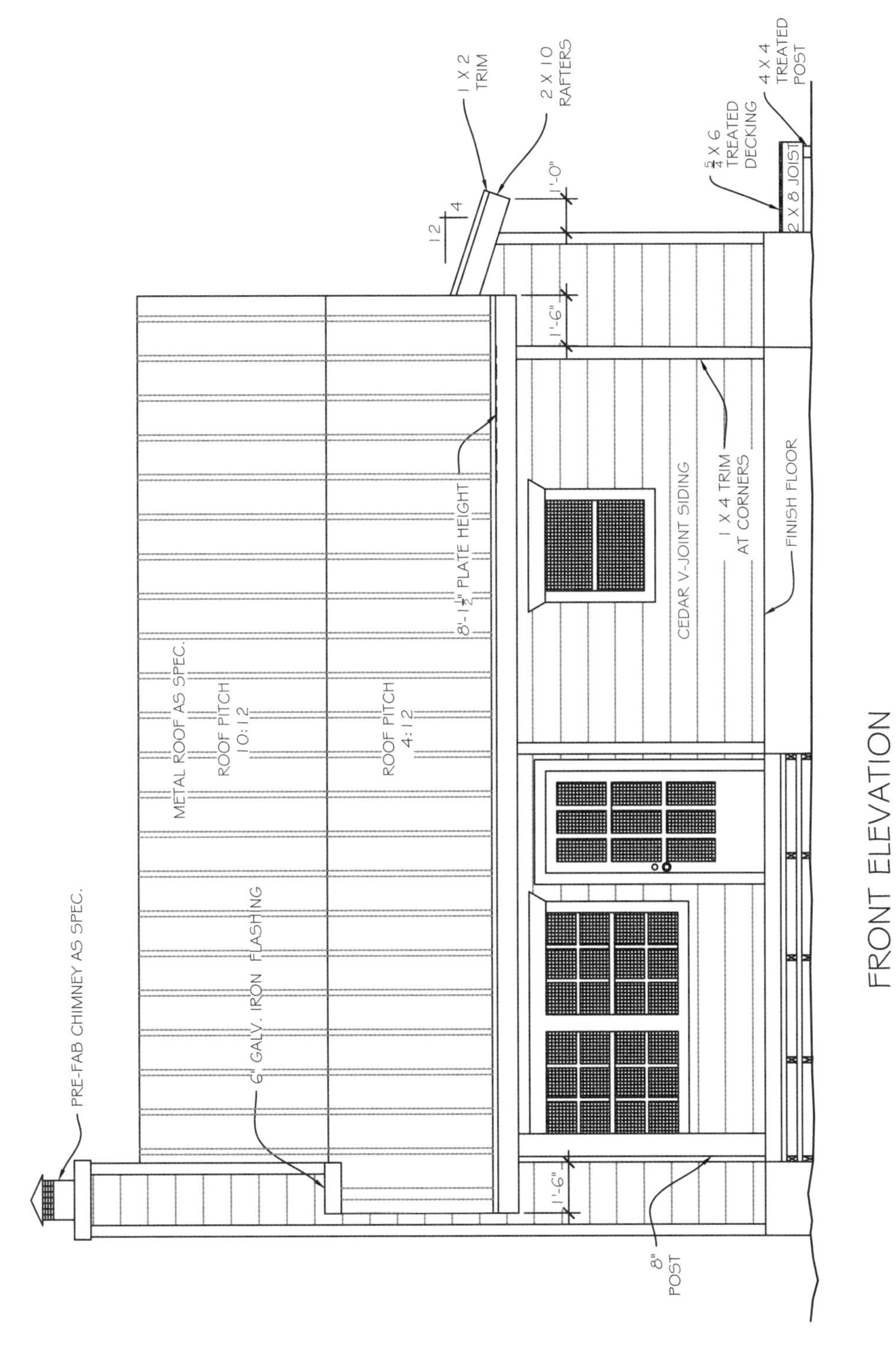
1 X 2
TRIM
2 X 10
RAFTERS
4 X 4
TREATED
POST
5/4 X 6
TREATED
DECKING
2 X 8 JOIST
12
4
1'-0"
1'-6"
8'-1 1/2" PLATE HEIGHT
CEDAR V-JOINT SIDING
1 X 4 TRIM
AT CORNERS
FINISH FLOOR
METAL ROOF AS SPEC.
ROOF PITCH
10:12
ROOF PITCH
4:12
PRE-FAB CHIMNEY AS SPEC.
6" GALV. IRON FLASHING
1'-6"
8"
POST
FRONT ELEVATION
SCALE: 3/16"= 1'-0"

CP2.43 Front Elevation Notes

ANSI/ASME STANDARDS

The American National Standards Institute (ANSI) oversees the creation of thousands of norms and guidelines that directly affect the manufacturing and development of products and services in the United States. ANSI is also the U.S. representative to ISO, the International Organization for Standardization *(see Appendix B)*.

A listing of all ANSI standards can be found at the ANSI website (www.ansi.org), but many of the ANSI/ASME standards that are relevant to the creation of engineering drawings are as follows:

- Y14.1-2012 Decimal Inch Drawing Sheet Size and Format
- Y14.1M-2012 Metric Drawing Sheet Size and Format
- Y14.2-2014 Line Conventions and Lettering
- Y14.3-2012 Orthographic and Pictorial Views
- Y14.31-2014 Undimensioned Drawings
- Y14.34-2013 Associated Lists
- Y14.35-2014 Revision of Engineering Drawings and Associated Documents
- Y14.5 Dimensioning and Tolerancing
- Y14.6-2001 Screw Thread Representation
- Y14.38-1999 Abbreviations and Acronyms

NOTES:

ISO STANDARDS

The International Organization for Standardization (ISO) is a consortium of the national standards institutes from more than 160 member countries. These institutes collaborate to establish standards for the development and manufacturing of products and services. The goal of ISO is not only to improve the quality of these products and processes, but also to make them safer and more efficient. Another goal of international standards is to make trade between countries easier and fairer and to provide governments with a technical base for health, safety, and environmental legislation. ISO standards also help safeguard consumers and end users of products and services. The United States is represented in the ISO network by the American National Standards Institute (ANSI).

The first attempts at international standardization began in the early twentieth century, but it was not until 1946 that delegates from 25 countries met and created an international organization. The purpose of this organization would be "to facilitate the international coordination and unification of industrial standards." The new organization, which was named ISO, officially began operations on February 23, 1947.

In English, the acronym ISO stands for the International Organization for Standardization; in French, ISO stands for *Organisation internationale de normalisation.* Because of the many countries and languages represented in the ISO network, it was decided to use a word derived from the Greek *isos*, meaning "equal" to describe the organization. For this reason, regardless of the language, the organization is referred to as ISO. ISO has published over 21,000 standards covering almost every product, system, or service.

A listing of most ISO standards can be found at the ISO website (www.iso.org), but a partial list of the ISO standards that relate specifically to the creation of technical drawings follows:

- *ISO 128-1- 2003 (R2015) Technical Drawings – General Principles*
- *ISO 129-1- 2004 (R2013) Indication of Dimensions and Tolerances*

There are multiple sub-categories of standards applying to technical drawings encompassed by the ISO 128 and 129 Standards. For example, ISO 128-2 covers basic conventions for lines and ISO 129-2 covers dimensions and tolerances on mechanical engineering drawings.

NOTES:

THE UNITED STATES NATIONAL CAD STANDARD

The United States National CAD Standard (NCS) coordinates CAD-related publications for the building design and construction industry in an attempt to make communication between design/construction teams and clients more consistent and direct.

Although adoption of the NCS by the building design and construction industry is voluntary, several government agencies have adopted the standard, and many public and private organizations are in the process of adopting it.

At its website (www.nationalcadstandard.org), the NCS lists the following benefits as the advantages of adopting the NCS standard:

BENEFITS TO CLIENTS AND OWNERS

- Consistent organization of data for all projects, from all sources
- Greater clarity of communication of design intent to the client
- Streamlined electronic data management of facility management data
- Enhanced potential for automated document storage and retrieval
- Streamlined construction document checking process

BENEFITS TO DESIGN PROFESSIONALS

- Consistent data classification for all projects, regardless of the project type or client
- Seamless transfer of information among architects, engineers, and other design team members
- Predictable file translation results between formats; reduced preparation time for translation
- Reduced file formatting and setup time when adopted by software application vendors
- Reduced staff training time to teach "office standards"
- Streamlined checking process for errors and omissions
- New opportunities for expanded services and revenue beyond building design
- Added value to design services; firms can feature compliance with the NCS

BENEFITS TO CONTRACTORS AND SUBCONTRACTORS

- Consistent drawing sheet order and sheet organization; information appears in the same place in all drawing sets

- Consistent detail reference system
- Reduction of discrepancies, reducing the potential for errors, change orders, and construction delays
- Enhanced potential for automated payment process
- Consistent organization of data for all projects, from all sources

INDUSTRY-WIDE BENEFITS FOR NCS ADOPTION

- Reduced in-office training time with "collective professional memory" of a drawing standard
- Improved training at undergraduate and graduate levels
- Enhanced potential for automated training and distance learning
- Substantial reduction of barriers to seamless exchange of building construction data, leading to greater efficiency and decreased costs

NOTES:

GEOMETRIC DIMENSIONING AND TOLERANCING BASICS

Applying plus/minus dimensioning to a drawing allows a mechanical designer to define the location and size of a part's features within a certain allowance, but applying the concepts of geometric dimensioning and tolerancing (GDT) allows the designer also to define the form (flatness, straightness, circularity, and cylindricity), orientation (perpendicularity, angularity, and parallelism), or position of a part's features.

Applying GDT increases the odds that a part will pass a quality control inspection, and fewer rejected parts will result in lower production costs.

On a drawing, GDT symbols are shown inside a rectangular box called a feature control frame. If the feature being defined by the GDT symbol is located relative to a datum feature (usually a surface or axis), the datum feature is defined with a datum feature symbol.

Figure D.1 shows a view of an object with GDT symbols added. Study this figure and note how the objects identified as the feature control frame and the datum feature symbol are represented.

INTERPRETING THE FEATURE CONTROL FRAME

The feature control frame contains the GDT characteristic symbol (parallelism, perpendicularity, etc.), the tolerance, and if the tolerance is referenced from a datum(s), the datum reference letter(s). Figure D.2 shows a feature control frame with its components identified.

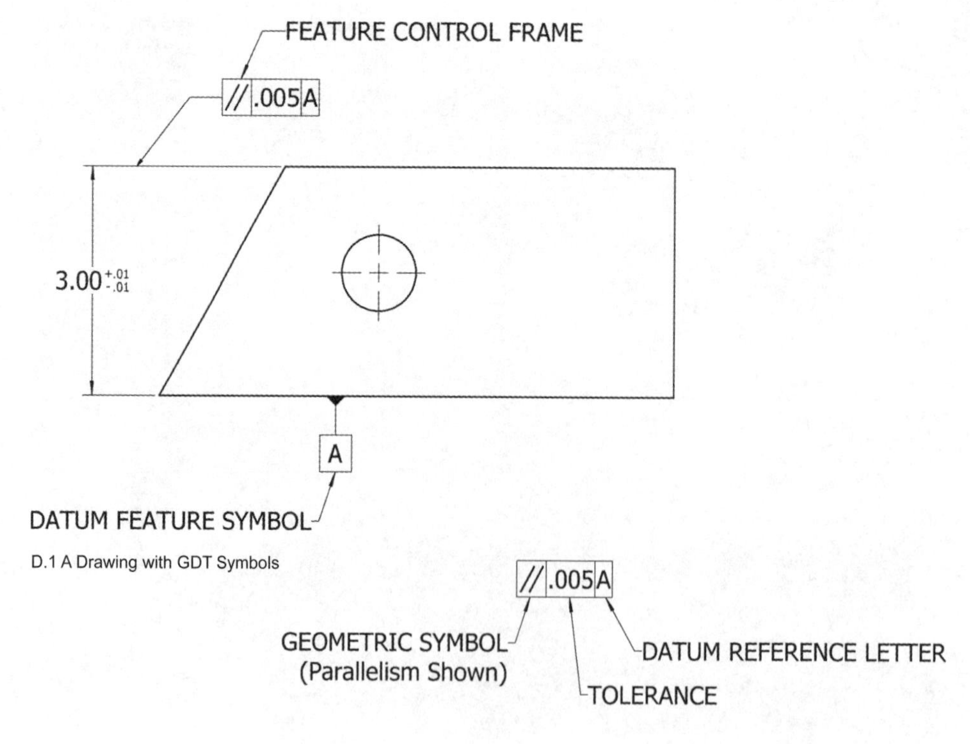

D.1 A Drawing with GDT Symbols

D.2 Interpreting a Feature Control Frame

GEOMETRIC CHARACTERISTIC SYMBOLS

The GDT characteristic symbols that may be included in a feature control frame are shown below. The symbols used to represent these characteristics are defined in the American Society of Mechanical Engineers standard for dimensioning and tolerancing, ASME Y14.5. These symbols are included in the Tolerance command found on AutoCAD's Dimension toolbar. See Figures 5.42(a) and Figures 5.42(b).

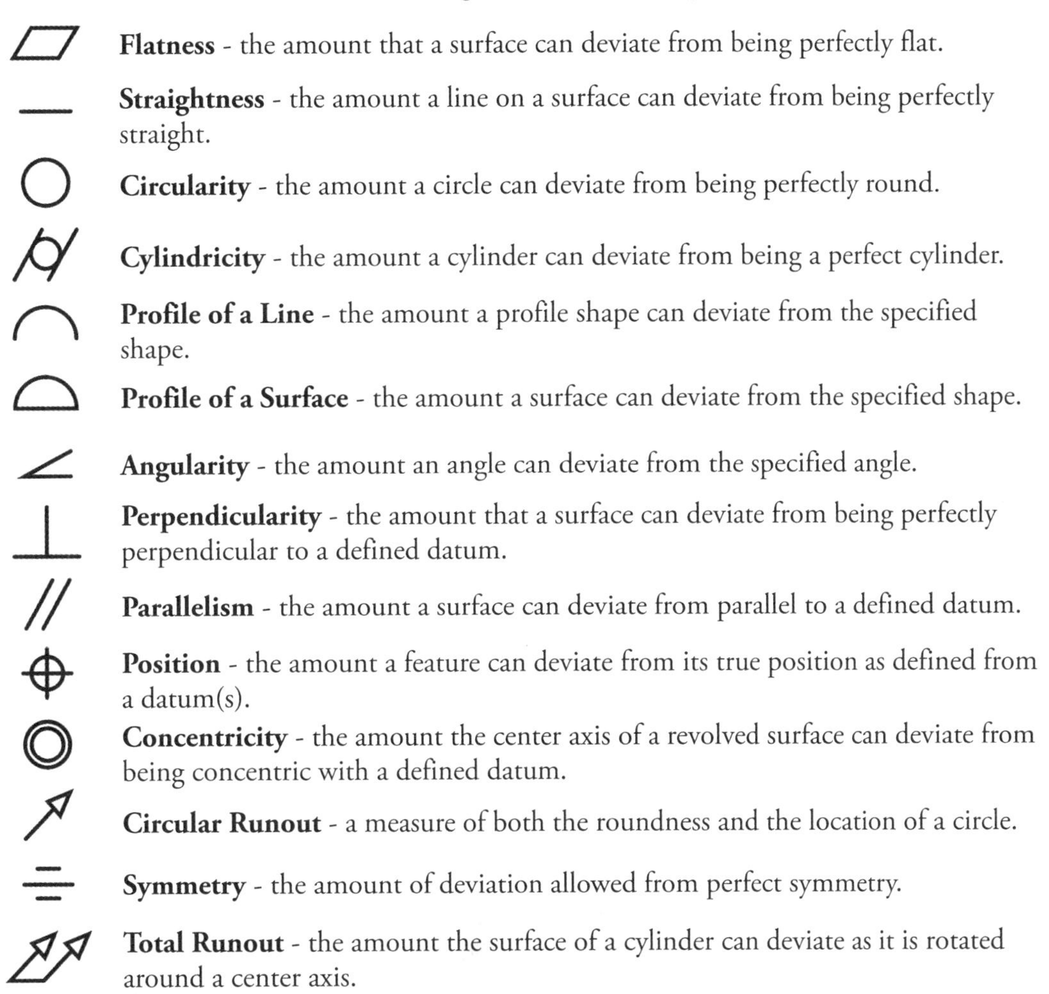

Flatness - the amount that a surface can deviate from being perfectly flat.

Straightness - the amount a line on a surface can deviate from being perfectly straight.

Circularity - the amount a circle can deviate from being perfectly round.

Cylindricity - the amount a cylinder can deviate from being a perfect cylinder.

Profile of a Line - the amount a profile shape can deviate from the specified shape.

Profile of a Surface - the amount a surface can deviate from the specified shape.

Angularity - the amount an angle can deviate from the specified angle.

Perpendicularity - the amount that a surface can deviate from being perfectly perpendicular to a defined datum.

Parallelism - the amount a surface can deviate from parallel to a defined datum.

Position - the amount a feature can deviate from its true position as defined from a datum(s).

Concentricity - the amount the center axis of a revolved surface can deviate from being concentric with a defined datum.

Circular Runout - a measure of both the roundness and the location of a circle.

Symmetry - the amount of deviation allowed from perfect symmetry.

Total Runout - the amount the surface of a cylinder can deviate as it is rotated around a center axis.

INTERPRETING GDT ON A DRAWING

Figure D.3 shows a drawing of an object with a feature control frame specifying the allowed deviation from perfect flatness for the bottom surface of the object. The distance between the two lines representing the .005 tolerance is exaggerated in this figure to better illustrate the concept.

INTERPRETING PARALLELISM

Figure D.4 shows a drawing of an object with a feature control frame specifying the allowed deviation that the top surface can have from being parallel to the bottom surface-identified with a datum feature symbol as datum A. The distance between the two lines representing the .005 tolerance is exaggerated in this figure to better illustrate the concept.

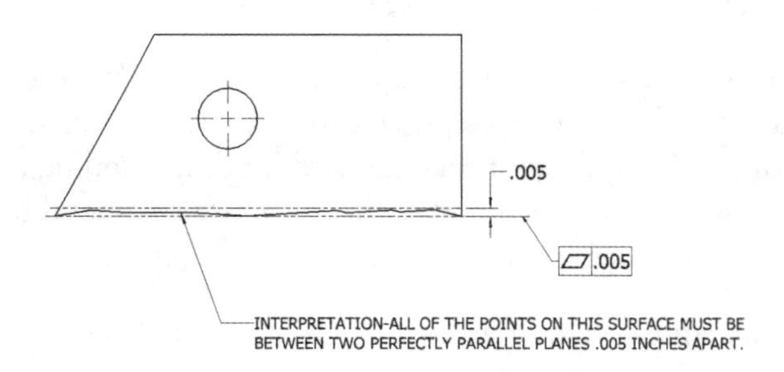

D.3 Interpreting the GDT Symbol for Flatness as Applied to the Bottom Surface on an Object

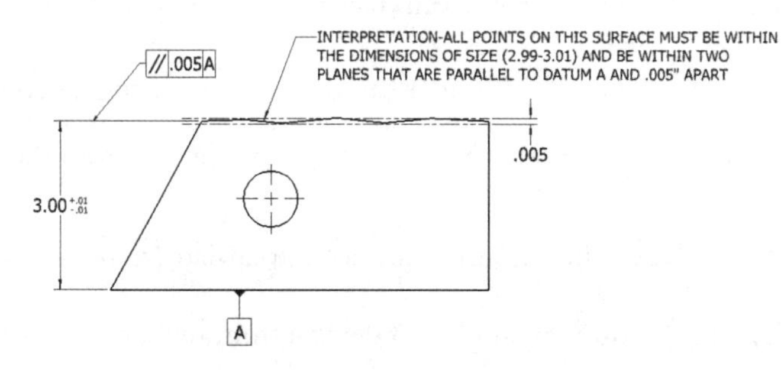

D.4 Interpreting the GDT Symbol for Parallelism as Applied to the Top Surface of an Object

ADDING GDT TO AN AUTOCAD DRAWING

Step 1. Select the **Tolerance** icon from AutoCAD's **Dimension** toolbar (see Figure D.5). This will open the **Geometric Tolerance** dialog box shown in Figure D.6.

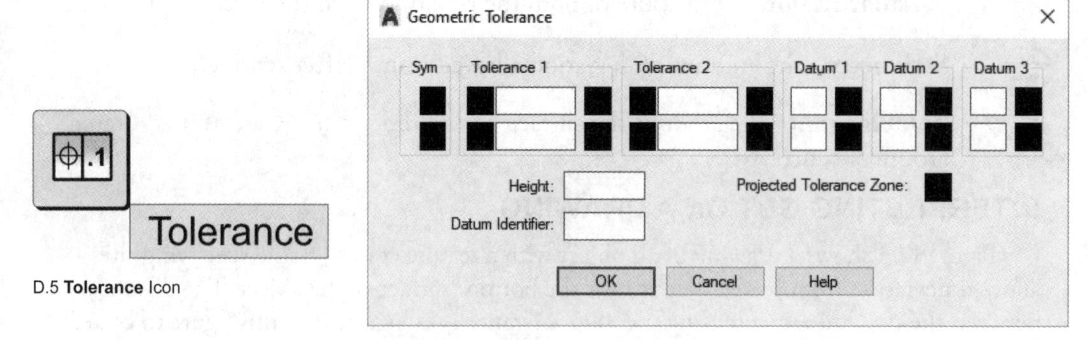

D.5 **Tolerance** Icon

D.6 **Geometric Tolerance** Dialog Box

Step 2. Select the first black box below the word *Sym* located in the **Geometric Tolerance** dialog box shown in Figure D.6. This will open the **Symbol** box shown in Figure D.7.

Step 3. Select the geometric characteristic symbol from the symbols available in the **Symbol** box (see Figure D.7). After you select a symbol, it will appear in the **Geometric Tolerance** dialog box in the window beneath *Sym*, as shown in Figure D.8.

Step 4. Fill in the values for the desired tolerance and datum(s) as shown in Figure D.8 and click **OK**. Then, use the mouse to move the feature control frame to its desired location in the drawing.

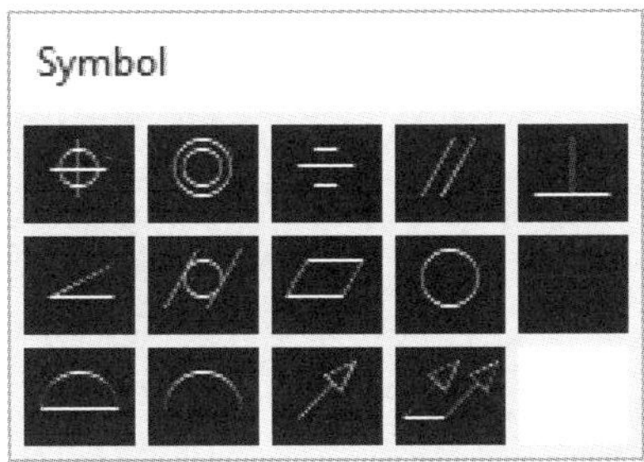

D.7 Symbol Box

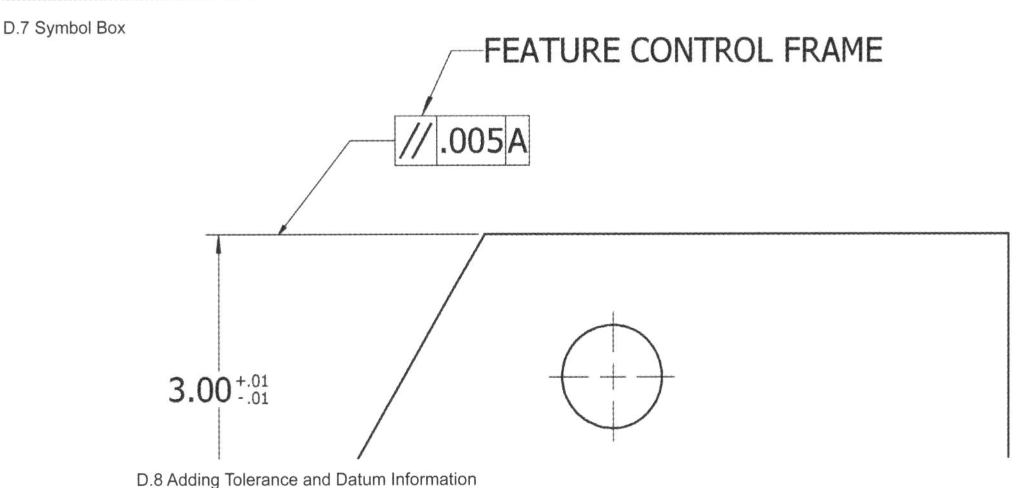

D.8 Adding Tolerance and Datum Information

ADDING A DATUM FEATURE SYMBOL

You can add a datum feature symbol by following Steps 1 and 2 in the preceding section, but instead of selecting the black box below the word **Sym**, type the desired datum identification character (an A, for example) in the **Datum Identifier** box and click **OK**. Then, use the mouse to move the datum feature symbol to its desired location in the drawing.

LEARNING TO APPLY GDT TO DRAWINGS

Learning to apply GDT to drawings is a process requiring instruction, study, and practice. GDT techniques are often included in the curriculum of advanced mechanical engineering drawing courses. Many organizations also offer continuing education GDT workshops for professionals. A good resource for GDT materials and certification is the ASME website: www.asme.org.

NOTES:

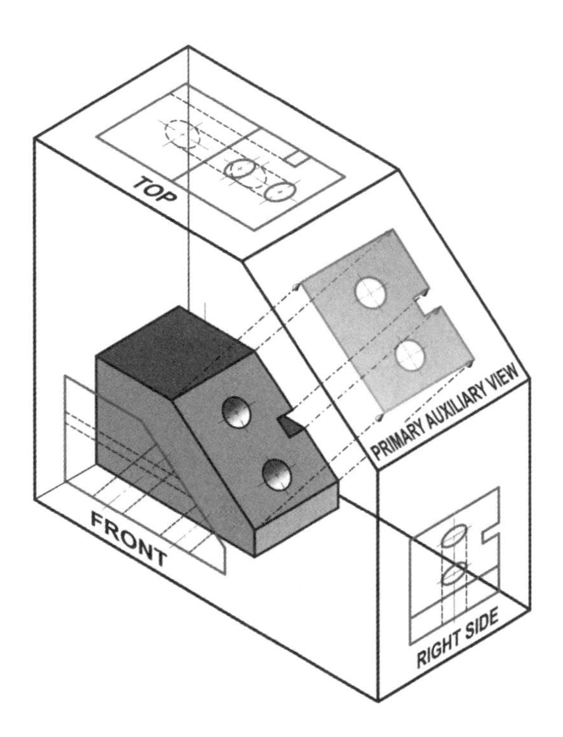

AUXILIARY VIEWS:

In some instances, such as when an object has features on an inclined plane, the regular multiviews may not describe these features in their true size or shape. In Figure E.1, the holes and slot located on the inclined plane labeled A are not shown in true shape in either the top or right-side view because the plane is foreshortened in both views. This situation may present problems when a drafter is attempting to dimension these features.

In such cases, the drafter may decide to draw an auxiliary view of the inclined plane. The auxiliary view is drawn as if the viewer's line of sight were perpendicular to the inclined plane. The features of the inclined plane will appear true size and shape in the auxiliary view.

If the drafter is working from a 3D CAD model of the object, preparing an auxiliary view is a relatively easy process of rotating the model until the inclined plane is parallel to the plane of projection.

If the drafter is working with 2D geometry, the process of creating an auxiliary view is more complicated. This appendix applies the techniques formulated in the 18th century by the founder of modern technical drawing, Gaspar Monge, to add an auxiliary view to a 2D multiview drawing.

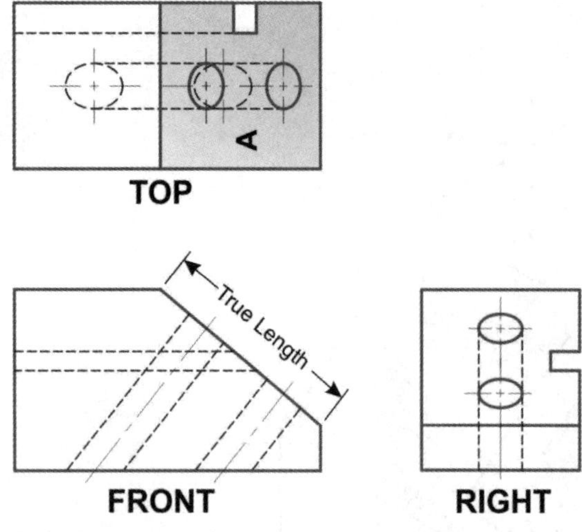

E.1 Multiview Drawing of an Object Including Inclined Plane A

The American Society of Mechanical Engineers standard governing the creation and labeling of auxiliary views is *ASME Y14.3-2012 Orthographic and Pictorial View Drawings.*

E.1 VISUALIZING AN AUXILIARY VIEW

In Figure E.2 the object shown in Figure E.1 has been placed inside a glass box. The box has a projection plane labeled ***primary auxiliary view***. This projection plane is parallel to the object's inclined plane.

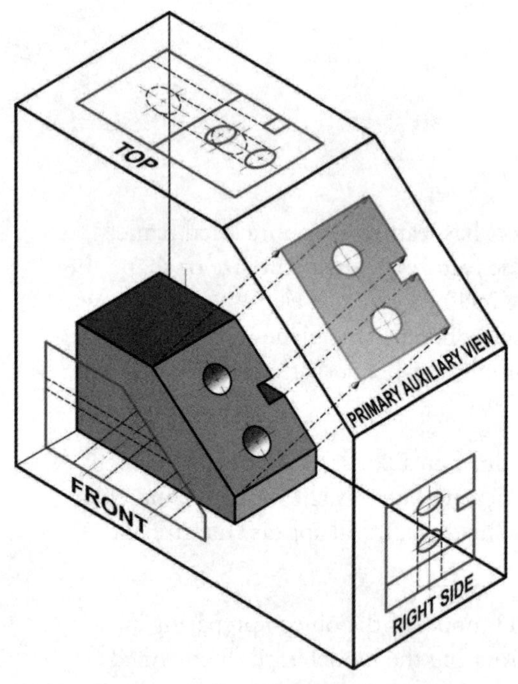

E.2 Glass Box with an Auxiliary View of *Plane A*

Viewed through this projection plane, the inclined surface is true size and shape. This is because its features are projected from the front view, perpendicular to the projection plane.

In Figure E.3 the projection planes of the glass box are unfolded. In the resulting views, you can see the position of the auxiliary view relative to the other views.

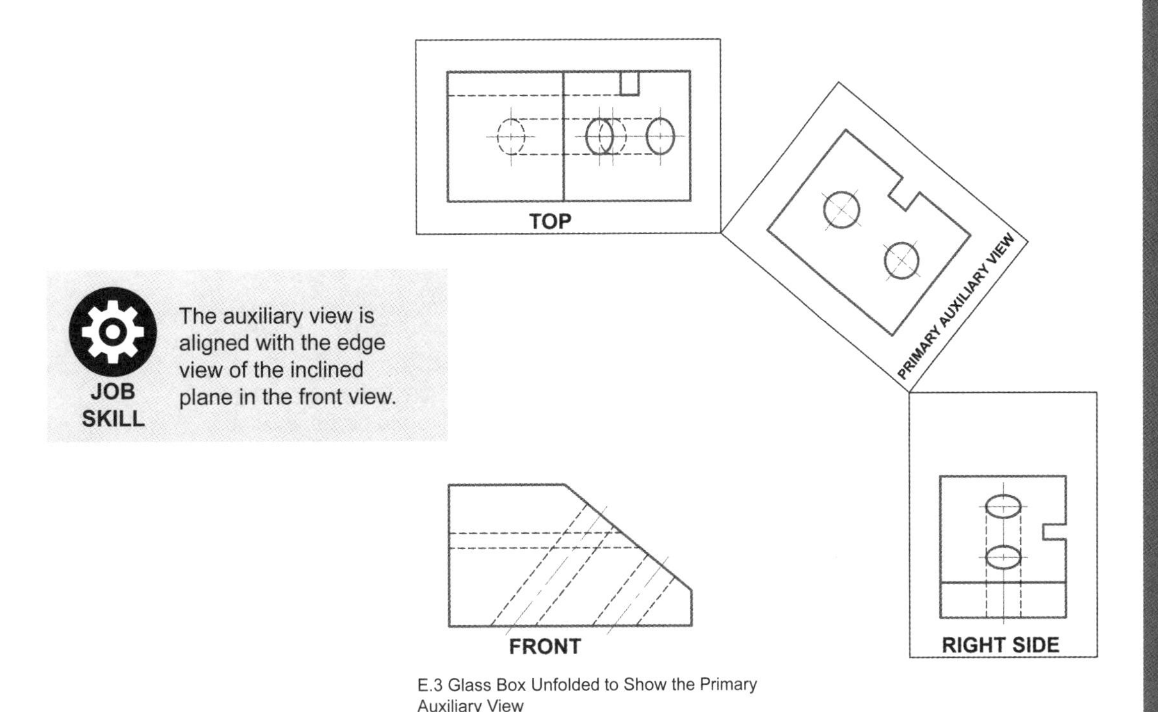

JOB SKILL

The auxiliary view is aligned with the edge view of the inclined plane in the front view.

E.3 Glass Box Unfolded to Show the Primary Auxiliary View

The view in Figure E.3 is called a primary auxiliary view because it is adjacent to, and aligned with, a principal view of the object.

A ***secondary auxiliary view*** would be adjacent to, and aligned with, a primary auxiliary view.

STEP–BY–STEP

TUTORIAL E.1: CONSTRUCTING A PRIMARY AUXILIARY VIEW – DESCRIPTIVE GEOMETRY METHOD

In this project you will use AutoCAD and the principles of descriptive geometry to create a primary auxiliary view for the object shown in **Figure E.3**.

1. Download the **Auxiliary View Tutorial Prototype** drawing file located in the *Prototype Drawing Files* folder associated with this book. Use **SAVE AS** to save the drawing to your **Home** directory, and rename the drawing **AUXILIARY VIEW TUTORIAL E.1**.
2. Follow Steps 1 through 33 to create a primary auxiliary view of the inclined plane.

Step 1. Create two new layers named **Reference** and **Projections**. Assign the **Phantom** linetype to the **Reference** layer. On the Reference layer, you will draw two reference lines from which distances are measured. You will use the Projections layer for laying out construction lines. The prototype drawing already contains layers 0, Center, Hidden, and Visible.

APPENDIX

STEP–BY–STEP

Step 2. At the command line, type **SNAP** and press **<Enter>**. When prompted with *Specify snap spacing or [ON/OFF/Aspect/Style/Type] <0.5000>:*, type **R** for **ROTATE**, and press **<Enter>**.

Step 3. When prompted with *Specify a base point:*, select the top end of the line in the front view labeled *Plane A* (see **Figure E.4**).

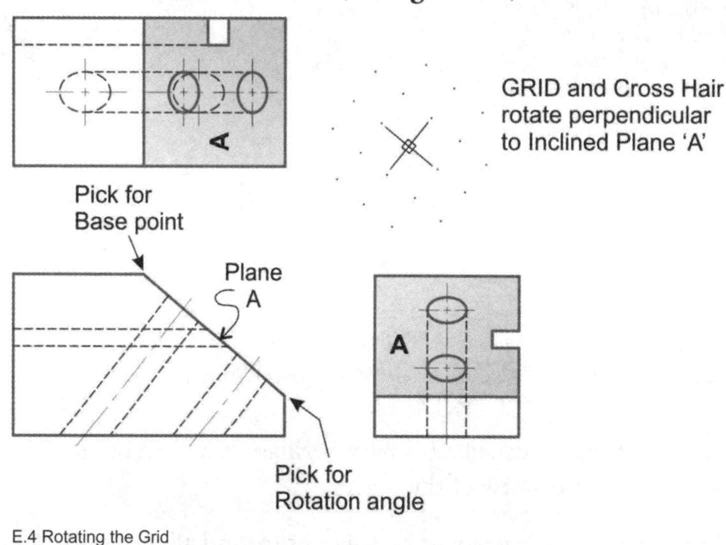

E.4 Rotating the Grid

Step 4. When prompted with *Specify rotation angle:*, select the bottom end of the line labeled *Plane A*. The grid and crosshairs will rotate and be perpendicular to *Plane A*. The grid and cursor should be turned as in **Figure E.4**.

Step 5. Set the **Projections** layer current, turn **Ortho** on, and draw two **12"** lines extending from each endpoint of inclined *Plane A* as shown in **Figure E.5**.

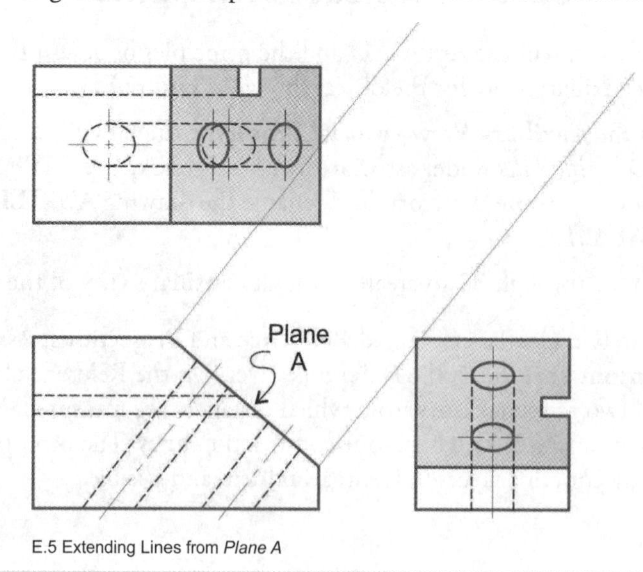

E.5 Extending Lines from *Plane A*

STEP–BY–STEP

Step 6. Set the **Reference** layer current. Draw a line parallel to inclined *Plane A* as shown in **Figure E.6**. An easy way to do this is by drawing a line from *midpoint* to *midpoint* of the two construction lines drawn in Step 5. This line will be referred to as *Reference line 1.*

Step 7. Turn **Ortho** off, and draw a second line on the left edge of the right-side view as shown in **Figure E.6**. This line will be referred to as *Reference line 2.*

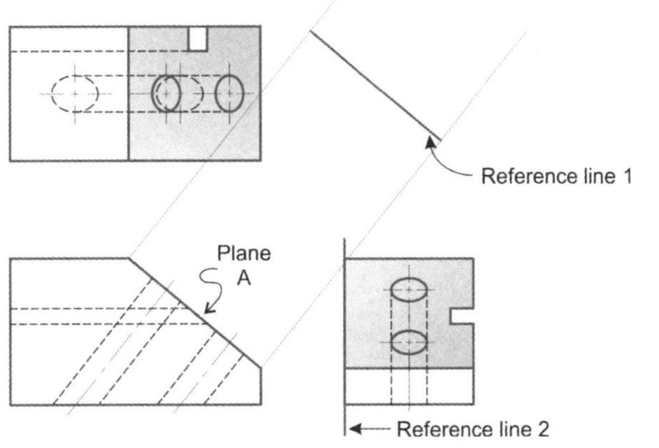

E.6 Adding Reference Line1 and Reference Line 2

Step 8. Use the **Distance** command to measure the distances from *Reference line 2* to the points labeled *a* and *b* on the right-side view in **Figure E.7** by selecting the **Distance** button from the Inquiry toolbar or typing **DI** at the command prompt. When prompted with *dist Specify first point:*, select point *1* on the right-side view where the object intersects *Reference line 2.*

Step 9. When prompted with *Specify second point:*, select point a on the right-side view. The **Distance** command determines that the distance between these points, measured along the *X*-axis, equals **2.75**.

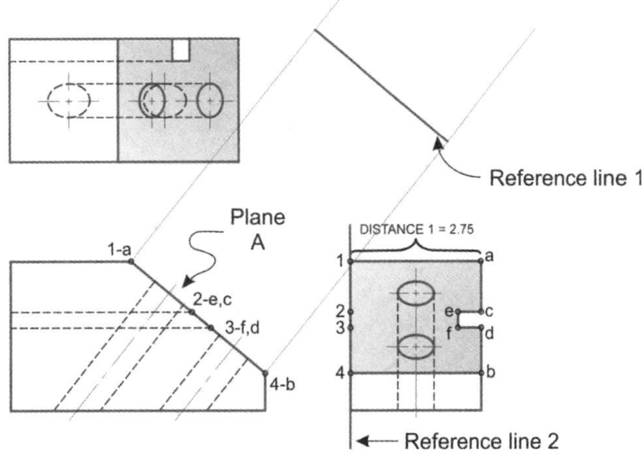

E.7 Measuring Distance between *Point 1* and *Point a*

APPENDIX

STEP–BY–STEP

Step 10. Use the **Distance** command to measure the distance between point *4* and point *b* in the right-side view, as you did in Step E.

Step 11. To add lines *1-a* and *4-b* to the auxiliary view as shown in **Figure E.8**, set the **Visible** layer current. Transfer the distance found between point *1* and point *a* in Step 8 by drawing a line **2.75"** in length from the top point of *Reference line 1* along the angled projection line constructed in Step 6.

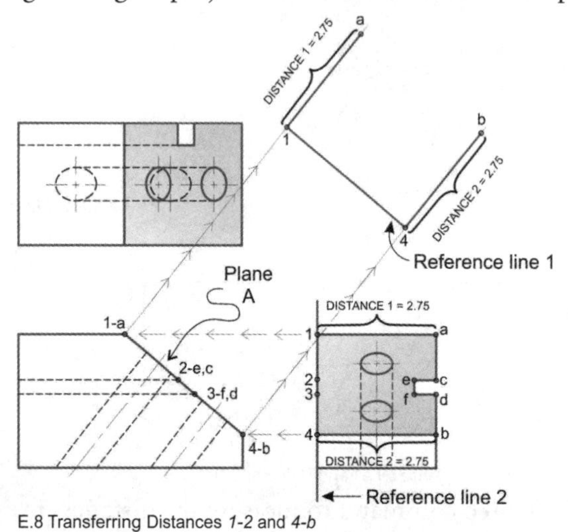

E.8 Transferring Distances *1-2* and *4-b*

Step 12. Draw another line **2.75"** in length from the bottom point of *Reference line 1* along the other angled projection line to represent the distance between point *4* and point *b* (see **Figure E.8**).

Step 13. To locate the top and bottom edges of the slot in the auxiliary view (see **Figure E.9**), set the **Projections** layer current, turn **Ortho** on, and project construction lines from the points in the front view labeled *2-e,c* and *3-f,d*, as described in Steps 14 and 15.

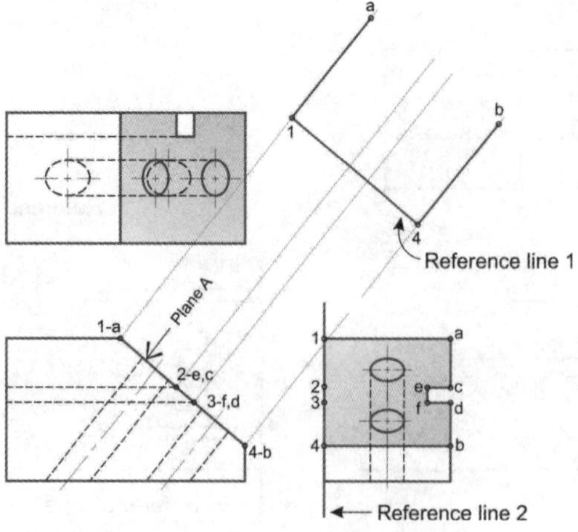

E.9 Extending Lines from *Point 2-e, c* and *Point 3-f, d*

STEP–BY–STEP

Step 14. Draw a construction line **12"** long from the point labeled *2-e,c* in the front view that is perpendicular to *Reference line 1*.

Step 15. Draw a construction line **12"** long from the point labeled *3-f,d* in the front view that is perpendicular to *Reference line 1*.

Step 16. Use the **POINT** command and the **Snap From** object snap setting to locate the points labeled *c* and *d* on the angled construction lines in **Figure E.10**. Set the **Reference** layer current, and follow Steps 17 through 24.

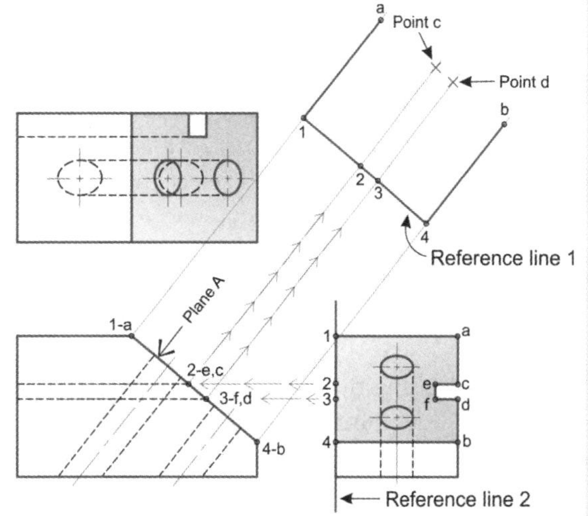

E.10 Locating *Point c* and *Point d* in Auxiliary View

Step 17. Select **Point Style** from the **Format** pull-down menu. When the **Point Style** dialog box opens, select the **X** point style (fourth from the right on the top row), select the **Set Size in Absolute Units** button, set the **Point Size** to **.125**, and click **OK**.

Step 18. Turn on the following **Osnaps**: **Endpoint, Midpoint, Center, Node,** and **Intersection**. **Dynamic Input** should also be on. Refer to **Figure E.10** and note that the distance in the right-side view from point *2* to point *c* is the same distance as measured earlier between point *1* and point *a* (**2.75"**). This is true of the distance between point *3* and point *d* as well.

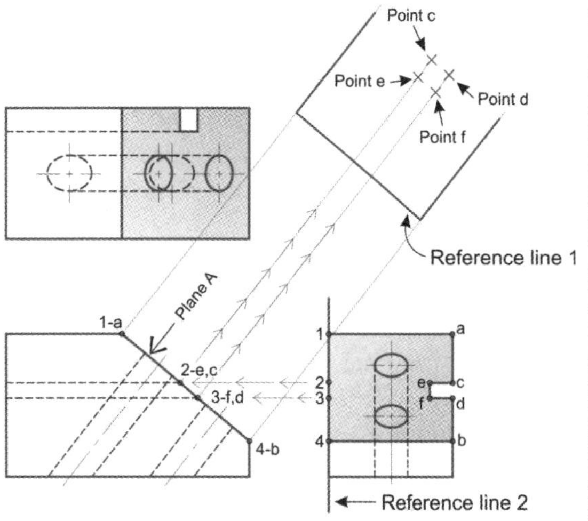

E.11 Locating *Point e* and *Point f* in Auxiliary View

APPENDIX

STEP–BY–STEP

Step 19. Open the **Object Snap** toolbar and select the **POINT** command from the **Draw** toolbar. When prompted with *Specify a point:*, select the **Snap From** icon (see **Figure 4.105**) on the **Object Snap** toolbar and select the point where the angled line drawn from point *2-e,c* in the front view intersects *Reference line 1*. When prompted with *from Base point:,* move the cursor to the top end of the angled projection line until the **Endpoint** object snap lights up (do not select the point), type **2.75**, and press **<Enter>**.

Step 20. Repeat the **POINT** command as in Step 19, except this time, when prompted with *Specify a point:*, select the **Snap From** icon on the **Object Snap** toolbar and select the point where the angled line drawn from point *3-f,d* in the front view intersects *Reference line 1*. When prompted with *from Base point:*, move the cursor to the top end of the angled projection line until the **Endpoint** object snap lights up (do not select the point), type **2.75**, and press **<Enter>**.

Step 21. Using the **POINT** command and the **Snap From** object snap setting, locate the points labeled *e* and *f* on the angled construction lines as shown in **Figure E.11**.

Step 22. Use the **Distance** command to find the distance between point *2* and point *e* in the right-side view.

Step 23. Select the **POINT** command from the **Draw** toolbar. When prompted with *Specify a point:*, select the **Snap From** icon on the **Object Snap** toolbar and select the point where the angled line drawn from point *2-e,c* in the front view intersects *Reference line 1*. When prompted with *from Base point:*, move the cursor to the top end of the angled projection line until the **Endpoint** object snap lights up (do not select the point), type the distance between point *2* and point *e* in the right-side view, and press **<Enter>**.

Step 24. Repeat the **Point** command as in Step 23, except this time, when prompted with *Specify a point:*, select the **Snap From** icon on the **Object Snap** toolbar and select the point where the angled line drawn from point *3-f,d* in the front view intersects *Reference line 1*. When prompted with *from Base point:*, move the cursor to the top end of the angled projection line until the **Endpoint** object snap lights up (do not select the point), type the distance between point *2* and point *e* in the right-side view, and press **<Enter>**.

STEP-BY-STEP

Step 25. Set the **Visible** layer current and use the **LINE** command to connect points *a,c,e,f,d,* and *b* in the auxiliary view as shown in **Figure E.12**. Change *Reference line 1* to the **Visible** layer. This completes the true shape profile of inclined *Plane A*.

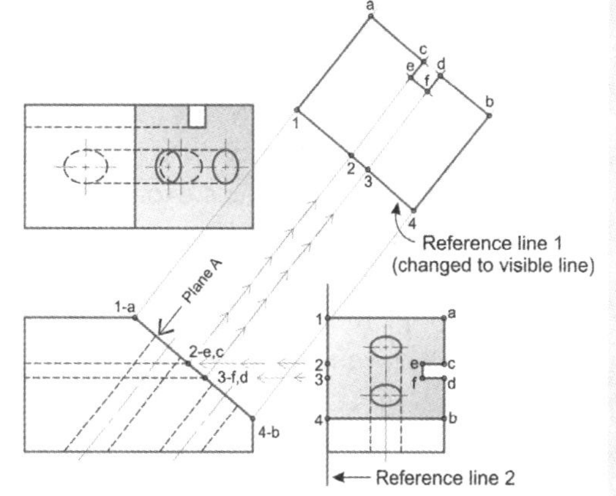

E.12 Connecting *Points a,c,e,f,d,* and *b* in the Auxiliary View

Step 26. To locate the centers of holes marked *g* and *h* on the right-side view in **Figure E.13** to their positions in the auxiliary view, set the **Projections** layer current.

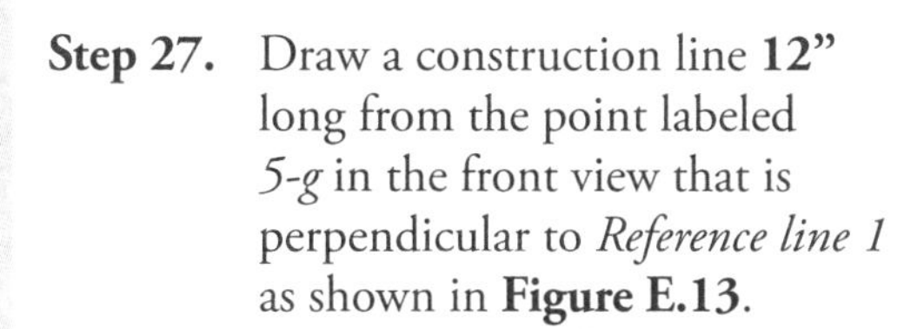

Step 27. Draw a construction line **12"** long from the point labeled *5-g* in the front view that is perpendicular to *Reference line 1* as shown in **Figure E.13**.

Step 28. Repeat Step 27, but this time, draw a construction line **12"** long from the point labeled *6-h* in the front view that is perpendicular to *Reference line 1* as shown in **Figure E.13**.

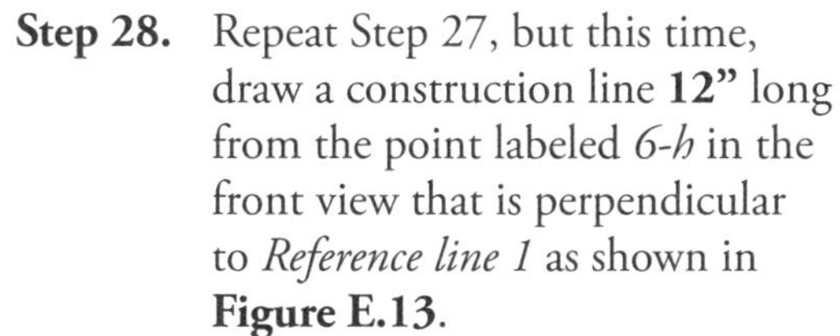

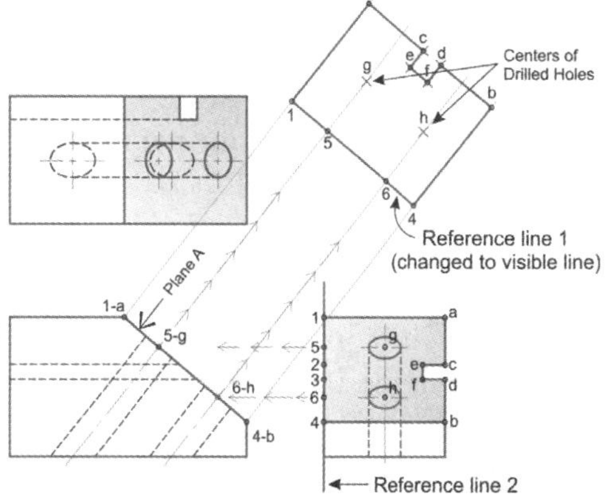

E.13 Locating *Points g* and *h* in the Auxiliary View

Step 29. Use the **Distance** command to find the horizontal distance between *Reference line 2* and point *g* in the right-side view.

Step 30. Select the **POINT** command from the **Draw** toolbar. When prompted with *Specify a point:,* select the **Snap From** icon on the **Object Snap** toolbar and select the point where the angled line drawn from point *5-g* in the front view intersects *Reference line 1*. When prompted with *from Base point:,* move the cursor to the top end of the angled projection line until the **Endpoint** object snap lights up (do not select the point), type the distance between *Reference line 2* and point *g* in the right-side view, and press **<Enter>**.

APPENDIX

STEP–BY–STEP

Step 31. Repeat the **POINT** command as in Step 30, except this time, when prompted with *Specify a point:*, select the **Snap From** icon on the **Object Snap** toolbar. Select the point where the angled line drawn from point *6-h* in the front view intersects *Reference line 1*. When prompted with *from Base point:*, move the cursor to the top end of the angled projection line until the **Endpoint** object snap lights up (do not select the point), type the horizontal distance between *Reference line 2* and point *h* in the right-side view (this distance is the same as between *Reference line 2* and point *g*, and press **<Enter>**.

E.14 Drawing Circles at *Points g* and *h* in the Auxiliary View

Step 32. Set the **Visible** layer current and draw two circles **.75"** in diameter at the points identified in Steps 26–31 for *g* and *h* as shown in **Figure E.14**.

Step 33. Turn off the **Reference** and **Projections** layers. Add centerlines to the holes in the auxiliary view. Complete the view by adding centerlines between the front view and the holes in the auxiliary view as shown in **Figure E.15**.

Technically, the view shown in **Figure E.15** would be considered a ***partial auxiliary view*** because the other planes of the object are not shown in this view. If the drafter wished, broken lines could be added to the view to show more of the object, as in **Figure E.16**.

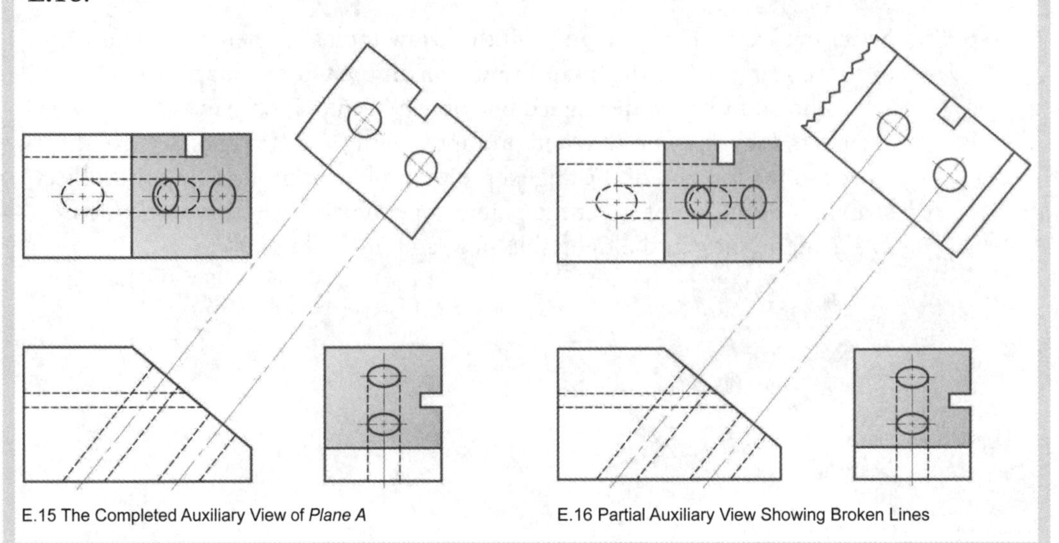

E.15 The Completed Auxiliary View of *Plane A*

E.16 Partial Auxiliary View Showing Broken Lines

E.2 CONSTRUCTING AN AUXILIARY VIEW WITH THE OFFSET COMMAND

A simpler, and quicker, way to construct an auxiliary view takes advantage of AutoCAD's **OFFSET** and **TRIM** commands. Also, instead of the **Distance** command, the **Linear** dimension command is used to measure distances in the side, and/or, top views.

The tutorial accompanying Project E.2 shows how to quickly create an auxiliary view using these CAD Commands.

STEP–BY–STEP

TUTORIAL E.2: CONSTRUCTING A PRIMARY AUXILIARY VIEW USING THE OFFSET COMMAND

In this project you will use AutoCAD's **OFFSET** command to create the primary auxiliary view for the object shown in **Figure E.3**.

1. Download the **Auxiliary View Tutorial Prototype** drawing file located in the *Prototype Drawing Files* folder associated with this book. Use **SAVE AS** to save the drawing to your **Home** directory, and rename the drawing **AUXILIARY VIEW TUTORIAL E.2**.
2. Follow Steps 1 through 10 to create a primary auxiliary view of the inclined plane.

Step 1. Create a new layer named **Projections** and set the **Visible** layer current. Select the **OFFSET** command and when prompted to *Specify offset distance:*, type **6.5** and press **<Enter>**. Select the line marked *Plane A* in the front view as the object to offset. Next, pick a point above the right-side view when *prompted to specify point on side to offset:*. The resulting line represents the left edge of the auxiliary view (see **Figure E.17**).

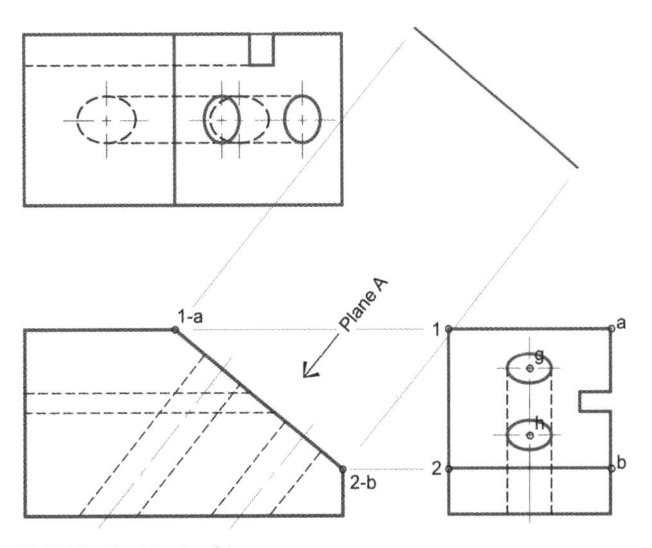

E.17 Offsetting Line *1-a, 2-b*

STEP-BY-STEP

Step 2. The width of the object in the right-side view is **2.75"**. Select the **OFFSET** command and type **2.75** as the offset distance. Select the line created in Step 1 and offset it to the right side. The resulting line represents the right edge of the auxiliary view as shown in **Figure E.18**.

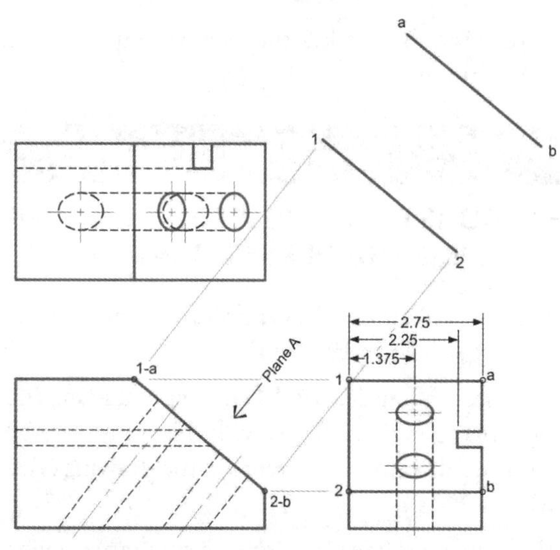

E.18 Offsetting Line *1,2* to Create Line *a,b*

Step 3. Draw a line connecting point *1* and point *a* in the auxiliary view as shown in **Figure E.19**. Draw another line connecting point *2* and point *b*.

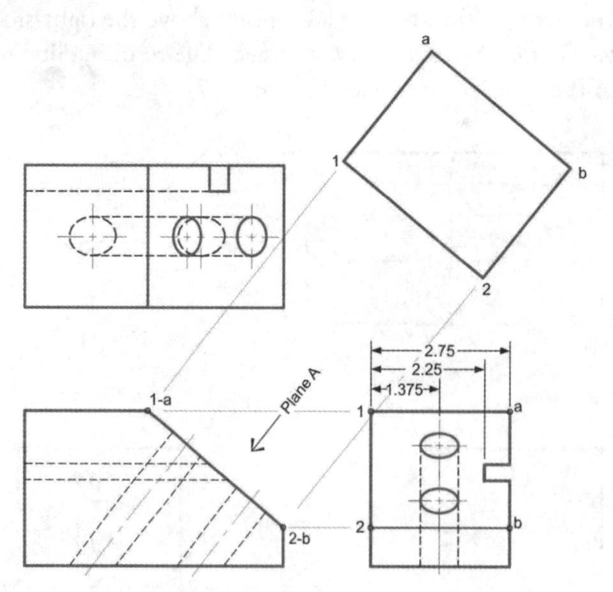

E.19 Connecting Point *1* to *a*, and Point *2* to *b*

STEP-BY-STEP

Step 4. Select the **OFFSET** command and pick line *1-2* in the auxiliary view as the line to offset. Enter **1.375** as the offset distance (the distance shown in the right-side view from the part's left edge to the center of the holes). Pick to the right of line *1-2* when prompted to *specify point on side to offset:* (see **Figure E.20**). This step creates line *x-y* in the auxiliary view as shown in **Figure E.20**.

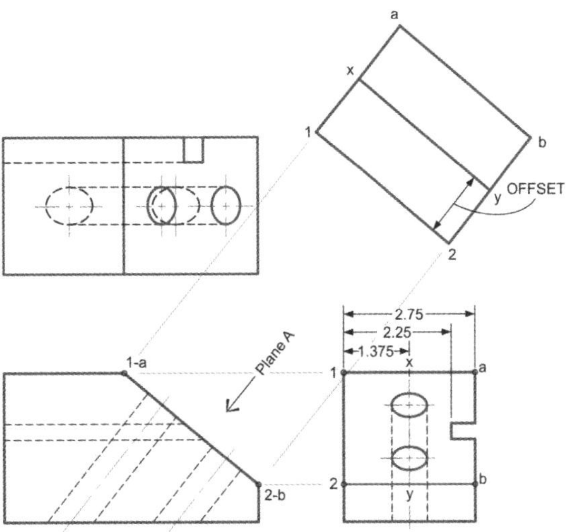

E.20 Offsetting Line *1-2* to Create Line *x-y*

Step 5. Set the **Projections** layer current and draw a line from point *3-g* on *Plane A* perpendicular to line *x-y* in the auxiliary view as shown in **Figure E.21**. Repeat this step, drawing a line from point *4-h* perpendicular to line *x-y*.

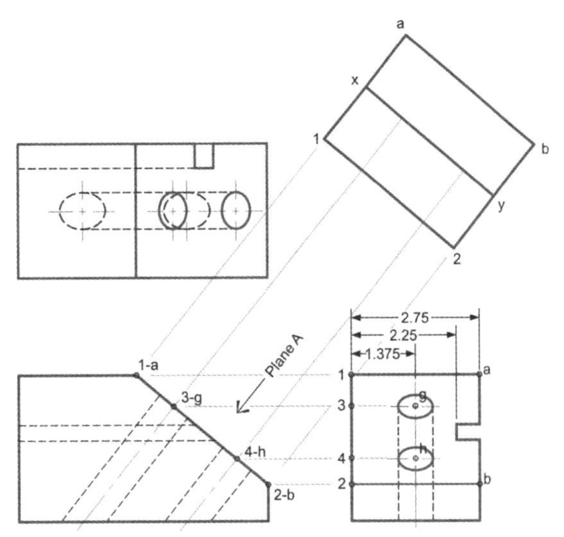

E.21 Projecting Points *3-g* and *4-h* to Line *x-y*

APPENDIX

STEP-BY-STEP

Step 6. The points where line *x-y* and the projection lines from points *3-g* and *4-h* intersect are the centers for the two drilled holes. Set the **Visible** layer current and draw a **.75"** diameter circle at each intersection as shown in **Figure E.22**.

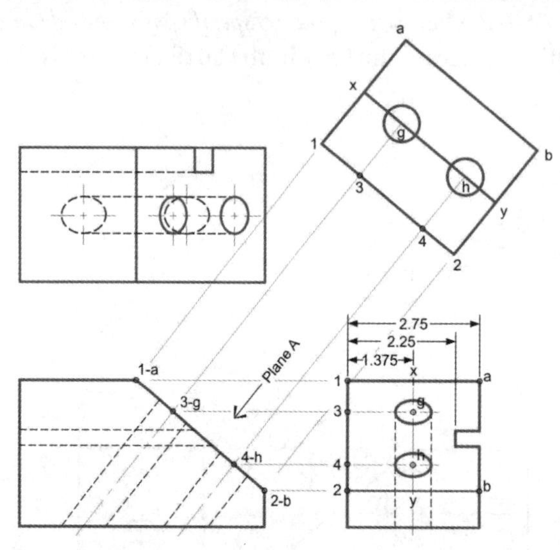

E.22 Adding Circles at Points *g* and *h*

Step 7. Select the **OFFSET** command and type **2.25** for the distance to offset (the distance shown in the dimension on the right-side view from the part's left edge to the inside edge of the slot). Select line *1-2* and offset it to the right of the line as shown in **Figure E.23**.

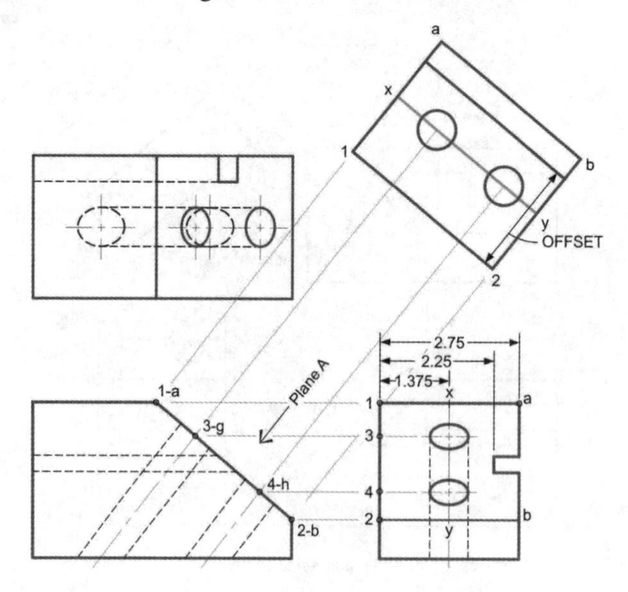

E.23 Offsetting Line *1-2* at 2.25"

STEP–BY–STEP

Step 8. Set the **Projections** layer current and project two construction lines from the intersections of points *5-e,c* and *6-f,d* perpendicular to line *a-b* in the auxiliary view as shown in **Figure E.24**.

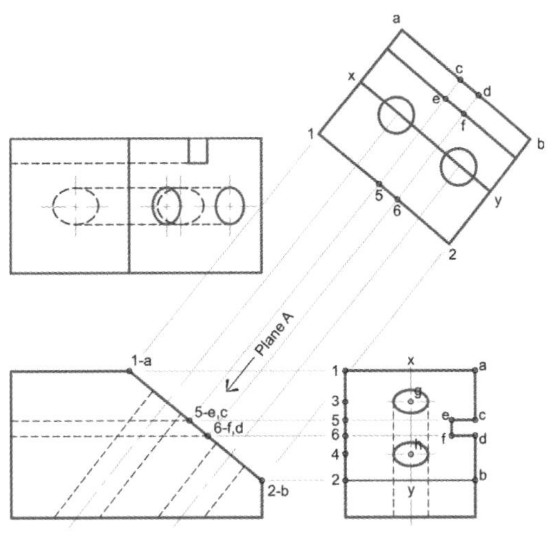

E.24 Projecting Points *5-e,c* and *6-f,d* to Locate Points *e* and *f*

Step 9. Use the **TRIM** command to trim the lines representing the slot in the auxiliary view as shown in **Figure E.25**. Set the **Visible** layer current and draw lines *c-e* and *d-f*. Erase line *x-y*.

Step 10. Turn off the **Projections** layer and complete the drawing by adding centerlines to the holes and between the front view and the holes in the auxiliary view as shown in **Figure E.26**.

This step completes the construction of the auxiliary view.

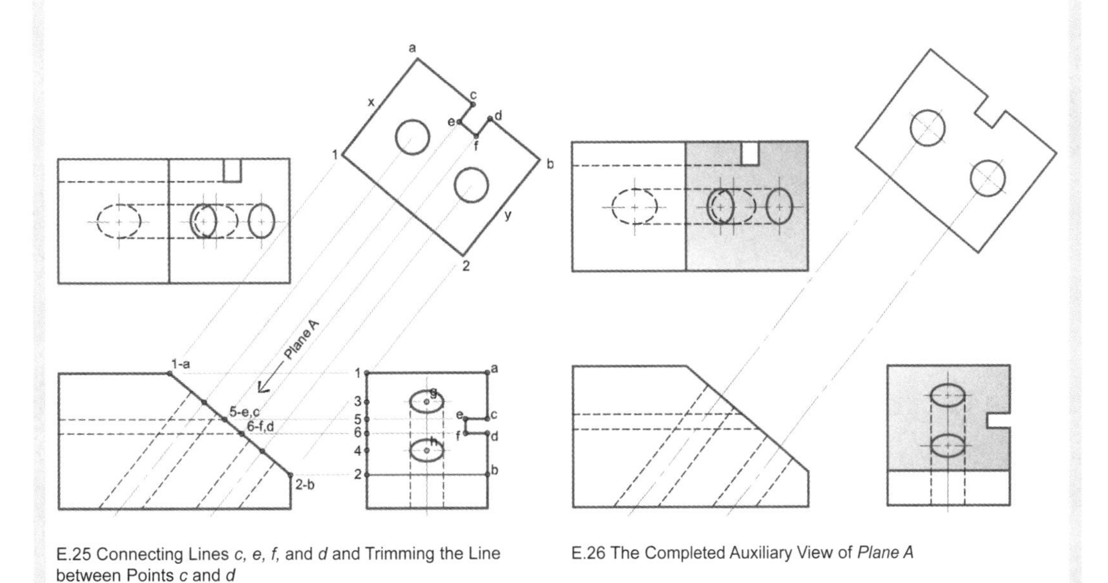

E.25 Connecting Lines *c, e, f,* and *d* and Trimming the Line between Points *c* and *d*

E.26 The Completed Auxiliary View of *Plane A*

The techniques presented in this appendix have their origins in Gaspard Monge's *Géométrie Descriptive*, written in the eighteenth century. As mentioned in Chapter 1, many of Monge's ideas later became the foundation of modern technical drawing. In modern times, with the advent of CAD software capable of 3D modeling, auxiliary views are more likely to be created by rotating a 3D model of the object until the inclined or oblique plane is parallel to the projection plane. The resulting view allows drafters to dimension the features of the plane where they appear in their *true shape* and *true size*.

Auxiliary Views
Partial Auxiliary View
Primary Auxiliary View
Secondary Auxiliary View

Multiple Choice

1. In an auxiliary view, the plane of projection is _____relative to the inclined plane.
2. Oblique
   a. Parallel
   b. Skewed
   c. Perpendicular
3. A primary auxiliary view is drawn adjacent to, and aligned with, a(n) ____ view.
   a. Principal
   b. End
   c. Secondary auxiliary
   d. Perpendicular
4. An auxiliary view may be needed if the features of an object do not appear true shape in:
   a. An oblique view
   b. An isometric view
   c. Any of the regular views
   d. All the above
5. A partial auxiliary view omits:
   a. Phantom lines
   b. Cutting plane lines
   c. Foreshortened features of the object
   d. None of the above
6. What linetype is assigned to the lines that are drawn between the principal view and an auxiliary view?
   a. Center
   b. Phantom
   c. Hidden
   d. Visible

AUXILIARY VIEW PROJECT
PROJECT E.

Download the **Auxiliary View Prototype E.1** drawing file located in the *Appendix E Prototypes* folder that is in the *Prototype Drawing Files* folder associated with this book. Use **SAVE AS** to save the drawing to your **Home** directory, and rename the drawing **AUXILIARY VIEW PROJECT E.1**.

Follow the steps shown in **Tutorial E.1** to create a partial, primary auxiliary view for the inclined plane represented by its edge in the front view as shown in **Figure E.27**. Project the auxiliary view from the front view of the object and transfer distances found in the top and side views.

When finished drawing the partial auxiliary view, left-click on the **Layout1** tab in the lower-left corner of the graphics window and edit the text in the title block by double-clicking on it to enable the text edit option. Save the file and plot the drawing as instructed by your instructor.

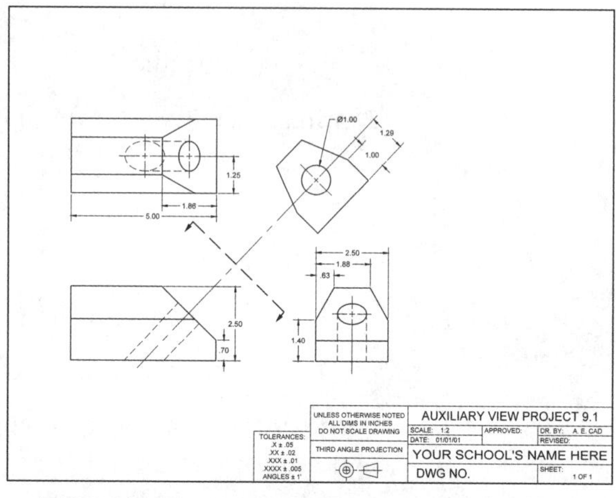

E.27 Partial Auxiliary View of Inclined Plane Projected from Front View

AUXILIARY VIEW PROJECT
PROJECT E.

Download the **Auxiliary View Prototype E.2** drawing file located in the *Appendix E Prototypes* folder that is in the *Prototype Drawing Files* folder associated with this book. Use **SAVE AS** to save the drawing to your **Home** directory, and rename the drawing **AUXILIARY VIEW PROJECT E.2**.

Follow the steps shown earlier in this appendix in **Tutorial E.2** to use the *Offset* command to create a partial, primary auxiliary view for the inclined plane represented by its edge in the front view that contains the .25" diameter hole as shown in **Figure E.28**. Project the auxiliary view from the front view of the object by offsetting distances found in the top and side views.

When finished drawing the partial auxiliary view, left-click on the **Layout1** tab in the lower-left corner of the graphics window and edit the text in the title block by double-clicking on it to enable the text edit option. Save the file and plot the drawing as instructed by your instructor.

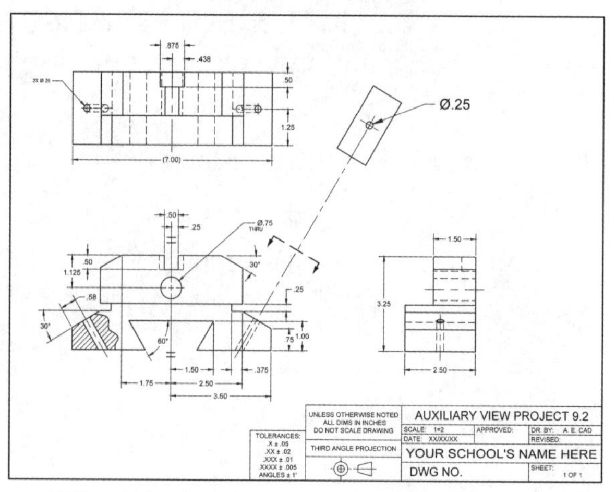

E.28 Partial Auxiliary View of Inclined Plane Projected from Front View

absolute coordinate Points that are located along the X-, Y-, and Z-axes that are relative to a point defined as 0,0,0. In an AutoCAD drawing, 0,0,0 is usually located in the lower left corner of the graphics window.

actual size The measured size of a finished part. This size determines whether the part passes a quality control inspection.

aeronautical or aerospace drafters Drafters who prepare engineering drawings detailing plans and specifications used in the manufacture of aircraft and related equipment.

aligned text Text placed on a technical drawing that faces the bottom and the right side of the sheet. This technique is common on architectural drawings but is not allowed on drawings employing the ASME dimensioning standard.

allowance The minimum clearance or maximum interference between parts.

American Institute of Architects (AIA) The accrediting body for architects.

American National Standards Institute (ANSI) The national organization for the development of standards in the United States. ANSI represents the United States as a member of the International Organization for Standardization.

American Society of Mechanical Engineers (ASME) Publisher of standards for the creation of technical drawing in the United States. ASME Y14.5 Dimensioning and Tolerancing and ASME Y14.2 Line Conventions and Lettering are two standards important to drafters.

architectural drafters Drafters who prepare the drawings used in construction industries.

assembly drawing A drawing that illustrates how the separate parts of an assembly are related to each other, for example, how mating parts fit together.

base point The point on a block that aligns to the insertion point in the field of the drawing defined when placing the block.

baseline dimensions A group of linear dimensions that are referenced from the same datum or baseline.

block library A group of block definitions stored in a drawing file. For example, an architectural firm might create a block library of symbols like doors and windows that are frequently used on floor plans. See block.

block AutoCAD term that refers to a predefined object or symbol stored in an AutoCAD drawing file that can be inserted into the drawing whenever it is needed.

broken-out section A view of an object that shows only a small area of the object as a section.

Building Information Modeling (BIM) A system of modeling and linking building system information into a digital, and increasingly 3D, database.

CAD (computer-aided design) A term often used to describe the creation of technical drawings. Also a term for the software used to create technical drawings. Some popular CAD programs include AutoCAD, SOLIDWORKS, Revit, Creo, and Inventor.

CAM (computer-aided manufacturing) Manufacturing processes in which manufacturing equipment and processes are controlled by computer commands.

Cartesian coordinate system A system of locating points along X- and Y-axes relative to a starting point representing "zero X" and "zero Y." Named for its originator, René Descartes.

checker An experienced designer/drafter with expertise in manufacturing, drafting techniques, and dimensioning conventions who is responsible for reviewing and approving the drawings prepared by other drafters.

civil drafters and design technicians Drafters who prepare construction drawings and topographical maps used in civil engineering projects.

construction documents (CDs) Drawings used in the construction of residential and commercial buildings. These drawings may include floor plans, elevations, foundations, wall sections, and roof framing plans.

continuous dimensions A dimensioning technique in which a linear dimension is placed using the second extension line origin of a selected dimension as its first extension line origin. This technique is also called chain dimensioning.

cutting plane An imaginary plane that slices through an object to reveal its interior features in a section view.

datum A theoretically perfect feature (plane, axis, or point) from which dimensions are referenced.

designer Individual who assists engineers or architects with the design process. Designers are often former drafters who have proven their ability to take on more responsibility and decision-making duties.

detail drawings Drawings that provide the information required to manufacture or purchase each part in an assembly including the necessary views, dimensions, notations, and specifications.

dimension commands The commands used to dimension an AutoCAD drawing. These commands are located on the Dimension toolbar and include Linear, Baseline, Continue, Angular, Diameter, and Radius.

dimension standards Dimensioning rules that have been created to standardize dimensioning styles and techniques. Dimensioning standards for mechanical drawings have been defined by the American Society of Mechanical Engineers (ASME) and the International Standards Organization (ISO). Dimensioning standards for construction documents are defined by the United States National CAD Standard.

dimensions Annotations that are added to a technical drawing specifying the size and location of the features of an object. There are two types of dimensions: size and location. For example, the diameter of a hole is a size dimension, whereas the dimensions that indicate the placement of the center of the hole are location dimensions. Dimensional information may include notes concerning the material from which the object is manufactured, special processes performed on the object (heat treating, polishing, etc.), and any other information needed during the manufacture of a part or the construction of a building.

drafter An individual with specialized training in the creation of technical drawings.

drafting A term often used to describe the creation of technical drawings.

draw commands The commands used to place geometry in an AutoCAD drawing. These commands are located on the Draw toolbar and Draw panel and include LINE, CIRCLE, ARC, and Multiline Text.

drawing limits The limits of an AutoCAD drawing define its drawing area; this is comparable to selecting the sheet size for the drawing. When setting limits, you are prompted to specify the lower left and upper right corners of the drawing area. In most cases, the lower left corner will default to 0,0, and you will define the upper right corners by typing the coordinates of the corresponding sheet size. For example, in decimal units, an A-size sheet limits would be 0,0 and 12,9; a B-size sheet limits would be 0,0 and 17,11; a C-size sheet limits would be 0,0 and 22,17; and a D-size sheet limits would be 0,0 and 34,22.

drawing units The units of measurement to be used in the creation of an AutoCAD drawing. For example, in an architectural drawing, one unit may equal one foot, whereas in a mechanical drawing, one unit may equal one inch or one millimeter. The drawing units should be set before beginning a drawing or defining the drawing limits.

electrical drafters Drafters who prepare diagrams used in the installation and repair of electrical equipment and building wiring.

electrical plans Drawings that provide electrical contractors information about the type, location, and installation of electrical components (switches, lamps, ceiling fans, electrical outlets, cable TV jacks, etc.) used in the project. All of the information needed by the electrical contractor to wire the building should be provided by this plan.

electro-mechanical drafters Drafters who split their duties between mechanical drafting and electrical/ electronics drafting.

electronics drafters Drafters who prepare schematic diagrams, printed circuit board artwork, integrated circuit layouts, and other graphics used in the design and maintenance of electronic (semiconductor) devices.

elevation drawings Drawings that provide information about the exterior details of a building. This information may include roof pitch, exterior materials and finishes, overall heights of features, and window and door styles. All the dimensions and notations required by workers on the jobsite should be included on this sheet.

engineer's, architect's, and metric scales Precision measurement instruments used to make measurements during the creation or interpretation of technical drawings.

engineering drawing A term often used to describe the creation of technical drawings.

engineering graphics A term often used to describe the creation of technical drawings.

features Geometric elements that are added to a base part. Features include holes, slots, arcs, fillets, rounds, angled planes, counterbored holes, and countersunk holes. Features are located on the object with location dimensions and are described with size dimensions.

fillet A rounded inside corner of a part.

finite element analysis (FEA) Software that is used in conjunction with CAD modeling software to perform advanced design analysis on 3D models. FEA allows engineers and designers to calculate such properties as an object's mass, center of gravity, strength, distribution of stresses, and moments of inertia.

first-angle projection A technique for arranging multiview drawings in which the left-side view is drawn to the right of the front view, and the top view is placed below the front view, and so on. Commonly used on drawings prepared outside North America.

floor plans Drawings that provide home builders and contractors with the necessary information to lay out the building, including the locations of features such as walls, doors, electrical components (switches, lamps, etc.), and plumbing fixtures (tubs, commodes, sinks, etc.). Floor plans usually include all of the dimensions and notations required by the workers on the job site. Doors and windows are dimensioned to their centers and continuous (also known as chain) dimensioning is typically employed on floor plans.

foreshortening A term used to describe the phenomenon that occurs when a feature on an inclined plane is not shown true size or true shape in a multiview drawing.

full section A view of an object that shows its interior detail as if it has been cut in half.

geometric dimensioning and tolerancing (GD&T) A dimensioning technique that is used to control the form (flatness, straightness, circularity, and cylindricity), orientation (perpendicularity, angularity, and parallelism), or position of a part's features.

graphic primitives Geometric shapes such as boxes, cylinders, cones, spheres, wedges, and prisms that can be combined (unioned) or removed from one another (subtracted) to create more complicated shapes.

half section A view of an object that shows its interior detail as if one fourth of it has been removed.

inclined plane A plane located on an object in a multiview drawing that is sloping and is not perpendicular to the line of sight of the viewer.

inquiry commands Commands used to display information about AutoCAD entities such as distance between two points, area of a closed figure like a rectangle or circle, or the volume of a 3D object. These commands are located on the Inquiry toolbar and Utilities panel.

insertion point The point defined in the field of the drawing that a block's base point will align to upon insertion.

International Organization for Standardization (ISO) The international organization for the development of standards including technical drawing and dimensioning standards. ANSI represents the United States as a member of ISO. The ISO dimensioning standard is almost identical to the ASME dimensioning standard.

isometric drawing A type of pictorial drawing in which receding lines are drawn at 30° relative to the horizon. Commonly used in the mechanical engineering field. See pictorial drawing.

layers In AutoCAD drawings, lines and other entities are drawn on layers. Think of layers as sheets of clear glass layered one on top of the other. A layer can have its own properties such as color, linetype, or lineweight.

lead hardness grade The scale that defines the hardness of graphite pencil leads. Soft leads range between 2B and 7B. Medium leads include 3H, 2H, H, F, HB, and B. Hard leads range between 4H and 9H.

least material condition (LMC) The condition of a part when it contains the least amount of material. The LMC of an external feature, such as a shaft, is the lower limit of size defined by the tolerance. The LMC of an internal feature, such as a hole, is the upper limit of size defined by the tolerance.

limits The maximum and minimum sizes of a feature as defined by its tolerances. For example, a feature with a nominal dimension of .50, with a tolerance of ±.02, has an upper limit of .52 and a lower limit of .48.

linetypes Include visible lines, which show the visible edges and features of an object; hidden lines, which represent features that would not be visible; and centerlines, which locate the centers of features such as holes and arcs. Standard linetypes have been established by the *American Society of Mechanical Engineers (ASME)* in ASME Y14.2.

lineweight Refers to the width of the lines in a technical drawing. Standard lineweights have been established by the *American Society of Mechanical Engineers (ASME)* in ASME Y14.2. In this standard, visible lines are drawn 0.6mm wide, and center and hidden lines are drawn .3mm wide.

maximum material condition (MMC) The condition of a part when it contains the greatest amount of material. The MMC of an external feature, such as a shaft, is the upper limit. The MMC of an internal feature, such as a hole, is the lower limit.

mechanical drafters Drafters who prepare detail and assembly drawings of machinery and mechanical devices.

mechanical working drawings Drawings used in the fabrication and assembly of machine parts.

miter line In drawings created with orthographic projection techniques, a construction line drawn at 45° that enables information to be projected from the top view to the side view, and from the side view to the top view. See orthographic projection.

modeling commands Commands used to create 3D models in an AutoCAD drawing. These commands are located on the Modeling toolbar and include UNION, SUBTRACT, 3DROTATE, and EXTRUDE.

modify commands Commands used to modify the geometry of an AutoCAD drawing. These commands are located on the Modify toolbar and Modify panel and include ERASE, MOVE, COPY, OFFSET, ROTATE, and SCALE.

multiview drawing A technique used by drafters and designers to depict a three-dimensional object (an object having height, width, and depth) as a group of related two-dimensional (having only width and height, or width and depth, or height and depth) views.

National Society of Professional Engineers The accrediting agency for engineers.

nominal size A dimension that describes the general size of a feature. Tolerances are applied to this dimension.

Object Snap settings A technique used in the creation and editing of AutoCAD drawings that allows the user to snap to exact points on an object. Common object snap settings include snap to endpoint, snap to midpoint, snap to intersection, snap to center, and snap to quadrant.

offset section A section that includes features that would not lie along the path of a straight cutting plane line. The cutting plane line is offset to take in these features.

orthographic projection The technique employed in the creation of multiview drawings to project geometric information (points, lines, planes, or other features) from one view to another.

parametric modeling A method of creating 3D CAD models in which the geometry of the model is driven by the dimensions associated with the geometry. This allows designers to modify the features of a model by simply editing its dimensions. When the parametric dimension is changed, the 3D model updates to reflect the new dimension value.

partial auxiliary view An auxiliary view that is simplified by omitting planes and other features not shown true shape in the view.

parts list A table placed on a technical drawing that itemizes all the parts in an assembly (sometimes referred to as a bill of materials or BOM). The parts list may include columns for part number, part name, description, quantity, and material.

perspective drawing A type of pictorial drawing in which receding lines appear to converge at a vanishing point. Commonly used in the architectural field. See pictorial drawing.

pictorial drawing A type of drawing in which an object appears to be three dimensional; that is, it appears to have width, height, and depth. But unlike an actual 3D model, a pictorial drawing is constructed using only X- and Y-coordinates. See isometric drawing and perspective drawing.

pipeline drafters and process piping drafters Drafters who prepare drawings used in the construction and maintenance of oil refineries, oil production and exploration industries, chemical plants, and process piping systems such as those used in the manufacture of semiconductor devices.

polar coordinates Coordinates defined by a length and an angle that are relative to the last point defined.

primary auxiliary view A view that is adjacent to, and aligned with, a principal view of the object showing the true shape of features that are not parallel to any of the principal projection planes (front, top, side, etc.).

professional engineer (P.E.) An engineer who is licensed by the National Society of Professional Engineers.

projection planes An imaginary two-dimensional plane, like a sheet of clear glass, placed parallel to a principal face of the object to be visualized. The object's features (points, lines, planes) are projected perpendicular to the projection plane when visualizing a multiview drawing of the object.

properties In an AutoCAD drawing, the properties of an object include its color, lineweight, layer, linetype, line-type scale, and so forth.

quality control inspection A step in the manufacturing cycle performed by a quality control (QC) inspector using precise measuring equipment to determine the actual size of the part. The QC inspector compares the actual size of the part with the dimensions noted on the technical drawing. Parts that measure within the allowable size limits will pass the QC inspection, whereas parts that measure outside the limits will be rejected.

reference dimension A dimension that is included on a technical drawing for information only and is not necessary to manufacture, or inspect, the part. No tolerances are applied to reference dimensions. Reference dimensions are enclosed in parentheses.

regular views In a multiview drawing, this term refers to an object's front, top, bottom, right, left, and back (or rear) views.

relative coordinates Points that are located along the X-, Y-, and Z-axes that are relative to the last point defined.

removed section A section view that is not drawn in its normal projected position but somewhere else on the sheet. Requires labeling of both the cutting plane line and the view it references.

revolved section A cross-sectional view of an object drawn on the object.

roof pitch The angle of a roof. A roof's pitch is expressed as a ratio of the vertical rise of the roof (measured in inches) to the horizontal run of the roof (measured in inches). Using this notation, a roof with a "four-twelve" pitch (labeled as 4/12 on the drawing) would rise 4" for every 12" of horizontal run. Roof pitch is noted on elevation drawings in a set of construction documents.

round A rounded outside corner of a part.

running object snaps Object Snap modes that are activated whenever the user is prompted to select the location of a point.

sans serif A typeface that does not include small marks, called serifs, at the ends of the main strokes of characters. Sometimes called gothic font.

secondary auxiliary view A view that is adjacent to, and aligned with, a primary auxiliary view.

section A drawing technique in which an object is drawn as if part of its exterior has been removed to reveal its interior features and details.

section lines Diagonal lines drawn that are placed on a section view to indicate the areas of the object that came in contact with the cutting plane line. On AutoCAD drawings, section lines are placed with the HATCH command.

sheet sizes Technical drawings are created on standardized sheet sizes. Sheet size varies with the type of drawing and/or the unit of measurement used to create the drawing.

technical drawing Term used to describe the process of creating the drawings used in the field of engineering and architecture.

technical lettering Freehand lettering that is added to a technical drawing or sketch. Technical lettering should be legible and consistent with regard to style.

text style The characteristics of text used in a drawing such as font name, height, width factor, and oblique angle. These values are determined by the values set in the Text Style dialog box.

third-angle projection A technique for arranging multiview drawings with the top view above the front view and the right side drawn to the right of the front view. Commonly used on drawings prepared in North America.

three-dimensional (3D) object An object having height, width, and depth.

tick mark A short diagonal line used to note the termination of dimensions in architectural drawings. Also referred to as a slash.

tolerances The total permissible variation in the size and/or shape of the object's features as defined by applying tolerances to the nominal size dimension. The difference between the upper and lower size limits of the feature.

traditional drafting tools The tools that were used to create technical drawings before CAD techniques became the standard. These tools include parallel straightedges, drafting machines, drafting boards, drafting triangles, protractors, circle and ellipse templates, and technical pens and pencils.

two-dimensional (2D) object An object having height and width, width and depth, or height and depth.

unidirectional text Text placed on a technical drawing that faces only the bottom of the sheet. This technique is required when the ASME text standard is applied to a drawing.

United States National CAD Standard (NCS) A standard developed to unify the preparation, interpretation, and formatting of electronic drawing files across the fields of building design and construction.

user coordinate system (UCS) The point in an AutoCAD drawing where the X-, Y-, and Z-axes intersect (0,0,0). This point is noted in the graphics window with the UCS icon.

viewport A *viewport* is an object found in *Paper Space* of a *Layout*. It is like a window in your *Layout* that displays geometry created in the *Model* tab. The scale of the objects displayed in the viewport can be adjusted for accuracy before plotting.

INDEX

Page numbers followed by ‡ indicate a table.

Page numbers followed by ‡ indicate a table.

Page numbers followed by ‡ indicate a table.

Page numbers followed by ‡ indicate a table.

Page numbers followed by ‡ indicate a table.

Page numbers followed by ‡ indicate a table.

Page numbers followed by ‡ indicate a table.

Page numbers followed by ‡ indicate a table.

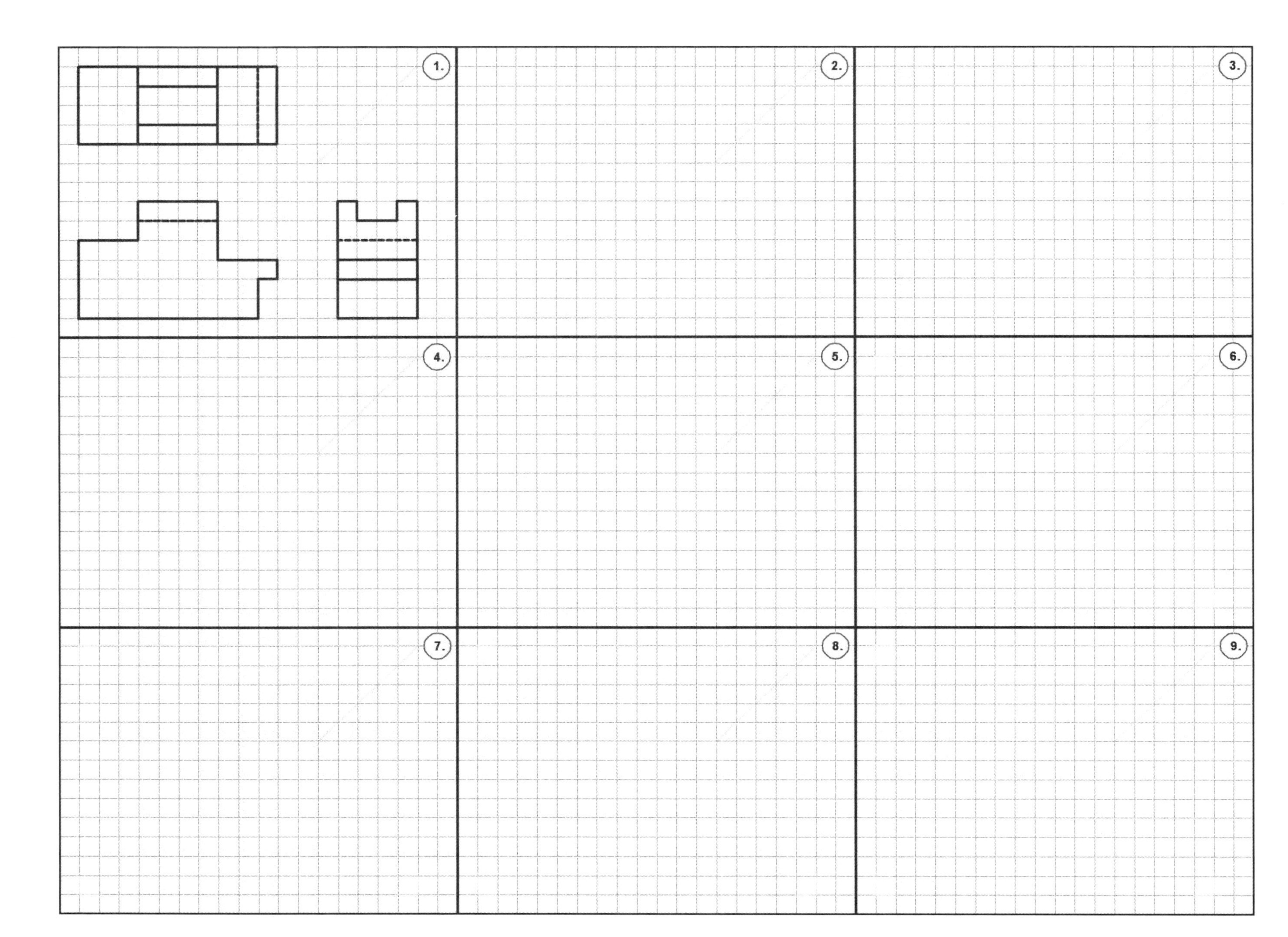

1.
2.
3.
4.
5.
6.
7.
8.
9.

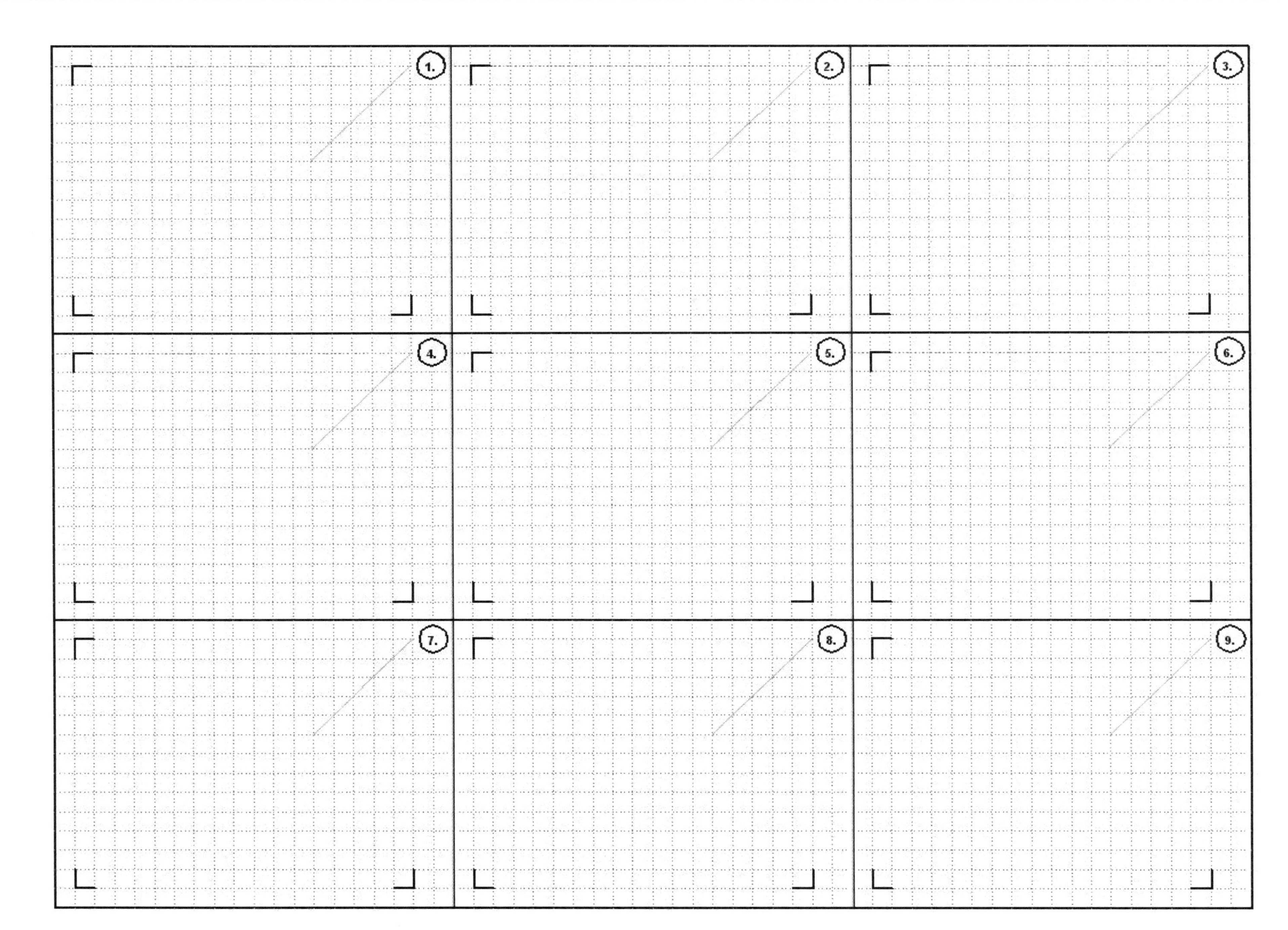

1.
2.
3.
4.
5.
6.
7.
8.
9.

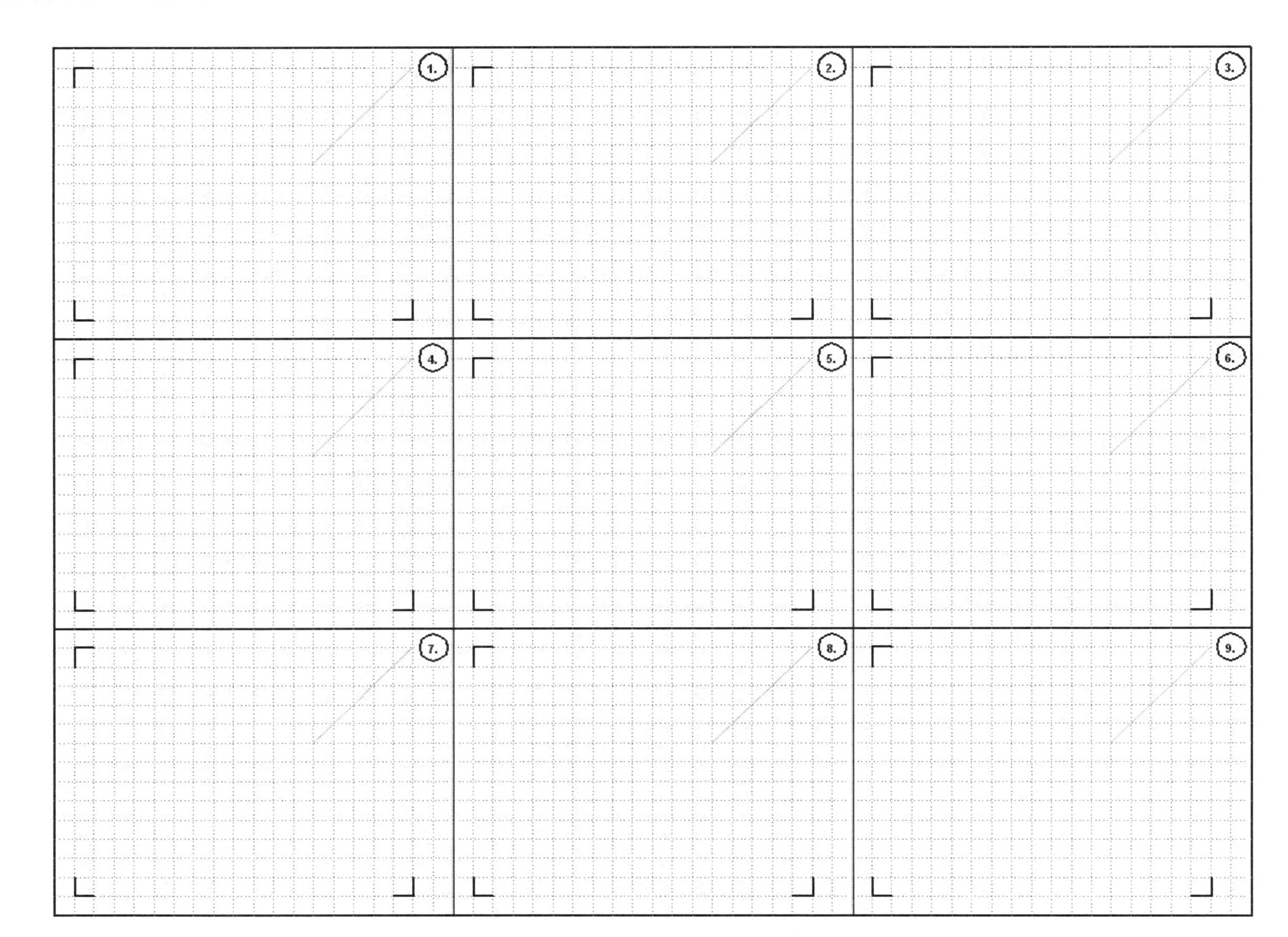
1.
2.
3.
4.
5.
6.
7.
8.
9.

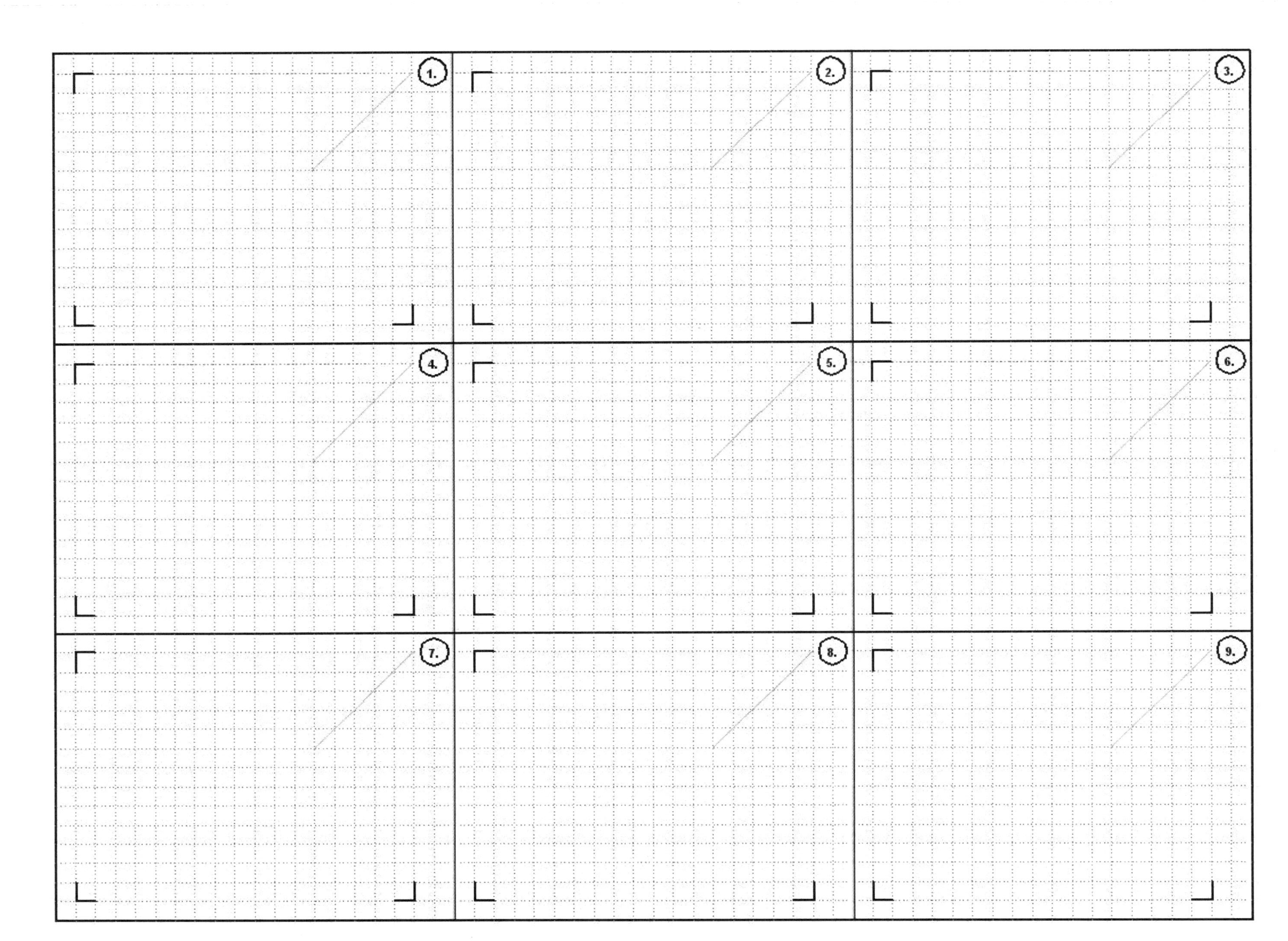
1.
2.
3.
4.
5.
6.
7.
8.
9.

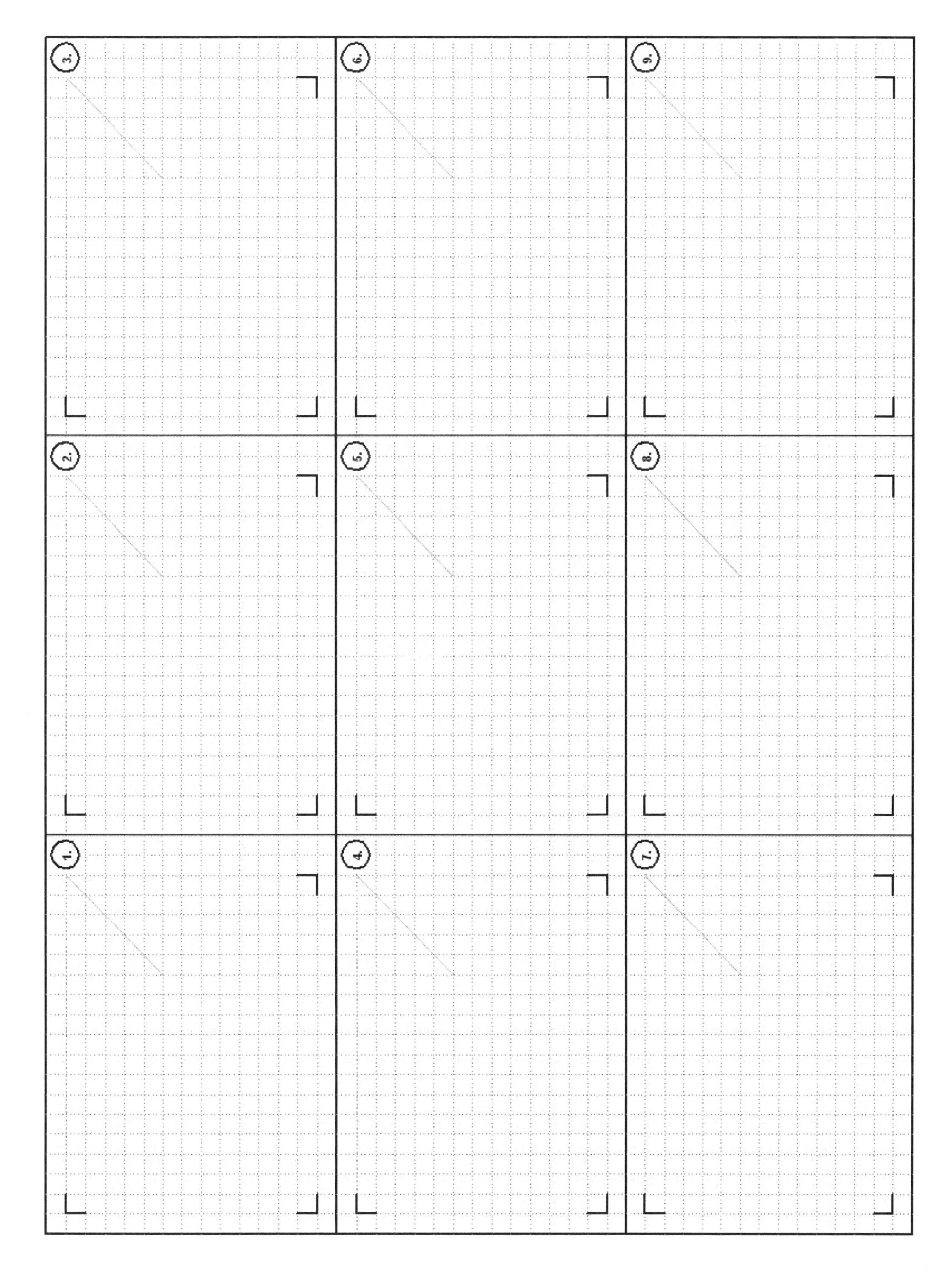

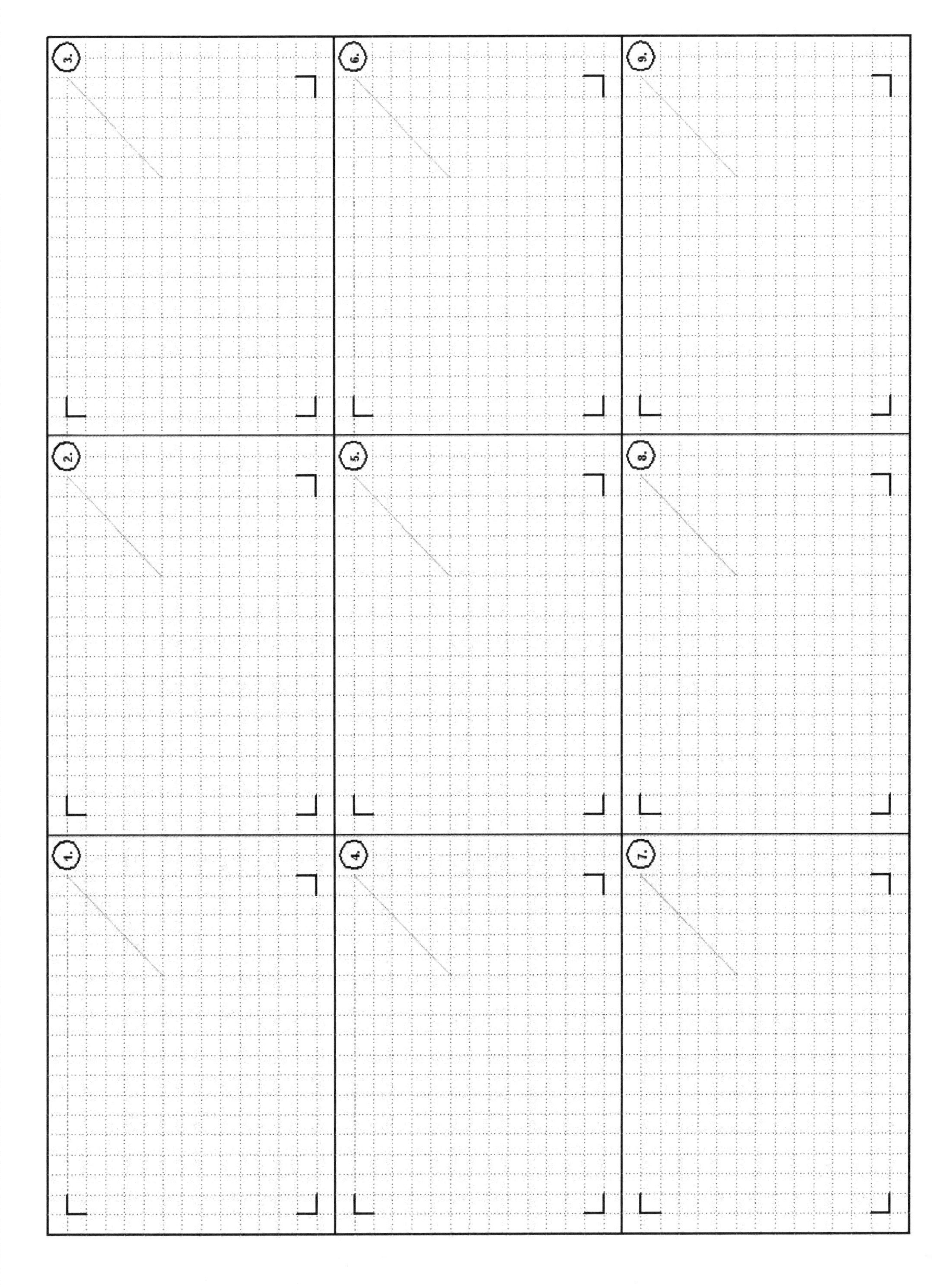

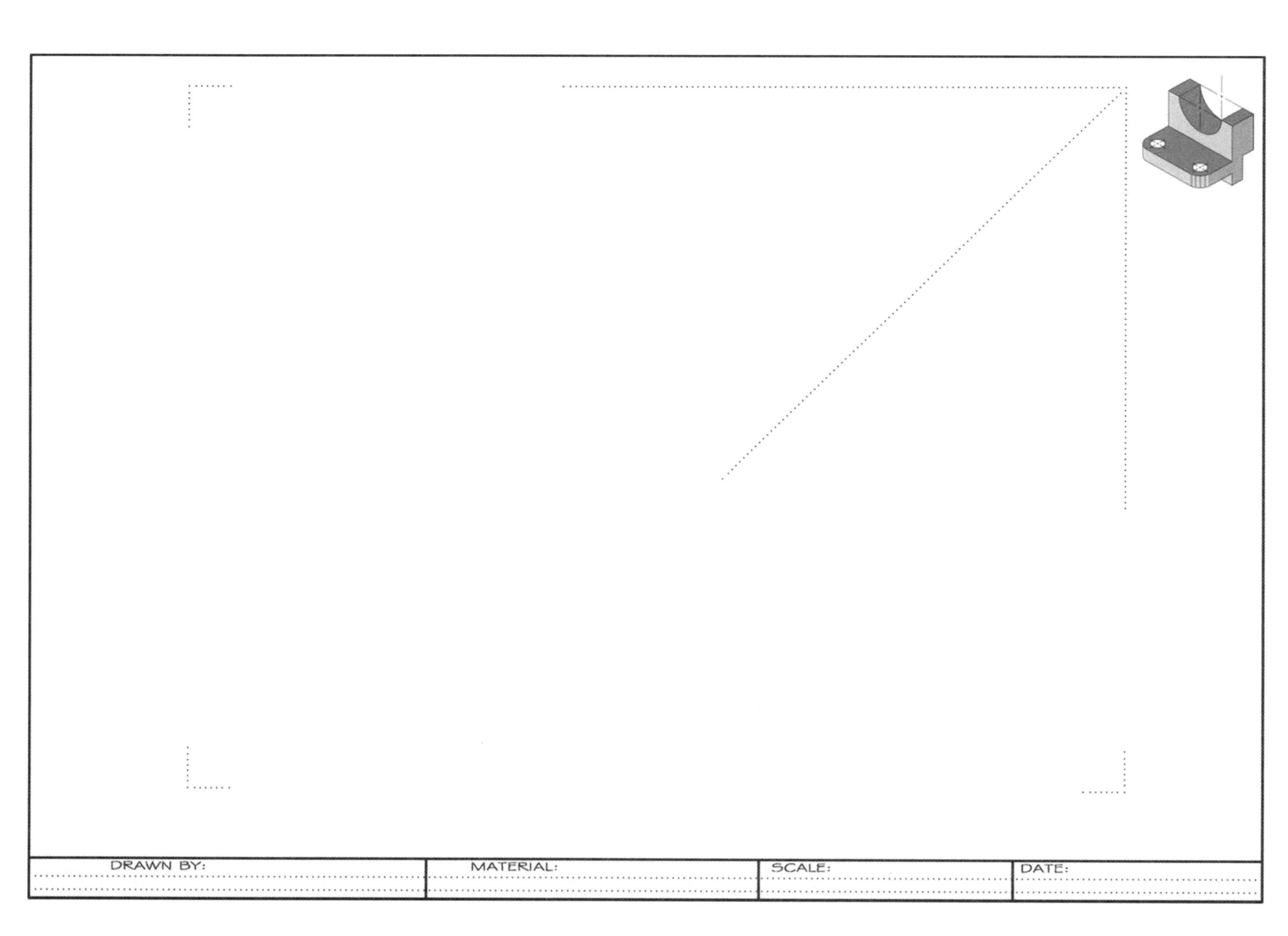
DRAWN BY:
MATERIAL:
SCALE:
DATE:

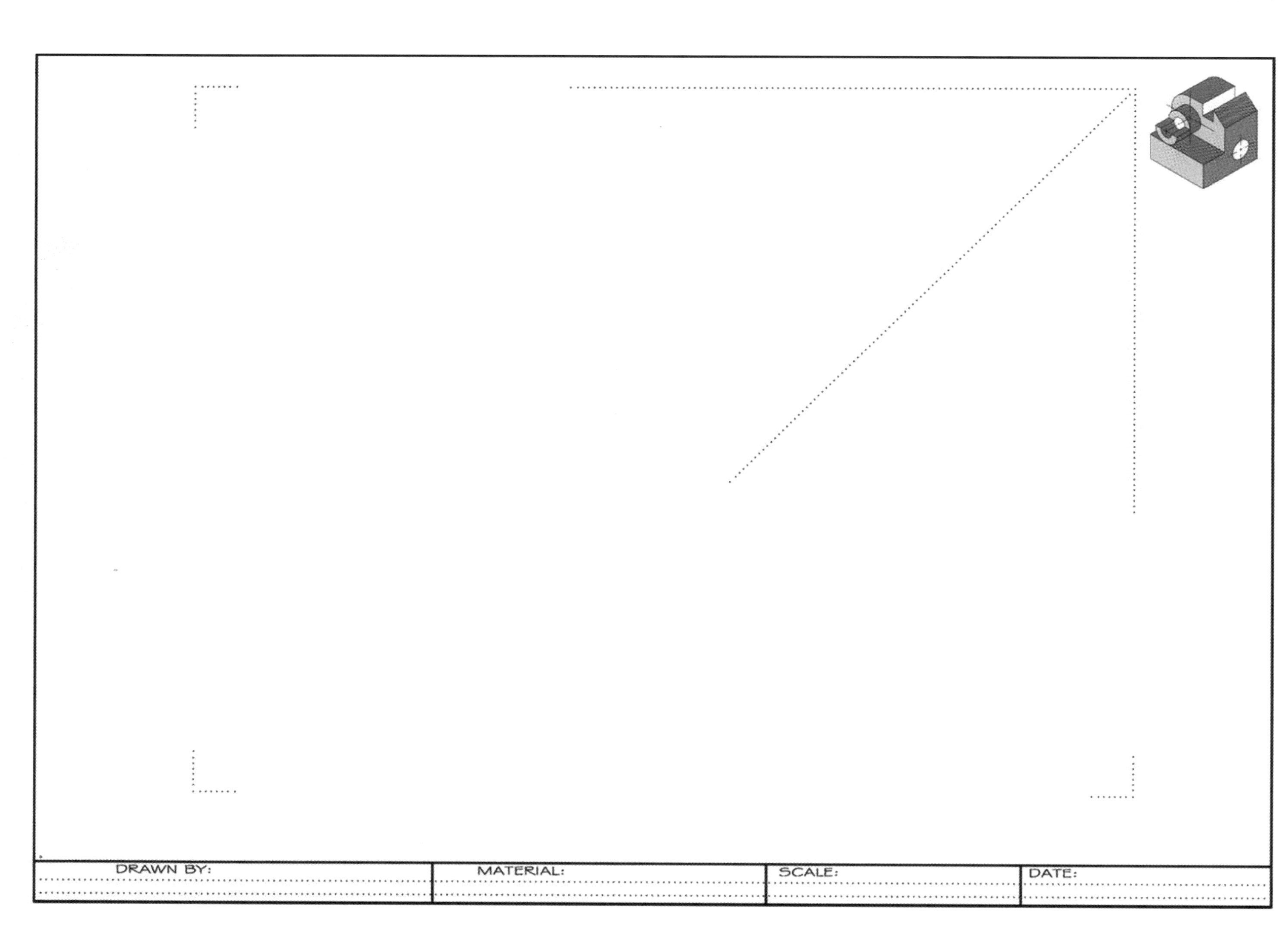
DRAWN BY:
MATERIAL:
SCALE:
DATE:

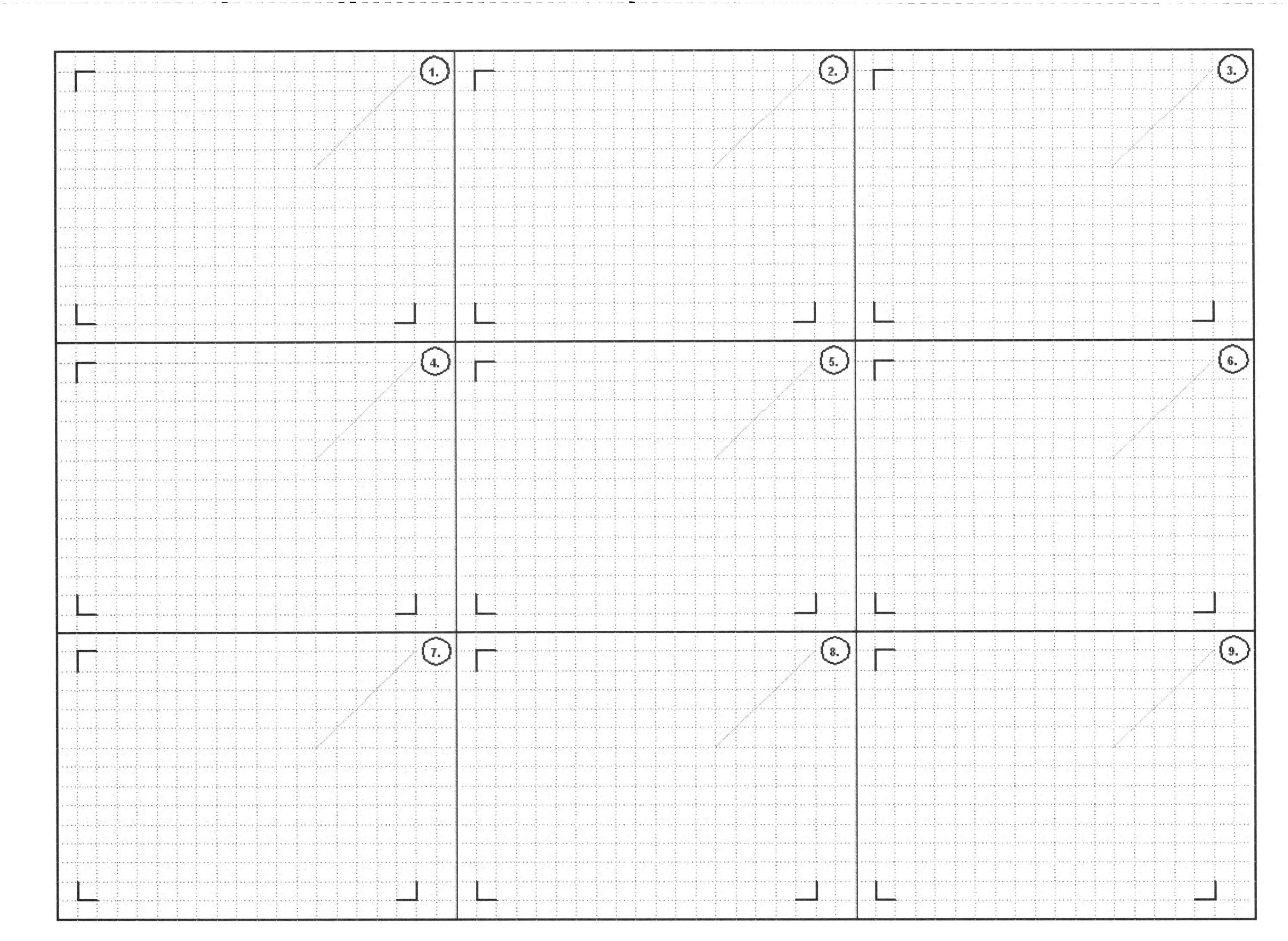
1.
2.
3.
4.
5.
6.
7.
8.
9.

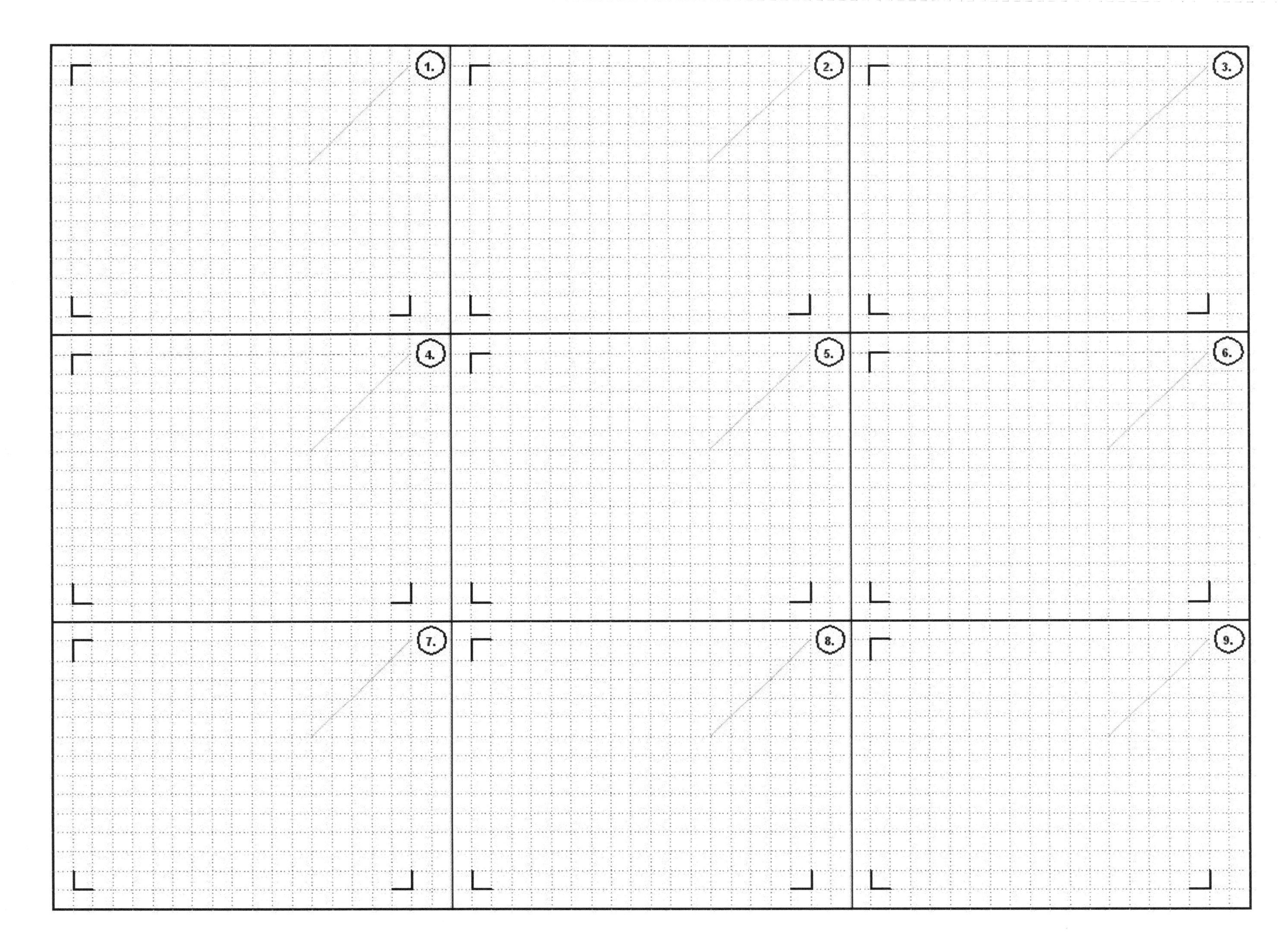
1.
2.
3.
4.
5.
6.
7.
8.
9.

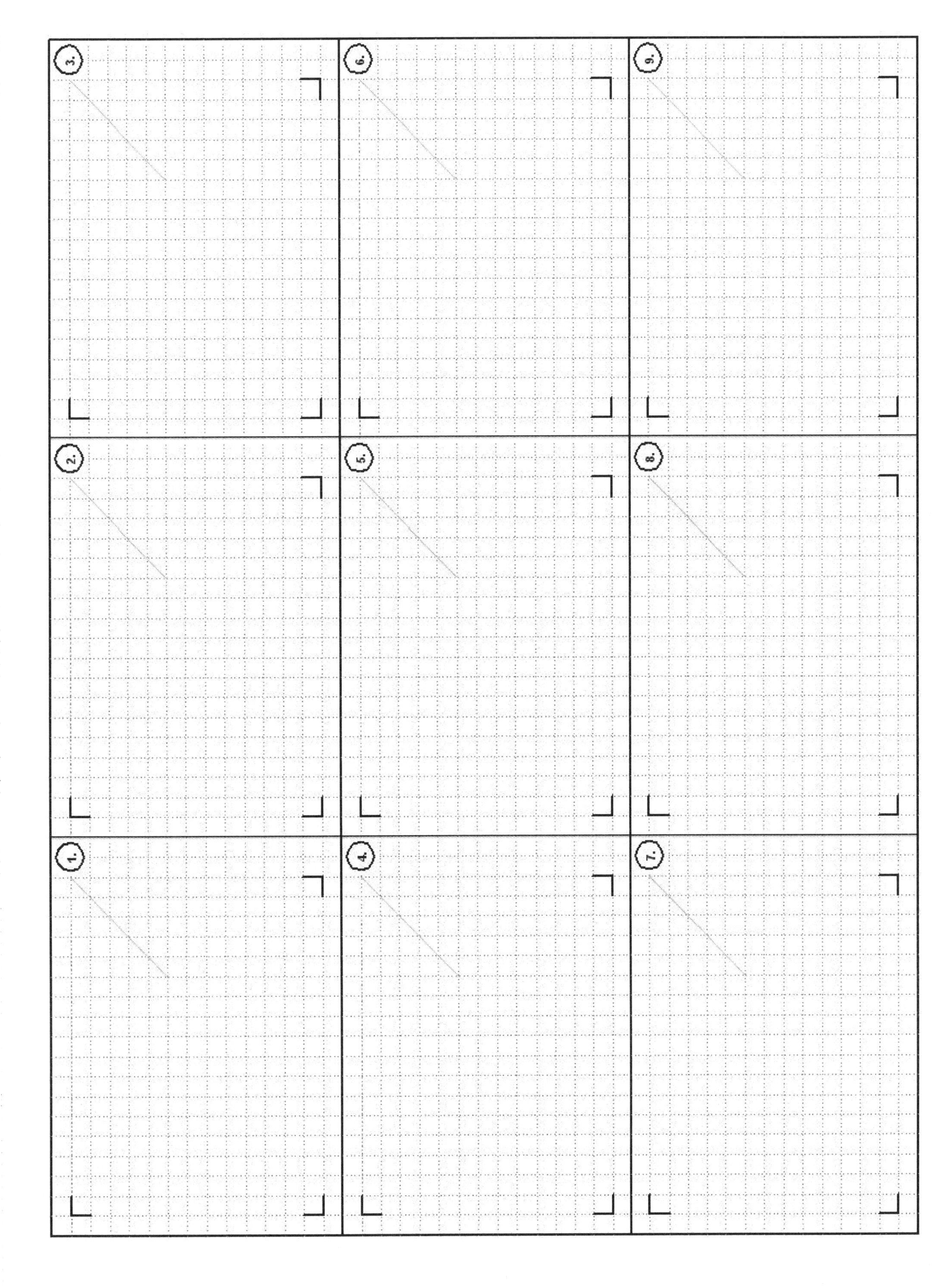

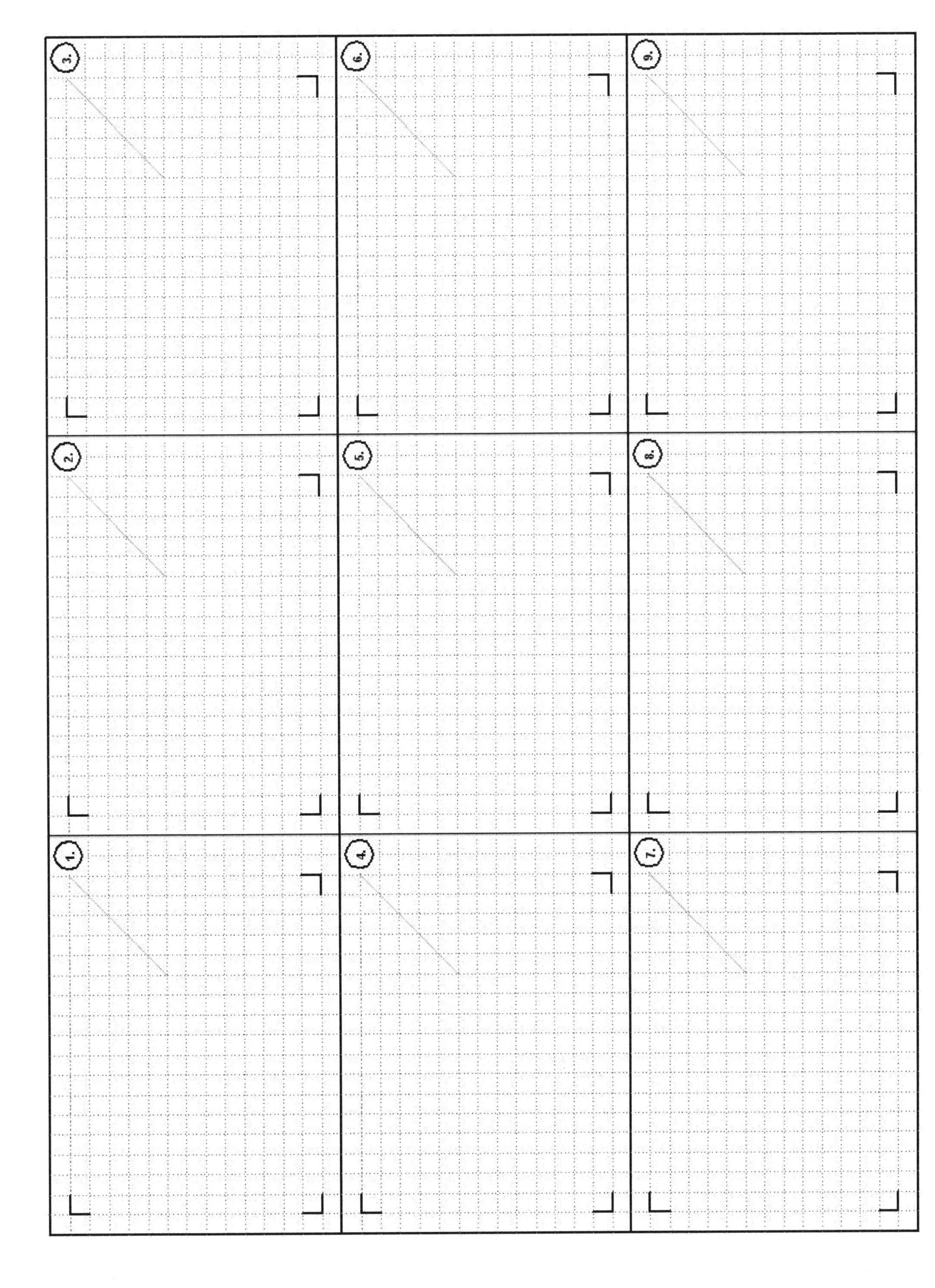

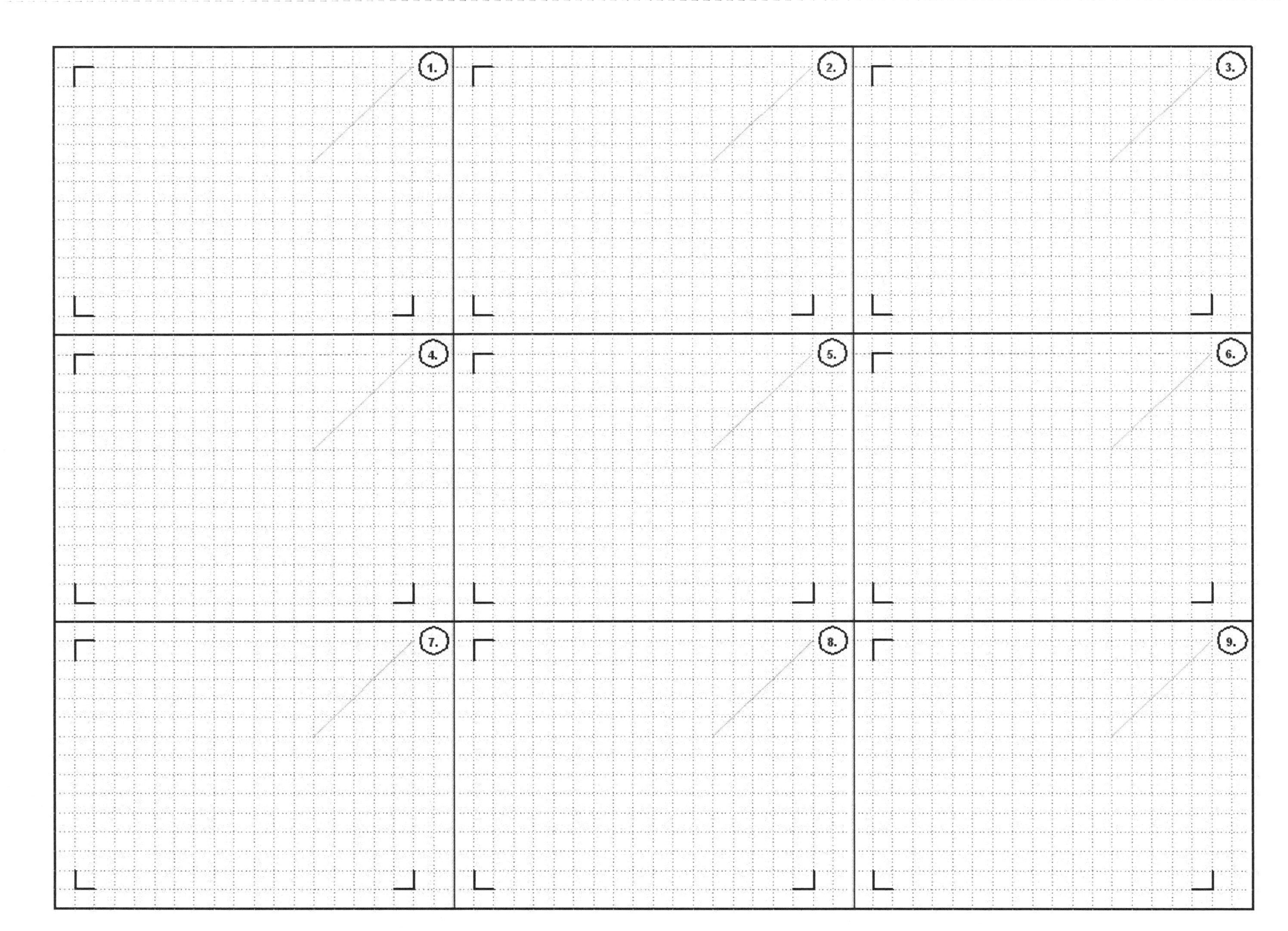
1.
2.
3.
4.
5.
6.
7.
8.
9.

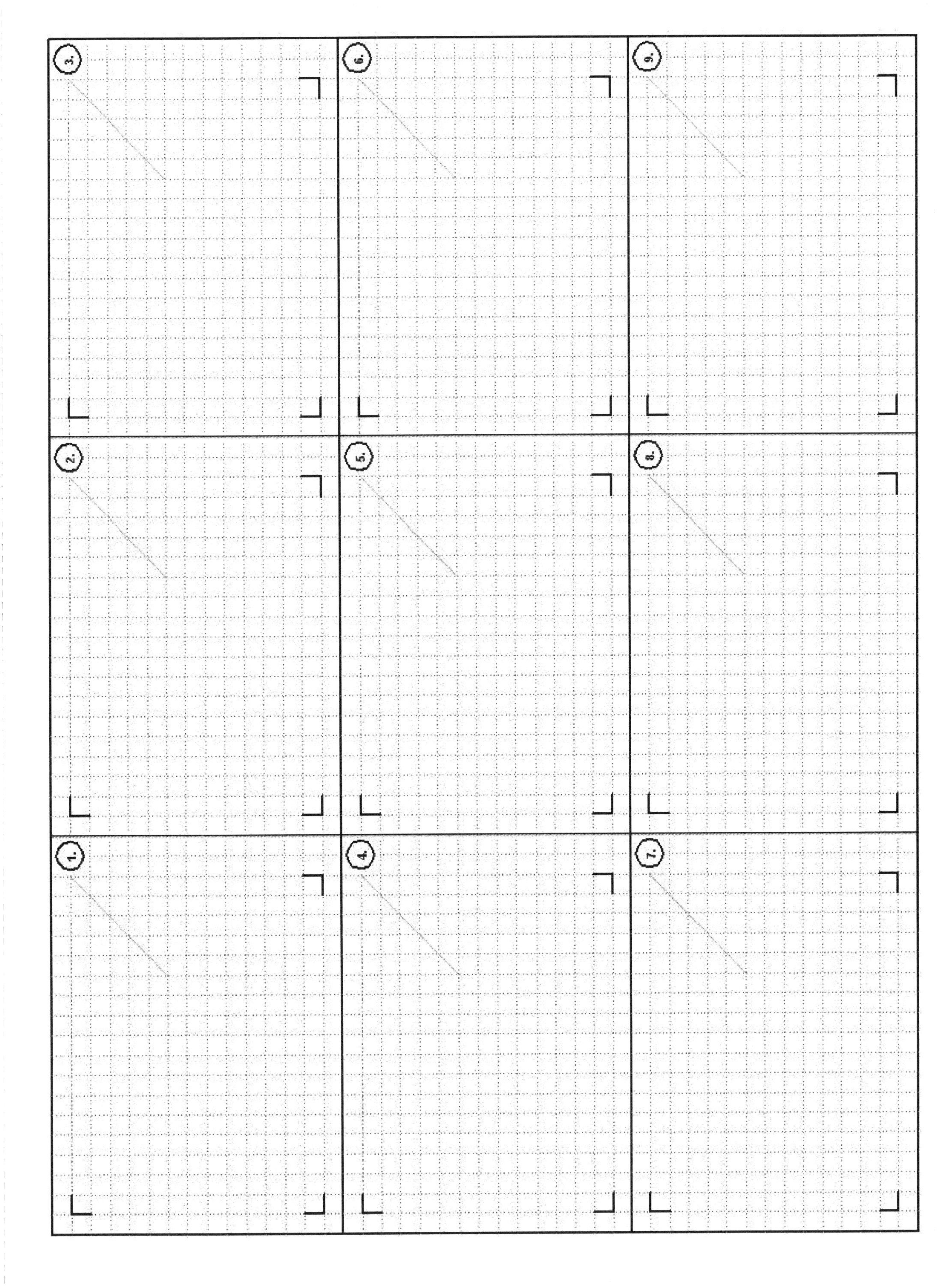